LIPIDS AND SYNDROMES OF INSULIN RESISTANCE
FROM MOLECULAR BIOLOGY TO CLINICAL MEDICINE

ANNALS OF THE NEW YORK ACADEMY OF SCIENCES

Volume 827

EDITORIAL STAFF

Executive Editor
BILL BOLAND

Managing Editor
JUSTINE CULLINAN

Associate Editor
STEFAN MALMOLI

The New York Academy of Sciences
2 East 63rd Street
New York, New York 10021

NEW YORK ACADEMY OF SCIENCES
(Founded in 1817)
BOARD OF GOVERNORS, October 1996–October 1997

MARTIN L. LEIBOWITZ, *Chairman of the Board*
RICHARD A. RIFKIND, *Vice Chairman of the Board*
RODNEY W. NICHOLS, *President and CEO* [ex officio]

Honorary Life Governors
WILLIAM T. GOLDEN   JOSHUA LEDERBERG

JOHN T. MORGAN, *Treasurer*

Governors
ELEANOR BAUM   D. ALLAN BROMLEY   LAWRENCE B. BUTTENWIESER
PRAVEEN CHAUDHARI   EDWARD COHEN   RONALD L. GRAHAM
BILL GREEN   JACQUELINE LEO   WILLIAM J. McDONOUGH
SANDRA PANEM   CHARLES RAMOND
WILLIAM C. STEERE, Jr.   TORSTEN WIESEL

HENRY M. GREENBERG, *Past Chairman of the Board*
HELENE L. KAPLAN, *Counsel* [ex officio]
CRAIG PURINTON, *Secretary* [ex officio]

ANNALS OF THE NEW YORK ACADEMY OF SCIENCES
Volume 827

LIPIDS AND SYNDROMES OF INSULIN RESISTANCE

FROM MOLECULAR BIOLOGY TO CLINICAL MEDICINE

Edited by I. Klimeš, S. M. Haffner, E. Šeböková, B. V. Howard, and L. H. Storlien

The New York Academy of Sciences
New York, New York
1997

Copyright © 1997 by the New York Academy of Sciences. All rights reserved. Under the provisions of the United States Copyright Act of 1976, individual readers of the Annals are permitted to make fair use of the material in them for teaching and research. Permission is granted to quote from the Annals provided that the customary acknowledgment is made of the source. Material in the Annals may be republished only by permission of the Academy. Address inquiries to the Executive Editor at the New York Academy of Sciences.

Copying fees: *For each copy of an article made beyond the free copying permitted under Section 107 or 108 of the 1976 Copyright Act, a fee should be paid through the Copyright Clearance Center, Inc., 222 Rosewood Drive, Danvers, MA 01923. The fee for copying an article is $3.00 for nonacademic use; for use in the classroom it is $0.07 per page.*

∞ *The paper used in this publication meets the minimum requirements of American National Standard for Information Sciences – Permanence of Paper for Printed Library Materials, ANSI Z39.48-1984.*

Library of Congress Cataloging-in-Publication Data

Lipids and syndromes of insulin resistance: from molecular biology to clinical medicine/edited by I. Klimeš . . . [et al.].
    p.   cm. – (Annals of the New York Academy of Sciences, ISSN 0077-8923; v. 827).
    The Third International Smolenice Insulin Symposium was an official satellite meeting of the 32nd Annual Meeting of the European Association for the Study of Diabetes held in Vienna, Austria.
    Includes bibliographical references and index.
    ISBN 1-57331-070-0 (cloth: alk. paper). – ISBN 1-57331-071-9 (paper: alk. paper).
    1. Insulin resistance – Congresses. 2. Lipids – Metabolism – Disorders – Congresses. 3. Hyperlipidemia – Congresses. 4. Obesity – Congresses. I. Klimeš, Iwar. II. European Association for the Study of Diabetes. Meeting (32nd: 1996: Vienna, Austria).
III. International Smolenice Insulin Symposium on "Lipids and Syndromes of Insulin Resistance: from Molecular Biology to Clinical Medicine" (1996). IV. Series.
    [DNLM: 1. Insulin Resistance – physiology – congresses. 2. Lipids – metabolism – congresses. 3. Insulin – metabolism – congresses. 4. Dietary Fats – metabolism – congresses. W1 AN626YL v.827 1997/ WK 820 L764 1997]
Q11.N5 vol. 827
[RC662.4]
500 s – dc21
[616.3'997]
DNLM/DLC
for Library of Congress   97-18285
                                                                                              CIP

CCP
Printed in the United States of America
ISBN 1-57331-070-0 (cloth)
ISBN 1-57331-071-9 (paper)
ISSN 0077-8923

ANNALS OF THE NEW YORK ACADEMY OF SCIENCES

Volume 827
September 20, 1997

LIPIDS AND SYNDROMES OF INSULIN RESISTANCE
FROM MOLECULAR BIOLOGY TO CLINICAL MEDICINE[a]

Editors and Conference Organizers
I. KLIMEŠ, S. M. HAFFNER, E. ŠEBÖKOVÁ, B. V. HOWARD, and L. H. STORLIEN

CONTENTS

Preface. By IWAR KLIMEŠ and ELENA ŠEBÖKOVÁ xi

Keynote Address: Progress in Population Analyses of the Insulin Resistance Syndrome. By STEVEN M. HAFFNER 1

Part I. Animal Models of the Insulin Resistance Syndrome

Hypertension and the Insulin Resistance Syndrome of Rats: Are They Related? By IWAR KLIMEŠ and ELENA ŠEBÖKOVÁ 13

Obesity Genes and Insulin Resistance Syndrome. By E. BOWIE KAHLE, RUDOLPH L. LEIBEL, DIRK W. DOMASCHKO, SAM G. RANEY, and K. TRACI MANN 35

Development of Obesity and Insulin Resistance in the Israeli Sand Rat (*Psammomys obesus*): Does Leptin Play a Role? By G. R. COLLIER, A. DE SILVA, A. SANIGORSKI, K. WALDER, A. YAMAMOTO, and P. ZIMMET 50

Diabetes and Hypertension in Rodent Models. By INGRID KLÖTING, SABINE BERG, PETER KOVÁCS, BIRGER VOIGT, LUTZ VOGT, and SIEGFRIED SCHMIDT 64

Insulin Resistance Syndrome in Mice Deficient in Insulin Receptor Substrate-1. By HIROYUKI TAMEMOTO, KAZUYUKI TOBE, TOSHIHISA YAMAUCHI, YASUO TERAUCHI, YASUSHI KABURAGI, and TAKASHI KADOWAKI 85

---

[a] The papers in this volume were presented at the Third International Smolenice Insulin Symposium on "Lipids and Syndromes of Insulin Resistance: From Molecular Biology to Clinical Medicine," which was held on August 28 to September 1, 1996, in Smolenice Castle, Slovak Republic, and organized by the Diabetes and Nutrition Research Group of the Institute of Experimental Endocrinology, Slovak Academy of Sciences, Bratislava, Slovak Republic, in collaboration with the Slovak Diabetes and Endocrine Societies; the Department of Medicine, Division of Clinical Epidemiology, University of Texas Health Science Center, San Antonio, Texas; the Department of Biomedical Science, University of Wollongong, New South Wales, Australia; and the Medlantic Research Institute, Washington, District of Columbia.

WOK.1W Rats: A Potential Animal Model of the Insulin Resistance Syndrome. *By* PETER KOVÁCS, BIRGER VOIGT, SABINE BERG, LUTZ VOGT, and INGRID KLÖTING 94

Part II. Molecular Biology of Syndromes of Insulin Resistance

Lipid Transport Genes and Their Relation to the Syndrome of Insulin Resistance. *By* DAVID J. GALTON, QIUPING ZHANG, ELISABETH CAVALLERO, JULIAN CAVANNA, ANDREA KAY, ALINE CHARLES, SYLVIE BRASCHI, LEON PERLEMUTER, and BERNARD JACOTOT 100

Progress in Determining the Genes for Hypertension, Insulin Resistance, and Dyslipidemia. *By* MARK CAULFIELD, PIERRE-MARC BOULOUX, and PATRICIA MUNROE 110

Hyperinsulinemia and Sympathoadrenal System Activity in the Rat. *By* RICHARD KVETŇANSKÝ, MILAN RUSNÁK, DANIELA GAŠPERÍKOVÁ, JANA JELOKOVÁ, ŠTEFAN ZÓRAD, ILJA VIETOR, KAREL PACÁK, ELENA ŠEBÖKOVÁ, LADISLAV MACHO, ESTHER L. SABBAN, and IWAR KLIMEŠ 118

Is a Mutation of the β_3-Adrenergic Receptor Gene Related to Non-Insulin-dependent Diabetes Mellitus and Juvenile Hypertension in the Czech Population? *By* B. BENDLOVÁ, I. MAZURA, J. VČELÁK, J. PERUŠIČOVÁ, D. PALYZOVÁ, I. KLIMEŠ, and E. ŠEBÖKOVÁ 135

Glucose Transport and Insulin Signaling in Rat Muscle and Adipose Tissue: Effect of Lipid Availability. *By* D. GAŠPERÍKOVÁ, I. KLIMEŠ, T. KOLTER, P. BOHOV, A. MAASSEN, J. ECKEL, M. T. CLANDININ, and E. ŠEBÖKOVÁ 144

Cafestol (a Coffee Lipid) Decreases Uptake of Low-Density Lipoprotein (LDL) in Human Skin Fibroblasts and Liver Cells. *By* A. C. RUSTAN, B. HALVORSEN, T. RANHEIM, and C. A. DREVON 158

High Fructose Feeding Enhances Erythrocyte Carbonic Anhydrase 1 mRNA Levels in Rat. *By* K. K. GAMBHIR, P. OATES, M. VERMA, S. TEMAM, and W. CHEATHAM 163

Pharmacological Treatment and Mechanisms of Insulin Resistance: Impact on Vascular Smooth Muscle Cells, Blood Pressure, and Lipids. *By* WILLA A. HSUEH and RONALD E. LAW 170

Part III. Role of Lipids in Diseases of Insulin Resistance

Fatty Acid Regulation of Gene Expression: Its Role in Fuel Partitioning and Insulin Resistance. *By* STEVEN D. CLARKE, REBECCA BAILLIE, DONALD B. JUMP, and MANABU T. NAKAMURA 178

Impact of Dietary Fatty Acid Composition on Insulin Action at the Nucleus. *By* NANA A. GLETSU and M. THOMAS CLANDININ 188

Molecular and Cellular Determinants of Triglyceride Availability.
By E. ŠEBÖKOVÁ and I. KLIMEŠ 200

Dietary Fatty Acids, Insulin Resistance, and Diabetes.
By BARBARA V. HOWARD 215

Lipid Abnormalities in Muscle of Insulin-resistant Rodents: The Malonyl CoA Hypothesis. By N. B. RUDERMAN, A. K. SAHA, D. VAVVAS, S. J. HEYDRICK, and T. G. KUROWSKI 221

Pharmacological Strategies for Reduction of Lipid Availability. By JAMES E. FOLEY, ROBERT C. ANDERSON, PHILIP A. BELL, BRYAN F. BURKEY, RHONDA O. DEEMS, CHRISTOPHER DE SOUZA, and BETH E. DUNNING ... 231

Effect of Oral Antidiabetics and Insulin on Lipids and Coronary Heart Disease in Non-Insulin-dependent Diabetes Mellitus. By M. HANEFELD, T. TEMELKOVA-KURKTSCHIEV, and C. KÖHLER 246

Relationships between Fatty Acid Composition and Insulin-induced Oxidizability of Low-Density Lipoproteins in Healthy Men. By T. PELIKÁNOVÁ, E. TVRZICKÁ, L. KAZDOVÁ, and A. ŽÁK 269

Small Dense Low-Density Lipoprotein (LDL) in Non-Insulin-dependent Diabetes Mellitus (NIDDM): Impact of Hypertriglyceridemia. By T. TEMELKOVA-KURKTSCHIEV, M. HANEFELD, and W. LEONHARDT 279

Part IV. Dietary Fats in Diseases of Insulin Resistance

Does Dietary Fat Influence Insulin Action? By L. H. STORLIEN, A. D. KRIKETOS, A. B. JENKINS, L. A. BAUR, D. A. PAN, L. C. TAPSELL, and G. D. CALVERT 287

Monounsaturated and Marine ω-3 Fatty Acids in NIDDM Patients.
By A. A. RIVELLESE 302

Omega-3 and Omega-6 Fatty Acids in the Insulin Resistance Syndrome: Lipid and Lipoprotein Metabolism and Atherosclerosis.
By A. C. RUSTAN, M. S. NENSETER, and C. A. DREVON 310

Omega-6/Omega-3 Fatty Acid Ratio and *Trans* Fatty Acids in Non-Insulin-dependent Diabetes Mellitus. By ARTEMIS P. SIMOPOULOS 327

Dietary Fats and Hypertension: Focus on Fish Oil. By PETER R. C. HOWE 339

Postprandial Triglyceride High Response and the Metabolic Syndrome.
By J. SCHREZENMEIR, S. FENSELAU, I. KEPPLER, J. ABEL, B. ORTH, CH. LAUE, W. STÜRMER, U. FAUTH, M. HALMAGYI, and W. MÄRZ 353

Evaluation of an Omega-3 Fatty Acid Supplement in Diabetics with Microalbuminuria. By YVONNE K. LUNGERSHAUSEN, PETER R. C. HOWE, PETER M. CLIFTON, CHRISTOPHER R. T. HUGHES, PAT PHILLIPS, JOHN J. GRAHAM, and DAVID W. THOMAS 369

Paleodiet and Its Relation to Atherosclerosis. By J. LIETAVA, M. THURZO, and A. DUKÁT 382

Part V. Energy Balance, Lipids, and Obesity

Passive Overconsumption: Fat Intake and Short-Term Energy Balance.
By JOHN E. BLUNDELL and JENNIE I. MACDIARMID 392

Dietary Fats and Thermogenesis. *By* A. TREMBLAY and C. BOUCHARD 408

Fat Metabolism in the Predisposition to Obesity. *By* ARNE ASTRUP,
ANNE RABEN, BENJAMIN BUEMANN, and SØREN TOUBRO 417

Fat and Energy Balance. *By* MICHAEL J. PAGLIASSOTTI, ELLIS C. GAYLES,
and JAMES O. HILL 431

Recent Advances in the Pharmacological Control of Energy Balance and
Body Weight. *By* RICHARD L. ATKINSON 449

Fat Substitutes and Energy Balance. *By* JOHN C. PETERS 461

Nonesterified Fatty Acid Regulation of Lipid and Glucose Oxidation in the
Obese. *By* SARAH E. KING, JANET M. BRYSON, LOUISE A. BAUR,
SOJI SWARAJ, and IAN D. CATERSON 476

Part VI. Poster Papers

Nuclear All-*trans* Retinoic Acid Receptors in Liver of Rats with
Diet-induced Insulin Resistance. *By* J. BRTKO, E. ŠEBÖKOVÁ,
D. GAŠPERÍKOVÁ, I. KLIMEŠ, S. HUDECOVÁ, and J. BRANSOVÁ 480

Diet-induced Insulin Resistance Is Associated with Decreased Activity of
Type I Iodothyronine 5'-Deiodinase in Rat Liver. *By* J. BRANSOVÁ,
A. BRTKOVÁ, E. ŠEBÖKOVÁ, I. KLIMEŠ, P. LANGER, and J. BRTKO 485

Insulin and Catecholamines Act at Different Stages of Rat Liver Regeneration.
By J. KNOPP, L. MACHO, M. FICKOVÁ, Š. ZÓRAD, R. KVETŇANSKÝ,
and I. JAROŠČÁKOVÁ 489

Fatty Acid Composition in Fractions of Structural and Storage Lipids in
Liver and Skeletal Muscle of Hereditary Hypertriglyceridemic Rats.
By PAVOL BOHOV, ELENA ŠEBÖKOVÁ, DANIELA GAŠPERÍKOVÁ,
PAVEL LANGER, and IWAR KLIMEŠ 494

Increased Adipose Tissue Lipolysis in a Hypertriglyceridemic Rat Line.
By A. VRÁNA and L. KAZDOVÁ 510

Structural Changes in the Aorta of the Hereditary Hypertriglyceridemic Rat.
By FRANTIŠEK KRISTEK, SOŇA EDELSTEINOVÁ, ELENA ŠEBÖKOVÁ,
JÁN KYSELOVIČ, and IWAR KLIMEŠ 514

Increased Lipoprotein Oxidability and Aortic Lipid Peroxidation in
an Experimental Model of Insulin Resistance Syndrome.
By LUDMILA KAZDOVÁ, ALEŠ ŽÁK, and ANTONÍN VRÁNA 521

Phenotype and Genotype Comparison of Hereditary Hypertriglyceridemic (hHTG) and Brown-Norway (BN) Rats: Identification of Quantitative Trait Loci (QTLs) for the Insulin Resistance Syndrome.
By PETER KOVÁCS, NILESH J. SAMANI, ELENA ŠEBÖKOVÁ, BIRGER VOIGT, DANIELA GAŠPERÍKOVÁ, DANIELA JEŽOVÁ, RICHARD KVETŇANSKÝ, DAVID LODWICK, INGRID KLÖTING, and IWAR KLIMEŠ 526

Insulin Resistance in Adipocytes of SHR Rats. *By* M. FICKOVÁ, Š. ZÓRAD, J. KUNEŠ, M. RUSNÁK, and L. MACHO 532

Disproportionate Increase of Fatty Acid Binding Proteins in the Livers of Obese Diabetic *Psammomys obesus*. *By* P. LEWANDOWSKI, D. CAMERON-SMITH, K. MOULTON, K. WALDER, A. SANIGORSKI, and G. R. COLLIER 536

Partial Characterization of Insulin Resistance in Adipose Tissue of Monosodium Glutamate–induced Obese Rats. *By* Š. ZÓRAD, L. MACHO, D. JEŽOVÁ, and M. FICKOVÁ 541

Biguanide Effects on Insulin Signaling. *By* EMMANUELLE MEUILLET, NICOLAS WIERNSPERGER, PIERRE HUBERT, and GÉRARD CRÉMEL 546

Effect of the High-Fat Diet on the Calcium Channels in Rat Myocardium. *By* I. MINAROVIČ, R. VOJTKO, E. ŠEBÖKOVÁ, I. KLIMEŠ, and I. ZAHRADNÍK 550

The Effect of Fasting and Vitamin E on Insulin Action in Obese Type 2 Diabetes Mellitus. *By* J. ŠKRHA, G. ŠINDELKA, and J. HILGERTOVÁ 556

The Effect of Hyperlipidemia on Serum Fatty Acid Composition in Type 2 Diabetics. *By* PAVOL BOHOV, VILIAM BALÁŽ, ELENA ŠEBÖKOVÁ, and IWAR KLIMEŠ 561

High Lipid Levels in Slovak Rural Population: Consequence of Thyroid Dysfunction or Nutritional Status? *By* P. LANGER, E. HANZEN, M. TAJTÁKOVÁ, Z. PUTZ, A. KREZE, E. ŠEBÖKOVÁ, P. BOHOV, D. GAŠPERÍKOVÁ, M. HUČKOVÁ, and I. KLIMEŠ 568

Activity of Antioxidant Enzymes during Hyperglycemia and Hypoglycemia in Healthy Subjects. *By* J. KOŠKA, D. SYROVÁ, P. BLAŽÍČEK, M. MARKO, D. J. GRŇA, R. KVETŇANSKÝ, and M. VIGAŠ 575

Index of Contributors 581

Financial assistance was received from:

Principal Sponsor
- NOVO NORDISK (DENMARK)

Sponsors
- BAYER (GERMANY)
- ELI LILLY (CZECH REPUBLIC)
- SCOTIA PHARMACEUTICALS (SCOTLAND, UNITED KINGDOM)
- SLOVAK DIABETES ASSOCIATION (SLOVAK REPUBLIC)

Associate Sponsors
- CIBA-GEIGY/NOVARTIS (SWITZERLAND)
- LABORATOIRES FOURNIER (SLOVAK REPUBLIC)
- L'UDOVÁ BANKA (SLOVAK REPUBLIC)
- MERCK LIPHA (SLOVAK REPUBLIC)
- MILEX-SCHÄRDINGER (SLOVAK REPUBLIC)
- OTČINA INSURANCE (SLOVAK REPUBLIC)

The New York Academy of Sciences believes it has a responsibility to provide an open forum for discussion of scientific questions. The positions taken by the participants in the reported conferences are their own and not necessarily those of the Academy. The Academy has no intent to influence legislation by providing such forums.

Preface

IWAR KLIMEŠ AND ELENA ŠEBÖKOVÁ
Diabetes and Nutrition Research Group
Institute of Experimental Endocrinology
Slovak Academy of Sciences
SK-83306 Bratislava, Slovak Republic

Lipids and Syndromes of Insulin Resistance: From Molecular Biology to Clinical Medicine arose from the Third International Smolenice Insulin Symposium held from August 28 to September 1, 1996, as an official satellite meeting of the Thirty-second Annual Meeting of the European Association for the Study of Diabetes in Vienna, Austria. The symposium took place at Smolenice Castle, the congress center of the Slovak Academy of Sciences at the southeastern foot of the Small Carpathian Mountains close to the capital of Slovakia, and was organized by the Diabetes and Nutrition Research Group of the Institute of Experimental Endocrinology, Slovak Academy of Sciences, Bratislava.

The past decade has seen the growing recognition that a cluster of prevalent conditions including non-insulin-dependent diabetes mellitus, obesity, dyslipidemias, hypertension, and cardiovascular disease, the "Insulin Resistance Syndrome," were linked through a common etiology. During the 1992 Smolenice Insulin Symposium, the focus was restricted to dietary lipids in relation particularly to insulin action. In the interim, evidence has firmed the linking of lipids in their multiple metabolic role (as diverse as major energy sources, to cell structural elements, to intracellular second messengers, to gene regulators) to many aspects of the diseases of the Insulin Resistance Syndrome. Therefore, the 1996 Smolenice Insulin Symposium attempted to bring together some of the world's best scientists and clinicians to coalesce the advances made in understanding the pivotal role that lipids play in the etiology and disease outcomes of the Insulin Resistance Syndrome. One hundred ten participants from 17 countries all over the world contributed to the conference. There were 30 invited lectures, 14 free communications, and 25 posters. This volume stands as a tribute to the excellence of their individual contributions as well as to their ability to link that work to endpoints of prevention and therapy for the diseases of the Insulin Resistance Syndrome.

We are delighted to have the chance to express appreciation to Barbara V. Howard, president of the Medlantic Research Institute in Washington, District of Columbia. As Honorary Chair, she significantly contributed again to the scientific and financial issues of the meeting. Due to her outstanding record of scientific achievement, her continuous support of all three Smolenice Insulin Symposia, and her scientific exchange with Slovakia, the president of the Slovak Academy of Sciences awarded the "Jan Jesenius" Golden Medal for "Outstanding Merit in Medical Sciences" to B. V. Howard.

Fond appreciation also goes to Leonard H. Storlien, head of the Department of Biomedical Science, University of Wollongong, NSW, Australia. L. H. Storlien, who also served on the organizing committee of the 1992 Smolenice Insulin Symposium, contributed immensely to all conceptional, editorial, and financial issues and even managed to "keep smiling." His merits were rewarded by nominating him as an "Honorary Member of the Slovak Diabetes Association."

Our organizing quartet, consisting of B. V. Howard, L. H. Storlien, and the two of us, was joined in 1995 by Steven M. Haffner, head of the Division of Clinical Epidemiology of the Department of Medicine, University of Texas Health Science Center, San Antonio, Texas. In addition to his keynote address, which was a remarkable overview on epidemiology of the Insulin Resistance Syndrome, thanks are due to S. M. Haffner for securing funds for several invited speakers from the United States and Canada. S. M. Haffner also received an "Honorary Membership in the Slovak Diabetes Association."

Thanks are due to all people who made this symposium possible: our honorary chair and both cochairs, who worked altruistically for the last two years; the members of the Local Organizing Committee; and, of course, the invited speakers who traveled so far to participate. This and other activities within the symposium would not have been imaginable without the significant support of the institutions listed at the end of the Table of Contents. Their support is greatly appreciated.

We also wish to acknowledge the generous support of the Institute of Experimental Endocrinology of the Slovak Academy of Sciences in Bratislava, led by Richard Kvetňanský, and of the president of the Slovak Academy of Sciences, Štefan Luby. Significant patronage was received for this symposium from the Slovak Diabetes Association and its chair, Juraj Vozár. Similarly, we wish to thank the Slovak Endocrine Society. Thanks are also extended to Alica Mitková for the extra hours given to her symposium-related tasks as the secretary of the Slovak Endocrine Society and as chief technician of our research group.

We are also indebted to our spouses, Zuzana Jezerská and Peter Zák, respectively, for their enormous support and understanding during our involvement in the organization of our symposium.

Finally, we express our sincere thanks to the New York Academy of Sciences, particularly Stefan Malmoli, for outstanding editorial expertise.

We hope that the Fourth International Smolenice Insulin Symposium, scheduled for the last summer or fall of the second millennium, will again be attended by so many outstanding colleagues from all over the world.

Progress in Population Analyses of the Insulin Resistance Syndrome

STEVEN M. HAFFNER

Department of Medicine
University of Texas Health Science Center at San Antonio
San Antonio, Texas 78284-7873

INTRODUCTION

Insulin resistance, that is, resistance to the insulin-mediated glucose disposal, has been long noted to be abnormal in non-insulin-dependent diabetes mellitus (NIDDM).[1] Attention over the last several years has also focused on possible effects of insulin resistance and/or hyperinsulinemia as a possible etiologic factor for other disorders related to cardiovascular disease. In 1988, Reaven suggested that insulin resistance may underlie a number of disorders including hypertension, dyslipidemia (especially high triglyceride and low high-density lipoprotein), impaired glucose tolerance, and coronary heart disease.[2]

Reaven used the term "Syndrome X" to describe these related disorders. We have proposed the name "Insulin Resistance Syndrome" (IRS) because it highlights the presumed etiology of the syndrome.[3] Although Reaven formulated his concept in terms of the harmful effects of insulin resistance ("low insulin-mediated glucose disposal"), many studies have used increased insulin concentrations as a surrogate for direct measurement of insulin resistance since the measurement of insulin resistance by the hyperinsulinemic euglycemic clamp, steady-state plasma glucose, or frequently sampled intravenous glucose tolerance test (FSIGT) is expensive and has limited patient acceptance. In nondiabetic subjects, increased insulin concentrations generally reflect insulin resistance.[4] In NIDDM subjects, fasting (but not postglucose insulin) may still be a reasonable surrogate for insulin resistance[4] since insulin secretion is decreased and thus insulin levels may be decreased following oral ingestion of glucose. Because of the expense and limited patient acceptance, insulin resistance has been assessed directly in only a few epidemiologic studies.[5-7] Since our review is emphasizing the role of population studies in assessing the insulin resistance, we will concentrate on studies measuring serum or plasma insulin. We also emphasize non-insulin-dependent diabetes, hypertension, and cardiovascular disease. Dyslipidemia will not be covered since this area has been well documented.[3,7-11] We will also not discuss proposed extensions to the IRS, including visceral fat,[12-16] plasminogen activator inhibitor-1 (PAI-1),[17-19] small dense LDL,[20,21] and decreased sex hormone binding globulin.[12,16,22,23]

1

PROSPECTIVE STUDIES OF INSULIN IN RELATION TO THE DEVELOPMENT OF CLUSTERING OF CARDIOVASCULAR RISK FACTORS

While a number of reports have confirmed that elevated insulin levels are associated "cross-sectionally" with increased triglyceride levels, decreased high-density lipoprotein, and hypertension,[8] few data are available on whether insulin concentrations predict the development of metabolic disorders. In the San Antonio Heart Study,[3] increased fasting insulin levels significantly predicted the development of NIDDM, low high-density lipoprotein (HDL), high triglyceride, and hypertension over an eight-year follow-up.

CLUSTERING OF CARDIOVASCULAR RISK FACTORS

In the San Antonio Heart Study,[3] subjects were more likely to develop multiple metabolic disorders than would be predicted from calculating the probability of developing only a single disorder, suggesting that these metabolic disorders do indeed cluster; subjects who developed multiple metabolic disorders had higher insulin concentrations than those who developed only a single disorder.[24-26] In another report, both "specific" or "true" insulin (which does not recognize proinsulin) and immunoreactive insulin (which does recognize proinsulin) were associated with certain cardiovascular risk factors (hypertension, low high-density lipoprotein cholesterol, high triglyceride, and glucose intolerance) and with clustering of these risk factors in both diabetic and nondiabetic subjects.[27] Thus, for the assessment of cardiovascular risk factors, specificity of the insulin assay may not be critical. (This is probably not true for prospective studies of NIDDM.)

PROINSULIN AND THE IRS

Most of the studies describing the association of hyperinsulinemia to hypertension have been limited by insulin assays that cross-react with proinsulin. Temple et al.[28] have suggested that proinsulin and split 32-33 proinsulin may comprise the majority of circulating immunoreactive insulin in subjects with NIDDM. Other studies have also suggested a disproportionate increase in proinsulin in subjects with NIDDM[29-31] and perhaps impaired glucose tolerance as well.[29,32] Proinsulin was more strongly associated with blood pressure than was insulin in diabetic subjects[33,34] and in nondiabetic subjects.[33,35] Fasting insulin (measured by a radioimmunoassay that does not recognize proinsulin), fasting proinsulin, and the fasting proinsulin/insulin ratio were all associated with hypertension in nondiabetic subjects.[33] The latter study suggests that the IRS may be associated not only with insulin resistance, but with a degree of β-cell dysfunction as well. However, because the amount of proinsulin is relatively low in subjects with normal glucose tolerance and even impaired glucose tolerance[29] (fasting proinsulin/true insulin: 0.07 and 0.11, respectively), the use of immunoreactive insulin assays to assess the insulin cardiovascular risk factor relationship is not likely to be a serious problem in nondiabetic subjects.

INSULIN CONCENTRATIONS, INSULIN RESISTANCE, AND HYPERTENSION

The relationship between insulin resistance and hypertension is the most-studied, but the most-controversial part of Syndrome X.[36] Michaela Modan initially suggested that insulin concentrations were associated with hypertension independent of glucose tolerance or obesity.[37] This concept was furthered by Ferrannini *et al.*, who showed a decrease in insulin sensitivity in normoglycemic subjects with hypertension relative to normotensive controls.[38] Multiple mechanisms have been suggested to explain the association of insulin resistance and blood pressure, including increased sympathetic nervous system activity, proliferation of vascular smooth muscle cells, altered cation transport, and increased sodium reabsorption.[39]

Cross-sectional studies have supported an association between insulin concentration and hypertension,[8,40-42] although other epidemiological studies have not found such a relationship.[43-45] Relatively few studies have examined the cross-sectional association between insulin resistance and hypertension.[38,46-51] These studies have been limited by selection by level of obesity,[38,47] diabetic status,[46] and race.[49,50]

Unfortunately, only in a few studies has the association between insulin concentrations and the incidence of hypertension been examined prospectively where the temporal relationship between insulin and hypertension can be elucidated. This is an important issue since Julius *et al.*[52] have suggested a hemodynamic basis for insulin resistance in hypertension and thus insulin resistance could develop subsequent to the occurrence of hypertensive changes. However, in four prospective studies insulin concentrations predicted the incidence of hypertension,[53-56] while in two studies insulin predicted the incidence of hypertension in lean subjects, but not in obese subjects.[57,58] Some of the issues that may obscure the relationship of insulin and hypertension include antihypertensive therapy, obesity, racial differences, and the use of insulin assays that cross-react with proinsulin.

Pharmacological therapy may affect insulin sensitivity. β-Blockers and thiazides worsen insulin resistance; α-blockers and angiotension-converting enzyme inhibitors may decrease insulin resistance.[59] For example, if an investigator included hypertensive subjects on thiazides, a spurious association between insulin and hypertension might be observed.

It is not surprising that obesity may be an important confounder of the hypertension/insulin resistance relationship since obesity both strongly predicts the development of hypertension[57,60] and is associated with hyperinsulinemia[37] and insulin resistance.[61] Some studies have found associations between insulin levels and blood pressure that disappeared after adjustment for obesity.[43,45] However, because obesity may be in the causal pathway, perhaps a case can be made for not adjusting for obesity in studies of insulin resistance and hypertension. More interesting is the possibility that the effect of insulin and blood pressure may differ by the level of obesity. Laakso et al.[46] have suggested that lean subjects with NIDDM and hypertension were more insulin-resistant than lean normotensive subjects, but obese hypertensive and nonhypertensive NIDDM subjects were equally insulin-resistant. Consistent with this outlook is a strong relationship between insulin resistance and hypertension in lean subjects,[38] but not in obese subjects.[47] However, in one report, a strong cross-sectional relationship between insulin resistance and hypertension was found in both lean and obese subjects.[48] In another study,[57] in lean subjects (BMI < 25 kg/m^2), the incidence of hypertension for those in the highest tertile of insulin concentrations compared with those in the lowest two tertiles was 10.0% versus 4.5%, respectively [relative risk (RR) = 2.24; p = 0.0321]. Conversely, in subjects with BMI between 25 and 30 kg/m^2, the incidence of hypertension was 15.0% versus 11.5%, respectively (RR = 1.32; p = NS). Finally, in obese subjects (BMI > 30 kg/m^2), the incidence of hypertension in corresponding categories was 12.2% versus 17.3%, respectively (RR = 0.70; p = NS). Similarly, Shetterly et al.[56] showed a stronger association between baseline insulin levels and incident hypertension in follow-up among lean subjects (as compared to their obese counterparts). Thus, there may exist a threshold of obesity above which increased insulin resistance may not affect the risk of hypertension.

Saad et al.[49] have reported an association between insulin resistance and blood pressure in Caucasians, but not in African-Americans or Pima Indians. In contrast, Falkner et al.[50] have reported that young lean hypertensive African-American males are more insulin-resistant than young lean normotensive African-American males. The discrepancy between these two studies might be explained by the greater obesity of African-Americans in the former study (BMI = 31 versus 24 kg/m^2, respectively). In a prospective study,[57] the relationship of fasting insulin to the incidence of hypertension was equally strong in Mexican Americans and non-Hispanic whites. However, in the San Luis Valley Diabetes Study, the association between baseline insulin concentrations and the incidence of hypertension was significant only in non-Hispanic whites.[56]

The area of insulin resistance in relation to the etiology of hypertension remains very controversial. We believe that the major confounding variable is the presence of obesity. If there is a strong relationship between insulin and blood pressure, it is most likely to be found in lean subjects. It has not been resolved whether the relation between insulin and blood pressure varies in different ethnic groups and whether the possible differences in ethnic groups in the association between blood pressure and insulin could be due to differences in adiposity among the ethnic groups.

DIABETES AND CARDIOVASCULAR DISEASE

Subjects with NIDDM have a twofold to fourfold increased risk of developing cardiovascular disease.[62-64] Unlike the situation with microvascular complications of diabetes (retinopathy and renal disease) where duration of diabetes and severity of glycemia have been strong and consistent risk factors,[65,66] cardiovascular disease has often not been associated with these traditional diabetic risk factors.[65,67-69] However, in two recent Finnish studies, duration of diabetes and glycosylated hemoglobin were statistically significant, although fairly weak, predictors of CHD in NIDDM subjects.[70,71] The relatively weak association between duration of diabetes and severity of glycemia and cardiovascular disease suggests that common antecedents might underlie both atherosclerotic heart disease and NIDDM.

PREDICTORS OF NIDDM: HYPERINSULINEMIA, INSULIN RESISTANCE, AND DECREASED INSULIN SECRETION

Previous prospective studies have consistently demonstrated that insulin resistance[72,73] and hyperinsulinemia[74-79] are strong predictors of NIDDM. Several studies have suggested that abnormal insulin secretion, as assessed by a low acute insulin response to intravenous glucose, a low increment of insulin to glucose over 30 minutes of an oral glucose tolerance test, or a low 2-h insulin postglucose load,[72,78-83] predicts the development of diabetes, especially in subjects with impaired glucose tolerance (IGT).

In some reports,[84,85] high proinsulin (but not specific insulin) predicted the development of NIDDM. These studies might call into question whether the older reports[74-79] that used immunoreactive insulin were correct. The newer reports dealt with Japanese Americans[85] (who are lean, but have increased visceral fat) and elderly Finnish subjects[84] and thus may not be typical of obese, high-risk populations such as Pima Indians or Mexican Americans. Moreover, these new studies only had a short follow-up (2.5-3.5 years) and thus it is likely that the majority of subjects who converted to NIDDM had impaired glucose tolerance at baseline and therefore already had pancreatic insufficiency. The use of an immunoreactive insulin assay would not invalidate these studies that showed that insulin resistance predicted NIDDM.[72,73] Additional studies of specific insulin and proinsulin need to be done particularly in obese, high-risk populations.

CARDIOVASCULAR RISK FACTORS IN PREDIABETIC SUBJECTS

As stated previously, Reaven has proposed that insulin resistance may underlie a cluster of risk factors including glucose tolerance, dyslipidemia, hypertension, and cardiovascular disease.[2] This is consistent with the finding of increased blood pressure, increased triglyceride, and decreased HDL cholesterol prior to the onset of marked hyperglycemia observed in a number of studies.[24,86-88] Increased cardiovascular risk factors were found even in subjects with normal glucose tolerance who later developed NIDDM.[88] Moreover, adjustment for fasting insulin concentrations abolished the difference between confirmed prediabetic subjects and those who remained normal at follow-up.[88]

HYPERINSULINEMIA, INSULIN RESISTANCE, AND CORONARY HEART DISEASE

Given that insulin resistance and hyperinsulinemia are strongly related to cardiovascular risk factors, it seems reasonable to believe that insulin resistance should be strongly related to cardiovascular disease. Surprisingly, there has been a marked controversy about this issue.[89]

Hyperinsulinemia has been identified as a risk factor for coronary heart disease (CHD) in several,[90-93] but not all, studies.[94-96] The studies of Pyörälä et al.[91] and Eschwege et al.[92] showed that insulin concentrations were significantly related to CHD in men. Welborn and Wearne,[90] who studied both men and women, found a significant relationship between insulin and cardiovascular disease (CHD) only in men. The recent study by Després et al.[93] provides the strongest evidence that hyperinsulinemia is associated prospectively with the development of CHD. These authors did a nested case control study in 91 male cases and 105 matched controls from the Quebec Cardiovascular Study. They reported more comprehensive lipid and lipoprotein, including HDL cholesterol and apoprotein B, than previous reports. (In earlier studies,[90-92] HDL cholesterol was not measured.) Furthermore, the investigators used an insulin assay that did not recognize the proinsulin. They found that adjustment for other risk factors did not diminish the predictive power of insulin for CHD and that both fasting insulin and apo B predicted CHD strongly.

The negative studies by Welin et al.[95] and Ferrara et al.[96] were done in elderly subjects and thus could have represented a survival bias in that subjects who might have had CHD due to increased insulin resistance might have already died. The Orchard et al.[94] study was done in high-risk subjects (Multiple Risk Factor Intervention Trial, MRFIT) and thus might have been enriched in subjects with insulin resistance. Furthermore, there was an interaction of apo E phenotype by insulin concentration in relation to CHD incidence in the MRFIT study.[94] At present, no study has prospectively shown the association between insulin concentrations and CHD in women. Furthermore, no prospective studies have examined the relationship of insulin resistance to coronary heart disease.

INSULIN RESISTANCE AND ATHEROSCLEROSIS

Only a few cross-sectional studies have found an association of insulin resistance (as determined by the hyperinsulinemic euglycemic clamp) with atherosclerosis as measured by carotid ultrasound[97] and coronary angiography.[98] These studies,[97,98] however, were small. The relationship between insulin resistance [by the frequently sampled intravenous glucose tolerance test (FSIGT)] and atherosclerosis (by β-mode ultrasound of the carotid) was recently reported in a large multiethnic population (398 blacks, 457 Hispanics, and 542 non-Hispanic whites).[99] In the Insulin Resistance Atherosclerosis Study (IRAS), there was a positive association between insulin resistance and the intimal medial thickness (IMT) of the carotid artery both in Hispanics and in non-Hispanic whites. This effect was reduced, but not totally explained by adjustment for traditional cardiovascular disease risk factors, glucose tolerance, measures of adiposity, and fasting insulin levels. However, there was no significant association between insulin resistance and IMT in blacks. Katz et al.[100] also reported that specific insulin and proinsulin are not related to atherosclerosis (as assessed by coronary angiography) in a predominantly black population.

SUMMARY

Insulin resistance is associated with a variety of cardiovascular risk factors including hypertension, dyslipidemia, and non-insulin-dependent diabetes. In blacks, the relation between insulin resistance, hypertension, and atherosclerosis has been questioned. Most data collected on the Insulin Resistance Syndrome have been collected in nondiabetic subjects; therefore, no inference can be drawn to exogenous insulin use in diabetic subjects where improved glycemic control is usually associated with improved cardiovascular risk factors (especially dyslipidemia) in the absence of weight gain.

REFERENCES

1. DEFRONZO, R. A. 1988. Lilly Lecture 1987—The triumvirate: β-cell, muscle, liver; a collusion responsible for NIDDM. Diabetes 37: 667–687.
2. REAVEN, G. M. 1988. Banting Lecture: Role of insulin resistance in human disease. Diabetes 37: 1595–1607.
3. HAFFNER, S. M., R. A. VALDEZ, H. P. HAZUDA, B. D. MITCHELL, P. A. MORALES & M. P. STERN. 1992. Prospective analyses of the insulin resistance syndrome (Syndrome X). Diabetes 41: 715–722.
4. LAAKSO, M. 1993. How good a marker is insulin level for insulin resistance? Am. J. Epidemiol. 137: 959–965.
5. WAGENKNECHT, L. E., E. J. MAYER, M. REWERS, S. M. HAFFNER, J. SELBY, G. M. BURKE, L. HENKIN, G. HOWARD, P. J. SAVAGE, M. F. SAAD, R. N. BERGMAN & R. HAMMAN. 1995. The Insulin Resistance Atherosclerosis Study (IRAS): objectives, design, and recruitment results. Ann. Epidemiol. 5: 464–471.
6. HAFFNER, S. M., R. D'AGOSTINO, M. F. SAAD, M. REWERS, L. MYKKÄNEN, J. SELBY, G. HOWARD, P. J. SAVAGE, R. F. HAMMAN, L. E. WAGENKNECHT & R. N. BERGMAN. 1996. Increased insulin resistance and insulin secretion in non-diabetic African-Americans and Hispanics compared to non-Hispanic whites: the Insulin Resistance Atherosclerosis Study. Diabetes 45: 742–748.
7. MYKKÄNEN, L., S. M. HAFFNER, T. RÖNNEMAA, R. BERGMAN & M. LAAKSO. 1994. Is there sex difference in the association of insulin levels and insulin sensitivity with lipids and lipoproteins? Metabolism 43: 523–528.
8. ZAVARONI, I., E. BONORA, M. PAGLIARA, E. DALL'AGLIO, L. LUCHETTI, G. BUONANNO, P. A. BONATI, M. BERGONZANI, L. GNUDI, M. PASSERI, & G. M. REAVEN. 1989. Risk factors for coronary artery disease in healthy persons with hyperinsulinemia and normal glucose tolerance. N. Engl. J. Med. 320: 702–706.
9. LAWS, A., A. C. KING, W. L. HASKELL & G. M. REAVEN. 1991. Relation of fasting plasma insulin concentrations to high density lipoprotein cholesterol and triglyceride concentrations in men. Arterioscler. Thromb. 11: 1636–1642.
10. GARG, A., J. H. HELDERMAN, M. KOFFLER, R. AYUSO, J. ROSENSTOCK & P. RASKIN. 1988. Relationship between lipoprotein levels and *in vivo* insulin action in normal young white men. Metabolism 37: 982–987.
11. MYKKÄNEN, L., J. KUUSISTO, S. M. HAFFNER, K. PYÖRÄLÄ & M. LAAKSO. 1994. High insulin levels predict multiple atherogenic changes in lipoproteins. Arteriosclerosis 14: 518–526.

12. KISSEBAH, A. H., N. VYDELINGUM, R. MURRAY, D. J. EVANS, A. J. HARTZ, R. K. KALKHOFF & P. W. ADAMS. 1982. Relation of body fat distribution to metabolic complications of obesity. J. Clin. Endocrinol. Metab. 54: 254-260.
13. OHLSON, L. O., B. LARSSON, K. SVARDSUDD, L. WELIN, H. ERIKSSON, L. WILHELMSEN, P. BJORNTORP & G. TIBBLIN. 1985. The influence of body fat distribution on the incidence of diabetes mellitus: 13.5 years of follow-up of the participants in the study of men born in 1913. Diabetes 34: 1055-1058.
14. DESPRÉS, J. P., S. MOORJANI, P. J. LUPIEN, A. TREMBLAY, A. NADEQU & C. BOUCHARD. 1990. Regional distribution of body fat, plasma lipoproteins, and cardiovascular disease. Arteriosclerosis 10: 497-511.
15. DESPRÉS, J., D. PRUD'HOMME, M. C. POULIOT, A. TREMBLAY & C. BOUCHARD. 1991. Estimation of deep abdominal adipose-tissue accumulation from simple anthropometric measurements in men. AJCN 54: 471-477.
16. HAFFNER, S. M., K. KARHAPÄÄ, L. MYKKÄNEN & M. LAAKSO. 1994. Insulin resistance, body fat distribution, and sex hormones in men. Diabetes 43: 212-219.
17. VAGUE, P., I. JUHAN-VAGUE, M. F. AILLAUD, C. BADIER, R. VIARD, M. C. ALESSI & D. COLLEN. 1986. Correlation between blood fibrinolytic activity, plasminogen activator inhibitor level, plasma insulin level, and relative body weight in normal and obese subjects. Metabolism 35: 250-253.
18. MYKKÄNEN, L., T. RÖNNEMAA, J. MARNIEMI, S. M. HAFFNER, R. BERGMAN & M. LAAKSO. 1994. Insulin sensitivity is not an independent determinant of plasma plasminogen activator inhibitor-1 activity. Arterioscler. Thromb. 14: 1264-1271.
19. JUHAN-VAGUE, I., M. C. ALESSI & P. VAGUE. 1991. Increased plasma plasminogen activator inhibitor-1 levels: a possible link between insulin resistance and atherothrombosis. Diabetologia 34: 457-462.
20. REAVEN, G. M., Y. D. I. CHEN, J. JEPPESEN, P. MAHEUX & R. M. KRAUSS. 1993. Insulin resistance and hyperinsulinemia in individuals with small dense, low density lipoprotein particles. J. Clin. Invest. 92: 141-146.
21. HAFFNER, S. M., L. MYKKÄNEN, M. PAIDI, R. VALDEZ, B. V. HOWARD & M. P. STERN. 1995. Small dense LDL is associated with the insulin resistance syndrome. Diabetologia 38: 1328-1336.
22. HAFFNER, S. M., M. S. KATZ, M. P. STERN, et al. 1989. Association of decreased sex hormone binding globulin and cardiovascular risk factors. Arteriosclerosis 9: 136-143.
23. NESTLER, J. E. 1993. Sex hormone binding globulin: a marker for hyperinsulinemia and/or insulin resistance? J. Clin. Endocrinol. Metab. 76: 273-274.
24. MYKKÄNEN, L., J. KUUSISTO, K. PYÖRÄLÄ & M. LAAKSO. 1993. Cardiovascular disease risk factors as predictors of type II (non-insulin-dependent) diabetes mellitus in elderly subjects. Diabetologia 36: 553-559.
25. DUNCAN, B., M. SCHMIDT, A. R. SHARRETT, R. WATSON, F. BRANCATI & G. HEISS. 1994. Association of fasting insulin with clustering of metabolic abnormalities in African-Americans and whites. Diabetes 43(suppl. 1): 150A.
26. MYKKÄNEN, L., T. RÖNNEMAA & M. LAAKSO. 1994. Low insulin sensitivity is associated with a clustering of cardiovascular risk factors. Diabetes 43(suppl.): 150A.
27. HAFFNER, S. M., L. MYKKÄNEN, R. VALDEZ & M. P. STERN. 1994. An evaluation of two insulin assays in the insulin resistance syndrome. Arterioscler. Thromb. 14: 1430-1437.
28. TEMPLE, R. C., C. A. CARRINGTON, S. D. LUZIO, D. R. OWENS, A. E. SCHNEIDER, W. J. SOBBEY & C. N. HALES. 1989. Insulin deficiency of non-insulin-dependent diabetes. Lancet 1: 293-295.
29. HAFFNER, S. M., R. R. BOWSHER, L. MYKKÄNEN, H. P. HAZUDA, B. D. MITCHELL, R. A. VALDEZ, R. GINGERICH, A. MONTEROSSA & M. P. STERN. 1994. Proinsulin and specific insulin concentrations in high and low risk populations for non-insulin-dependent diabetes mellitus. Diabetes 43: 1490-1493.
30. WARD, W. K., E. C. LACAVA, T. L. PAQUETTE, J. C. BEARD, B. J. WALLUM & D. PORTE. 1987. Disproportionate elevation of immunoreactive proinsulin in type II (non-insulin-dependent) diabetes mellitus and in experimental insulin resistance. Diabetologia 30: 698-702.
31. SAAD, M. F., S. E. KAHN, R. G. NELSON, D. J. PETTITT, W. C. KNOWLER, M. W. SCHWARTZ, S. KOWALYK, P. H. BENNETT & D. PORTE, JR. 1990. Disproportionately elevated proinsulin

in Pima Indians with non-insulin-dependent diabetes mellitus. J. Clin. Endocrinol. Metab. **70:** 1247–1253.
32. DAVIES, M., G. RAYMAN, I. P. GRAY, J. L. DAY & C. N. HALES. 1993. Insulin deficiency and increased plasma concentrations of intact and 32/33 split proinsulin in subjects with impaired glucose tolerance. Diabetes Med. **10:** 313–320.
33. HAFFNER, S. M., L. MYKKÄNEN, R. A. VALDEZ, M. P. STERN, D. L. HOLLOWAY, A. MONTEROSSA & R. R. BOWSHER. 1994. Disproportionately increased proinsulin levels are associated with the insulin resistance syndrome. J. Clin. Endocrinol. Metab. **79:** 1806–1810.
34. NAGI, D. K., T. J. HENDRA, A. J. RYLE, T. M. COOPER, R. C. TEMPLE, P. M. S. CLARK, A. E. SCHNEIDER, C. N. HALES & J. S. YUDKIN. 1990. The relationship of concentrations of insulin, intact proinsulin, and 32-33 split proinsulin with cardiovascular risk factors in type II (non-insulin-dependent) diabetic subjects. Diabetologia **33:** 532–537.
35. HAFFNER, S. M., L. MYKKÄNEN, M. P. STERN, R. A. VALDEZ, J. A. HEISSERMAN & R. R. BOWSHER. 1993. Relationship of proinsulin and insulin to cardiovascular risk factors in nondiabetic subjects. Diabetes **42:** 1297–1302.
36. HAFFNER, S. M. 1993. Editorial: insulin and blood pressure–fact or fantasy? J. Clin. Endocrinol. Metab. **76:** 541–543.
37. MODAN, M., H. HALKIN, S. ALMOG, A. LUSKY, A. ESHKOL, M. SHEFI, A. SHITRIT & Z. FUCHS. 1985. Hyperinsulinemia: a link between hypertension, obesity, and glucose intolerance. J. Clin. Invest. **75:** 809–817.
38. FERRANNINI, E., G. BUZZIGOLI, R. BONADONNA, M. A. GIORICO, M. OLEGGINI, L. GRAZIADEI, R. PEDRINELLI, L. BRANDI & S. BEVILACQUA. 1987. Insulin resistance in essential hypertension. N. Engl. J. Med. **317:** 350–357.
39. DEFRONZO, R. A. & E. FERRANNINI. 1991. Insulin resistance: a multifaceted syndrome responsible for NIDDM, obesity, hypertension, dyslipidemia, and atherosclerotic cardiovascular disease. Diabetes Care **14:** 173–194.
40. FERRANNINI, E., S. M. HAFFNER, M. P. STERN, B. D. MITCHELL, A. NATALI, H. P. HAZUDA & J. K. PATTERSON. 1991. High blood pressure and insulin resistance: influence of ethnic background. Eur. J. Clin. Invest. **21:** 280–287.
41. WING, R. R., C. H. BUNKER, L. H. KULLER & K. A. MATTHEWS. 1989. Insulin, body mass index, and cardiovascular risk factors in premenopausal women. Arteriosclerosis **9:** 479–484.
42. EVERY, N. R., E. J. BOYKO, E. M. KEANE, J. A. MARSHALL, M. REWERS & R. F. HAMMAN. 1993. Blood pressure, insulin, and C-peptide in San Luis Valley, Colorado. Diabetes Care **16:** 1543–1550.
43. CAMBIEN, F., J. WARNET, E. ESCHWEGE, A. JACQUESON, J. L. RICHARD & G. ROSSELIN. 1987. Body mass, blood pressure, glucose, and lipids: does plasma insulin explain their relationships? Arteriosclerosis **7:** 197–202.
44. DOWSE, G. K., V. R. COLLINS, K. G. M. M. ALBERTI, P. Z. ZIMMET, J. TUOMILEHTO, P. CHITSON & H. GAREEBOO. 1993. Insulin and blood pressure levels are not independently related in Mauritians of Asian Indian, Creole, or Chinese origin. J. Hypertens. **11:** 297–307.
45. MULLER, D. C., D. ELAHI, R. E. PRATLEY, J. D. TOBIN & R. ANDRES. 1993. An epidemiological test of the hyperinsulinemia-hypertension hypothesis. J. Clin. Endocrinol. Metab. **76:** 544–548.
46. LAAKSO, M., H. SARLUND & L. MYKKÄNEN. 1989. Essential hypertension and insulin resistance in non-insulin-dependent diabetes. Eur. J. Clin. Invest. **19:** 518–526.
47. BONORA, E., P. MOGHETTI, M. ZENERE, F. TOSI, D. TRAVIA & M. MUGGEO. 1990. β-Cell secretion and insulin sensitivity in hypertensive and normotensive obese subjects. Int. J. Obes. **14:** 735–742.
48. POLLARE, T., H. LITHELL & C. BERNE. 1990. Insulin resistance is a characteristic feature of primary hypertension, independent of obesity. Metabolism **39:** 167–174.
49. SAAD, M. F., S. LILLIOJA, B. L. NYOMBA, C. CASTILLO, R. FERRARO, M. DE GREGORIO, E. RAVUSSIN, W. C. KNOWLER, P. H. BENNETT, B. V. HOWARD & C. BOGARDUS. 1991. Racial differences in the relation between blood pressure and insulin resistance. N. Engl. J. Med. **324:** 733–739.
50. FALKNER, B., S. HULMAN, J. TANNENBAUM & H. KUSHNER. 1990. Insulin resistance and blood pressure in young black men. Hypertension **16:** 706–711.

51. MYKKÄNEN, L., S. M. HAFFNER, T. RÖNNEMAA, R. WATANABE & M. LAAKSO. 1996. Relationship of plasma insulin concentration and insulin sensitivity to blood pressure: is it modified by obesity? J. Hypertens. **14:** 399-405.
52. JULIUS, S., T. GUDBRANDSSON, K. JAMERSON, S. T. SHAHAB & O. ANDERSSON. 1991. The hemodynamic link between insulin resistance and hypertension. J. Hypertens. **9:** 983-986.
53. NISKANEN, L. K., M. I. UUSITUPA & K. PYÖRÄLÄ. 1991. The relationship of hyperinsulinemia to the development of hypertension in type 2 diabetic patients and in non-diabetic subjects. J. Hum. Hypertens. **5:** 155-159.
54. SKARFORS, E. T., H. O. LITHELL & I. SELINUS. 1991. Risk factors for the development of hypertension: a 10-year longitudinal study in middle-aged men. J. Hypertens. **9:** 217-223.
55. LISSNER, L., C. BENGTSSON, L. LAPIDUS, K. KRISTJANSSON & H. WEDEL. 1992. Fasting insulin in relation to subsequent blood pressure changes and hypertension in women. Hypertension **20:** 797-801.
56. SHETTERLY, S. M., M. REWERS, R. HAMMAN & J. A. MARSHALL. 1994. Patterns and predictors of hypertension incidence among Hispanics and non-Hispanic whites: the San Luis Valley Diabetes Study. J. Hypertens. **12:** 1095-1102.
57. HAFFNER, S. M., E. FERRANNINI, H. P. HAZUDA & M. P. STERN. 1992. Clustering of cardiovascular risk factors in confirmed prehypertensive individuals. Hypertension **20:** 38-45.
58. MYKKÄNEN, L., J. KUUSISTO, S. M. HAFFNER, K. PYÖRÄLÄ & M. LAAKSO. 1994. Fasting plasma insulin in relation to the development of hypertension in the elderly subjects. Circulation **90**(suppl. 2): 665.
59. LITHELL, H. O. L. 1991. Effect of antihypertensive drugs on insulin, glucose, and lipid metabolism. Diabetes Care **14:** 203-209.
60. DYER, A. R., J. STAMLER, R. B. SHEKELLE, J. A. SCHOENBERGER, R. STAMLER, S. SHEKELLE, D. M. BERKSON, O. PAUL, M. K. LEPPER & H. A. LINDBERG. 1982. Relative weight and blood pressure in four Chicago epidemiologic studies. J. Chronic Dis. **35:** 897-908.
61. OLEFSKY, J. M., O. G. KOLTERMAN & J. A. SCARLETT. 1982. Insulin action and resistance in obesity and non-insulin-dependent type II diabetes mellitus. Am. J. Physiol. **243:** E15-E30.
62. GARCIA, M. J., P. M. MCNAMARA, T. GORDON & W. B. KANNELL. 1974. Morbidity and mortality in diabetics in the Framingham population: sixteen-year follow-up. Diabetes **23:** 105-111.
63. STAMLER, J., O. VACCARO, J. D. NEATON & D. WENTWORTH. 1993. Multiple Risk Factor Intervention Trial Research Group: diabetes, other risk factors, and 12-year cardiovascular mortality for men screened in the Multiple Risk Factor Intervention Trial. Diabetes Care **16:** 434-444.
64. ASSMANN, G. & H. SCHULTE. 1988. The Prospective Cardiovascular Munster (PROCAM) Study: prevalence of hyperlipidemia in persons with hypertension and/or diabetes mellitus and the relationship to coronary heart disease. Am. Heart J. **116:** 1713-1724.
65. DIABETES DRAFTING GROUP. 1985. Prevalence of small vessel and large vessel disease in diabetic patients from 14 centres: The World Health Organization Multinational Study of vascular disease in diabetics. Diabetologia **28:** 615-640.
66. KLEIN, R., B. E. K. KLEIN, S. E. MOSS, M. D. DAVIS & D. L. DE METS. 1988. Glycosylated hemoglobin predicts the incidence and progression of diabetic retinopathy. JAMA **260:** 2864-2871.
67. FULLER, J. H., M. J. SHIPLEY, G. ROSE, R. J. JARRETT & H. KEEN. 1980. Coronary heart disease risk and impaired glucose tolerance: The Whitehall Study. Lancet **1:** 1373-1376.
68. JARRETT, R. J. 1984. Type II (non-insulin-dependent) diabetes mellitus and coronary heart disease—chicken, egg, or neither? Diabetologia **26:** 99-102.
69. HERMAN, J. B., J. H. MEDALIE & U. GOLDBOURT. 1977. Differences in cardiovascular morbidity and mortality between previously known and newly diagnosed adult diabetics. Diabetologia **13:** 229-234.
70. LAAKSO, M., S. LEHTO, I. PENTTILÄ & K. PYÖRÄLÄ. 1993. Lipids and lipoproteins predicting coronary heart disease mortality and morbidity in patients with non-insulin-dependent diabetes. Circulation **88:** 1421-1430.
71. KUUSISTO, J., L. MYKKÄNEN, K. PYÖRÄLÄ & M. LAAKSO. 1994. NIDDM and its metabolic control predict coronary heart disease in the elderly subjects. Diabetes **43:** 960-967.
72. LILLIOJA, S., D. M. MOTT, M. SPRAUL, R. FERRARO, J. E. FOLEY, E. RAVUSSIN, W. C. KNOWLER, P. H. BENNETT & C. BOGARDUS. 1993. Insulin resistance and insulin secretory

dysfunction as precursors of non-insulin-dependent diabetes mellitus: prospective studies of Pima Indians. N. Engl. J. Med. 329: 1988-1992.
73. WARRAM, J. H., B. C. MARTIN, A. S. KROLEWSKI, J. S. SOELDNER & C. R. KAHN. 1990. Slow glucose removal rate and hyperinsulinemia precede the development of type II diabetes in the offspring of diabetic parents. Ann. Intern. Med. 113: 903-915.
74. SICREE, R. A., P. Z. ZIMMET, H. O. M. KING & J. S. COVENTRY. 1987. Plasma insulin response among Nauruans: prediction of deterioration in glucose tolerance over 6 years. Diabetes 36: 179-186.
75. HAFFNER, S. M., M. P. STERN, B. D. MITCHELL, H. P. HAZUDA & J. K. PATTERSON. 1990. Incidence of type II diabetes in Mexican Americans predicted by fasting insulin and glucose levels, obesity, and body fat distribution. Diabetes 39: 283-288.
76. FUJIMOTO, W. Y. 1990. Association of elevated fasting C-peptide level and increased intraabdominal fat distribution with development of NIDDM in Japanese American men. Diabetes 39: 104-111.
77. CHARLES, M. A., A. FONTBONNE, N. THIBULT, J. M. WARNET, G. E. ROSSELIN & E. ESCHWEGE. 1991. Risk factors for NIDDM in the white population: Paris Prospective Study. Diabetes 40: 796-799.
78. SAAD, M. F., W. C. KNOWLER, D. J. PETTITT, R. G. NELSON, D. M. MOTT & P. H. BENNETT. 1988. The natural history of impaired glucose tolerance in the Pima Indians. N. Engl. J. Med. 319: 1500-1506.
79. LUNDGREN, H., C. BENGTSSON, G. BLOHME, L. LAPIDUS & J. WALDENSTRÖM. 1990. Fasting serum insulin concentration and early insulin response as risk determinants for developing diabetes. Diabetes Med. 7: 407-413.
80. EFENDIC, S., R. LUFT & A. WAJNGOT. 1984. Aspects of the pathogenesis of type 2 diabetes. Endocr. Rev. 5: 395-410.
81. KOSAKA, K., R. HAGURA & T. KUZUYA. 1977. Insulin responses in equivocal and definite diabetes with special reference to subjects who had mild glucose intolerance but later developed definite diabetes. Diabetes 26: 944-952.
82. KADOWAKI, T., Y. MIYAKE, R. HAGURA, Y. AKANUMA, H. KAJINUMA, T. KUZUYA, F. TAKAKU & K. KOSAKA. 1984. Risk factors for worsening to diabetes in subjects with impaired glucose tolerance. Diabetologia 26: 44-49.
83. HAFFNER, S. M., H. MIETTINEN, S. P. GASKILL & M. P. STERN. 1995. Decreased insulin secretion and increased insulin resistance are independently related to the 7-year risk of non-insulin-dependent diabetes mellitus. Diabetes 44: 1386-1391.
84. MYKKÄNEN, L., S. M. HAFFNER, C. N. HALES, J. KUUSISTO & M. LAAKSO. 1995. Serum proinsulin levels are disproportionately elevated in elderly prediabetic subjects. Diabetologia 38: 1176-1182.
85. KAHN, S. E., D. L. LEONETTI, R. L. PRIGEON, E. J. BOYKO, R. W. BERGSTROM & W. Y. FUJIMOTO. 1995. Proinsulin as a marker for the development of NIDDM in Japanese-American men. Diabetes 44: 173-179.
86. MCPHILLIPS, J. B., E. BARRETT-CONNOR & D. L. WINGARD. 1990. Cardiovascular disease risk factors prior to the diagnosis of impaired glucose tolerance and non-insulin-dependent diabetes mellitus in a community of older adults. Am. J. Epidemiol. 131: 443-453.
87. MEDALIE, J. H., C. M. PAPIER, U. GOLDBOURT & J. B. HERMAN. 1975. Major factors in the development of diabetes mellitus in 10,000 men. Arch. Intern. Med. 135: 811-817.
88. HAFFNER, S. M., M. P. STERN, H. P. HAZUDA, B. D. MITCHELL & J. K. PATTERSON. 1990. Cardiovascular risk factors in confirmed prediabetic individuals: does the clock for coronary heart disease start ticking before the onset of clinical diabetes? JAMA 263: 2893-2898.
89. STERN, M. P. 1996. Do non-insulin-dependent diabetes mellitus and cardiovascular disease share common antecedents? Ann. Intern. Med. 124: 110-116.
90. WELBORN, T. A. & K. WEARNE. 1979. Coronary heart disease incidence and cardiovascular mortality in Busselton with reference to glucose and insulin concentrations. Diabetes Care 2: 154-160.
91. PYÖRÄLÄ, K., E. SAVOLAINEN, S. KAUKOLA & J. HAAPAKOSKI. 1985. Plasma insulin as coronary heart disease risk factor: relationship to other risk factors and predictive during 9½

year follow-up of the Helsinki Policeman Study population. Acta Med. Scand. 701(suppl.): 38–52.
92. ESCHWEGE, E., J. L. RICHARD, N. THIBULT, P. DUCIMETIERE, J. M. WARNET, J. R. CLAUDE & G. E. ROSSELIN. 1985. Coronary heart disease mortality in relation with diabetes, blood glucose, and plasma insulin levels: the Paris Prospective Study 10 years later. Horm. Metab. Res. 15(suppl.): 41–46.
93. DESPRÉS, J. P., B. LAMARCHE, P. MAURIÉGE, B. CANTIN, G. R. DAGENAIS, S. MOORJANI & P. J. LUPIEN. 1996. Hyperinsulinemia as an independent risk factor for ischemic heart disease. N. Engl. J. Med. **334**: 952–957.
94. ORCHARD, T. J., J. EICHNER, L. H. KULLER, D. J. BECKER, L. M. McCALLUM & G. A. GRANDITS. 1994. Insulin as a predictor of coronary heart disease—interaction with apolipoprotein E phenotype: a report from the Multiple Risk Factor Intervention Trial. Ann. Epidemiol. **4**: 40–45.
95. WELIN, L., H. ERIKSSON, B. LARSSON, L. O. OHLSON, K. SVARDSUDD & G. TIBBLIN. 1992. Hyperinsulinemia is not a major coronary risk factor in elderly men: the study of men born in 1913. Diabetologia **35**: 766–770.
96. FERRARA, A., E. L. BARRETT-CONNOR & S. L. EDELSTEIN. 1994. Hyperinsulinemia does not increase the risk of fatal cardiovascular disease in elderly men or women without diabetes: the Rancho Bernardo Study, 1984–1991. Am. J. Epidemiol. **140**: 857–869.
97. LAAKSO, M., H. SARLUND, R. SALONEN, M. SUHONEN, K. PYÖRÄLÄ, J. T. SALONEN & P. KARHAPÄÄ. 1991. Asymptomatic atherosclerosis and insulin resistance. Arterioscler. Thromb. **11**: 1068–1076.
98. BRESSLER, P., S. BAILEY, M. SAAD & R. A. DEFRONZO. 1992. Insulin resistance and coronary artery disease (CAD): the missing link. Diabetes 41(suppl. 1): 24A.
99. HOWARD, G., D. H. O'LEARY, D. ZACCARO, S. M. HAFFNER, M. REWERS, R. HAMMAN, J. V. SELBY, M. F. SAAD, P. SAVAGE & R. BERGMAN. 1996. Insulin sensitivity and atherosclerosis. Circulation **93**: 1809–1817.
100. KATZ, R. J., R. E. RATNER, R. M. COHEN, E. EISENHOWER & D. VERME. 1996. Are insulin and proinsulin independent risk markers for premature coronary artery disease? Diabetes **45**: 736–741.

Hypertension and the Insulin Resistance Syndrome of Rats

Are They Related?[a]

IWAR KLIMEŠ AND ELENA ŠEBÖKOVÁ

Diabetes and Nutrition Research Group
Institute of Experimental Endocrinology
Slovak Academy of Sciences
SK-83306 Bratislava, Slovak Republic

In the mid-sixties, Welborn *et al.*[1] described hyperinsulinemia in normoglycemic patients with essential hypertension and suggested that these subjects could be insulin-resistant. This observation passed unnoticed for many years until the early eighties when several groups started to promote the idea of a possible central role of insulin resistance (IR) and/or hyperinsulinemia in the pathogenesis of hypertension (for review, see reference 2). In 1988, Reaven and colleagues advanced the concept of the insulin resistance syndrome and a possible causal relationship between metabolic disturbances (such as hypertriglyceridemia, hyperinsulinemia, insulin resistance, glucose intolerance, and obesity) and hypertension.[3] This stimulated further interest in this field and led to numerous publications examining this theory in humans and in experimental animals.

A number of reviews emphasizing various aspects of the association of insulin resistance with hypertension have covered the issue until now.[2,4-16] Nevertheless, a number of issues (including the impact of existing genetic background) remain open, and a clear demarcation between rat and human studies has not always been made. The latter is particularly important in light of the different pressor action of insulin observed in rats as compared to humans and dogs.[14]

Therefore, the goal of this article is (a) to review the results obtained from rat models of hypertension and/or of the insulin resistance syndrome with a special emphasis on the pathogenesis of hypertension and (b) to examine the insulin resistance-hyperinsulinemia-hypertension hypothesis on the basis of data obtained in the hereditary hypertriglyceridemic, insulin-resistant, and hypertensive rat.[13,17-19]

---

[a] This work was supported by research grants of the Slovak Grant Agency for Science of the Slovak Republic (Nos. GAV 425/1991-1992 and 2/544/1993-1996) and by a research grant of the Slovak Grant Agency for Technic, Slovak Republic (No. GAT PEKO 931 004) within the frame of EURHYPGEN Concerted Action of the European Community [Contract No. ERBBMH1CT920869 (Bratislava, Slovak Republic)].

INSULIN RESISTANCE IN RAT MODELS OF HYPERTENSION

The most frequently used genetic model of human hypertension, the spontaneously hypertensive rat (SHR), has been intensely examined for the presence of IR and related abnormalities. It has been shown that SHR have fasting hyperglycemia, impaired oral glucose tolerance, hyperinsulinemia, and hypertriglyceridemia in comparison with the normotensive Wistar-Kyoto (WKY) rats.[20-22] Moreover, Reaven's group repeatedly described *in vitro* resistance to insulin-stimulated glucose uptake in adipocytes isolated from SHR,[22,23] and this was confirmed by others as well.[24] On the other hand, Bader *et al.*,[25] who measured insulin receptor tyrosine kinase activity and glucose transporter (GLUT4) number in skeletal muscle of SHR, did not find any difference in comparison to Lewis or Wistar rats.

Mondon and Reaven[21] and Hulman *et al.*[26] established that there is *in vivo* IR in SHR using the islet suppression test and the euglycemic hyperinsulinemic clamp (EHC), respectively. On the contrary, other groups found enhanced glucose disposal in SHR using the intravenous glucose tolerance test and the EHC.[27-29] The higher *in vivo* insulin action in the SHR was mainly due to an increase in nonoxidative disposal and glycogen synthase activity.[30]

As pointed out recently by Meehan *et al.*[2] in their critical review, a plausible explanation for this discrepancy could be that the former experiments were performed in stressed anesthetized or restrained rats, whereas the animals were unstressed, conscious, and unrestrained in the latter studies. SHR show a more enhanced response of the sympathetic nervous system (SNS) to stress than WKY,[31] which could lead to a greater impairment of glucose metabolism in the former strain when they are anesthetized or restrained. We have made a similar observation with hereditary hypertriglyceridemic (hHTg), insulin-resistant, and hypertensive rats.[32]

Thus, SHR may not be a true model for a genetic or a causal association between IR and hypertension. Nevertheless, Buchanan *et al.*[28] demonstrated an exaggerated insulin response to glucose in SHR and suggested that nutrient-stimulated hyperinsulinemia could contribute to the pathogenesis of hypertension in these rats.

Other genetic models of hypertension have also been studied. Fasting hypertriglyceridemia[33] as well as fasting and postglucose load hyperinsulinemia that were suggestive of IR[33,34] have been described in Dahl salt-sensitive (DS) rats—the most extensively studied genetic model of salt-sensitive hypertension.[35] It was also reported that maximal insulin-stimulated glucose transport was significantly lower in adipocytes isolated from DS rats than in those from Sprague-Dawley rats.[33] It is of interest that these metabolic changes in Dahl rats are not dependent on salt intake. Some recent experiments, based on EHC studies in conscious DS rats, indicate that chronic treatment with pioglitazone (an agent that increases insulin sensitivity) increased the glucose clearance rate and attenuated the development of hypertension in these rats.[36]

The Milan hypertensive strain (MHS) of rats develops hypertension because of a defect in renal sodium reabsorption. Thus, these rats can be a very useful animal model for studying some types of renal hypertensive mechanisms.[37] Abnormalities of insulin and lipid metabolism were also studied in these rats. In particular, fasting plasma insulin, plasma triglycerides (TG), as well as total cholesterol (Ch) levels were described to be higher in the MHS than in the Milan normotensive (MNS) rats.[38] No differences in *in vivo* insulin action were found, however, between MHS and MNS using the EHC technique.[29]

Plasma TG and insulin levels, as well as the insulin/glucose ratio, are higher in Lyon hypertensive than in Lyon normotensive rats, and these parameters cosegregated with blood pressure (BP) in F_2 hybrids.[39]

HYPERTENSION IN RAT MODELS OF INSULIN RESISTANCE

The obese Zucker rat (OZR) is a model of obesity, IR, and hyperinsulinemia. IR is present both *in vitro* and *in vivo*. EHC studies showed that sensitivity to insulin in adult rats with pronounced obesity is decreased in all insulin target tissues.[40] An increase in protein-tyrosine phosphatase (PTPase) enzyme activity and protein in the muscle of the OZR has been recently proposed to be a major cause of IR in these rats.[41] A special characteristic of OZR is their hyperlipidemia, which involves all lipoprotein classes, especially VLDL. Finally, muscle lipoprotein lipase activity is decreased, resembling the IR of the Zucker rats.[40] The presence of hypertension in the OZR and its relation to hyperinsulinemia and IR has been recently reviewed by Meehan *et al.*[2] While some investigators[42-45] reported no difference in blood pressure between the OZR and the normal lean Zucker rat, others[46-50] found the OZR hypertensive. However, the studies differed in a number of variables (e.g., techniques for BP measurement, anesthesia, sex, and age of the rats). Zemel *et al.*[51] found elevated BP and raised vascular reactivity in young OZR even after ganglionic blockade, and they propose that a lower vasodilator action of insulin on vascular smooth muscles may contribute to hypertension in the OZR. Hypertension in OZR seems to be potentiated by high salt intake,[52] but unaffected by salt restriction,[50] and is associated with abnormal calcium metabolism.[53] It is evident that increased vascular reactivity to various pressor agents[54] is due to impaired recovery of vascular smooth muscle intracellular calcium following agonist stimulation in OZR.[55] Insulin attenuates vasoconstrictor response of vascular smooth muscle to phenylephrine[56] as well as vasopressin-induced calcium response in arterial smooth muscle cells isolated from OZR.[57] Conversely, some investigators think that raised BP in the OZR is related neither to insulinemia nor to IR. One proposal is that high BP occurs only in old rats secondary to renal injury.[58]

In this context, it is interesting that another obese and hypertensive rat model, the spontaneously hypertensive/NIH-corpulent rat, is also significantly hyperinsulinemic and insulin-resistant, and yet has lower blood pressure than its lean littermates.[40,59,60] On the other hand, obese hyperinsulinemic and hyperlipidemic Dahl salt-sensitive rats have higher BP than their lean littermates[61] and the sensitivity of BP to salt was also higher in obese versus lean animals.

The SHHF/Mcc-*fa*^{cp} (abbreviated SHHF) rat is a relatively new genetic model of non-insulin-dependent diabetes mellitus (NIDDM), obesity, hypertension, and heart failure.[62] It has severe cardiovascular lesions in both lean and obese rats and is the only colony to develop spontaneous congestive heart failure (CHF). The obese males have overt NIDDM, whereas the obese females have normal fasting blood glucose, but an abnormal glucose tolerance. Obese female SHHF rats have IR and, like many obese hypertensive women, develop CHF in middle to late middle age.

Of the number of rat strains that spontaneously develop glucose intolerance and IR, most either are of the IDDM type or are obese with NIDDM. Few rat models exist for the study of nonobese NIDDM-like syndromes. The nonobese BHE/cdb rat is characterized by impaired glucose tolerance, fasting hypertriglyceridemia, and abnormalities in hepatic metabolism, plus numerous changes in whole-body glucose turnover, diabetic nephropathy, and hypertension.[63]

Another nonobese model is the hereditary hypertriglyceridemic rat (hHTg). These rats were originally developed as a genetic model of hypertriglyceridemia.[17] The strain has been produced by selective inbreeding from Wistar rats according to the rise of plasma TG induced by high-sucrose diet. Although hHTg rats display hypertriglyceridemia, impaired glucose tolerance, IR, and increased BP even without nutritional stimuli,[18] high-sucrose feeding aggravates these symptoms.[13] Systolic, diastolic, and mean arterial BP are elevated by about 20–40 mmHg above the values seen in control rats.[18,32] It is notable that BP of the hHTg rats correlates positively with their plasma TG levels.[18,64] Moreover, plasma TG levels correlate negatively with insulin action *in vivo*.[65] The latter inverse association has been recently described by a number of investigators in other experimental situations as well (for review, see reference 66). Thus, all the aforementioned findings imply a certain association of IR and hypertension in the hHTg rats.

Very recently, Klöting's group from Karlsburg, Germany, presented for the first time a further potential nonobese animal model of the IR syndrome—the WOK.1W rats.[67] These animals were selected from an outbred Wistar strain, and homozygous RT1a and RT1u (diabetes susceptible haplotype) were inbred separately and selected for "healthy fertility," named WOK.1A (RT1a) and WOK.1W (RT1u). After more than 35 inbred generations, fertility and survival of pups dropped drastically in the WOK.1W line. During studies aimed at elucidation of the aforementioned problems, nonfasting hypertriglyceridemia, hyperinsulinemia, altered glucose tolerance, and moderate hypertension were identified in these rats. Interestingly, male rats had more serious derangements of their metabolic parameters and of BP as well.

We have examined other inbred rat strains with either hypertriglyceridemia and IR or hypertension.[68] The Choco rats were hypertensive without any other metabolic abnormality typical for the IR syndrome. Conversely, PD rats (strain of Wistar origin carrying polydactyly-luxate syndrome) displayed hypertriglyceridemia, impaired glucose tolerance, hyperinsulinemia, and IR, but no hypertension. Thus, these data provide evidence against a causal relation between IR and hypertension, at least for the rat models studied. A similar dissociation of metabolic and hemodynamic changes has been observed in Fischer 344 rats in which high-carbohydrate diet induced pronounced elevation of fasting plasma insulin and TG levels without significant BP changes.[69,70] Nevertheless, either longer exposure to high-sucrose diet[71] or a combination of sucrose with salt[69] was able to elevate BP in these animals.

BLOOD PRESSURE CHANGES IN DIET-INDUCED INSULIN RESISTANCE

In animal studies, there is almost a consensus that feeding greater amounts of sucrose or fructose is associated with fasting hypertriglyceridemia, hyperinsulinemia, impaired insulin action *in vitro* and *in vivo*, and mild impairment of glucose tolerance.[72-75] In addition, diets rich in sucrose or fructose also raise BP.[76] Since the seventies, several investigations examining the effect of sucrose and other sugars on BP in normal rats[77,78] or in the SHR[78,79] have been published. Because sucrose feeding stimulates sympathetic nervous system (SNS) activity,[78,80] the increase in BP could be due to SNS-mediated effects in various tissues, including blood vessels, heart, and the kidneys.[81,82] This view has been further supported by studies of Hwang *et al.*,[83] who lowered BP, but not insulin levels, in fructose-fed Sprague-Dawley rats by clonidine.

On the other hand, infusion of somatostatin, which suppresses insulin levels, caused a reduction in BP and plasma TG levels in fructose-fed rats.[84] Furthermore, exercise training of rats, which enhances insulin sensitivity and reduces fructose-induced hyperinsulinemia,[85] also attenuated the increase of BP seen in fructose-fed rats.[86] Very recently, Donelly *et al.*[70] have shown that the relationship of hyperinsulinemia and BP in various strains of rats fed diets enriched with simple sugars is not sodium-dependent. These results are compatible with the earlier data of Brands *et al.*,[87] who demonstrated that hypertension during chronic hyperinsulinemia in rats is not salt-sensitive. It seems likely, therefore, that mechanisms other than sodium and/or fluid retention may contribute to carbohydrate-induced hypertension. Nevertheless, this series of studies provided solid evidence for the suggestion that hyperinsulinemia and IR might play a role in the pathogenesis of hypertension. In contrast, Reed *et al.*[69] demonstrated that, although BP did not increase in response to a fructose diet alone in Wistar or Fischer 344 rats, there was a significant increase of BP in both these groups when they ate the diet containing fructose in combination with salt.

Several studies in rats demonstrated *in vitro* and *in vivo* IR after feeding diets high in fats.[66,88,89] Further, the IR after fat feeding has been linked to increased TG storage in skeletal muscle,[90] skeletal muscle being the most-important tissue for insulin-stimulated glucose disposal. These results have been interpreted within the framework of the glucose/fatty acid cycle of Randle; that is, increased free fatty acid oxidation competitively inhibits glucose oxidation.[66] Finally, Kaufman's group[91,92] has shown that a high-fat (mostly saturated fat) diet increases BP in rats that were allowed to overeat and in which hyperinsulinemia was observed.

Although fat feeding stimulates SNS activity,[91,93] which in turn regulates blood pressure through cardiac, vascular, and renal effects,[94,95] the role of hyperinsulinemia and/or IR cannot be neglected. In addition, Clandinin's group from Edmonton, Canada, recently observed simultaneous insulin-like growth factor-I (IGF-I) resistance and insulin resistance in skeletal muscles of rats being fed the high-fat diet.[96] The high-fat diet reduced both the responsiveness and sensitivity of IGF-I-stimulated glucose transport.

It must be stressed that the BP response to feeding high-fat diets has also a genetic component. This has been clearly shown by Maher *et al.*,[97] who did not see any additional effect of high-fat feeding on BP in obese Zucker rats. Moreover, they obtained two groups of lean animals, one of which was more phenotypically (BP, insulinemia, and glycemia) sensitive to high-fat feeding than the other.

One of the obvious proofs for accepting the role of hyperinsulinemia and/or IR in the pathogenesis of hypertension is the necessity to normalize BP by means that affect plasma insulin levels and/or insulin action. A very clear message has been recently published by two groups in describing the effects of troglitazone.[49,98] Troglitazone belongs to the thiazolidinediones, which are a new class of orally active hypoglycemic compounds that appear to work by either mimicking or enhancing insulin action without stimulation of beta-cell insulin secretion.[99-102] Yoshioka et al.[49] studied the effect of troglitazone on individual components of the IR syndrome in the obese Zucker rat. After four weeks of treatment, plasma insulin, TG, Ch, and systolic BP decreased in a dose-dependent manner. Lee et al.[98] found that troglitazone treatment completely restored in vivo insulin action (EHC) and also normalized the elevated plasma TG and systolic BP levels in normal rats fed high-fructose diet. Other members of the thiazolidinedione family (i.e., ciglitazone and pioglitazone or CS-045) were shown to have similar positive effects on BP and pertinent metabolic variables in the obese Zucker rats and Dahl rats.[10,14,101,103]

Nevertheless, the above-mentioned observations do not prove cause-and-effect relationships because the mechanisms by which these drugs lower BP have not been clearly established. Dubey et al.[104] found that pioglitazone reduced arterial BP in renal hypertensive rats, which are not insulin-resistant or hyperinsulinemic, suggesting that the antihypertensive effect of these agents may be independent of changes in insulin sensitivity. Also, the thiazolidinediones may inhibit calcium channels, which may account for their BP-lowering actions independently of increased insulin sensitivity.[105]

Another interesting idea has been proposed by Storlien's group. They claim a pivotal role in increased lipid availability for the features (including BP changes) of the IR syndrome in rats.[66] Therefore, they examined whether an agent that lowers the TG supply—benfluorex[106]—could ameliorate the range of abnormalities that develop with fructose, sucrose, or fat feeding in rats.[107] They found that fructose and fat feeding significantly impaired insulin action at the whole-body level and increased the availability of circulating (fructose diet) and muscle (fat diet) TG. Feeding fructose, but not fat, significantly increased mean arterial BP, an effect prevented by benfluorex. All the other components of the IR syndrome were normalized with benfluorex treatment as well.

Further supportive evidence for the above-mentioned hypothesis of Storlien comes from a different model of IR, that is, from tumor necrosis factor-alpha (TNF-alpha) infused rats.[108] They are hypertriglyceridemic and markedly hyperinsulinemic at euglycemia. The IR is likely to be due to suppression of the insulin-mediated phosphorylation of the insulin receptor by TNF-alpha. Interestingly, marine fish oil [rich in omega-3 polyunsaturated fatty acids (n-3 PUFA)] feeding corrects almost all of the described pathology to normal levels.[109]

Marine fish oils lower the TG supply, which occurs through reduction of hepatic TG output.[110] Several investigators (including ourselves) published data showing that dietary supplementation of rats with n-3 PUFA is associated with improvement of insulin action both *in vitro* and *in vivo*, decrease of plasma TG and insulin levels, amelioration of TG storage in skeletal muscles,[90,109-111] and reduction in BP.[112,113] This hypotensive action of n-3 PUFA has been well documented in rats with genetic hypertension, for example, in SHR or stroke-prone SH animals[114,115] or in the hereditary hypertriglyceridemic, IR, and hypertensive rats.[116]

The parallel effect of n-3 PUFA on IR, lipids, and hypertension provides further evidence for the association of IR with raised BP. This does not further elucidate the mechanisms responsible because of the very complex effects of n-3 PUFA on the regulation of BP.[113,117] On the basis of data obtained in SHR, at least two mechanisms for the hypotensive action of fish oil have been proposed. The first one is an increase in endothelial-dependent relaxation in resistance arteries,[115,118] which could be due to suppression of the endothelial-dependent constrictor factors, thromboxane A_2 or its precursors. The second mechanism could be the decreased reactivity of resistance vessels to sympathetic nerve stimulation and noradrenaline.[119] The latter effect is most likely due to changes in membrane and intracellular phospholipids by incorporation of n-3 PUFA, influencing vascular and myocardial adrenoreceptor function and intracellular signal/transduction mechanisms.[117]

The increased incorporation of n-3 PUFA into membrane phospholipids has been repeatedly reported to be accompanied by increases in membrane fluidity.[120,121] This in turn influences the activities of membrane-bound enzymes, including insulin receptor tyrosine kinase,[121] and those of ion transport systems,[11] which appear to participate in the pathogenesis of hypertension.[11]

Another link between insulin and hypertension has been proposed by Axelrod.[122] In his exhaustive review, he collected a body of evidence showing that insulin in physiological concentrations decreases catecholamine-induced production of PGI_2 and PGE_2, two potent vasodilators, in adipose tissue. Thus, hyperinsulinemia could increase peripheral vascular resistance and BP by inhibiting the stimulatory effect of catecholamines on the production of PGI_2 and PGE_2. In this respect, it is of interest that n-3 PUFA decrease insulin release both *in vivo* and *in vitro*.[111,123] Although all the aforementioned mechanisms overlap significantly, a certain role of hyperinsulinemia and/or IR in the pathogenesis of hypertension cannot be ignored.

EFFECTS OF EXOGENOUS INSULIN ADMINISTRATION ON BLOOD PRESSURE

Hyperinsulinemia during an EHC led in normal rats to a dose-dependent increase in BP and heart rate that was sympathetically mediated.[124] Moreover, insulin enhanced the pressor response to infused norepinephrine in rat mesenteric vasculature.[125] Brands *et al.*[126] demonstrated that 5 days of insulin infusion, which caused a moderate hyperinsulinemia (similar to fasting insulinemia of obese individuals), produced an increase in mean arterial BP in conscious normal rats. Similarly, administration of insulin with subcutaneous minipumps caused an increase of BP.[127,128] Plasma noradrenaline levels rose slightly in 12-day insulin-treated rats and were positively correlated with mean arterial BP.[128] Štolba *et al.*[129] have demonstrated that long-term insulin administration

(10 weeks) increased BP in Wistar rats. Moreover, Brands et al.[87] showed that hypertension during chronic hyperinsulinemia in rats is not salt-sensitive. This pressor effect of exogenous hyperinsulinemia has not been confirmed in other animal species, in particular in the dog.[130] The pressor action of exogenous insulin is, however, supported by recent findings of Bhanot et al.,[131] who demonstrated that long-term vanadyl sulfate treatment led to a sustained decrease of plasma insulin and BP in SHR as compared to nontreated SHR. Conversely, restoration of plasma insulin in the vanadyl-treated SHR to pretreatment levels by subcutaneous administration of insulin reversed the effects of vanadyl on BP; that is, BP rose again. Thus, although the exact mechanisms remain yet unclear, at least in the rat insulin seems to have pressor effects.

POSSIBLE INSULIN-DEPENDENT BLOOD PRESSURE-REGULATING FACTORS

Several main mechanisms of insulin's effects on BP regulation have been recently intensely discussed in a number of reviews.[2,9,10,132-134]

Sympathetic Activity and Catecholamines

Clearly, insulin administration provokes an adrenergic response if hypoglycemia is not prevented.[135-137] However, when euglycemia is maintained, chronic administration of insulin is still able to activate the SNS of rats.[128,129] Fischer 344 rats fed a high-fat diet develop obesity, hypertension, hyperinsulinemia, and IR. During EHC studies, BP increased, and this BP response was reversible by combined alpha- and beta-blockade. This suggests a role for increased SNS activity in these insulin-mediated BP changes.[132]

Indeed, dietary-induced (high-fat, high-sucrose, or fructose diets) states of IR, hyperinsulinemia, and raised BP in rats have been repeatedly reported to be associated with upregulation of the SNS activity.[9,78,91,93,138] Conversely, fasting or caloric restriction was shown to suppress the SNS activity.[139] Evidence has accumulated indicating that insulin-mediated metabolism in certain neurons related to the ventromedial portion of the hypothalamus is involved in this process. It is also well established that dietary-induced changes in SNS activity contribute to changes in dietary thermogenesis, that is, to an adaptive change in metabolic rate taking place in response to food intake. This stimulated Landsberg[9] to put forward the hypothesis that insulin-mediated sympathetic stimulation could be a mechanism recruited in the obese to increase metabolic rate and restore energy balance; concomitant increases in sympathetic stimulation of the heart, vasculature, and kidney result in hypertension, which, according to this hypothesis, would be an unfortunate by-product of the cardiovascular response to a metabolic adaptation.

However, in rats, two weeks of sustained hyperinsulinemia in conjunction with sucrose supplements was no more stimulatory to the SNS of rats than sucrose alone.[140] Thus, whether chronic hyperinsulinemia can effectively influence BP via enhancement of the SNS activity remains to be yet established.

Direct measurements of catecholamines in blood obtained via arterial catheters in various rat models of hypertension under conscious, resting conditions produced variable results. In SHR compared often to WKY, findings ranged from no differences[141] to an elevation of plasma noradrenaline in SHR,[142] an elevation of both noradrenaline and adrenaline in young SHR,[142] or an elevation of adrenaline only in old SHR[143] and stroke-prone SHR.[144] Finally, Jablonskis and Howe[145] found that adrenaline was elevated in both young and old SHR at rest. In the hereditary hypertriglyceridemic, IR, and hypertensive Wistar rat (hHTg), a statistically significant elevation of plasma adrenaline and noradrenaline levels was found in comparison with Wistar controls,[18] as well as an elevation of plasma adrenaline only in comparison with Brown-Norway rats.[32] In response to immobilization stress, both plasma adrenaline and noradrenaline levels of hHTg rats were much higher than those of the Brown-Norway ones.[32]

How the above observations in rat models of hypertension and IR syndrome, respectively, relate to the action of insulin on BP is not clear. Yet, they provide additional evidence for the simultaneous appearance of hypertension, increased SNS activity, and impaired insulin action.

Sodium-Fluid Retention

Renal Na^+ retention leading to hypervolemia has been suggested as one of the possible mechanisms of insulin-induced hypertension.[5,7] In normal rats, in which serum insulin levels were raised by infusion for one week, hypertension developed without sodium retention.[14] Conversely, a sodium-retaining effect of insulin was demonstrated earlier using the EHC technique in rats.[146] Moreover, there was no difference in the antinatriuretic effect of insulin in SHR and in normotensive rats.[147]

The exact mechanisms of insulin's antinatriuretic effect are not entirely clear. Evidence exists showing that insulin acts in the rat kidney at the loop of Henle where an increased chloride reabsorption has been reported.[146] Other studies that used micropuncture and microperfusion techniques have shown that the antinatriuretic effect of insulin can be exerted on both the proximal and distal parts of the nephron.[148,149] Insulin might also indirectly enhance sodium reabsorption via its effects on the SNS activity and on the renin-angiotensin-aldosterone system. Both the increase in circulating catecholamines and the stimulation of renal nerves increase sodium reabsorption,[150,151] whereas angiotensin II may mediate the sodium-retaining action by an effect on the proximal tubular cells.[152] However, the latter mechanisms are probably of minor importance.[153]

If chronic hyperinsulinemia induces hypertension by its sodium-retaining effect, this would imply that, at least to some extent, the resulting increase in BP is volume-dependent and salt-sensitive. This might be true for the Dahl salt-sensitive rats,[154] which represent the most extensively studied genetic model of salt-sensitive hypertension associated with fasting and postglucose load hyperinsulinemia.[33,34] Data on sodium retention during administration of exogenous insulin to rats are contradictory.[146,155] Information on salt sensitivity of another rat model of hypertension, insulin resistance, and hyperinsulinemia — obese Zucker rats — is also rather contradictory.[45,50,52]

Modulation of Cation Transport

Insulin is known to affect several transmembrane cation transport systems, thereby influencing intracellular sodium and calcium homeostasis.[11,135,156,157] In particular, insulin stimulates Na^+/K^+ ATPase, Ca^{++} ATPase, and the Na^+/H^+ antiport, as well as potential- and receptor-mediated Ca^{++} channels.[158,159] The net effect of insulin would thus be the reduction of intracellular Na^+, Ca^{++}, and H^+ concentrations and the elevation of Mg^{++} concentration. Vascular smooth muscle cells show attenuated responses to agonists in the presence of insulin. In the syndrome of IR and hypertension, the contrary should be found, with a consequent increase of intracellular Ca^{++}. The latter is a major determinant of vascular smooth muscle contractility.[160,161]

The Na^+/H^+ antiport is an insulin-sensitive cell membrane system involved in the regulation of intracellular Na^+ concentration and pH, cell growth, proximal renal tubular sodium reabsorption, and also Ca^{++} exchange.[157,162] Overactivity of this system as a result of hyperinsulinemia could lead to hypertension through various mechanisms as summarized recently by Meehan et al.[2] First, it leads to increased intracellular Na^+ in vascular smooth muscle cells, making them more sensitive to pressor effects of catecholamines and angiotensin II. Second, increased intracellular Na^+ is usually associated with increased intracellular Ca^{++}, resulting in increased vascular tone. Finally, it causes alkalinization of the cytoplasm, which stimulates cellular growth leading to hyperplasia and hypertrophy of vascular smooth muscle cells. All these effects could cause increased peripheral vascular resistance and a rise of BP.

Hyperinsulinemia (and hypertriglyceridemia) has been reported in several rat strains with genetic hypertension (e.g., in SHR,[22] Dahl rats,[33] Milan hypertensive rats,[38] and hHTg rats[110]) in which alterations of erythrocyte Na^+ and K^+ transport were described.[64,163-166] In the hereditary HTg rats, in which BP is substantially elevated above control levels,[13,18,32] increased plasma TG levels exhibit a clear-cut association with an elevated red cell Na^+ content. The latter was caused by an augmented ouabain-resistant Na^+ influx.[64] It is of interest that ouabain-resistant Na^+ influx and Na^+ leak cosegregated with blood pressure in recombinant inbred strains.[167]

It should be noted that there is considerable epidemiologic evidence on the association of red cell Na^+ and K^+ transport alterations with abnormalities of lipid metabolism in normotensive and hypertensive humans.[168-171] Particular changes in plasma lipids and especially in the lipid composition of the erythrocyte membrane were reported to affect kinetic parameters of various transport systems, including Na^+-Li^+ countertransport, Na^+-K^+ cotransport, and Na^+-K^+ pump.[172-174] Modification of fatty acid composition of phosphatidylcholine causes profound modification of Na^+-Li^+ exchange and Na^+-K^+ cotransport in human erythrocytes.[175] Subsequent studies revealed altered membrane phospholipid composition due to an abnormal plasma lipoprotein pattern, which is responsible for kinetic alterations of major Na^+ and K^+ transport systems.[176,177] This might explain some mechanisms responsible for cosegregation of plasma lipids, red cell Na^+ content, ion transport abnormalities, and BP in our studies on hHTg rats and their F_2 hybrids.

This is consistent with other results obtained in hHTg rats. We found that plasma triglycerides correlate with a number of variables, that is, with elevated BP and red cell

Na+ content,[64] and IR *in vivo* and fasting insulinemia.[65] Moreover, insulin sensitivity *in vivo* correlated positively with the percentage of long-chain fatty acids (Σ20–22 PUFA) in skeletal muscle phospholipids of hHTg rats. In contrast, a negative correlation was found between *in vivo* insulin action and the ratio of n-6/n-3 fatty acids in membrane phospholipids (unpublished data). Thus, changes in fatty acid composition of membrane phospholipid could, in turn, affect not only insulin action, but also membrane ion transport systems through modulation of local protein-lipid interactions.

Vasculopathy

This can be defined as dysfunction in the form of inappropriate vasoconstriction and/or structural alteration expressed as arterial narrowing and stiffening due to muscle cell proliferation.[10]

Interestingly, most studies in humans and in dogs showed that insulin appears to have a vasodilatory effect in some vascular beds (for review, see references 14 and 134). In particular, insulin infusions during euglycemia caused vasodilation in skeletal muscles[178] and in the coronary circulation.[179] The vasodilator effect of insulin has been confirmed also *in vitro* using the rat tail artery[180] and rabbit femoral artery and vein.[181] Information derived from a study of obese Zucker rats supports this view. These IR rats exhibit an increased vascular reactivity (vascular smooth muscle preparations) to various vasoactive substances. Exogenous insulin has been shown to attenuate this vascular hyperactivity, but to a significantly lesser extent in the obese as compared to lean Zucker rats. This suggests that the increase in vascular reactivity in obese Zucker rats may be due to a reduced ability of insulin to attenuate vasoconstrictor responses.[53,56] Similarly, in a rat model of insulin-deficient IR, that is, in streptozotocin diabetes, administration of exogenous insulin restored normal vascular responsiveness.[182]

Vascular tone in terms of contractility and/or relaxation is also significantly controlled by multiple vasoactive substances that are released from the endothelium, with dominant relaxing properties.[183] Endothelium-derived relaxing factor, identified as nitric oxide (NO), is a potent vasodilator that regulates arterial tone by increasing cyclic guanosine monophosphate (cGMP) production in vascular smooth muscle.[184] Acute inhibition of NO synthase with increasing doses of L-NAME in normal rats, who were kept either acutely hyperinsulinemic during an EHC[185] or chronically (for 10 days) hyperinsulinemic (at euglycemia), led to a higher and dose-dependent BP increase in comparison to rats treated with placebo.[186] These data suggest that chronic hyperinsulinemia, which is associated with IR, increases the dependency of the vasculature to NO formation. Such mechanisms may be important in attenuating the increase in BP that could result from the activation of other pressor mechanisms by elevated insulin levels.

However, this effect of insulin on NO formation may be altered in hypertensive states since decreased endothelial vasodilation has been observed in the hypertensive and hyperinsulinemic Dahl rats.[187] Conversely, administration of L-arginine, the main substrate for NO synthesis,[188] to Dahl/JR rats completely prevented their salt-sensitive hypertension.[189] On the other hand, no effect of L-arginine on hypertension was found in the SHR.[190]

Interesting data were recently published by Baron et al.,[185] who found that acute induction of hypertension with L-NAME caused marked IR in awake unstressed normal rats. Moreover, the degree of induced IR was related to the height of BP elevation. Given the known dependence of skeletal muscle blood flow on endothelium-derived NO in rats,[191] and the findings that insulin vasodilates skeletal muscle vasculature in rats,[180,192] Baron et al.[185] hypothesize that administration of the NO synthase inhibitor L-NAME leads to a marked reduction in perfusion to that tissue and causes IR via reduction in insulin and glucose delivery. Moreover, their data indicate that L-NAME-induced IR occurred largely (but not exclusively) via nonadrenergic mechanisms. On the other hand, the available data on the effect of L-NAME administration to rats point to an early suppression of the SNS activity followed by a late activation (at least in renal sympathetic activity) as a result of central disinhibition of sympathetic tone by the blockade of neuronal nitric oxide synthesis.[193] Therefore, IR resulting from NO inhibition in the rat could be due to activation of the SNS and/or reduction in peripheral perfusion. Interestingly, very recently, Baron[194] published human results supporting the validity of the aforementioned experimental data.

Another piece to the mosaic of the vasodilator hypothesis was recently published by Kohlman et al.[195] They found that BP increased significantly during EHC in rats submitted to inhibition of bradykinin, prostaglandin, or NO, suggesting that these vasodilator systems tend to counteract the pressoric effect of hyperinsulinemia. Bradykinin seems to play the most-important homeostatic role under these conditions because its inhibition significantly reduces insulin sensitivity and allows BP to rise.

Hence, hyperinsulinemia causing an increased sympathetic drive (and possibly a sodium retention) in the setting of blunted insulin-dependent vasodilation and catecholamine antagonism seems to provide a reasonable scenario in which hypertension might result, at least partially, from defects in insulin and glucose metabolism. However, a bidirectional relation is not excluded either.

Low concentrations of insulin have been shown to stimulate proliferation of smooth muscle cells in tissue culture.[196,197] Since insulin effects on glucose uptake and cellular growth are mediated by different receptors and intracellular pathways,[198,199] and growth is apparently unaltered despite resistance in the glucoregulatory action of insulin, chronic hyperinsulinemia might result in medial hyperplasia, arterial stiffness, and luminal narrowing and thus could lead to increased peripheral vascular resistance. The mitogenic effect of insulin has been studied in several vascular cell types from a number of species. A distinctive insulin sensitivity was observed in microvascular endothelium.[200]

On the other hand, Hall et al.[14] in their recent review wrote that there is little direct evidence that hyperinsulinemia, similar to that found in insulin-resistant states, is capable of causing significant stimulation of vascular smooth muscle growth *in vivo*.

POTENTIAL EFFECTS OF HYPERTENSION ON INSULIN SENSITIVITY

With increasing age, the rise in BP of normal rats and rat models of hypertension[201,202] is not only associated with a decrease in insulin sensitivity,[203-205] but muscle capillary density declines as well.[206,207] Moreover, arteriolar rarefaction (disappearance of microvessels) has also been documented in skeletal muscle of the SHR.[208] Apart from contributing to the increase in total peripheral resistance in hypertension, the decrease in capillary density may increase the diffusion distance between the capillary and muscle cell, thus impeding glucose metabolism.[209] High BP might therefore contribute to IR in the process of vessel rarefaction.

Capillary density (capillary rarefaction) is proposed to be causally related with redifferentiation of skeletal muscle fibers of SHR,[208,210] with the decrease in perfusion suggested as the stimulus for redifferentiation. This fiber redifferentiation leads to a relative increase in type IIb fibers (white, fast-twitch, glycolytic fibers) and a comparable decrease in type I fibers (red, slow-twitch, oxidative fibers).

There are substantial differences in the rates of glucose uptake in normal rat muscles composed of predominantly different fiber types,[211,212] which are further modulated by the functional status of the muscle (resting versus exercising) and by the insulin concentration present (no insulin versus low and high insulin concentrations).[213,214] In particular, at high insulin doses, the highest glucose uptake rates have been reported in muscles composed of predominantly slow-twitch, oxidative, type I fibers.[213,214] Moreover, Storlien et al.[215] recently showed that in sedentary Wistar rats increased amounts of unsaturated fatty acids were found in the more oxidative, insulin-sensitive red quadriceps and soleus muscles (containing more type I fibers), whereas reduced levels of PUFA were found in primarily glycolytic white quadriceps muscle (containing more type IIb fibers).[215] These data are very compatible with previous results on the relation between fatty acid composition of the skeletal muscle cell membrane and insulin action in rats.[66,90]

Thus, it may be expected that a reduction of type I fibers in skeletal muscles of SHR and possibly in other rat models of hypertension could contribute to a reduction of insulin-induced glucose disposal in skeletal muscles. This could then be followed by hyperinsulinemia with its deleterious effect on insulin action and BP regulation.

At least in the SHR animals, not all researchers have observed signs of capillary rarefaction and fiber redifferentiation in skeletal muscles.[206,216] In addition to other reasons, these discrepancies could be due to genetic differences in the subcolonies of SHR and their "control" WKY rats around the world.[208] Moreover, gene loci responsible for the rise in BP may be slightly different in different colonies and may not be linked to loci controlling skeletal muscle properties.

GENETICS OF HYPERTENSION AND INSULIN RESISTANCE

Recently, Mark and Anderson[217] suggested the concept that there is a "sensitivity or resistance to the BP effects of insulin resistance" and that genetic factors could be important. There is convincing evidence that genetic factors can play a critical role in salt sensitivity or resistance, and it is possible that genetic factors also significantly regulate the BP response to IR and hyperinsulinemia. Differences in IR observed among various rat strains[69,154] strongly suggest that this phenotype has a genetic component. It has also been demonstrated that genes responsible for "BP sensitivity to IR" may be distinct from those responsible for salt sensitivity.[69] Thus, significant new knowledge on the hyperinsulinemia-IR-hypertension relationship can be expected to come from rigorous genetic analyses of rat models of hypertension and/or of the IR syndrome as well.

The first results employing the tools of molecular biology and signaling the importance of changes at the gene level for BP regulation were available in the late eighties and early nineties. Several studies have shown cosegregation of renin (Ren) alleles with BP. These included crosses between the SS/Jr rat and the inbred Dahl salt-resistant (SR/Jr) rat,[218] the SHR and the Lewis rat,[219] the SHR and the BN rat,[220] the Lyon hypertensive (LH) rat and the Lyon normotensive (LN) rat,[221] and the New Zealand genetically hypertensive (GH) rat and the BN rat.[222]

Two studies in 1991—Jacob et al.[223] and Hilbert et al.[224]—identified a polymorphism on chromosome 10 that cosegregated with raised BP in a cross between SHR stroke-prone (SP) and Wistar-Kyoto (WKY) rats. The particular region was shown to contain the angiotensin-converting enzyme (ACE) gene. ACE alleles have subsequently been shown to cosegregate with BP in a further independent cross between SHR-SP and WKY rats[225] and in a cross between the inbred Dahl salt-sensitive rat (SS/Jr) and the Milan normotensive (MNS) rat.[226] More studies using other crosses followed and brought further supportive evidence for the significance of the ACE with regard to BP.[222,227]

Cosegregation with BP was also shown for alleles of the guanyl cyclase A (GCA) gene that codes for the atrial natriuretic factor receptor. In particular, the aforementioned was observed in crosses between the SS/Jr rat and the WKY rat and between the SS/Jr rat and the MNS rat.[226]

Crossbreeding studies have implicated other genes in the pathogenesis of hypertension. These include genes for the kallikrein hormonal family,[228] adducin,[229] NO synthase,[230] angiotensinogen,[231] 11β-hydroxylase and aldosterone synthase,[232] and a novel gene of unknown function—the Sa gene.[233]

The above-mentioned results obtained with molecular biology tools seem to be substantiated also by "classical" physiology. For example, feeding losartan, an antagonist of the angiotensin AT_1 receptors, to normal rats with the fructose-induced syndrome of IR prevented both BP elevation and hyperinsulinemia, but not elevation of plasma TG.[234] BP in a model of genetic hypertension, that is, in the SHR, was shown to be reduced in a dose-dependent manner by both benazepril (an ACE inhibitor) and valsartan (an angiotensin AT_1 inhibitor).[235] Benazepril also significantly increased the insulin-stimulated glucose disposal and both of the inhibitors used were without effect on plasma TG.

Treatment of Zucker rats (IR rats that are also hypertensive) and SHR (genetically hypertensive animals with moderate IR) with ACE inhibitors (captopril and trandolapril, respectively) showed that these drugs stimulated the *in vitro* insulin action in skeletal muscle—the 2-deoxyglucose uptake in Zucker rats[236] and glycogen synthesis in the SHR.[237] Both groups claim that the aforementioned actions of ACE inhibitors on insulin action are related to bradykinin.

Nevertheless, the present data cannot provide conclusive evidence that modification of one or more genes is responsible for the development of hypertension, which might negatively influence insulin action. Indeed, other nearby, closely linked genes may be the determining genes. For example, the region of rat chromosome 10 containing the ACE gene also contains the genes for nerve growth factor receptor[224] and growth hormone,[223,224] and the syntenous region of the human genome contains the gene for phenylethanolamine N-methyltransferase.[238] Similarly, rat chromosome 2 contains the genes for GCA and the alpha-1 subunit of the Na,K-ATPase, which has also been implicated in the etiology of hypertension.[227]

Finally, the existing results also demonstrate the importance of studying animals of both sexes from reciprocal F_0 crosses. Many reported differences occur only in one sex or, indeed, in only one lineage of particular sex.[222,239]

As far as genetics of IR is concerned, there have been a few limited studies on genes that might affect insulin sensitivity or lipid profiles (for review, see Caulfield et al.[240] in this volume). There are some preliminary data suggesting linkage and association of the apolipoprotein B locus with dyslipidemia in hypertension, and of the lipoprotein lipase gene at chromosome 12 with hypertension. Further IR genes are in the "pipeline."[241] Among others, an interesting possibility could be the beta3-adrenergic receptor.[242] The beta3-adrenergic receptor has been shown to be the principal mediator of catecholamine-mediated thermogenesis and fatty acid beta-oxidation in brown fat tissue and it is thought to be an important stimulator of lipolysis in white adipose tissue.[243] It was hypothesized that defective receptors could contribute to the pathogenesis of obesity and IR.

CONCLUSIONS

There is overwhelming evidence from rat studies for a link between IR/hyperinsulinemia and hypertension. The exact interactions and sequence of events in various insulin-resistant states seem to vary and they are still not yet completely understood. Studies with dietary-induced hyperinsulinemia or with administration of exogenous insulin to rats suggest that, in the rat, insulin seems to have a pressor effect that may not be salt-sensitive. Mechanisms seem to involve the stimulatory action of insulin on the activity of the sympathetic nervous system, increased Na^+ reabsorption (which may not apply for all rat models of hypertension and IR), stimulation of the cation transport with consequent changes in vascular reactivity, and growth-promoting effects on arteriolar smooth muscle cells. In rats, insulin may also have a vasodilator action, which is related to endothelial production of nitric oxide and seems to be attenuated in IR/hypertension states. Increased availability of triglycerides (circulating or accumulated in skeletal muscles) correlates with BP, cation transport, and IR. More detailed studies indicate a crucial role of fatty acid composition of membrane phospholipids as one of the hypothetical links between IR and BP. Differences in the degree and appearance of individual traits of IR, hypertriglyceridemia, BP, etc., among various rat models of hypertension and/or of the IR syndrome indicate a strong genetic influence on the phenotypic expression of each marker evaluated. Thus, caution is required before generalizations are made from any rat model of the IR, hypertriglyceridemia, and blood pressure triad.

ACKNOWLEDGMENTS

We thank J. Zicha and J. Kuneš, Institute of Physiology, Academy of Sciences of the Czech Republic in Prague, for their suggestions and critical reading of some parts of this manuscript.

REFERENCES

1. WELBORN, T. A., A. BRECKENRIDGE, A. H. RUBINSTEIN, C. T. DOLLERY & T. R. FRASER. 1986. Lancet **1**: 1336–1337.
2. MEEHAN, W. P., CH. DARWIN, N. B. MAALOUF, T. A. BUCHANAN & M. F. SAAD. 1993. Steroids **58**: 621–634.
3. REAVEN, G. M. 1988. Diabetes **37**: 1595–1607.
4. MODAN, M., H. HALKIN, S. ALMOG, A. LUSKY, A. ESHKOL, M. SHEFI, A. SHITRIT & Z. FUCHS. 1985. J. Clin. Invest. **75**: 809–817.
5. FERRANNINI, E., G. BUZZIGOLI, R. BONADONNA, M. A. GIORICO, M. OLEGGINI, L. GRAZIADEI, R. PEDRINELLIA, L. BRANDI & L. BEVILACQUA. 1987. N. Engl. J. Med. **317**: 350–356.
6. IZZO, J. L. & A. L. M. SWISLOCKI. 1991. Am. J. Med. **90**(suppl. 2A): 26S–31S.
7. REAVEN, G. M. 1991. Diabetes Care **14**: 195–202.
8. SOWERS, J. R., P. R. STANDLEY, J. L. RAM, M. B. ZEMEL & L. M. RESNICK. 1991. Am. J. Hypertens. **4**: 466S–472S.
9. LANDSBERG, L. 1992. Hypertension **19**(suppl. I): I61–I66.
10. WEIDMANN, P., M. DE COURTEN & L. BÖHLEN. 1993. J. Hypertens. **11**(suppl. 5): S27–S38.
11. RESNICK, L. M. 1993. Am. J. Hypertens. **6**: 123S–134S.
12. VETTOR, R., I. CUSIN, D. GANTEN, F. ROHNER-JEANRENAUD, E. FERRANNINI & B. JEANRENAUD. 1994. Am. J. Physiol. **267**: R1503–R1509.
13. KLIMEŠ, I., A. VRÁNA, J. KUNEŠ, E. ŠEBÖKOVÁ, Z. DOBEŠOVÁ, P. ŠTOLBA & J. ZICHA. 1995. Blood Pressure **4**: 137–142.
14. HALL, J. E., M. W. BRANDS, D. H. ZAPPE & M. ALONSO-GALICIA. 1995. Proc. Soc. Exp. Biol. Med. **208**: 317–329.
15. MEDIRATTA, S., A. FOZAILOFF & W. H. FRISHMAN. 1995. J. Clin. Pharmacol. **35**: 943–956.
16. REAVEN, G. M., H. LITHELL & L. LANDSBERG. 1996. N. Engl. J. Med. **334**: 374–381.
17. VRÁNA, A. & L. KAZDOVÁ. 1990. Transplant. Proc. **22**: 2579.
18. ŠTOLBA, P., Z. DOBEŠOVÁ, P. HUŠEK, H. OPLTOVÁ, J. ZICHA, A. VRÁNA & J. KUNEŠ. 1992. Life Sci. **51**: 733–740.
19. KLIMEŠ, I., E. ŠEBÖKOVÁ, A. VRÁNA, P. ŠTOLBA, L. KAZDOVÁ, J. KUNEŠ, P. BOHOV, M. FICKOVÁ, J. ZICHA, D. RAUČINOVÁ & V. KŘEN. 1995. *In* Lessons from Animal Diabetes, p. 271–282. Smith-Gordon. London.
20. YAMORI, Y., M. OHTAKA, H. UESHIMA, Y. NARA, R. HORIE, T. SHIMAMOTO & Y. KOMACHI. 1978. Jpn. Circ. J. **42**: 841–847.
21. MONDON, C. E. & G. M. REAVEN. 1988. Metabolism **37**: 303–305.
22. REAVEN, G. M. & H. CHANG. 1991. Am. J. Hypertens. **4**: 34–38.
23. REAVEN, G. M., H. CHANG, B. B. HOFFMAN & S. AZHAR. 1989. Diabetes **38**: 1155–1160.
24. BURSZTYN, M., D. BEN-ISHAY & A. GUTMAN. 1992. J. Hypertens. **10**: 137–142.
25. BADER, S., R. SCHOLZ, M. KELLERER, S. TIPPMER, K. RETT, S. MATHAEI, P. FREUND & H. U. HÄRING. 1992. Diabetologia **35**: 712–718.
26. HULMAN, S., B. FLAKNER & Y. Q. CHEN. 1991. Metabolism **40**: 359–361.
27. TSUTSU, N., Y. TAKATA, K. NUNOI, M. KIKUCHI, S. TAKASHITI, S. SADOSHIMA & M. FUJISHIMA. 1989. Metabolism **38**: 63–66.
28. BUCHANAN, T. A., J. H. YOUNG, V. M. CAMPESE & G. F. SIPOS. 1992. Diabetes **41**: 872–878.
29. FRONTONI, S., L. OHMAN, J. R. HAYWOOD, R. A. DEFRONZO & L. ROSETTI. 1992. Am. J. Physiol. **262**: E191–E196.
30. FARRACE, S., S. FRONTONI, S. GAMBARDELLA, G. MENZINGER & L. ROSETTI. 1995. Am. J. Hypertens. **8**: 949–953.
31. MCMURTY, J. P. & B. C. WEXLER. 1981. Endocrinology **108**: 1730–1736.

32. KLIMEŠ, I., E. ŠEBÖKOVÁ, D. GAŠPERÍKOVÁ, I. JAROŠČÁKOVÁ, E. VISKUPIČ, J. MALINOVÁ, D. JEŽOVÁ, E. SABBAN & R. KVETŇANSKÝ. 1996. *In* Catecholamines and Other Neurotransmitters in Stress: Molecular Genetic and Neurobiological Advances, p. 343-352. Gordon & Breach. New York.
33. REAVEN, G. M., J. TWERSKY & H. CHANG. 1991. Hypertension **18**: 630-635.
34. KOTCHEN, T. A., H. Y. ZHANG, M. COVELLI & N. BLEHSCHMIDT. 1992. Am. J. Physiol. **262**: E692-E697.
35. DAHL, L. K., M. HEINE & L. TASSINARI. 1962. J. Exp. Med. **115**: 1173-1190.
36. ZHANG, H. Y., S. R. REDDY & T. A. KOTCHEN. 1994. Hypertension **24**: 106-110.
37. BIANCHI, G., P. FERRARI & B. R. BARBER. 1984. *In* Handbook of Hypertension. Vol. 4, p. 328-349. Elsevier. Amsterdam/New York.
38. DALL-AGLIO, E., P. TOSINI, P. FERRARI, I. ZAVARONI, M. PASSERI & G. M. REAVEN. 1991. Am. J. Hypertens. **4**: 773-775.
39. VINCENT, M., E. H. BOUSSAIRI, R. CARTIER, M. LO, A. SASSOLAS, C. CERUTTI, C. BARRES, M.-P. GUSTIN, G. CUISINAUD, N. J. SAMANI, G. M. LATHROP & J. SASSARD. 1993. J. Hypertens. **11**: 1179-1185.
40. SHAFRIR, E. 1990. *In* Diabetes Mellitus, p. 299-340. Elsevier. Amsterdam/New York.
41. AHMAD, F. & B. J. GOLDSTEIN. 1995. Metabolism **44**: 1175-1184.
42. O'DONNELL, M. P., B. L. KASISKE, M. P. CLEARY & W. F. KEANE. 1985. J. Lab. Clin. Med. **106**: 605-610.
43. BUNAG, R. D. & D. L. BARRINGER. 1988. Clin. Exp. Hypertens. **10**(suppl. 1): 257-262.
44. AUGUET, M., S. DEFLOTTE & P. BRAQUET. 1989. J. Pharm. Pharmacol. **41**: 861-864.
45. PAWLOSKI, C. M., N. L. KANAGY, L. H. MORTENSON & G. D. FINK. 1992. Hypertension **19**: 190-195.
46. KASISKE, B. L., M. P. CLEARY, M. P. O'DONNELL & W. F. KEANE. 1985. J. Lab. Clin. Med. **106**: 598-604.
47. KURTZ, T. W., C. MORRIS & H. A. PERSHADSINGH. 1989. Hypertension **13**: 896-901.
48. ZEIGLER, D. W. & K. P. PATEL. 1991. Am. J. Physiol. **261**: R712-R718.
49. YOSHIOKA, S., H. NISHINO, T. SHIRATI, K. IKEDA, H. KOIKE, A. OKUNO, M. WADA, T. FUJIWARA & H. HORIKOSHI. 1993. Metabolism **42**: 75-80.
50. KIM, J. H., T. SNIDER, M. ABEL & M. B. ZEMEL. 1994. J. Nutr. **124**: 713-716.
51. ZEMEL, M. B., J. D. PEUKER, J. R. SOWERS & L. SIMPSON. 1992. Am. J. Physiol. **262**: E368-E371.
52. REDDY, S. R. & T. A. KOTCHEN. 1992. Hypertension **20**: 389-393.
53. ZEMEL, M. B., S. REDDY, S. E. SHEHIN, W. LOCKETTE & J. R. SOWERS. 1990. J. Vasc. Med. Biol. **2**: 82-85.
54. AMBROZY, S. L., S. E. SHEHIN, C. Y. CHIOU, J. R. SOWERS & M. B. ZEMEL. 1991. Am. J. Hypertens. **4**: 592-596.
55. ABEL, M. A. & M. B. ZEMEL. 1993. Am. J. Hypertens. **6**: 500-504.
56. ZEMEL, M. B., S. REDDY & J. R. SOWERS. 1991. Am. J. Hypertens. **4**: 537-540.
57. STANDLEY, P. R., J. L. RAM & J. R. SOWERS. 1993. Endocrinology **133**: 1693-1699.
58. KASISKE, B. L., M. P. O'DONNELL & W. F. KEANE. 1992. Hypertension **19**(suppl. I): I110-I115.
59. MICHAELIS, O. E., IV, D. H. PATRICK, C. T. HANSEN, J. J. CANARY, R. M. WERNER & N. CARSWELL. 1986. Am. J. Pathol. **123**: 398-400.
60. BALY, D. L., M. J. ZARNOWSKI, N. CARSWELL & O. E. MICHAELIS. 1989. J. Nutr. **119**: 628-632.
61. MICHAELIS, O. E., IV, M. T. VELASQUEZ, A. A. ABRAHAM, D. A. SERVETNICK, D. J. SCHOFIELD & C. T. HANSEN. 1995. Am. J. Hypertens. **8**: 467-473.
62. MCCUNE, S. A., M. J. RADIN, J. E. JENKINS, Y. CHU, S. PARK & R. G. PETERSON. 1995. *In* Lessons from Animal Diabetes, p. 255-270. Smith-Gordon. London.
63. BERDANIER, C. D. 1995. *In* Lessons from Animal Diabetes, p. 231-246. Smith-Gordon. London.
64. KUNEŠ, J., K. H. BIN TALIB, Z. DOBEŠOVÁ & A. VRÁNA. 1994. Clin. Sci. **86**: 11-13.
65. RAUČINOVÁ-GAŠPERÍKOVÁ, D., E. ŠEBÖKOVÁ, I. KLIMEŠ & P. LANGER. 1995. Physiol. Res. **44**: P15 (abstract).
66. STORLIEN, L. H., D. A. PAN, A. D. KRIKETOS & L. A. BAUR. 1993. Ann. N.Y. Acad. Sci. **683**: 82-90.
67. KOVÁCS, P., N. J. SAMANI, E. ŠEBÖKOVÁ, B. VOIGT, D. GAŠPERÍKOVÁ, J. JEŽOVÁ, R. KVETŇANSKÝ, D. LODWICK, I. KLÖTING & I. KLIMEŠ. 1997. This volume.

68. VRÁNA, A., L. KAZDOVÁ, Z. DOBEŠOVÁ, J. KUNEŠ, V. KŘEN, V. BÍLÁ, P. ŠTOLBA & I. KLIMEŠ. 1993. Ann. N.Y. Acad. Sci. **683:** 57–68.
69. REED, M. J., H. HO, R. DONNELLY & G. M. REAVEN. 1994. Blood Pressure **3:** 197–201.
70. DONNELLY, R., H. HO & G. M. REAVEN. 1995. Blood Pressure **4:** 164–169.
71. PREUSS, H. G., J. J. KNAPKA, P. MACARTHY, A. K. YOUSUFI, S. G. SABNIS & T. T. ANTONOVYCH. 1992. Am. J. Hypertens. **5:** 585–591.
72. AHRENS, E. H., J. HIRSCH, K. OETTE, J. W. FARQUHAR & Y. STEIN. 1961. Trans. Assoc. Am. Physicians **74:** 134–145.
73. VRÁNA, A. & P. FÁBRY. 1983. World Rev. Nutr. Diet. **42:** 55–101.
74. STORLIEN, L. H., E. W. KRAEGEN, A. B. JENKINS & D. J. CHISHOLM. 1988. Am. J. Clin. Nutr. **47:** 420–427.
75. KLIMEŠ, I., E. ŠEBÖKOVÁ, A. VRÁNA & L. KAZDOVÁ. 1993. Ann. N.Y. Acad. Sci. **683:** 69–81.
76. AHRENS, R. A. 1974. Am. J. Clin. Nutr. **27:** 403–422.
77. PREUSS, M. B. & H. G. PREUSS. 1980. Lab. Invest. **143:** 101–107.
78. FOURNIER, R. D., C. C. CHIUEH, I. J. KOPIN, J. J. KNAPKA, D. DIPETTE & H. G. PREUSS. 1986. Am. J. Physiol. **250:** E381–E385.
79. YOUNG, J. B. & L. LANDSBERG. 1981. Metabolism **30:** 421–424.
80. YOUNG, J. B. & L. LANDSBERG. 1977. Nature **269:** 615–617.
81. KOPIN, I. J., D. S. GOLDSTEIN & G. Z. FEUERSTEIN. 1981. *In* Frontiers in Hypertension Research, p. 283–289. Springer-Verlag. New York/Berlin.
82. KRIEGER, D. R. & L. LANDSBERG. 1988. Am. J. Hypertens. **1:** 84–90.
83. HWANG, I. S., H. HO, B. B. HOFFMAN & G. M. REAVEN. 1987. Hypertension **10:** 512–516.
84. REAVEN, G. M., H. HO & B. HOFFMAN. 1989. Hypertension **14:** 117–120.
85. ZAVARONI, I., Y. CHEN, C. E. MONDON & G. M. REAVEN. 1981. Metabolism **30:** 476–480.
86. REAVEN, G. M., H. HO & B. B. HOFFMAN. 1988. Hypertension **12:** 129–132.
87. BRANDS, M. W., D. A. HILDEBRANDT, H. L. MIZELLE & J. E. HALL. 1992. Hypertension **19**(suppl. I): 83–89.
88. GRUNDLEGER, M. L. & S. W. THENEN. 1982. Diabetes **31:** 232–237.
89. STORLIEN, L. H., D. E. JAMES, K. M. BURLEIGH, D. J. CHISHOLM & E. W. KRAEGEN. 1986. Am. J. Physiol. **251:** E576–E583.
90. STORLIEN, L. H., A. B. JENKINS, D. J. CHISHOLM, W. S. PASCOE, S. KHOURI & E. W. KRAEGEN. 1991. Diabetes **40:** 280–289.
91. KAUFMAN, L. N., M. M. PETERSON & S. M. SMITH. 1991. Am. J. Physiol. **260:** E95–E100.
92. KAUFMAN, L. N., M. M. PETERSON & S. M. SMITH. 1994. Metabolism **43:** 1–3.
93. SCHWARTZ, J. H., J. B. YOUNG & L. LANDSBERG. 1983. J. Clin. Invest. **72:** 361–370.
94. KRIEGER, D. R. & L. LANDSBERG. 1988. Am. J. Hypertens. **1:** 84–90.
95. DIBONA, G. F. 1986. Fed. Proc. **45:** 2871–2884.
96. LIU, S., V. E. BARACOS, H. A. QUINNEY, T. LE BRICON & M. T. CLANDININ. 1995. Endocrinology **136:** 3318–3324.
97. MAHER, M. A., W. J. BANZ & M. B. ZEMEL. 1995. J. Nutr. **125:** 2618–2622.
98. LEE, M. K., P. D. G. MILES, M. KHOURSHEED, K. M. GAO, A. R. MOOSSA & J. M. OLEFSKY. 1994. Diabetes **43:** 1435–1439.
99. KRAEGEN, E. W., D. E. JAMES, A. B. JENKINS, D. J. CHISHOLM & L. H. STORLIEN. 1989. Metabolism **38:** 1089–1093.
100. IWANISHI, M. & M. KOBAYASHI. 1993. Metabolism **42:** 1017–1021.
101. STEVENSON, R. W., E. M. GIBBS, J. C. PARKER, G. I. SHULMAN, M. J. KESSEDJIAN, J. A. REYNOLDS & D. K. KREUTTER. 1995. *In* Lessons from Animal Diabetes, p. 181–195. Smith-Gordon. London.
102. HÄRING, H. U., S. TIPPMER, K. KELLERER, L. MOSTHAF, G. KRODER, B. BOSSENMAIER & L. BERTI. 1996. Diabetes **45**(suppl. 1): S115–S119.
103. APWEILER, R., H. F. KÜHNLE, G. RITTER, R. SCHELL & P. FREUND. 1995. Metabolism **44:** 577–583.
104. DUBEY, R. K., H. Y. ZHANG, S. R. REDDY, M. A. BOEGEHOLD & T. A. KOTCHEN. 1993. Am. J. Physiol. **265:** R726–R732.
105. PERSHADSINGH, H. A., J. SZOLLOSI, S. BENSON, W. C. HYUN, B. G. FEUERSTEIN & T. W. KURTZ. 1993. Hypertension **21:** 1020–1023.
106. GEELEN, M. J. H. 1983. Biochem. Pharmacol. **32:** 1765–1772.

107. STORLIEN, L. H., N. D. OAKES, D. A. PAN, M. KUSUNOKI & A. B. JENKINS. 1993. Diabetes **42:** 457-462.
108. RAINA, N., J. MATSUI, S. C. CUNNANE & K. N. JEEJEEBHOY. 1995. Lipids **30:** 713-718.
109. SIERRA, P., P. R. LING, N. W. ISTFAN & B. R. BISTRIAN. 1995. Metabolism **44:** 1365-1370.
110. KLIMEŠ, I., M. FICKOVÁ, E. ŠVÁBOVÁ, A. VRÁNA, P. BOHOV, Š. ZÓRAD, L. ŠIKUROVÁ, L. KAZDOVÁ, V. BALÁŽ & L. MACHO. 1989. In Biomembranes and Nutrition. Vol. 195, p. 429-440. Colloque INSERM. Paris.
111. VRÁNA, A., L. KAZDOVÁ, I. KLIMEŠ, M. FICKOVÁ & A. ŽÁK. 1990. In Insulin and the Cell Membrane, p. 397-413. Harwood. New York.
112. ZIEMLANSKI, S., B. PANCZENKO-KRESOWSKA, G. OKOLSKA, E. WIELGUS-SERAFINSKA & K. ZELAKIEWICZ. 1985. Ann. Nutr. Metab. **29:** 223-231.
113. IACONO, J. M. & R. M. DOUGHERTY. 1993. Annu. Rev. Nutr. **13:** 243-260.
114. HOWE, P. R. C., P. F. ROGER & Y. LUNGERSHAUSEN. 1991. Prostaglandins Leukotrienes Essent. Fatty Acids **44:** 113-117.
115. BEXIS, S., Y. K. LUNGERSHAUSEN, M. T. MANO, P. R. C. HOWE, J. Q. KONG, D. L. BIRKLE, D. A. TAYLOER & R. J. HEAD. 1994. Blood Pressure **3:** 120-126.
116. EDELSTEINOVÁ, S., J. KYSELOVIČ, I. KLIMEŠ, E. ŠEBÖKOVÁ, B. KOVÁCSOVÁ, F. KRISTEK, A. MITKOVÁ, A. VRÁNA & P. ŠVEC. 1993. Ann. N.Y. Acad. Sci. **683:** 353-356.
117. BEILIN, L. J. 1993. Ann. N.Y. Acad. Sci. **683:** 35-45.
118. YIN, K., Z. M. CHU & L. J. BEILIN. 1991. Br. J. Pharmacol. **15:** 991-997.
119. MANO, M. T., S. BEXIS, Y. ABEYWARDENA, E. J. MCMURCHIE, R. A. KING, R. M. SMITH & R. J. HEAD. 1995. Blood Pressure **4:** 177-186.
120. GINSBERG, B. H., P. CHATTERJEE & M. YOREK. 1990. In Insulin and the Cell Membrane, p. 313-330. Harwood. New York.
121. CRÉMEL, G., M. FICKOVÁ, I. KLIMEŠ, C. LERAY, V. LERAY, E. MEUILLET, M. ROQUES, C. STAEDEL & P. HUBERT. 1993. Ann. N.Y. Acad. Sci. **683:** 164-171.
122. AXELROD, L. 1991. Diabetes **40:** 1223-1227.
123. KLIMEŠ, I., E. ŠEBÖKOVÁ, A. MINČENKO, A. VRÁNA, M. FICKOVÁ, M. HROMADOVÁ, E. ŠVÁBOVÁ, P. BOHOV & L. KAZDOVÁ. 1991. In Lipoproteins and Atherosclerosis, p. 55-62. Fischer. Jena, Germany.
124. EDWARDS, J. G. & C. M. TIPTON. 1989. J. Appl. Physiol. **67:** 2335-2342.
125. TOWNSEND, R. R., R. YAMAMOTO, M. NICKOLS, D. J. DIPETTE & G. A. NICKOLS. 1992. Hypertension 19(suppl. II): 105-110.
126. BRANDS, M. W., D. A. HILDEBRANDT, H. L. MIZELLE & J. E. HALL. 1991. Am. J. Physiol. **260:** R764-R768.
127. YOKOTA, M., K. SHIMODA, Y. KOBAYASHI & N. KAJIWARA. 1989. Nippon Jinzo Gakkai Shi **31:** 875-881.
128. MOREAU, P., L. LAMARCHE, A. K. LAFLAMME, A. CALDERONE, N. YAMAGUCHI & J. DE CHAMPLAIN. 1995. J. Hypertens. **13:** 333-340.
129. ŠTOLBA, P., P. HUŠEK & J. KUNEŠ. 1994. Physiol. Res. **43:** 329-334.
130. BRANDS, M. W. & J. E. HALL. 1992. J. Am. Soc. Nephrol. **3:** 1064-1077.
131. BHANOT, S., J. H. MCNEILL & M. BRYER-ASH. 1994. Hypertension **23:** 308-312.
132. DALY, P. A. & L. LANDSBERG. 1991. Diabetes Care **14:** 240-248.
133. GANS, R. O. B. & A. J. M. DONKER. 1991. J. Intern. Med. 45(suppl. 2): 49-64.
134. HALL, J. E., M. W. BRANDS, D. H. ZAPPE & M. ALONSO-GALICIA. 1995. Clin. Exp. Pharmacol. Physiol. **22:** 689-700.
135. JEŽOVÁ, D., R. KVETŇANSKÝ, K. KOVÁCS, Z. OPRŠALOVÁ, M. VIGAŠ & G. B. MAKARA. 1987. Endocrinology **120:** 409-415.
136. YAMAGUCHI, N. 1991. Can. J. Physiol. Pharmacol. **70:** 167-206.
137. MACHO, L., Š. ZÓRAD, P. BLAŽÍČEK, M. FICKOVÁ & R. KVETŇANSKÝ. 1996. In Catecholamines and Other Neurotransmitters in Stress: Molecular Genetic and Neurobiological Advances. Gordon & Breach. New York.
138. RETT, K., M. WICKLMAYR & H. MEHNERT. 1994. Eur. Heart J. 15(suppl. C): 78-81.
139. LANDSBERG, L. & J. B. YOUNG. 1986. In Monitoring Neurotransmitter Release during Behaviour, p. 33-47. Ellis Horwood. Chichester.
140. YOUNG, J. B. 1988. Life Sci. **43:** 193-200.

141. FERRARI, P., G. B. PICOTTI, E. MINOTTI, G. B. BONDIOLOTTI, A. M. CARAVAGGI & G. BIANCHI. 1981. Clin. Sci. **61:** 199-202.
142. PAK, CH. 1981. Jpn. Heart J. **22:** 987-995.
143. BORKOWSKI, K. R. & P. QUINN. 1983. J. Auton. Pharmacol. **3:** 157-160.
144. UNGER, T., H. BECKER & R. DIETZ. 1984. Circ. Res. **54:** 30-37.
145. JABLONSKI, L. T. & P. R. C. HOWE. 1994. Blood Pressure **3:** 106-111.
146. KIRCHNER, K. A. 1988. Am. J. Physiol. **255:** F1206-F1213.
147. FINCH, D., G. DAVIS, J. BOWER & K. KIRCHNER. 1990. Hypertension **15:** 514-518.
148. DEFRONZO, R. A., M. GOLDBERG & Z. S. AGUS. 1976. J. Clin. Invest. **58:** 83-90.
149. BAUM, M. 1987. J. Clin. Invest. **79:** 104-109.
150. BELLO-REUSS, E., R. E. COLINDRES, E. PASTORIZA-MUNOZ, R. A. MUELLER & C. W. GOTTSCHALK. 1975. J. Clin. Invest. **56:** 200-217.
151. BESAROB, A., P. SILVA, L. LANDSBERG & F. EPSTEIN. 1977. Am. J. Physiol. **233:** F39-F45.
152. HARRIS, P. G. & L. G. NAVAR. 1985. Am. J. Physiol. **248:** F621-F628.
153. GANS, R. O. B. 1991. J. Intern. Med. **45**(suppl. 2): 49-64.
154. TOMIYAMA, H., T. KUSHIRO, H. ABETA, H. KURUMATANI, H. TAGUCHI, N. KUGA, F. SAITO, F. KOBAYASHI, Y. OTSUKA, K. KANMATSUSE & N. KAJIWARA. 1992. Hypertension **20:** 596-600.
155. HALL, J. E., M. W. BRANDS, D. A. HILDEBRANDT & H. L. MIZELLE. 1992. Hypertension **19**(suppl. I): I45-I55.
156. MOORE, R. D. 1983. Biochim. Biophys. Acta **737:** 1-49.
157. HUOT, S. J. & P. S. ARONSON. 1991. Diabetes Care **14:** 521-535.
158. FERRARI, P. & P. WEIDMANN. 1990. J. Hypertension **8:** 491-500.
159. CANESA, M. 1994. Curr. Opin. Nephrol. Hypertens. **3:** 511-517.
160. BORLE, A. B. 1981. Rev. Physiol. Biochem. Pharmacol. **90:** 113-153.
161. LEVY, J., M. B. ZEMEL & J. R. SOWERS. 1989. Am. J. Med. **87**(suppl. 6A): 7-16.
162. MAHNENSMITH, R. L. & P. S. ARONSON. 1985. Circ. Res. **56:** 773-788.
163. DUHM, J., B. O. GÖBEL & F. X. BECK. 1983. Hypertension **5:** 642-652.
164. BIANCHI, G., P. FERRARI & D. TRIZIO. 1985. Hypertension **7:** 319-325.
165. FEIG, P. U., P. P. MITCHELL & J. W. BOYLAN. 1985. Hypertension **7:** 423-429.
166. ZICHA, J. & J. DUHM. 1990. Hypertension **15:** 612-627.
167. BIN TALIB, K. H., Z. DOBEŠOVÁ, P. KLÍR, V. KŘEN, J. KUNEŠ, M. PRAVENEC & J. ZICHA. 1992. Hypertension **20:** 575-582.
168. HUNT, S. C., R. R. WILLIAMS, J. D. SMITH & K. O. ASH. 1986. Hypertension **8:** 30-36.
169. DUHM, J. & J. BEHR. 1986. Scand. J. Lab. Invest. **46**(suppl. 180): 82-95.
170. HAJEM, S., T. MOREAU, P. HANNAERT, J. LELOUCH, G. ORSSAUD, G. HUEL, J. R. CLAUDE & R. P. GARAY. 1990. J. Hypertens. **8:** 891-896.
171. LIJNEN, P., R. FAGARD, J. STAESSEN, L. THIJS & A. AMERY. 1992. J. Hypertens. **10:** 1205-1211.
172. PAGNAN, A., R. CORROCHE, G. B. AMBROSIO, S. FERRARI, P. GUARINI, D. PICCOLO, A. OPPORTUNO, A. BASSI, O. OLIVIERI & G. BAGGIO. 1989. Clin. Sci. **76:** 87-93.
173. KELLY, R. A., M. L. CANESSA, T. I. STEINMAN & W. E. MITCH. 1989. Kidney Int. **35:** 595-603.
174. RUTHERFORD, P. AA., T. H. THOMAS, M. F. LAKER & R. WILKINSON. 1992. Eur. J. Clin. Invest. **22:** 719-724.
175. ENGELMANN, B., J. A. OP DEN KAMP & B. ROELOFSON. 1990. Am. J. Physiol. **258:** C682-C691.
176. DUHM, J. & B. ENGELMANN. 1992. Pharm. Pharmacol. Lett. **2:** 32-35.
177. DUHM, J., B. ENGELMANN, U. M. SCHÖNTHIER & S. STREICH. 1993. Biochim. Biophys. Acta **749:** 185-188.
178. LIANG, C. S., J. U. DOHERTY, R. FAILLACE, K. MAEKAWA, S. ARNOLD, H. GARRAS & W. B. HOOD, JR. 1982. J. Clin. Invest. **69:** 1321-1336.
179. LAAKSO, M., S. V. EDELMAN, G. BRECHTEL & A. D. BARON. 1990. J. Clin. Invest. **85:** 1844-1852.
180. ALEXANDER, W. D. & R. J. OAKE. 1977. Diabetes **26:** 611-614.
181. YAGI, S., K. TAKATA, H. HIYOKAWA, M. YAMAMOTO, Y. NOTO, T. IKEDA & N. HATTORI. 1988. Diabetes **37:** 1064-1067.
182. REDDY, S., S. SHEHIN & J. R. SOWERS. 1990. J. Vasc. Med. Biol. **2:** 47-51.
183. VANHOUTTE, P. M. 1989. Hypertension **13:** 658-667.
184. SANDERS, P. W. 1992. J. Nephrol. **5:** 23-30.
185. BARON, A. D., J. ZHU, S. MARSHALL, O. IRSULA, G. BRECHTEL & C. KEECH. 1995. Am. J. Physiol. **269:** E709-E715.

186. MOREAU, P., N. YAMAGUCHI & J. DE CHAMPLAIN. 1993. J. Hypertens. 11(suppl. 5): S270–S271.
187. BOEGEHOLD, M. A. 1992. Hypertension 19: 290–295.
188. PALMER, R. M. J., D. S. ASHTON & S. MONCADA. 1988. Nature 333: 664–666.
189. CHEN, P. Y. & P. W. SANDERS. 1993. Hypertension 22: 812–818.
190. CHEN, P. Y. & P. W. SANDERS. 1991. J. Clin. Invest. 88: 1559–1567.
191. TAKAHISHI, H., T. NAKANISHI, M. NISHIMURA, H. TANAKA & M. YOSHIMURA. 1992. J. Cardiovasc. Pharmacol. 20: S176–S178.
192. PITRE, M. & H. BACHELARD. 1994. Hypertension 24: 376 (abstract).
193. TOGASHI, H., I. SAKUMA, M. YOSHIOKA, T. KOBAYASHI, H. YASUDA, A. KITABATAKE, H. SAITO, S. S. GROSS & R. LEVI. 1992. J. Pharmacol. Exp. Ther. 262: 343–347.
194. BARON, A. D. 1996. Diabetes 45(suppl. 1): S105–S109.
195. KOHLMAN, O., F. DE ASSIS ROCHA NEVES, M. GINOZA, A. TAVARES, M. L. CEZARETTI, A. B. RIBEIRO, I. GAVRAS & H. GAVRAS. 1995. Hypertension 25: 1003–1007.
196. STOUT, R. W. 1990. Diabetes Care 13: 631–654.
197. PFEIFLE, B. & H. DITSCHUNEIT. 1981. Diabetologia 20: 155–158.
198. KING, G. L., D. GOODMAN, S. BUZNEY, A. MOSES & C. R. KAHN. 1985. J. Clin. Invest. 75: 1028–1036.
199. KAHN, C. R. 1994. Diabetes 43: 1066–1084.
200. VINTERS, H. V. & J. A. BEREINER. 1987. Diabetes Metab. 13: 294–300.
201. KIHARA, M., R. HORIE, W. LOVENBERG & Y. YAMORI. 1993. Heart Vessels 8: 7–15.
202. SCHORK, N. J., P. JOKELAINEN, E. J. GRANT, M. A. SCHORK & A. B. WEDER. 1994. Am. J. Physiol. 266: R702–R708.
203. MACHO, L., M. FICKOVÁ & A. HUBKOVÁ. 1977. Endocrinol. Exp. 11: 75–83.
204. CONET, CH., J. DELARUE, T. CONSTANS & F. LAMISSE. 1992. Horm. Res. 38: 46–50.
205. BARZILAI, N. & L. ROSETTI. 1995. Am. J. Physiol. 269: E591–E597.
206. GRAY, S. D. 1988. Microvasc. Res. 36: 228–238.
207. SMITH, D., H. GREEN, J. THOMPSON & M. SHARATT. 1989. Am. J. Physiol. 256: C50–C58.
208. BACHIR-LAMRINI, L. B., B. DEMPORE, M. H. MAYET & R. J. FAVIER. 1990. Am. J. Physiol. 258: R352–R357.
209. LILLIOJA, S., A. A. YOUNG, C. CULTER, J. L. IVY, W. G. H. ABBOTT, J. K. ZAWADZKI, H. YKI-JARVINEN, L. CHRISTIN, T. W. SECOMB & C. BOGARDUS. 1987. J. Clin. Invest. 80: 415–424.
210. LEWIS, D. M., A. J. LEVI, P. BROOKSBY & J. V. JONES. 1994. Exp. Physiol. 79: 377–386.
211. BONEN, A., M. H. TAN & W. M. WATSON-WRIGHT. 1981. Diabetes 30: 702–704.
212. JAMES, D. E., E. W. KRAEGEN & D. J. CHISHOLM. 1985. Am. J. Physiol. 248: E575–E580.
213. JAMES, D. E., E. W. KRAEGEN & D. J. CHISHOLM. 1985. J. Clin. Invest. 76: 657–666.
214. PLOUGH, T., H. GALBO, J. VINTEN, M. JORGENSEN & E. A. RICHTER. 1987. Am. J. Physiol. 253: E12–E20.
215. KRIKETOS, A. D., D. A. PAN, J. R. SUTTON, J. F. Y. HOH, L. A. BAUR, G. J. COONEY, A. B. JENKINS & L. H. STORLIEN. 1995. Am. J. Physiol. 269: R1154–R1162.
216. ATRAKCHI, A., S. D. GRAY & R. C. CARLSEN. 1994. Am. J. Physiol. 267: C827–C835.
217. MARK, A. L. & E. A. ANDERSON. 1995. Proc. Soc. Exp. Biol. Med. 208: 330–336.
218. RAPP, J. P., S. M. WANG & H. DENE. 1989. Science 243: 542–544.
219. KURTZ, T. W., L. SIMONET, P. M. KABRA, S. WOLFE, L. CHAN & B. L. HJELLE. 1990. J. Clin. Invest. 85: 1328–1332.
220. PRAVENEC, M., L. SIMONET, V. KŘEN, J. KUNEŠ, G. LEVEN & J. SPRIZER. 1991. Genomics 9: 466–472.
221. DUBAY, C., M. VINCENT, N. SAMANI, P. HILBERT, M. A. KAISER & J. P. BERESSI. 1993. Nat. Genet. 3: 354–357.
222. HARRIS, E. L., E. LINTON PHELAN, C. M. THOMPSON, J. A. MILLAR & M. R. GRIGOR. 1995. J. Hypertens. 13: 397–404.
223. JACOB, H. J., K. LINDPAINTNER, S. E. LINCOLN, K. KUSUMI, R. K. BUNKER & Y. P. MAO. 1991. Cell 67: 213–224.
224. HILBERT, P., K. LINDPAINTNER, I. S. BECKMANN, T. SERIKAWA, F. SOUBRIER & C. DUBAY. 1991. Nature 353: 521–529.
225. NARA, Y., T. NABIKA, K. IKEDA, M. SAWAMURA, J. ENDO & Y. YAMORI. 1991. Biochem. Biophys. Res. Commun. 181: 941–946.
226. DENG, A. Y. & J. P. RAPP. 1991. Nat. Genet. 1: 267–272.

227. DENG, A. Y., H. DENE & J. P. RAPP. 1994. J. Clin. Invest. **94:** 431–436.
228. PRAVENEC, M., V. KŘEN, J. KUNEŠ, G. A. SCICLI, O. A. CARRETERO, L. SIMONET & T. W. KURTZ. 1991. Hypertension **17:** 242–246.
229. BIANCHI, G., G. TRIPODI, G. CASARI, S. SALARDI, B. R. BARBER, R. GARCIA, P. LEONI, L. TORIELLI, D. CUSI, M. FERRANDI, L. A. PINNA, F. E. BARALLE & P. FERRARI. 1994. Proc. Natl. Acad. Sci. U.S.A. **91:** 3999–4003.
230. DENG, A. Y. & J. P. RAPP. 1995. J. Clin. Invest. **95:** 2170–2177.
231. LODWICK, D., M. A. KAISER, J. HARRIS, F. CUMIN, M. VINCENT & N. J. SAMANI. 1995. Hypertension **25:** 1245–1251.
232. INGLIS, G. C., C. J. KENYON, C. SZPIRER, K. KLINGA-LEVAN, R. G. SUTCLIFFE & J. CONNELL. 1995. J. Mol. Endocrinol. **14:** 303–311.
233. KURTZ, T. 1994. Lancet **344:** 167–168.
234. NAVARRO-CID, J., R. MAESO, F. PEREZ-VIZCAINO, V. CACHOFEIRO, L. M. RUILOPE, J. TAMARGO & V. LAHERA. 1995. Hypertension **26:** 1074–1078.
235. CHOW, L., M. DE GASPARO & N. LEVENS. 1995. Eur. J. Pharmacol. **282:** 77–86.
236. HENRIKSEN, E. J. & S. JACOB. 1995. Metabolism **44:** 267–272.
237. LEIGHTON, B., A. L. SANDERSON, M. E. YOUNG, G. K. RADDA, E. A. BOEHM & J. F. CLARK. 1996. Diabetes **45**(suppl. 1): S120–S124.
238. HOEHE, M. R., R. PLEATKE, B. OTTERUD, D. STAUFFER, J. HOLIK & W. F. BYERLEY. 1992. Hum. Mol. Genet. **1:** 175–178.
239. DAVIDSON, A. O., N. SCHORK, B. C. JAQUES, A. W. KELMAN, R. G. SUTCLIFFE, J. L. REID & A. F. DOMINICZAK. 1995. Hypertension **26:** 452–459.
240. CAULFIELD, M., P.-M. BOULOUX & P. MUNROE. 1997. This volume.
241. BENNETT, P. H. 1996. Abstract. This conference.
242. GIACOBINO, J.-P. 1995. Eur. J. Endocrinol. **132:** 377–385.
243. MEIER, CH. A. 1995. Eur. J. Endocrinol. **133:** 762–763.

Obesity Genes and Insulin Resistance Syndrome

E. BOWIE KAHLE,[a] RUDOLPH L. LEIBEL,[b]
DIRK W. DOMASCHKO,[a] SAM G. RANEY,[a]
AND K. TRACI MANN[a]

[a] *Department of Biology*
Marshall University
Huntington, West Virginia 25755

[b] *Rockefeller University*
New York, New York 10021

INTRODUCTION

Obese rodents are widely studied experimental models for human obesity. Their reproduction rate, offspring numbers, and degree of inbreeding are advantages over other mammals as research models for human obesity. Six single-gene obesity mutations have been identified in mice. Obese (*ob*), diabetes (*db*), fat (*fat*), and tubby (*tub*) are inherited as autosomal recessive syndromes. The agouti (yellow) (A^y) and Adipose (*Ad*) (now extinct) obese phenotypes are inherited as dominant traits. The obese phenotypes of all are associated with some degree of insulin resistance. Each mutation is associated with complex neuroendocrine disturbances.[1] Homologous genes, but not mutations, have been identified in humans. Their genetic locations have been determined based upon synteny relationships between the mouse and human genomes.[2]

This review will emphasize *ob* and *db* early-onset obese phenotypes which in all strains exhibit elevated insulin secretion and insulin resistance. Also, reviews of the less insulin-resistant, later-onset obesity phenotypes of *fat*, *agouti yellow* (A^y), and *tub* (a brief description) will be made. The mechanisms for obesity and related insulin resistance differ among the mutant genes. However, each includes elements of similarity at two levels: the overstimulation of insulin secretion and the apparent effects on hypothalamic centers influencing energy expenditure and food intake.

MUTATIONS IN THE *ob* AND *db* GENES

The most studied rodent single-gene obesity mutations are *obese* (*ob*)[3] and *diabetes* (*db*).[4] All strains of mice segregating for mutations in *ob* or *db*, or rats with the *db* homologue, *fa* (Zucker fatty)[5] and *fa*^{k},[6] exhibit insulin resistance at some point during development of obesity. A distinguishing characteristic among the strains is susceptibility to beta-cell failure (overt diabetes).[3,7,8]

The Mouse ob Mutant

The *ob* mutation identified in 1950 was the first recessive obesity mutation described in mice.[9] The *ob* mutation results in profound obesity and insulin resistance as part of a syndrome that resembles morbid obesity in humans.[10] Cross-circulation experiments between mutant and wild-type mice suggested that *ob* mice are deficient for a circulating factor that suppresses food intake.[3] The *ob* mutation is associated with myriad hormonal and metabolic alterations,[6] including effects on thermoregulation, fertility, adrenal, and thyroid function.[11]

Zhang *et al.*[12] reported the cloning and sequencing of the mouse *ob* gene and its human homologue. The *ob* gene encodes an adipose tissue messenger RNA for a 167-amino-acid peptide which is secreted from adipose tissue. Its amino-acid sequence is highly conserved (84% identical) between human and mouse. The gene product is apparently part of a signaling pathway from adipose tissue that acts to regulate the size of body fat[13] stores by effects on food intake and energy expenditure. Because a principal action of the protein is to make an animal thinner, Halaas *et al.*[14] proposed that this 16-kDa protein be called "leptin," derived from the Greek root, leptos, meaning thin.

Leptin reduces food intake, body weight, and blood glucose concentrations and also affects fertility in mice with leptin deficiency due to mutation of *ob* (*ob/ob* mice).[14-18] The hyperphagia and excessive body weight gain characteristics of the complex *ob* phenotype are consistent with an imbalance in the activity of the autonomic nervous system (ANS) and with low sympathetic and high parasympathetic tone.[11]

The Mouse db Mutant

The autosomal recessive diabetes mutation (*db*) was first detected in progeny of the C57BL/KsJ strain at the Jackson Laboratory[4] and mapped to the middle of chromosome 4. Subsequently, the mutation has been detected at least four more times in other mouse strains, as well as in two rat strains.[19,20] The phenotype of *db/db* mice, which includes severe early-onset obesity, extreme insulin resistance, and strain-specific susceptibility to diabetes, is identical to that of *ob/ob* mice on the same strain background.[3] The *fatty* (*fa*) gene in rats[5] was thought to be a homologue of *db* because *fa/fa* mutants have a similar behavioral and metabolic phenotype to *db/db* mutants and because *fa* and *db* map to syntenic chromosomal regions.[15] Definitive evidence of this correspondence has recently been obtained with the demonstration that *db* and *fa* are mutations in the same gene, the leptin receptor.[21]

The weight loss of *ob/ob* mice that are joined by parabiosis to *db/db* mice,[3] the failure of *db/db* mice to respond to centrally administered leptin,[17,18] the high levels of *ob* mRNA in adipose tissue of *db/db* mice,[22,23] and the high serum concentrations of leptin in *db/db* and *fa/fa* animals[24] together provide compelling evidence that the *db* gene product acts distally to the *ob* gene product in the same regulatory pathway for body fat. Coleman[3] postulated that *db* encodes a receptor for *ob*. The molecular cloning of a transmembrane receptor[25,26] for leptin[12] made it possible to test the hypothesis that the mouse *diabetes* and rat *fatty* phenotypes are due to mutations in the receptor for leptin.

Chua et al.[21] showed by genetic mapping and restriction analysis that mouse *db*, rat *fa*, and the gene encoding the leptin receptor (*Lepr*) are the same gene. The precise mechanism or mechanisms of the mutations described by Chua et al.[21] and those that are present in other rodents (*db*, *db*^{Pas}, and *fa*^{k}) with apparent loss of *Lepr* function are now known. The leptin receptor gene encodes at least five splice variants of the receptor.[27,28] Different organs express certain isoforms of the receptor.[27,28] The short form, which may be a transmembrane transporter for leptin, is highly expressed in many tissues (including brain). The long form, which includes intracellular signaling motifs, is expressed mainly in the hypothalamus.

Recently, Lee et al.[27] and Tartaglia et al.[25] reported that *db* is due to a G→T substitution that introduces an extraneous splice donor site in the short-form terminal exon. The long form of the leptin receptor is truncated and nonfunctional. The *fa* mutation[29] is a single amino-acid transversion which results in reduced intracytoplasmic transport of the receptor, resulting in reduced, but perhaps not totally absent function; *fa*^{k} is a null (stop) mutation.[29] The *db*^{Pas} (Pasteur) mutation is a duplication which eliminates all isoforms of the receptor. The fine structure of the human leptin gene[30] and its genetic mapping have been reported.

The leptin receptor is homologous to members of the cytokine receptor superfamily.[25,31] In such receptors, binding of ligand to the receptor activates a *janus* (JAK) kinase[32] and leads to phosphorylation of cytoplasmic target proteins. Among the cytoplasmic target proteins activated by the *db/db* leptin receptor are a class of cytoplasmic transcription factors called STAT proteins (signal transducers and activators of transcription).[33] Ghilardi et al.[28] have demonstrated that the short-form *Lepr* is unable to activate STAT proteins,[33] thereby providing further evidence that reduced expression of the long form is sufficient to cause the *db/db* phenotype. STAT 3 is the probable mediator of the leptin signal.[34]

The STAT pathway may play an important role in leptin-mediated signaling. The loss of STAT pathway activation in *db/db* mice with the loss of leptin's putative lipostatic function may be analogous to the effect of targeted disruption of the *Stat 1* gene in mice resulting in a complete deficiency in interferon-mediated biologic responses.[35,36] Increasing evidence now exists showing that the activation of the STAT proteins might be essential for most, if not all, of the specific biological effects of the cytokine receptors.

Overstimulation of Neuropeptide Y (NPY) Secretion and Obese Phenotype ob/ob *(Leptin-deficient) and* db/db *(Leptin-unresponsive) Mice*

NPY cell bodies in the hypothalamic arcuate nucleus project to the paraventricular nucleus (PVN), where NPY release lowers energy expenditure and potently stimulates food intake.[37,38] Overexpression of NPY mRNA in the hypothalamus of *ob/ob* and *db/db* mice has been reported.[12,39,40] Campfield et al.[17] and Smith et al.[41] have reported that leptin administration into the left ventricular cavity of the cerebrum inhibits feeding in *ob/ob* mice following administration of NPY. Thus, leptin may determine the sensitivity of the feeding response to exogenous NPY. However, the total dependence of leptin function on an interaction with NPY-mediated effects of leptin has recently been questioned.[42] Erickson, Clegg, and Palmiter[42] generated an NPY knockout mouse that transiently responds to recombinant leptin administration with decreased food intake and weight loss. Later, in the five-day leptin-treatment period of this NPY knockout model,[42] food intake resembled that of controls. These results demonstrate that leptin can suppress feeding and promote weight loss via signaling pathways that are independent of those using NPY.

Weight loss with leptin administration in ob/ob mice reduces elevated concentrations of both blood glucose and NPY mRNA.[14–18] To determine whether these responses are due to a specific action of leptin or to the reversal of the obese state, Schwartz et al.[43] investigated the specificity of the effect of systemic leptin administration to ob/ob mice on concentrations of plasma glucose and insulin and on hypothalamic expression of NPY mRNA. Their experiments controlled for changes of food intake induced by leptin with saline-treated controls either fed *ad libitum* or pair-fed to the intake of the leptin-treated group. They found the systemic administration of leptin to inhibit NPY gene overexpression through a specific action in the arcuate nucleus and to exert a glucose-lowering action that is partly independent of its weight-reducing effects.

The mechanism that best explains the observations of Schwartz et al.[43] would suggest that leptin increases insulin sensitivity of peripheral tissues in ob/ob mice, resulting in an increased rate of glucose uptake. A second possibility is that the metabolic rate increased during leptin administration similar to that reported by Pelleymounter et al.,[18] thus accelerating glucose utilization through an insulin-independent mechanism. In either scenario, decreased glucose stimulation of the pancreatic islets would be expected to lower the circulating insulin level.

Rat fa and fa^k (Corpulent), Homologues of db

Two single-gene obesity mutations, fa^5 and $f,^6$ have been identified in the rat. The gene symbol *corpulent* or *cp* has also been used as a designation for the fa^k mutation.[44]

The fatty (*fa*) mutation in the rat arose spontaneously in the Merck strain in 1961 and shows strong phenotypic similarities to the *db* and *ob* mouse mutations.[5] The Zucker fatty is hyperinsulinemic with normal blood glucose levels[45] and develops adipocyte hypertrophy and hyperplasia.[46] During the intervening quarter of a century, the *fa* animal has been among the most-studied experimental models for human obesities. The *fa* obese mutation is due to a single nucleotide substitution, Gln269Pro, in the *fa* allele of the leptin receptor, *Lepr* (*Lepr*^{fa}).[29,47,48]

The *f* mutation arose spontaneously in Koletsky's hypertensive rat strain[6] and the mutation was introgressed into the LA/N and SHR/N strains developed at the NIH by Hansen.[44] In complementation studies, we have produced obese, apparent 13M-LA/N *fa/f* compound mutants,[49] confirming Yen's earlier findings with Zucker-Koletsky *fa/f* compound mutants.[50] Recently, in genetic mapping studies, we have placed the *f* gene in an interval between (*Pgm1*) and (*Glut1*) on rat chromosome (Chr) 5,[51] which also contains *fa*.[19,52] The locus designation has not been modified to recognize the fact that *fa* and *fa^k* are allelic. In this review, subsequent reference to *f* or *cp* will be made as *fa^k* ("*corpulent*"). Based on three pieces of evidence, namely, the production of *fa/f* compound mutants,[49,50] mutation sequencing identification, and molecular mapping,[51] it is now clear that *fa* and *f* are mutations in the same gene, the leptin receptor (*Lepr*).[21,25,26,29,30] The Koletsky mutation is due to a Tyr763Stop mutation in the leptin receptor.[53,54] The human homologous region for *Lepr* is 1p31-p22.[30] Linkage of a microsatellite in this region to acute insulin release in humans has been described.[55] Rat strains carrying the *fa^k* ("*corpulent*") mutation serve as a useful experimental model for disorders of carbohydrate utilization.[49,56-58] The obese LA/N-*fa^k* ("*corpulent*") rat exhibits metabolic characteristics associated with type IV hyperlipoproteinemia (carbohydrate-sensitivity) in humans,[59,60] including normoglycemia or mild hyperglycemia and basal hyperinsulinemia.

The fa^k ("Corpulent") *Rat as a Model for the Role of Muscle in Obesity Development*

The obese *fa^k* ("*corpulent*") rat may serve as a model of whole body lipid production in obesity. This is particularly important in light of the significant role of skeletal muscle in the regulation of energy balance.[61-64]

Incorporation of ^{3}H into fatty acids following ^{3}H$_2$O administration shows that the rates of *de novo* fatty acid synthesis in liver and adipose depots account for only 22% and 7%, respectively, of the total fatty acid synthesis during 60-minute posttreatment of lean (normal weight) mice.[65] Surprisingly, 21% of the total *de novo* fatty acid was localized in the postural ("back") skeletal muscles, which include the predominantly red muscles psoas major and iliacus.[65] The majority of mature, differentiated skeletal muscle cells in the postural muscles had large numbers of mitochondria.[65] We[66] and others[67,68] have found enzymatic markers for tricarboxylic acid (TCA) cycle flux in hindlimb, predominately red muscles of obese *fa^k* ("*corpulent*") and *fa* rats, respectively, to be higher than those of lean littermates. This is surprising because one of the main aberrations in the obese Zucker *fa* rat is its decreased metabolic rate coupled with an increased efficiency of energy storage.[69] Enzymatic markers for tricarboxylic flux,[64,70-72] as well as fatty acid synthesis and NADPH production,[71] were elevated in the selected muscles of obese LA/N-*fa^k* ("*corpulent*") animals compared to their lean littermates. With the apparently higher nutrient substrate supply to skeletal muscle attributed to greatly elevated food intake by the obese *fa^k* ("*corpulent*"),[56] it can be speculated that mitochondrial metabolism may meet ATP production requirements and that excess TCA cycle intermediates could be shuttled to the cytosol with greater availability to fatty acid synthetic pathways. These data are supported by our recent finding of elevated *in vivo* incorporation of tritium from ^{3}H$_2$O into *de novo* fatty acid synthesis in the predominantly red fiber muscles (FIGURE 1) [hindlimb−gastrocnemius, soleus, and vastus medius; and postural−psoas and iliacus] in obese compared with lean *fa^k* ("*corpulent*") littermates (unpublished data). Presently, none of the derangement of skeletal muscle metabolism of rodent obesity models described here can be

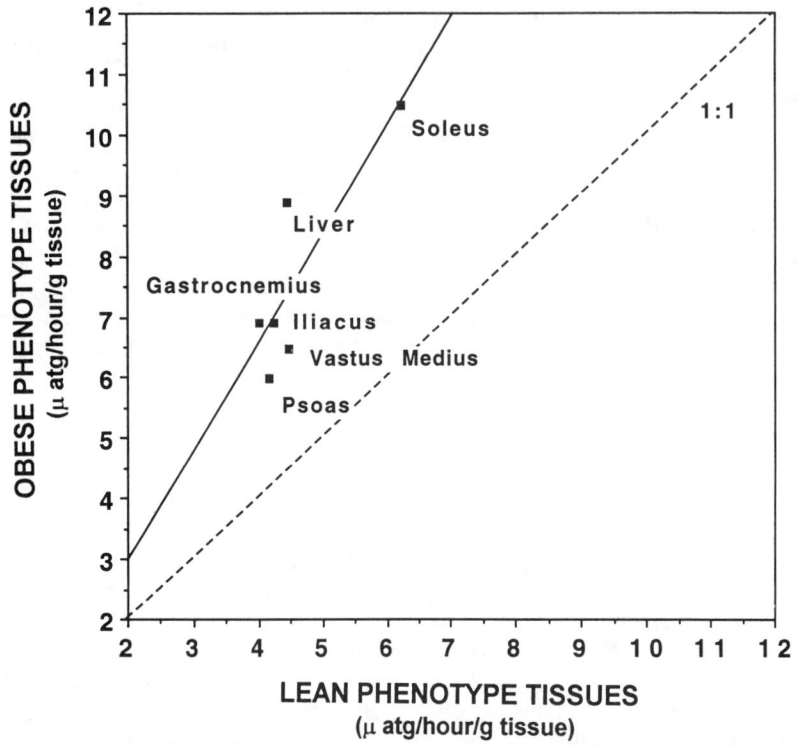

FIGURE 1. ^{3}H-labeled fatty acid concentrations in selected skeletal muscles and liver of fa^k ("corpulent") rats at 60 minutes following intraperitoneal administration of ^{3}H$_2$O. Obese fa^k ("corpulent") rats exhibited higher concentrations compared with lean littermates ($P < 0.05$). Data are presented with values for obese rat tissue types on the y-axis and those of lean littermates on the x-axis. The solid line represents the following equation: $y = -0.060631 + 1.7888x$, $r^2 = 0.722$, $\alpha = 0.01$. The broken line represents a null hypothesis (1:1 relationship) that there is no difference between obese and lean phenotypes for fatty acid synthesis.

linked to specific obesity mutations. Each metabolic aberration in obese compared with lean counterparts may be secondary to the obese condition per se.

Current investigations of the ultrastructure of the soleus, gastrocnemius, and psoas major muscles show distinctive differences in undifferentiated myoblasts or satellite cells between obese and lean muscle (unpublished data). Skeletal muscle satellite cells are single cells that are intimately associated with myotubes.[73] As myoblasts fuse to form multinucleated myotubes and undergo terminal differentiation,[74,75] undifferentiated myoblasts become encased in the basement membrane of mature myotubes and are termed satellite cells.[75] At least 30% of the myofiber nuclei of young (juvenile) muscle are satellite cells.[74] This cell population is greater in predominantly oxidative muscles compared with the numbers in predominantly glycolytic muscles. Satellite cells in mature, uninjured lean animal skeletal muscle show properties of atrophy with highly condensed chromatin, reduced cytoplasm, and absent or poorly developed organelles.[74-76] The satellite cells of the LA/N-fa^k ("corpulent") lean phenotype display "normal" (atrophic) morphology (FIGURES 2A and 2C; FIGURE 3A). However, those from obese animals (FIGURES 2B and 2D; FIGURES 3B-D) exhibit well-developed organelles, retention of cytoplasmic volume, and highly fenestrated cell membranes. The mature myotube ultrastructures of these obese and lean LA/N-fa^k ("corpulent") rats were similar.

Analyses of the ultrastructures from electron micrographs of satellite cells from obese animals (18 micrographs representing 4 rats) compared with lean animals (16 micrographs representing 4 rats) demonstrate a higher cytoplasm-to-nuclear ratio ($P < 0.01$), presence (obese) versus absence (lean) of endoplasmic reticulum ($P < 0.01$), elevated numbers of fenestrations ($P < 0.01$), and higher numbers of vesicles ($P < 0.01$) and of mitochondria ($P < 0.01$). These structural characteristics suggest that the skeletal muscle satellite cells from these obese rats are capable of extensive metabolic activity.[74,75] Skeletal muscle satellite cell morphology or metabolism has not been reported for other obese animal models. Well-developed organelles and retention of cytoplasmic volume in the obese satellite cells pose the intriguing prospect

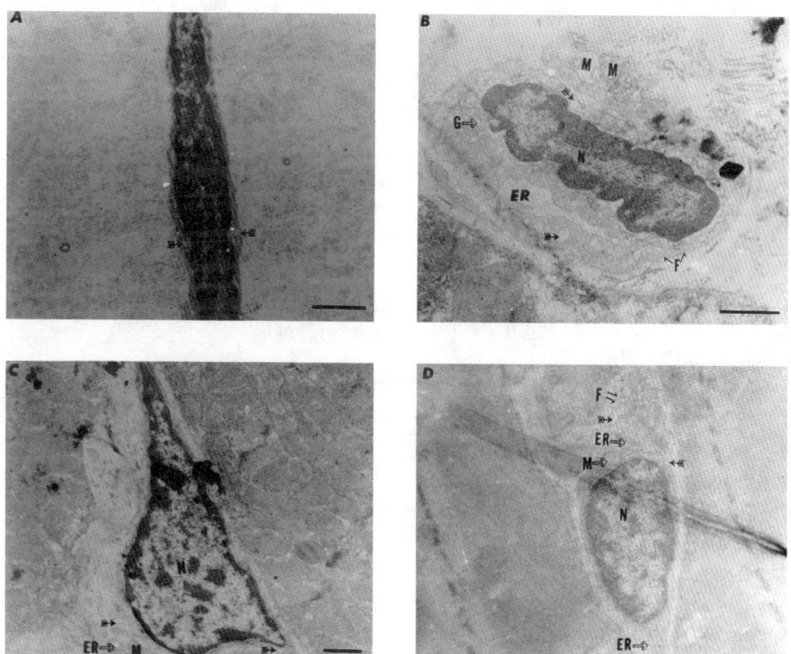

FIGURE 2. Ultrastructure of fa^k ("corpulent") rat soleus and gastrocnemius muscle satellite cells. Lean phenotype: (A) soleus and (C) gastrocnemius exhibit underdeveloped mitochondria (M) and endoplasmic reticulum (ER). Nuclei (N) are very condensed and possess few vesicles (arrows). Obese phenotype: (B) soleus and (D) gastrocnemius have well-developed endoplasmic reticulum (ER), Golgi apparatus (G), and multiple mitochondria (M). Nuclei (N) are less condensed than lean littermates. Multiple vesicles (arrows) and fenestrations (F) are present. Bar = 1 mm.

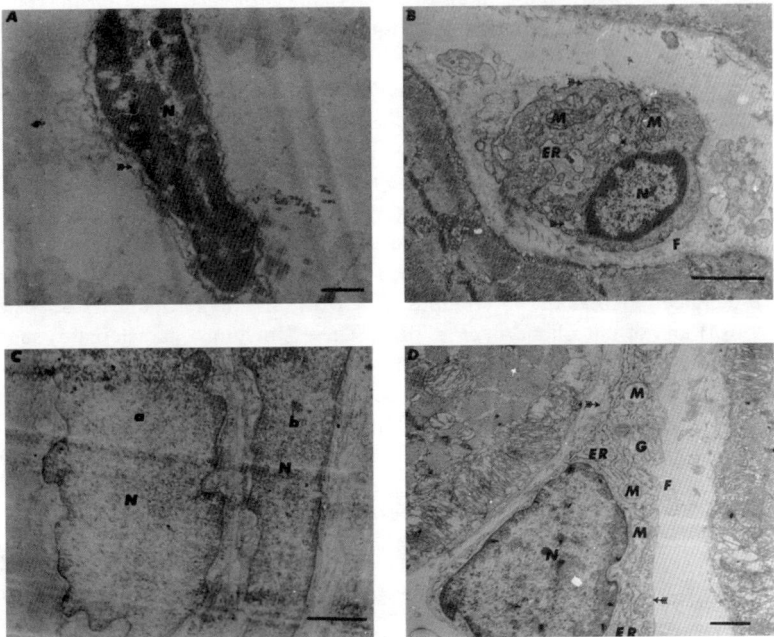

FIGURE 3. Ultrastructure of fa^k *("corpulent")* rat psoas muscle satellite cells. Lean phenotype: (*A*) satellite cell lacks organelles, has a very condensed nucleus (N), and possesses few vesicles (arrows). Obese phenotype: panels *B* and *D* have well-developed endoplasmic reticulum (ER), Golgi apparatus (G), and multiple mitochondria (M). Nuclei (N) are less condensed than lean littermates. Multiple vesicles (arrows) and marked fenestrations (F) are present. Obese phenotype: (*C*) adjacent myofiber (*a*) and satellite (*b*) cell nuclei exhibit similar loosely packed nuclei. Bar = 1 mm.

that, in "obese animals," the metabolic activity of skeletal muscle may be enhanced by its satellite cell content.

The idea that skeletal muscle cells have the capacity to synthesize large quantities of fatty acids is consonant with the microanatomic observation of fat deposits within muscle. The cell type within whole muscle that is involved, whether undifferentiated muscle cells or intermuscular connective tissue adipocytes, may play an important role in providing free fatty acids for oxidation by adjacent mature skeletal muscle myotubes. There is a growing body of evidence to support the idea of such a function. We have experiments with satellite cell primary cultures under way to test this hypothesis.

Characterization of the muscle satellite cell as a novel cell type in genetically obese rodents may help to elucidate the role of skeletal muscle in the regulation of energy balance (partitioning of nutrients between storage and oxidation), possibly via effects on metabolic fuel (glucose versus free fatty acids) utilized.[64] Again, there is no evidence, at present, to link the novel structural characteristics of skeletal muscle satellite cells of this rodent obesity model to specific obesity mutations. Each structural aberration in obese compared with lean counterparts may be secondary to the obese condition itself.

MUTATIONS OF THE Cpe^{fat} AND tub GENES

The Cpe^{fat} mutation (Chr 8) occurred spontaneously on the inbred HRS/J (hairless) strain at the Jackson Laboratory in 1973.[77] The fat mutation maps to mouse Chr 8.[78] The tub mutation (Chr 7)[79] was first noted in a C57BL/6J (FI25) male breeder also at the Jackson Laboratory. Both mutations were initially characterized phenotypically by Coleman and Eicher.[77]

The homozygous obese Cpe^{fat} animal exhibits elevated plasma proinsulin concentrations due to the virtual absence of CPE activity in the islet beta-cell. The Cpe^{fat} mutant becomes obese much later than ob and db animals. Cpe^{fat}/Cpe^{fat} mutants are not grossly distinguishable from lean littermates until 8–12 weeks of age.[80]

In normal animals, CPE is present at high concentrations in secretory granules and is required for excision of paired dibasic residues remaining at the C-terminus of various peptide prohormone intermediates, including those derived from proinsulins.[81] Naggert et al.[80] reported a missense mutation in Cpe(Ser202Pro) of fat mice resulting in the virtual absence of enzyme activity in pancreatic islets and the pituitary. This loss of exopeptidase function correlates with an aberrant increase in proinsulin and partially processed intermediates in both β-cells and serum. Hyperproinsulinemia is apparent well before the development of overt obesity and is independent of the glycemic status of mutant fat/fat mice. The fat mutation represents the first demonstration of an obesity-diabetes syndrome produced by a genetic defect in a prohormone processing pathway.

The causal relationship, if any, between the hyperproinsulinemia and the appearance of a later-developing obesity has not been determined. The observation that the Cpe defect is not limited to islet tissue, but also is present in the pituitary as well as the brain,[80] suggests that the obesity may develop as a result of widespread defects in the exopeptidase processing step required for full maturation of a number of neuroendocrine/endocrine prohormones.[82] A defect in NPY processing might be expected to decrease food intake; thus, it is more likely that a defect in processing of CRF, GLP, CCK, etc. (i.e., neuropeptides that normally decrease food intake), is involved. Thus, the later-developing obesity in fat/fat mutants may reflect defects in processing of prohormone forms of a variety of neuropeptides associated with the control of energy expenditure, nutrient partitioning, and satiety.

Most physiological characteristics of tub mutant and fat mutant mice are similar.[77] Differences between fat and tub mutants include slower progression to obesity in the tub homozygote.[77] Also, average body weight is less in tub compared with fat mutants.[77] Finally, the tub mutant is hyperinsulinemic, mild at weaning, but more pronounced in mature animals, and transiently hypoglycemic (between 12–20 weeks), a phenotypic characteristic not seen in fat/fat animals.[77] There is anecdotal evidence (Les Kozak, Jackson Laboratory) that neither tubby nor fat animals are hyperphagic, suggesting that their obesity may be the result of partitioning of ingested calories to fat.

The tub/tub animal develops hyperinsulinemic postpubertal obesity associated with progressive degeneration of retinal and cochlear ganglion cells. In this aspect, the syndrome resembles somewhat that seen in the Bardet-Biedl syndrome.[83] The tub mutation was recently cloned by two separate groups[84,85] and encodes a novel phosphadiesterase that may play a role in apoptosis. The gene order in the mouse Hbb-*tub*-Pth-Calc is conserved on human 11p15, making it likely that the human homologue of *tub* resides on 11p15.[79]

MUTATIONS AT THE *agouti* LOCUS

Obese phenotypes in association with yellow coat hair are characteristic of certain dominant mutations at the agouti (*a*) locus.[86] Lethal yellow (A^y) and viable yellow (A^{vy}) are insulin-resistant,[87] exhibit hypertrophic and hyperplastic pancreatic islets,[88] and are glucose-intolerant.[87] Regulation of hair pigmentation resulting in the *agouti* coat color (intermittent yellow and black pigment bands) is the only established function of the *agouti* gene in the wild-type mouse.[86]

The agouti gene (*a*) encodes agouti signaling protein (ASP), which blocks the action of MSH on hair follicle pigment-producing cells. With MSH action blocked, the animal switches from production of the black hair pigment (melanin) to production of the yellow pigment (eumelanin). The A^y mutation[89] results in ectopic overexpression of ASP in skin (yellow coat) as well as in other tissues in which the gene is not normally expressed (e.g., brain). The mechanism by which ASP causes insulin resistance may be related to the induction of increased intramyocyte free calcium.[90]

ASP action in skeletal muscle may alter the intracellular free calcium, $[Ca^{2+}]_i$. Zemel et al.[91] reported elevated levels of $[Ca^{2+}]_i$ in the soleus muscle of A^{vy}/a mice and found that recombinant agouti protein induced elevated $[Ca^{2+}]_i$ levels in cultured skeletal muscle myocytes. Because skeletal muscle is the primary site of peripheral glucose disposal[92] and because increased skeletal muscle $[Ca^{2+}]_i$ can produce insulin resistance and hyperinsulinemia,[93] ectopic expression of the protein in skeletal muscle could be responsible for these characteristics. These data are important because they suggest that nutrient partitioning can lead to obesity (excess fat) without any need for a dramatic increase of food intake. However, A^y animals have increased lean body mass as well as increased fat, suggesting that insulin may mediate both effects.

Ectopic overexpression of ASP in adipocytes may alter lipolysis. Adipocytes of A^{vy}/a mice have a depressed basal lipolytic rate, which may be due to an ASP-mediated alteration in the signal transduction pathway of lipolysis at the level of the production or maintenance of intracellular cAMP.[94] This hypothesis is compatible with the fact that ASP affects hair pigmentation by modulating intracellular levels of cAMP in the melanocyte. While this effect appears to occur in the melanocyte through ASP's antagonism of α-MSH activation of its receptor (MC1R),[95] it may occur in adipocytes by ASP antagonism of an as yet unidentified receptor (or receptors) in the melanocortin receptor family.[95] Obesity-promoting effects of ASP in adipocytes could be independent of its possible hyperinsulinemia-inducing effects in muscle or could result from synergistic effects in the two tissue types.

The effects on energy balance of ectopic expression of ASP in the brain must also be considered. Obesity and even hyperinsulinemia indirectly may be the result of primary ASP effects on areas of the brain that control weight, body fat, and insulin production. Consistent with this possibility is the recent demonstration that recombinant agouti protein antagonizes α-MSH activation of the melanocortin receptor MC4R,[95] which is expressed in brain nuclei involved in neuroendocrine and sympathetic control.[95,96] It is conceivable that this action in the brain may trigger the hyperphagia, increased efficiency of food utilization, and reduced sympathetic tone observed in obese yellow mice.[94,97,98] Decreased adrenergic tone may lead to decreased lipolysis and to increased lipogenesis and insulin production.[94]

REFERENCES

1. BRAY, G. A., J. FISLER & D. A. YORK. 1990. Neuroendocrine control of the development of obesity: understanding gained from studies of experimental animal models. Front. Obes. **11:** 128–181.
2. LEIBEL, R. L., N. BAHARY & J. M. FRIEDMAN. 1990. Genetic variation and nutrition in obesity: approaches to the molecular genetics of obesity. World Rev. Nutr. Diet. **63:** 90–101.
3. COLEMAN, D. L. 1978. Obese and diabetes: two mutant genes causing diabetes-obesity syndromes in mice. Diabetologia **14:** 141–148.
4. HUMMEL, K. P., M. M. DICKIE & D. L. COLEMAN. 1966. Diabetes, a new mutation in the mouse. Science **153:** 1127–1128.
5. ZUCKER, L. M. & T. F. ZUCKER. 1961. Fatty, a new mutation in the rat. J. Hered. **52:** 275–278.
6. KOLETSKY, S. 1973. Obese spontaneously hypertensive rats: a model for study of atherosclerosis. Exp. Mol. Pathol. **19:** 53–60.
7. HUMMEL, K. P., D. L. COLEMAN & P. W. LANE. 1972. The influence of genetic background on expression of mutations at the diabetes locus in the mouse: I. C57BL.KsJ and C57BL/6J strains. Biochem. Genet. **7:** 1–13.
8. MICHAELIS, O. E., K. C. ELLWOOD, J. M. JUDGE, N. W. SCHOENE & C. T. HANSEN. 1984. Effect of dietary sucrose on the SHR/N-corpulent rat: a model for insulin-independent diabetes. Am. J. Clin. Nutr. **39:** 612–618.
9. INGALLS, A. M., M. M. DICKIE & G. D. SNELL. 1950. Obese, a new mutation in the house mouse. J. Hered. **41:** 317–318.
10. FRIEDMAN, J., R. L. LEIBEL, D. A. SIEGEL, J. WALSH & N. BAHARY. 1991. Molecular mapping of the mouse ob mutation. Genomics **11:** 1054–1062.
11. BRAY, G. A. 1991. Obesity, a disorder of nutrient partitioning: the Mona Lisa hypothesis. Nutrition **121:** 1146–1162.
12. ZHANG, Y., R. PROENCA, M. MAFFEI, M. BARONE, L. LEOPOLD & J. M. FRIEDMAN. 1994. Positional cloning of the mouse obese gene and its human homologue. Nature **372:** 425–432.
13. KENNEDY, G. C. 1953. The role of depot fat in the hypothalamic control of food intake in the rat. Proc. R. Soc. Lond. **140:** 578–596.
14. HALAAS, J. L., K. S. GAJIWALA, M. MAFFEI, S. L. COHEN, B. T. CHAIT, D. RABINOWITZ, R. L. LALLONE, S. K. BURLEY & J. M. FRIEDMAN. 1995. Weight-reducing effects of the plasma protein encoded by the obese gene. Science **269:** 543–546.
15. STEPHENS, T. W., M. BASINSKI, P. K. BRISTOW, J. M. BUE-VALLESKEY, S. G. BURGETT, L. CRAFT, J. HALE, J. HOFFMANN, H. M. HSIUNG, A. KRIAUCIUNAS, W. MACKELLAR, P. R. ROSTECK, B. SCHONER, D. SMITH, F. C. TINSLEY, X.-Y. ZHANG & M. HEIMAN. 1995. The role of neuropeptide Y in the antiobesity action of the obese gene product. Nature **377:** 530–532.
16. WEIGLE, D. S., T. R. BUKOWSKI, D. C. FOSTER, S. HOLDERMAN, J. M. KRAMER, G. LASSER, C. E. LOFTON-DAY, D. E. PUNKARD, C. RAYMOND & J. L. KUIJPER. 1995. Recombinant ob protein reduces feeding and body weight in the ob/ob mouse. J. Clin. Invest. **96:** 2065–2070.
17. CAMPFIELD, L. A., F. J. SMITH, Y. GUISEZ, R. DEVOS & P. BURN. 1995. Recombinant mouse OB protein: evidence for a peripheral signal linking adiposity and central neural networks. Science **269:** 546–549.
18. PELLEYMOUNTER, M. A., M. J. CULLEN, M. B. BAKER, R. HECHT, D. WINTERS, T. BOONE & F. COLLINS. 1995. Effects of the obese gene product on body weight regulation in ob/ob mice. Science **269:** 540–543.
19. TRUETT, G. E., N. BAHARY, J. M. FRIEDMAN & R. L. LEIBEL. 1991. Rat obesity gene fatty (fa)

maps to chromosome 5: evidence for homology with the mouse gene diabetes (db). Proc. Natl. Acad. Sci. U.S.A. **88**: 7806-7809.
20. TRUETT, G. E., H. J. JACOB, J. MILLER, G. DROUIN, N. BAHARY, J. W. SMOLLER, E. S. LANDER & R. L. LEIBEL. 1995. Genetic map of rat chromosome 5 including the fatty (fa) locus. Mamm. Genome **6**: 25-30.
21. CHUA, S. C., W. K. CHUNG, X. S. WU-PENG, Y. ZHANG, S-M. LIU, L. TARTAGLIA & R. L. LEIBEL. 1996. Phenotypes of mouse diabetes and rat fatty due to mutations in the OB (leptin) receptor. Science **271**: 994-996.
22. MAFFEI, M., H. FEI, G. H. LEE, C. DANI, P. LEROY, Y. ZHANG, R. PROENCA, R. NEGREL, G. AILHAUD & J. M. FRIEDMAN. 1995. Increased expression in adipocytes of ob RNA in mice with lesions of the hypothalamus and with mutations at the db locus. Proc. Natl. Acad. Sci. U.S.A. **92**: 6957-6960.
23. MAFFEI, M., J. HALAAS, E. RAVUSSIN, R. E. PRATLEY, G. H. LEE, Y. ZHANG, H. FEI, S. KIM, R. LALLONE, S. RANGANATHAN & J. M. FRIEDMAN. 1995. Leptin levels in human and rodent: measurements of plasma leptin and ob RNA in obese and weight-reduced subjects. Nat. Med. **1**: 1-7.
24. FREDERICH, R. C., B. LOLLMANN, A. HAMANN, A. NAPOLITANO-ROSEN, B. B. KAHN, B. B. LOWELL & J. S. FLIER. 1995. Expression of OB mRNA and its encoded protein in rodents. J. Clin. Invest. **96**: 1658-1663.
25. TARTAGLIA, L. A., M. DEMBSKI, X. WENG, N. DENG, J. CULPEPPER, R. DEVOS, G. J. RICHARDS, L. A. CAMPFIELD, F. T. CLARK, J. DEEDS, C. MUIR, S. SANKER, A. MORIARTY, K. J. MOORE, J. S. SMUTKO, G. G. MAYS, E. A. WOOLF, C. A. MONROE & R. I. TEPPER. 1995. Identification and expression cloning of a leptin receptor, OB-R. Cell **83**: 1263-1271.
26. CHEN, H., O. CHARLAT, L. A. TARTAGLIA, E. A. WOOLF, X. WENG, S. J. ELLIS, N. D. LAKEY, J. CULPEPPER, K. J. MOORE, R. E. BREITBART, G. M. DUYK, R. I. TEPPER & J. P. MORGENSTERN. 1996. Evidence that the diabetes gene encodes the leptin receptor: identification of a mutation in the leptin receptor gene in db/db mice. Cell **84**: 491-495.
27. LEE, G. H., R. PROENCA, J. M. MONTEZ, K. M. CARROLL, J. G. DARVISHZADETH, J. I. LEE & J. M. FRIEDMAN. 1996. Abnormal splicing of the leptin receptor in diabetic mice. Nature (Lond.) **379**: 632-655.
28. GHILARDI, N., S. ZIEGLER, A. WIESTNER, R. STOFFEL, M. H. HEIM & R. C. SKODA. 1996. Defective STAT signaling by the leptin receptor in diabetic mice. Proc. Natl. Acad. Sci. U.S.A. **93**: 6231-6235.
29. CHUA, S. C., D. W. WHITE, X. S. WU-PENG, S-M. LIU, N. OKADA, E. E. KERSHAW, W. K. CHUNG, L. POWER-KEHOE, M. CHUA, L. A. TARTAGLIA & R. L. LEIBEL. 1996. Phenotype of fatty due to Gln269Pro mutation in the leptin receptor (Lepr). Diabetes **45**: 1141-1143.
30. CHUNG, W. K., L. POWER-KEHOE, M. CHUA & R. L. LEIBEL. 1996. Mapping of the OB receptor (OBR) to 1p in the region of non-conserved gene order from mouse and rat to human. Genome Res. **6**: 431-438.
31. BAZAN, J. F. 1990. Structural design and molecular evolution of a cytokine receptor superfamily. Proc. Natl. Acad. Sci. U.S.A. **87**: 6934-6938.
32. IHLE, J. N. 1995. Cytokine receptor signaling. Nature (Lond.) **377**: 591-594.
33. DARNELL, J. E., I. M. KERR & G. R. STARK. 1994. Jak-STAT pathways and transcriptional activation in response to IFNs and other extracellular signaling proteins. Science **264**: 1415-1421.
34. VAISSE, C., J. L. HALAAS, C. M. HORVATH, J. D. DARNELL, M. STOFFEL & J. M. FRIEDMAN. 1996. Leptin activation of Stat3 in the hypothalamus of wild-type and ob/ob mice, but not db/db mice. Nat. Genet. **14**: 95-97.
35. MERZA, M. A., J. M. WHITE, K. C. F. SHEEHAN, E. A. BACH, S. J. RODIG, A. S. DIGHE, D. H. KAPLAN, J. K. RILEY, A. C. GREENLUND, D. CAMPBELL, K. CARVER-MOORE, R. N. DUBOIS, R. CLARK, M. AGUET & R. D. SCHREIBER. 1996. Targeted disruption of the STAT1 gene in mice reveals unexpected physiologic specificity in the JAK-STAT signaling pathway. Cell **84**: 431-442.
36. DURBIN, J. E., R. HACKENMILLER, M. C. SIMON & D. E. LEVY. 1996. Targeted disruption of the mouse STAT1 gene results in compromised innate immunity to viral disease. Cell **84**: 443-450.
37. STANLEY, B. G. 1993. Neuropeptide Y in multiple hypothalamic sites controls eating behavior,

endocrine, and autonomic systems for energy balance. *In* The Biology of Neuropeptide Y and Related Peptides, p. 457–509. Humana. Totowa, New Jersey.
38. BILLINGTON, C. J., J. E. BRIGGS, M. GRACE & A. S. LEVINE. 1991. Effects of intracerebroventricular injection of neuropeptide Y on energy metabolism. Am. J. Physiol. **260:** R321–R327.
39. WILDING, J. P. H., S. G. GILBEY, C. J. BAILEY, R. A. L. BATT, G. WILLIAMS, M. A. GHATEI & S. R. BLOOM. 1993. Increased neuropeptide-Y messenger ribonucleic acid (mRNA) and decreased neurotensin mRNA in the hypothalamus of the obese (ob/ob) mouse. Endocrinology **132:** 1939–1944.
40. CHUA, S. C., A. W. BROWN, J. KIM, K. L. HENNESSY, R. L. LEIBEL & J. HIRSCH. 1991. Food deprivation and hypothalamic neuropeptide gene expression effects of strain background and the diabetes mutation. Mol. Brain Res. **11:** 291–299.
41. SMITH, F. J., L. A. CAMPFIELD, J. A. MOSCHERA, P. S. BAILON & P. BURN. 1996. Feeding inhibition by neuropeptide Y. Nature **382:** 307.
42. ERICKSON, J. C., K. E. CLEGG & R. D. PALMITER. 1996. Sensitivity to leptin and susceptibility to seizures of mice lacking neuropeptide Y. Nature **381:** 415–421.
43. SCHWARTZ, M. W., D. G. BASKIN, T. R. BUKOWSKI, J. L. KUIJPER, D. FOSTER, G. LASSER, D. E. PRUNKARD, D. PORTE, S. C. WOODS, R. J. SEELEY & D. S. WEIGLE. 1996. Specificity of leptin action on elevated blood glucose levels and hypothalamic neuropeptide Y gene expression in ob/ob mice. Diabetes **45:** 531–535.
44. HANSEN, C. T. 1983. Two new congenic rat strains for nutrition and obesity research. Fed. Proc. **42:** 573 (abstract).
45. ZUCKER, L. M. & H. N. ANTONIADES. 1972. Insulin and obesity in the Zucker genetically obese rat "fatty." Endocrinology **90:** 1320–1330.
46. JOHNSON, P. R., L. M. ZUCKER, J. A. CRUCE & J. HIRSCH. 1972. Cellularity of adipose depots in the genetically obese Zucker rat. J. Lipid Res. **12:** 706–714.
47. PHILLIPS, M. S., Q. LIU, H. A. HAMMOND, V. DUGAN, P. J. HEY, C. J. CASKEY & J. F. HESS. 1996. Leptin receptor missense mutation in the fatty Zucker rat [letter]. Nat. Genet. **13:** 18–19.
48. IIDA, M., T. MURAKAMI, K. ISHIDA, A. MIZUNO, M. KUWAJ & K. SHIMA. 1996. Phenotype-linked amino acid alteration in leptin receptor cDNA for Zucker fatty (fa/fa) rat. Biochem. Biophys. Res. Commun. **222:** 19–26.
49. KAHLE, E. B., K. G. BUTZ, O. E. MICHAELIS, R. L. LEIBEL & C. T. HANSEN. 1994. The first reproduction documentation of a genetically obese female fa/cp rat. ILAR News **36:** 75–77.
50. YEN, T. T., W. N. SHAW & P. L. YU. 1997. Genetics of obesity in Zucker rats and Koletsky rats. Heredity **38:** 373–377.
51. KAHLE, E. B., K. G. BUTZ, S. C. CHUA, E. E. KERSHAW, R. L. LEIBEL, T. W. FENGER, C. T. HANSEN & O. E. MICHAELIS. 1996. The corpulent (cp) mutation maps to the same interval on (Pgm1-Glut1) rat chromosome 5 as the fatty (fa) mutation. Obes. Res. **5:** 142–145.
52. KERSHAW, E. E., S. C. CHUA, J. A. WILLIAMS, E. M. MURPHY & R. L. LEIBEL. 1995. Molecular mapping of SSRs for Pgm1 and C8b in the vicinity of the rat fatty locus. Genomics **27:** 149–154.
53. WU-PENG, X. S., S. C. CHUA, N. OKADA, S-M. LIU, M. NICOLSON & R. L. LEIBEL. 1997. Phenotype of the obese Koletsky (f) rat due to Tyr763Stop mutation in the extracellular domain of the leptin receptor (Lepr): evidence for deficient plasma-to-CSF transport of leptin in both the Zucker and Koletsky obese rat. Diabetes **46:** 513–518.
54. TAKAYA, K., Y. OGAWA, J. HIRAOKA, Y. YAMORI, K. NAKAO & R. J. KOLETSKY. 1996. Nonsense mutation of leptin receptor in the obese spontaneously hypertensive Koletsky rat. Nat. Genet. **14:** 130–131.
55. THOMPSON, D. B., R. C. JANSSEN, V. M. OSSOWSKI, M. PROCHAZKA, W. C. KNOWLER & C. BOGARDUS. 1995. Evidence for linkage between a region on chromosome 1p and the acute insulin response in Pima Indians. Diabetes **44:** 478–481.
56. MICHAELIS, O. E., K. C. ELLWOOD, J. HALLFRISCH & C. T. HANSEN. 1983. Effect of dietary sucrose and genotype on metabolic parameters of a new strain of genetically obese rat: LA/N-corpulent. Nutr. Res. **3:** 217–228.
57. ELLWOOD, K. C., O. E. MICHAELIS, J. J. EMBERLAND & S. J. BHATHENA. 1985. Hormonal and

lipogenic and gluconeogenic enzyme responses in LA/N-corpulent rats. Proc. Soc. Exp. Biol. Med. **179:** 163-167.
58. KAHLE, E. B., K. G. BUTZ, R. L. LEIBEL, C. E. HANSEN, S. J. BHATHENA & O. E. MICHAELIS. 1996. Glucose homeostasis in 3 interstrains (LA/N-BN/Crl cp/cp; Zucl3M-BN/Crl fa/fa; and Zucl3M-LA/N fa^{cp}) of genetically obese rats. *In* Lessons from Animal Diabetes. Volume 6, p. 411-418. Birkhäuser. Basel.
59. REISER, S., M. C. BICKARD, J. HALLFRISCH, O. E. MICHAELIS IV & E. S. PRATHER. 1981. Blood lipids and their distribution in lipoproteins in hyperinsulinemic subjects fed three different levels of sucrose. J. Nutr. **111:** 1045-1057.
60. REISER, S., E. BOHN, J. HALLFRISCH, O. E. MICHAELIS IV, M. KEENEY & E. S. PRATHER. 1981. Serum insulin and glucose in hyperinsulinemic subjects fed three different levels of sucrose. Am. J. Clin. Nutr. **34:** 2348-2358.
61. SWINBURN, B. & E. RAVUSSIN. 1993. Energy balance or fat balance. Am. J. Clin. Nutr. **57** (suppl.): 766S-771S.
62. ABOU MRAD, J., F. YAKUBU, D. LIN, J. C. PETERS, J. B. ATKINSON & J. O. HILL. 1992. Skeletal muscle composition in dietary obesity-susceptible and dietary obesity-resistant rats. Am. J. Physiol. **262:** R684-R688.
63. WADE, A. J., M. M. MARBUT & J. M. ROUND. 1990. Muscle fibre type and aetiology of obesity. Lancet **335:** 805-808.
64. PAGLIASSOTTI, M. J., D. A. PAN, P. A. PRACH, T. A. KOPPENHAFER, L. H. STORLIEN & J. O. HILL. 1995. Tissue oxidative capacity, fuel stores, and skeletal muscle fatty acid composition in obesity-prone and obesity-resistant rats. Obes. Res. **3:** 459-464.
65. HOLLANDS, M. A. & J. A. CAWTHORNE. 1981. Important sites of lipogenesis in the mouse other than liver and white adipose tissue. Biochem. J. **196:** 645-647.
66. KAHLE, E. B., J. M. DADGARI, G. A. DUDLEY, C. T. HANSEN, O. L. TULP & O. E. MICHAELIS. 1988. Adaptive response of enzymes of carbohydrate and lipid metabolism to exercise training. *In* New Models of Genetically Obese Rats for Studies in Diabetes, Heart Disease, and Complications of Obesity, p. 143-148. National Institutes of Health. Bethesda, Maryland.
67. WARDLAW, G. M., M. L. KAPLAN & S. LANZA-JACOBY. 1986. Effects of treadmill training on muscle oxidative capacity and accretion in young male obese and non-obese Zucker rats. J. Nutr. **116:** 1841-1852.
68. TROGAN, C. E., J. T. BROZINICK, G. M. KASTELLO & J. L. IVY. 1989. Muscle morphological and biochemical adaptations to training in obese Zucker rats. J. Appl. Physiol. **67:** 1807-1813.
69. BRAY, G. A. & D. A. YORK. 1979. Hypothalamic and genetic obesity in experimental animals: an autonomic and endocrine hypothesis. Physiol. Rev. **59:** 719-809.
70. NEWSHOLME, E. A., B. CRABTREE & V. A. ZAMMIT. 1980. Use of enzyme activities as indices of maximum rates of fuel utilization. *In* Excerpta Medica: Trends in Enzyme Histochemistry and Cytochemistry. Volume 7, p. 245-258. Ciba Foundation Symposium. Amsterdam.
71. WAKIL, S. J., J. K. STOOPS & V. C. JOSHI. 1983. Fatty acid synthesis and its regulation. Annu. Rev. Biochem. **52:** 537-579.
72. SIMONEAU, J. A. & C. BOUCHARD. 1989. Human variation in skeletal muscle fiber-type proportion and enzyme activities. Am. J. Physiol. **257:** E567-E572.
73. MAURO, A. 1961. Satellite cell of skeletal muscle fibers. J. Biophys. Biochem. Cytol. **9:** 493-495.
74. SCHULTZ, E. & K. M. MCCORMICK. 1993. Cell biology of the satellite cell. *In* Molecular and Cell Biology of Muscular Dystrophy, p. 190-209. Chapman & Hall. London.
75. SCHULTZ, E. & K. M. MCCORMICK. 1994. Skeletal muscle satellite cells. Rev. Physiol. Biochem. Pharmacol. **123:** 213-257.
76. SCHULTZ, E. 1989. Satellite cell behavior during skeletal muscle growth and regeneration. Med. Sci. Sports Exercise 21(suppl.): 5181-5186.
77. COLEMAN, D. L. & E. M. EICHER. 1990. Fat (fat) and tubby (tub): two autosomal recessive mutations causing obesity syndromes in the mouse. J. Hered. **81:** 424-427.
78. PAIGEN, B. J. & D. L. COLEMAN. 1990. Linkage of fat to esterase-1. Mouse Genome **86:** 240.
79. CHUNG, W. K., L. POWER-KEHOE & R. L. LEIBEL. 1996. Detailed molecular mapping of the mouse fat mutation: fat is non-recombinant with carboxypeptidase E. Genomics **32:** 210-217.
80. NAGGERT, J. K., L. D. FRICKER, O. VARLAMOV, P. M. NISHINA, Y. ROUILLE, D. F. STEINER, R. J. CARROL, B. J. PAIGEN & E. H. LEITER. 1995. Hyper-proinsulinaemia in obese fat/fat

mice associated with a carboxypeptidase-E mutation which reduces enzyme activity. Nat. Genet. **10:** 135-142.
81. FRICKER, L. D. 1991. *In* Peptide Biosynthesis and Processing, p. 199-228. CRC Press. Boca Raton, Florida.
82. O'RAHILLY, S., H. GRAY, P. J. HUMPHREYS, A. KROOK, K. S. POLONSKY, A. WHITE, S. GIBSON, K. TAYLOR & C. CARR. 1995. Brief report: impaired processing of prohormones associated with abnormalities of glucose homeostasis and adrenal function. N. Engl. J. Med. **333:** 1386-1390.
83. BRAY, G. A. 1995. Laurence, Moon, Bardet, and Biedl: reflections on a syndrome [comment]. Obes. Res. **3:** 383-403.
84. KLEYN, P. W., W. FAN, S. G. KOVATS, J. J. LEE, J. C. PULIDO, Y. WU, L. R. BERKEMEIER, D. J. MISUMI, L. HOLMGREN, O. CHARLAT, E. A. WOOLF, O. TAYBER, T. BRODY, P. SU, F. HAWKINS, B. KENNEDY, L. BALDINI, C. EBELING, G. D. ALPERIN, J. DEEDS, N. D. LAKEY, J. CULPEPPER & K. J. MOORE. 1996. Identification and characterization of the mouse obesity gene tubby: a member of a novel gene family. Cell **85:** 281-290.
85. NOBEN-TRAUTH, K., J. K. NAGGERT, M. A. NORTH & P. M. NISHINA. 1996. A candidate gene for the mouse mutation tubby. Nature **380:** 534-538.
86. SILVERS, W. K. 1979. The agouti and extension series of alleles, umbrous and sable. *In* The Coat Colors of Mice, p. 6-44. Springer-Verlag. New York/Berlin.
87. FRIGERI, L. G., G. L. WOLFF & G. ROBEL. 1983. Impairment of glucose tolerance in yellow (A^{vy}/A) (BALB/c × VY) F-1 hybrid mice by hyperglycemic peptide(s) from human pituitary glands. Endocrinology **113:** 2097-2105.
88. WARBRITTON, A., A. M. GILL, T. T. YEN, T. BUCCI & G. L. WOLFF. 1994. Pancreatic islet cells in preobese yellow $A^{vy}/-$ mice: relation to adult hyperinsulinemia and obesity. Proc. Soc. Exp. Biol. Med. **206:** 145-151.
89. BULTMAN, S. J., E. J. MICHAUD & R. P. WOYCHIK. 1992. Molecular characterization of the mouse agouti locus. Cell **71:** 1195-1204.
90. KLEBIG, M. L., J. E. WILKINSON, J. G. GEISLER & R. P. WOYCHIK. 1995. Ectopic expression of the agouti gene in transgenic mice causes obesity, features of type II diabetes, and yellow fur. Proc. Natl. Acad. Sci. U.S.A. **92:** 4728-4732.
91. ZEMEL, M. B., J. H. KIM, R. P. WOYCHIK, E. J. MICHAUD, S. H. KADWELL, I. R. PATEL & W. O. WILKISON. 1995. Agouti regulation of intracellular calcium: role in the insulin resistance of viable yellow mice. Proc. Natl. Acad. Sci. U.S.A. **92:** 4733-4737.
92. MOLLER, D. E., Ed. 1993. *In* Insulin Resistance. Wiley. New York.
93. DRAZNIN, B., K. SUSSMAN, M. KAO, D. LEWIS & N. SHERMAN. 1987. The existence of an optimal range of cytosolic free calcium for insulin-stimulated glucose transport in rat adipocytes. J. Biol. Chem. **262:** 14385-14388.
94. YEN, T. T., A. M. GILL, L. G. FRIGERI, G. S. BARSH & G. L. WOLFF. 1994. Obesity, diabetes, and neoplasia in yellow $A^{vy}/-$ mice: ectopic expression of the agouti gene. FASEB J. **8:** 479-488.
95. LU, D., D. WILLARD, I. R. PATEL, S. KADWELL, L. OVERTON, T. KOST, M. LUTHER, W. CHEN, R. P. WOYCHIK, W. O. WILKISON & R. D. CONE. 1994. Agouti protein is an antagonist of the melanocyte-stimulating-hormone receptor. Nature (Lond.) **371:** 799-802.
96. GANTZ, I., H. MIWA, Y. KONDA, Y. SHIMOTO, T. TASHIRO, S. J. WATSON, J. DELVALLE & T. YAMADA. 1993. Molecular cloning, expression, and gene localization of a fourth melanocortin receptor. J. Biol. Chem. **268:** 15174-15179.
97. FRIGERI, L. G., G. L. WOLFF & C. TEGUH. 1988. Differential responses of yellow A^y/A and agouti A/a (BALB/c × VY) F-1 hybrid mice to the same diets: glucose tolerance, weight gain, and adipocyte cellularity. Int. J. Obes. **12:** 305-320.
98. DICKERSON, G. E. & J. W. GOWEN. 1947. Hereditary obesity and efficient food utilization in mice. Science **105:** 496-498.

Development of Obesity and Insulin Resistance in the Israeli Sand Rat (*Psammomys obesus*)

Does Leptin Play a Role?

G. R. COLLIER,[a] A. DE SILVA,[a] A. SANIGORSKI,[a] K. WALDER,[a] A. YAMAMOTO,[b] AND P. ZIMMET[c]

[a]*School of Nutrition and Public Health*
Deakin University
Geelong, Victoria 3217, Australia

[b]*Faculty of Education*
Shizuoka University
Shizuoka-shi 422, Japan

[c]*International Diabetes Institute*
Caulfield, Melbourne, Victoria, Australia

INTRODUCTION

Obesity has emerged as the most common metabolic disorder in affluent countries. It is a major health problem, increasing the risk of a number of chronic diseases including cardiovascular disease (CVD) and non-insulin-dependent diabetes mellitus (NIDDM). Despite enormous research efforts, our understanding of the molecular mechanisms underlying the etiology of obesity remains incomplete.

In an effort to dissect the various metabolic pathways and defects that occur during the development of obesity, animal models have been used extensively. A number of single gene mutation animal models are available and they include *db, ob, fat, tub, Ad,* and A^y mice and *fa* rats. However, these models may have limited value when used to examine the development of obesity in humans, which is a multigenic disease with a heterogeneous phenotype.

In contrast to single gene mutation animal models, the Israeli Sand Rat (*Psammomys obesus*) is a polygenic animal model that responds to dietary manipulation and displays a heterogeneous pattern of disease development, mimicking the human condition. The sand rat would appear to be an ideal animal model to examine the metabolic pathways involved in the development of obesity and NIDDM.

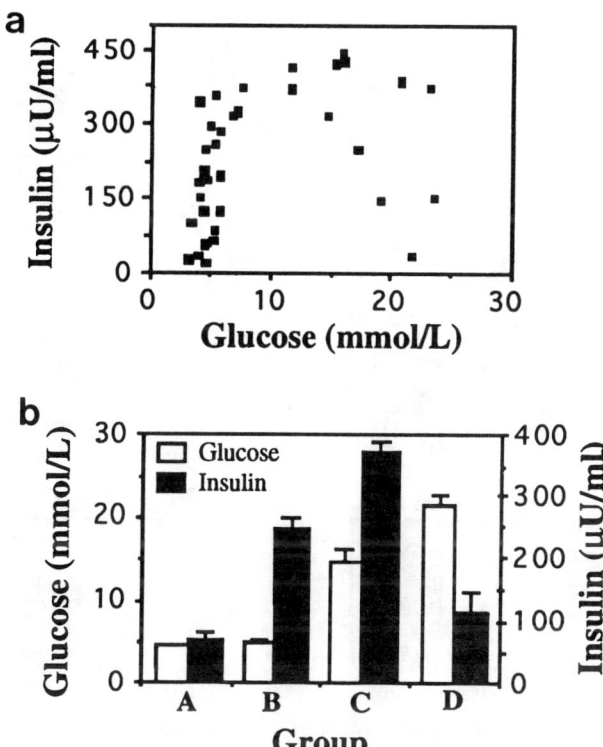

FIGURE 1. (a) The inverted U-shaped curve relationship describing the plasma glucose and insulin concentration of Israeli Sand Rats (*Psammomys obesus*) in the fed state ($n = 37$). (b) Plasma glucose (□) and plasma insulin (■) concentrations of four separate groups of Israeli Sand Rats in the fed state. Results are expressed as mean ± SEM. Group A was normoglycemic and normoinsulinemic ($n = 10$). Group B was normoglycemic and hyperinsulinemic ($n = 13$). Group C was hyperglycemic and hyperinsulinemic ($n = 11$). Group D was hyperglycemic and normoinsulinemic ($n = 7$). Adapted from reference 3.

ISRAELI SAND RAT (*PSAMMOMYS OBESUS*)

Sand rats are native to the sandy areas of North Africa and are diurnally active above ground.[1] Their staple diet consists of stems and leaves of saltbush. However, when fed a diet of standard laboratory chow (a substantial increase in energy density from their native diet), sand rats can develop diabetes.[1] Interestingly, a range of metabolic responses has been demonstrated in the transition from a diet of low-energy density to a diet of high-energy density, with some animals remaining lean and healthy, while others develop obesity and moderate to severe diabetes.[2] This range of response is similar to that observed in human populations.

We have established a breeding colony of Israeli Sand Rats at Deakin University,

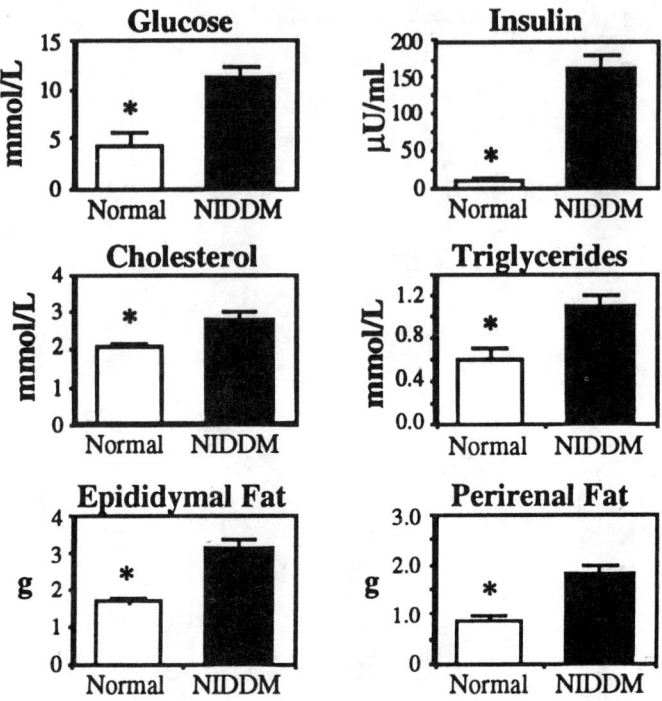

FIGURE 2. Data collected from 12-week-old Israeli Sand Rats: normal, group A animals (□) ($n = 10$); diabetic, group C animals (■) ($n = 10$). All values are expressed as mean ± SEM. All differences were considered significant at $p < 0.05$ (*). Adapted from reference 4.

Geelong, Australia. Over the past few years, we have characterized the metabolic defects that occur.[3-7] As can be seen in FIG. 1a, an inverted U-shaped curve describes a relationship in this model between fed glucose and insulin levels. This pattern is similar to that reported from cross-sectional studies in human populations. DeFronzo[8] has coined this relationship as "Starling's curve of the pancreas."

It is also possible in age-matched animals (see FIG. 1b) to divide sand rats into four separate groups based on glucose and insulin levels in the fed state. The characteristics of these groups were then investigated. When compared with normoglycemic and normoinsulinemic animals in group A, animals in group C developed a number of abnormalities including hyperglycemia, hyperinsulinemia, elevated triglyceride and cholesterol levels, increased fat stores, and increased body weight (FIG. 2). These metabolic changes constitute some of the major features of the so-called Insulin Resistance Syndrome or the "Metabolic Syndrome."[9]

Further studies in our laboratory were carried out in an effort to characterize the defects in glucose metabolism that develop in these animals. In these studies, we have demonstrated a direct correlation between the levels of fasting plasma glucose and the rates of hepatic glucose production.[5] In addition, there was an inverse correlation between fasting plasma glucose and the metabolic clearance rates of glucose, sup-

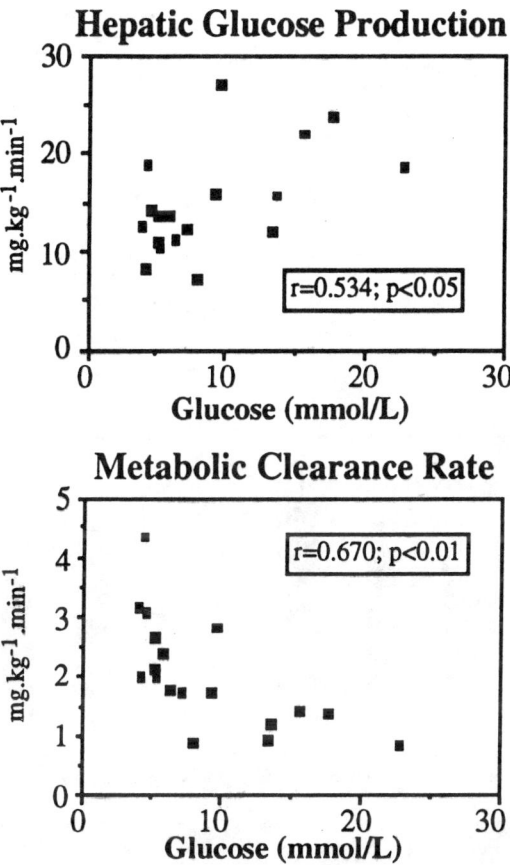

FIGURE 3. (Top) The correlation between hepatic glucose production (HGP) and fasting plasma glucose in Israeli Sand Rats ($n = 19$, $r = 0.534$, $p < 0.05$). (Bottom) The inverse correlation between the metabolic clearance rate of glucose and fasting plasma glucose ($n = 19$, $r = 0.670$, $p < 0.01$). Adapted from reference 5.

porting earlier observations that the pattern of diabetes development in this animal model is similar to that seen in the human condition (FIG. 3).

The short time span required to reach maturity and to develop obesity and diabetes in the Israeli Sand Rat makes it ideal for longitudinal investigations. Further results from our laboratory suggested that fasting hyperglycemia in group C animals was primarily due to elevated basal glucose turnover, evident at the whole body and tissue level.[5] Results of our studies inhibiting free fatty acid oxidation implied that elevated fatty acid oxidation was at least partly responsible for the increased turnover via increased hepatic glucose production.[6] Other metabolic defects demonstrated in group C rats include impaired oral and intravenous glucose tolerance, as well as reduced insulin sensitivity after an insulin bolus challenge (FIG. 4). Interestingly, in this study, glucose uptake into individual muscles was not impaired in the insulin-insensitive

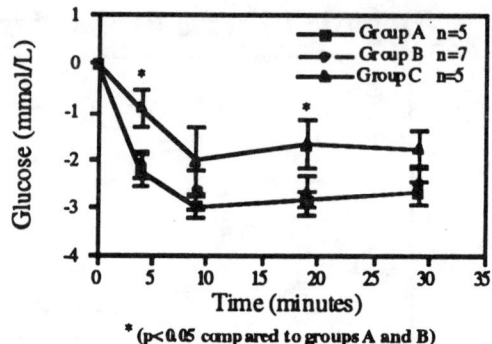
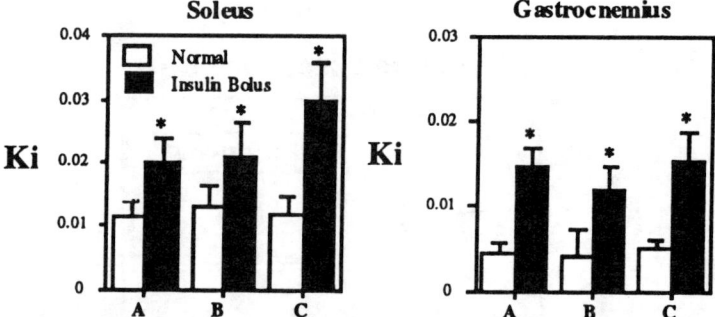

FIGURE 4. (Top) An insulin bolus (5 U/kg), together with 3 μg 2-deoxyglucose, was administered to group A, group B, and group C Israeli Sand Rats, and blood glucose was measured for 30 minutes. (Bottom) The mean fractional rate of 2-deoxyglucose uptake (Ki) in soleus and gastrocnemius muscles for the three groups of animals with (■) and without (□) an insulin bolus. All values are expressed as mean ± SEM. All differences were considered significant at $p < 0.05$ (*).

group C animals, suggesting that insulin resistance at the level of hepatic glucose production was critical in the development of hyperglycemia in Israeli Sand Rats (unpublished observations).

OBESITY DEVELOPMENT

In the Israeli Sand Rat, as in susceptible human populations,[10] it is the transition from the traditional low-energy-dense diets to high-energy-dense diets that triggers the development of impaired glucose tolerance or NIDDM. Why do some animals remain lean and free from diabetes, whereas others develop obesity and associated metabolic disturbances?

At weaning, animals that go on to develop obesity and hyperinsulinemia are indistinguishable, based on metabolic characteristics such as body weight, glucose, insulin, and triglycerides, from those that remain lean throughout life. Interestingly,

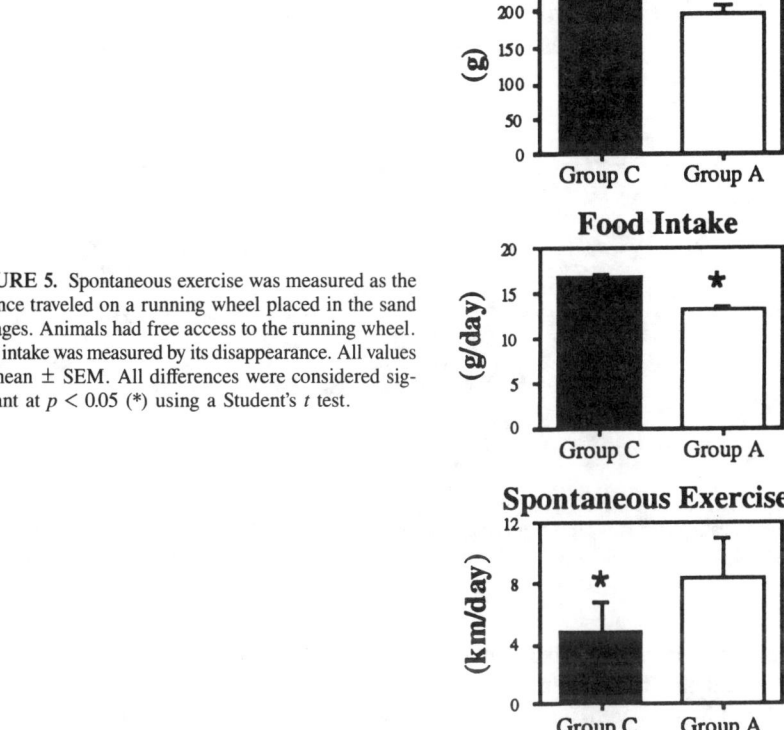

FIGURE 5. Spontaneous exercise was measured as the distance traveled on a running wheel placed in the sand rat cages. Animals had free access to the running wheel. Food intake was measured by its disappearance. All values are mean ± SEM. All differences were considered significant at $p < 0.05$ (*) using a Student's t test.

when we compared mature, lean (group A) animals with those that developed obesity and metabolic disturbances (group C), group C animals demonstrated hyperphagia and decreased spontaneous exercise levels (FIG. 5; unpublished observations). These results highlight a major energy imbalance in these animals. It appears that, at an early stage, some animals will overeat and develop the subsequent metabolic disturbances. Why do some animals choose to overeat when all animals are presented with identical environmental conditions?

We cannot answer this question from the studies performed to date; however, the Israeli Sand Rat provides an excellent opportunity to further our investigation of the possible underlying mechanisms. One explanation of the molecular changes underlying the development of hyperphagia and metabolic disturbances could be found in the recent discovery of the *ob* gene and its circulating protein product, leptin.

Recently, leptin has been suggested as the "missing circulating factor" that could help explain the early "lipostat hypothesis."[11] This theory stemmed from the pioneering work of Kennedy in the 1950s[11] and was strengthened by the parabiosis studies of Coleman and coworkers in the 1970s.[12,13] In the parabiosis studies, Coleman utilized lean mice and two animal models of obesity, the *ob/ob* and *db/db* mice. It was es-

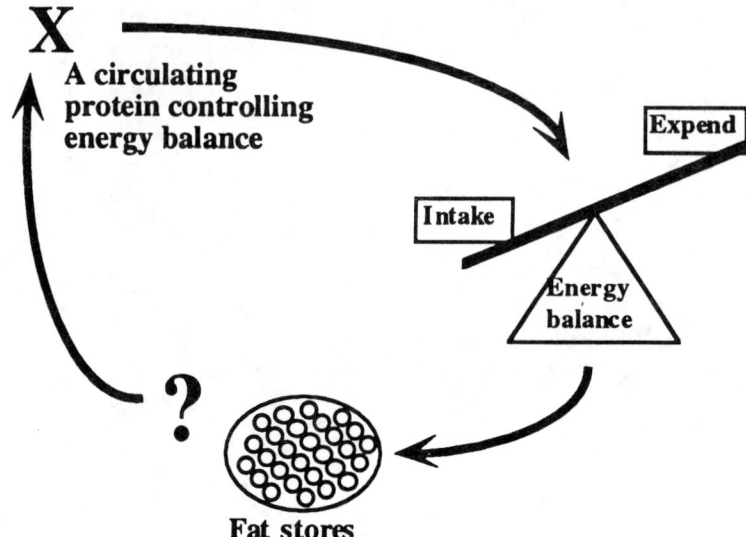

FIGURE 6. Schematic representation of the "lipostat hypothesis." Energy balance is regulated by a circulating factor secreted from adipose tissue that interacts with the hypothalamus to alter appetite and energy expenditure.

tablished from these studies that energy balance was regulated in the obese mice by a circulating factor from the lean mice and it was suggested that a satiety factor, produced in adipose tissue, circulated in the blood and affected appetite through interaction with the hypothalamus. It was further suggested that *ob/ob* mice lacked the circulating satiety factor.[12] This concept is represented diagrammatically in FIG. 6.

OBESE GENE AND LEPTIN

The suggestion that a circulating satiety factor or "lipostat" was important in controlling energy balance remained in the literature for some time without much scientific support, until Friedman and coworkers at Rockefeller University, New York, cloned the *ob* gene.[14] The *ob* gene encodes a 4.5-kb adipose tissue mRNA with a highly conserved 167-amino-acid open reading frame. The amino-acid sequence is 84% identical between human and mouse.[14] When the *ob* structure was examined in obese mice, it was discovered that they had a premature stop codon, resulting in a truncated protein, and expression of the mutated gene was 20 times greater.[14] Subsequently, the *ob* gene was expressed in *E. coli* and purified and then the protein product was administered intraperitoneally (5 µg/g/day) to *ob/ob* mice. *Ob* protein administration dramatically reduced food intake and body weight and was called "leptin" (derived from the Greek *leptos*, meaning "thin").[15]

In subsequent studies, leptin administration to *ob/ob* mice decreased body weight, percentage body fat, food intake, serum glucose, and insulin, and increased metabolic rate, body temperature, and exercise levels.[15-17] These studies highlighted the potential of leptin as a regulator of energy balance and stimulated a major research focus worldwide on the regulation and actions of leptin and *ob* gene expression.

In our own studies with the Israeli Sand Rat,[18] we isolated the *ob* gene and demonstrated exclusive expression of the gene in adipose tissue, consistent with previous results. We sequenced the gene and demonstrated 90% homology between the sand rat and the mouse, and 78% homology with the human *ob* gene sequence. These results confirmed that the *ob* gene is highly conserved between species. Further analysis of *ob* gene expression clearly demonstrated that obese sand rats had elevated levels of *ob* gene expression when compared with lean animals; however, they did not have the same *ob* gene mutation as the *ob/ob* mice and, in fact, had elevated circulating leptin levels.[18] Elevated leptin levels in obesity have also been demonstrated in animal models of obesity, including *fa/fa* rats and *db/db*, A^y, VMH-lesioned, GTG, and high-fat-fed mice.[19-24] This is consistent with our own human data[25,26] where elevated leptin levels are significantly associated with increased insulin levels and body weight.

LEPTIN AND OBESITY DEVELOPMENT

Interestingly, in animal models of obesity, elevated leptin levels do not result in reduced energy intake. These data imply that resistance to the action of leptin has developed, resulting in a compensatory overproduction of leptin. To further understand the relationship between leptin and obesity development, we designed a study to examine the longitudinal changes in leptin levels during weight gain.[27] It is clear from these studies that glucose, insulin, and leptin levels remained unchanged in lean animals from 4 weeks (weaning) until 12 weeks of age, despite increased body weight from 67 g to 170 g. In contrast, animals that developed obesity (body weight, 65 g to 210 g), hyperglycemia, and hyperinsulinemia also developed elevated circulating leptin levels.[27]

It appears from these studies that leptin is not simply a marker of adipose tissue mass. In fact, only animals that developed a disproportionate increase in body fat mass, hyperinsulinemia, and hyperglycemia had elevated leptin levels. This is further supported by our recent unpublished findings that high leptin levels are associated with insulin resistance independent of body weight (FIG. 7). In these studies, we separated Israeli Sand Rats into two groups matched for body weight: group 1 were insulin-sensitive (normoglycemic and normoinsulinemic) and group 2 were insulin-resistant (normoglycemic and hyperinsulinemic). Interestingly, group 2 animals had significantly higher leptin levels than the insulin-sensitive group 1 animals.[18] Although we have no supporting data from this study, it is interesting to speculate that leptin directly inhibits insulin action as a protective mechanism limiting the accumulation of adipose tissue mass.

It is also possible that our animals are leptin-resistant. If this is the case, what will be the effect of leptin administration? Few studies have examined the effect of leptin administration in animal models of obesity that display elevated circulating leptin

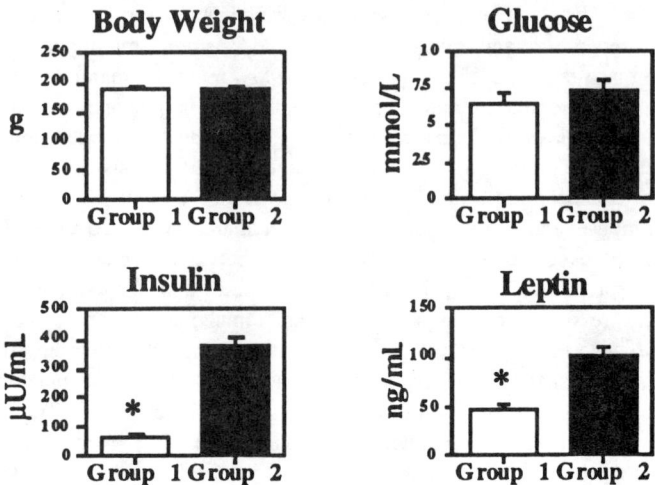

FIGURE 7. Body weight, blood glucose, plasma insulin, and leptin levels in two groups of *Psammomys obesus* matched for body weight: group 1 ($n = 33$), normoinsulinemic; group 2 ($n = 36$), hyperinsulinemic. All values are expressed as mean ± SEM. All differences were considered significant at $p < 0.05$ (*).

levels. In a recent study,[28] we examined the effect of daily administration of 5 mg leptin/kg body weight to obese, diabetic Israeli Sand Rats over a two-week period. The animals were paired according to age, sex, body weight, blood glucose, and plasma insulin concentrations and divided into two groups ($n = 10$): leptin-treated and control. Throughout the study period, animals were given a single daily intraperitoneal injection of leptin (kindly supplied by Margery Nicolson, Amgen Incorporated, California) or phosphate-buffered saline. Our results from this study[28] showed that leptin administration (5 mg/kg/day) had no significant effect on food intake, body weight, percentage body fat, blood glucose, or plasma insulin concentrations. This dosage was sufficient to cause significant reductions in all of these parameters in *ob/ob* mice.[15-17] The results from these studies examining the longitudinal changes in leptin levels and leptin treatment suggest that these sand rats develop resistance to leptin's action in controlling energy balance.

It is interesting to speculate that development of leptin resistance is, in fact, a survival advantage for the animals in their natural habitat when food is scarce; that is, leptin resistance at the level of the CNS may be an expression of the "thrifty gene,"[10] a condition well described in this animal model.[2] When food was available, animals needed to consume large quantities and store the excess energy as fat for use during periods of famine. The ability to develop leptin resistance would be a survival advantage to the animal allowing consumption of large amounts of food without leptin-suppressing appetite signals, thus enabling storage of excess energy as fat. However, when the animal is placed in an environment where food is plentiful, the development of leptin resistance could result in hyperphagia, hyperinsulinemia, and obesity, and later NIDDM.

LEPTIN RECEPTORS

The suggestion that leptin resistance could be an underlying feature of obesity has concentrated research efforts on the mechanism of leptin action. This research effort resulted in the recent discovery of the leptin receptor.[29] The leptin receptor is a single membrane-spanning receptor most closely related to class 1 cytokine receptors and is expressed in choroid plexus. In subsequent studies, Friedman and coworkers[30] cloned the leptin receptor from the choroid plexus and demonstrated that it has at least six alternatively spliced forms. One of these splice variants is expressed at high levels in the hypothalamus and is abnormally spliced in *db/db* mice.[30] The mutant protein is missing the cytoplasmic region and is likely to be defective in signal transduction.[30] These studies suggest that the weight-reducing effect of leptin may be mediated by signal transduction through leptin receptors in the hypothalamus. Genetic mapping studies have determined that the mouse *db* and rat *fatty* genes are the same as the gene encoding the leptin receptor,[31,32] implying that the obesity in these models is due to defective leptin receptor activity. It has been proposed that the leptin receptor on the choroid plexus acts as a transporter allowing leptin to cross the blood-brain barrier into the hypothalamus where it can interact with neurotransmitters. It is likely that leptin alters energy balance through interaction with a number of neurotransmitters, including neuropeptide Y (NPY), glucagon-like peptide-1 (GLP-1), CCK8, melanin-concentrating hormone (MCH), and others. NPY and MCH have been shown to increase food intake,[33,34] whereas CCK8 and GLP-1 are potent inhibitors of food intake[35,36] when administered directly into the CNS of animal models. How leptin may coordinate the action of these neurotransmitters (and others) to regulate the complex control of energy balance is unknown and remains purely speculative.

Recent studies by Stephens *et al.*[37] demonstrated that leptin administration in *ob/ob* mice resulted in a significant reduction in NPY expression in the hypothalamus, possibly mediating the effects of leptin on energy balance. However, this control is certainly more complex and experimental data from Smith *et al.*[38] suggest that leptin determines the sensitivity of the feeding response to exogenous NPY, perhaps by altering NPY binding. Also, studies have shown that normal-weight "gene knockout" mice, which lack NPY, still responded to peripheral administration of leptin.[35] Further research is needed before we can begin to unravel the complex control mechanisms of energy balance in the hypothalamus.

However, is leptin's only site of action the hypothalamus? This is an important question if obesity therapies are to be developed, including chronic leptin administration. As discussed above, our studies have shown that leptin administration (under the same conditions that were effective in the *ob/ob* mouse) was ineffective in a hyperleptinemic model of obesity. Will large doses of leptin overcome leptin resistance at the level of the hypothalamus? Will this leptin treatment have any other possible actions? To answer these questions, we need to further understand the physiological action of leptin.

Recent evidence has emerged demonstrating a number of effects of leptin independent of its role in control of food intake. These effects include reduction of blood glucose and insulin levels,[39] increased thermogenesis and reduced body weight,[40] and

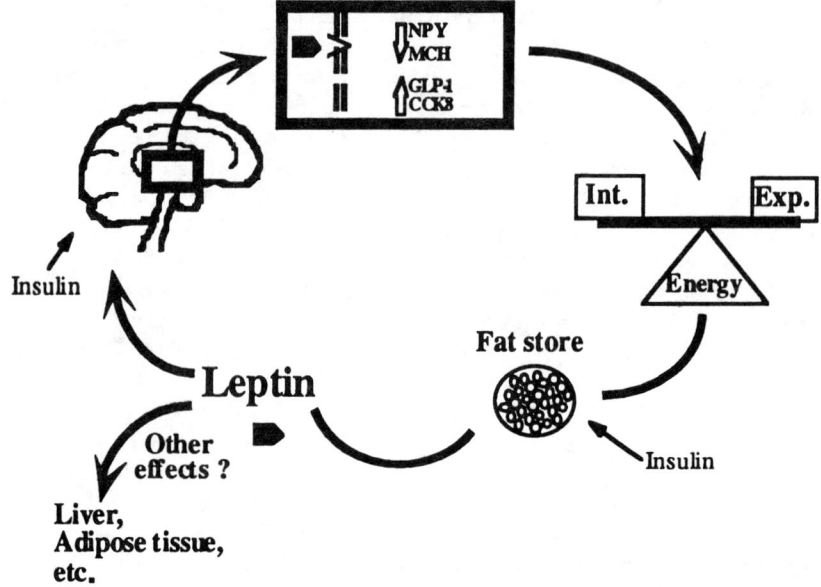

FIGURE 8. Leptin and energy balance.

(most recently) effects of leptin in the neuroendocrine response to fasting.[41] In this recent study, leptin administration reduced (or prevented) the changes in gonadal, adrenal, and thyroid axes in male mice following food deprivation for 48 hours and prevented the starvation-induced delay in ovulation in female mice.

Is then the major role of leptin in the regulation of food intake or does it exert more-important actions as a peripheral endocrine hormone? We have unpublished results examining leptin receptors in the Israeli Sand Rat and these preliminary studies have uncovered leptin receptors in a range of tissues, with the highest number of receptors found in liver, kidney, heart, muscle, hypothalamus, and adipose tissue. These studies raise unanswered questions regarding leptin and its potentially diverse actions (FIG. 8).

SUMMARY

The Israeli Sand Rat (*Psammomys obesus*) is an excellent polygenic model for the study of obesity and diabetes. The metabolic characteristics and the heterogeneous development of these defects, including elevated leptin levels, mimic those found in susceptible human populations. Interestingly, only animals that develop metabolic abnormalities demonstrate hyperleptinemia and, in these animals, leptin administration at the same dose that is effective in *ob/ob* mice is ineffective in reducing food intake or body weight. Perhaps leptin resistance needs to develop in Israeli Sand Rats to allow the development of obesity and, in fact, leptin resistance may be the "thrifty gene" that predisposes individuals to the development of obesity and subsequent metabolic abnormalities. However, there remain many unanswered questions about the physiological actions of leptin. The widespread tissue location of receptors and the actions of leptin independent of food intake highlight the need for further research aimed at determining the major physiological action of this newly discovered and exciting hormone.

REFERENCES

1. SHAFRIR, E. & A. GUTMAN. 1993. *Psammomys obesus* of the Jerusalem colony: a model for nutritionally induced, non-insulin-dependent diabetes. J. Basic Clin. Physiol. Pharmacol. **4:** 83-99.
2. KALDERON, B., A. GUTMAN, E. LEVY, E. SHAFRIR & J. H. ADLER. 1986. Characterization of stages of development of obesity-diabetes syndrome in sand rat (*Psammomys obesus*). Diabetes **35:** 717-724.
3. BARNETT, M., G. R. COLLIER, F. M. COLLIER, P. ZIMMET & K. O'DEA. 1994. A cross-sectional and short-term longitudinal characterization of NIDDM in *Psammomys obesus*. Diabetologia **37:** 671-676.
4. BARNETT, M., G. R. COLLIER, P. ZIMMET & K. O'DEA. 1995. Energy intake with respect to the development of diabetes mellitus in *Psammomys obesus*. Diabetes Nutr. Metab. **8:** 42-47.
5. HABITO, R., M. BARNETT, A. YAMAMOTO, D. CAMERON-SMITH, K. O'DEA, P. ZIMMET & G. R. COLLIER. 1995. Basal glucose turnover in *Psammomys obesus*: an animal model of type 2 (non-insulin-dependent) diabetes mellitus. Acta Diabetol. **32:** 187-192.
6. BARNETT, M., R. HABITO, D. CAMERON-SMITH, A. YAMAMOTO & G. R. COLLIER. 1996. The effect of inhibiting fatty acid oxidation on basal glucose metabolism in *Psammomys obesus*. Horm. Metab. Res. **28:** 165-170.
7. BARNETT, M., G. R. COLLIER, P. ZIMMET & K. O'DEA. 1994. The effect of restricting energy intake on diabetes in *Psammomys obesus*. Int. J. Obes. **18:** 789-794.
8. DEFRONZO, R. A. 1988. The triumvirate: b-cell, muscle, liver–a collusion responsible for NIDDM. Diabetes **37:** 667-687.
9. REAVEN, G. M. 1988. Role of insulin resistance in human disease. Diabetes **37:** 1595-1607.
10. ZIMMET, P. Z. 1992. Challenges in diabetes epidemiology–from west to the rest. Diabetes Care **15:** 232-247.
11. KENNEDY, G. 1953. The role of depot fat in the hypothalamic control of food intake in the rat. Proc. R. Soc. Lond. **140:** 578-592.
12. COLEMAN, D. L. 1978. Obesity and diabetes: two mutant genes causing diabetes-obesity syndromes in mice. Diabetologia **14:** 141-148.
13. COLEMAN, D. L. 1973. Effects of parabiosis of obese with diabetes and normal mice. Diabetologia **9:** 294-298.
14. ZHANG, Y., R. PROENCA, M. MAFFEI, M. BARONE, L. LEOPOLD & J. M. FRIEDMAN. 1994. Positional cloning of the mouse *obese* gene and its human homologue. Nature **372:** 425-432.
15. HALAAS, J. L., K. S. GAJIWALA, M. MAFFEI, S. L. COHEN, B. T. CHAIT, D. RABINOWITZ, R. L. LALLONE, S. K. BURLEY & J. M. FRIEDMAN. 1995. Weight-reducing effects of the plasma protein encoded by the *obese* gene. Science **269:** 543-546.
16. CAMPFIELD, L. A., F. J. SMITH, Y. GUISEZ, R. DEVOS & P. BURN. 1995. Recombinant mouse OB protein: evidence for a peripheral signal linking adiposity and central neural networks. Science **269:** 546-549.
17. PELLEYMOUNTER, M. A., M. J. CULLEN, M. B. BAKER, R. HECHT, D. WINTERS, T. BOONE & F. COLLINS. 1995. Effects of the *obese* gene product on body weight regulation in *ob/ob* mice. Science **269:** 540-543.
18. WALDER, K., M. WILLET, P. ZIMMET & G. R. COLLIER. 1996. Ob gene expression and leptin levels in Israeli Sand Rats (*Psammomys obesus*). Biochim. Biophys. Acta. In press.
19. FREDERICH, R. C., A. HAMANN, S. ANDERSON, B. LÖLLMANN, B. B. LOWELL & J. S. FLIER. 1995. Leptin levels reflect body lipid content in mice: evidence for diet-induced resistance to leptin action. Nat. Med. **1:** 1311-1314.

20. FREDERICH, R. C., B. LÖLLMANN, A. HAMANN, A. NAPOLITANO-ROSEN, B. B. KAHN, B. B. LOWELL & J. S. FLIER. 1995. Expression of ob mRNA and its enclosed protein in rodents: impact of nutrition and obesity. J. Clin. Invest. 96: 1658-1663.
21. FUNAHASHI, T., I. SHIMOMURA, H. HIRAOKA, T. ARAI, M. TAKAHASHI, T. NAKAMURA et al. 1995. Enhanced expression of rat obese (ob) gene in adipose tissues of ventromedial hypothalamus (VMH)-lesioned rats. Biochem. Biophys. Res. Commun. 211: 469-475.
22. MAFFEI, M., H. FEI, G. H. LEE, C. DANI, P. LEROY, Y. ZHANG, et al. 1995. Increased expression in adipocytes of ob RNA in mice with lesions of the hypothalamus and with mutations at the db locus. Proc. Natl. Acad. Sci. U.S.A. 92: 6957-6960.
23. MAFFEI, M., J. HALAAS, E. RAVUSSIN, R. E. PRATLEY, G. H. LEE, Y. ZHANG, H. FEI, R. LALLONE, S. RANGANATHAN, P. A. KERN & J. M. FRIEDMAN. 1995. Leptin levels in human and rodent: measurement of plasma leptin and ob RNA in obese and weight-reduced subjects. Nat. Med. 1: 1155-1161.
24. OGAWA, Y., H. MASUZAKI, N. ISSE, T. OKAZAKI, K. MORI, M. SHIGEMOTO, et al. 1995. Molecular cloning of rat obese cDNA and augmented gene expression in genetically obese Zucker fatty (fa/fa) rats. J. Clin. Invest. 96: 1647-1652.
25. ZIMMET, P., A. HODGE, M. NICHOLSON, M. STATEN, M. DE COURTEN, J. MOORE, A. MORAWIECZKI, J. LUBINA, G. COLLIER, G. ALBERTI & G. DOWSE. 1996. Serum leptin concentration, obesity, and insulin resistance in Western Samoans: the expression of the "thrifty gene"? Br. Med. J. 313: 965-969.
26. ZIMMET, P. Z. & G. R. COLLIER. 1996. Of mice and (wo)men: the obesity (ob) gene, its product, leptin, and obesity. Med. J. Aust. 164: 393-394.
27. COLLIER, G. R., K. WALDER, P. LEWANDOWSKI, A. SANIGORSKI & P. ZIMMET. 1996. Does the "thrifty gene" cause leptin resistance? Submitted.
28. WALDER, K., P. LEWANDOWSKI, A. SANIGORSKI, S. LEE, P. ZIMMET & G. R. COLLIER. 1996. Leptin administration has no effect on food intake, body weight, or diabetic status in a multigenic model of obesity: Psammomys obesus. Submitted.
29. TARTAGLIA, L. A., M. DEMBSKI, X. WENG, N. DENG, J. CULPEPPER, R. DEVOS, G. J. RICHARDS, L. A. CAMPFIELD, F. T. CLARK, J. DEEDS, C. MUIR, S. SANKER, A. MORIARTY, K. J. MOORE, J. S. SMUTKO, G. G. MAYS, E. A. WOOLF, C. A. MONROE & R. I. TEPPER. 1995. Identification and expression cloning of a leptin receptor, OB-R. Cell 83: 1263-1271.
30. LEE, G.-H., R. PROENCA, J. M. MONTEZ, K. M. CARROLL, J. G. DARVISHZADEH, J. I. LEE & J. M. FRIEDMAN. 1996. Abnormal splicing of the leptin receptor in diabetic mice. Nature 379: 632-635.
31. CHUA, S. C., JR., W. K. CHUNG, X. S. WU-PENG, Y. ZHANG, S.-M. LIU, L. TARTAGLIA & R. L. LEIBEL. 1996. Phenotypes of mouse diabetes and rat fatty due to mutations in the OB (leptin) receptor. Science 271: 994-996.
32. PHILLIPS, M. S., Q. LU, H. A. HAMMOND, V. DUGAN, P. J. HEY, C. T. CASKEY & J. F. HESS. 1996. Leptin receptor missense mutation in the fatty Zucker rat. Nat. Genet. 13: 18-19.
33. ZARJEVSKI, N., I. CUSIN, R. VETTOR, F. ROHNER-JEANRENAUD & B. JEANRENAUD. 1993. Chronic intracerebroventricular neuropeptide-Y administration to normal rats mimics hormonal and metabolic changes of obesity. Endocrinology 133: 1753-1758.
34. QU, D., D. S. LUDWIG, S. GAMMELTOFT, M. PIPER, M. A. PELLEYMOUNTER, M. J. CULLEN, W. F. MATTHER, J. PRZYPEX, R. KANAREK & E. MARTALOS-FLIER. 1996. A role for melanin concentrating hormone in the central regulation of feeding behavior. Nature 380: 243-246.
35. ERICKSON, J. C., K. E. CLEGG & R. D. PALMITER. 1996. Sensitivity to leptin and susceptibility to seizures of mice lacking neuropeptide Y. Nature 381: 415-418.
36. TURTON, M. D., D. O'SHEA, I. GUNN, S. A. BEAK, C. M. B. EDWARDS, K. MEERAN, S. J. CHOI, G. M. TAYLOR & S. R. BLOOM. 1996. A role for glucagon-like peptide-1 in the central regulation of feeding. Nature 389: 69-72.
37. STEPHENS, T. W., M. BASINSKI, P. K. BRISTOW, J. M. BUE-VALLESKEY, S. G. BURGETT, L. CRAFT, J. HALE, J. HOFFMANN, H. M. HSLUNG, A. KRIAUCIUNAS, W. MACKELLAR, P. R. ROSTECK, JR., B. SCHONER, D. SMITH, F. C. TINSLEY, X-Y. ZHANG & M. HEIMAN. 1995. The role of neuropeptide Y in the antiobesity action of the obese gene product. Nature 377: 530-532.
38. SMITH, F. J., L. S. CAMPFIELD, J. A. MOSCHERA, P. S. BAILON & P. BURN. 1996. Feeding inhibition by neuropeptide Y. Nature 382: 307.
39. SCHWARTZ, M. W., D. G. BASKIN, T. R. BUKOWSKI, J. L. KUIJPER, D. FOSTER, G. LASSER, D. E. PRUNKARD, D. PORTE, JR., S. C. WOODS, R. J. SEELEY & D. S. WEIGLE. 1996. Spec-

ificity of leptin action on elevated blood glucose levels and hypothalamic neuropeptide Y gene expression in *ob/ob* mice. Diabetes **45:** 531–535.
40. LEVIN, N., C. NELSON, A. GURNEY, R. VANDLEN & F. DE SAUVAGE. 1996. Decreased food intake does not completely account for adiposity reduction after ob protein infusion. Proc. Natl. Acad. Sci. U.S.A. **93:** 1726–1730.
41. AHIMA, R. S., D. PRABAKARAN, C. MANTZOROS, D. QU, B. BLOWELL, E. MARATOS-FLIER & J. S. FLIER. 1996. Role of leptin in the neuroendocrine response to fasting. Nature **382:** 250–252.

Diabetes and Hypertension in Rodent Models[a]

INGRID KLÖTING, SABINE BERG, PETER KOVÁCS,
BIRGER VOIGT, LUTZ VOGT,
AND SIEGFRIED SCHMIDT

"Gerhardt Katsch" Institute of Diabetes
University of Greifswald
17495 Karlsburg, Germany

INTRODUCTION

Complex diseases like diabetes, hypertension, cardiovascular disease, or cancer are quantitative multifactorial traits with both environmental and genetic determinants. While much is known about the environmental factors that predispose to the development of complex diseases, the nature of the genetic susceptibility factors remains largely obscure in human beings. Genetic heterogeneity in an outbred population and varying environmental conditions are the main causes for this dilemma in humans. Because of this, geneticists have searched for alternative systems to define genes that could be involved in complex diseases. The use of animal models closely resembling complex human diseases can overcome the problem of genetic heterogeneity by inbreeding, and the experiments can be carried out in a standardized environment. Thus, using genetically defined inbred animal models, the predisposition to complex disease can be dissected into discrete genetic factors, each of which makes a contribution to a threshold trait. In this way, one can study hundreds of meioses from a single set of parents, which is impossible in human families. Of course, the applicability of the results from animal studies to human diseases must always be carefully evaluated. However, animal studies can help to identify key genes acting in the same biochemical pathway or physiological system and can help to probe more complex genetic interactions than is possible in humans.

To identify discrete genetic factors contributing to a complex disease, phenotypic and genetic information must be available. For this purpose, disease-prone and disease-resistant inbred strains are normally crossed to generate F1 hybrids that either are intercrossed to generate F2 hybrids or are backcrossed onto disease-prone and disease-resistant strains to generate first backcross hybrids (BC1). The phenotypes and genotypes of the F2 and BC1 hybrids are determined and the data are used for linkage analysis. A major prerequisite for the identification of quantitative trait loci markers (QTLs) is the availability of a dense genetic linkage map. A suitable analytical ap-

[a] This work was supported by Grant No. Kl 771/2-2 from Deutsche-Forschungsgemeinschaft.

proach for whole-genome analysis is also required. In mice and rats, both criteria are met in these species. Most models of complex diseases have been established in mice and rats. Up to now, most QTLs have been detected in mouse models, for example, for insulin-dependent diabetes mellitus,[1-3] alcoholism,[4,5] stress vulnerability,[6] and dietary obesity,[7] and in rat models for non-insulin-dependent diabetes mellitus[8,9] and hypertension.[10-17] The number of identified QTLs is increasing from year to year as can be seen in insulin-dependent diabetes of the non-obese diabetes mouse (NOD) and in hypertension of several rat models such as the spontaneously hypertensive rat (SHR), the stroke-prone SHR (SHRSP), the Milan hypertensive rat, or the Dahl salt-sensitive rat. For both NOD diabetes and hypertension, more than 10 QTLs on different chromosomes have been described. However, the identification of a QTL is one side of the coin; the demonstration of its physiologic effect is the other.

Another important point in QTL mapping is the fact that mapping results need not be consistent among different crosses. Linkage analysis reveals only those trait-causing genes that differ between the two animal models used for crossing. This means that a QTL may be detected in an A × B cross, but not necessarily in an A × C cross demonstrated in several crossing studies. For instance, the analysis of diabetes in the spontaneously diabetic BioBreeding (BB) rat as the model for the insulin-dependent diabetes mellitus points to one, two, or three genes required to produce disease, depending on the disease-resistant strain used.[18-21] Also, crosses with SHR or SHRSP rats as models for essential hypertension have indicated that the genetic determination of blood pressure is strain-dependent, caused by different allele effects between disease-prone and disease-resistant strains.[22-28] This fact has consequences not only for the calculation of the number of QTLs involved in the genetic control of a disease, but also for the odds of identifying a disease-relevant locus. Despite this, the use of animal models of human diseases remains an alternative in the search for key genes and for the physiologic function of their products. Crossing studies, QTL mapping, and the analysis of physiologic effects by creating congenic strains that differ in a chromosomal region of interest will be described to demonstrate the dissection of complex disease traits.

BACKGROUND OF CROSSING STUDIES

It is well known that diabetic subjects develop hypertension more frequently than nondiabetic persons. Although each disease alone is serious, the combination of the two gives the patient with diabetes an increased risk of developing diabetic complications and cardiovascular disease. Whereas the clinical association of hypertension and diabetes has long been appreciated, the underlying biological mechanisms that initiate and sustain the elevation of blood pressure in diabetic subjects remain poorly understood. However, considering that more than 50% of all diabetics do not develop hypertension despite a long duration of diabetes, obesity, insulin resistance, and genetic factors must surely play a crucial role. This phenomenon is also seen in most animal models used in diabetes research. Although most animal models develop hyperglycemia or impaired glucose tolerance with obesity, hyperinsulinemia, and/or hyper-

TABLE 1. Widely Used Animal Models of Non-Insulin-dependent (NIDDM) and Insulin-dependent Diabetes Mellitus (IDDM)[a]

Animal Models	Hyper-glycemia	Hyper-insulinemia	iGT	Obesity	Hyper-lipi-demia	Hyper-tension
NIDDM						
Diabetes mouse (db)	+++	+		+	−	?
Obese mouse (ob)	+	+		+	−	?
Fatty mouse (fat)	−	+		+	−	?
Adipose mouse (ad)	+	+		+	−	?
Yellow obese mouse (A^y)	+			+	−	?
NZO mouse	±	+		+	−	?
Zucker rat (fa)	−	+	+	+	+	±
Koletzky rat (cp)	−	+	+	+	+	+
OLET rat	+ males	+ age		+	−	?
hHTG rat	−	±	±	−	+	+
WOK.1W/K rat	−	+	+	−	+	+
GK rat	+	+	+	−	−	?
SHR/N-cp rat	+ diet	+		+	−	±
DSS/N-cp rat	−	+		+	+	+
IDDM						
NOD mouse	+++	−		−	−	?
BB rat	+++	−		−	−	± diabetics
LETL rat	+++	−		−	−	?

[a] iGT, impaired glucose tolerance; ±, inconsistent findings are described; ?, up to now not studied (for details, cf. Klimeš et al. in this volume).

lipidemia, only a few develop hypertension (TABLE 1; details in Klimeš et al. in this volume). In an attempt to overcome this problem, the corpulent allele (cp) that causes obesity and hyperinsulinemia in the Koletzky rat was transferred onto the genetic background of SHR (SHR/N-cp) and Dahl salt-sensitive rats (DSS/N-cp) in the hope of obtaining hybrids with obesity, hyperinsulinemia, and hypertension. However, these crosses did not result in progeny showing a combination of diseases found in the parental strains. Nevertheless, we attempted to generate animal models that develop diabetes on the one hand and hypertension on the other since some of the genes causing hypertension in nondiabetics may also be involved in causing susceptibility to hypertension in diabetics. This prompted us to use for crossing studies normotensive, but diabetes-prone BB/OK rats bred in our own animal facility (described in detail in reference 29) and commercially available hypertensive, but nondiabetes-prone SHR/Mol rats.

CHARACTERISTICS OF BB/OK AND SHR/MOL RATS

About 50% of BB/OK rats spontaneously develop insulin-dependent diabetes at a mean age of 18.6 ± 4.3 weeks. All nondiabetic BB/OK rats are normotensive with a mean systolic blood pressure of 121.4 ± 6.0 mmHg in males and 122.3 ± 2.3 mmHg in females, as determined using the tail cuff method described previously.[30] Diabetes occurrence in BB/OK rats can be explained by three recessively acting diabetogenic genes, named *Iddm1*, *Iddm2*, and *Iddm3*. The *Iddm1* gene causes lymphopenia and was mapped between loci *D4Mit6* and *Npy* on chromosome 4.[21,31] The class-II genes of the major histocompatibility complex (MHC) of the RT1^u haplotype are the second genetic factor, *Iddm2*, required for diabetes development in BB rats. The third gene, *Iddm3*, was recently located on chromosome 18 between the *Olf* and *D18Mit9* loci.[31]

All SHR/Mol rats descended from the National Institutes of Health (NIH) population (abbreviation N) develop hypertension at an age of about 6 weeks. The systolic blood pressure changes from 160 to 280 mmHg. In contrast to the BB rat, the development of hypertension in SHR and SHRSP rats requires the action of at least five genes mapped by QTL analysis. They have been located near the angiotensin-I-converting enzyme on chromosome 10 (*Bp1*[10,11]), around the *D18Mit1* corresponding to the *Ttr* locus on chromosome 18 (*Bp2*[11]), between the loci *DXCep2* and *DXCep4* on chromosome X (*Bp3*[10]), near the renin gene on chromosome 13 (*Bp5*[14]), and at the so-called *Sa* gene near the *Lsn* locus on chromosome 1 (*Sa*[12,13]). Further QTLs involved in the development of hypertension were mapped on chromosomes 2 (*Bp6*[14]), 5 (*Bp7*[17]), and 17 (*Bp8*[17]); however, spontaneously hypertensive rat strains other than SHR or SHRSP were used for this analysis.

Both rat strains differ not only in blood pressure and diabetes occurrence, but also in other phenotypic traits. Because the diabetic state of BB rats markedly influences such traits, 38 nondiabetic BB/OK rats, of which 18 (7M:11F) remained nondiabetic up to an age of 32 weeks, and 22 SHR/Mol rats (12M:10F) were studied for glucose tolerance and serum concentrations of triglycerides, cholesterol, total protein, creatinine, urea, calcium, and phosphate. The animals were checked two times between 14 and 18 weeks (cf. TABLE 2). After diabetes onset (>32 weeks), 24-h excretion of urine total protein, albumin, creatinine, urea, sodium, potassium, calcium, and phosphate was determined using metabolic cages. As shown in TABLE 2, the two rat strains differed not only in blood pressure, but also in the majority of traits studied. Most differences were observed in 24-h excretion, which could be expected because of the relationship of blood pressure to kidney function. SHR rats had significantly higher values in total protein, creatinine, urea, and calcium than BB/OK rats. In the other urine constituents, either female or male SHR rats indicated significantly higher values than BB/OK rats. Sex differences were also observed in serum and urine total protein and in urine creatinine and potassium within BB and SHR rats. Because of these differences, both rat strains are suitable for crossing studies to dissect complex traits like blood pressure and serum and urine constituents. These phenotypic differences were supplemented by a search for genetic differences between the parental strains. Both rat strains were studied for polymorphisms using 112 PCR-analyzed microsatellite markers on 21 chromosomes of the rat genome. Eighty-six out of 112 markers (76.8%) were polymorphic between BB/OK and SHR/Mol rats. This relatively high polymorphism is a good basis for mapping of QTLs in some of the traits studied in the cross hybrids.

TABLE 2. Phenotypic Characteristics of Nondiabetic BB/OK and Hypertensive SHR/Mol Rats[a]

		Serum Constituents					24-h Urine Excretion		
Trait	Sex	BB/OK	Sig.	SHR/Mol		Trait	Sex	BB/OK	Sig.	SHR/Mol
No. of animals		18 (7M:11F)		22 (12M:10F)		No. of animals		15 (7M:8F)		22 (12M:10F)
Mean age of animals (weeks)	M/F	14.8 ± 2.3		14.9 ± 2.3		Mean age of animals (weeks)	M/F	35.0 ± 2.5		34.8 ± 2.3
No. of values		15M:19F		24M:22F		No. of values		7M:8F		12M:10F
Glucose tolerance (mmol × min)	M	540 ± 65.8		516 ± 78.1 ++		Total protein (g/day)	M	0.030 ± 0.005 ++	**	0.049 ± 0.009 ++
	F	554 ± 67.3	**	441 ± 98.2			F	0.013 ± 0.004	**	0.019 ± 0.004
Triglycerides (mmol/L)	M	1.5 ± 0.4	**	1.1 ± 0.3		Albumin (μg/day)	M	12.4 ± 6.5 +		23.6 ± 18.3
	F	1.3 ± 0.6	*	1.7 ± 0.3 ++			F	4.5 ± 6.1	**	37.4 ± 18.5
Cholesterol (mmol/L)	M	2.3 ± 0.3 ++	**	1.8 ± 0.3 ++		Creatinine (mmol/day)	M	0.120 ± 0.013 ++	**	0.100 ± 0.009 ++
	F	2.5 ± 0.3		2.4 ± 0.2			F	0.063 ± 0.005		0.071 ± 0.005
Total protein (g/L)	M	60.6 ± 14.8	**	78.4 ± 5.5		Urea (mmol/day)	M	8.7 ± 1.4	**	10.7 ± 1.3
	F	61.4 ± 11.6	**	78.1 ± 5.6			F	8.1 ± 1.4	**	11.2 ± 1.0
Creatinine (mmol/L)	M	45.1 ± 8.2	**	35.0 ± 4.5 ++		Na (mmol/day)	M	1.5 ± 0.8		1.2 ± 0.3
	F	43.0 ± 6.3		40.0 ± 5.7			F	1.8 ± 1.1		1.5 ± 0.3
Urea (mmol/L)	M	7.1 ± 0.6		7.3 ± 0.5		K (mmol/day)	M	3.4 ± 0.7		3.7 ± 0.7
	F	6.7 ± 0.7	**	8.0 ± 0.7			F	3.3 ± 0.4	**	4.4 ± 0.6
Ca (mmol/L)	M	2.7 ± 0.11		2.6 ± 0.09		Ca (mmol/day)	M	0.043 ± 0.01	**	0.140 ± 0.06
	F	2.6 ± 0.13		2.6 ± 0.13			F	0.115 ± 0.04	**	0.019 ± 0.004
P (mmol/L)	M	1.9 ± 0.3		2.0 ± 0.3		P (mmol/day)	M	0.090 ± 0.044	**	0.021 ± 0.015 +
	F	1.8 ± 0.3		1.7 ± 0.3			F	0.042 ± 0.043		0.048 ± 0.037

[a] Data are given as mean ± SD. Significantly different (Sig.) at 5% (*) or 1% (**) level between BB/OK and SHR rats. Significantly different at 5% (+) or 1% (++) level between males and females.

CROSSING OF BB/OK AND SHR/MOL RATS AND PHENOTYPES

Because various sex differences were found in both BB/OK and SHR/Mol rats, male and female diabetic BB/OK rats were reciprocally crossed with hypertensive SHR/Mol rats to generate F1 hybrids. All F1 hybrids were normoglycemic, but hypertensive with a systolic blood pressure of 192 ± 10 mmHg, comparable with that of hypertensive SHR rats, indicating a dominant effect of blood pressure genes of SHR rats. F1 hybrids were backcrossed onto male and female BB/OK and SHR rats. Progeny were designated BC1BB-M(ale), BC1BB-F(emale), BC1SHR-M, and BC1SHR-F, respectively. From these backcrosses, we recovered 117 BC1BB-M (62M:55F), 99 BC1BB-F (45M:54F), 97 BC1SHR-M (43M:54F), and 135 BC1SHR-F (68M:67F) hybrids that were also characterized for diabetes- and hypertension-related traits. Because diabetes occurrence in BB/OK rats can be explained by three autosomal recessively acting genes, diabetic BC1 hybrids were only observed in BC1BB hybrids. Fifteen out of 117 BC1BB-M hybrids (12.8%) and 7 out of 99 BC1BB-F hybrids (7.1%) were diabetic with an age of onset of 129.9 ± 41.3 and 147.9 ± 57.9 days, respectively.

Turning to the phenotypes of BC1 hybrids, the results are summarized in TABLE 3 for BC1BB and in TABLE 4 for BC1SHR hybrids. No significant differences were found between BC1BB-M and BC1BB-F hybrids in the serum constituents studied, indicating that the sex of the parent backcrossed has no effect on these traits in the progeny. In contrast to the serum constituents, significant differences were detected in blood pressure and in some of the urine constituents. Significantly higher blood pressure and phosphate excretion were observed in male BC1BB-F versus male BC1BB-M hybrids, whereas significantly higher values were found for sodium excretion in male BC1BB-M and for potassium as well as calcium excretion in female BC1BB-M versus female BC1BB-F hybrids.

More-pronounced reciprocal cross effects were found in BC1SHR hybrids (cf. TABLE 4). Whereas in BC1BB hybrids no reciprocal cross effect was found in serum constituents, in BC1SHR only serum calcium was without cross effect. Reciprocal cross effects were found in triglycerides, cholesterol, and serum urea in both sexes. In addition, cross effects were seen in blood pressure and urine potassium. Male and female BC1SHR-F hybrids were characterized by significantly higher blood pressure and significantly lower potassium excretion versus BC1SHR-M animals. The other reciprocal cross effects were mainly limited to one sex. Significantly higher total protein, albumin, and sodium excretion was observed in male BC1SHR-F hybrids. For phosphate excretion, significantly lower values were found in female BC1SHR-F hybrids.

These findings demonstrate clearly that BC1BB and BC1SHR hybrids are characterized by evidently different trait behavior in both males and females. These effects must be considered in linkage analysis to identify QTLs in the appropriate traits because such effects can be caused by different genetic factors, for example, X- and Y-linked, mitochondrial, or imprinted genes. Nevertheless, these BC1 hybrid populations will be a useful tool to dissect the quantitative traits studied.

TABLE 3. Phenotypic Characteristics of Nondiabetic BCIBB Hybrids[a]

Serum Constituents

Trait	Sex	BC1BB-M	BC1BB-F
No. of animals	M/F	102	92
Mean age of animals (weeks)	M/F	15.3 ± 3.2	14.9 ± 2.3
No. of values		126M:103F	90M:109F
Glucose tolerance (mmol × min)	M	551 ± 126.3 (61)	546 ± 88.1 (45)
	F	590 ± 124.1 (49) ++	574 ± 112.3 (54)
Triglycerides (mmol/L)	M	2.0 ± 0.9	1.9 ± 0.7
	F	1.6 ± 0.7 ++	1.7 ± 0.7
Cholesterol (mmol/L)	M	2.6 ± 0.6	2.5 ± 0.6
	F	2.9 ± 0.6 ++	2.7 ± 0.4 ++
Total protein (g/L)	M	66.7 ± 13.4	65.9 ± 14.9
	F	71.4 ± 8.8 ++	69.4 ± 11.4
Creatinine (mmol/L)	M	41.3 ± 8.1	42.5 ± 8.8
	F	42.8 ± 8.3	42.9 ± 6.5
Urea (mmol/L)	M	7.6 ± 1.3 ++	7.6 ± 1.1 ++
	F	8.2 ± 1.2	8.1 ± 1.1
Ca (mmol/L)	M	2.6 ± 0.4	2.6 ± 0.2
	F	2.6 ± 0.4	2.6 ± 0.2
P (mmol/L)	M	2.1 ± 0.5 ++	2.2 ± 0.3 ++
	F	1.8 ± 0.4	1.9 ± 0.3

Blood Pressure and 24-h Urine Excretion

Trait	Sex	BC1BB-M	Sig.	BC1BB-F
No. of animals	M/F	102		92
Mean age of animals (weeks)	M/F	35.6 ± 2.1		35.5 ± 2.2
No. of values		56M:46F		42M:50F
Blood pressure (mmHg)	M	135.8 ± 17.9 (56)	**	146.4 ± 14.4 (42)
	F	143.5 ± 16.4 (42) +		148.2 ± 10.8 (56)
Total protein (g/day)	M	0.033 ± 0.012		0.031 ± 0.017 ++
Albumin (μg/day)	F	0.026 ± 0.034		0.017 ± 0.021 ++
Creatinine (mmol/day)	M	24.5 ± 24.4		25.4 ± 27.4
	F	33.3 ± 23.8		37.8 ± 33.4
Urea (mmol/day)	M	0.130 ± 0.036		0.133 ± 0.027 ++
	F	0.106 ± 0.116		0.080 ± 0.024 ++
Na (mmol/day)	M	11.7 ± 2.8		11.7 ± 2.5
	F	12.8 ± 5.6		11.9 ± 3.1
K (mmol/day)	M	1.7 ± 0.6	*	1.4 ± 0.5
	F	1.7 ± 0.5		1.5 ± 1.7
Ca (mmol/day)	M	4.7 ± 1.0		5.0 ± 0.9
	F	5.0 ± 1.0	**	4.5 ± 0.8
	F	0.053 ± 0.021 ++		0.045 ± 0.025 ++
P (mmol/day)	F	0.127 ± 0.047 ++	**	0.099 ± 0.052 ++
	M	0.022 ± 0.035 ++	*	0.043 ± 0.046 ++
	F	0.048 ± 0.112		0.057 ± 0.060

[a] Data are given as mean ± SD. Significantly different (Sig.) at 5% (*) or 1% (**) level between BC1BB-M and BC1BB-F. Significantly different at 5% (+) or 1% (++) level between males and females.

TABLE 4. Phenotypic Characteristics of Nondiabetic BC1SHR Hybrids[a]

Trait	Sex	Serum Constituents BC1SHR-M	Sig.	BC1SHR-F	Trait	Sex	Blood Pressure and 24-h Urine Excretion BC1SHR-M	Sig.	BC1SHR-F
No. of animals	M/F	97		135	No. of animals	M/F	97		135
Mean age of animals (weeks)	M/F	15.0 ± 2.2		14.9 ± 2.2	Mean age of animals (weeks)	M/F	35.0 ± 2.4		35.1 ± 2.2
No. of values		89M:108F		102M:110F	No. of values		43M:54F		68M:67F
Glucose tolerance (mmol × min)	M	549 ± 76.9 (45)	*	609 ± 158.5 (50)	Blood pressure (mmHg)	M	170 ± 13.7	**	195 ± 23.4
	F	621 ± 125.9 (54) ++		641 ± 177.1 (55)		F	172 ± 13.3	**	205 ± 23.7
Triglycerides (mmol/L)	M	1.4 ± 0.5	**	1.9 ± 0.6	Total protein (g/day)	M	0.038 ± 0.017 ++	*	0.052 ± 0.035 ++
	F	1.5 ± 0.6		2.0 ± 0.7		F	0.017 ± 0.018		0.018 ± 0.017
Cholesterol (mmol/L)	M	2.1 ± 0.3 ++	**	2.6 ± 0.7	Albumin (μg/day)	M	25.3 ± 9.6	**	44.4 ± 37.4
	F	2.4 ± 0.5		2.7 ± 0.6		F	33.3 ± 14.0		40.4 ± 42.2
Total protein (g/L)	M	71.2 ± 5.6	**	76.2 ± 8.1	Creatinine (mmol/day)	M	0.116 ± 0.042 ++		0.122 ± 0.058 ++
	F	71.0 ± 5.8	**	75.2 ± 5.8		F	0.073 ± 0.025		0.071 ± 0.026
Creatinine (mmol/L)	M	39.2 ± 5.3 ++	**	38.0 ± 6.1 ++	Urea (mmol/day)	M	13.3 ± 4.9		14.3 ± 5.8 ++
	F	44.1 ± 7.6	**	40.3 ± 4.9		F	13.9 ± 7.8		12.3 ± 2.8
Urea (mmol/L)	M	7.7 ± 0.8	**	8.9 ± 1.2	Na (mmol/day)	M	1.5 ± 0.5	*	1.7 ± 0.4
	F	7.8 ± 1.5	**	9.0 ± 1.7		F	1.4 ± 0.8		1.6 ± 0.4
Ca (mmol/L)	M	2.6 ± 0.2		2.7 ± 0.4	K (mmol/day)	M	4.9 ± 1.1	**	4.4 ± 0.8
	F	2.6 ± 0.2		2.7 ± 0.2		F	5.1 ± 1.4	**	4.0 ± 0.8
P (mmol/L)	M	2.1 ± 0.3 ++		2.2 ± 0.4 ++	Ca (mmol/day)	M	0.071 ± 0.035		0.084 ± 0.042
	F	1.7 ± 0.4	**	1.9 ± 0.4		F	0.159 ± 0.075		0.147 ± 0.055
					P (mmol/day)	M	0.047 ± 0.042		0.040 ± 0.043
						F	0.075 ± 0.054	**	0.052 ± 0.033

[a] Data are given as mean ± SD. Significantly different (Sig.) at 5% (*) or 1% (**) level between BC1SHR-M and BC1SHR-F. Significantly different at 5% (+) or 1% (++) level between males and females.

QTLs FOR HYPERTENSION IN BC1BB AND BC1SHR HYBRIDS

As mentioned above, QTL mapping results need not be consistent among different crosses. Linkage analysis reveals only those trait-causing genes that differ between the two animal models used for crossing. To dissect hypertension in rats, different hypertensive models and nonhypertensive strains were used for linkage analysis. With the (SHRSP × WKY) F2 cross, blood pressure loci were detected on chromosomes 1 (*Sa*), 10 (*Bp1*), 18 (*Bp2*), and X (*Bp3*). Furthermore, the heat stress protein *hsp70* gene located within the MHC on chromosome 20 was also found to be associated with hypertension using congenic BN.1K (SHR), SHR.1N (BN), and BN × SHR recombinant inbred rat strains.[32]

To search for QTLs involved in determining blood pressure in BC1BB and BC1SHR hybrids, we selected 26 microsatellite markers for genetic analysis[33] that were polymorphic between BB/OK and SHR/Mol rats and located on chromosomes 1 (*Sa, Lsn, Igf2, Secr, Tnt, D1Mgh12, D1Mgh13*), 10 (*IL4, Abp, Aep, Ppy, Ace, Gh*), 18 (*Ttr, D18Mit1, Grl, Tilp, Gjal, Olf, D18Mit9*), 20 (*RT1.A, Tnfα*), and X (*Ar, Mycs, Pfkb1, DXMgh3*). Data were analyzed with the MAPMAKER/QTL computer package.[34] A threshold lod of >4.8 for F2 crosses and of >3.0 for first backcrosses, corresponding to a P value of 5×10^{-5}, are required for this analysis to ensure that the probability is at most 5% for a peak occurring by chance.[35]

At first, each backcross population was separately analyzed. There was no region on any chromosome indicating an lod score of >3.0. Because the number of backcross progeny that must be genotyped to map a QTL is an important factor,[35] the low number of BC1 hybrids could be the reason for this negative result. Therefore, we combined the data of BC1BB-M and BC1SHR-M as well as BC1BB-F and BC1SHR-F and in this way constructed two F2 crosses named F2M and F2F. Using these F2 populations, all loci on all chromosomes studied showed highly statistically significant QTLs of blood pressure with lod scores of >5, as shown in FIGURE 1. Higher lod scores were found in the F2F versus the F2M populations.

These findings confirm published data using SHR or SHRSP cross hybrids demonstrating that blood pressure loci map on chromosomes 1, 10, 18, 20, and X. However, with the mapping of a QTL, nobody knows the physiologic effect of this QTL. To elucidate the physiologic effect of a QTL found in a defined chromosomal region, there are several methods—the generation of congenic strains is one of them.

BACKGROUND OF CONGENIC BB.SHR RATS

Strains that are genetically identical except for a single chromosome segment are said to be congenic. Congenic strains are usually derived by backcross breeding and genomic selection techniques, by which a specific chromosome segment is transferred from the strain A onto the genetic background of a recipient strain B designated B.A. If this congenic B.A strain shows a difference from the progenitor strain B, one can conclude that there is a locus within the transferred segment affecting the appropriate trait. If a library of congenics (B.A1, B.A2, B.A3, etc.) exists, one can cross the con-

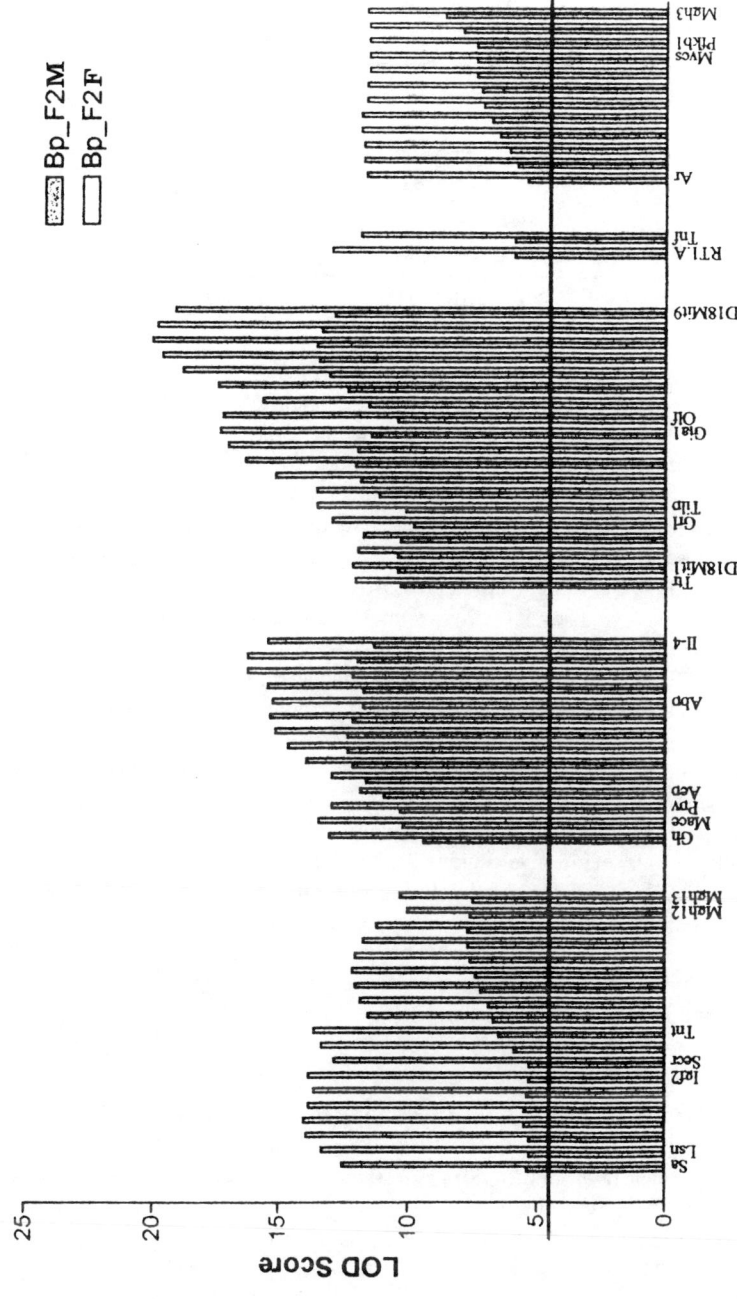

FIGURE 1. Location scores for loci affecting blood pressure in F2M and F2F hybrids.

TABLE 5. Genetic Characteristics of Congenic BB.SHR Rats

Characteristics	Designation of Congenics				
	BB.Sa	BB.Bp1	BB.Bp2	BB.1K	BB.Xs
Chromosome	1	10	18	20	X
Blood pressure locus	Sa	Bp1 (Ace)	Bp2	Hsp	Bp3
Region of SHR rat	Sa-Igf2	Gh-Aep	Ttr-Tilp	RT1.A-Tnf	Ar-DXMgh3
Region size ($\approx$cM)^a	16	3	12	2	36
Status of congenics	established	intercross	established	established	established

^{a} Mapped in BC1BB and BC1SHR hybrids.

genics to study the interactions of two (B.A1 × B.A2, B.A1 × B.A3, etc.) or more transferred genes (A1 × A2 × A3). However, one should take into account that different congenic strains (B.A, C.A, D.A, etc.) carrying the same chromosomal region of the strain A on different genetic backgrounds need not show the same phenotypic effect. The cause is the genetic difference of recipient strains, leading to different gene interactions between the genetic background of the recipient strain and the transferred segment of the strain A.

To study the role of a single blood pressure QTL found in the BC1BB and BC1SHR hybrids, we transferred several chromosome segments of the SHR/Mol rat onto the genetic background of diabetes-prone BB/OK rats using BC1BB-M hybrids for further backcrosses. To accelerate the generation of congenics, we selected animals that were heterozygous for the region of interest and homozygous for BB alleles at most of the 72 background loci examined. After four backcross generations, the animals were already homozygous at background loci and were intercrossed. Animals homozygous for SHR alleles at the loci of interest were selected and established the appropriate congenic BB.SHR strain.

As shown in TABLE 5, five regions of the SHR chromosomes 1, 10, 18, 20, and X, each carrying a QTL for blood pressure, were transferred onto BB/OK rats. The congenics were named after the blood pressure locus found to be involved in hypertension, except for the BB.Xs strain where a large region of more than 30 cM was transferred. First phenotypic results of the newly established congenic BB.Sa, BB.Bp2, BB.Bp3, and BB.1K lines are shown.

DIABETES AND BLOOD PRESSURE IN CONGENIC BB.SHR

Three complete litters of BB/OK rats and their congenic derivatives BB.Sa, BB.Bp2, BB.1K, and BB.Xs were studied for diabetes occurrence up to an age of 30 weeks as described elsewhere.[29] All animals were kept in the same animal room as described.[29] As shown in TABLE 6, BB.Sa, BB.Bp2, and BB.Xs rats developed diabetes at a frequency and age comparable with those of BB rats. As expected, no diabetes was detected in BB.1K rats since the *Iddm2* allele that is essential for diabetes development in BB rats is missing in this rat strain.[18-21] This finding indicates that con-

TABLE 6. Diabetes Frequency in BB/OK Rats and Their Congenic Derivatives

Strain	Diabetes Frequency	Age at Onset (days) (mean ± SD)
BB/OK	18 (11:7) 50% (9/18)	121 ± 22.0
BB.Sa	14 (8:6) 64% (9/14)	125 ± 31.8
BB.Bp2	21 (10:11) 43% (9/21)	125 ± 26.1
BB.Xs	26 (9:17) 42% (11/26)	112 ± 38.4
BB.1K	17 (10:7) 0%	

genic BB.Sa, BB.Bp2, and BB.Xs rats possess all diabetogenic genes of BB/OK rats on the one hand and that blood pressure regions of SHR/Mol rats do not influence diabetes occurrence on the other.

To elucidate the effect of different blood pressure regions of the SHR/Mol rat, nondiabetic (nd) and diabetic (d) males of BB.Sa (nd = 3, d = 4), BB.Bp2 (nd = 3, d = 1), BB.Xs (nd = 3, d = 3), BB.1K (nd = 2), and BB/OK rats (nd = 5, d = 3) with a mean age of 15.2 ± 1.2 weeks were implanted with sensors (Data Sciences International, St. Paul, Minnesota) measuring telemetrically systolic (SBP, mmHg) and diastolic blood pressure (DBP, mmHg), heart rate (HR, beats/min), and motor activity (MA, movements/5 min). One week after implantation of the sensors, the measurements were carried out every 5 min over a period of at least 5 days per animal.[36] In diabetic animals, the measurements were carried out one week after diabetes diagnosis. All diabetic animals were treated with insulin (Lente®, Novo Industry, Denmark) using a daily injection of 7-10 U/kg body weight as described in detail elsewhere.[37]

The mean values of the traits studied are summarized in TABLE 7. In comparison with nondiabetic BB/OK rats, all nondiabetic congenics have significantly elevated SBP, which is more evident in the day profiles shown as the mean values of animals studied in FIGURES 2 and 3. In contrast to all other strains, a remarkable feature was observed in BB.Sa rats (cf. FIG. 2). As shown in FIGURES 2 and 4, not only the SBP, but also DBP as well as HR are characterized by rapid fluctuations not seen in BB/OK rats and in the other congenics. This phenomenon is also reflected in the difference between SBP and DBP changing most in BB.Sa rats (cf. FIG. 5). For all rat strains shown in FIGURE 5, the lowest blood pressure amplitude was found in BB.Sa rats followed by BB/OK, BB.Xs, and BB.1K rats, with the latter two being almost indistinguishable. The highest blood pressure amplitude was observed in BB.Bp2 animals. BB.Sa rats had not only the lowest blood pressure difference, but also the lowest HR and MA values (cf. TABLE 7) of all rats studied. Comparable HRs were observed in BB/OK and BB.1K rats, which were significantly lower than those of BB.Xs and higher

TABLE 7. Systolic (SBP) and Diastolic Blood Pressure (DBP), Heart Rate (HR), and Motor Activity (MA) in Nondiabetic BB Rats and Their Nondiabetic Congenic Derivatives[a]

Trait	BB	d BB/OK	BB.Sa	d BB.Sa	BB.Bp2	d BB.Bp2	BB.Xs	d BB.Xs	BB.1K
No. of males	5	3	3	4	3	1	3	3	3
Age of animals (weeks)	13.6 ± 1.6	15.0	15.6 ± 0.5	19.8 ± 5.1	14.4 ± 2.8	20.0	14.6 ± 1.1	18.6 ± 2.1	14.4 ± 0.5
No. of values	7467	4550	4233	5522	4937	1551	4647	3866	5763
SBP (mmHg)	118.7 ± 9.9	127 ± 14.9[b]	124.0 ± 11.6[c]	124.3 ± 15.0	126.3 ± 10.1[c]	122.1 ± 10.1[b]	126.9 ± 10.7[c]	137.7 ± 11.3[b]	122.7 ± 7.0[c]
DBP (mmHg)	88.8 ± 10.6	90.0 ± 12.0[b]	107.0 ± 12.1[c]	101.8 ± 13.1[b]	87.2 ± 7.8[c]	80.7 ± 7.9[b]	92.1 ± 9.2[c]	99.8 ± 9.8[b]	87.4 ± 7.2[c]
SBP − DBP (mmHg)	29.1 ± 9.6	37.9 ± 5.8[b]	17.1 ± 2.2[c]	22.8 ± 10.9[b]	39.0 ± 3.5[c]	41.4 ± 3.3[b]	34.8 ± 5.3[c]	38.0 ± 4.6[b]	35.1 ± 2.3[c]
HR (beats/min)	349.9 ± 57.5	323.3 ± 59.4[b]	330.2 ± 60.8[c]	305.8 ± 51.9[b]	341.2 ± 48.1[c]	305.1 ± 38.1[b]	358.3 ± 55.7[c]	302.1 ± 59.9[b]	346.4 ± 57.2[c]
MA (movements/ 5 min)	17.4 ± 35.5	19.1 ± 40.0[b]	14.8 ± 30.3[c]	10.0 ± 21.2[b]	24.7 ± 41.6[c]	20.8 ± 31.8[b]	17.6 ± 34.2	27.0 ± 57.5[b]	17.5 ± 34.6

[a] Data are given as mean ± SD.
[b] Significant difference between diabetic (d) and nondiabetic rats at 1% level.
[c] Significant difference between BB/OK rats and their appropriate congenics at 1% level.

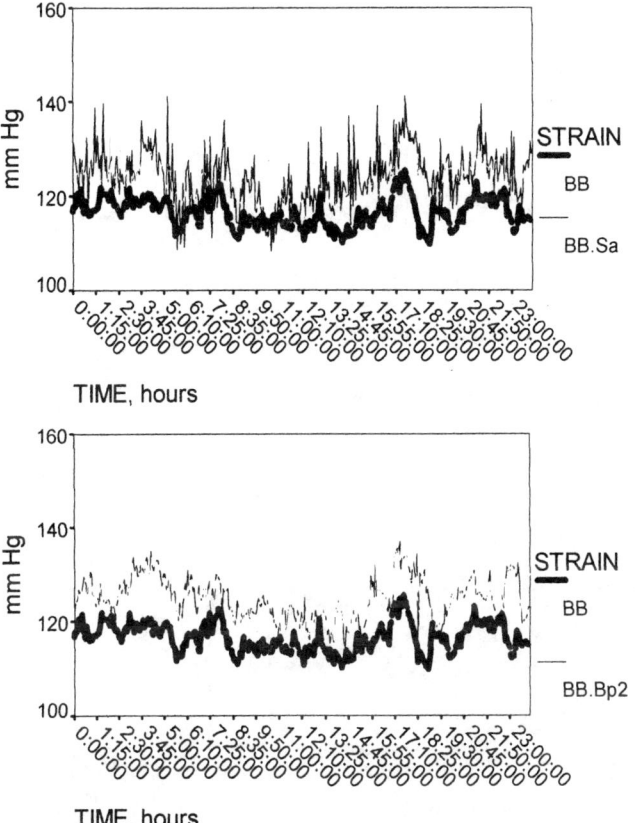

FIGURE 2. Day profile of systolic blood pressure in nondiabetic BB.Sa and BB.Bp2 rats compared with nondiabetic BB/OK rats (mean of males studied).

than those of BB.Bp2 rats. The MA was highest in BB.Bp2 rats followed by BB/OK, BB.Xs, and BB.1K animals.

The SBP increased evidently in diabetic BB/OK and BB.Xs rats, but remained at a level similar to that of the nondiabetics in BB.Sa animals (TABLE 7). In BB.Bp2 rats, the SBP decreased one week after diabetes onset. The DBP increased in diabetic BB/OK and BB.Xs rats and decreased in BB.Sa and BB.Bp2 rats, even though the SBP was unchanged in BB.Sa. An analogous behavior was seen in MA. However, in all diabetics, the behavior of HR was comparable. All diabetics of each strain showed a decrease in HR. Comparing the HR day profiles of nondiabetics and diabetics, changes were especially evident in both BB.Bp2 and BB.Xs rats (FIG. 6), being more pronounced in BB.Xs. In BB.Xs rats also, the changes of HR and MA were more pronounced (cf. FIG. 7).

Although other traits were also studied in BB/OK rats and their congenic deriv-

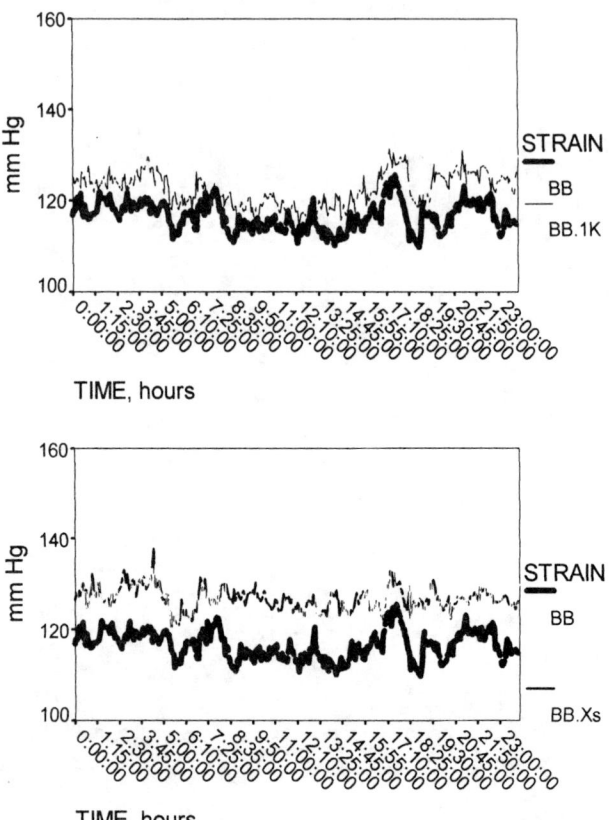

FIGURE 3. Day profile of systolic blood pressure in nondiabetic BB.1K and BB.Xs rats compared with nondiabetic BB/OK rats (mean of males studied).

atives, showing differences between them in several serum and urine constituents, the findings of blood pressure, HR, and MA demonstrated clearly that a single blood pressure QTL of SHR rats transferred onto the genetic background of BB/OK rats causes a significant increase of the systolic blood pressure and influences differently DBP, HR, and MA. This is a finding that has never been described up to now for there are no congenic rat strains carrying genetically defined QTL regions of the SHR rat.

SUMMARY AND CONCLUSIONS

As shown by ourselves and others, animal models closely resembling human complex diseases like IDDM in BB/OK and hypertension in SHR/Mol rats can be used to dissect a complex disease into discrete genetic factors as has been done for hypertension in (BB/OK × SHR/Mol) cross hybrids. Discrete genetic factors, so-called

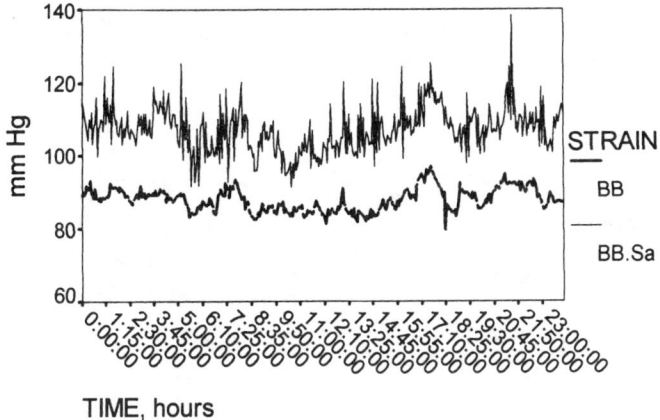

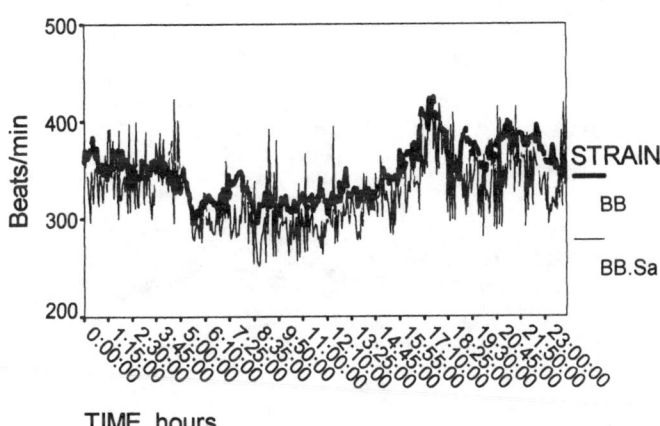

FIGURE 4. Day profile of diastolic blood pressure and heart rate in nondiabetic BB.Sa rats compared with nondiabetic BB/OK rats (mean of males studied).

QTLs, were detected on chromosomes 1, 10, 18, 20, and X. To gain additional information about the physiologic effect of the mapped blood pressure QTLs, genetically defined regions of the SHR rat were transferred onto the genetic background of diabetes-prone BB/OK rats. Four new congenic BB.SHR rats named BB.Sa, BB.Bp2, BB.1K, and BB.Xs were generated and characterized telemetrically for blood pressure, heart rate, and motor activity. The data demonstrate clearly that each single blood pressure QTL of the SHR rat causes a significant increase of the systolic blood pressure and has a different influence on diastolic blood pressure, heart rate, and motor activity. The effects were modified differently by the diabetic state in BB.Sa, BB.Bp2, and BB.Xs rats carrying all diabetogenic genes of the BB/OK rats. The results demonstrate that

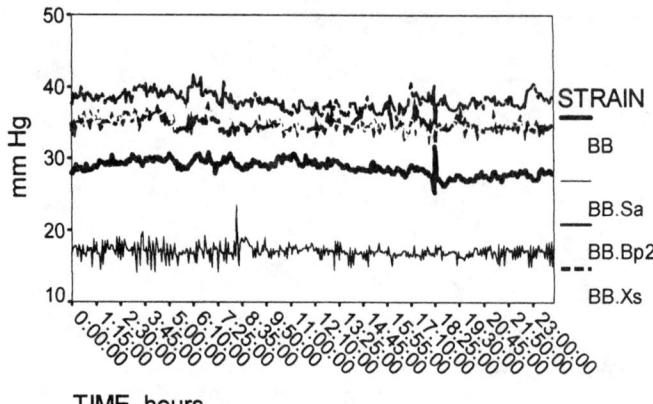

FIGURE 5. Day profile of blood pressure amplitude in nondiabetic BB/OK, BB.Sa, BB.Bp2, and BB.Xs rats (mean of males studied).

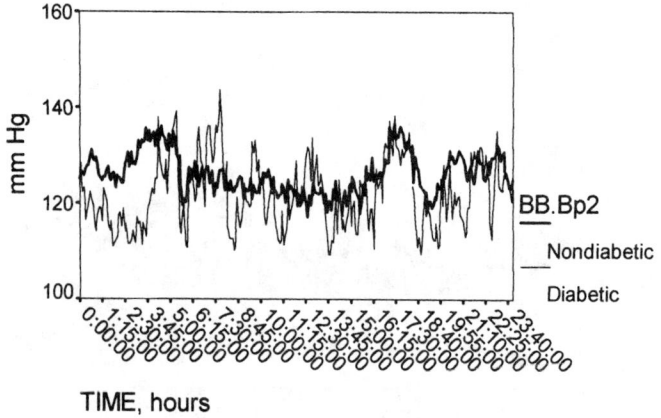

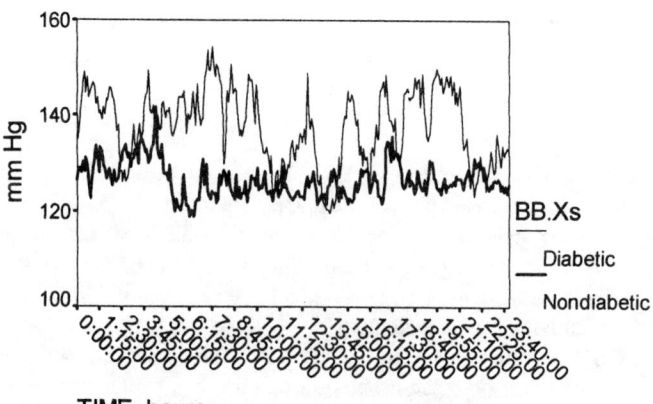

FIGURE 6. Day profile of systolic blood pressure in nondiabetic and diabetic BB.Bp2 and BB.Xs rats (mean of males studied).

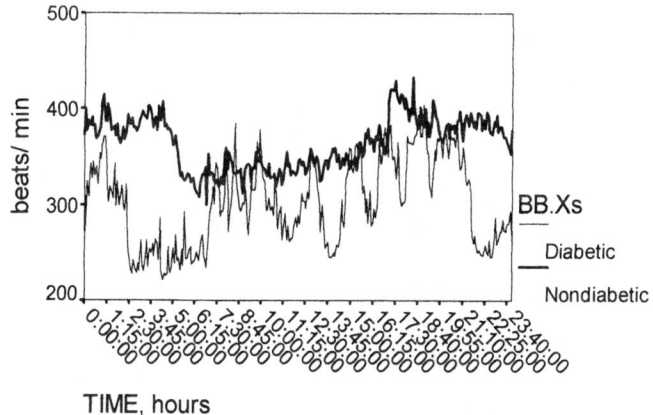

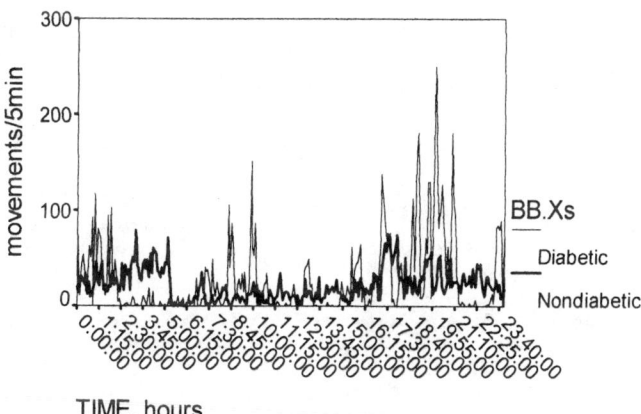

FIGURE 7. Day profile of heart rate and motor activity in diabetic and nondiabetic BB.Xs rats (mean of males studied).

these newly established congenic strains are a unique tool to study the physiological control of blood pressure by a single blood pressure QTL on the one hand and their interaction with hyperglycemia on the other. It is well within the bounds of possibility that diabetic congenics reflect the diabetic hypertension seen in diabetic patients. Because of the synteny conservation in gene order between different mammals, genes of the appropriate human region could therefore be candidate genes for hypertension in diabetics. Furthermore, these congenic strains can also be used to study interactions between a blood pressure QTL and various selected environmental conditions. In this way, one could learn which QTL can be influenced by environmental factors and to what extent. Another point is the study of gene interactions. Because congenics are genetically identical except for the defined transferred region, congenics can be crossed to investigate the interaction between two or three blood pressure QTLs selected by the investigator and not by nature. These QTL combinations can be studied in the nondiabetic as well as diabetic state.

Although the advantage of congenic strains has been shown, the transferred chromosomal regions are too large to pinpoint the gene responsible for the phenotypic change. Therefore, regions on each chromosome must be systematically whittled down, which can be done by crossing the congenics with BB/OK rats and intercrossing their progeny to generate recombinants. These can then be used for the creation of new congenic lines carrying a much smaller region of the SHR/Mol rat. This has been started for the region on chromosome 1 spanning a 16-cM region from the *Sa* to the *Igf2* gene. BB.Sa rats were therefore backcrossed onto BB/OK rats and the resulting progeny were intercrossed. The aim will be to create at least three new congenic BB.Sa rat strains homozygous for the SHR alleles of *Sa*, *Lsn*, or *Igf2* genes. However, new problems will emerge with these new congenics. To genetically define small regions requires more dense polymorphic markers than are currently available. Dense polymorphic markers will also be necessary to split the other regions on chromosomes 10, 18, 20, and X. We expect that in the near future it will be possible using this approach to define small regions of <0.5 cM. The recent progress in gene mapping in the rat gives hope that the use of such congenic lines will allow the identification and recovery of the blood pressure genes in the near future.

REFERENCES

1. TODD, J. A., T. J. AITMAN, R. J. CORNALL, S. GHOSH, J. R. S. HALL, C. M. HEARNE, A. W. KNIGHT, J. M. LOVE, M. A. MCALEER, J. B. PRINS, N. RODRIGUES, M. LATHROP, A. PRESSEY, N. H. DELAROTE, L. B. PETERSON & L. S. WICKER. 1991. Genetic analysis of autoimmune type 1 diabetes mellitus in mice. Nature 351: 542–547.
2. GHOSH, P., S. M. PALMER, N. R. RODRIGUES, H. J. CORDELL, C. M. HEARNE, R. J. CORNALL, J. B. PRINS, P. MCSHANE, G. M. LATHROP & L. B. PETERSON. 1993. Polygenic control of autoimmune diabetes in nonobese diabetic mice. Nat. Genet. 4: 404–409.
3. WICKER, L. S., J. A. TODD & L. B. PETERSON. 1995. Genetic control of autoimmune diabetes in the NOD mouse. Annu. Rev. Immunol. 13: 179–200.
4. PHILLIPS, T. J., J. C. CRABBE, P. METTEN & J. K. BELKNAP. 1994. Localization of genes affecting alcohol drinking in mice. Alcohol Clin. Exp. Res. 18: 931–941.
5. RODRIGUES, L. A., R. PLOMIN, D. A. BLIZARD, B. C. JONES & G. E. MCCLEARN. 1995. Alcohol acceptance, preference, and sensitivity in mice. II. Quantitative trait loci mapping analysis using B × D recombinant inbred strains. Alcohol Clin. Exp. Res. 19: 367–373.
6. TARRICONE, B. J., J. N. HINGTGEN, J. K. BELKNAP, S. R. MITCHEL & J. I. NURNBERGER. 1995. Quantitative trait loci associated with behavioral response of B × D recombinant inbred mice to restraint stress: a preliminary communication. Behav. Genet. 25: 489–495.
7. WEST, D. B., L. J. GOUDEY, B. YORK & G. E. TRUETT. 1994. Dietary obesity linked to genetic loci on chromosome 9 and 15 in a polygenic mouse model. J. Clin. Invest. 94: 1410–1416.
8. GALLI, J., L. S. LI, A. GLASER, C. G. ÖSTENSON, H. JIAO, H. FAKHRAI, H. J. JACOB, E. S. LANDER & H. LUTHMAN. 1996. Genetic analysis of non-insulin-dependent diabetes mellitus in the GK rat. Nat. Genet. 12: 31–37.
9. GAUGUIER, D., P. FRGUEL, V. PARENT, C. BERNARD, M. T. BIHOREA, B. PORTHA, M. R. JAMES, L. PENICAUD, M. LATHROP & A. KTORZA. 1996. Chromosomal mapping of genetic loci associated with non-insulin-dependent diabetes in the GK rat. Nat. Genet. 12: 38–43.
10. HILBERT, P., K. LINDPAINTNER, J. S. BECKMANN, T. SERIKAWA, F. SOUBRIER, C. DUBAY, P. CARTWRIGHT, B. DE GOUYON, C. JULIER, S. TAKAHASI, M. VINCENT, D. GANTEN, M. GEORGES & G. M. LATHROP. 1991. Chromosomal mapping of two loci associated with blood-pressure regulation in hereditary hypertensive rats. Nature 353: 521–529.

11. JACOB, H. J., K. LINDPAINTNER, S. E. LINCOLN, K. KUSUMI, R. K. BUNKER, Y. P. MAO, D. GANTEN, V. J. DZAU & E. S. LANDER. 1991. Genetic mapping of a gene causing hypertension in the stroke-prone spontaneously hypertensive rat. Cell 67: 213-224.
12. IWAI, N. & T. INAGAMI. 1992. Identification of a candidate gene responsible for high blood pressure of spontaneously hypertensive rats. J. Hypertens. 10: 1155-1157.
13. LINDPAINTNER, K., P. HILBERT, D. GANTEN, B. NADAL-GINARD, T. INAGAMI & N. IWAI. 1993. Molecular genetics of the Sa-gene: cosegregation with hypertension and mapping to rat chromosome 1. J. Hypertens. 11: 19-23.
14. DUBAY, C., M. VINCENT, N. J. SAMANI, P. HILBERT, M. A. KAISER, J. P. BERESSI, Y. KOTELEVTSEV, J. S. BECKMANN, F. SOUBRIER, J. SASSARD & G. M. LATHROP. 1993. Genetic determinants of diastolic and pulse pressure map to different loci in Lyon hypertensive rats. Nat. Genet. 3: 354-357.
15. SAMANI, N. J., D. LODWICK, M. VINCENT, C. DUBAY, M. A. KAISER, M. P. KELLY, M. LO, J. HARRIS, J. SASSARD, M. LATHROP & J. D. SWALES. 1993. A gene differentially expressed in the kidney of the spontaneously hypertensive rat cosegregates with increased blood pressure. J. Clin. Invest. 92: 1099-1103.
16. BIANCHI, G., G. TRIPODI, G. CASARI, S. SALARDI, B. R. BARBER, R. GARCIA, P. LEONI, L. TORIELLI, D. CUSI, M. FERRANDI, L. A. PINNA, F. E. BARALLE & P. FERRARI. 1994. Two point mutations within the adducin genes are involved in blood pressure variation. Proc. Natl. Acad. Sci. U.S.A. 91: 3999-4003.
17. DENG, A. Y., H. DENE, M. PRAVENECE & J. P. RAPP. 1994. Genetic mapping of two new blood pressure quantitative trait loci in the rat by genotyping endothelin system genes. J. Clin. Invest. 93: 2701-2709.
18. JACKSON, R. A., J. B. BUSE, R. RIFAI, D. PELLETIER, E. L. MILFORD, C. B. CARPENTER, G. S. EISENBARTH & R. M. WILLIAMS. 1984. Two genes required for diabetes in BB rats: evidence from cyclical intercrosses and backcrosses. J. Exp. Med. 159: 1629-1636.
19. KLÖTING, I. & O. STARK. 1987. Genetic studies of IDDM in BB rats: the incidence of diabetes in F2 and first backcross hybrids allows rejection of a recessive hypothesis. Exp. Clin. Endocrinol. 98: 312-318.
20. KLÖTING, I. & L. VOGT. 1990. Coat colour phenotype, leucopenia, and insulin-dependent diabetes mellitus (IDDM) in BB rats. Diabetes Res. 15: 37-39.
21. JACOB, H. J., A. PETTERSSON, D. WILSON, Y. MAO, A. LERNMARK & E. S. LANDER. 1992. Genetic dissection of autoimmune type I diabetes in the BB rat. Nat. Genet. 2: 56-60.
22. LOUIS, W. J., R. TABEI, A. SJOERDSMA & S. SPECTOR. 1969. Inheritance of high blood pressure in the spontaneously hypertensive rat. Lancet 1: 1035-1036.
23. KNUDSEN, K. D., L. K. DAHL, K. THOMPSON, J. IWAI, M. HEINE & G. LEITL. 1970. Effects of chronic excess salt ingestion: inheritance of hypertension in the rat. J. Exp. Med. 132: 976-1000.
24. TANASE, H., Y. SUZUKI, A. OOSHIMA, Y. YAMORI & K. OKAMOTO. 1970. Genetic analysis of blood pressure in spontaneously hypertensive rats. Jpn. Circ. J. 34: 1197-1212.
25. TANASE, H. 1979. Genetic control of blood pressure in spontaneously hypertensive rats (SHR). Jikken Dobutsu 28: 519-530.
26. OKAMOTO, K. 1972. Spontaneous Hypertension, Ist Pathogenesis und Comüplications. Springer-Verlag. Berlin/New York.
27. YEN, T. T., P. L. YU, H. ROEDER & P. W. WILLARD. 1974. A genetic study of hypertension in Okamoto-Aoki spontaneously hypertensive rats. Heredity 33: 309-316.
28. SCHLAGER, G. 1972. Spontaneous hypertension in laboratory animals. J. Hered. 63: 35-38.
29. KLÖTING, I. & L. VOGT. 1991. BB/O(ttawa)K(arlsburg) rats: features of a subline of diabetes-prone BB rats. Diabetes Res. 18: 79-87.
30. KLÖTING, I., M. STIELOW & L. VOGT. 1995. Development of new animal models in diabetes research: spontaneously hypertensive-diabetic rats. Diabetes Res. 29: 127-138.
31. KLÖTING, I., L. VOGT & T. SERIKAWA. 1995. Locus on chromosome 18 cosegregates with diabetes in the BB/OK subline. Diabetes Metab. 21: 338-344.
32. HAMET, P., Y. L. SUN, D. MALO, D. KONG, V. KREN, M. PRAVENEC, J. KUNES, P. DUMAS, L. RICHARD, F. GAGNON & J. TREMBLAY. 1994. Genes of stress in experimental hypertension. Clin. Exp. Pharmacol. Physiol. 21: 907-911.
33. KLÖTING, I., B. VOIGT & L. VOGT. 1995. Forty-seven polymorphic microsatellite loci in different inbred rat strains. J. Exp. Anim. Sci. 37: 42-47.

34. LANDER, E. S., P. GREEN, J. ABRAHAMSON, A. BARLOW, M. J. DALY, S. E. LINCOLN & L. NEWBURG. 1987. MAPMAKER: an interactive computer package for constructing genetic linkage maps of experimental and natural populations. Genomics 1: 174–181.
35. LANDER, E. S. & D. BOTSTEIN. 1989. Mapping Mendelian factors underlying quantitative traits using RFLP linkage maps. Genetics 121: 185–199.
36. BERG, S., A. DUNGER, L. VOGT, I. KLÖTING & S. SCHMIDT. 1996. Circadian variations in blood pressure and heart rate in diabetes-prone and resistant rat strains compared with spontaneously hypertensive rats. Exp. Clin. Endocrinol. Diabetes. In press.
37. KLÖTING, I. & K. REIHER. 1985. Einige Aspekte zur Haltung und Reproduktion spontandiabetischer BB-Ratten. Z. Verstierkd. 27: 5–11.

Insulin Resistance Syndrome in Mice Deficient in Insulin Receptor Substrate-1

HIROYUKI TAMEMOTO, KAZUYUKI TOBE,
TOSHIHISA YAMAUCHI, YASUO TERAUCHI,
YASUSHI KABURAGI, AND TAKASHI KADOWAKI

Third Department of Internal Medicine
Faculty of Medicine
University of Tokyo
Tokyo 113, Japan

INTRODUCTION

It is well established that the insulin receptor is a heterotetramer containing two alpha and two beta subunits and that the tyrosine kinase domain in the beta subunit plays the key role in signal transmission across the plasma membrane. The insulin receptor substrate-1, or IRS-1, is the major substrate of both the insulin receptor and the type I IGF-1 receptor kinase. Its molecular weight is 170 to 185 kDa on SDS-polyacrylamide gel electrophoresis. After activation of insulin receptor kinase, IRS-1 is phosphorylated on many tyrosine phosphorylation sites and these tyrosine residues are binding sites for several signaling molecules containing the SH2 domain.[1] A lipid kinase phosphatidylinositol-3 kinase, or PI3 kinase, binds to tyrosine in sequences Tyr-x-x-Met. An adapter protein Grb2 binds to Tyr-Val-Asn-Ile, and tyrosine phosphatase Syp binds to Tyr-Ile-Asp-Leu.[2] There are several other substrates of the insulin receptor kinase. An adapter protein Shc is also phosphorylated on tyrosine in insulin-stimulated cells. In adipocytes, a 60-kDa protein called pp60 is also tyrosine-phosphorylated.[3] There are differences in their ability to bind signaling molecules. IRS-1 binds PI3 kinase, Grb2, and Syp. Shc binds Grb2, but not PI3 kinase or Syp. The pp60 binds PI3 kinase, but not Grb2 or Syp.

Insulin has both metabolic and growth-promoting actions, which seem to be mediated through distinct pathways. There is increasing support for the idea that Grb2 and also Syp are involved in the pathways through activation of p21ras, Raf-1, MAP kinase cascade, and growth promotion.[2,4] PI3 kinase is believed to be important in the translocation of GLUT4 and glucose transport, and some data suggest that it is also important in the activation of p70 S6 kinase,[5] protein synthesis,[6] and glycogen synthase.[7] To clarify the physiological role of IRS-1, we have made IRS-1 knockout mice.

ESTABLISHMENT OF IRS-1 KNOCKOUT MICE

The IRS-1 gene has no intron in the coding sequence of IRS-1. In our targeting vector, a neomycin resistance gene driven by a phosphoglycerokinase promoter was inserted in the EcoRI site, which is located just 3' to the site coding for the tyrosine for Grb2 binding. Several homologous recombinant ES clones were obtained, and from two of them the mutated IRS-1 gene was transmitted to the offspring.[8] The IRS-1 knockout mice were born alive and apparently healthy. They were also fertile and bore healthy offspring. However, they were smaller compared with their heterozygous and wild-type littermates. After 10 weeks of age, the heterozygous and wild-type littermates weighed about 24 grams, while the IRS-1 knockout mice weighed about 16 grams. The blood glucose levels of the IRS-1 knockout mice were not elevated compared with their heterozygous and wild-type littermates when challenged with an oral glucose load. However, the serum insulin levels were significantly elevated. The fasting serum insulin levels of the IRS-1 knockout mice were about twice the levels of heterozygous and wild-type mice (FIG. 1). Therefore, the insulin resistance of the IRS-1 knockout mice is milder than that of the insulin receptor mutant homozygotes, who show much higher serum insulin levels.

When the glucose transport activity was analyzed using isolated fat cells, the maximum level was reduced to about 50% of that of control and also the dose response curve was shifted to the right (data not shown). The growth retardation of the IRS-1 knockout mice is also milder than that of the IGF-I receptor knockout mice.

These phenotypes of IRS-1 knockout mice can be explained by the existence of alternative pathways. This idea was supported by the observation that, in the liver of IRS-1 knockout mice, a PI3 kinase activity that immunoprecipitated with antiphosphotyrosine antibody still remained in the absence of IRS-1. Although the tyrosine phosphorylation of Shc was intact in the knockout mice, Shc cannot bind PI3 kinase. Therefore, these data suggested that there are some unidentified substrates that mediate the activation of PI3 kinase and glucose transport in response to insulin.

ALTERNATIVE PATHWAYS THAT SUBSTITUTE FOR THE LOSS OF IRS-1

To identify the alternative substrate of insulin receptor, we injected insulin into the portal vein of IRS-1 knockout mice, and tyrosine-phosphorylated proteins were extracted from the liver by immunoprecipation with antiphosphotyrosine antibody. Western blotting with antiphosphotyrosine antibody revealed a new band with slightly higher molecular weight than IRS-1.[9] This protein, pp190, was not detected with anti-IRS-1 antibody (FIG. 2). The important function of IRS-1 is to bind several signaling molecules. Therefore, we tested whether the pp190 can bind the p85 subunit of PI3 kinase. In the anti-p85 immunoprecipitates, a 190-kDa band was detected by antiphosphotyrosine antibody (FIG. 2). Similarly, in the anti-Grb2 antibody immunoprecipitates, a 190-kDa band was detected by antiphosphotyrosine antibody (data not shown).

The next question is whether this 190-kDa protein is a single molecule that binds

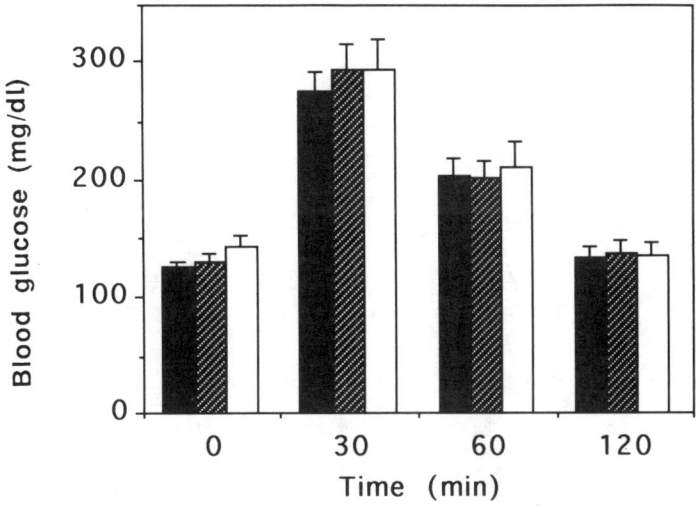

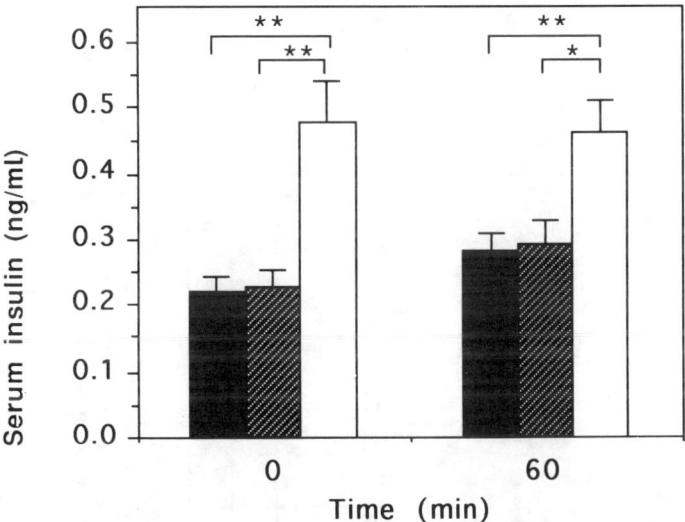

FIGURE 1. The blood glucose and serum insulin levels in the oral glucose tolerance test. The filled, hatched, and open bars represent wild-type, heterozygous, and knockout mice, respectively.

both PI3 kinase and Grb2 as IRS-1 does. An anti-Grb2 antibody immunoprecipitated PI3 kinase activity (data not shown). This result shows that both PI3 kinase and Grb2 coincide in the same immunoprecipitates. Since PI3 kinase cannot bind Grb2 directly, there should be a bridge between the two. These data strongly suggest that pp190 is an IRS-1-like protein that can bind both PI3 kinase and Grb2. Kahn's group also no-

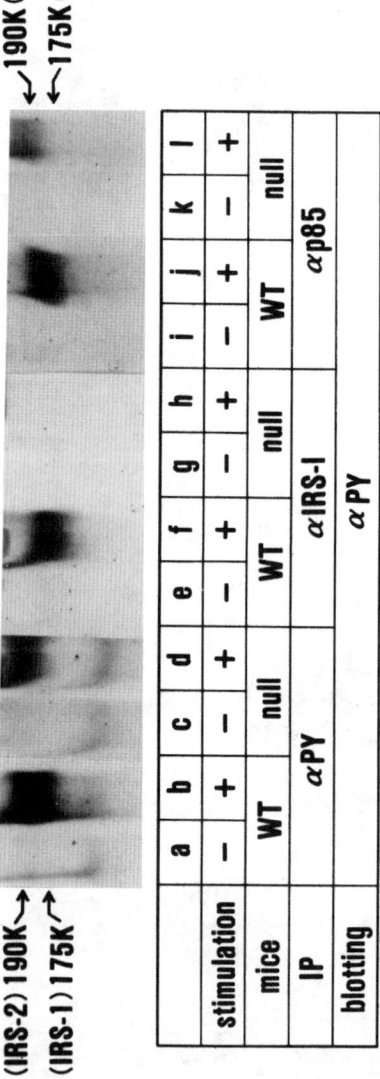

FIGURE 2. Identification of IRS-2 in the liver of IRS-1 knockout mice. Mice were anesthetized, insulin was injected through the portal vein, and the liver was removed and homogenized. The proteins precipitated with one of the three possible antibodies were blotted with an antiphotyrosine antibody (RC20). Lanes a, b, c, d: immunoprecipitated with antiphosphotyrosine antibody; lanes e, f, g, h: with anti-IRS-1 antibody; lanes i, j, k, l: with anti-PI3 kinase antibody. WT: wild-type mice; null: IRS-1 knockout mice.

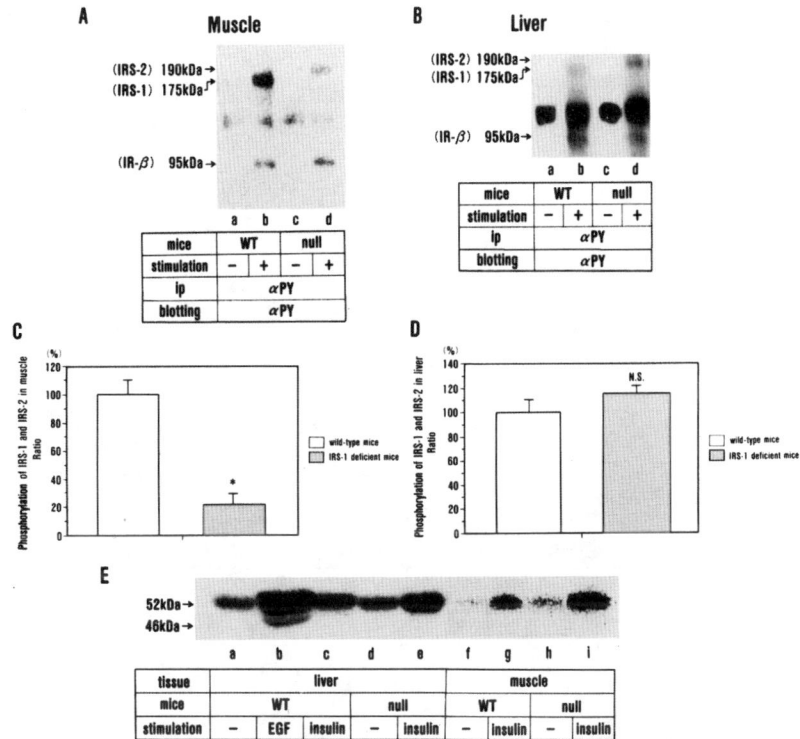

FIGURE 3. The tyrosine phosphorylation of IRS-2 and Shc in the liver and muscle of IRS-1 knockout mice. Mice were anesthetized and insulin was injected through the portal vein (for liver) or inferior vena cava (for muscle). The tyrosine-phosphorylated proteins were immunoprecipitated with an antiphosphotyrosine antibody (PY20) (for A, B, C, D) or an anti-Shc antibody (for E), and then blotted with another antiphosphotyrosine antibody (RC20).

ticed this protein and called it IRS-2. IRS-2 was recently cloned by White's group.[10] The published amino acid sequence showed high similarity throughout the protein. Most notably, the important phosphorylation sites were completely conserved.

INSULIN ACTIONS IN THE MUSCLE AND LIVER OF IRS-1 KNOCKOUT MICE

The tyrosine phosphorylation of IRS-2 was analyzed in the muscle and liver of IRS-1 knockout mice.[11] For stimulation of muscle with insulin, insulin was injected through the inferior vena cava of the anesthetized animals. In the muscle, tyrosine phosphorylation of IRS-2 was detected. However, the intensity of the band was about 20–30% of that of IRS-1 of the wild-type mice (FIGS. 3A and 3C). Therefore, IRS-2 may be insufficient to compensate for the absence of IRS-1. In the liver, the tyrosine phosphorylation of IRS-2 is almost comparable to that of IRS-1 in the wild-type mice

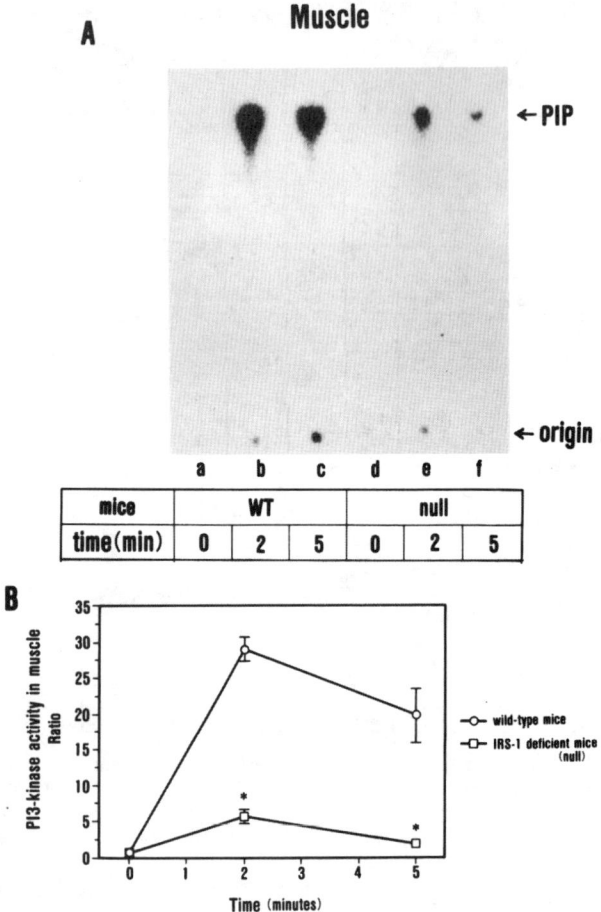

FIGURE 4. The activation of PI3 kinase in muscle by insulin. The hind leg muscle of mice was stimulated with insulin, and PI3 kinase activity immunoprecipitated with antiphosphotyrosine antibody was analyzed. (A) Representative data showing the phosphorylated lipid product. (B) The radioactivity of the phosphorylated lipid was quantitated and expressed in arbitrary units.

(FIGS. 3B and 3D). There was no difference in the tyrosine phosphorylation of Shc between the wild-type and the IRS-1 knockout mice (FIG. 3E).

We analyzed the insulin-stimulated PI3 kinase activity in hind limb muscle. Without insulin stimulation, the PI3 kinase activity that associates with the antiphosphotyrosine antibody was almost undetectable. Injection of insulin elevated the PI3 kinase activity in the muscles of both the wild-type and IRS-1 knockout mice. However, the intensity of the phosphorylated lipid product was about one-sixth of that of wild-type mice (FIG. 4).

The activation of PI3 kinase is believed to trigger the translocation of GLUT4 and the activation of p70 S6 kinase. Therefore, we measured the 2-deoxyglucose uptake of the isolated soleus. The 2-deoxyglucose uptake of the muscle of the wild-type mice

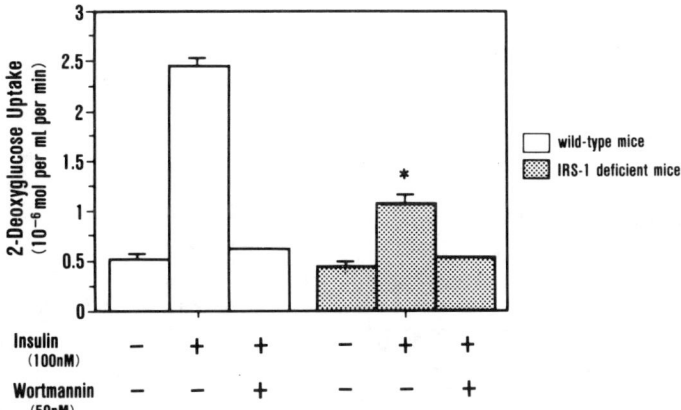

FIGURE 5. The 2-deoxyglucose uptake in the soleus muscle. The isolated soleus muscle was preincubated with insulin *in vitro*, with or without wortmannin, and washed, and then 2-deoxyglucose uptake was analyzed.

was elevated by 5-fold, while that of the IRS-1 knockout mice was elevated only by 2-fold (FIG. 5). This stimulation of 2-deoxyglucose was completely inhibited with 50 nM of wortmannin, an inhibitor of PI3 kinase. We also analyzed the activation of p70 S6 kinase in the muscle. In the muscle of the wild-type mice, the p70 S6 kinase activity was elevated by about 3-fold, while that of the IRS-1 knockout mice was elevated only by 1.5-fold (FIG. 6). This stimulation of p70 S6 kinase was also sensitive to wortmannin treatment.

Hyperphosphorylation of PHAS-I triggers the transcription of mRNA. Thus, protein synthesis in the isolated soleus muscle was analyzed. Tritium-labeled tyrosine

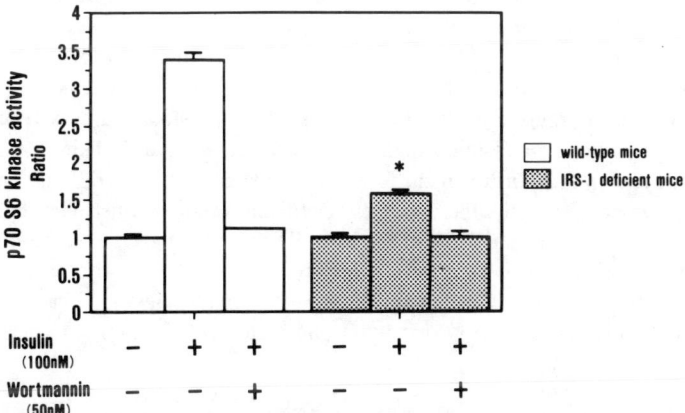

FIGURE 6. The activation of p70 S6 kinase in the soleus muscle. The isolated soleus muscle was preincubated as in the legend of FIGURE 5. The immune complex kinase assay for p70 S6 kinase was performed as described in reference 5.

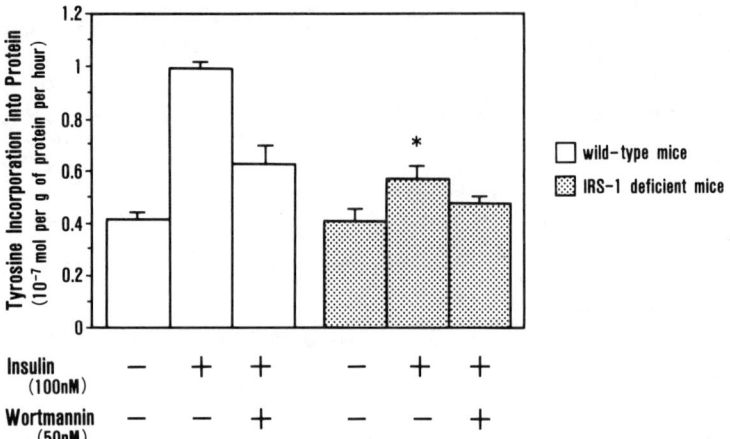

FIGURE 7. The activation of protein synthesis in the soleus muscle. The pretreatment of the muscle is described in the legend of FIGURE 5. The radioactivity of the labeled tyrosine incorporated into the newly synthesized protein was quantitated.

incorporated into newly synthesized protein was measured. The basal incorporation of radiolabeled tyrosine was similar in wild-type and IRS-1 knockout mice. Incubation with insulin stimulated the radiolabeled tyrosine incorporation by 2.5-fold, while the stimulation was increased only by 1.2-fold in the IRS-1 knockout mice (FIG. 7). This stimulation was again wortmannin-sensitive.

In summary, PI3 kinase activation by insulin was severely impaired in IRS-1 knockout mice. The 2-deoxyglucose uptake and protein synthesis that are supposed to be downstream of PI3 kinase were also impaired in the IRS-1 knockout mice. MAP kinase activity was also analyzed in the muscle, by an immune complex kinase assay using myelin basic protein as substrate. By stimulation with EGF, MAP kinase was activated in both IRS-1 knockout and wild-type mice to similar levels. However, when stimulated with insulin, the MAP kinase in the muscle was activated by 3.5-fold in the wild-type mice, while that of the IRS-1 knockout mice was activated only by 1.5-fold. Therefore, in the muscle of the IRS-1 knockout mice, MAP kinase activation was impaired specifically in insulin stimulation. On the other hand, PI3 kinase in the liver was activated by insulin treatment to a similar extent in both wild-type and IRS-1 knockout mice. Similarly, there was no significant difference between the insulin-stimulated MAP kinase activity in the liver. These data suggest that the insulin resistance of the IRS-1 knockout mice is largely explained by the insulin resistance in muscle. The difference in insulin sensitivity between muscle and liver is correlated with the relative amount of tyrosine-phosphorylated IRS-2 in these tissues.

DISCUSSION

We have made mice deficient in IRS-1. In these animals, insulin's actions were substituted in part by the presence of IRS-2. However, because of the relative higher abundance of IRS-2 in the liver compared to muscle, several actions of insulin were more severely impaired in muscle. Another animal model that has insulin resistance primarily in the muscle has been made by Moller's group.[12] They made transgenic mice that expressed a dominant negative mutant of the insulin receptor under the control of creatinine kinase. Their mice showed mild hyperinsulinemia and reduced sensitivity to insulin in the insulin tolerance test. This phenotype is very similar to that of IRS-1 knockout mice, except that the growth of their mice was normal. Interestingly, they recently reported that their mice developed diabetes later and dyslipidemia and also that the content of body fat was increased in these animals.[12] It is an important question if IRS-1 knockout mice also develop dyslipidemia. Although it is a preliminary result, Abe et al. observed that the serum triglyceride levels of wild-type B6 mice became lower with age, while those of IRS-1 knockout mice remained almost the same; also, the systolic blood pressure of the mice was higher in IRS-1 knockout mice (Hideki Abe, personal communication).[13]

These data suggest that insulin resistance in muscle is one factor in the development of Syndrome X. In conclusion, IRS-1 knockout mice will serve as a tool to study the mechanism of insulin resistance and related syndromes.

REFERENCES

1. SUN, X. J. et al. 1991. The structure of the insulin receptor substrate-1 defines a unique signal transduction protein. Nature 352: 73–77.
2. WHITE, M. F. & C. R. KAHN. 1994. The insulin signaling system. J. Biol. Chem. 269: 1–4.
3. LAVAN, B. E. & G. E. LIENHARD. 1993. The insulin-elicited 60-kDa phosphotyrosine protein in rat adipocytes is associated with phosphatidylinositol 3-kinase. J. Biol. Chem. 268: 5921–5928.
4. XIAO, S., D. W. ROSE, T. SASAOKA, H. MAEGAWA, T. R. BURKE, JR., P. P. ROLLER, S. E. SHOELSON & J. M. OLEFSKY. 1994. Syp (SH-PTP2) is a positive mediator of growth factor–stimulated mitogenic signal transduction. J. Biol. Chem. 269: 21244–21248.
5. CHEATHAM, B., C. J. VLAHOS, L. CHEATHAM, L. WANG, J. BLENIS & C. R. KAHN. 1994. Phosphatidylinositol 3-kinase activation is required for insulin stimulation of pp70 S6 kinase, DNA synthesis, and glucose transporter translocation. Mol. Cell. Biol. 14: 4902–4911.
6. MENDEZ, R., M. G. MYERS, JR., M. F. WHITE & R. E. RHOADS. 1996. Stimulation of protein synthesis, eukaryotic translation initiation factor 4E phosphorylation, and PHAS-I phosphorylation by insulin requires insulin receptor substrate-1 and phosphatidylinositol 3-kinase. Mol. Cell. Biol. 16: 2857–2864.
7. YAMAMOTO-HONDA, R. et al. 1995. Upstream mechanism of glycogen synthase activation by insulin and insulin-like growth factor-I. J. Biol. Chem. 270: 2729–2734.
8. TAMEMOTO, H. et al. 1994. Insulin resistance and growth retardation in mice lacking insulin receptor substrate-1. Nature 372: 182–186.
9. TOBE, K. et al. 1995. Identification of a 190-kDa protein as a novel substrate for the insulin receptor kinase functionally similar to insulin receptor substrate-1. J. Biol. Chem. 270: 5698–5701.
10. SUN, X. J. et al. 1995. Role of IRS-2 in insulin and cytokine signaling. Nature 377: 173–177.
11. YAMAUCHI, T. et al. 1996. Insulin signalling and insulin actions in the muscles and liver of insulin-resistant, insulin receptor substrate-1–deficient mice. Mol. Cell. Biol. 16: 3074–3084.
12. MOLLER, D. et al. 1996. Transgenic mice with muscle-specific insulin resistance develop increased adiposity, impaired glucose tolerance, and dyslipidemia. Endocrinology 137: 2397–2405.
13. ABE, H. et al. 1996. J. Jpn. Diabetes Soc. 34: suppl. 207.

WOK.1W Rats
A Potential Animal Model of the Insulin Resistance Syndrome[a]

PETER KOVÁCS, BIRGER VOIGT, SABINE BERG,
LUTZ VOGT, AND INGRID KLÖTING

"Gerhardt Katsch" Institute of Diabetes
University of Greifswald
17495 Karlsburg, Germany

INTRODUCTION

In 1981, diabetic BB rats and animals from the founder outbred Wistar rat stock were transferred from the Biobreeding Laboratories in Ottawa, Canada, to our animal facility.[1] Genetic studies of the BB rats showed that they were homozygous for the RT1^u haplotype of the major histocompatibility complex (MHC), while the animals from the outbred stock were heterozygous for the haplotypes RT1^a and RT1^u.[2,3] Subsequent crossing studies indicated that the RT1^u haplotype is a predisposing factor for development of diabetes in these rats.[2,4] These findings prompted us to select RT1^a and RT1^u homozygous animals of the outbred Wistar rat stock to generate two inbred lines carrying either the diabetes-susceptible RT1^u or the diabetes-resistant RT1^a haplotype. The lines were designated as W(istar) O(ttawa) K(arlsburg).1A (RT1^a) and WOK.1W (RT1^u).[1] During the inbreeding process of more than 30 generations, no animals manifested diabetes, although WOK.1W rats are characterized by the same diabetes-susceptible RT1^u haplotype as BB rats. The WOK.1W and WOK.1A rats were inbred separately and artificially selected only for "healthy reproduction." After more than 35 generations, fertility and survival of pups decreased rapidly in WOK.1W rats. Because infections could be excluded as the cause, metabolic traits and blood pressure of a few animals were investigated in order to examine the reason for the low rate of reproduction. The findings indicated that, in comparison to nondiabetic BB/OK rats, all WOK.1W rats studied were normoglycemic, but might be hypertriglyceridemic and hypertensive. This fact prompted us to characterize phenotypically WOK.1W as well as nondiabetic, normotriglyceridemic, and normotensive BB/OK rats, both types descended from the same progenitor outbred stock, for body weight gain, glucose tolerance, blood and urine constituents, and blood pressure to obtain information about the metabolic state of WOK.1W rats that might be related to their low rate of reproduction.

---

[a] This work was supported by Grant No. Kl 771/1-3 of Deutsche-Forschungsgemeinschaft.

EXPERIMENTAL METHODS

Nine male and 12 female WOK.1W rats (F37) and 8 male and 10 female nondiabetic BB/OK rats (F35) were studied from the 3d until the 28th weeks of life. Males and females were kept separately in groups of 3 in Macrolon cages under strict hygienic conditions and were free of major pathogens as described previously.[5] They had free access to food (Sniff R, Soest, Germany) and acidulated water.

Body weight was assessed at 3, 6, 8, 10, 12, 14, 16, 20, 24, and 28 weeks. Serum triglycerides and blood glucose were determined at 18, 22, and 26 weeks. Since glucose tolerance was disturbed in WOK.1W rats, plasma insulin was additionally determined at an age of 26 and 28 weeks in WOK.1W rats and at an age of 28 weeks in nondiabetic BB/OK rats. A glucose tolerance test was performed by injecting glucose (2 g/kg as 20% solution) intraperitoneally in nonfasting animals between 7:00 and 9:00 A.M. at 8, 12, 16, 20, 24, and 28 weeks. Blood samples were obtained from tail vein before (0) and 10, 30, and 60 min after the glucose load, and glucose was determined by a glucose analyzer (Medingen ESAT 6660-2). To measure the urine excretion of total protein, albumin, creatinine, urea, sodium, potassium, calcium, and phosphate, the animals were kept in metabolic cages for 24 hours at 18, 22, and 26 weeks. At the same time intervals, the systolic blood pressure was measured indirectly with the tail-cuff method described previously[6] (Kent Scientific Corporation, Kent, England).

The values are shown as mean ± SD. Significant differences of mean values were checked by ANOVA and assumed to be significant at $p < 0.05$.

RESULTS AND DISCUSSION

As shown in TABLE 1, there were significant differences in body weight between WOK.1W and nondiabetic BB/OK rats. The body weight was significantly higher after 3, 6, 8, 10, and 12 weeks of life in WOK.1W males, but there were no differences

TABLE 1. Body Weight Gain (g) of WOK.1W and Nondiabetic BB/OK Rats[a]

	Males		Females	
Weeks	WOK.1W (9)	BB/OK (8)	WOK.1W (12)	BB/OK (10)
3	45 ± 5^A	35 ± 6^C	39 ± 6^B	33 ± 8^C
6	146 ± 12^A	129 ± 17^B	111 ± 21^C	100 ± 21^C
8	231 ± 12^A	183 ± 26^B	151 ± 9^C	134 ± 20^D
10	289 ± 16^A	258 ± 31^B	168 ± 28^C	171 ± 23^C
12	338 ± 17^A	310 ± 20^B	199 ± 10^C	199 ± 26^C
14	355 ± 30^A	342 ± 32^A	213 ± 11^B	214 ± 24^B
16	373 ± 49^A	377 ± 38^A	224 ± 4^B	229 ± 24^B
20	411 ± 35^A	413 ± 36^A	234 ± 10^B	251 ± 29^B
24	433 ± 43^A	443 ± 35^A	240 ± 9^C	263 ± 26^B
28	463 ± 24^A	448 ± 40^A	243 ± 24^C	267 ± 27^B

[a] Data are means ± SD; values without a common superscript (A, B, C, D) within a row are significantly different ($p < 0.05$). Number of animals tested is given in the parentheses.

TABLE 2. Blood Pressure (BP), Levels of Blood Constituents [Serum Triglycerides (TG), Plasma Glucose (PG), Plasma Insulin], and Urine Constituents per Day (Total Protein, Albumin, Creatinine, Urea, Sodium, Potassium, Calcium, Phosphate) in WOK.1W and Nondiabetic BB/OK Rat Strains[a]

Trait	Males		Females	
	WOK.1W	BB/OK	WOK.1W	BB/OK
No. of values	27	24	36	30
BP (mmHg)	143 ± 10.9^B	123 ± 5.4^C	153 ± 14.0^A	125 ± 5.16^C
TG (mmol/L)	3.33 ± 0.91^A	1.54 ± 0.43^B	3.50 ± 0.92^A	1.38 ± 0.48^B
PG (mmol/L)	5.00 ± 0.49^B	6.22 ± 0.73^A	5.16 ± 0.78^B	6.11 ± 0.24^A
Insulin (ng/mL)[b]	$5.86 \pm 1.90\ (16)^A$	$3.04 \pm 1.83\ (8)^B$	$2.37 \pm 0.55\ (24)^B$	$1.81 \pm 1.08\ (9)^B$
Total protein (mg/day)	29.1 ± 13.5^A	27.8 ± 20.8^A	8.5 ± 6.1^C	20.8 ± 16.7^B
Albumin (µg/day)	30.9 ± 25.3^A	9.4 ± 8.0^B	34.9 ± 23.1^A	7.3 ± 7.7^B
Creatinine (mmol/day)	0.111 ± 0.037^A	0.102 ± 0.039^A	0.071 ± 0.031^B	0.080 ± 0.030^B
Urea (mmol/day)	12.06 ± 3.44^A	9.09 ± 3.17^C	9.92 ± 3.87^B	8.35 ± 2.05^C
Sodium (mmol/day)	1.35 ± 0.319^B	1.67 ± 0.747^A	0.96 ± 0.53^C	1.74 ± 0.99^A
Potassium (mmol/day)	4.32 ± 0.78^A	3.59 ± 0.80^B	3.42 ± 1.12^B	3.40 ± 0.85^B
Calcium (mmol/day)	0.038 ± 0.021^B	0.100 ± 0.099^A	0.094 ± 0.028^A	0.089 ± 0.053^A
Phosphate (mmol/day)	0.150 ± 0.092^A	0.0385 ± 0.0518^B	0.150 ± 0.086^A	0.0457 ± 0.0469^B

[a] Data are means ± SD; values without a common superscript (A, B, C) within a row are significantly different ($p < 0.05$).
[b] Number of values measured for insulin is given in the parentheses (cf. EXPERIMENTAL METHODS).

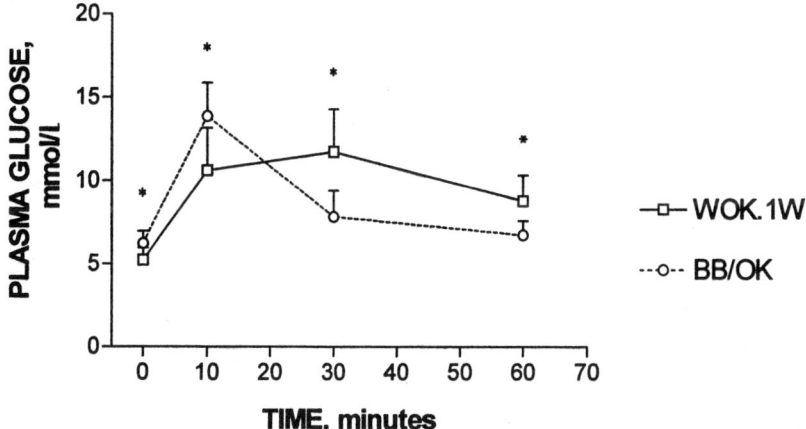

FIGURE 1. Glucose tolerance in male WOK.1W and nondiabetic BB/OK rats. Plasma glucose levels were measured before (0) and 10, 30, and 60 min after glucose load. (*) Significantly different at $p < 0.05$.

at subsequent intervals until the 28th week. Females of WOK.1W had a higher body weight gain than those of BB/OK at the 3d and 8th weeks of life, but lower ones in the 24th and 28th weeks. These findings suggest that the WOK.1W rats may have a more pronounced weight gain in early life than BB/OK rats. However, in the later age intervals, comparable body weights were observed, indicating that WOK.1W rats, like BB/OK animals, are not obese.

Significant differences between WOK.1W and nondiabetic BB/OK rats were observed not only in triglycerides and blood pressure, but also in the majority of traits studied as summarized in TABLE 2. The blood pressure was significantly increased in male (+18 mmHg) and female WOK.1W rats (+22 mmHg) compared to BB/OK rats. Triglycerides were more than two times higher in male and female WOK.1W rats than in BB/OK rats. On the other hand, the plasma glucose levels were significantly lower in WOK.1W rats than in BB/OK rats. Plasma insulin levels were significantly higher only in male WOK.1W rats compared with male BB/OK rats, whereas with total protein excretion significantly decreased values were observed only in female WOK.1W rats. For albumin, urea, sodium, and phosphate excretion, significant differences were found between male WOK.1W and BB/OK and between female WOK.1W and BB/OK, respectively. Whereas most of these urine values were significantly higher in WOK.1W, there was an exception for sodium excretion. Significantly lower values were observed in WOK.1W rats. Significant differences were also found in urine potassium and calcium, but only between male WOK.1W and BB/OK rats. In WOK.1W males, higher values were obtained for potassium, whereas lower values were obtained for calcium.

Compared with nondiabetic BB/OK rats, glucose tolerance was disturbed in WOK.1W rats, as indicated by significantly different plasma glucose values before and 10 and 30 min after glucose load. As demonstrated in FIGURES 1 and 2, male and female WOK.1W rats are characterized by a different glucose response than BB

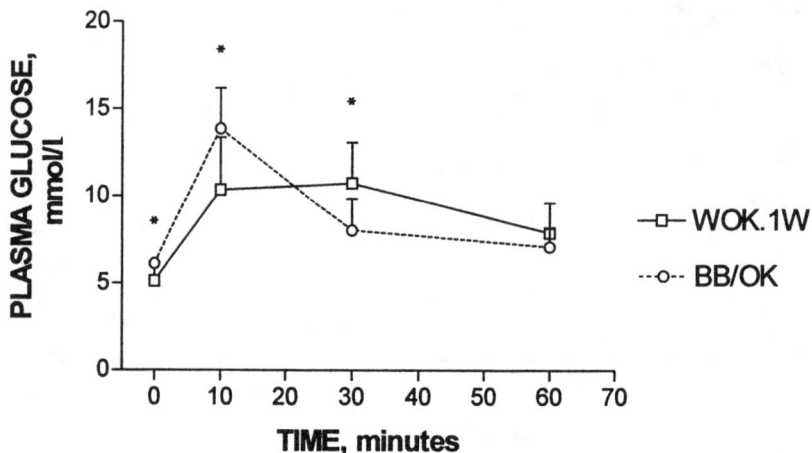

FIGURE 2. Glucose tolerance in female WOK.1W and nondiabetic BB/OK rats. Plasma glucose levels were measured before (0) and 10, 30, and 60 min after glucose load. (*) Significantly different at $p < 0.05$.

rats. Whereas BB rats show a rapid increase of blood glucose followed by a decrease at 10 min after glucose load, the response in WOK.1W rats is characterized by significantly lower blood glucose values at 10 min after glucose load and by significantly higher values at 30 min after glucose load.

Our findings indicate that WOK.1W rats are characterized by nonfasting hypertriglyceridemia, hyperinsulinemia, and moderate hypertension, which could cause the low reproductive performance in this strain. These characteristics resemble the symptoms of the insulin resistance syndrome, which has been addressed in recent years in many investigations.[7] In considering the relationship between hyperinsulinemia and hypertension, which is still not fully explained, the hypertension in the hereditary hypertriglyceridemic (hHTG) rat, which is the only animal model manifesting all the above-mentioned phenotypic traits of the WOK.1W rat, seems to be influenced by increased activity of the sympathetic nervous system.[8] The lower levels of urine sodium in WOK.1W rats found in our study could be due to increased renal reabsorption, which is also considered to be a factor contributing to hypertension.[9] However, the high values of albumin excretion, urea, and phosphate, indicating kidney injury, do not let us exclude other mechanisms leading to the higher systolic blood pressures observed in WOK.1W rats. Further studies of the activity of the sympathetic nervous system as well as studies investigating the origin of hypertriglyceridemia in the WOK.1W rat should be carried out.

Despite the low reproductive performance, WOK.1W may become an interesting animal model not only for metabolic, but also for genetic studies of the human insulin resistance syndrome. While the hHTG rat seems to show a nutritionally induced genetic predisposition for the insulin resistance syndrome,[10] the WOK.1W rat might be added to the list of animals with the genetically determined characteristics of insulin resistance published by Shafrir.[11] The fact that WOK.1W rats are not obese might make this strain convenient for studies of the insulin resistance syndrome because obesity is currently a complicating factor in its further investigation.[12] In addition, the already known genetic profile of WOK.1W rats comprising 97 marker loci[13] could be an advantage for future searches of quantitative trait loci (QTLs) for this syndrome.

REFERENCES

1. KLÖTING, I. & L. VOGT. 1989. Rat News Lett. 21: 20–21.
2. KLÖTING, I., O. STARK, M. KOHOUTOVÁ & H. J. HAHN. 1981. Diabetologia 21: A277.
3. STARK, O., I. KLÖTING, K. REIHER & K. D. KOHNERT. 1982. Acta Biol. Med. Ger. 41: 1129–1133.
4. COLLE, E., R. D. GUTTMANN & T. SEEMAYER. 1981. J. Exp. Med. 154: 1237–1242.
5. KLÖTING, I. & L. VOGT. 1991. Diabetes Res. 18: 79–87.
6. KLÖTING, I., M. STIELOW & L. VOGT. 1995. Diabetes Res. 29: 127–138.
7. REAVEN, G. M. 1988. Diabetes 37: 1595–1607.
8. KLIMEŠ, I., A. VRÁNA, J. KUNEŠ, E. ŠEBÖKOVÁ, Z. DOBEŠOVÁ, P. ŠTOLBA & J. ZICHA. 1995. Blood Pressure 4: 137–142.
9. FINCH, D., G. DAVIS, J. BOWER & K. KIRCHNER. 1990. Hypertension 15: 514–518.
10. KLIMEŠ, I., E. ŠEBÖKOVÁ, A. VRÁNA, P. ŠTOLBA, L. KAZDOVÁ, J. KUNEŠ, P. BOHOV, M. FICKOVÁ, J. ZICHA, D. RAUCINOVÁ & V. KŘEN. 1994. The hereditary hypertriglyceridemic rat, a new model of the insulin resistance syndrome. In Lessons from Animal Diabetes, p. 271–283. Smith–Gordon. London.
11. SHAFRIR, E. 1993. The plurimetabolic syndrome in animals: models for experimentation, their implication, and relation to NIDDM. In Diabetes, Obesity, and Hyperlipidemia: The Plurimetabolic Syndrome, p. 45–55. Excerpta Medica. Amsterdam.
12. SHAFRIR, E. 1990. Diabetes in animals. In Diabetes Mellitus, p. 299–340. Elsevier. Amsterdam/New York.
13. KLÖTING, I., B. VOIGT & L. VOGT. 1995. Diabetes Res. 29: 65–71.

Lipid Transport Genes and Their Relation to the Syndrome of Insulin Resistance[a]

DAVID J. GALTON,[b] QIUPING ZHANG,[b]
ELISABETH CAVALLERO,[c] JULIAN CAVANNA,[b]
ANDREA KAY,[b]
ALINE CHARLES,[c] SYLVIE BRASCHI,[c]
LEON PERLEMUTER,[c] AND BERNARD JACOTOT[c]

[b]Department of Human Metabolism and Genetics
St. Bartholomew's Hospital
London EC1M 6BQ, United Kingdom
[c]Department of Internal Medicine
and
Department of Endocrinology
Hôpital Henri Mondor
Creteil, France

INTRODUCTION

The metabolic syndrome (or insulin resistance syndrome) is usually defined as the clustering of obesity, NIDDM, hyperlipemia, and (in some cases) hypertension. All of these are established risk factors for the development of premature atherosclerosis, which would then count as one of the important complications of the syndrome.

With regard to the pathogenesis, the simplest explanation is that this cluster represents a clinical association of several common conditions linked by environmental and genetic factors. This would make it somewhat analogous to the concurrence of bronchitis, asthma, and emphysema, which are loosely connected by environmental factors such as repeated bronchial infections, cigarette smoking, air pollution, and exposure to allergens, together with genetic factors such as variants or deficiencies of alpha-1 antitrypsin. For the metabolic syndrome, such factors could include calorie and dietary fat intake, the level of physical exercise, alcohol and salt intake, and a family history of diabetes mellitus, lipemia, or hypertension.

However, there may exist a defined syndrome with a single etiological determinant such as resistance to the action of insulin. The main proponents of this hypothesis[1-3]

---

[a] This work was supported by Grant No. PL 931211 of the Commission of the European Communities; the Joint Research Board of St. Bartholomew's Hospital, London; and an Overseas Research Fellowship (British Universities) to Q. Zhang.

point out that many of the features of the syndrome can be deduced from insulin resistance. For example, there is impaired glucose uptake, leading to hyperglycemia and diabetes mellitus; there is failure of the antilipolytic action of insulin, leading to an enhanced flux of plasma free fatty acids that drives hepatic synthesis and secretion of VLDL-triglycerides; there is impaired action of insulin on renal tubular cells, leading to salt and water retention and possibly hypertension; and the occurrence of obesity further aggravates the state of peripheral insulin resistance.

However, one can reproduce many of the clinical features of the metabolic syndrome by postulating a primary defect in the transport of plasma lipids and free fatty acids (FFA). Thus, an increase in the levels or flux of plasma FFA impairs glucose utilization by the operation of the Randle glucose–fatty acid cycle; it stimulates hepatic synthesis and release of the VLDL-triglycerides, and obesity provides a large tissue source for the plasma FFA. However, it is difficult to account for the associated hypertension unless one invokes the recent report of a genetic locus at or near the lipoprotein lipase (LPL) gene that contributes 52–73% to the interindividual variation in systolic pressure in 125 Taiwanese sib-pairs.[4]

Perhaps a more-likely explanation for the metabolic syndrome is multifactorial, including environmental factors such as obesity, dietary fat, and alcohol intake, and with a genetic predisposition involving variation at several candidate genes underlying insulin resistance (insulin receptor, glucose transporters, glucokinase 1 and 2, and signal transduction proteins), lipid transport (apolipoproteins, lipoprotein lipase, and cholesterol ester transfer protein), and hypertension (angiotensin-converting enzyme, renin, and angiotensinogen).

Part of the problem of analyzing a complex trait involving several common conditions (obesity, diabetes, lipemia, hypertension) is to distinguish primary metabolic determinants from secondary effects. This is important to do for the development of new therapies to treat causes rather than complications of the syndrome. One possible approach is to attempt to define the genetic determinants of the condition that can directly lead to the environmental factors with which they interact. We have therefore examined 18 subjects with NIDDM, obesity, and lipemia for various candidate genes[5] and here report the extent of genetic variation found at the lipoprotein lipase gene locus.

PATIENTS

Subjects under Study

Eighteen obese subjects (TABLE 1) with known or recently diagnosed NIDDM (type II diabetes mellitus) and severe hyperlipemia (fasting TG > 1000 mg/dL or 11.4 mmol/L) were referred for admission to the Endocrinology or Lipid Units of the H. Mondor Hospital between March 1993 and February 1994 and were included in this study. There were 15 men (mean age, 45.6; range, 27–72 years; mean BMI, 31 kg/m²; mean fasting blood glucose, 15.2 mmol/L; and mean alcohol intake, 191 g/week). The 3 women had a mean age of 43 years, a mean BMI of 37 kg/m², a mean

TABLE 1. Clinical Data of the Patients at Admission

Patient	Age	BMI (kg/m^2)	Sex	Plasma Cholesterol (mM)	Plasma TG (mM)	HDL-C (mM) at Entry	HDL-C (mM)	Fasting Glucose at Discharge	Alcohol Intake (g/week)	Complications
1	42	29.3	M	9.3	24.1	0.67	0.93	17.5	1050	liver steatosis
2	39	36.3	M	6.3	13.3	0.49	0.67	9.5	0	liver steatosis
3	41	39.7	M	11.3	75.0	0.36	0.49	11.7	0	eruptive xanthomatosis
4	43	27.0	M	19.3	41.2	—	0.65	19.5	175	eruptive xanthomatosis
5	36	32.1	M	7.61	12.5	—	0.60	16.7	0	liver steatosis, pancreatitis
6	45	33.5	M	14.7	14.7	—	—	7.2	350	liver steatosis
7	28	32.0	F	23.0	76.0	0.39	0.85	11.7	40	liver steatosis, pancreatitis
8	42	42.9	F	9.0	18.9	—	0.60	13.7	0	—
9	58	40.1	M	8.2	30.8	0.36	0.60	27.9	518	liver steatosis
10	39	30.7	M	16.6	33.3	—	1.06	12.2	56	liver steatosis
11	46	25.7	M	5.6	44.9	0.26	0.31	12.8	0	liver steatosis
12	58	30.9	M	14.0	—	0.39	—	19.5		pancreatitis
13	72	28.0	M	3.6	20.5	0.75	0.57	17.2		liver steatosis, eruptive xanthomatosis, history of possible pancreatitis
14	56	27.8	M	4.0	26.2	—	0.70	11.0	280	liver steatosis
15	36	28.1	M	8.0	28.0	0.52	—	19.5	230	liver steatosis, eruptive xanthomatosis
16	47	29.4	M	8.0	79.0	—	0.49	16.1	140	liver steatosis, eruptive xanthomatosis, history of possible pancreatitis
17	27	27.7	M	4.5	28.0	0.41	0.67	11.1	70	pancreatitis
18	58	32.1	F	44.2	33.0	—	0.80	12.2	0	—

fasting blood glucose level of 12.5 mmol/L, and a mean alcohol intake of 13 g/week. Five patients had eruptive xanthomatosis and 6 had symptoms of pancreatitis. They were selected for this degree of severity because it made the chances of finding a primary defect in lipid transport more likely. The other main clinical and metabolic features are outlined in TABLE 1. Eleven patients were of French origin; 3 were of North-African origin; and the other 4 patients were of Spanish, Italian, Chinese, and Haitian origin, respectively. All patients received strict hypocaloric nutritional therapy on admission combined with several hypoglycemic and/or hypolipidemic drugs, depending on clinical features. Additional therapies were the insulin pump for patients 1, 4, 5, 7, 9, 10, 11, 12, and 17, usually for 48 hours; combined oral hypoglycemic therapy with metformin, sulfonylureas, and/or fibrates for subjects 3, 6, 8, 13, and 14; and diet alone for subject 2 whose dietary record showed a daily caloric intake of 7200 kcal, the highest of all subjects studied.

The different treatments that the patients thus received during hospitalization do not allow reliable conclusions about lipid responses to therapies. The diagnosis of NIDDM and hypertriglyceridemia occurring together was responsible for the hospital admission of patients 4, 5, 7, 10, 12, 15, 16, and 17. A family history of NIDDM was present in 11 out of the 18 patients (1, 2, 3, 5, 6, 7, 8, 9, 10, 16, and 18); patients 6 and 14 had a family history of dyslipidemia.

Eighty-one control subjects were recruited randomly from the H. Mondor Hospital with diabetes mellitus and no overt hyperlipidemia. These were used to analyze the frequencies of the three new mutants that we found in this study.

METHODS

Total cholesterol, triglyceride, and HDL-cholesterol (after precipitation of B-containing lipoproteins by phosphotungstic acid and magnesium chloride) were determined with an Abbott Diagnostics VP analyzer using enzymatic reagents (Boehringer Mannheim, Germany). Apolipoproteins A1 and apo-B were measured by immunoturbidimetric methods using Daichi kits (Tokyo, Japan). Lipoprotein lipase activity measurements were performed and none of the patients had an absolute deficiency of this enzyme.

DNA Methods

DNA was prepared from peripheral leukocytes using a Nucleon II kit (Scotlabs Limited, United Kingdom) and resuspended in TE buffer.

Amplification of LPL Gene Exons

Oligonucleotide primers were synthesized (IGI Limited, United Kingdom) to allow the amplification of individual exons of the LPL gene. For each exon, flanking intronic

DNA (sequence courtesy of K. Oka) was examined for potential PCR primer sites using PRIMER software, and primer pairs were identified. In each pair, one primer was biotinylated at the 5' end. Typical PCR reaction conditions were as follows: 50 → 100 ng DNA, 50 mM KCl, 10 mM Tris-HCl (pH 8.3), 2.0 mM $MgCl_2$, 0.001% (w/v) gelatin, 200 µM dNTPs, 100 nM of each primer, and 2 units of *Taq* polymerase (Life Sciences, United Kingdom) in a 100-µL reaction. Amplification was achieved in a Perkin Elmer Cetus 480 thermal cycler using an initial denaturing step of 94°C for 1 min and then 35 cycles of 94°C for 30 s, 60°C for 1 min, and 72°C for 1 min, followed by a final elongation step of 72°C for 10 min. Small aliquots were removed to test the PCR reactions on 1.5% agarose gels.

Sequencing of PCR Products

PCR products were sequenced using the dideoxy chain termination method. The PCR products were pretreated by one of two methods:

1. PCR products were incubated with streptavidin-coated magnetic beads (Dynal Limited, United Kingdom). After washing, the bound PCR product was denatured and the biotinylated strand was recovered. The bound single-stranded product was then sequenced using the Sequenase II kit (Amersham Life Sciences, United Kingdom) and the nonbiotinylated primer.[7]
2. The PCR product was sequenced directly after treatment with Exonuclease I and Shrimp Alkaline Phosphatase (Sequenase PCR product sequencing kit, Amersham Life Sciences, United Kingdom).

The reaction products from sequencing reactions were electrophoresed on 6% acrylamide-bisacrylamide (19:1)/7 M urea gels using a glycerol-tolerant gel buffer. The gels were prewarmed and samples were electrophoresed at 50 V/cm. Subsequent to electrophoresis, the gels were fixed in 10% acetic acid and 10% methanol and vacuum-dried. Dry gels were autoradiographed for 18 → 72 hours with Hyperfilm (Amersham Life Sciences, United Kingdom) and the sequence was read off the developed autoradiograph.

RESULTS

Mutations Detected

The mutations in the lipoprotein lipase gene revealed by DNA sequence analysis are shown in TABLE 2, and 8 of the 18 subjects carried mutants in the heterozygous state. Three previously unpublished mutations were found. No mutations were found in the catalytic triad (residues Asp^{156}-His^{241}-Ser^{132}), but three of the mutations occurred in exons 3 and 5, which code for protein domains flanking the catalytic cleft of the enzyme. With regard to the most-severely lipemic subjects at presentation (patients 16, 3, and 11, with plasma triglycerides of 79, 76, and 45 mmol/L, respectively), no

TABLE 2. Lipoprotein Lipase Gene Mutations in French Diabetic Hyperlipidemic Individuals

Exon	Nucleotide	αα Residue	Individuals	Publication Information
3	G^{579}-A	Val^{108}-Val GTG→GTA	17, 18	in literature[a]
5	G^{818}-A	Gly^{188}-Glu GGG→GAG	13, 17	in literature[a]
	C^{829}-T	Arg^{192}-Ter CGA→TGA	12	this study
6	A^{1127}-G	Asn^{291}-Ser AAT→AGT	6, 9	in literature[a]
7	C^{1308}-G	Phe^{351}-Leu TTC→TTG	8	this study
8	C^{1338}-A	Thr^{361}-Thr ACC→ACA	9, 14	this study
9	C^{1595}-G	Ser^{447}-Ter TCA→TGA	12, 13	in literature[a]

[a] See reference 3 for further details.

mutations were detected at the LPL locus. Possibly regulatory sequences of the LPL gene or other genetic loci may be involved in their predisposition to develop severe dyslipidemia. The mean plasma triglyceride level of the patients carrying the LPL mutants after treatment was 5.0 ± 2.3 ($n = 8$) mmol/L compared to 7.9 ± 5.5 ($n = 10$) mmol/L of the others. The numbers studied are too small to draw conclusions regarding the impact of these mutants on the effects of therapy.

The New Mutations

The DNA sequence gels showing the Arg^{192} and Phe^{351} mutations are presented in FIGURES 1A and 1B. Sequence analysis reveals that both subjects (numbers 8 and 12) are heterozygous for the C^{1308}-G and C^{829}-T substitutions, respectively. The first mutation results in an amino-acid substitution of phenylalanine (TTC) for leucine (TTG) at residue 351; the second mutation results in a substitution of arginine (CGA) at residue 192 for a stop codon (TGA). PCR amplifications of exons 5 and 7 and subsequent DNA sequencing on several occasions for both coding and noncoding strands were performed to confirm the presence of these two new mutations.

The silent codon mutation of C^{1338}-A at threonine 361 (sequence gels not presented) would not be expected to have any functional effects on the activity of lipoprotein lipase, but could be used as a biallelic disease marker.

The Allele Frequencies

The Arg^{192} variant was subsequently analyzed by PCR amplification of exon 5 and digestion with Taq 1. This produces bands for the common allele of 123 bp and 140

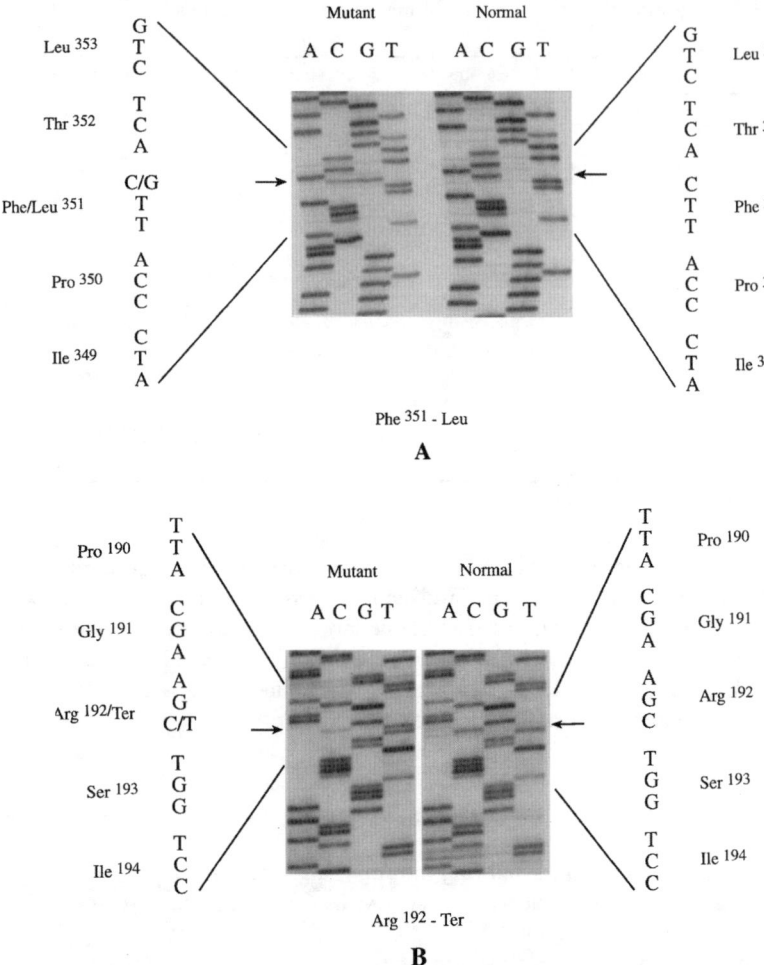

FIGURE 1. Autoradiographs of DNA gel sequences. (A) Nucleotide sequence of the noncoding strand for the mutant and normal exon 7 showing a C-G transversion at nucleotide 1308 that results in the substitution of phenylalanine for leucine at codon 351. (B) Nucleotide sequence of the noncoding strand for the mutant and normal exon 5 showing a C-T transversion at nucleotide 829 that results in the substitution of arginine for a stop codon (Ter) at position 192.

bp, and the variant abolishes the Taq 1 site to produce a 263-bp band on gel electrophoresis. The Phe[351] variant was further analyzed by using a modified amplimer that introduces an Hinc 11 site in the presence of the mutant allele.

The Thr[361] mutation was analyzed by using a modified 5' amplimer (-GCC TGA AGT TTC CAC AAA TAA GGC-) where an additional mutation (A-G) was introduced at nucleotide 1336 into the amplification product. Thus, a novel Hae 111 restriction site was created as the enzyme has the recognition sequence of GGCC. Am-

TABLE 3. Frequencies of Novel Mutations of Lipoprotein Lipase in Diabetic Hyperlipidemic and Normolipidemic Subjects

Location in LPL Gene	Nucleotide Change	Codon Change	Allele Frequency[a]	
			Controls	Subjects
exon 5	C^{829}-T	Arg^{192}-Ter	0 (162)	0.027 (36)
exon 7	C^{1308}-G	Phe^{351}-Leu	0 (162)	0.027 (36)
exon 8	C^{1338}-A	Thr^{361}-Thr	0.12 (162)	0.055 (36)

[a] Numbers in parentheses (n) represent the chromosomes scored.

plification followed by digestion with Hae 111 yielded bands of 214 and 23 bp for the common allele, whereas the Thr 361 mutation abolished the Hae 111 site, yielding three bands of 237, 214, and 23 bp for the rare heterozygote. The frequencies of these three new variants in the 81 subjects tested are presented in TABLE 3.

DISCUSSION

It is surprising to be able to identify three new mutations of the lipoprotein lipase gene by DNA analysis of 18 patients with marked diabetes and lipemia. One of these in exon 5 predicts the termination of the enzyme protein at amino acid 192. This would be expected to abolish the catalytic activity of the enzyme since it is known that carboxy-terminal truncation at residue 381 reduces the catalytic activity by about 85%.[8] The heterozygous state may be unaffected or may show a mild lipemia; however, when associated with NIDDM, a condition that predisposes to hypertriglyceridemia, the mutant could lead to severe lipemia. The other newly described variant (Phe^{351}-Leu) may have some effect on enzyme function due to steric differences between the two amino acids; however, this mutant will need further evaluation since both residues are hydrophobic. We are currently investigating the properties of such a variant lipase by site-directed mutagenesis, as well as its expression by transfection into HEK 293 cells.[9] It is of interest, however, that the severity of the lipemia in this case (fasting plasma triglyceride level of 19 mmol/L at presentation) was similar to the lipemia with the truncated mutant (18.9 mmol/L). The severity of the diabetes at presentation was also similar (fasting blood glucose levels of 13 mmol/L and 19.5 mmol/L, respectively), so this enzyme (Phe^{351}-Leu) may be expected to show reduced activity.

The third newly described mutation (C^{1338}-A) is conservative and does not change the amino-acid sequence of the protein product. Other than its possible use as a biallelic marker for the locus, we are not investigating this further. Another recent report[10] has shown that, of 9 patients presenting with primary hypertriglyceridemia, 4 were found to possess mutations at the lipoprotein lipase. These mutants were Gly^{210}-Asp, Tyr^{302}-stop, Asp^{250}-Asn, and Ser^{447}-stop. By contrast, 20 patients were examined for mutations at the hepatic lipase and no coding variants were found; only a conservative codon variant was detected.[11]

More than 32 disease-related mutations of lipoprotein lipase have been documented since the initial report in 1989.[12,13] The majority of these appear to be rare, occurring at frequencies of less than 1:1000. However, four common population polymorphisms (frequencies of >1%) at the LPL locus have been documented that show disease relationships. The first to be published was a biallelic marker in intron 8 at an Hind 111 restriction site[13] whose alleles showed asymmetric frequencies in subjects with dyslipidemias compared to controls[13-15] and in subjects with premature coronary artery disease.[16] These effects were subsequently attributed to linkage disequilibrium of a common exon mutation located 635 bp downstream at a Ser^{447}-Ter mutation[17] in exon 9.[18] This variant truncates the mature protein by two amino acids and has been found to show disease associations with dyslipidemia.[19,20] Other common polymorphic variants (population frequencies of >1%) at LPL have been found at Asp^9-Asn (see reference 21) and at Asn^{291}-Ser (see reference 22). These have been reported to have disease associations with dyslipidemia.[23] In our 18 diabetic lipemic subjects, we found the Asn^{291} and Ser^{447} variants, but not the Asp^9 mutant. However, three new variants have come to light, suggesting that there is much more genetic diversity at this locus than was previously recognized. Possible explanations for the accumulation of mutations at this locus would have to include (i) a possible absence of selective pressures at a locus predisposing to a metabolic disorder that only manifests in middle age after the reproductive period is over or that (ii) heterozygotes provide some selective advantage possibly under conditions of prolonged famine where genetic variants that spare the utilization of plasma fuels may have some survival value. This may be analogous to the "thrifty" genotype hypothesis to explain the high incidence of type II diabetes mellitus in many populations.

REFERENCES

1. REAVEN, G. M. 1988. Role of insulin resistance in human disease. Diabetes 37: 1595-1607.
2. CREPALDI, G., A. TIENGO & E. MANZATO, Eds. 1993. Diabetes, Obesity, and Hyperlipidaemia: The Plurimetabolic Syndrome. Excerpta Medica. Amsterdam.
3. HAFFNER, S. M., M. P. STERN, B. D. MITCHELL, H. P. HAZUDA & J. K. PATTERSON. 1990. Incidence of type II diabetes in Mexican Americans predicted by fasting insulin and glucose levels, obesity, and body fat distribution. Diabetes 39: 283-288.
4. WU, D. A., X. BU, C. H. WARDEN, D. D. C. SHEN, C-Y. JENG, et al. 1996. Quantitative trait locus mapping of human blood pressure to a genetic region at or near the lipoprotein lipase gene locus on chromosome 8p22. J. Clin. Invest. 97: 2111-2118.
5. ZHANG, Q., E. CAVALLERO, J. CAVANNA, A. KAY, A. CHARLES, L. PERLEMUTER, B. JACOTOT & D. J. GALTON. 1996. Diabetes, obesity, and lipaemia: mutations at the lipoprotein lipase gene locus. Submitted.
6. MA, Y., M. S. LIU, D. GINZINGER, J. FROHLICH, J. D. BRUNZELL & M. R. HAYDEN. 1993. Gene-environment interaction in the conversion of mild to severe phenotype in a patient homozygous for Ser172Cys mutation in the lipoprotein lipase gene. J. Clin. Invest. 91: 1953-1958.
7. HULTMAN, T., S. STAHL, E. HORNES & M. UHLEN. 1989. Direct solid-phase sequencing of genomic and plasmid DNA using magnetic beads as solid support. Nucleic Acids Res. 17: 4937-4946.
8. KOZAKI, K., T. GOTOVA, M. KAWAMURA, H. SHIMANO, Y. YAZAKI, Y. OUCHI, H. ORIMO & N. YAMADA. 1993. Mutation analysis of human lipoprotein lipase by carboxy-terminal truncation. J. Lipid Res. 34: 1765-1772.
9. HOFFMAN, M. M. & W. STOFFEL. 1996. Functional characterization of a fusion protein genetically engineered from human lipoprotein lipase and human apolipoprotein CII. Submitted.
10. CALANDRA, S., M. L. SIMONE & S. BERTOLINI. 1996. New mutations of the lipoprotein lipase gene in Italian patients with hyperchylomicronemia. In European Atherosclerosis Society 66th Congress, Florence, p. 19.

11. JANSEN, H., A. J. M. VERHOEVEN, D. J. J. HALLEY & J. KASTELEIN. 1996. C-T substitution at -480 of the hepatic lipase promoter associated with a lowered lipase activity. *In* European Atherosclerosis Society 66th Congress, Florence, p. 20.
12. LANGLOIS, S., S. DEEB, J. D. BRUNZELL, J. J. P. KASTELEIN & M. R. HAYDEN. 1989. A major insertion accounts for a significant proportion of mutations underlying human lipoprotein lipase deficiency. Proc. Natl. Acad. Sci. U.S.A. **86:** 948–952.
13. CHAMBERLAIN, J. C., J. A. THORN, K. OKA, D. J. GALTON & J. STOCKS. 1989. DNA polymorphisms at the lipoprotein lipase gene locus: associations in normal and hypertriglyceridemic subjects. Atherosclerosis **79:** 85–89.
14. HEINZMANN, C., T. KIRCHGESSNER, P. O. KWITEROVICH, J. A. LEDIAS, G. E. ANTONIARAKIS & A. J. LUSIS. 1991. DNA polymorphism haplotypes of the human lipoprotein lipase gene: possible association with high density lipoprotein levels. Hum. Genet. **86:** 578–584.
15. PEACOCK, R. E., A. HAMSTEN, P. NILSSON EHLE & S. E. HUMPHRIES. 1992. Associations between lipoprotein lipase gene polymorphisms and plasma correlations of lipids, lipoproteins, and lipase activity in young myocardial infarction survivors and age-matched healthy individuals from Sweden. Atherosclerosis **97:** 171–181.
16. THORN, J. A., J. C. CHAMBERLAIN, J. C. ALCOLADO, K. OKA, L. CHAN, J. STOCKS & D. J. GALTON. 1990. Lipoprotein and hepatic lipase gene variants in coronary atherosclerosis. Atherosclerosis **85:** 55–60.
17. HATA, A., M. ROBERTSON, M. EMI & J. M. LALOUEL. 1990. Direct detection and automated sequencing of individual alleles after electrophoretic strand separation. Nucleic Acids Res. **18:** 5407–5410.
18. STOCKS, J., J. A. THORN & D. J. GALTON. 1992. Lipoprotein lipase genotypes for a common premature termination codon mutation detected by PCR-mediated site-directed mutagenesis and restriction enzyme analysis. J. Lipid Res. **33:** 853–857.
19. MATTU, R. K., E. W. A. NEEDHAM, R. MORGAN, A. REES, A. K. HACKSHAW, J. STOCKS, P. C. ELWOOD & D. J. GALTON. 1994. DNA variants at the LPL gene locus associate with angiographically defined severity of atherosclerosis and serum lipoprotein levels in a Welsh population. Arterioscler. Thromb. **14:** 1090–1097.
20. ZHANG, Q., J. CAVANNA, B. R. WINKELMANN, B. SHINE, W. CROSS, W. MARZ & D. J. GALTON. 1995. Common genetic variants of lipoprotein lipase that relate to lipid transport in patients with premature coronary artery disease. Clin. Genet. **48:** 293–298.
21. GAGNE, E., J. GENEST, H. ZHANG, L. A. CLARKE & M. R. HAYDEN. 1994. Analysis of DNA changes in the LPL gene in patients with familial combined hyperlipidemia. Arterioscler. Thromb. **14:** 1250–1257.
22. REYMER, P., E. GAGNE, B. E. GROENEMEYER, H. ZHANG, J. FORSYTH, H. JANSEN, J. C. SEIDELL, D. KROMHOUT, K. E. LIE, J. J. P. KASTELEIN & M. R. HAYDEN. 1995. A lipoprotein lipase mutation (Arg^{291}-Ser) is associated with reduced high density lipoprotein cholesterol (HDL) levels in patients with premature atherosclerosis. Nat. Genet. **10(1):** 28–34.
23. PIMSTONE, S. N., E. GAGNE, S. M. CLEE, E. A. STEIN & M. R. HAYDEN. 1995. Postprandial retinyl palmitate response supports evidence for a functional effect of the Arg^{291}-Ser mutation in the lipoprotein lipase gene. Circulation **92:** no. 8, abstract 2350.

Progress in Determining the Genes for Hypertension, Insulin Resistance, and Dyslipidemia[a]

MARK CAULFIELD,[b] PIERRE-MARC BOULOUX,[c]
AND PATRICIA MUNROE[d]

[b]Department of Clinical Pharmacology
St. Bartholomew's Hospital
London EC1A 7BE, United Kingdom

[c]Department of Endocrinology
Royal Free Hospital School of Medicine
London NW3 2QG, United Kingdom

[d]Department of Pediatrics
University College London Medical School
The Rayne Institute
London WC1E 6JJ, United Kingdom

ESSENTIAL HYPERTENSION

In the general population, blood pressure adopts a normal distribution, with essential hypertension being defined on the basis of thresholds for pharmacological intervention to reduce blood pressure and thus risk of stroke and heart attack. Family studies imply that approximately 30% of blood pressure variation is due to genes, with the remainder arising from environmental influences such as sodium, alcohol, and obesity.[1] Since populations cannot be completely demarcated into hypertensives or normotensives, it is likely that several genes are involved in the genetic basis of this trait.

GENETIC STRATEGIES FOR HYPERTENSION

During the early nineties, there have been considerable advances in molecular approaches available to investigate the genetic basis of complex diseases such as human essential hypertension. The strategy that has been most widely employed thus far to identify the major genes for hypertension involves investigation of candidate genes arising from systems that are physiologically implicated in blood pressure regulation.

---

[a] This work was supported by the Joint Research Board of St. Bartholomew's Hospital and by funds from the Medical Research Council of Great Britain to M. Caulfield and P. Munroe.

However, there are many potential candidates that could alter blood pressure levels and evaluation of the role of all of these genetic loci could be extremely time-consuming.

An alternative and complimentary strategy is to screen the whole genome utilizing a dense map of highly polymorphic markers spread throughout the human genome to test for linkage of regions to blood pressure. Using this approach, it may be possible to identify chromosomal regions that harbor susceptibility genes that we do not even suspect contribute to the hypertensive phenotype. However, this still leaves the task of refining the region down to a specific causative gene for which no function may yet have been described and this may be quite laborious.

LINKAGE STUDIES IN HYPERTENSION

The most-powerful approach to understanding the genetics of complex traits will probably prove to be family-based linkage studies.[2] In many cases, these studies are based on affected sibling pairs from nuclear families because raised blood pressure may have a variable age of onset. In addition, it may be possible for apparently unaffected family members to possess a susceptibility variant for hypertension, but not manifest the trait because of the absence of an environmental stimulus.[2] Equally, the manifestation of elevated blood pressure may rely on gene–gene interaction and may require the inheritance of genetic variants at more than one locus.[2]

With linkage studies, the aim is to observe the tracking of a particular allele at a genetic locus with hypertension through a family. This can be based on computing a logarithmic likelihood ratio (Lod score), which simply expresses the likelihood of linkage divided by the likelihood of nonlinkage.[2] It has previously been accepted that support for linkage has been demonstrated if Lod scores are greater than 3.0, which means that data at the genetic locus under evaluation are 1000 to 1 in favor of linkage.[2] However, in complex traits, large data sets may be required to offer adequate power to generate such support for linkage. Recently, some workers have proposed that the stringency accepted for such Lod scores should be even greater in genomewide screens.[2] This would require identification of large numbers of families to offer any chance of being able to refute or accept involvement of a region in raised blood pressure.

A complimentary statistical method of linkage analysis employed in hypertension genetics research is to ask the question of whether hypertensive affected siblings share alleles at a genetic locus more often than would be expected by chance in the population.[2] This allele-sharing method uses a t statistic to measure excess allele sharing in affected sibling pairs.[2] An advantage of such methods is that they can take account of variable penetrance of susceptibility variants and do not require specification of Mendelian modes of inheritance, which makes them particularly useful in complex traits like hypertension.[2]

POPULATION ASSOCIATION STUDIES

This study design has been widely employed to compare the distribution of genetic variants of a known candidate gene in unrelated cases and controls. A positive association implies that an allele of the variant studied exhibits linkage disequilibrium with the complex trait under investigation.[2] However, this does not mean that the associated variant causes hypertension. Since linkage disequilibrium only operates over short chromosomal distances, it is possible that the association may be merely highlighting the presence of a causative mutation close by, but probably within, the same gene.[2] When evaluating reports of association studies, it is important to check that there has not been selection bias in the ascertainment of cases and controls[2] — for example, by comparing people of different ethnic origin or recruiting study groups from highly selected clinic populations rather than the general population.[2]

PROGRESS TOWARD THE GENETIC BASIS OF HYPERTENSION

"Simple Mendelian Forms of Hypertension"

Lifton and Shimkets have approached the genetic investigation of hypertension in a novel way by investigating rare Mendelian hypertensive syndromes and demonstrating genetic influences that may have extremely large effects on blood pressure.[3,4] This approach could provide important insights into mechanisms that may hold wider importance when investigated in the entire hypertensive population or in defined homogenous subsets.

Glucocorticoid-suppressible Hyperaldosteronism

This syndrome is characterized by autosomal dominant transmission of hypertension with variably elevated aldosterone levels, suppressed plasma renin activity, and high levels of abnormal adrenal steroids, 18-hydroxycortisol and 18-oxocortisol.[5] It appears that the aberrant steroids and the aldosterone produced in this syndrome are under direct adrenocorticotrophic hormone control since they are suppressible by exogenous glucocorticoids such as dexamethasone. Families with this syndrome have early onset hypertension, which may be severe, and there appears to be an excess of cerebral hemorrhage in some families.

The genes encoding 11β-hydroxylase, which is normally expressed in the zona fasiculata and glomerulosa, and aldosterone synthase, which is normally only expressed in the zona glomerulosa, have 95% homology in nucleotide sequence and lie in close physical proximity on chromosome 8.[6]

In 1992, Lifton et al. described a chimeric gene duplication arising from unequal crossing over between 11β-hydroxylase regulatory sequences that became fused to the aldosterone synthase gene.[3] This results in ectopic aldosterone production under adrenocorticotrophic hormone control. Subsequently, several crossover sites ranging from intron 2 to 4 have been described.[7] Although these appear to be independent mutations, there is certainly a preponderance of Celtic ancestry in the families with this form of hypertension.[3] Interestingly, some families with this "Simple Mendelian Form of Hypertension" have individuals who possess the chimeric aldosterone synthase gene, but are not hypertensive. It is possible that this incomplete penetrance of the chimeric gene is due to the requirement of coexisting environmental factors such as sodium ingestion or indeed other genes to permit the hypertensive phenotype to be manifest. While this is a rare hypertensive trait, it illustrates the benefit of specific intermediate phenotypes in demarcating hypertensive subgroups and focusing research for genetic markers. It also underscores the complexity of apparently simple hypertensive traits.

Liddle's Syndrome

Liddle's syndrome is a rare form of hypertension described initially in 1963 within a single kindred with multiple siblings having severe hypertension.[8] Some of these siblings manifest hypokalemia with suppressed plasma renin activity and low aldosterone levels even when challenged by a low sodium diet.[8] Absence of response of these features to an aldosterone antagonist, but reduction in blood pressure and correction of hypokalemia by triamterene suggested that an abnormality of the distal epithelial sodium channel might underlie this condition.[8]

Recently, mutations within subunits of the epithelial sodium channel have been described in the original Liddle's kindred.[4] Functional studies reveal that these mutations could lead to constitutive overactivity of this channel with excess reabsorption of sodium in the kidney.[4]

Hypertension and Brachydactyly

An autosomal dominant form of severe hypertension associated with brachydactyly has been documented in a large Turkish kindred.[9] In studying this form of hypertension, there was no specific phenotypic clue as to chromosomal localization of the causative gene or genes and therefore a genomewide screen was used to link a region on 12p to this trait.[9] While the precise nature of the defect remains to be determined, this demonstrates the potential value of genome screening.

STUDIES IN HUMAN ESSENTIAL HYPERTENSION

The Renin Angiotensin System

The close involvement of this system in blood pressure regulation and cardiovascular disease has prompted researchers to focus on candidate genes from within the system.[10] Cleavage of angiotensinogen by renin is a key rate determinant of this cascade.[10] The observation that plasma angiotensinogen may be elevated with increased blood pressure and indeed track with hypertension through families makes this an attractive candidate gene for hypertension.[11] Indeed, three studies have reported support for the linkage of the angiotensinogen gene to hypertension in affected sibling pairs of white European origin and African-Caribbeans.[12-14] In addition, in one of these studies, two genetic variants within the angiotensinogen molecule were associated with hypertension.[12] Furthermore, one of the variants, encoding threonine instead of methionine at position 235 in the final molecule, was associated with raised plasma angiotensinogen levels.[12] However, there have been a number of studies that have not found an association of these variants of angiotensinogen with hypertension and therefore it is important to remain cautious about the role of this gene in hypertension.[13,14]

Hypertension as a Metabolic Syndrome

Hypertension commonly clusters with dyslipidemia and insulin resistance, which raises the possibility that genetic factors predisposing to these coexistent features might play a primary role in the development of raised blood pressure.[15] From the San Antonio Heart Study, it has been reported that glucose intolerance and raised insulin levels may predate hypertension by eight years.[16] However, it could be that insulin sensitivity is altered by a general membrane defect as part of the metabolic syndrome of hypertension.[15] Evidence from strong familial aggregation of non-insulin-dependent diabetes mellitus (NIDDM) where insulin resistance is prevalent suggests that, in some phenotypes, sensitivity to insulin may be affected by genes in some individuals. Recently, a preliminary report from a genome screen in Mexican-American families with NIDDM suggests that there is a potential locus on chromosome 2.[17] However, this will require confirmation in further studies and, interestingly, the locus linked to NIDDM is not near any potential candidates for this trait.[17]

Insulin Receptor Gene in Hypertension

The insulin receptor mediates the actions of insulin, enhances glucose transport, and may represent an end-organ site for insulin resistance, but studies investigating a role for this receptor in NIDDM have reported conflicting results.[18,19] An association study on Australian hypertensives reported an association of an *Rsa* 1 restriction fragment length polymorphism with high blood pressure.[19] Subsequently, a linkage study on 31 affected hypertensive pairs reported very borderline support for linkage using *Rsa* 1 and *Sst* 1 restriction fragment length polymorphisms with analysis using an allele-sharing method.[20] However, this was not confirmed by the Lod score, which was zero, suggesting there is no support for linkage in this data set.[20] It must be emphasized that none of the studies on the insulin receptor gene in hypertension have characterized the subject's status with regard to insulin sensitivity nor offered adequate power to refute a role for this gene in hypertension.

Dyslipidemia and the Genetics of Hypertension

Several reports have noted the association between hypertension and dyslipidemia in the form of raised cholesterol and triglycerides and low levels of high-density lipoprotein.[16] Indeed, familial clustering of lipid anomalies and hypertension has been documented in families from Utah and has led to the proposal that familial dyslipidemic hypertension may occur in as many as 12–20% of hypertensives.[21] It has been suggested that dyslipidemia may affect endothelial function and consequently may contribute to impaired vasodilatation and thus to elevated blood pressure.

The Lipoprotein Lipase Genetic Locus and Systolic Blood Pressure

Recently, a study on 48 Taiwanese families selected by a proband with NIDDM rather than hypertension were analyzed using quantitative trait locus analysis for the relationship of a variety of loci to blood pressure.[22] There was support for linkage of the region chromosome 8, which contains lipoprotein lipase, to systolic blood pressure, which was also supported by data from flanking markers.[22] This does not mean that lipoprotein lipase causes raised systolic blood pressure since other loci in the area may be important. It is unclear how an enzyme like lipoprotein lipase, which hydrolyzes triglyceride to free fatty acids and transfers particles to high-density lipoprotein, might influence blood pressure. Further studies to define the role of this locus in blood pressure are required and it remains possible that this locus is really implicated in NIDDM.

Apolipoprotein B Gene and Hypertension

Apolipoprotein B is the predominant protein of low-density lipoprotein and a component of lipoprotein (a) constituting atherogenic lipid fractions.[23] Genetic associations of restriction fragment length polymorphisms and cardiovascular disease and plasma apolipoprotein B levels have been observed.[24,25] In a study on 56 hypertensive sibling pairs of white European origin, there was borderline support for linkage of a highly polymorphic marker in the flanking region of the apolipoprotein B gene.[26] This was strengthened if the data were partitioned according to cholesterol level, and 20 hypertensive sibling pairs had cholesterol levels greater than 6.5 mmol/L.[26] There was support for linkage in these families with elevated cholesterol, but not among families with low cholesterol, even though both sibships were similarly hypertensive.[26] These data need confirmation and must be treated with caution. However, there is some suggestion that the apolipoprotein B gene may influence dyslipidemia in hypertensives.

SUMMARY

For the first time, we have techniques available that may enable us to determine cause and effect in common cardiovascular diseases. The observations presented herein require cautious interpretation until replicated. There are currently several large programs under way that seek to establish substantial family resources for investigation of the genetic basis of hypertension. When interpreting the results of genome screens, it will be necessary to consider that we may link genetic loci for coexistent features such as dyslipidemia and insulin resistance to hypertension. Therefore, careful phenotypic data will be necessary to dissect out the causes of hypertension.

REFERENCES

1. WARD, R. 1990. Familial aggregation and genetic epidemiology of blood pressure. *In* Hypertension: Pathophysiology, Diagnosis, and Management, p. 81-100. Raven Press. New York.
2. LANDER, E. S. & N. J. SCHORK. 1994. Genetic dissection of complex traits. Science **265**: 2037-2048.
3. LIFTON, R. P., R. G. DLUHY, M. POWERS, G. M. RICH, S. COOK, S. ULICK & J. M. LALOUEL. 1992. A chimaeric 11-beta-hydroxylase/aldosterone synthase gene causes glucocorticoid-remediable aldosteronism and human hypertension. Nature **355**: 262-265.
4. SHIMKETS, R., D. G. WARNOCK, C. M. BOSITIS, C. NELSON-WILLIAMS, J. H. HANSSON, M. SCHAMBELAN, J. R. GILL, S. ULICK, R. V. MILORA, J. W. FINDLING, C. M. CANESSA, B. C. ROSSIER & R. P. LIFTON. 1994. Liddle's syndrome: heritable human hypertension caused by mutations in the beta subunit of the epithelial sodium channel. Cell **79**: 407-414.
5. SUTHERLAND, D. J., J. L. RUSE & J. C. LAIDLAW. 1966. Hypertension, increased aldosterone secretion, and low plasma renin activity relieved by dexamethasone. Can. Med. Assoc. J. **95**: 1109-1119.
6. OGISHIMA, T., H. SHIBATA, H. SHIMADA, F. MITANI, H. SUZUKI & T. I. Y. SARUTA. 1991. Aldosterone synthase cytochrome P-450 expressed in the adrenals of patients with primary aldosteronism. J. Biol. Chem. **266**: 10731-10734.
7. LIFTON, R. P., R. G. DLUHY, M. POWERS, G. M. RICH, M. GUTKIN, F. FALLO, J. R. J. GILL, L. FELD, A. GANGULY, J. C. LAIDLAW, *et al.* 1992. Hereditary hypertension caused by chimeric gene duplications and ectopic expression of aldosterone synthase. Nat. Genet. **2**: 66-74.
8. LIDDLE, G. W., T. BLEDSOE & W. S. COPPAGE. 1963. A familial renal disorder stimulating primary aldosteronism, but with negligible aldosterone secretion. Trans. Assoc. Am. Physicians **76**: 199-213.
9. SCHUSTER, H., T. F. WIENKER, S. BAHRING, N. BILGINTURAN, H. R. TOKA, E. J. NEITZEL, O. TOKA, D. GILBERT, A. LOWE, J. OTT, H. HALLER & F. C. LUFT. 1996. Severe autosomal dominant hypertension and brachydactyly in a unique Turkish kindred maps to human chromosome 12. Nat. Genet. **13**: 98-100.
10. MACGREGOR, G. A., N. D. MARKANDU, J. E. ROULSTON, J. C. JONES & J. J. MORTON. 1981. Maintenance of blood pressure by the renin angiotensin system in normal man. Nature **291**: 329-331.
11. WATT, G. C., S. B. HARRAP, C. J. FOY, D. W. HOLTON, H. V. EDWARDS, H. R. C. DAVIDSON, A. F. LEVER & R. FRASER. 1992. Abnormalities of glucocorticoid metabolism and the renin-angiotensin system: a four-corners approach to the identification of genetic determinants of blood pressure. J. Hypertens. **10**: 473-482.
12. JEUNEMAITRE, X., F. SOUBRIER, Y. V. KOTELEVTSEV, R. P. LIFTON, C. S. WILLIAMS, S. C. HUNT, P. N. HOPKINS, R. R. WILLIAMS, J. M. LALOUEL, *et al.* 1992. Molecular basis of human hypertension: role of angiotensinogen. Cell **71**: 169-180.
13. CAULFIELD, M., P. LAVENDER, M. FARRALL, P. MUNROE, M. LAWSON, P. C. TURNER & A. J. L. CLARK. 1994. Linkage of the angiotensinogen gene to essential hypertension. N. Engl. J. Med. **330**: 1629-1633.
14. CAULFIELD, M., P. LAVENDER, J. NEWELL-PRICE, M. FARRALL, S. KAMDAR, H. DANIEL, M. LAWSON, P. DE FREITAS, P. FOGARTY & A. J. L. CLARK. 1995. Linkage of the angiotensinogen gene locus to human essential hypertension in African Caribbeans. J. Clin. Invest. **95**: 687-692.
15. REAVEN, G. M. & B. B. HOFFMAN. 1987. A role for insulin in the aetiology and course of hypertension. Lancet **ii**: 435-437.
16. HAFFNER, S. M., M. W. BRANDS, H. P. HAZUDA & M. P. STERN. 1992. Clustering of risk factors in confirmed prehypertensive individuals. Hypertension **20**(1): 38-55.
17. HARRIS, C. L., E. BOERWINKLE, R. CHAKRABORTY, *et al.* 1996. A genome-wide search for human non-insulin-dependent (type 2) diabetes genes reveals a major susceptibility locus on chromosome 2. Nat. Genet. **13**(2): 161-166.
18. OELBAUM, R. S., P. M. G. BOULOUX, S. R. LI, M. G. BARONI, J. STOCKS & D. J. GALTON.

1991. Insulin receptor gene polymorphisms in type 2 (non-insulin-dependent) diabetes mellitus. Diabetologia **34:** 260–264.
19. YING, L-H., R. Y. L. ZEE, L. R. GRIFFITHS & B. J. MORRIS. 1991. Association of a RFLP for the insulin receptor gene, but not insulin with essential hypertension. Biochem. Biophys. Res. Commun. **181:** 486–492.
20. MUNROE, P. B., H. I. DANIEL, M. FARRALL, M. LAWSON, P. M. BOULOUX & M. J. CAULFIELD. 1995. Absence of genetic linkage between polymorphisms of the insulin receptor gene and essential hypertension. J. Hum. Hypertens. **9:** 669–670.
21. WILLIAMS, R. R., S. C. HUNT, P. N. HOPKINS, B. M. STULTS, L. L. WU, S. J. HASSTEDT, G. K. BARLOW, M. C. SCHUMACHER, R. P. LIFTON & J. M. LALOUEL. 1988. Familial dyslipidemic hypertension: evidence from 58 Utah families for a syndrome present in approximately 12% of patients with essential hypertension. JAMA **259**(24): 3579–3586.
22. WU, D-A., X. BU, C. H. WARDEN, et al. 1996. Quantitative trait locus mapping of human blood pressure to a genetic region at or near the lipoprotein lipase gene locus on chromosome 8p22. J. Clin. Invest. **97:** 2111–2118.
23. DAMMERMAN, M. & J. L. BRESLOW. 1995. Genetic basis of lipoprotein disorders. Circulation **91:** 505–512.
24. HEGELE, R. A., L-S. HUANG, P. N. HERBERT, B. B. CONRAD, J. E. BURING, C. H. HENNEKENS & J. L. BRESLOW. 1986. Apolipoprotein B gene DNA polymorphisms associated with myocardial infarction. N. Engl. J. Med. **315:** 1509–1515.
25. MENDIS, S., J. SHEPARD, C. J. PACKARD & D. GAFFNEY. 1991. Restriction fragment length polymorphisms in the apo B gene in relation to coronary heart disease in a southern Asian population. Clin. Chim. Acta **196:** 107–117.
26. MUNROE, P. B., A. JOHNSTON, H. DANIEL, M. LAWSON, P. BOULOUX & M. CAULFIELD. 1995. Linkage of the apolipoprotein B gene to hypercholesterolemia in patients with essential hypertension. Abstract: Proc. Eur. Soc. Hypertens. J. Hypertens.

Hyperinsulinemia and Sympathoadrenal System Activity in the Rat[a]

RICHARD KVETŇANSKÝ,[b] MILAN RUSNÁK,[b]
DANIELA GAŠPERÍKOVÁ,[b] JANA JELOKOVÁ,[b]
ŠTEFAN ZÓRAD,[b] ILJA VIETOR,[b] KAREL PACÁK,[c]
ELENA ŠEBÖKOVÁ,[b] LADISLAV MACHO,[b]
ESTHER L. SABBAN,[d] AND IWAR KLIMEŠ[b]

[b]Institute of Experimental Endocrinology
Slovak Academy of Sciences
833 06 Bratislava, Slovak Republic

[c]Department of Medicine
Washington Hospital Center
Washington, District of Columbia

[d]Department of Biochemistry and Molecular Biology
New York Medical College
Valhalla, New York 10595

A relationship between hyperinsulinemia, insulin resistance, and hypertension has been described; however, it is still not completely clear whether these phenomena are causally related (for review, see references 1–5).

In rats, both acute and chronic hyperinsulinemia lead to an increase in blood pressure (BP).[6–8] The mechanisms of insulin-induced changes in BP regulation could involve several factors—activation of catecholaminergic systems, sodium-fluid retention, modulation of cation transport, vasculopathy, etc.[5] Moreover, several investigators have indicated a role for the sympathoneural system in insulin-mediated BP changes.[7,9–13] Thus, one of the important ways by which hyperinsulinemia could increase BP is the activation of the sympathoadrenal system. However, the mechanism by which hyperinsulinemia could affect sympathoadrenal system activity remains unclear. Therefore, the aim of this contribution, which represents a review with new data, was to study the following:

---

[a] This study was supported by the Slovak Grant Agency for Science (Grant Nos. 95/5305/272 and 95/5305/043) and the United States–Slovak Science and Technology Joint Fund (Project No. 93 024), and in part by a Public Health Service Award (No. NS-32166) and a grant of the European Communities, EURHYPGEN (No. ERBBMHICT 920869).

(i) the effects of hyperinsulinemia on sympathoadrenal system (SAS) activity in rats evaluated by measurements of (a) catecholamine (CA) secretion and (b) CA synthesis and CA biosynthetic enzyme activity, protein concentration, and gene expression;
(ii) the SAS activity in insulin-resistant and hypertensive hereditary hypertriglyceridemic rats (hHTg) at basal conditions and in stress;
(iii) the mechanism(s) by which hyperinsulinemia could induce the activation of the SAS: is the insulin effect solely induced by hypoglycemia and mediated via sympathetic neurons or is it also mediated by a direct insulin action at the level of adrenal medullary and/or sympathetic neurons?

HYPERINSULINEMIA

Effects of Hyperinsulinemia on Catecholamine Secretion

Hyperinsulinemia leads to a dose-dependent BP increase in rats.[5,9] Euglycemic insulin infusion also has been shown to increase BP, an effect that is reversed by alpha- and beta-adrenergic blockade. This suggests an important role for increased SAS activity in insulin-mediated BP changes.[11] Nevertheless, the pressor effect of exogenous hyperinsulinemia, shown in rats, has not been confirmed in other animal species.[3,14]

In our experiments, we studied the question of how different levels of hyperinsulinemia could affect plasma levels of CA. Conscious, 20-h-fasted, tail artery–cannulated male Sprague-Dawley rats (Taconic Farm), weighing 300–350 g, were exposed to insulin-induced hypoglycemia. Porcine regular insulin (Eli Lilly) was injected iv (0.3, 1.0, 3.0 IU/kg). Blood was collected via the cannula at 0, 15, 45, 75, and 105 min after insulin injection. Plasma levels of norepinephrine (NE), epinephrine (EPI), and dopamine (DA) were assessed using the HPLC method described previously.[15]

Insulin-induced hypoglycemia produced a dose-dependent adrenomedullary activation (plasma EPI levels reached extreme values over 5000 pg/mL – more than 60-fold increases). In contrast, plasma NE levels increased by a maximum of about 2.5-fold. High insulin doses also induced increases in plasma DA levels.

The data show that insulin induces marked increases in plasma EPI levels. Since almost all of the total circulating EPI is secreted from the adrenal medulla, these increases indicate that insulin predominantly activates the adrenomedullary function.

Plasma NE levels changed only slightly in insulin-treated animals, suggesting that the activity of the sympathoneural system may not be markedly affected by the insulin administration. Since the adrenal medullary contribution to circulating NE in stressed rats is only about 30% of the total increase,[16] plasma NE levels found in our experiments might predominantly reflect the NE release from the sympathetic nerves. The evidence for drastic adrenomedullary stimulation and only moderate sympathoneural stimulation during insulin-induced hypoglycemia has also been shown by others.[17–21]

Effects of Hyperinsulinemia on Catecholamine Synthetic Enzyme Activity, Protein Concentration, and Gene Expression

The effects of hyperinsulinemia on adrenomedullary activity were also studied by measurement of the activity,[22,23] protein, and mRNA[24-26] levels of the catecholamine biosynthetic enzymes—tyrosine hydroxylase (TH) and phenylethanolamine N-methyltransferase (PNMT). Male Sprague-Dawley rats (400–450 g, Charles River Wiga, Germany) were used in these experiments. Insulin (Actrapid, Novo Nordisk, Denmark) or saline was administered ip at a dose of 5 IU/kg of body weight. Animals were injected one or seven times (once a day, in the morning) and sacrificed 5 hours after the drug administration. Changes in the adrenal medullary TH and PNMT activity and mRNA levels are shown in FIGURES 1A and 1B. Significant elevations of TH and PNMT mRNA levels were observed in rats exposed to both the first and the seventh insulin administrations, whereas activities of these enzymes were significantly elevated only after repeated injections (FIG. 1A). A typical Northern blot of the insulin-induced increase in TH mRNA levels is shown in FIGURE 1B. Data on increased adrenal TH and PNMT activities in rats after insulin-induced hypoglycemia have been already reported more than 20 years ago.[27-29] However, no data on gene expression of these enzymes (besides the data on TH mRNA from our group[26]) have been found in the available literature.

These data demonstrate that hyperinsulinemia produces not only increased CA secretion from the adrenal medulla, but also increased activity and gene expression of the enzymes involved in CA biosynthesis. TH protein levels were also found to be increased in the adrenal medulla of rats repeatedly injected with insulin.[26] Insulin-induced adrenal medullary TH gene expression is mainly dependent on neural regulation since cutting the splanchnic nerve innervating the adrenals completely prevents the increase in adrenal TH mRNA levels.[26] Similar increases of adrenal TH and PNMT mRNA levels were described after exposure of rats to different stressors that highly activate SAS activity.[24,25,30-35]

Thus, our data suggest that hyperinsulinemia markedly activates the adrenomedullary system and partly the sympathoneural system, demonstrated by increases in plasma levels of catecholamines, CA metabolites, and precursor DOPA, as well as by increases in the activity and mRNA levels of the CA biosynthetic enzymes, tyrosine hydroxylase and phenylethanolamine N-methyltransferase, in the adrenal medulla.

SYMPATHOADRENAL ACTIVITY IN INSULIN-RESISTANT AND HYPERTENSIVE HEREDITARY HYPERTRIGLYCERIDEMIC RATS

The hereditary hypertriglyceridemic (hHTg) rat represents a unique animal model of the human insulin resistance syndrome where hypertension is one of its prominent features.[5,36,37] In this study, we confirmed earlier findings[38,39] that both systolic BP (157 ± 5.5 versus 115 ± 3.4 mmHg) and diastolic BP (131 ± 3.9 versus 95 ± 3.9 mmHg) were highly significantly increased in hHTg rats. This time the normotensive Brown-Norway (BN) rats served as controls. The BN rats were chosen for comparison

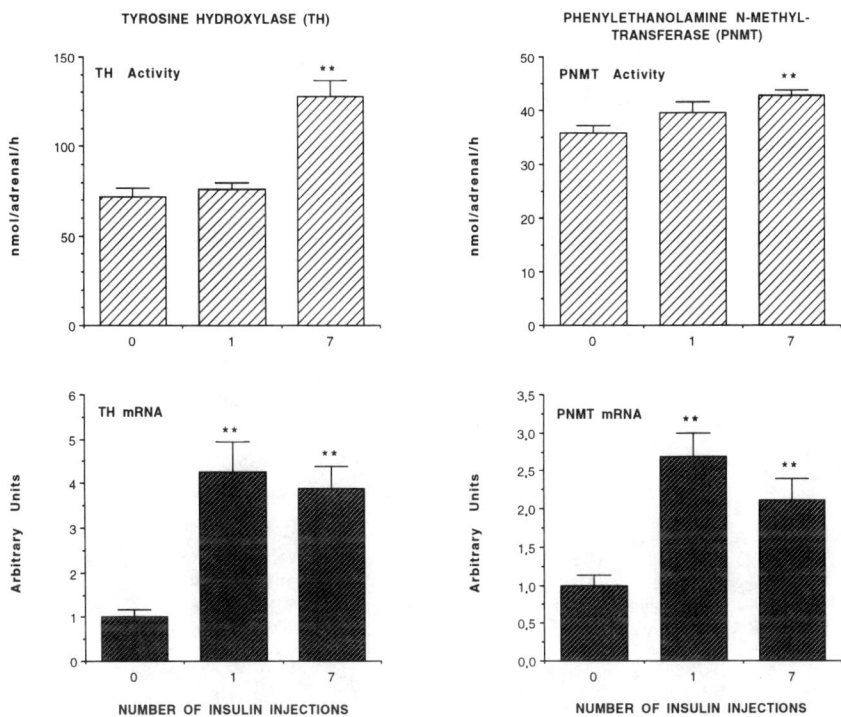

FIGURE 1A. Effects of insulin (Actrapid, Novo Nordisk) administration (5 IU/kg, ip) on TH and PNMT activities and mRNA levels in the adrenals of rats as measured by Northern blotting. Animals were injected one or seven times (once a day) and sacrificed by decapitation 5 hours after the last drug administration. Mean values ± SEM for 6–8 rats per group are shown. Statistical significance calculated by one-factor ANOVA followed by Scheffe's post-hoc test: (**)$p < 0.01$ versus control groups.

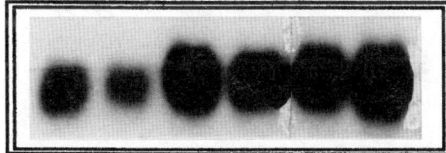

FIGURE 1B. A representative Northern blot of TH mRNA levels in the adrenals of rats treated with insulin once or seven times.

since this strain is completely normal as far as the pertinent symptoms of insulin resistance syndrome are concerned.

The aim of this part of the study was to ascertain whether hHTg rats might show changes in SAS activity and whether these changes would participate in the etiology of hypertension in the animals with insulin resistance syndrome. Four-month-old hHTg Wistar male rats were obtained from our colony (35th generation). BN rats of the

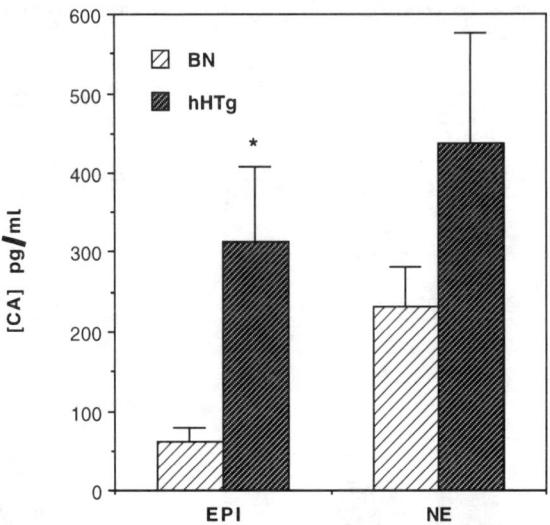

FIGURE 2. Plasma epinephrine (EPI) and norepinephrine (NE) levels in normotensive Brown-Norway (BN) and hypertensive hereditary hypertriglyceridemic (hHTg) rats at rest. Blood was obtained via a chronic carotid catheter at strictly stress-free conditions. Mean values ± SEM for 6-8 rats. One-factor ANOVA followed by Scheffe's post-hoc test was used: (∗) $p < 0.05$ versus BN rats.

same age were purchased from Charles River Farm in France. Plasma catecholamines were measured in blood samples of conscious hHTg and BN rats with indwelling catheters in the carotid artery (4 days after the surgery) at rest and during immobilization stress[40] by a modification of the radioenzymatic assay.[41]

As shown in FIGURE 2, the resting plasma EPI levels were significantly elevated in hHTg rats, whereas the NE levels were not. However, during exposure to immobilization stress, the responses of both plasma EPI and NE levels were significantly elevated in hHTg rats compared to the control BN animals (FIG. 3). To the best of our knowledge, these are the first data showing an increase in CA secretion in hHTg rats at rest and during stress in well-controlled studies. Plasma catecholamines were previously measured only in blood collected after decapitation of rats,[42] a procedure known to increase CA baseline levels more than 50-fold.[43] An increase in plasma CA levels in cannulated rats after exposure to different stressors is a well-established phenomenon.[44]

The relationships between elevated BP, insulin resistance, hyperinsulinemia, and the activity of the SAS are still not entirely clear. Several rat models of hypertension also exhibit the symptomology of insulin resistance syndrome[5] and are like the hHTg rats. For example, spontaneously hypertensive rats (SHR) show a more enhanced response of the sympathoneural system to stressors than WKY rats.[45,46] Buchanan and others[47] demonstrated in SHR an exaggerated insulin response to glucose and suggested that nutrient-stimulated hyperinsulinemia could contribute to the pathogenesis of hypertension in these rats. In contrast, the presence of hypertension in Zucker obese rats has not been consistently demonstrated. Since plasma CA and CA metabolite

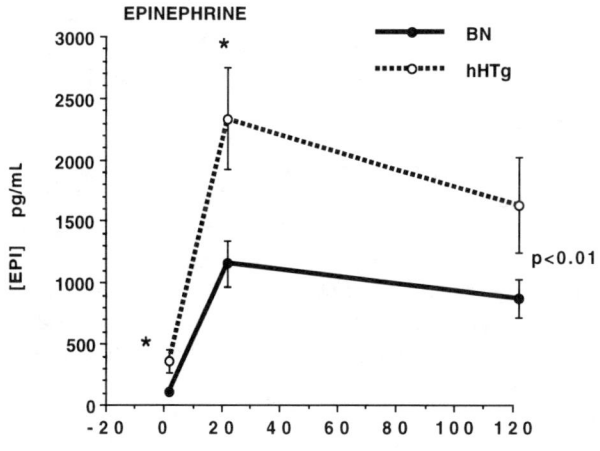

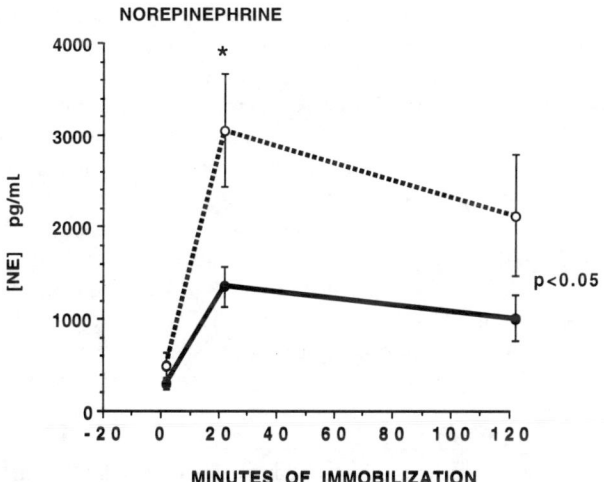

FIGURE 3. Effects of immobilization stress on plasma levels of epinephrine (EPI) and norepinephrine (NE) in normotensive Brown-Norway (BN) and hypertensive hereditary hypertriglyceridemic (hHTg) rats. Blood was obtained via a chronic carotid catheter, the procedure excluding additional stressors. Data are expressed as mean values ± SEM for 6–8 rats. Two-factor ANOVA for repeated measures followed by Scheffe's post-hoc test was used for statistical evaluation between strain groups: (∗) $p < 0.05$ compared to BN group. Significance between the time courses in hHTg and BN rats is marked in the figure. Values in all immobilization intervals are significantly elevated compared to controls.

responses have been shown to be significantly lower in Zucker obese rats during immobilization stress[48] and BP and vascular reactivity in such animals stayed elevated even after ganglionic blockade,[49] the involvement of the SAS in regulation of BP in these rats is not likely.

To look for the origin of the increased plasma CA levels in hHTg rats, the enzyme activity and gene expression of the rate-limiting enzyme in CA biosynthesis, tyrosine hydroxylase, and the enzyme converting NE to EPI, phenylethanolamine N-methyltransferase, in the adrenals were measured. Adrenal TH activity was measured according to Nagatsu and others,[22] and adrenal mRNA levels were measured using Northern blots as described in reference 24. Adrenal TH activity in hHTg rats was not significantly different from BN rats both at rest and after immobilization stress (data not shown). Adrenal TH and PNMT mRNA levels in hHTg and BN control and immobilized rats are shown in FIGURE 4A. A representative Northern blot of TH mRNA levels is demonstrated in FIGURE 4B. Control TH and PNMT mRNA levels revealed no differences between these two strains. Two-hour immobilization induced an approximately 3-fold increase of adrenal TH mRNA levels in BN rats, and the response was significantly exaggerated in hHTg rats (about 5-fold increase) (FIG. 4A). Immobilization increased the adrenal PNMT mRNA levels to about the same extent in both strains.

These data have shown that not only CA secretion, but also the gene expression of the rate-limiting enzyme in CA synthesis are increased in hHTg rats during stress. On the other hand, the TH activity was not significantly changed and even mRNA levels were not affected at control conditions. How thus to explain this difference?

An increase in adrenal TH activity is a time-dependent process that needs a longer time than an increase in TH mRNA levels.[50] During a single 2-hour immobilization, adrenal TH mRNA levels were elevated, but 24 hours after immobilization the levels returned to baseline.[31] This stress interval, however, was not long enough for an increase in TH activity and protein.[32] Only long-term increased TH mRNA levels are linked to increases in TH activity or protein level.[32] In hHTg rats, neither TH activity nor rest TH mRNA levels were increased, suggesting that the mechanism of CA synthesis may not be changed in control hHTg rats. A dissociation between regulation of CA secretion and adrenal TH gene expression has been described.[24] More important, however, is our finding of exaggerated responsiveness in expression of the TH gene in adrenals of hHTg rats during stress, which is in good agreement with exaggerated plasma CA levels in these animals.

Thus, in resting hHTg rats, an increase in CA secretion has been demonstrated without an increase in gene expression of enzymes involved in CA synthesis. During stress, the increased CA secretion positively correlated with the increased TH mRNA levels, strongly suggesting an elevation in SAS activity in hHTg rats.

POSSIBLE MECHANISMS OF HYPERINSULINEMIA-INDUCED ACTIVATION OF THE SAS

As shown in the literature, as well as in the studies described above, insulin administration results in an increase in plasma CA levels in rats. The majority of literary sources explain insulin-induced activation of the SAS by a counterregulatory effect of hypoglycemia mediated via central mechanisms.[21,51-53] The newest findings suggest that catecholamines are in the very first line of defense against hypoglycemia, even in humans.[54] However, besides central nervous mechanisms involved in the

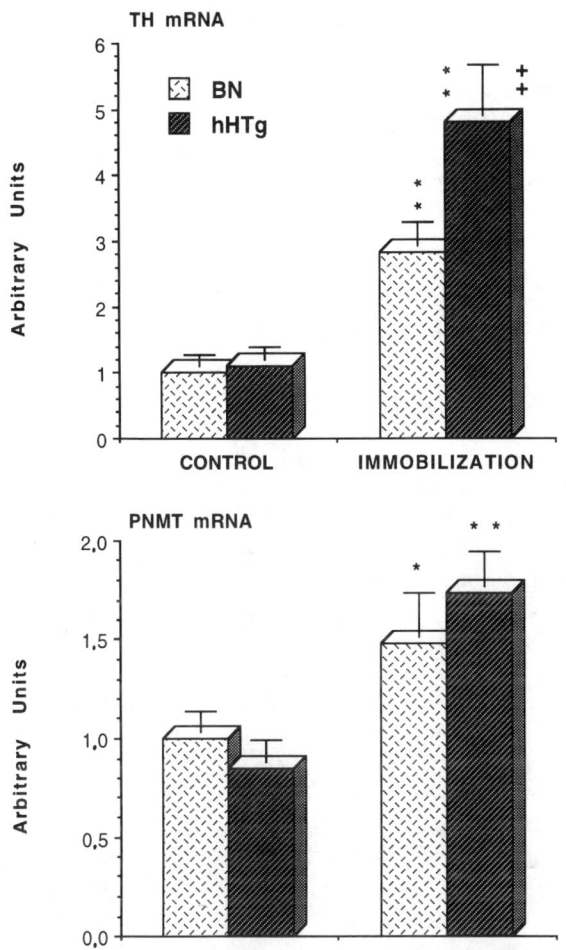

FIGURE 4A. Effects of immobilization stress on adrenal tyrosine hydroxylase (TH) and phenylethanolamine N-methyltransferase (PNMT) mRNA levels in normotensive Brown-Norway (BN) and hypertensive hereditary hypertriglyceridemic (hHTg) rats. The animals were sacrificed by decapitation at rest conditions (control) or after 2-hour immobilization. Mean values ± SEM for 6–10 rats per group are shown. Statistical significance calculated by one-factor ANOVA followed by Scheffe's post-hoc test: (∗) $p < 0.05$, (∗∗) $p < 0.01$ versus control groups; (++) $p < 0.01$ versus immobilized BN group.

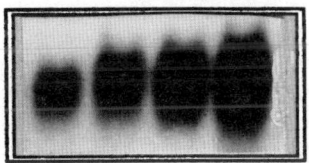

FIGURE 4B. A representative Northern blot of adrenal TH mRNA levels in BN and hHTg rats at rest and after 2-hour immobilization.

hypoglycemia-induced increase of plasma CA levels, the participation of nonneurogenic factors that could directly stimulate adrenal catecholamine secretion has also been suggested.[21,55,56] Therefore, an indirect and/or direct influence of insulin on CA release in rats was investigated in this study.

Hypoglycemic or Euglycemic Hyperinsulinemic Clamp

Hypoglycemic (HHC) or euglycemic (EHC) hyperinsulinemic clamps were used in conscious, chronically cannulated, unstressed Wistar male rats. Insulin was infused into the jugular vein in a dose of 6.4 mU/kg/min and blood was collected in 10-min intervals from the carotid artery.

Plasma glucose concentration was reduced to about 2.8 mmol/L during HHC, whereas glucose levels during EHC were not different from the basal levels (FIG. 5). In rats with HHC, plasma EPI levels were highly increased (about 10 times) and plasma NE levels were less, but also significantly elevated (less than 2 times). During euglycemic hyperinsulinemia, this increase in plasma CA levels was almost completely prevented (FIG. 5). Moreover, there was a negative correlation between plasma EPI and glucose levels ($p < 0.01$; data not shown). These data support the idea that insulin-induced elevation of plasma CA levels in the rat is not a consequence of a direct insulin action, but rather an indirect effect mediated by hypoglycemia.

There is no doubt that the adrenal medulla is the origin of the circulating EPI levels during hypoglycemia. To establish the origin of plasma NE during hypoglycemia is much more complicated. Conclusions of previous studies are confusing. Some investigators suggested the adrenal medulla as the exclusive source of circulating NE, whereas others found an increased NE turnover in the pancreas, suggesting an increased activation of the sympathoneural system during insulin-induced hypoglycemia. Even if our summary data shown in FIGURE 5 suggest a clear-cut response to the question under discussion, it has to be mentioned that not all animals uniformly responded to euglycemia. Some rats with hyperinsulinemia, in spite of euglycemia, showed a small (about 2-fold), but significant increase in plasma CA levels, mainly in EPI levels (unpublished data). During euglycemic hyperinsulinemia, BP still increased and was reversed by alpha- and beta-adrenergic blockade.[11] An increase in muscle nerve sympathetic activity and minor, but significant, increases in forearm venous plasma NE levels were found after euglycemic hyperinsulinemia in healthy humans.[57] Thus, even if hypoglycemia is the main factor responsible for the CA release, an existence of other factors cannot be excluded.

In agreement with this conclusion is our finding of elevated TH and PNMT gene expression in the adrenals of rats exposed to euglycemic hyperinsulinemic clamp. These animals were decapitated at the end of 90 minutes of hyperinsulinemia and euglycemia started; they were then compared to saline-treated animals. As shown in FIGURE 6, the administration of insulin even to rats with regulated euglycemia produced a significant elevation of adrenal TH and PNMT mRNA levels. These findings suggest that hyperinsulinemia might regulate gene expression of CA biosynthetic enzymes directly without involvement of hypoglycemic counterregulation.

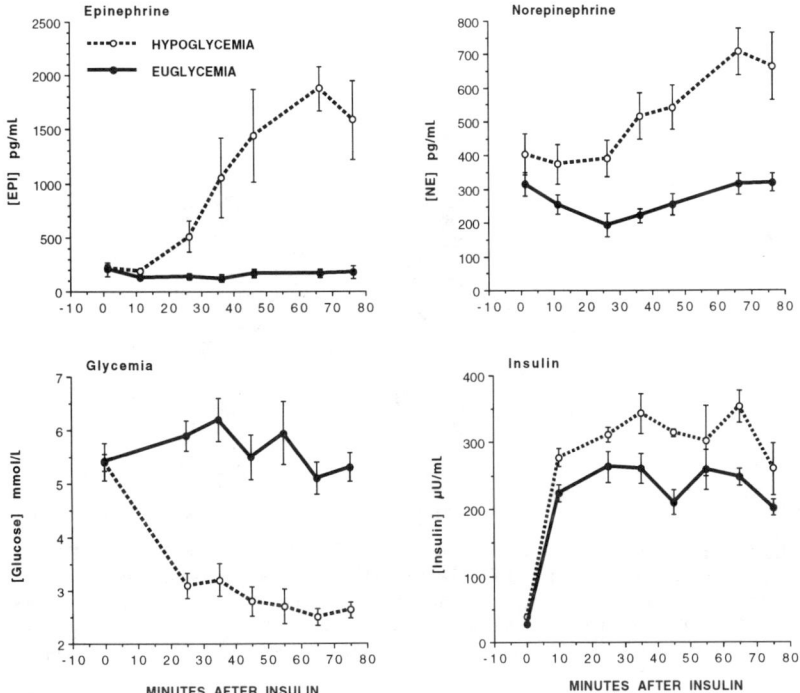

FIGURE 5. Effects of hypoglycemic or euglycemic hyperinsulinemic clamp on plasma epinephrine (EPI) and norepinephrine (NE) levels in conscious Wistar male rats. Insulin was infused into the jugular vein in a dose of 6.4 mU/kg/min and blood was collected in 10-min intervals from the carotid artery. Glycemia and plasma insulin levels are also shown. Data are expressed as mean values ± SEM for 6 rats.

Direct Effect of Insulin on Adrenal Medullary Catecholamine Release

The data shown earlier in this paper suggest a possible direct effect of insulin on catecholamine release from the adrenal medullary cells in rats. In support of this, binding of insulin to the membrane fraction of whole adrenals has been observed in different species, including rats[58] and human beings.[59] In the present paper, insulin binding to isolated plasma membranes from the rat adrenal medulla was determined. Plasma membranes were prepared by differential centrifugation after the tissue was homogenized in 50 mM Tris-HCl/1 mM EDTA buffer (pH 7.4). ^{125}I-(Tyr-A14) insulin binding was performed in tubes containing 200 µg of membrane proteins, for 5 hours at 10°C. FIGURE 7 shows the presence of insulin specific binding sites in the rat adrenal medulla. The specific binding of insulin was also found in the adrenal cortex. The pattern of displacement of insulin binding by unlabeled insulin was similar in the adrenals, liver, and fat tissue, demonstrating the presence of specific insulin receptors in rat adrenals. Preliminary data on the presence of insulin receptors in the adrenal medulla and their changes during stress have been reported in abstract form.[60] High-

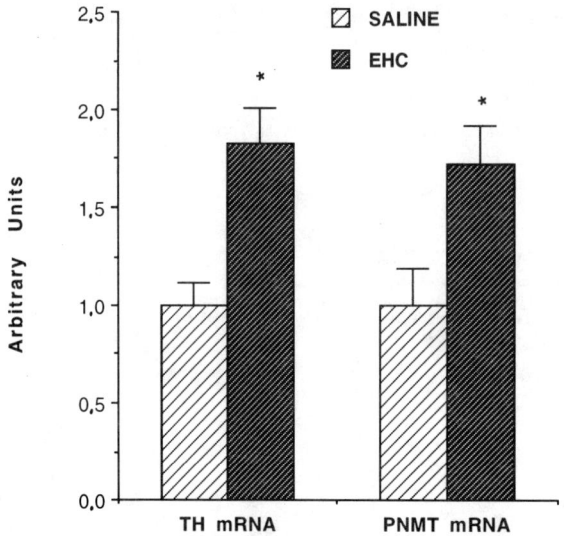

FIGURE 6. Effects of euglycemic hyperinsulinemic clamp (EHC) on adrenal TH and PNMT mRNA levels in conscious Wistar male rats. Saline or insulin in a dose of 6.4 mU/kg/min was infused into the jugular vein and the animals were decapitated 90 minutes after the commencement of insulin infusion. Mean values ± SEM for 5–8 rats per group are shown. Statistical significance calculated by one-factor ANOVA followed by Scheffe's post-hoc test: (∗) $p < 0.05$ versus saline-treated group.

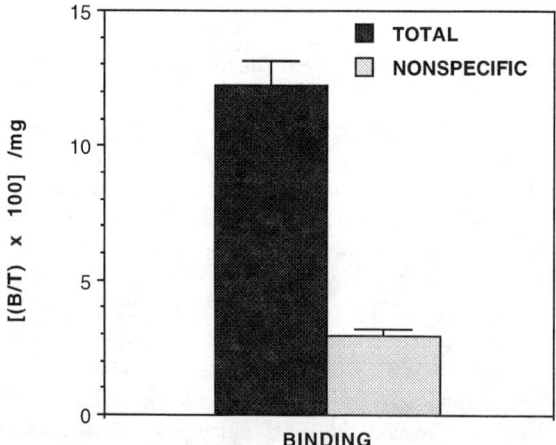

FIGURE 7. Presence of specific insulin binding sites in isolated plasma membranes from the rat adrenal medulla. Mean values ± SEM for 8 rats are shown. Specific insulin binding in the adrenal medulla represents 76.4% of the total insulin binding.

affinity insulin receptors together with specific receptors for IGF-1 were demonstrated in isolated chromaffin cells from bovine adrenal medulla.[61,62] The authors suggested that binding of insulin to its receptors on chromaffin cells can modulate the function of these cells.

To confirm the presence of insulin receptors in the adrenal medulla, it was essential to detect gene expression of insulin receptors in that tissue as well as specific insulin binding. Our preliminary experiments using PCR amplification of fragments comprising the alternatively spliced exon 11 of the insulin receptor gene[63] have shown the presence of mRNAs of both insulin receptor isomers (IR-A and IR-B) in the adrenal medulla (Zórad et al., unpublished data).

The presence of insulin receptors in the rat adrenal medulla suggests a direct insulin role in the regulation of adrenal medullary function. To confirm this hypothesis, adrenal medullary tissue was incubated with different concentrations of insulin (0.02–10 µg/mL, porcine insulin, Sigma). The incubation medium was collected after 10 or 120 min of incubation time and analyzed for EPI and NE concentration by the HPLC method with electrochemical detection described earlier.[15] The level of CA release in each incubation experiment is expressed as a percentage of its own controls (tissue incubated without adding of insulin). The methodological details have been described elsewhere.[64]

Incubation of the adrenal medullary tissue with insulin in doses from 0.02 to 1.00 µg/mL showed no changes in EPI and NE levels. However, insulin in doses of 5 and 10 µg/mL markedly stimulated EPI and also NE release from the adrenal medulla at both incubation intervals (FIG. 8). It has been reported that insulin has no effect on CA content in the adrenal medulla. However, both insulin and IGF-1 enhance CA secretion, with insulin being less potent than IGF-1.[62,65] Since high insulin concentrations were used for stimulation of CA release in those as well as in our studies, a possible interaction and involvement of IGF-1 receptors in this process cannot be excluded.

Thus, our data show that insulin has a direct effect on the release of catecholamines from the rat adrenal medulla *in vitro*, suggesting a possible role of insulin in regulation of catecholamine secretion from the adrenal medulla *in vivo*.

CONCLUSIONS

In summary, our data suggest the following in rats:

(i) Hyperinsulinemia-induced hypoglycemia markedly activates the adrenomedullary system and, to a smaller extent, also the sympathoneural system, as demonstrated by increases in plasma levels of epinephrine, norepinephrine, and dopamine and by increases in the activity and gene expression of the catecholamine biosynthetic enzymes, tyrosine hydroxylase and phenylethanolamine *N*-methyltransferase, in the adrenal medulla.

(ii) Insulin-resistant and hypertensive rats with hereditary hypertriglyceridemia (hHTg) have significantly increased basal levels of plasma CA and significantly exaggerated responses to immobilization stress. Adrenal tyrosine hydroxylase mRNA levels were not significantly changed in hHTg rats at rest, but an exposure to immobilization stress exaggerated the gene expression of this enzyme in comparison to normotensive BN rats. Thus, a relationship among increased blood pressure, plasma catecholamine levels, and gene expression

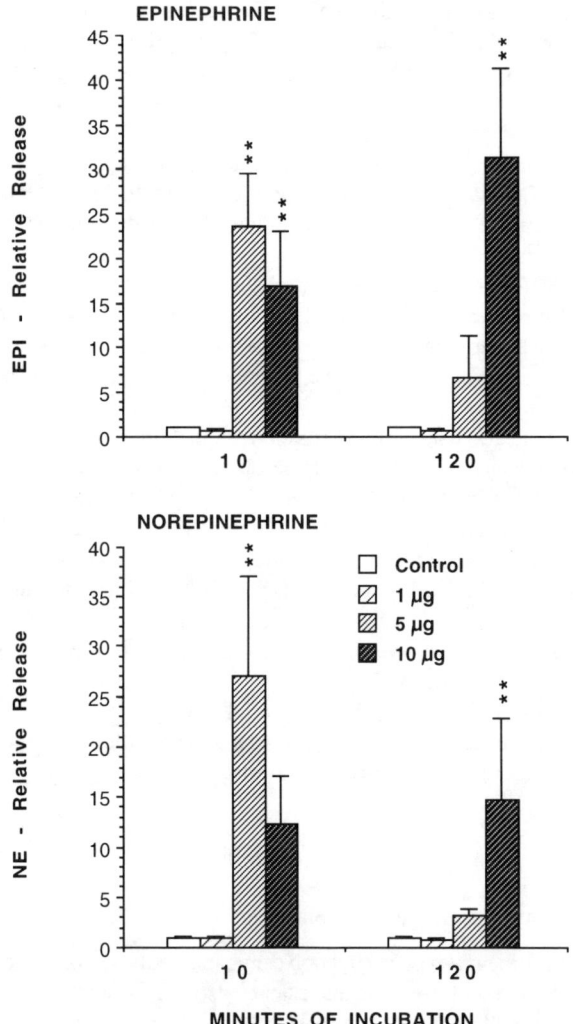

FIGURE 8. Effects of insulin (1–10 μg/mL) on epinephrine (EPI) and norepinephrine (NE) *in vitro* release from slices of isolated adrenal medulla. One half of the adrenal medulla was incubated with insulin for 10 or 120 minutes and the other half without insulin (control–basal release). Data are expressed as the relative release of basal catecholamine in each experiment. Basal EPI release was 36.3 ± 13.5 pmol/mg/min and NE release was 15.1 ± 4.0 pmol/mg/min. Mean values ± SEM for 5–13 samples per group are shown. Statistical significance calculated by one-factor ANOVA followed by Scheffe's post-hoc test: (∗∗) $p < 0.01$ versus control groups.

of tyrosine hydroxylase, the rate-limiting enzyme in catecholamine biosynthesis, has been found, especially under stress conditions. The increased activity of the sympathoadrenal system is suggested as one of the important factors involved in the mechanism maintaining hypertension in hHTg rats.

(iii) Euglycemic hyperinsulinemia failed to produce increases in plasma epinephrine and norepinephrine levels in rats. These data indicate that the mechanism by which insulin activates the sympathoadrenal system is mainly mediated by hypoglycemia via preganglionic neurons. However, insulin via insulin receptors localized in the adrenal medulla could play a role in regulation of adrenal TH and PNMT gene expression, as well as in regulation of catecholamine release from the adrenal medulla.

Thus, our data suggest that a direct *in vivo* effect of insulin on adrenal medullary cells should be considered.

REFERENCES

1. REAVEN, G. M. 1988. Role of insulin resistance in human disease. Diabetes **37**: 1595–1607.
2. MEEHAN, W. P., CH. DARWIN, N. B. MAALOUF, T. A. BUCHANAN & M. F. SAAD. 1993. Insulin and hypertension: are they related? Steroids **58**: 621–634.
3. HALL, J. E., M. W. BRANDS, D. H. ZAPPE & M. ALONSO-GALICIA. 1995. Insulin resistance, hyperinsulinemia, and hypertension: causes, consequences, or merely correlations? Proc. Soc. Exp. Biol. Med. **208**: 317–329.
4. REAVEN, G. M., H. LITHELL & L. LANDSBERG. 1996. Hypertension and associated metabolic abnormalities—the role of insulin resistance and the sympathoadrenal system. N. Engl. J. Med. **334**: 374–381.
5. KLIMEŠ, I. & E. ŠEBÖKOVÁ. 1997. Hypertension and the insulin resistance syndrome of rats: are they related? This volume.
6. BRANDS, M. W., D. A. HILDEBRANDT, H. L. MIZELLE & J. E. HALL. 1991. Sustained hyperinsulinemia increases arterial pressure in conscious rats. Am. J. Physiol. **260**: R764–R768.
7. MOREAU, P., L. LAMARCHE, A. K. LAFLAMME, A. CALDERONE, N. YAMAGUCHI & J. DE CHAMPLAIN. 1995. Chronic hyperinsulinemia and hypertension: the role of the sympathetic nervous system. J. Hypertens. **13**: 333–340.
8. ŠTOLBA, P., P. HUŠEK & J. KUNEŠ. 1994. Blood pressure changes induced by chronic insulin treatment in Wistar rats. Physiol. Res. **43**: 329–334.
9. EDWARDS, J. G. & C. M. TIPTON. 1989. Influences of exogenous insulin on arterial blood pressure measurements of the rat. J. Appl. Physiol. **67**: 2335–2342.
10. TOWNSEND, R. R., R. YAMAMOTO, M. NICKOLS, D. J. DiPETTE & G. A. NICKOLS. 1992. Insulin enhances pressor responses to norepinephrine in rat mesenteric vasculature. Hypertension **19**(suppl. II): 105–110.
11. DALY, P. A. & L. LANDSBERG. 1991. Hypertension in obesity and NIDDM: role of insulin and sympathetic nervous system. Diabetes Care **14**: 240–248.
12. LANDSBERG, L. 1992. Hyperinsulinemia: possible role in obesity-induced hypertension. Hypertension **19**(suppl. I): 61–66.
13. FOURNIER, R. D., C. C. CHIUEH, I. J. KOPIN, J. J. KNAPKA, D. DiPETTE & H. G. PREUSS. 1986. Refined carbohydrate increases blood pressure and catecholamine excretion in SHR and WKY. Am. J. Physiol. **250**: E381–E385.
14. HALL, J. E., M. W. BRANDS, D. H. ZAPPE & M. ALONSO-GALICIA. 1995. Cardiovascular actions of insulin: are they important in long-term blood pressure regulation? Clin. Exp. Pharmacol. Physiol. **22**: 689–700.
15. KVETŇANSKÝ, R., D. S. GOLDSTEIN, V. K. WEISE, C. HOLMES, K. SZEMEREDI, G. BAGDY & I. J. KOPIN. 1992. Effects of handling or immobilization on plasma levels of 3,4-dihydroxyphenylalanine, catecholamines, and metabolites in rats. J. Neurochem. **58**: 2296–2302.
16. KVETŇANSKÝ, R., V. K. WEISE, N. B. THOA & I. J. KOPIN. 1979. Effects of chronic guanethidine

treatment and adrenal medullectomy on plasma levels of catecholamines and corticosterone in forcibly immobilized rats. J. Pharmacol. Exp. Ther. **209:** 287–291.
17. GOLDSTEIN, D. S., M. GARTY, G. BAGDY, K. SZEMEREDI, E. M. STERNBERG, S. LISTWAK, K. PACÁK, A. DEKA-STAROSTA, A. HOFFMAN, P. C. CHANG, R. STULL, P. W. GOLD & I. J. KOPIN. 1993. Role of CRH in glucopenia-induced adrenomedullary activation in rats. J. Neuroendocrinol. **5:** 475–486.
18. KELLER-WOOD, M. E., C. E. WADE, J. SHINSAKO, L. C. KEIL, G. R. VAN LOON & M. F. DALLMAN. 1983. Insulin-induced hypoglycemia in conscious dogs: effect of maintaining carotid arterial glucose levels on the adrenocorticotropin, epinephrine, and vasopressin responses. Endocrinology **112:** 624–632.
19. KERR, D., I. A. MACDONALD & R. B. TATTERSALL. 1989. Influence of duration of hypoglycemia on the hormonal counterregulatory response in normal subjects. J. Clin. Endocrinol. Metab. **68:** 1118–1122.
20. GARBER, A. J., P. E. CRYER, J. V. SANTIAGO, M. W. HAYMOND, A. S. PAGLIARA & D. M. KIPNIS. 1976. The role of adrenergic mechanisms in the substrate and hormonal response to insulin-induced hypoglycemia in man. J. Clin. Invest. **58:** 7–15.
21. YAMAGUCHI, N. 1992. Sympathoadrenal system in neuroendocrine control of glucose: mechanisms involved in the liver, pancreas, and adrenal gland under hemorrhagic and hypoglycemic stress. Can. J. Physiol. Pharmacol. **70:** 167–206.
22. NAGATSU, T., M. LEVITT & S. UDENFRIEND. 1964. A rapid and simple radioassay for tyrosine hydroxylase activity. Anal. Biochem. **9:** 122–126.
23. AXELROD, J. 1962. Purification and properties of phenylethanolamine N-methyltransferase. J. Biol. Chem. **237:** 1657–1660.
24. KVETŇANSKÝ, R., B. NANKOVÁ, B. HIREMAGALUR, E. VISKUPIČ, I. VIETOR, M. RUSNÁK, A. MCMAHON, I. J. KOPIN & E. L. SABBAN. 1996. Induction of adrenal tyrosine hydroxylase mRNA by single immobilization stress occurs even after splanchnic transection and in the presence of cholinergic antagonists. J. Neurochem. **66:** 138–146.
25. VISKUPIČ, E., R. KVETŇANSKÝ, E. L. SABBAN, K. FUKUHARA, V. K. WEISE, I. J. KOPIN & J. P. SCHWARTZ. 1994. Increase in rat adrenal phenylethanolamine N-methyltransferase mRNA level caused by immobilization stress depends on intact pituitary-adrenocortical axis. J. Neurochem. **63:** 808–814.
26. VIETOR, I., M. RUSNÁK, E. VISKUPIČ, P. BLAŽIČEK, E. L. SABBAN & R. KVETŇANSKÝ. 1996. Glucoprivation by insulin leads to trans-synaptic increase in rat adrenal tyrosine hydroxylase mRNA levels. Eur. J. Pharmacol. **313:** 119–127.
27. SILBERGELD, S., R. KVETŇANSKÝ, V. K. WEISE & I. J. KOPIN. 1971. Effect of repeated administration of 2-deoxy-D-glucose or insulin on catecholamine-synthesizing enzymes in rat adrenals. Biochem. Pharmacol. **20:** 1763–1768.
28. KVETŇANSKÝ, R., S. SILBERGELD, V. K. WEISE & I. J. KOPIN. 1971. Effects of restraint on rat adrenomedullary response to 2-deoxy-D-glucose. Psychopharmacologia (Berl.) **20:** 22–31.
29. FLUHARTY, S. J., G. L. SNYDER, E. M. STRICKER & M. J. ZIGMOND. 1983. Short- and long-term changes in adrenal tyrosine hydroxylase activity during insulin-induced hypoglycemia and cold stress. Brain Res. **267:** 384–387.
30. KVETŇANSKÝ, R. & E. L. SABBAN. 1993. Stress-induced changes in tyrosine hydroxylase and other catecholamine biosynthetic enzymes. *In* Tyrosine Hydroxylase: From Discovery to Cloning, pp. 253–281. VSP Press. Utrecht.
31. MCMAHON, A., R. KVETŇANSKÝ, K. FUKUHARA, V. K. WEISE, I. J. KOPIN & E. L. SABBAN. 1992. Regulation of tyrosine hydroxylase and dopamine-β-hydroxylase mRNA levels in rat adrenals by a single and repeated immobilization stress. J. Neurochem. **58:** 2124–2130.
32. NANKOVÁ, B., R. KVETŇANSKÝ, A. MCMAHON, E. VISKUPIČ, B. HIREMAGALUR, G. FRANKLE, K. FUKUHARA, I. J. KOPIN & E. L. SABBAN. 1994. Induction of tyrosine hydroxylase gene expression by a non-neuronal non-pituitary mediated mechanism in immobilization stress. Proc. Natl. Acad. Sci. U.S.A. **91:** 5937–5941.
33. SABBAN, E. L., B. HIREMAGALUR, B. NANKOVÁ & R. KVETŇANSKÝ. 1995. Molecular biology of stress-elicited induction of catecholamine biosynthetic enzymes. Ann. N.Y. Acad. Sci. **771:** 327–338.
34. BARUCHIN, A. M., E. P. WEISBERG, L. L. MINER, D. ENNIS, L. K. NISENBAUM, E. NAYLOR, E. M. STRICKER, M. J. ZIGMOND & B. B. KAPLAN. 1990. Effects of cold exposure on rat

adrenal tyrosine hydroxylase: an analysis of RNA, protein, enzyme activity, and cofactor levels. J. Neurochem. **54:** 1769–1775.
35. TUMER, N., C. HALE, J. LAWLER & R. STRONG. 1992. Modulation of tyrosine hydroxylase gene expression in the rat adrenal gland by exercise: effects of age. Mol. Brain Res. **14:** 51.
36. VRÁNA, A. & L. KAZDOVÁ. 1990. The hereditary hypertriglyceridemic nonobese rats: an experimental model of human hypertriglyceridemia. Transplant. Proc. **22:** 2579.
37. KLIMEŠ, I., E. ŠEBÖKOVÁ, A. VRÁNA, P. ŠTOLBA, L. KAZDOVÁ, J. KUNEŠ, P. BOHOV, M. FICKOVÁ, J. ZICHA, D. RAUČINOVÁ & V. KŘEN. 1994. The hereditary hypertriglyceridemic rat, a new animal model of the insulin resistance syndrome. *In* Lessons from Animal Diabetes, pp. 271–283. Smith-Gordon. London.
38. ŠTOLBA, P., Z. DOBEŠOVÁ, P. HUŠEK, H. OPLTOVÁ, J. ZICHA, A. VRÁNA & J. KUNEŠ. 1992. The hypertriglyceridemic rat as a genetic model of hypertension and diabetes. Life Sci. **51:** 733–740.
39. EDELSTEINOVÁ, S., J. KYSELOVIČ, I. KLIMEŠ, E. ŠEBÖKOVÁ, B. KOVÁCSOVÁ, F. KRISTEK, A. MITKOVÁ, A. VRÁNA & P. ŠVEC. 1993. Effects of marine fish oil on blood pressure and vascular reactivity in the hereditary hypertriglyceridemic rat. Ann. N.Y. Acad. Sci. **683:** 353–356.
40. KVETŇANSKÝ, R. & L. MIKULAJ. 1970. Adrenal and urinary catecholamines in rat during adaptation to repeated immobilization stress. Endocrinology **87:** 738–743.
41. PEULER, J. D. & G. A. JOHNSON. 1977. Simultaneous single isotope radioenzymatic assay of plasma norepinephrine, epinephrine, and dopamine. Life Sci. **21:** 625–636.
42. ŠTOLBA, P., H. OPLTOVÁ, P. HUŠEK, J. NEDVÍDKOVÁ, J. KUNEŠ, Z. DOBEŠOVÁ, J. NEDVÍDEK & A. VRÁNA. 1993. Adrenergic overactivity and insulin resistance in nonobese hereditary hypertriglyceridemic rats. Ann. N.Y. Acad. Sci. **683:** 281–288.
43. KVETŇANSKÝ, R., C. SUN, C. R. LAKE, N. B. THOA, T. TORDA & I. J. KOPIN. 1978. Effect of handling and forced immobilization on rat plasma levels of epinephrine, norepinephrine, and dopamine-β-hydroxylase. Endocrinology **103:** 1868–1874.
44. KVETŇANSKÝ, R., K. PACÁK, E. L. SABBAN, I. J. KOPIN & D. S. GOLDSTEIN. 1997. Stressor-specificity of peripheral catecholaminergic activation. *In* Catecholamines: Bridging Basic Science with Clinical Medicine. Academic Press. New York. In press.
45. KVETŇANSKÝ, R., C. MCCARTY, N. B. THOA, C. R. LAKE & I. J. KOPIN. 1979. Sympatho-adrenal responses of spontaneously hypertensive rats to immobilization stress. Am. J. Physiol. **236:** H457–H462.
46. MCMURTY, J. P. & B. C. WEXLER. 1981. Hypersensitivity of spontaneously hypertensive rats (SHR) to heat, ether, and immobilization. Endocrinology **108:** 1730–1736.
47. BUCHANAN, T. A., J. H. YOUNG, V. M. CAMPESE & G. F. SIPOS. 1992. Enhanced glucose tolerance in spontaneously hypertensive rats: pancreatic beta cell hyperfunction with normal insulin sensitivity. Diabetes **41:** 872–878.
48. PACÁK, K., R. MCCARTY, M. PALKOVITS, G. CIZZA, I. J. KOPIN, D. S. GOLDSTEIN & G. P. CHROUSOS. 1995. Decreased central and peripheral catecholaminergic activation in obese Zucker rats. Endocrinology **136:** 4360–4367.
49. ZEMEL, M. B., J. D. PEUKER, J. R. SOWERS & L. SIMPSON. 1992. Hypertension in insulin-resistant Zucker rats is independent of sympathetic neural support. Am. J. Physiol. **262:** E368–E371.
50. MINER, L. L., A. BARUCHIN & B. B. KAPLAN. 1992. Trans-synaptic modulation of rat adrenal tyrosine hydroxylase gene expression during cold stress. *In* Stress: Neuroendocrine and Molecular Approaches, pp. 313–324. Gordon & Breach. New York.
51. CANNON, W. B., M. A. MCIVER & S. W. BLISS. 1924. Studies on the conditions of activity in endocrine glands. XIII. A sympathetic and adrenal mechanism for mobilizing sugar in hypoglycemia. Am. J. Physiol. **69:** 46–66.
52. YOSHIMATSU, H., Y. OOMURA, T. KATAFUCHI & A. NIIJIMA. 1987. Effects of hypothalamic stimulation and lesion on adrenal nerve activity. Am. J. Physiol. **253:** R418–R424.
53. SMYTHE, G. A., W. S. PASCOE & L. H. STORLIEN. 1989. Hypothalamic noradrenergic and sympathoadrenal control of glycemia after stress. Am. J. Physiol. **256:** E231–E235.
54. BOLLI, G. B. 1997. Importance of catecholamines in defense against insulin hypoglycemia in humans. *In* Catecholamines: Bridging Basic Science with Clinical Medicine. Academic Press. New York. In press.
55. KHALIL, Z., P. D. MARLEY & B. LIVETT. 1986. Elevation in plasma catecholamines in response to insulin stress is under both neuronal and nonneuronal control. Endocrinology **119:** 159–167.

56. KHALIL, Z., B. LIVETT & P. D. MARLEY. 1986. The role of sensory fibers in the rat splanchnic nerve in the regulation of adrenal medullary secretion during stress. J. Physiol. 370: 201-215.
57. BERNE, C., J. FAGIUS, T. POLLARE & P. HJEMDAHL. 1992. The sympathetic response to euglycaemic hyperinsulinemia: evidence from microelectrode nerve recording in healthy subjects. Diabetologia 35: 873-879.
58. BERGERON, J. J., R. RACHUBINSKI, N. SEARLE, D. BORTS, R. SIKSTORM & B. I. POSNER. 1980. Polypeptide hormone receptors *in vivo*: demonstration of insulin binding to adrenal gland and gastrointestinal epithelium by quantitative radioautography. J. Histochem. Cytochem. 28: 824-835.
59. PILLION, D. J., P. ARNOLD, M. YANG, C. R. STOCKARD & W. E. GRIZZLE. 1989. Receptors for insulin and insulin-like growth factor-I in the human adrenal gland. Biochem. Biophys. Res. Commun. 165: 204-211.
60. ZÓRAD, S., M. RUSNÁK, M. FICKOVÁ, R. KVETŇANSKÝ & L. MACHO. 1994. Insulin receptors in adrenal gland—characterization and effect of starvation and immobilization stress. Neuroendocrinology 60(S1): 99.
61. DELICADO, E. G. & M. M. PORTUGAL. 1987. Glucose transporters in isolated chromaffin cells. Biochem. J. 243: 541-547.
62. DAHMER, K. M. & R. L. PERLMAN. 1988. Bovine chromaffin cells have insulin-like growth factor-I (IGF-I) receptors: IGF-I enhances catecholamine secretion. J. Neurochem. 51: 321-323.
63. BREINER, M., M. WIELEND, W. BECKER, D. MULLER-WIELEND, R. STREICHER, M. FABRY & H. G. JOOST. 1993. Heterogeneity of insulin receptors in rat tissues as detected with the partial agonist B29, B29′-suberoyl-insulin. Mol. Pharmacol. 44: 271-276.
64. MACHO, L., S. ZÓRAD, P. BLAŽIČEK, M. FICKOVÁ & R. KVETŇANSKÝ. 1996. Direct effect of insulin and corticosterone on catecholamine release from adrenal medulla of rat. *In* Stress: Molecular Genetic and Neurobiological Advances. Vol. 1, pp. 333-342. Harwood. Amsterdam.
65. WILSON, S. P., O. H. VIVEROS & N. KIRSCHNER. 1985. Relationship between regulation of enkephalin-containing peptide and dopamine-β-hydroxylase levels in cultured adrenal chromaffin cells. J. Neurochem. 45: 1363-1370.

Is a Mutation of the β₃-Adrenergic Receptor Gene Related to Non-Insulin-dependent Diabetes Mellitus and Juvenile Hypertension in the Czech Population?[a]

B. BENDLOVÁ,[b,c] I. MAZURA,[b] J. VČELÁK,[b]
J. PERUŠIČOVÁ,[d] D. PALYZOVÁ,[e] I. KLIMEŠ,[f]
AND E. ŠEBÖKOVÁ[f]

[b]Institute of Endocrinology
Prague, Czech Republic

[d]General Teaching Hospital
First Medical Faculty
[e]Clinic for Children and Adolescents
Third Medical Faculty
Charles University
Prague, Czech Republic

[f]Institute of Experimental Endocrinology
Slovak Academy of Sciences
Bratislava, Slovak Republic

Glucose intolerance leading to non-insulin-dependent diabetes mellitus (NIDDM), hypertension, and dyslipidemia are the most-frequently observed metabolic and clinical manifestations of the metabolic syndrome described by Reaven,[1,2] which is evident in a substantial fraction (5–25%) of adults in Western societies. The central feature of this syndrome is most probably insulin resistance with ensuing hyperinsulinemia.[3,4] An association between obesity, particularly abdominal visceral obesity, and the insulin resistance syndrome has been noted by numerous investigators,[2,5,6] as well as the hypothesis that insulin-mediated sympathetic stimulation can contribute to the development of hypertension and metabolic disturbances in obese patients.[7]

The link between abdominal visceral obesity and several insulin resistance and hyperinsulinemia-induced pathological states could be associated with an abnormality in the β₃-adrenergic receptor (β₃-AR),[8,9] which is expressed especially in visceral fat in

---

[a] This study was supported by grants from the Czech Ministry of Health – Nos. IGA 3414-3, IGA 1065-4, and COST B5-20.

[c] Address for correspondence: Department of Endocrine Biochemistry, Institute of Endocrinology, Národní 8, 116 94 Prague 1, Czech Republic.

humans.[10] This adrenergic receptor is the main receptor involved in the catecholamine-induced regulation of thermogenesis in brown adipose tissue and lipolysis in white adipose tissue, and thus it is very important for the regulation of peripheral energy expenditure.[11] It is characterized by a lower affinity to catecholamines (activated by high concentrations of catecholamines as they are found in noradrenergic synapses, for example, after meals or during cold exposure) and by resistance to desensitization as compared to the β_1 and β_2-AR.[9] Impaired function of β_3-AR may predispose subjects to obesity and insulin resistance by decreasing energy expenditure.[12] This possibility is supported by findings that several β_3-adrenergic agonists have antiobesity effects[13,14] and improve the metabolic profile of obese subjects. The increased function of the β_3-AR is probably responsible for the elevated catecholamine-induced rate of lipolysis in obesity and for the subsequent release of excess free fatty acids into the liver through the portal system.[15,16]

In 1995, a missense mutation was identified in the gene for the β_3-adrenergic receptor that results in the replacement of tryptophan for arginine (Trp64Arg),[17] and its occurrence was subsequently associated with a tendency for obesity and insulin resistance,[18-20] a lower metabolic rate, and an earlier onset of NIDDM.[17,18]

The aim of our study was to search for the Trp64Arg mutation in Czech NIDDM patients, juvenile hypertensives, and juvenile controls and to establish its possible role in the pathogenesis of NIDDM and hypertension.

METHODS

Study Subjects

We studied the frequency of the Trp64Arg mutation in the Czech population—in patients with non-insulin-dependent diabetes mellitus (D, $n = 59$), among them 9 sibling pairs; in probands with juvenile hypertension (H, $n = 68$); and in juvenile controls (C, $n = 81$). The age of subjects, BMI (calculated as the weight in kilograms divided by the square of the height in meters), and blood pressure characteristics are given in TABLE 1.

NIDDM patients (20 men, 39 women) were volunteers recruited from Prague diabetes centers and diagnosed by criteria of the World Health Organization.[21]

All juvenile hypertensives were asymptomatic individuals and their high pressure was not the cause of their examination. The cardiogenic, nephrogenic, renovascular, endocrine, and neurogenic causes of hypertension were excluded. All blood pressure values were evaluated in compliance with the criteria for classification and blood pressure distribution defined in the Report of the Second Task Force on Blood Pressure Control.[22] Permanent hypertension was found in 16 patients; the others showed, during long-term blood pressure control, fluctuating values around the 90th and 95th percentiles for age, sex, and with regard to present height and weight. Also, 13% of hyper-

TABLE 1. Characteristics of the Study Subjects

	Group	Age (Years)	BMI (kg/m^2)	Blood Pressure[a]
D	NIDDM (n = 59)	52.6 ± 8.7	29.4 ± 4.7	S: 140 ± 23 D: 85 ± 9 [mmHg]
H	juvenile hypertensives (n = 68)	17.7 ± 2.4	25.0 ± 4.1	S and/or D > 95th percentile[22]
C	controls (n = 81)	18.4 ± 2.9	23.3 ± 4.7	S and/or D < 90th percentile[22]

[a] S: systolic; D: diastolic.

tensives used continuous antihypertension therapy. The average duration of hypertension was 28 months.

The study was approved by the Ethic Committee, and all subjects gave their informed consent to participate in the study.

Clinical and Metabolic Characterization

Physical measures of the participants were obtained in the fasting state, after voiding. Waist circumference at the umbilicus and maximum hip circumference were measured (W/H ratio). Sitting systolic and diastolic blood pressure were determined after 20 min of rest and were measured with a mercury sphygmomanometer.

After an overnight fast, venous blood samples were obtained for the determination of a number of laboratory parameters. Fasting blood glucose levels detected by the glucose oxidase method (Beckman Glucose Analyzer 2), glycated hemoglobin (Abbott IMx Glycated Hemoglobin Kit, Abbott Laboratories), and glycated proteins (spectrophotometric redox reaction using nitroblue tetrazolium as a sensitive redox indicator for the specific quantification of fructosamine in alkaline solution) were determined in the NIDDM patients. Immunoreactive insulin (IRI) was assayed in patients not on insulin therapy using a radioimmunoassay (RIA) kit (RIA-Sax-Insulin 100, Laboratorium Saxoniae GmbH, Germany) with guinea pig anti-insulin antiserum having 40% cross-reactivity to human proinsulin. Proinsulin and split-proinsulins were extracted using our monoclonal antibody to human C-peptide, C-PEP-01. The quantity of proinsulinlike compounds was expressed as the % of IRI. Serum and urine levels of C-peptide were evaluated by the RIA kit using guinea pig anti-C-peptide antibody with 5.3% cross-reactivity with human proinsulin in weight units (Immunotech, Czech Republic). Radioimmunoassays were used for the determination of growth hormone (HUMA-LAB, Slovak Republic), plasma renin activity (Angio I, Immunotech, Czech Republic), cortisol (our RIA method), SHBG (IRMA, Orion, Finland), and DHEA-sulfate (Immunotech, Czech Republic). Serum concentrations of total cholesterol

(Merckotest, CHOD-PAP-Method), high-density lipoprotein (HDL) cholesterol (Merck System Cholesterin, CHOD-PAP-Method), and triacylglycerols (Merck System, GPO-PAP-Method) were measured using an automatic analyzer (Merck, Vitalab Eclipse). The status of thyroid hormones was determined in NIDDM patients. TSH (Liana, Immunotech, Czech Republic), free T3, and free T4 (RIA, Immunotech, Czech Republic) were measured.

The levels of norepinephrine, epinephrine, dopamine, vanillylmandelic acid (VMA), homovanillic acid (HVA), and 5'-hydroxyindole acetic acid (HIAA) were determined by reversed-phase HPLC in urine (8-hour overnight collection). We measured the urine levels of N-acetyl-β-D-glucosaminidase (fluorometric method using the fluorogenic methylumbelliferyl N-acetyl-β-D-glucosaminide as a substrate), microalbuminuria (radial immunodiffusion on agarose gel using our own rabbit antiserum and human albumin as a standard), and β_2-microglobulin (RIA kit, Immunotech, Czech Republic).

The juvenile hypertensives and the age- and BMI-matched controls underwent an oral glucose tolerance test (oral administration of 75 g of glucose after an overnight fast). Venous blood samples were drawn at time 0 and at 30, 60, 90, and 120 min after glucose administration for determination of blood glucose and serum insulin and C-peptide concentrations. At the times 0 and 90 min, we also determined plasma renin activity, somatostatin (after acetone, petrol ether extraction by RIA using our own rabbit antiserum and standard; ^{125}I-somatostatin, Amersham), glucagon (RIA using antiglucagon antibody given by Pffeifer, Ulm, Germany; ^{125}I-glucagon, Amersham; standard, Novo Nordisk), and plasma norepinephrine and epinephrine levels. At time 0, glycated hemoglobin and proteins, triacylglycerols, total cholesterol and HDL cholesterol, C-peptide, N-acetyl-β-D-glucosaminidase, β_2-microglobulin, norepinephrine, epinephrine, dopamine, VMA, HVA, and HIAA in urine (8-hour overnight collection) were determined using the methods described above.

Detection of the Trp64Arg Polymorphism of the β_3-Adrenergic Receptor Using PCR and Mva I Restriction Endonuclease

DNA was isolated from peripheral blood leukocytes using common methods. Polymerase chain reaction (PCR) was carried out in a volume of 50 mL using the primers BSTNUP (5'CGCCCAATACCGCCAACAC) and BSTNDOWN (5'CCACCAGGAGTCCCATCACC). The reaction was performed under similar conditions as described by Widén et al.[18] The amplified PCR products (210 bp) were digested with Mva I restriction endonuclease (Boehringer Mannheim) overnight at 37°C. Mva I is an isoschizomer to BstO I and BstN I and is specific for the sequence CC(A/T)GG. The digested PCR products were analyzed by 4% agarose gel electrophoresis (80 V, 20.5 hours). The Trp64Arg mutation was visualized by the occurrence of the 161-bp fragment instead of the 99- and 62-bp fragments that are typical for the wild type of the β_3-adrenergic receptor.

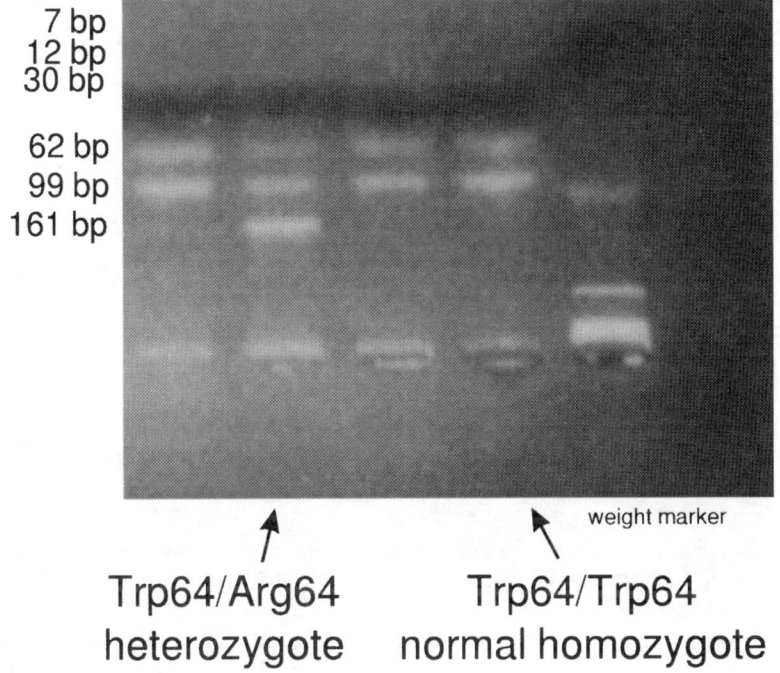

FIGURE 1. Detection of the Trp64Arg mutation of the β_3-adrenergic receptor using PCR and Mva I restriction endonuclease.[18]

Statistical Analysis

Values are given as the mean ± SD. Student's paired t test was used to evaluate the differences between the means, and differences were considered significant when $p < 0.05$.

RESULTS

Frequency of the Trp64Arg Polymorphism in Czech NIDDM Patients, Juvenile Hypertensives, and Controls

Detection of the Trp64Arg mutation using Mva I restriction endonuclease is shown in FIGURE 1.

None of our studied subjects was homozygous for the Trp64Arg allele. The Trp64Arg mutation was most-frequently detected in NIDDM patients—18.6% of them were heterozygotes. Among 9 sibling pairs, only in 1 pair were both siblings with the Trp64/Arg64 genotype; and in 1 pair, one sibling was with the mutation and the other without.

TABLE 2. Frequency of the Trp64Arg Mutation in the β_3-Adrenergic Receptor in Czech NIDDM Patients, Juvenile Hypertensives, and Controls

Group	Number	Trp64/Trp64 Homozygotes	Trp64/Arg64 Heterozygotes
D	59	48	11 (18.6%)
H	68	58	10 (14.7%)
C	81	75	6 (7.4%)

Fewer Trp64/Arg64 heterozygotes than among NIDDM patients were found in the group of juvenile hypertensives—14.7%; and this number was nearly double in comparison with juvenile controls—7.4%. The frequency of Trp64/Arg64 heterozygotes was similar in controls with or without a family risk of hypertension (TABLE 2).

Comparison of Phenotypic Features in Trp64/Trp64 and Trp64/Arg64 Genotypes in NIDDM Patients, Juvenile Hypertensives, and Controls

The occurrence of the Trp64Arg mutation in our NIDDM patients (TABLE 3) did not correlate with the age of onset of the disease. NIDDM patients with the Trp64Arg mutation had lower BMI ($p = 0.042$), lower serum triacylglycerol levels ($p = 0.045$), and lower TSH levels ($p = 0.03$); furthermore, men had significantly lower diastolic blood pressure ($p = 0.02$) and women had lower free T3 levels ($p = 0.002$) compared to those without the mutation. No other variables differed significantly as to genotype in our NIDDM patients.

The juvenile hypertensives with the Trp64Arg mutation showed significantly higher fasting serum blood glucose levels ($p = 0.002$), higher fasting and stimulated C-peptide levels ($p = 0.0004$, $p < 0.05$), higher serum insulin levels at 90 min ($p = 0.031$), and much lower levels of β_2-microglobulin in urine ($p = 0.004$) than the hypertensive homozygotes Trp64/Trp64, but they did not differ in their BMI nor in any other measured parameter. Trp64/Arg64 controls had higher fasting glucose levels ($p = 0.025$), but their fasting and stimulated C-peptide and insulin levels were on the other hand significantly lower ($p < 0.01$) than in Trp64/Trp64 controls (TABLE 4); in fact, this persisted even after elimination of 15 hyperinsulinemic Trp64/Trp64 controls.

When we compared all Trp64/Trp64 homozygotes (recruited from hypertensives as well as controls) versus all Trp64/Arg64 heterozygotes, significant differences between those genotypes occurred in fasting blood glucose levels ($p = 0.0002$) and urine β_2-microglobulin ($p = 0.0038$).

DISCUSSION

Our results suggest that the β_3-adrenergic receptor gene could be one of the candidate genes whose defect may play a pathogenic role in the development of impaired glucose tolerance and the ensuing hyperinsulinemia and insulin resistance.

There appear to be ethnic differences in the frequency of the Trp64Arg mutation.

TABLE 3. Some Clinical Characteristics of NIDDM Patients in Relation to the Trp64Arg Mutation of the β_3-Adrenergic Receptor

Characteristics	NIDDM ($n = 59$)		p
	Trp64/Trp64	Trp64/Arg64	
Number	48	11	
Age at diagnosis	48.5 ± 8.2	47.7 ± 10.7	ns[a]
Sex (M/F)	15/33	5/6	
BMI (kg/m^2) (total)	30.7 ± 5.8	28.2 ± 3.8	0.04
W/H ratio (total)	0.91 ± 0.8	0.89 ± 0.7	ns
Blood pressure (mmHg)			
systolic	139.9 ± 24.4	140.9 ± 22.3	ns
diastolic	87.4 ± 11.3	83.6 ± 6.06	0.064
Serum blood glucose−fasting (mmol/L)	9.1 ± 3.8	8.04 ± 2.88	0.15, ns
Serum insulin−fasting (mIU/L)	13.4 ± 9.0	11.9 ± 7.1	ns
Serum C-peptide−fasting (nmol/L)	0.68 ± 0.28	0.64 ± 0.21	ns
Glycosylated hemoglobin (%Hb)	9.98 ± 2.81	8.83 ± 2.17	0.07
Total cholesterol (mmol/L)	6.60 ± 1.68	6.44 ± 1.00	ns
HDL cholesterol (mmol/L)	1.17 ± 0.34	1.24 ± 0.37	ns
Triacylglycerols (mmol/L)	2.87 ± 2.91	2.01 ± 0.74	0.045
TSH (mIU/L)	2.34 ± 1.36	1.83 ± 0.57	0.03
free T3 (pmol/L)	3.71 ± 0.68	3.23 ± 1.01	0.07
free T4 (pmol/L)	17.02 ± 2.34	15.67 ± 3.85	ns

[a] ns: not significant.

The frequency of the Trp64Arg allele in Czech NIDDM patients (9.3%) was similar to the frequency in Finnish patients (11% and 8%),[18] but lower than that in Japanese NIDDM patients (21%)[20] and in Pima Indians with NIDDM (31%).[17] There was a higher frequency of the Trp64Arg allele in juvenile hypertensives (7.4%) compared to controls (3.7%). The frequency of this mutant allele in our juvenile controls is the lowest in comparison with other control groups (American white controls−8%; Finnish nondiabetics−12% and 14%; Japanese controls−19%; French controls−10%).[17-20]

We did not find (similar to Kadowaki et al.[20]) the correlation with the time of onset of NIDDM that was observed in Pima Indians and Finns.[17,18] Our subjects with Trp64Arg did not differ in their waist/hip ratio and, in fact, we observed lower BMI in those with the Trp64Arg mutation. We found some relations between the occurrence of the mutation and lower TSH and free T3. It is well known that brown adipose tissue is the site of a complex interaction between the sympathetic nervous system and the thyroid hormone. The responses of brown adipose tissue to cold or exogenous norepinephrine as well as the lipolytic response in white adipose tissue are blunted in the absence of thyroid hormone; on the other hand, thyroid hormone stimulates the expression of β-adrenergic receptors including β_3-AR in brown and white adipose tissue.[23] However, an increase in T3 was observed in the transgenic mice with decreased brown adipose tissue and fed with Western diets.[24] The connection of the mutated β_3-AR with the decreased production of the thyroid hormone is unclear.

TABLE 4. Some Clinical Characteristics of Juvenile Hypertensives and Controls according to the Presence of the Trp64Arg Mutation of the β_3-Adrenergic Receptor

Characteristics	Juvenile Hypertensives			All Controls		
	Trp64/Trp64	Trp64/Arg64	p	Trp64/Trp64	Trp64/Arg64	p
Number	58	10		75	6	
Sex (M/F)	54/4	8/2		50/25	6/0	
BMI (kg/m^2)	24.6 ± 3.8	25.8 ± 3.6	ns[a]	23.5 ± 4.4	21.8 ± 3.0	ns
Serum blood glucose (mmol/L)						
fasting	4.67 ± 0.55	5.15 ± 0.45	0.002	4.42 ± 0.72	4.64 ± 0.18	0.025
60 min	7.34 ± 1.78	7.80 ± 1.77	ns	6.81 ± 1.39	6.40 ± 2.12	ns
120 min	6.05 ± 1.45	6.05 ± 1.28	ns	5.86 ± 1.05	5.13 ± 0.54	0.0022
Serum insulin (mIU/L)						
fasting	11.6 ± 13.0	16.0 ± 10.6	ns	11.9 ± 9.34	10.7 ± 9.5	ns
60 min	56.9 ± 37.6$^{90'}$	94.5 ± 60.5$^{90'}$	0.031	69.4 ± 43.8	39.7 ± 28.8	0.011
120 min	54.1 ± 39.8	75.8 ± 52.6	ns	52.6 ± 34.9	22.0 ± 16.4	0.0002
Serum C-peptide (nmol/L)						
fasting	0.48 ± 0.13	0.64 ± 0.13	0.0004	0.48 ± 0.21	0.32 ± 0.04	0.0003
60 min	1.97 ± 0.77	2.28 ± 0.42	0.035	1.94 ± 0.71	1.35 ± 0.44	0.0016
120 min	1.76 ± 0.60$^{90'}$	2.35 ± 0.68$^{90'}$	0.007	1.79 ± 0.65	1.15 ± 0.20	0.0003
Total cholesterol (mmol/L)	4.33 ± 0.73	4.11 ± 0.98	ns	4.23 ± 0.76$^{(n=11)}$	—	—
HDL cholesterol (mmol/L)	1.31 ± 0.34	1.17 ± 0.2	0.053	1.29 ± 0.28$^{(n=6)}$	—	—
Triacylglycerols (mmol/L)	1.17 ± 0.61	1.27 ± 0.54	ns	1.25 ± 0.38$^{(n=10)}$	—	—
β_2-Microglobulin (μg/L)	0.108 ± 0.104	0.05 ± 0.037	0.004	0.187 ± 0.109	0.15 ± 0.08	ns

[a] ns: not significant.

There was an unexpected trend toward lower triacylglycerols and diastolic blood pressure in men in our NIDDM patients with the mutation compared to those without the mutation. No other differences were observed in our NIDDM patients, but they could be masked by the concomitant hyperglycemia and its treatment as proposed by Widén et al.[18] and Kadowaki et al.[20]

The association of the Trp64Arg mutation of the β_3-AR with several characteristics of the insulin resistance syndrome was apparent in our juvenile hypertensives. The juvenile hypertensives with the Trp64Arg mutation had higher fasting blood glucose levels and higher fasting and stimulated C-peptide and insulin levels compared to those without the mutation. Surprisingly, the juvenile controls with the mutant allele had significantly higher fasting blood glucose levels, but insulin and C-peptide levels were significantly lower than in those without the mutation.

These interesting and in part controversial results obtained after the evaluation of the Trp64Arg mutation in groups of Czech juvenile hypertensives and normotensives in correlation to their phenotypical features await further study.

REFERENCES

1. REAVEN, G. M. 1988. Diabetes **37**: 1595–1607.
2. DEFRONZO, R. A. & E. FERRANNINI. 1991. Diabetes Care **14**: 173–194.
3. FERRANNINI, E., G. BUZZIGOLI, R. BONADONNA, M. A. GIORICO, M. OLEGGINI, L. GRAZIDEI, R. PADRINELLI, L. S. BRANDI & S. BEVILACQUA. 1987. N. Engl. J. Med. **317**: 350–357.
4. FERRANNINI, E., S. M. HAFFNER, B. D. MITCHELL & M. P. STERN. 1991. Diabetologia **34**: 416–422.
5. CARO, J. F. 1991. J. Clin. Endocrinol. Metab. **73**: 691–695.
6. POULIOT, M-C., J. P. DEPRES, A. NADAU et al. 1992. Diabetes **41**: 826–834.
7. LANDSBERG, L. 1992. Hypertension **19**(suppl. 1): 161–166.
8. EMORINE, L. J., S. MARULLO, M-M. BRIEND-SUTREN et al. 1989. Science **245**: 1118–1121.
9. GIACOBINO, J-P. 1995. Eur. J. Endocrinol. **132**: 377–385.
10. KRIEF, S., F. LÖNNQVIST, S. RAIMBAULT et al. 1993. J. Clin. Invest. **91**: 344–349.
11. EMORINE, L., N. BLIN & A. D. STROSBERG. 1994. Trends Pharmacol. Sci. **15**: 3–7.
12. LOWELL, B. B. & J. S. FLIER. 1995. J. Clin. Invest. **95**: 923.
13. ARCH, J. R. S., A. T. AINSWORTH, M. A. CAWTHORNE, V. PIERCY, M. V. SENNITT & V. E. THODY. 1984. Nature **309**: 163–165.
14. HOWE, R. 1993. Drugs Future **18**: 529–549.
15. LÖNNQVIST, F., A. THÖRNE, K. NILLSEN, J. HOFFSTEDT & P. ARNER. 1995. J. Clin. Invest. **95**: 1109–1116.
16. ARNER, P. 1995. Ann. Med. **27**: 435–438.
17. WALSTON, J., K. SILVER, C. BOGARDUS, W. C. KNOWLER, F. S. CELI, S. AUSTIN, B. MANNING, A. D. STROSBERG, M. P. STERN, N. RABEN, J. D. SORKIN, J. ROTH & A. R. SHULDINER. 1995. N. Engl. J. Med. **333**(6): 343–347.
18. WIDÉN, E., M. LEHTO, T. KANNINEN, J. WALSTON, A. R. SHULDINER & L. C. GROOP. 1995. N. Engl. J. Med. **333**(6): 348–351.
19. CLÉMENT, K., CH. VAISSE, B. MANNING, A. BASDEVANT, B. G. GRAND, J. RUIZ, K. D. SILVER, A. R. SHULDINER, P. FROGUEL & A. D. STROSBERG. 1995. N. Engl. J. Med. **333**(6): 352–354.
20. KADOWAKI, H., K. YASUDA, K. IWAMOTO, S. OTABE, K. SHIMOKAWA, K. SILVER, J. WALSTON, H. YOSHINAGA, K. KOSAKA, N. YAMADA, Y. SAITO, R. HAGURA, Y. AKANUMA, A. SHULDINER, Y. YAZAKI & T. KADOWAKI. 1995. Biochem. Biophys. Res. Commun. **215**(2): 555–560.
21. WHO. 1985. Diabetes mellitus: report of a WHO study group. WHO Tech. Rep. Ser. **727**: 7–113.
22. SECOND TASK FORCE. 1987. Report of the second task force on blood pressure control in children. Pediatrics **79**(1): 1–25.
23. RUBIO, A., A. RAASMAJA & J. E. SILVA. 1995. Endocrinology **136**(8): 3277–3284.
24. HAMANN, A., J. S. FLIER & B. B. LOWELL. 1996. Endocrinology **137**(1): 21–29.

Glucose Transport and Insulin Signaling in Rat Muscle and Adipose Tissue

Effect of Lipid Availability[a]

D. GAŠPERÍKOVÁ,[b] I. KLIMEŠ,[b] T. KOLTER,[c]
P. BOHOV,[b] A. MAASSEN,[d] J. ECKEL,[c]
M. T. CLANDININ,[e] AND E. ŠEBÖKOVÁ[b,f]

[b]*Diabetes and Nutrition Research Group*
Institute of Experimental Endocrinology
Slovak Academy of Sciences
833 06 Bratislava, Slovak Republic

[c]*Diabetes Research Institute*
Düsseldorf, Germany

[d]*University of Leiden*
Leiden, the Netherlands

[e]*Nutrition and Metabolism Research Group*
University of Alberta
Edmonton, Alberta, Canada

INTRODUCTION

Studies *in vivo* and *in vitro* have accumulated substantial evidence showing that hypertriglyceridemias of exogenous origin (diets high in fat and/or simple sugar) and endogenous origin (inborn) are frequently linked to the impairment of insulin action in various peripheral tissues.[1,2] Several studies have indicated that raised plasma triglyceride levels in insulin-resistant states are accompanied by accumulation of triglycerides in skeletal muscle.[3-5]

Skeletal muscle is the main site of insulin-induced glucose disposal.[6] This process has been shown to be related to the fatty acid composition of skeletal muscle membrane structural lipids (i.e., phospholipids).[7,8] Increasing the percentage of long-chain polyunsaturated fatty acids (PUFA) of the n-3 series in muscle membrane phospholipids improves insulin action.[7,9,10] Saturated fatty acids have an opposite effect.[11]

---

[a] This work was supported by research grants from the Slovak Grant Agency for Sciences (GAV) (No. 2/543/95), Deutsche Forschungsgemeinschaft (No. SFB 351 C2), the Netherlands Diabetes Foundation, the Natural Sciences and Engineering Research Council of Canada, and COST B5 and the Slovak Diabetes Association (No. 1/95).

[f] To whom all correspondence should be addressed.

Glucose transport is a rate-limiting step for glucose utilization in muscle.[12] Thus, its impairment is one of the possible candidates for a defect in insulin action. Indeed, lately we have described a 50% reduction in the GLUT4 protein levels in the quadriceps femoris muscle in a nonobese rat model of the insulin resistance syndrome, that is, in the hHTg rat.[13] The hHTg strain was selected about 10 years ago by A. Vrána[14] in Prague, Czech Republic, from an outbred Wistar population. In this model, the endogenous hyperproduction of triglycerides and moderately increased blood pressure are accompanied by insulin resistance both *in vitro* and *in vivo*.[15,16]

Insulin-induced glucose transport is mediated by a process involving translocation of the GLUT4 glucose transporter from the cytoplasm to the plasma membrane.[17] Insulin receptor substrate (IRS-1) and phosphatidylinositol 3-kinase (PI 3-kinase), signaling molecules of the intracellular insulin action cascade, have been proposed to play a major role in activation of the glucose transport machinery.[18,19] However, mainly *in vitro* systems have been used to study the effect of insulin on activation of the IRS-1 and PI 3-kinase chain.[20,21] These studies have shown that insulin activates PI 3-kinase maximally for about 10 minutes. If this enzyme plays a role in the regulation of glucose transport under physiological conditions, it can be speculated that the PI 3-kinase should be activated by insulin on a long-term basis, which could be achieved by continuous insulin infusion at euglycemia.

There were three major aims to these studies: first, to use dietary regimens, known to increase or to reduce availability of circulating plasma levels and/or tissue stores of triglycerides, to verify the role of fatty acid composition of muscle membrane phospholipids for the *in vivo* insulin action in hHTg rats; second, to study the regulation of expression of the gene for GLUT4, the major glucose transporter isoform in muscle, in the hHTg rat line; third, to increase the understanding of the regulation of phosphatidylinositol 3-kinase activity in normal rats and in our animal model of the insulin resistance syndrome.

MATERIALS AND METHODS

Animals

Adult male hHTg rats (250–300 g), taken from the colony at our institute, were housed in a temperature and light–controlled room (12-h light:dark cycle; lights off at 18:00 h) with free access to food and water. Control normotriglyceridemic (NTg) Wistar rats, the progenitor strain of the hHTg rats, were purchased from VELAZ, Prague, Czech Republic. The animals were fed experimental diets for 2 weeks and randomly divided into 4 dietary groups (8 rats per group). The diets contained the same amount of fat (10 wt%). The only difference was that beef tallow (Palma, Bratislava, Slovak Republic) in basal diet (B) or in high sucrose diet (HS) was replaced with fish oil (Activepa 30, Tg, Martens, Norway) in the fish oil (FO) or high sucrose + fish oil (HS + FO) diet. Details on diet composition were published previously.[22]

Euglycemic Hyperinsulinemic Clamp

On the 11th day of feeding, rats were anesthetized by injection of xylazine hydrochloride (10 mg/kg) plus ketamine hydrochloride (75 mg/kg), and fitted with chronic carotid artery and jugular vein cannulae. Three days later after overnight fasting (16–18 h), conscious and freely moving animals in metabolic cages were subjected to a 90-min euglycemic hyperinsulinemic clamp (EHC, 6.4 mU/kg/min).[23] After completion of the clamp, rats were sacrificed and soleus, heart, and white epididymal adipose tissue were immediately removed and frozen until assayed. Control Wistar rats underwent the same protocol.

Biochemical and Lipid Analyses

Plasma obtained from blood sampled before, during, and after the clamp of overnight fasted rats was used for assays of insulin (RIA kit from Novo Nordisk, Copenhagen, Denmark). The plasma glucose was measured using a Beckman glucose analyzer (Fullerton, California). Plasma triglycerides were examined in the fed state one day before application of the clamp using a commercial kit (Lachema, Brno, Czech Republic). The composition of fatty acids in tissues was analyzed by gas chromatography. For details, see Bohov et al. in this volume.[24]

Quantitation of GLUT4 mRNA in hHTg Rats

Total RNA from tissue samples was extracted by TRIZOL reagent (GIBCO BRL, Life Technologies, Maryland). RNA (10 µg) was analyzed by Northern blotting after electrophoresis on 1% agarose/0.66 M formaldehyde gels and transferred onto MSI Nitropure membrane (Micron Separation Incorporated, Westborough, Massachusetts). The cDNA probe for GLUT4 was a 2.95-kb plasmid p-bluescript/Eco RI fragment corresponding to a region of the gene encoding the adipose/muscle glucose transporter obtained from D. E. James[25] (Washington University, St. Louis, Missouri). Probes were labeled with ^{32}P using the Random Primer DNA Labeling System (GIBCO BRL, Life Technologies). Following the prehybridization for 30 min at 68 °C, the filters were hybridized with GLUT4-labeled probes in QuikHyb solution (Stratagene, La Jolla, California) and denaturated salmon sperm DNA (0.15 mg/mL) at 68 °C for 1 hour. Then, the membranes were washed two times for 15 min at room temperature in 2× SSC and 0.1% (w/v) SDS and once for 20–30 min at 68 °C in 0.1× SSC and 0.1% (w/v) SDS. Thereafter, the filters were exposed to Kodak X-OMAT AR film for 20–40 hours at −80 °C with intensifying screens. The intensity of the mRNA bands on the blots was quantified by scanning densitometry using the Ultroscan XL (LKB, Bromma, Sweden).

Analysis of GLUT4 Protein Level

Total Membrane Preparation

For the preparation of total membrane (TM) from soleus, heart, and white adipose tissue, a modification of the method described by Kahn et al.[26] was used. Muscle homogenate in buffer A (250 mM sucrose, 10 mM Hepes, 10 mM EDTA, 0.1 mM phenylmethylsulfonyl fluoride, 2 µg/mL leupeptin, pH 7.4) was centrifuged at $1200g_{max}$ for 10 min and $9000g_{max}$ for 10 min, followed by $200,000g_{max}$ centrifugation for 80 min. The final pellet (TM) was resuspended in water. Protein of the total membrane preparation was determined according to Bradford's method[27] using BSA (Sigma, St. Louis, Missouri) as standard.

Electrophoresis and Western Blotting

The proteins (10 µg/lane) from the total membrane fraction in Laemmli[28] sample buffer were resolved by 12% sodium dodecyl sulfate–polyacrylamide gel electrophoresis (SDS-PAGE) under denaturating conditions (200 V for 45 min) in Tris/glycine/SDS buffer (25 mM/192 mM/0.5%; pH 8.3) using the Bio-Rad Mini Protean II system. After electrophoresis, proteins were transferred to HYBOND-C extra membrane (Amersham, 0.45-µm pore size) in the blotting buffer (25 mM Tris, 192 mM glycine, 20% v/v methanol, pH 8.3; at 100 V for 1 hour) using the Bio-Rad Mini Trans-Blot system.

The Western blotting protocol was performed exactly according to the ECL kit, RPN 2108 (Amersham, London, United Kingdom). The polyclonal GLUT4 antibody (used in dilution 1:500) was obtained from East Acres Biologicals (catalogue no. RaIRGT, Southbridge, Massachusetts). After ECL detection, the blots were exposed to X-ray sensitive film (Fuji, Tokyo, Japan) and analyzed by scanning densitometry using a MACINTOSH II computer-based image analysis system with IMAGE software (Wayna Rasband, NIMH, Bethesda, Maryland).

Immunoprecipitation and Measurement of PI 3-Kinase Activity

Polyclonal antisera against the 85-kDa subunit of PI 3-kinase and IRS-1 were generated by immunizing rabbits with particular recombinant proteins, as described in detail by Ouwens et al.[29] Tissue homogenate (10 wt%) in lysis buffer [30 mM Tris (pH 7.5), 150 mM NaCl, 2 mM EDTA, 5 mM $MgCl_2$, 10% glycerol, 0.5 mM sodium orthovanadate, 5 mM sodium fluoride, 0.1 mM PMSF, 1 µg/mL aprotinin, and 1 µg/mL leupeptin] containing 0.5% NP40 was cleared by centrifugation for 10 min at 14,000g at 4 °C and then incubated for 4 h at 4 °C with either the polyclonal anti-P85 antiserum (5 µL, to examine the total PI 3-kinase activity) or the IRS-1 antiserum (5 µL, to measure the IRS-1-associated PI 3-kinase activity) coupled to protein-A Sepharose beads. Immunoprecipitates were washed twice with lysis buffer containing 0.1%

TABLE 1. Effect of Diet on the Body Weight and Plasma (Fed State) Triglycerides (Tg) in Hereditary Hypertriglyceridemic Rats (hHTg) and Control Animals (NTg)

		B	FO	HS	HS + FO
NTg	body weight (g)	394 ± 9.7^a	375 ± 7.6^a	382 ± 11.3^a	382 ± 11.3^a
	plasma Tg (mmol/L)	2.56 ± 0.3^a	0.58 ± 0.2^b	3.63 ± 0.3^c	0.89 ± 0.2^b
hHTg	body weight (g)	283 ± 9.4^b	275 ± 9.8^b	277 ± 9.8^b	279 ± 15.7^b
	plasma Tg (mmol/L)	3.95 ± 0.6^c	0.78 ± 0.2^b	10.7 ± 0.3^d	1.64 ± 0.4^b

Note: Data are expressed as mean ± SEM, $n = 8$. Values without a common superscript (a–d) within a row are significantly different ($p < 0.05$). B = basal diet; B + FO = basal diet supplemented with fish oil; HS = high sucrose diet; HS + FO = high sucrose diet supplemented with fish oil.

NP40 and twice with 30 M Hepes (pH 7.5). After that, PI 3-kinase activity was determined by incorporation of ^{32}P-labeled gamma-ATP into phosphatidylinositol exactly as described by Burgering et al.[30] Incorporated ^{32}P was quantitated using a Phosphor Bio Imager System (Molecular Dynamics, United States). Intensity of the spots represents the relative values for the PI 3-kinase activity. It is expressed in percentage; the PI 3-kinase activity of control (NTg) rats before the clamp commencement is considered as 100%.

Statistical Evaluation

All data are presented as mean ± SEM. The effects of dietary treatment, strain, and euglycemic hyperinsulinemic clamp were evaluated by an analysis of variance (ANOVA) procedure, followed by an appropriate post hoc test. Treatments without a common superscript are significantly different.

RESULTS

Body Weight and Plasma Triglycerides

hHTg rats weighed approximately 20% less in comparison with age-matched control animals. Body weight of both animal strains did not differ significantly among the groups of rats (TABLE 1). hHTg rats had significantly elevated plasma levels of triglycerides when compared to control animals (TABLE 1). Feeding a high sucrose diet markedly increased plasma triglyceride levels in both control and hHTg rats. The addition of fish oil to basal or high sucrose diet prevented development of hypertriglyceridemia in hHTg rats and decreased plasma triglyceride concentration below initial values (TABLE 1). Similar results were also obtained in control rats.

Euglycemic Hyperinsulinemic Clamp

Plasma glucose levels before and during the clamp did not differ significantly between the dietary groups in hHTg rats (TABLE 2). Insulin values during the clamp

TABLE 2. Plasma Glucose and Insulin Levels before ("Basal") and after Euglycemic Hyperinsulinemic Clamp ("Clamp"), and Glucose Infusion Rate (GIR) in Hereditary Hypertriglyceridemic (hHTg) Rats

	B	FO	HS	HS + FO
Basal				
plasma insulin (μU/mL)	40.7 ± 4.8^a	28.6 ± 2.4^a	43.9 ± 12.7^a	28.0 ± 9.3^a
plasma glucose (mmol/L)	6.8 ± 0.6^a	6.3 ± 0.4^a	6.7 ± 0.3^a	6.0 ± 0.2^a
Clamp				
plasma insulin (μU/mL)	319.0 ± 35^b	317.8 ± 27.2^b	308.9 ± 3.9^b	366.6 ± 23.3^b
plasma glucose (mmol/L)	5.9 ± 0.2^a	5.7 ± 0.2^a	5.4 ± 0.5^a	5.2 ± 0.1^a
GIR (mg/kg/min)	25.3 ± 1.1^a	28.5 ± 0.1^b	21.6 ± 0.7^c	30.5 ± 0.4^d

Note: Data are expressed as mean ± SEM, $n = 8$. Values without a common superscript (a–d) within a row are significantly different ($p < 0.05$). B = basal diet; B + FO = basal diet supplemented with fish oil; HS = high sucrose diet; HS + FO = high sucrose diet supplemented with fish oil.

were approximately 10-fold higher than basal fasting insulin levels (TABLE 2). The glucose infusion rate (= GIR; an index of insulin sensitivity) was significantly decreased in high sucrose–fed rats. In contrast, diets supplemented with n-3 PUFA increased the *in vivo* insulin action (glucose infusion rate) in hHTg rats fed FO or HS + FO (TABLE 2).

Influence of Triglyceride Availability and Skeletal Muscle Phospholipid Fatty Acid Composition on Insulin Action in hHTg Rats

The glucose infusion rate correlated negatively with triglyceride levels of hHTg rats fed various diets. This relationship was more pronounced for the plasma triglyceride levels ($n = 16$, $r = -0.83$, $p < 0.001$) (FIG. 1a) than for the muscle triglyceride content ($n = 16$, $r = -0.57$, $p < 0.05$; data not shown).

Correlation analysis of fatty acid composition of skeletal muscle phospholipids with insulin action *in vivo* in hHTg rats showed a tight positive relationship between the percentage of long-chain PUFA and the glucose infusion rate (GIR) ($n = 16$, $r = 0.93$, $p < 0.001$) in hHTg rats (FIG. 1b) and a negative correlation between the ratio of n-6/n-3 PUFA and GIR ($n = 16$, $r = -0.89$, $p < 0.001$) (FIG. 1c). For further details on fatty acid composition of skeletal muscle phospholipids, see Bohov *et al.* in this volume.[24]

Glucose Transport in hHTg Rats: Relation to GLUT4 Protein and Gene Expression

GLUT4 protein levels were found to be reduced in soleus muscle (hHTg: 76 ± 14 versus NTg: 137 ± 5.1 arbitrary units [AU]; $p < 0.05$) and white adipose tissue

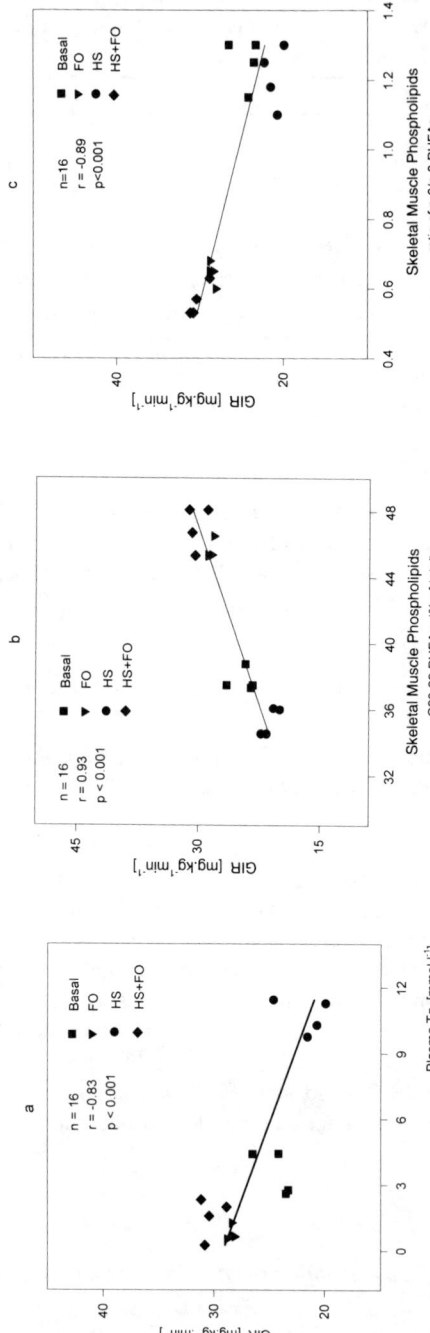

FIGURE 1. Correlation analysis between glucose infusion rate (GIR) and (a) plasma triglyceride level, (b) percentage of C20-22 polyunsaturated fatty acids, and (c) ratio of n-6/n-3 PUFA in skeletal muscle phospholipids of the hereditary hypertriglyceridemic rat. B = basal diet; B + FO = basal diet supplemented with fish oil; HS = high sucrose diet; HS + FO = high sucrose diet supplemented with fish oil; C20-22 PUFAs = sum of the percentages of the following polyunsaturated fatty acids—20:3 n-6 + 20:4 n-6 + 20:5 n-3 + 22:4 n-6 + 22:5 n-3 + 22:5 n-6 + 22:6 n-3.

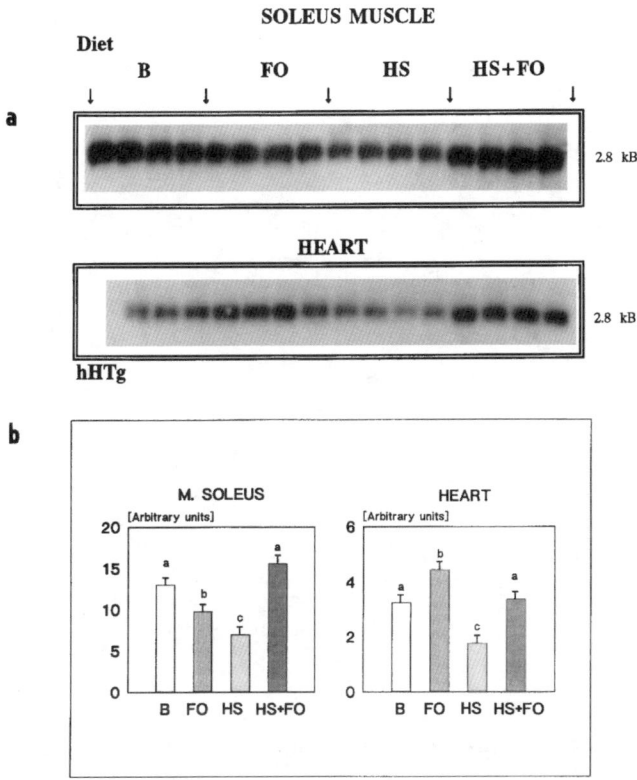

FIGURE 2. Autoradiogram of Northern blot (a) and mRNA levels for GLUT4 (b) in soleus and heart of hereditary hypertriglyceridemic (hHTg) rats fed various types of diet. Values without a common superscript (*a, b, c*) are significantly different ($p < 0.05$). B = basal diet; B + FO = basal diet supplemented with fish oil; HS = high sucrose diet; HS + FO = high sucrose diet supplemented with fish oil.

(hHTg: 26 ± 5.3 versus NTg: 103 ± 9.9 AU; $p < 0.05$) of hHTg rats when compared to data of the control animals. However, there were no changes of GLUT4 protein level in the heart of hHTg rats (hHTg: 151 ± 8.0 versus NTg: 135 ± 2.0 AU; not significant).

The high sucrose feeding reduced GLUT4 mRNA measured after the euglycemic hyperinsulinemic clamp, in soleus muscle and heart of hHTg rats. Conversely, feeding the HS diet enriched with n-3 PUFA restored the GLUT4 mRNA level in both tissues (FIG. 2). Quantitation of GLUT4 mRNA was completed only for hHTg rats.

Interestingly, GLUT4 gene expression correlated negatively with plasma triglycerides in soleus muscle ($n = 16$, $r = -0.61$, $p < 0.05$) (FIG. 3a) of hHTg rats. Finally, the GLUT4 gene expression correlated positively with the long-chain fatty acid (C20-22 PUFA) content in the skeletal muscles ($n = 16$, $r = 0.55$, $p < 0.05$) (FIG. 3b). Comparable results were found also in the heart of hHTg rats (data not shown).

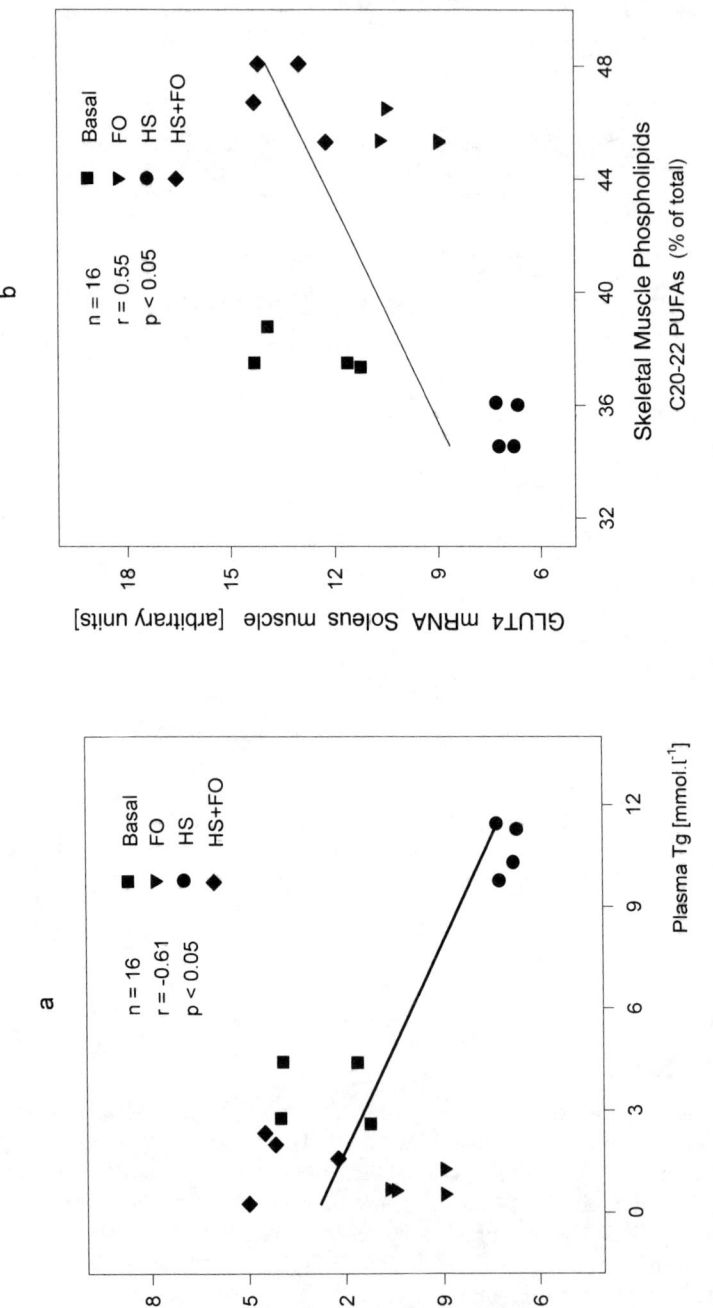

FIGURE 3. Relation between GLUT4 gene expression in skeletal muscle and (a) plasma triglyceride level and (b) skeletal muscle phospholipid C20-22 PUFAs in the hereditary hypertriglyceridemic rat. B = basal diet; B + FO = basal diet supplemented with fish oil; HS = high sucrose diet; HS + FO = high sucrose diet supplemented with fish oil; C20-22 PUFAs = sum of the percentages of the following polyunsaturated fatty acids—20:3 n-6 + 20:4 n-6 + 20:5 n-3 + 22:4 n-6 + 22:5 n-3 + 22:5 n-6 + 22:6 n-3.

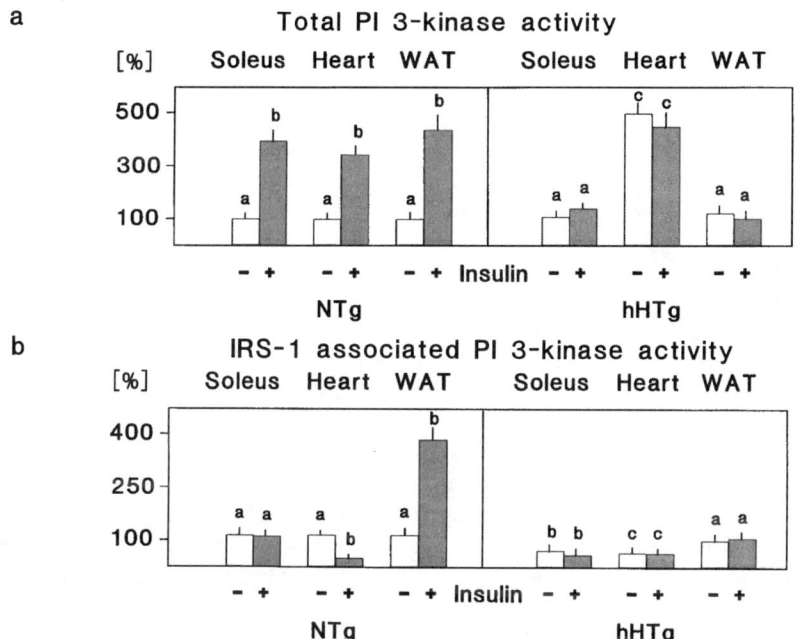

FIGURE 4. Total (a) and IRS-1-associated (b) activity of PI 3-kinase in the soleus and white adipose tissue of control and hHTg rats after 90 min of hyperinsulinemia at euglycemia (expressed as "+ Insulin"). Values are expressed as percentage, where the PI 3-kinase activity of control rats (NTg) before the clamp commencement is considered as 100%. Values without a common superscript (a, b, c) are significantly different ($p < 0.05$). Terms: NTg = control normotriglyceridemic rats; hHTg = hereditary hypertriglyceridemic rats; WAT = white adipose tissue.

Effect of Euglycemic Hyperinsulinemia on Total and IRS-1-associated PI 3-Kinase Activity

Ninety min of euglycemic hyperinsulinemia stimulated the total PI 3-kinase activity in both investigated tissues (i.e., in the soleus muscle and the white adipose tissue) of the control rats. This was not the case in the insulin-resistant hHTg rats, where no change in the total PI 3-kinase activity was seen in response to hyperinsulinemia (FIG. 4a).

A fraction of the total PI 3-kinase that is activated by insulin, the IRS-1-associated PI 3-kinase activity, was increased by euglycemic hyperinsulinemia only in the white adipose tissue of the control rats (FIG. 4b). In contrast, hyperinsulinemia failed to activate the IRS-1-associated PI 3-kinase activity in skeletal muscle of control rats. No effect of hyperinsulinemia was observed in any of the tissues obtained from insulin-resistant hHTg animals.

DISCUSSION

The present study verified that increased circulating triglyceride availability (regardless of whether of endogenous or exogenous origin) is associated with lower *in vivo* insulin action in our nonobese rat model of the insulin resistance syndrome, the hHTg rat. Simultaneously, insulin action was shown to increase with an increasing proportion of the long-chain PUFAs in muscle phospholipids induced by feeding the animals diets supplemented with marine fish oil rich in n-3 PUFAs.

Studies both *in vitro* and *in vivo* have shown that feeding high sucrose (fructose) or high fat diets to normal rats is followed by raised plasma triglyceride and/or triglyceride content in skeletal muscle and heart. A causal relation to the impairment of insulin action seen under these circumstances has been proposed.[2,3,5] In contrast, insulin action improves in situations where increased triglyceride availability is reduced by raised dietary intake of n-3 PUFA.[2,5] Feeding the insulin-resistant hHTg rats the high sucrose diet further impaired whole body insulin action (clamp GIR). Supplementation of the high sucrose diet with marine fish oil brought back the glucose infusion rate to values usually seen in normal rats. The most-insulin-resistant hHTg rats had the highest plasma triglyceride levels. This is consistent with the basic concept of the glucose–fatty acid cycle of Randle *et al.*[31,32] and has parallels and supportive evidence in the literature.[5,33,34]

The next variable highlighted by our study is the positive relation of the glucose infusion rate during the hyperinsulinemic clamp to the percentage of long-chain (20 and 22 carbon) n-3 PUFAs in the total skeletal muscle phospholipids of the hHTg rat. These data are again in harmony with earlier observations obtained both in rat[7] and in humans.[35,36] It is well known that the fatty acid composition of muscle membrane phospholipids can influence a range of cellular functions, including insulin action, possibly via the modulation of local protein-lipid interactions and/or by yielding potent intracellular second messengers.[37-39] Alternately, it has been proposed that increased unsaturation of membrane lipids increases the permeability of membranes to ion fluxes, which in turn raises energy-dependent ion pumping into the cell in order to maintain proper gradients.[40] Differences in metabolic rate could account then for the different amounts of triglycerides stored in the muscle and, thus, for the differences in insulin action via the glucose–fatty acid cycle.[5]

In searching for potential mechanisms responsible for the impaired insulin action of the hHTg rats, we confirmed our earlier observation of a 50% reduction of the GLUT4 protein content in skeletal muscle of the hHTg rat.[13] The lower GLUT4 protein levels in skeletal muscles have been seen in a number of our animal models;[13,41,42] however, it is not a universal finding.[42-44] The results in white adipose tissue are generally a consistent finding in insulin-resistant rat models.[43-47] hHTg rats fed the high sucrose diet had about 50% lower GLUT4 mRNA levels in the soleus muscle in comparison to data obtained from hHTg rats fed the basal diet. Supplementation of the high sucrose diet with n-3 PUFA brought back the gene expression to initial levels. Comparable data were obtained in the heart. As the pattern of changes in the mRNA levels corresponded to plasma triglyceride levels, we constructed a correlation plot and found a significant negative correlation between GLUT4 mRNA and plasma triglyceride levels in soleus muscle and heart of the hHTg rats. The possible implication of the aforementioned is twofold: (i) the reduction of GLUT4 protein levels in muscle and adipose tissue of the insulin-resistant hHTg rat is an intrinsic feature since it does not relate to changes in GLUT4 mRNA tissue content induced by various diets (unpublished observation); (ii) the long-chain PUFAs of the n-3 family may have their own pathway for regulating the gene expression,[48] including the GLUT4 gene, at least in skeletal muscle and heart. This is evidenced in these data also by finding the lowest percentage of C20-22 PUFA in skeletal muscle phospholipids of rats fed the high sucrose diet, which had the lowest gene expression for GLUT4, and vice versa.

There is no doubt that the major regulation of GLUT4 (and hence of muscle glucose uptake) by insulin involves the translocation of the protein from intracellular vesicles to the plasma membrane, and this process is mediated via the activation of the phosphatidylinositol 3-kinase (PI 3-kinase).[18,49] It has been so far generally accepted that this insulin-induced activation of the PI 3-kinase lasts up to 10 minutes.[20,21] However, if the PI 3-kinase–induced translocation of GLUT4 has a role in physiology, then this enzyme should be activated by insulin on a long-term rather than a temporary basis.

We demonstrated in control rats that total PI 3-kinase activity in the soleus muscle and in white adipose tissue undergoes insulin stimulation after 90 min of hyperinsulinemia at euglycemia. Furthermore, after insulin stimulation, a significant fraction of the PI 3-kinase activity was associated with IRS-1 as determined by immunoprecipitation with antibody directed against IRS-1. This phenomenon was observed only in white adipose tissue and not in skeletal muscle of the control rats. Moreover, we have presented evidence that the activation of PI 3-kinase by insulin is impaired in skeletal muscle and adipose tissue of hHTg rats in comparison to control animals.

Our data are in harmony with several recent studies showing a defect in the stimulation of PI 3-kinase in muscle of various animal models of insulin resistance.[50,51] Our results, however, expand the recent knowledge by reporting on a failure of insulin to activate the PI 3-kinase/IRS-1 system of an insulin-resistant animal under *in vivo* conditions.

In summary, (1) *in vivo* insulin action of hHTg rats changes in relation to the degree of triglyceride supply; (2) *in vivo* insulin resistance of the hHTg rats decreases with increasing proportion of the long-chain PUFAs in muscle phospholipids; (3) the reduced numbers of GLUT4 transporters in skeletal muscle, heart, and white adipose tissue of hHTg rats together with the defect in insulin activation of the PI 3-kinase signaling machinery may contribute to the insulin resistance of this animal model; and (4) the data on GLUT4 mRNA tissue levels point also to a possible regulatory effect of n-3 PUFA on gene expression of GLUT4.

ACKNOWLEDGMENTS

The skillful technical assistance of Alica Mitková and Silvia Kuklová is greatly appreciated.

REFERENCES

1. STORLIEN, L. H., D. E. JAMES, K. M. BURLEIGH, D. J. CHISHOLM & E. W. KRAEGEN. 1986. Am. J. Physiol. **251**: E576–E583.
2. KLIMEŠ, I., E. ŠEBÖKOVÁ, A. VRÁNA & L. KAZDOVÁ. 1993. Ann. N.Y. Acad. Sci. **683**: 69–81.
3. VRÁNA, A. & L. KAZDOVÁ. 1986. Prog. Biochem. Pharmacol. **21**: 59–73.
4. KLIMEŠ, I., E. ŠEBÖKOVÁ, A. MINCHENKO, A. VRÁNA, M. FICKOVÁ, M. HROMADOVÁ, E. ŠVÁBOVÁ, P. BOHOV & L. KAZDOVÁ. 1991. In Advances in Lipoprotein and Atherosclerosis Research, Diagnostics, and Treatment, p. 56–62. Fischer. Jena.
5. STORLIEN, L. H., D. A. PAN, A. D. KRIKETOS & L. A. BAUR. 1993. Ann. N.Y. Acad. Sci. **683**: 82–90.
6. DEFRONZO, R. A., R. C. BONADONNA & E. FERRANNINI. 1992. Diabetes Care **15**: 318–368.
7. STORLIEN, L. H., A. B. JEMLIS, D. J. CHISHOLM, W. S. PASCOE, S. KHOURI & E. W. KRAEGEN. 1991. Diabetes **40**: 280–289.
8. STORLIEN, L. H., L. A. BAUR, A. D. KRIKETOS, D. A. PAN, G. J. COONEY, A. B. JENKINS, G. D. CALVERT & L. V. CAMPBELL. 1996. Diabetologia **39**: 621–631.
9. SIMOPOULOS, A. P. 1994. Nutr. Today January/February: 12–16.
10. LIU, S., V. E. BARACOS, H. A. QUINNEY & M. T. CLANDININ. 1994. Biochem. J. **299**: 831–837.
11. LIU, S., V. E. BARACOS, H. A. QUINNEY, T. BRICON & M. T. CLANDININ. 1995. Endocrinology **136**: 3318–3324.
12. ZIEL, F. H., N. VENKATESAN & M. B. DAVIDSON. 1988. Diabetes **37**: 885–890.
13. ŠEBÖKOVÁ, E., I. KLIMEŠ, R. MOSS, M. WIERSMA & P. BOHOV. 1995. Physiol. Res. **44**: 87–92.
14. VRÁNA, A. & L. KAZDOVÁ. 1990. Transplant. Proc. **22**: 2579.
15. ŠTOLBA, P., H. OPLTOVÁ, P. HUŠEK, J. NEVÍDKOVÁ, J. KUNEŠ, Z. DOBEŠOVÁ, J. NEDVÍDEK & A. VRÁNA. 1993. Ann. N.Y. Acad. Sci. **683**: 281–288.
16. KLIMEŠ, I., A. VRÁNA, J. KUNEŠ, E. ŠEBÖKOVÁ, Z. DOBEŠOVÁ, P. ŠTOLBA & J. ZICHA. 1995. Blood Pressure **4**: 137–142.
17. GOULD, G. W. & G. D. HOLMAN. 1993. Biochem. J. **295**: 329–341.
18. KAHN, C. R. 1994. Diabetes **43**: 1066–1084.
19. CHEATHAM, B. & C. R. KAHN. 1995. Endocr. Rev. **16**: 117–142.
20. FOLLI, F., M. J. A. SAAD, J. M. BACKER & C. R. KAHN. 1992. J. Biol. Chem. **267**: 22171–22177.
21. KELLY, K. L., N. B. RUDERMAN & K. S. CHEN. 1992. J. Biol. Chem. **267**: 3423–3428.
22. ŠEBÖKOVÁ, E., I. KLIMEŠ, M. HERMANN, M. HROMADOVÁ, P. BOHOV, A. MITKOVÁ & M. HUETTINGER. 1992. Diabetes Nutr. Metab. **5**: 249–257.
23. KOOPMANS, S. J., A. MAASSEN, J. K. RADDRER, FRÖHLICH & H. M. KRANS. 1991. Biochim. Biophys. Acta **34**: 218–224.
24. BOHOV, P., E. ŠEBÖKOVÁ, D. GAŠPERÍKOVÁ, P. LANGER & I. KLIMEŠ. 1997. This volume.
25. JAMES, D. E., M. STRUBE & M. MUECKLER. 1989. Nature (London) **338**: 83–87.
26. KAHN, B. B., L. ROSSETTI, H. F. LODISH & M. J. CHARRON. 1991. J. Clin. Invest. **87**: 2197–2206.
27. BRADFORD, M. M. 1976. Anal. Biochem. **72**: 248–254.
28. LAEMMLI, U. K. 1970. Nature (London) **227**: 680–685.
29. OUWENS, D. M., G. C. M. VAN DER ZON, G. J. PRONK, J. L. BOS, W. MOLLER, B. CHEATHAM, C. R. KAHN & A. J. MAASSEN. 1994. J. Biol. Chem. **269**: 3316–3322.
30. BURGERING, B. M. TH., R. H. MEDEMA, A. J. MAASSEN, M. L. VAN DER WETERING, A. J. VAN DER EB, F. MCCORMICK & J. L. BOS. 1991. EMBO J. **10**: 1103–1109.
31. RANDLE, P. J., P. B. GARLAND, C. N. HALES & E. A. NEWSHOLM. 1963. Lancet I: 785–789.
32. RANDLE, P. J., A. L. KERBEY & J. ESPINAL. 1988. Diabetes Metab. Rev. **4**: 623–638.
33. JENKINS, A. B., L. H. STORLIEN, D. J. CHISHOLM & E. W. KRAEGEN. 1988. J. Clin. Invest. **82**: 293–299.
34. NUUTILA, P., V. A. KOIVISTO, J. KNUUTI, U. RUOTSALAINEN, M. TERAS, M. HAAPARANATA, J. BERGMAN, O. SOLIN, L. VOIPIO-PULKKI, U. WEGELIUS & H. YKI-JARVINEN. 1992. J. Clin. Invest. **89**: 1767–1774.
35. BORKMAN, M., L. H. STORLIEN, D. A. PAN, A. B. JENKINS, D. J. CHISHOLM & L. V. CAMPBELL. 1993. N. Engl. J. Med. **328**: 238–244.
36. VESBY, B., S. TENGBLAT & H. LITHELL. 1994. Diabetologia **37**: 1044–1050.
37. GINSBERG, B. H., P. CHATTERJEE & M. YOREK. 1990. In Insulin and the Cell Membrane, p. 413–427. Harwood. Chur.

38. CLANDININ, M. T., S. CHEEMA, C. J. FIELD & V. A. BARACOS. 1993. Ann. N.Y. Acad. Sci. **683**: 151-163.
39. CRÉMEL, G., M. FICKOVÁ, I. KLIMEŠ, C. LERAY, V. LERAY, E. MEUILLET, M. ROQUES, C. STAEDEL & P. HUBERT. 1993. Ann. N.Y. Acad. Sci. **683**: 164-171.
40. BRANDT, M. D., P. COUTURE, P. L. ELSE, K. W. WITHERS & A. J. HULBERT. 1991. Biochem. J. **275**: 81-86.
41. ŠEBÖKOVÁ, E., I. KLIMEŠ, R. MOSS, P. ŠTOLBA, M. WIERSMA & A. MITKOVÁ. 1993. Ann. N.Y. Acad. Sci. **683**: 218-227.
42. ŠEBÖKOVÁ, E., I. KLIMEŠ & D. RAUČINOVÁ. 1994. In Nutrition in Sustainable Environment, p. 442-446. Smith-Gordon. London.
43. KLIP, A., T. TSAKIRIDIS, A. MARETTE & P. A. ORTIZ. 1994. FASEB J. **8**: 43-53.
44. TSAKIRIDIS, T., A. MARETTE & A. KLIP. 1995. In Lessons from Animal Diabetes, p. 141-159. Smith-Gordon. London.
45. PEDERSEN, O., C. R. KAHN, J. S. FLIER & B. B. KAHN. 1991. Endocrinology **129**: 771-777.
46. CAMPS, M., A. CASTELLÓ, P. MUNOZ, M. MONFAR, X. TESTAR & M. PALACÍN. 1992. Biochem. J. **282**: 765-772.
47. KLIMEŠ, I., E. ŠEBÖKOVÁ, A. MINCHENKO, A. MAASSEN, R. MOSS, A. MITKOVÁ, M. WIERSMA & P. BOHOV. 1994. J. Nutr. Biochem. **5**: 389-396.
48. CLARKE, S. D., R. BAILLIE, D. B. JUMP & M. T. NAKAMURA. 1997. This volume.
49. CHEATHAM, B., C. J. VLAHOS, L. CHEATHAM, L. WANG, J. BLENIS & C. R. KAHN. 1994. Mol. Cell. Biol. **14**: 4902-4911.
50. FOLLI, F., M. J. A. SAAD, J. M. BACKER & C. R. KAHN. 1993. J. Clin. Invest. **92**: 1787-1794.
51. SAAD, M. J. A., F. FOLLI, J. A. KAHN & C. R. KAHN. 1993. J. Clin. Invest. **92**: 2065-2072.

Cafestol (a Coffee Lipid) Decreases Uptake of Low-Density Lipoprotein (LDL) in Human Skin Fibroblasts and Liver Cells[a]

A. C. RUSTAN,[b] B. HALVORSEN,[c] T. RANHEIM,[c]
AND C. A. DREVON[c]

[b]*Department of Pharmacology
School of Pharmacy*
[c]*Institute for Nutrition Research
University of Oslo
N-0316 Oslo, Norway*

INTRODUCTION

Coffee consumption is associated with an increased risk of coronary heart disease.[1,2] In 1983, Thelle *et al.*[3] reported a significant positive correlation between the amount of coffee ingested and serum cholesterol levels in a population-based study in northern Norway. These findings have been confirmed in several subsequent studies.[4-9] It was later shown that the method of brewing ("boiled coffee," that is, coffee prepared by boiling ground coffee beans with water and decanting the fluid into a cup without filtration) was crucial for the hypercholesterolemic effect of coffee. In contrast, consumption of filtered coffee had little or no effect on serum cholesterol level.[4-9]

Zock *et al.*[10] prepared a lipid-rich fraction of boiled coffee by centrifugation and administered the coffee lipids to 10 volunteers for six weeks. This treatment resulted in raised serum LDL cholesterol and triacylglycerol levels. Recently, Weusten–Van der Wouw *et al.*[11] showed that a mixture of cafestol and kahweol (FIGURE 1), two coffee-specific diterpenes present in the lipid fraction of boiled coffee, raised serum concentrations of cholesterol and triacylglycerol in healthy subjects. A daily intake of 73 mg cafestol and 58 mg kahweol for six weeks gave an increase of approximately 35% in total plasma cholesterol, 75% of which could be attributed to a rise in LDL cholesterol. Coffee oil without cafestol and kahweol caused no effects on serum cholesterol.[11]

The mechanisms by which these coffee lipids raise serum cholesterol are unknown, but a concurrent alteration in liver function enzymes is observed.[11] Moreover, the

---

[a] This work was supported by the Norwegian Cancer Society, the Norwegian Council on Cardiovascular Disease, the Norwegian Research Council, the Anders Jahres Foundation, Freia Chocoladefabrik's Medical Foundation, Pronova AS, Rachel and Otto Bruun's Legacy, and the Nordic Insulin Foundation.

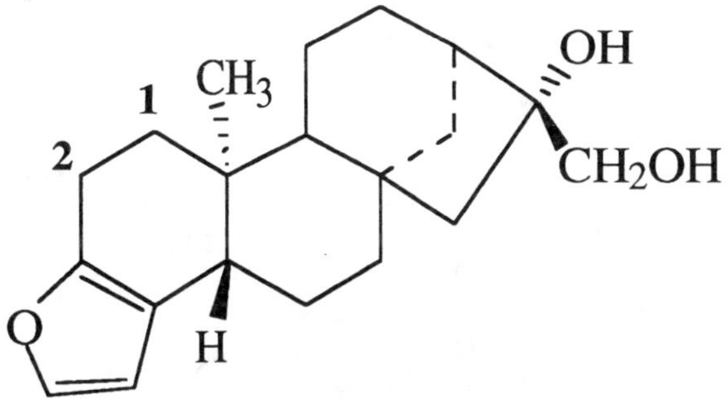

Cafestol, Mw = 314.4

Kahweol, Δ^{1,2}

FIGURE 1. Chemical structure of cafestol. Kahweol has an additional double bond between C1 and C2.

cholesterol-raising effect of the diterpenes from coffee oil seems to be specific for human primates.[12]

A potential site of action for cafestol and kahweol may be the low-density lipoprotein (LDL) receptor, which is involved in the endocytic process of apo B- and apo E-containing lipoproteins (FIGURE 2). One important way to regulate the cholesterol content of cells is via feedback repression of the gene for the LDL receptor.[13,14] When cells are depleted of cholesterol, the LDL receptor gene is transcribed actively and LDL is cleared from plasma and taken up by cells expressing this receptor on the cell surface. However, when cholesterol accumulates within cells, the number of LDL receptors is downregulated. By regulating the number of cell-surface LDL receptors, cells are able to control the rate of entry of cholesterol, thereby ensuring an optimal supply of the sterol.[13] The liver is responsible for 70–90% of the LDL clearance from plasma, approximately two-thirds of which is mediated via the LDL receptor.[15]

CELL CULTURE STUDIES

We have studied the effects of pure cafestol upon cholesterol metabolism in human skin fibroblasts (HSF)[16] and human hepatoma cells (HepG2)[17] to explore the possible mechanisms behind the plasma cholesterol-raising effect observed in humans. HSF regulate their intracellular cholesterol content by regulating the activity of the LDL receptor and have been extensively used to study the regulation of intracellular cholesterol homeostasis.[18] HepG2 cells express several of the normal biochemical functions of liver parenchymal cells, including high-affinity receptors for the uptake and degradation of LDL.[19] This cell line is therefore considered to be a reliable and useful model for studies on the regulation of hepatic LDL catabolism.[20] Moreover, the effects

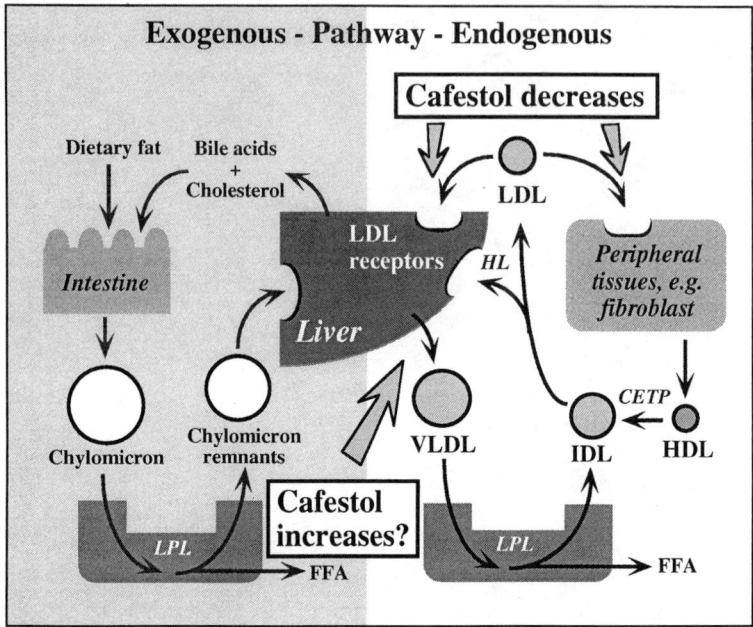

FIGURE 2. Possible sites of action of cafestol on lipoprotein metabolism. Terms: FFA, free fatty acids; VLDL, very-low-density lipoprotein; IDL, intermediate-density lipoprotein; LDL, low-density lipoprotein; HDL, high-density lipoprotein; LPL, lipoprotein lipase; HL, hepatic lipase; CETP, cholesterol ester transfer protein.

of cafestol and kahweol on cholesterol metabolism have been compared with 25-hydroxycholesterol. This oxysterol, which suppresses both the number of LDL receptors and endogenous cholesterol synthesis in HSF and HepG2 cells, exerts similar effects upon cellular cholesterol metabolism as cholesterol derived from LDL within the culture medium.[21,22]

Cafestol (20 µg/mL) decreased the uptake of [^{125}I]tyramine cellobiose–labeled LDL ([^{125}I]TC-LDL) by about 50% in HSF and by 20–25% in HepG2 cells after 18-h preincubation, whereas 25-hydroxycholesterol (5 µg/mL) reduced uptake by 70–80%. No cytotoxic effect was observed, as evaluated by measurement of lactate dehydrogenase (LDH) leakage. Furthermore, specific binding of radiolabeled LDL to the cells at 4°C was also reduced by cafestol in both cell types. Cafestol also decreased LDL metabolism in a new immortalized normal human liver epithelial cell line (THLE).[23]

HSF and HepG2 cells were also transfected with a highly active promoter, sterol regulatory element-1 (SRE-1), for the LDL receptor gene bound to the reporter gene chloramphenicol acetyltransferase (CAT).[24] Cells transfected with the active promoter did not respond to incubations with cafestol, whereas 25-hydroxycholesterol reduced the expression of CAT activity by 40–50%. Moreover, LDL receptor mRNA level was also unchanged after treatment with cafestol in these cells.

Cafestol caused a 2.3-fold higher incorporation of radiolabeled [^{14}C]oleic acid into

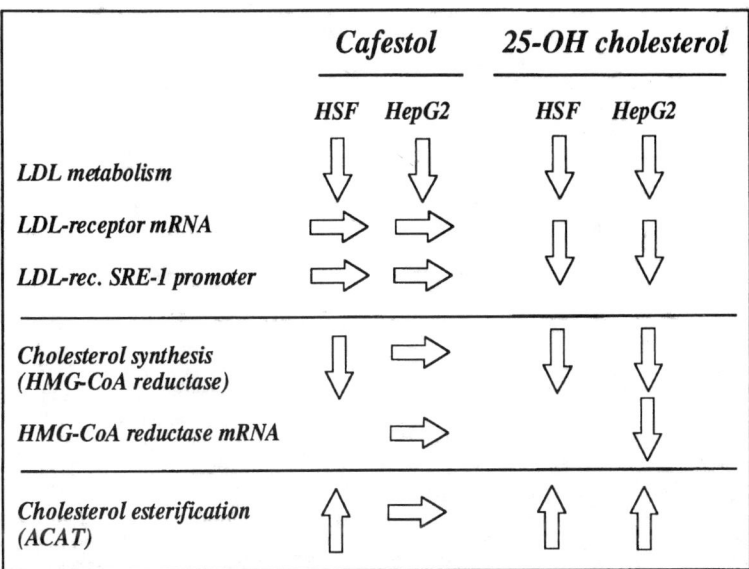

FIGURE 3. Summary of the effects of cafestol and 25-hydroxycholesterol (25-OH cholesterol) on cholesterol metabolism in human skin fibroblasts (HSF) and HepG2 cells. Terms: LDL, low-density lipoprotein; SRE-1, sterol regulatory element-1; HMG-CoA, 3-hydroxy-3-methyl-glutaryl CoA; ACAT, acyl-CoA:cholesterol acyltransferase.

cholesteryl ester in HSF, indicating an increased acyl-CoA:cholesterol acyltransferase (ACAT) activity. Incorporation of [^{14}C]acetate into cholesterol was reduced by approximately 40% with cafestol (20 μg/mL) as compared to control medium after 24-h preincubation, indicating a decreased 3-hydroxy-3-methyl-glutaryl CoA (HMG-CoA) reductase activity. For HepG2 cells, the effects of cafestol upon cholesterol synthesis and upon HMG-CoA reductase activity and mRNA level were similar to cells incubated with control medium. In addition, cholesterol esterification was unchanged in the presence of cafestol in HepG2 cells.

Our results summarized in FIGURE 3 demonstrate that cafestol reduces the LDL receptor activity in HSF and HepG2 cells. Thus, dietary intake of cafestol might cause increased concentration of plasma cholesterol in humans via downregulation of LDL receptors, probably by posttranscriptional mechanisms. Cafestol may provide a tool to increase our understanding of the regulation of cholesterol (and triacylglycerol) metabolism.

REFERENCES

1. THELLE, D. S. 1995. Coffee, tea, and coronary heart disease. Curr. Opin. Lipidol. 6: 25–27.
2. TVERDAL, A., I. STENSVOLD, K. SOLVOLL, O. P. FOSS, P. LUND-LARSEN & K. BJARTVEIT. 1990. Coffee consumption and death from coronary heart disease in middle-aged Norwegian men and women. Br. Med. J. 300: 566–569.

3. THELLE, D. S., E. ARNESEN & O. H. FØRDE. 1983. The Tromsø heart study: does coffee raise serum cholesterol? N. Engl. J. Med. **308:** 1454-1457.
4. ARO, A., J. TUOMILEHTO, E. KOSTIAINEN, U. UUSITALO & P. PIETINEN. 1987. Boiled coffee increases serum low density lipoprotein concentration. Metabolism **36:** 1027-1030.
5. BØNAA, K., E. ARNESEN, D. S. THELLE & O. H. FØRDE. 1988. Coffee and cholesterol: is it all in the brewing? The Tromsø study. Br. Med. J. **297:** 1103-1104.
6. BAK, A. A. & D. E. GROBBEE. 1989. The effect on serum cholesterol levels of coffee brewed by filtering or boiling. N. Engl. J. Med. **321:** 1432-1437.
7. ARO, A., J. TEIRILA & C. G. GREF. 1990. Dose-dependent effect on serum cholesterol and apoprotein B concentrations by consumption of boiled, non-filtered coffee. Atherosclerosis **83:** 257-261.
8. VAN DUSSELDORP, M., M. B. KATAN, T. VAN VLIET, P. N. DEMACKER & A. F. STALENHOEF. 1991. Cholesterol-raising factor from boiled coffee does not pass a paper filter. Arterioscler. Thromb. **11:** 586-593.
9. RATNAYAKE, W. M., R. HOLLYWOOD, E. O'GRADY & B. STAVRIC. 1993. Lipid content and composition of coffee brews prepared by different methods. Food Chem. Toxicol. **31:** 263-269.
10. ZOCK, P. L., M. B. KATAN, M. P. MERKUS, M. VAN DUSSELDORP & J. L. HARRYVAN. 1990. Effect of a lipid-rich fraction from boiled coffee on serum cholesterol. Lancet **335:** 1235-1237.
11. WEUSTEN-VAN DER WOUW, M. P., M. B. KATAN, R. VIANI, A. C. HUGGETT, R. LIARDON, P. G. LUND-LARSEN, D. S. THELLE, I. AHOLA, A. ARO, S. MEYBOOM & A. C. BEYNEN. 1994. Identity of the cholesterol-raising factor from boiled coffee and its effects on liver function enzymes. J. Lipid Res. **35:** 721-733.
12. TERPSTRA, A. H., M. B. KATAN, M. P. WEUSTEN-VAN DER WOUW, R. J. NICOLOSI & A. C. BEYNEN. 1994. Coffee oil consumption does not affect serum cholesterol in rhesus and cebus monkeys. J. Nutr. **125:** 2301-2306.
13. GOLDSTEIN, J. L. & M. S. BROWN. 1977. The low-density lipoprotein pathway and its relation to atherosclerosis. Annu. Rev. Biochem. **46:** 897-930.
14. GOLDSTEIN, J. L. & M. S. BROWN. 1990. Regulation of the mevalonate pathway. Nature **343:** 425-430.
15. STEINBERG, D., S. PARTHASARATHY, T. E. CAREW, J. C. KHOO & J. L. WITZTUM. 1989. Beyond cholesterol: modifications of low-density lipoprotein that increase its atherogenicity. N. Engl. J. Med. **320:** 915-924.
16. HALVORSEN, B., T. RANHEIM, M. S. NENSETER, A. C. HUGGETT & C. A. DREVON. 1997. Effect of a coffee lipid (cafestol) on cholesterol metabolism in human skin fibroblasts. Submitted.
17. RUSTAN, A. C., B. HALVORSEN, A. C. HUGGETT, T. RANHEIM & C. A. DREVON. 1997. Effect of coffee lipids (cafestol and kahweol) on regulation of cholesterol metabolism in HepG2 cells. Arterioscler. Thromb. Vasc. Biol. In press.
18. BROWN, M. S. & J. L. GOLDSTEIN. 1975. Regulation of the activity of the low density lipoprotein receptor in human fibroblasts. Cell **6:** 307-316.
19. LEICHTNER, A. M., M. KRIEGER & A. L. SCHWARTZ. 1984. Regulation of low density lipoprotein receptor function in a human hepatoma cell line. Hepatology **4:** 897-901.
20. GIBBONS, G. F. 1994. A comparison of in vitro models to study hepatic lipid and lipoprotein metabolism. Curr. Opin. Lipidol. **5:** 191-199.
21. DASHTI, N. 1992. The effect of low density lipoproteins, cholesterol, and 25-hydroxycholesterol on apolipoprotein B gene expression in HepG2 cells. J. Biol. Chem. **267:** 7160-7169.
22. CARLSON, T. L. & B. A. KOTTKE. 1989. Effect of 25-hydroxycholesterol and bile acids on the regulation of cholesterol metabolism in HepG2 cells. Biochem. J. **264:** 241-247.
23. PFEIFER, A. M. A., K. MACE, Y. TROMVOUKIS & M. M. LIPSKY. 1995. Highly efficient establishment of immortalized cells from adult human liver. Methods Cell Sci. **17:** 83-89.
24. YOKOYAMA, C., X. WANG, M. R. BRIGGS, A. ADMON, J. WU, X. HUA, J. L. GOLDSTEIN & M. S. BROWN. 1993. SREBP-1, a basic-helix-loop-helix-leucine zipper protein that controls transcription of the low density lipoprotein receptor gene. Cell **75:** 187-197.

High Fructose Feeding Enhances Erythrocyte Carbonic Anhydrase 1 mRNA Levels in Rat

K. K. GAMBHIR,[a] P. OATES,[a] M. VERMA,[b]
S. TEMAM,[a] AND W. CHEATHAM[a]

[a]Molecular Endocrinology Laboratory
Department of Medicine
College of Medicine
Howard University
Washington, District of Columbia 20059

[b]Department of Biochemistry
Georgetown University
Washington, District of Columbia 20007

INTRODUCTION

Many investigations have demonstrated that high fructose and sucrose feeding can cause hyperinsulinemia[1-3] and insulin resistance.[4-6] An increase in the concentration of insulin has been associated with hypertension, dyslipidemia, and glucose intolerance.[7-10] Insulin resistance is associated not only with quantitative changes in lipoproteins (decreased HDL and elevated triglycerides), but also with compositional changes in lipoproteins.[11] However, the biological mechanism by which high fructose or sucrose feeding leads to insulin resistance is not fully understood. Hwang et al.[1] assumed that insulin resistance could develop as a result of catecholamines that oppose insulin action. These researchers[1] also proposed that insulin resistance and hyperinsulinemia develop after high-carbohydrate diets because of sodium imbalance. The availability of extracellular Na^+ plays a role in Na^+/H^+ exchange. Thus, the efflux of Na^+ and the influx of K^+ through the cell membrane rely on an active transport mechanism.

These transport mechanisms may depend on insulin to regulate Na^+ transport.[12] Insulin also has an effect on the pH regulator, carbonic anhydrase (CA). Parui et al. showed that erythrocyte receptor-bound insulin increased the CA activity[13] and proposed that the insulin induced CA activity and increased proton production. Recently, we have observed an increase in intercellular CA activity and elevated $^{22}Na^+$ uptake in RBC.[12] The level of CA in various diseases has been studied and found to be different.[11] However, there are no reports on the level of CA1 mRNA in erythrocytes (E) of high fructose-fed rodents. In the present investigation, we therefore studied the effect of high fructose feeding on the level of CA1 mRNA in RBC.

MATERIALS AND METHODS

Materials

Rats: Male, Fischer Norway (F1) rats (Charles River), weighing 280–300 g, were acclimatized for a week and maintained on a standard rat chow diet *ad libitum*. All animals were maintained on a 12-hour light and dark cycle. The experimental diet was composed of fructose, protein, and fat [65%, 18%, and 12% (w/w), respectively].

Ficoll, Percoll, and other chemicals were purchased from Sigma Chemical (St. Louis, Missouri). All chemicals were of analytical company grade. Hypaque was obtained from Winthrop Laboratories, a division of Sterling Drug, Incorporated. All reagents of CA1 mRNA isolation and RT-PCR were purchased from Life Technologies (Gaithersburg, Maryland) and Sigma Chemical (St. Louis, Missouri).

Experimental Design

After acclimatization for a week, 28 rats were divided into two groups of 14 rats each. Rats were caged individually. The control rats were fed standard rat chow diet, whereas the experimental group of rats were fed high fructose diet *ad libitum*. Every 2 weeks, 2 rats from each group were sacrificed and blood was obtained. The study was carried out for a period of 14 weeks.

Isolation and Age Fractionation of Erythrocytes

Erythrocytes were isolated and purified as described previously.[13] The purified erythrocytes were then layered over a Percoll gradient consisting of 35%, 40%, 45%, 50%, 55%, 65%, 80%, and 100% of Percoll mixed with Dulbecco's modified Eagle medium as schematized in FIGURE 1.[14] This mixture was centrifuged at 500g for 30 minutes at 22°C. Three subfractions of Percoll density (g/mL)—1.07–1.09 as young (y), 1.09–1.11 as mid-age (m), and 1.11–1.12 as old (o)—were separated. All these subfractions were frozen at −75°C until CA1 mRNA analyses.

CA1 mRNA Level Estimation

Isolation of RNA

Purified RBC and subpopulations y, m, and o were lysed by Trizol and incubated for 5 minutes. Then, 0.2 mL of chloroform per 1 mL of Trizol was added and mixed by hand. After incubating at room temperature for 2–3 minutes, the samples were centrifuged at 12,000g for 15 minutes at 4°C. The sample mixture was fractionated

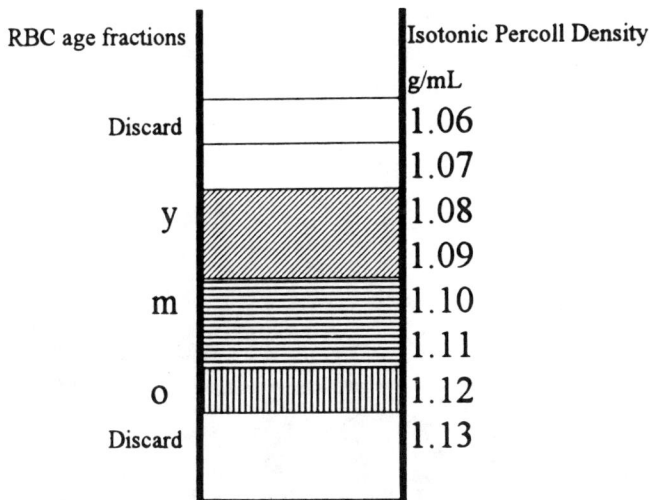

FIGURE 1. Centrifugation pattern of erythrocytes. Schematic representation of the density base erythrocyte age fractions: young (y) = 1.07–1.09, mid-age (m) = 1.09–1.11, old (o) = 1.11–1.12 g/mL Percoll.

into a lower red phenol–chloroform phase, an interphase, and a colorless upper aqueous phase (RNA). The aqueous phase was collected and precipitated using 0.5 mL of isopropyl alcohol per 1 mL of Trizol reagent used initially. The samples were incubated at 25°C for 10 minutes and centrifuged at 12,000g for 10 minutes at 4°C. The supernatant was discarded and the gel-like pellet on the side and bottom of the tube (RNA) was washed with 1 mL of 75% ethanol per 1 mL of Trizol used initially. The samples were vortexed and centrifuged at 12,000g for 5 minutes at 4°C.

The RNA pellet was dried (air) and dissolved in 20 µL of RNase-free water plus 1 µL of RNasin. After RNA was completely dissolved, it was stored at −70°C. Extracted RNAs were quantified by absorbance measurement at 260 nm, where 1 OD was taken as 40 µg RNA/mL. RNA polymerase was used for RT-PCR analysis using 1 µg RNA from each sample according to the supplier's conditions (Perkin-Elmer). RNA was analyzed on agarose gel.[15,16]

Northern Blot Analysis

The RNAs (20 µg) were adjusted to 50% formamide, 0.1 M MOPS [3-(N-morpholino)propanesulfonic acid] (pH 7.0), 50 mM sodium acetate, 5 mM EDTA, and 2.2 M formaldehyde. Quantification of the data was done by densitometric analysis. Densities were normalized with respect to GAPDH (glyceraldehyde phosphate dehydrogenase) hybridization and standardization for each gene.[15]

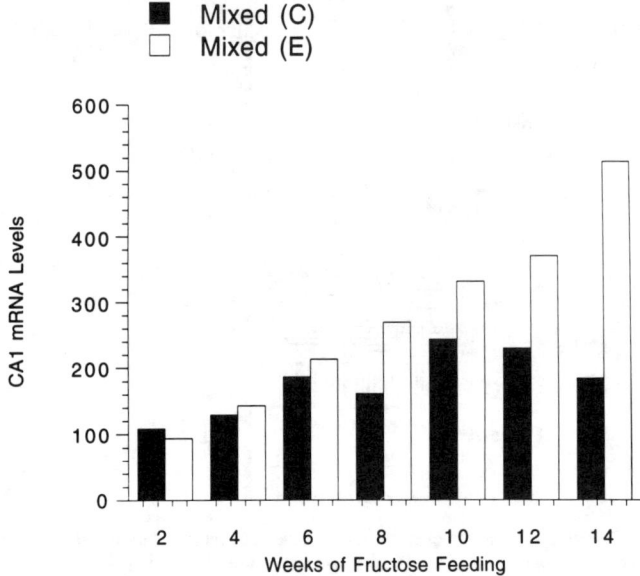

FIGURE 2. Carbonic anhydrase 1 mRNA levels in the purified erythrocytes (mixed age fraction) (C = control; E = experimental). Purified erythrocytes were lysed, and RNA was isolated and subjected to RT-PCR. The GAPDH gene was used as a control in the experiments.

RESULTS AND DISCUSSION

FIGURE 2 shows the CA1 mRNA levels in the purified E for the control and experimental rats fed high fructose for a period of 14 weeks. These results clearly indicate that high fructose feeding has a significant effect on the levels of CA1 mRNA in E. In chow-fed rats as compared to rats that consumed high fructose diet, the level of CA1 mRNA began to increase slightly for the first 6 weeks. Afterwards, the level of CA1 mRNA slowly started to decrease. On the other hand, in the experimental rats, the CA1 mRNA level was elevated 446% within 14 weeks.

FIGURE 3 summarizes the CA1 mRNA levels in the young subpopulation of erythrocytes (y). This figure illustrates a steady increase of CA1 mRNA in animals fed high fructose and standard chow diet. The values in control-fed animals were always less than half of those of fructose-fed animals. The difference between the 2d and 14th week was 253%.

The CA1 mRNA levels isolated from the mid-age erythrocytes (m) from both the control and experimental rats are shown in FIGURE 4. mRNA levels were higher in experimental rats than in control rats, except for the 2d week. Rats fed fructose showed 335% greater CA1 mRNA levels over the experimental period.

FIGURE 5 illustrates the CA1 mRNA levels in the old cell subpopulation of erythrocytes (o) from both the experimental and control rats. The CA1 mRNA levels of experimental rats were again higher than those of controls.

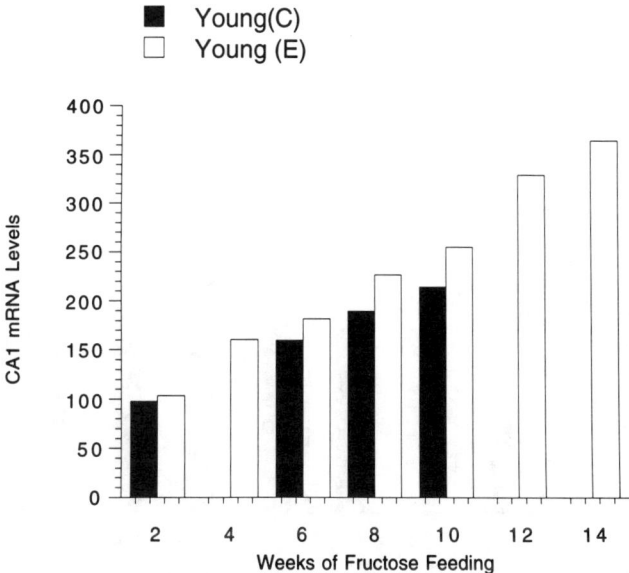

FIGURE 3. Carbonic anhydrase 1 mRNA levels in the young (y) subpopulation of purified erythrocytes. The experimental conditions were exactly the same as in FIGURE 2.

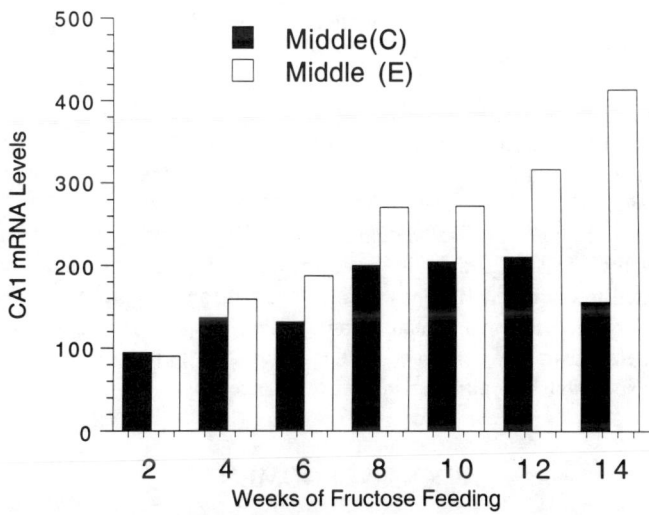

FIGURE 4. Carbonic anhydrase 1 mRNA levels in the mid-age (m) subpopulation of purified erythrocytes. The experimental conditions are described in FIGURE 2.

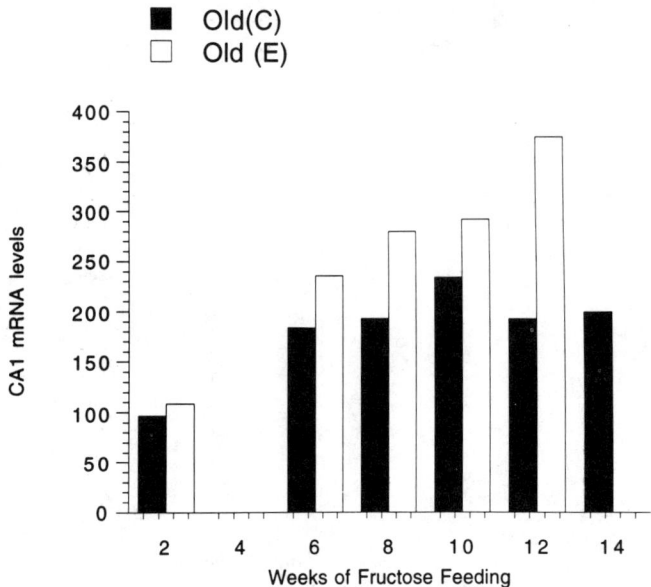

FIGURE 5. Carbonic anhydrase 1 mRNA levels in the old (o) subpopulation of purified erythrocytes. The experimental conditions were exactly the same as in FIGURE 2.

The results presented in this study demonstrate that high fructose feeding over 14 weeks increases CA1 mRNA progressively and that the magnitude of difference is highest in the oldest cells. The differences in CA1 mRNA levels were obvious in all age subfractions of erythrocytes.

Kazumi et al. reported that high fructose feeding increased plasma triglycerides in diabetic[17-19] and nondiabetic rodents.[19,20] In a recent study, these investigators concluded that Wistar fatty rats, an animal model of NIDDM, became hypertriglyceridemic due to increased triglyceride production as well as because of impaired removal of triglycerides.[21]

In conclusion, high fructose feeding in these rodents enhanced both CA1 mRNA levels as well as total CA activity in the erythrocytes. Further, this enhanced CA activity resulted in increased H^+ production, which enhanced intracellular $^{22}Na^+$ uptake in the erythrocytes, that is, cation imbalance. We speculate that this imbalance in cation metabolism may be triggering the perturbation in intracellular pH and metabolism responsible for initiating insulin resistance.

ACKNOWLEDGMENTS

We would like to thank Deborah King at the Office of Research Administration (ORA), Howard University, for her excellent help in the preparation of this manuscript.

REFERENCES

1. HWANG, I. S., H. HO, B. B. HOFFMAN & G. M. REAVEN. 1987. Fructose induced insulin resistance and hypertension in rats. Hypertension **10:** 512-516.
2. ZAVARONI, I., S. SANDERS, S. SCOTT & G. M. REAVEN. 1980. Effect of fructose feeding on insulin secretion and action in the rat. Metabolism **29:** 970-973.
3. TOBEY, J. A., C. E. MONDON, I. ZAVARONI & G. M. REAVEN. 1982. The chronism of insulin resistance in fructose-fed rats. Metabolism **31:** 608-612.
4. REAVEN, G. M., T. R. RISSER, Y-DI. CHEN & E. P. REAVEN. 1979. Characterization of a model of dietary-induced hypertriglyceridemia in young, nonobese rats. J. Lipid Res. **20:** 371-378.
5. WRIGHT, D. W., R. I. HANSEN, C. E. MONDON & G. M. REAVEN. 1983. Sucrose induced insulin resistance in the rat: modulation by exercise and diet. Am. J. Clin. Nutr. **38:** 879-883.
6. HAFFNER, S. M. 1996. The insulin resistance syndrome revisited. Diabetes Care **19:** 275-277.
7. MODAN, M., H. HALKIN, S. ALMOG, A. LUSKI, A. ESHKOL, M. SHEFI, A. SHITRIT & Z. FUCHS. 1985. Hyperinsulinemia: a link between hypertension, obesity, and glucose tolerance. J. Clin. Invest. **75:** 809-812.
8. ZAVARONI, I., E. BONORA, M. PAGLIVRA, E. DALLAGLIO, L. LUCHETTI, G. BUONANNO, P. A. BONATI, M. BERGONZANI, L. GUNDI, M. PASSERI & G. REAVEN. 1989. Risk factors for coronary artery disease in healthy persons with hyperinsulinemia and normal glucose tolerance. N. Engl. J. Med. **320:** 702-706.
9. HAFFNER, S. M., D. FONG, H. P. HAZUDA, J. A. PUGH & J. K. PATTERSON. 1988. Hyperinsulinemia, upper body adiposity, and cardiovascular risk factors in nondiabetics. Metabolism **37:** 338-345.
10. ORCHARD, T. J., D. J. BECHER, M. BATES, L. H. KULLER & A. L. DRASH. 1983. Plasma insulin and lipoprotein concentrations: an atherogenic association. Am. J. Epidemiol. **118:** 326-337.
11. YAMAKIDO, M., N. YORIOKA, K. GORIKI, K. WADA, J. HATA & Y. NISHIMOTO. 1980. Carbonic anhydrase I and II levels in erythrocyte K chronic renal disease patients. Hiroshima J. Med. Sci. **31:** 1-16.
12. GAMBHIR, K. K., R. PARUI, J. L. TOWNSEND & W. L. HENRY. 1990. Insulin may regulate ATPase-independent sodium transport. *In* Proc. Int. Symp.: Insulin and Cell Membrane, p. 185-193.
13. PARUI, R., K. K. GAMBHIR & P. MEHROTA. 1991. Changes in carbonic anhydrase may be the initial step of altered metabolism in hypertension. Biochem. Int. **23:** 779-798.
14. ALDERMAN, E. M., J. H. FUDENBERG & E. R. LOUINS. 1980. Binding of immunoglobulin classes to subpopulations of human red blood cells separated by density-gradient centrifugation. Blood **55:** 817-822.
15. BAILEY, J. M. & M. VERMA. 1991. Analytical procedures for a cryptic messenger RNA that mediates translational control of prostaglandin synthase by glucocorticoids. Anal. Biochem. **196:** 11-18.
16. VERMA, M. & E. A. DAVIDSON. 1993. Molecular cloning and sequencing of a canine tracheobronchial mucin cDNA containing a cysteine-rich domain. Proc. Natl. Acad. Sci. U.S.A. **90:** 7144-7148.
17. YOSHINO, G., M. MATSUSHITA, E. MAEDA, Y. NAKA, K. NAGATA, M. MORITA, K. MATSUBA, T. KAZUMI & M. KASUGA. 1992. Effect of long-term deficiency and insulin treatment on the composition of triglyceride-rich lipoproteins and triglyceride turnover in rats. Atherosclerosis **92:** 243-250.
18. YOSHINO, G., M. MATSUSHITA, M. IWAI, M. MORITA, K. MATSUBA, K. NAGATA, S. FURUKAWA, T. HIRANO & T. KAZUMI. 1992. Effect of mild diabetes and dietary fructose on VLDL triglyceride turnover in rats. Metabolism **41:** 236-240.
19. KAZUMI, T., M. VRANIC & G. STEINER. 1986. Triglyceride kinetics: effect of dietary fructose, sucrose, or glucose alone or with hyperinsulinemia. Am. J. Physiol. **250:** E325-E330.
20. KAZUMI, T., G. YOSHINO, K. MATSUBA, M. IWAI, I. IWATANI, M. MATSUSHITA, T. KASAMA, T. HOSOKAWA, F. NUMANO & S. BABA. 1991. Effect of dietary glucose or fructose on secretion and particle size of triglyceride-rich lipoproteins in Zucker fatty rats. Metabolism **40:** 962-966.
21. KAZUMI, T., T. HIRANO, H. ODAKA, T. EBARA, N. AMANO, T. HOZUMI, Y. ISHIDA & G. YOSHINO. 1996. VLDL triglyceride kinetics in Wistar fatty rats, an animal model of NIDDM: effects of dietary fructose alone or in combination with pioglitazone. Diabetes **45:** 806-811.

Pharmacological Treatment and Mechanisms of Insulin Resistance

Impact on Vascular Smooth Muscle Cells, Blood Pressure, and Lipids

WILLA A. HSUEH AND RONALD E. LAW

Division of Endocrinology, Diabetes, and Hypertension
University of Southern California
School of Medicine
Los Angeles, California 90033

THE INSULIN RESISTANCE SYNDROME AND CARDIOVASCULAR DISEASE

Insulin resistance is defined as a defect in the ability of skeletal muscle to take up glucose in response to insulin.[1] Skeletal muscle is the major target tissue into which insulin promotes the transport of glucose; this action of insulin is regulated by genetic factors, environmental factors, blood flow, circulating substances, and insulin signaling pathways. Obesity, non-insulin-dependent diabetes mellitus (NIDDM), and hypertension are associated with insulin resistance.[2-4] Decreased insulin-mediated glucose uptake into skeletal muscle occurs in adiposity; in nearly all populations, insulin resistance directly correlates with body mass index.[5] Visceral adiposity is particularly associated with insulin resistance.[6] Substances from fat tissue appear to suppress insulin-mediated glucose uptake; these include free fatty acids, tumor necrosis factor-alpha (TNF-α), and possibly leptins.[7,8] These substances are elevated in the circulation in some obese animals and humans. Infusion of these factors into normal animals suppresses insulin-mediated glucose uptake, and infusion of antibodies to TNF-α into obese animals improves insulin-mediated glucose uptake.[9] Nonobese subjects with the essential hypertension have been demonstrated to be insulin-resistant, but the mechanism of this association is unknown.[3] Impaired vasodilatory responses to insulin have been demonstrated to be associated with insulin resistance.[10] Increased blood flow into skeletal muscle increases delivery of insulin and glucose to this target tissue and can account for as much as 40% of insulin-mediated glucose uptake.[11] Insulin induces vasodilation by stimulation of nitric oxide from endothelial cells;[12] this action of insulin is blunted in obese and NIDDM subjects and may also be altered in essential hypertension.[10,13] Insulin resistance is a major contributing factor to NIDDM.[2] The pancreas increases insulin production in insulin resistance subjects to overcome the defect in insulin action; however, if pancreatic insulin production does not remain elevated to overcome this defect, hyperglycemia results. Additionally, there is also increased hepatic glucose production in NIDDM.[14] Subjects with a high propensity to develop NIDDM, such as children of diabetics, display defects in insulin-mediated glucose uptake prior to the development of diabetes.[15] Recent studies demonstrate that hyperinsulinemia, a marker of insulin resistance, is heritable in Caucasians, Pima Indians, and Mexican Americans, suggesting that genes influence insulin action.[16-18] Polymorphisms of several genes have been associated with insulin resistance; however, none of these have been confirmed.[15] Only rarely have abnormalities of the insulin receptor, itself, been associated with insulin resistance.[19] A major contribution to defects in insulin action in the skeletal muscle likely involves insulin signaling steps that are activated following the binding of insulin to its receptor.[1] The specific site of the defect is unknown, but it results in decreased mobilization of the glucose transporter, GLUT-4, from cytoplasmic vesicles to the cell surface to transport glucose into the cell.

In addition to glucose intolerance and hypertension, insulin resistance and hyperinsulinemia are associated with other characteristic features that are themselves risk factors for increased cardiovascular disease. These include a dyslipidemia that consists of low high-density lipoprotein (HDL) cholesterol, elevated triglycerides, and increased small low-density lipoprotein (LDL) cholesterol, which represents the moiety of LDL cholesterol that has a propensity to oxidation.[20,21] More recently, an enhanced tendency to thrombus formation has been described in the insulin resistance syndrome. Subjects with insulin resistance have been shown to have elevated circulating plasminogen activator inhibitor-1 (PAI-1) levels and increased fibrinogen levels.[22] A genetic polymorphism associated with the PAI-1 gene has been associated with PAI-1 expression and correlates with cardiovascular disease risk in young myocardial infarction patients.[23] In that study, PAI-1 alleles appeared to significantly influence the correlation between plasma PAI-1 levels and circulating levels of both insulin and triglycerides, suggesting that genetic polymorphisms in the PAI-1 gene may influence its responsiveness to insulin resistance.[23] Another polymorphism of the PAI-1 gene has also been correlated with increased frequency of myocardial infarction and has been linked to increased plasma PAI-1 levels in NIDDM.[24]

Patients with the insulin resistance syndrome are clearly at increased risk for coronary artery disease. This increased risk is due to the constellation of factors associated with the insulin resistance syndrome, although insulin resistance and hyperinsulinemia themselves may increase the risk for coronary artery disease. Four large prospective studies have demonstrated that hyperinsulinemia is an independent risk factor for coronary artery disease in Caucasian males; however, similar findings have not been demonstrated in women and have not been addressed in other ethnic groups.[25-30] Circulating insulin levels have also been found to correlate with other markers of cardiovascular disease, such as left ventricular mass index and carotid intimal-medial wall thickness.[31,32] Although insulin has direct effects on the vasculature as discussed below, its relationship in the pathogenesis and progression of atherosclerosis remains controversial.

TABLE 1. Effects of Insulin on the Vasculature

Endothelial Cells
transcytosis
nitric oxide production—aorta and peripheral resistance arteries
vasoconstricting cyclooxygenase—mesenteric artery
growth
glycogen synthesis

Vascular smooth muscle cells (VSMCs)
promotes growth—potentiates effects of other growth factors
enhances PAI-1 production—potentiates effects of angiotensin II
inhibits I-CAM production
promotes migration
stimulates extracellular matrix production

INSULIN ACTION IN THE VASCULATURE

Vascular responses affect insulin action and insulin itself has direct effects on the vascular wall. The effects of insulin on vascular cells are summarized in TABLE 1. Insulin has little effect to stimulate glucose uptake into vascular cells,[33] but it has a variety of other actions. An important action in endothelial cells is stimulation of nitric oxide release, which promotes vasodilation.[12] This vasodilatory response contributes to insulin's ability to enhance glucose uptake into skeletal muscle since inhibitors of nitric oxide synthase inhibit the vasodilatory response and blunt the ability of insulin to transport glucose into skeletal muscle.[34] Nitric oxide inhibits growth factor effects on the vasculature, such as VSMC migration and proliferation, and it also inhibits clotting.[35,36] Thus, nitric oxide is considered to be vascular-protective, inhibiting mechanisms involved in the atherosclerotic process.

In contrast, insulin has actions in VSMCs that promote the atherosclerotic process. Insulin itself has a modest effect to induce proliferation of VSMCs; however, insulin significantly potentiates the effect of platelet-derived growth factor (PDGF) and other growth factors on VSMC proliferation.[37] A number of different signaling pathways may be involved in regulation of VSMC growth. One likely candidate appears to be the RAS → RAF → MAP kinase → *c-fos* cascade that has been linked to proliferation in many other cell types.[38] Inhibition of this pathway with a selective inhibitor, PD098059, of MAP kinase (MEK), which activates MAP kinase by phosphorylation of serine and tyrosine residues, inhibited insulin-stimulated growth of quiescent VSMCs in a dose-dependent fashion and also inhibited to a lesser extent serum-induced growth.[39] Thus, the MAP kinase pathway appears to be important for insulin mitogenic signal transduction in VSMCs. This compound also inhibited the activation of the serum response element (SRE) attached to the chloramphenicol acetyltransferase (CAT) reporter gene, which is dependent on phosphorylation of the transcription factor Elk-1 by MAP kinase.[39] Activation of *c-fos* gene expression by insulin is mediated through a 5'-flanking SRE.[40] Thus, inhibition of the MAP kinase pathway by PD098059 likewise resulted in a near-total suppression of insulin-induced *c-fos* mRNA expression.[39] In combination, these findings suggest that the mitogenic action of insulin on VSMCs is involved in the induction of early growth response genes, such as *c-fos*, through the MAP kinase pathway.

In addition to its effects on growth, insulin has a modest effect to stimulate PAI-1 production in VSMCs, but markedly enhances the effect of angiotensin II. Insulin and angiotensin II are additive in the induction of MAP kinase in both VSMCs and cultured mesangial cells, which are modified smooth cells.[41] Insulin also stimulates extracellular matrix production by VSMCs.[42] A variety of VSMC behaviors including migration, proliferation, PAI-1 production, and extracellular matrix production all contribute to the atherosclerotic process and appear to be regulated in conjunction with other growth factors by insulin. Thus, insulin has dual effects to stimulate endothelial cell production of nitric oxide that may protect the vasculature from atherosclerosis, while promoting VSMC behaviors that enhance the atherosclerotic process. In the presence of endothelial cell damage that occurs due to hypertension (with high circulating LDL cholesterol), smoking, and other well-known risk factors for coronary artery disease, the nitric oxide response to the insulin response may be blunted. Ultimately, this could lead to a balance in insulin action such that the nitric oxide pathway would be impaired, while the action of VSMCs would proceed. Therefore, in the presence of endothelial cell damage, insulin appears to be atherogenic.

EFFECTS OF THIAZOLIDINEDIONE ON THE VASCULATURE

Thiazolidinedione analogues are novel insulin-sensitizing agents that are currently being tested clinically for both prevention and treatment of NIDDM.[43,44] The mechanism by which they enhance insulin action is unknown, although they likely affect insulin signaling pathways in skeletal muscle. In skeletal muscle, there is evidence that thiazolidinediones enhance insulin receptor tyrosine kinase phosphorylation and PI-kinase activities.[45–48] In patients with NIDDM, troglitazone (a member of the thiazolidinedione class) has been shown to improve glucose utilization and to lower fasting insulin and blood sugar levels.[43] In insulin-resistant prediabetic subjects, troglitazone has also been demonstrated to improve insulin-mediated glucose uptake and to lower triglycerides, circulating insulin levels, and blood pressure.[44] In a group of gestational diabetics who were insulin-resistant and at high risk for NIDDM, troglitazone improved sensitivity by 80% after six months of treatment.[49] Thus, these agents hold great promise in the treatment of insulin-resistant states.

In animal models, thiazolidinedione analogues lower blood pressure even in the absence of insulin resistance. Fructose feeding in the rat mimics several aspects of the insulin resistance syndrome in humans, as it decreases insulin action, increases triglycerides, and increases blood pressure. Thiazolidinedione analogues prevent or reverse these effects.[50,51] However, these agents also lower blood pressure in rats with renal vascular hypertension, which is not an insulin resistance state, and in normal rats.[51,52] Further studies in isolated vessels have demonstrated that thiazolidinedione analogues antagonize the action of vasoconstrictors on VSMC and that at higher doses

TABLE 2. Thiazolidinedione Action on Vascular Smooth Muscle Cells

Action	Reference
Inhibits Ca^{++} channels	51, 60, 61
Inhibits insulin, bFGF, EGF, and serum-induced mitogenesis	52, 55
Inhibits MAP kinase action in the nucleus	55
Inhibits c-fos induction by bFGF and insulin	55
Inhibits PDGF and A_{II}-induced migration	58, 59

(20 µM) these agents act as calcium channel blockers.[51] This effect likely contributes to their blood pressure–lowering capability.

A major concern regarding the use of these agents is their ability to stimulate insulin action to promote VSMC migration and proliferation.[53,54] Surprisingly, we found that troglitazone inhibited c-fos induction and DNA synthesis in VSMCs induced by a variety of growth factors, including platelet-derived growth factor (PDGF), basic fibroblast growth factor (bFGF), insulin, and angiotensin II (A_{II}).[55] Although troglitazone did not inhibit the increase in MAP kinase activity associated with administration of these growth factors to VSMCs, it did inhibit activation of Elk-1, an important transcription factor regulating c-fos expression.[55] Since Elk-1 is activated by MAP kinase,[56,57] these results suggest that troglitazone inhibits MAP kinase action in the nucleus. Further studies are under way to elucidate these mechanisms. Troglitazone also inhibited migration of VSMCs induced by PDGF and A_{II}.[58] Indeed, the MAP kinase signaling pathway is also involved in this VSMC function since the MEK inhibitor, PD098059, which inhibits MEK activation of MAP kinase, and antisense to MAP kinase both inhibited migration.[59] A summary of the actions of troglitazone in VSMCs is provided in TABLE 2. These observations were highly relevant *in vivo*. Treatment with troglitazone prior to aortic balloon angioplasty substantially prevented the intimal hyperplasia that results from this procedure.[55] Thus, the thiazolidinedione analogues may have dual effects to improve insulin action into skeletal muscle as well as to protect the vasculature.

REFERENCES

1. CHEATHAM, B. & C. R. KAHN. 1995. Insulin action and the insulin signaling network. Endocr. Rev. **16:** 117–142.
2. OLEFSKY, J. M., O. G. KOLTERMAN & J. A. SCARLETT. 1982. Insulin action and resistance in obesity and non-insulin-dependent type II diabetes mellitus. Am. J. Physiol. **243:** E15–E30.
3. MARTIN, B. C., J. H. WARRAM, A. S. KROLEWSKI, R. N. BERGMAN, J. S. SOELDNER & C. R. KAHN. 1992. Role of glucose and insulin resistance in the development of type 2 diabetes mellitus: results of a 25-year follow-up study. Lancet **340:** 925–929.
4. POLLARE, T., H. LITHELL & C. BERN. 1990. Insulin resistance is a characteristic feature of primary hypertension independent of obesity. Metabolism **39:** 167–174.
5. HAFFNER, S. M., R. D'AGOSTINO, JR., M. F. SAAD, M. REWERS, L. MYKKANEN, J. SELBY, G. HOWARD, P. J. SAVAGE, R. F. HAMMAN, L. WAGENKNECKT & R. N. BERGMAN. 1997. Increased insulin resistance and insulin secretion in non-diabetic African-Americans and Hispanics compared to non-Hispanic whites: the Insulin Resistance Atherosclerosis Study. Diabetes **46:** 63–69.

6. KISSELBAH, A., A. N. PEIRIS & D. J. EVANS. 1988. Mechanisms associating body fat distribution to glucose intolerance and diabetes mellitus: window with a view. Acta Med. Scand. **723**: 79–89.
7. HABILTON, B. S., D. PAGLIA, A. Y. M. KWAN & M. DEITEL. 1995. Increased obese mRNA expression in omental fat cells from massively obese humans. Nat. Med. **1**: 953–956.
8. HOTAMISLIGIL, G. S., P. ARNER, J. F. CARO, R. L. ATKINSON & B. M. SPIEGELMAN. 1995. Increased adipose tissue expression of tumor necrosis factor-alpha in human obesity and insulin resistance. J. Clin. Invest. **95**(5): 2409–2415.
9. HOTAMISLIGIL, G. S., N. S. SHARGILL & B. M. SPIEGELMAN. 1993. Adipose expression of tumor necrosis factor-alpha: direct role in obesity-linked insulin resistance. Science **259**(5091): 87–91.
10. LAAKSO, M., S. V. EDELMAN, G. BRECHTEL & A. D. BROWN. 1990. Decreased effect of insulin to stimulate skeletal muscle blood flow in obese men. J. Clin. Invest. **85**: 1844–1852.
11. BARON, A. D., H. STEINBERG, G. BRECHTEL & A. JOHNSON. 1994. Skeletal muscle blood flow independently modulates insulin-mediated glucose uptake. Am. J. Physiol. **266**: E248–253.
12. WU, H-Y., Y. Y. JENG, C-J. YUE, K-Y. CHYU, W. A. HSUEH & T. M. CHAN. 1994. Endothelial-dependent vascular effects of insulin and insulin-like growth factor-1 in the perfused rat mesenteric artery and aortic ring. Diabetes **43**: 1027–1031.
13. BARON, A. D., G. BRECHTEL-HOOK, A. JOHNSON & D. HARDIN. 1993. Skeletal muscle blood flow: a possible link between insulin resistance and blood pressure. Hypertension (Dallas) **21**: 129–135.
14. REAVEN, G. M. 1988. Role of insulin resistance in human disease. Diabetes **37**: 1595–1607.
15. HENNING, B. N. & L. C. GROOP. 1994. Metabolic and genetic characterization of prediabetic states. J. Clin. Invest. **94**: 1714–1721.
16. BAIER, L. J., J. C. SACCHETTINI, W. C. KNOWLER, J. EADS, G. PAOLISSO, P. A. TATARANNI, H. MOCHIZUKI, P. H. BENNETT, C. GOGARDUS & M. PROCHAZKA. 1995. An amino acid substitution in the human intestinal fatty acid binding protein is associated with increased fatty acid binding, increased fat oxidation, and insulin resistance. J. Clin. Invest. **95**(3): 1281–1287.
17. MITCHELL, B. D., C. M. KAMMERER, J. E. HIXSON, L. D. ATWOOD, S. HACKLEMAN, J. BLANGERO, S. M. HAFFNER, M. P. STERN & J. W. MACCLUER. 1995. Evidence for a major gene affecting postchallenge insulin levels in Mexican-Americans. Diabetes **44**(3): 284–289.
18. WILLIAMS, R. R., S. C. HUNT, P. N. HOPKINS, L. L. WU & J. M. LALOUEL. 1994. Evidence for single gene contributions to hypertension and lipid disturbances: definition, genetics, and clinical significance. Clin. Genet. **46**: 80–87.
19. TAYLOR, S. I. & A. ACCILI. 1991. Molecular genetics of insulin-resistant diabetes mellitus. J. Clin. Endocrinol. Metab. **73**: 1158–1163.
20. ZAVARONI, I., E. BONORA, M. PAGLIARA et al. 1989. Risk factors for coronary artery disease in healthy persons with hyperinsulinemia and normal glucose tolerance. N. Engl. J. Med. **320**: 702–706.
21. REAVEN, G. M., Y. D. I. CHEN, J. JAPPASEN, P. MAJEUS & R. M. KRAUSS. 1993. Insulin resistance and hyperinsulinemia in individuals with small, dense, low density lipoprotein particles. J. Clin. Invest. **92**: 141–146.
22. LANDIN, K., L. STIGENDAL, E. ERIKSSON, M. KROTKIEWSKI, B. RISBERG, L. TENGBORN & U. SMIT. 1990. Abdominal obesity is associated with an impaired fibrinolytic activity and elevated plasminogen activator inhibitor-1. Metabolism **39**: 1044–1048.
23. DAWSON, S. J., B. WIMAN, A. HAMSTEIN, F. GREEN, S. HUMPHRIES & A. M. HENNEY. 1993. The two allele sequences of a common polymorphism in the promoter of the plasminogen activator inhibitor-1 (PAI-1) gene respond differently to interleukin-1 in HepG2 cells. J. Biol. Chem. **268**: 10739–10745.
24. PANAHLOO, A., V. MOHAMED-ALI, A. LANE, F. GREEN, S. E. HUMPHRIES & J. S. YUDKIN. 1995. Determinants of plasminogen activator inhibitor 1 activity in noninsulin dependent diabetic patients—relationship with plasma insulin. Thromb. Haemostasis **61**: 370–373.
25. WELBORN, T. A. & K. WEARNE. 1979. Coronary heart disease, incidence, and cardiovascular mortality in Busselton with references to glucose and insulin concentrations. Diabetes Care **2**: 154–160.
26. PYORALA, K. 1979. Relationship of glucose tolerance and plasma insulin to the incidence of coronary heart disease: results from two population studies in Finland. Diabetes Care **2**: 131–141.
27. CUCIMETIERO, P., E. ESCHWEGE, L. PAPOZ, J. L. RICHARD, J. R. CLAUDE & G. ROSSELIN. 1980. Relationship of plasma insulin levels to the incidence of myocardial infarction and coronary heart disease mortality in a middle-aged population. Diabetologia **19**: 205–210.
28. DEAPREE, J. P., B. LAMARCHE, P. MAURIEGA, B. CANTIN, G. R. DAGENALA, S. MOORJANI &

P. J. LUPION. 1996. Hyperinsulinemia as an independent risk factor for ischemic heart disease. N. Engl. J. Med. **334:** 952-957.
29. WINGARD, D. L., E. L. BARRETT-CONNOR & A. FERRARA. 1995. Is insulin really a heart disease risk factor? Diabetes Care **16:** 1299-1304.
30. HSUEH, W. A., R. E. LAW, M. SAAD, J. DY, E. FEENER & G. KING. 1996. Insulin resistance and macrovascular disease. Curr. Opin. Endocrinol. Diabetes **3:** 346-354.
31. SASSON, Z., Y. RASOOLY, T. BHESANIA & I. RASOOLY. 1993. Insulin resistance is an important determinant of left ventricular mass in the obese. Circulation **88:** 1431-1436.
32. HOWARD, G., D. H. O'LEARY, D. ZACCARO, S. HAFFNER, M. ROWERS, R. HAMMAN, J. V. SELBY, M. F. SAAD, P. SAVAGE & R. BERGMAN. 1996. Insulin sensitivity and atherosclerosis. Circulation **93:** 1809-1817.
33. CORKEY, R. F., B. E. CORKEY & M. A. GIMBRONE, JR. 1978. Hexose transport and sorbitol accumulation in cultured human endothelial cells. Diabetes **27:** 446.
34. STEINBERG, H. O., G. GRECTEL, A. JOHNSON, N. FINEBERG & A. D. BARON. 1994. Insulin-mediated skeletal muscle vasodilation is nitric oxide dependent. J. Clin. Invest. **94:** 1172-1179.
35. RAGHVEBDRA, D. K., E. K. JACKSON & T. F. LUSCHER. 1995. Nitric oxide inhibits angiotensin II induced migration of rat aortic smooth muscle cell. J. Clin. Invest. **96:** 141-149.
36. GARG, U. C. & A. HASSID. 1989. Nitric-oxide generating vasodilators and 8-bromo-cyclic guanosine monophosphate inhibit mitogenesis and proliferation of cultured rat vascular SMCs. J. Clin. Invest. **83:** 1774-1777.
37. BANSKOTA, N. K., R. TAUB, K. ZELLNER & G. L. KING. 1989. Insulin, insulin-like growth factor I, and platelet-derived growth factor interact additively in the induction of the protooncogene c-myc and cellular proliferation in cultured bovine aortic smooth cells. Mol. Endocrinol. **89:** 1182-1190.
38. SEGER, R. & E. G. KREBS. 1995. The MAPK signaling cascade. FASEB J. **9:** 726-735.
39. GRAF, K., X-P. XI, D. A. WUTHRICH, W. A. HSUEH & R. E. LAW. 1996. Inhibition of MAP kinase activation blocks insulin-mediated growth and transcriptional activation of c-fos by Elk-1 in VSMCs. Circulation 94(suppl. I): I410.
40. STUMPO, D. J., T. N. STEWART, M. Z. GILMAN & P. J. BLACKSHEAR. 1988. Identification of c-fos sequences involved in induction by insulin and phorbol esters. J. Biol. Chem. **263:** 1611-1614.
41. ANDERSON, P. W., X-Y. ZHANG, J. TIAN, J. M. CORREALE, X-P. XI, D. YANG, K. GRAF, R. E. LAW & W. A. HSUEH. 1996. Insulin and angiotensin II are additive in stimulating TGF-β1 and matrix mRNAs in mesangial cells. Kidney Int. **50:** 745-753.
42. TAMAROGLIO, T. A. & C. S. LO. 1994. Regulation of fibronectin by insulin-like growth factor-1 in cultured rat thoracic aortic smooth muscle cells and glomerular mesangial cells. Exp. Cell Res. **215:** 338-346.
43. IWAMATO, I., T. KUZUYA, A. MATSUDA, S. KUMAKURA, G. INOOKA & I. SHIRAISHI. 1991. Effect of new oral antidiabetic agent CS-045 on glucose tolerance and insulin secretion in patients with NIDDM. Diabetes Care **14:** 1083-1086.
44. NOLAN, J. J., B. LUDVIK, R. BEERDSEN, M. JOYCE & J. OLEFSKY. 1994. Improvement in glucose tolerance and insulin resistance in obese subjects treated with troglitazone. N. Engl. J. Med. **331:** 1188-1193.
45. KELLERER, M., G. KRODER, S. TIPPMER, L. BERTI, R. KIEHN, L. MOSTHAF & H. HARING. 1994. Troglitazone prevents glucose-induced insulin resistance of insulin receptors in rat-1 fibroblasts. Diabetes **43:** 447-453.
46. IWANISHI, M. & M. KOBAYASHI. 1993. Effect of pioglitazone on insulin receptors of skeletal muscles from high-fat-fed rats. Metabolism **42(8):** 1017-1021.
47. KOBAYASHI, M., M. IWANISHI, K. EWAGA & Y. SHIGETA. 1992. Pioglitazone increases insulin sensitivity by activating insulin receptor kinase. Diabetes **41:** 476-483.
48. QON, M. J., H. CHEN, B. L. ING, M. L. LUI, M. J. ZARNOWSKI, K. YONEZAWA, M. KASUGA, S. W. CHUSHMAN & S. I. TAYLOR. 1995. Roles of 1-phosphatidylinositol 3-kinase in regulating translocation of GLUT4 in transfected rat adipose cell. Mol. Cell. Biol. **15:** 5403-5411.
49. BERKOWITZ, K., R. PETERS, S. KJOS, M. DUNN, A. XIANG, J. GOICO, A. MARROQUIN, S. AZEN & T. BUCHANAN. 1996. Effects of troglitazone on insulin sensitivity and B-cell function in women with prior gestational diabetes. Diabetes **45:** 57A.
50. LEE, M. K., D. G. PHILLIP, M. KHOURSHEED, K. M. GAO, A. R. MOOSSA & J. M. OLEFSKY. 1994. Metabolic effects of troglitazone on fructose-induced insulin resistance in the rat. Diabetes **43:** 1435-1439.

51. BUCHANAN, T. A., W. P. MEEHAN, Y. Y. JENG, D. YANG, T. M. CHAN, J. L. NADLER, S. SCOTT, R. K. RUDE & W. A. HSUEH. 1995. Blood pressure lowering by pioglitazone: evidence for a direct vascular effect. J. Clin. Invest. **96:** 354–360.
52. DUBEY, R. K., H. Y. ZHANG, S. R. REDDY, M. A. BOEGEHOLD & T. A. KOTCHEN. 1993. Pioglitazone attenuates hypertension and inhibits growth of renal arteriolar smooth muscle in rats. Am. J. Physiol. **265:** R726–R732.
53. BORNFELDT, K. E., E. W. RAINES, T. NAKANO, L. M. GRAVES, E. G. KREBS & R. ROSS. 1994. Insulin-like growth factor-1 and platelet-derived growth factor-BB induce directed migration of human arterial smooth muscle cells via signaling pathways that are distinct from those of proliferation. J. Clin. Invest. **93:** 1266–1274.
54. BORNFELDT, K. E., H. J. ARNQVIST & L. CAPRON. 1992. In vivo proliferation of vascular smooth muscle in relation to diabetes mellitus, insulin-like growth factor 1, and insulin. Diabetologia **35:** 104–108.
55. LAW, R. E., W. P. MEEHAN, X-P. XI, K. GRAF, D. A. WUTHRICH, W. COATS, D. FAXON & W. A. HSUEH. 1996. Troglitazone inhibits vascular smooth muscle cell growth and intimal hyperplasia. J. Clin. Invest. **98:** 1897–1905.
56. GILLE, H., M. KORTENJANN, O. THOMAE, C. MOOMAW, C. SLAUGHTER, M. H. COBB & P. E. SHAW. 1995. ERK phosphorylation potentiates Elk-1-mediated ternary complex formation and transactivation. EMBO J. **14:** 951–962.
57. GILLE, H., M. KORTENJANN, T. STRAHL & P. E. SHAW. 1996. Phosphorylation-dependent formation of a quaternary complex at the *c-fos* SRE. Mol. Cell. Biol. **16:** 1094–1102.
58. GRAF, K., X-P. XI, J. TIAN, R. E. LAW & W. A. HSUEH. 1996. MAP kinase–activation is involved in angiotensin II and PDGF-directed migration by vascular smooth muscle cells. *In* 50th Scientific Sessions, Council for High Blood Pressure (abstract).
59. GRAF, K., X-P. XI, D. YANG, E. FLECK, R. E. LAW & W. A. HSUEH. 1997. Map kinase-activation is involved in PDGF-directed migration by vascular smooth muscle cells. Hypertension **29:** 334–339.
60. PERSHADSINGH, H. A., J. SZOLLOSI, S. BENSON, W. C. HYUN, B. G. FEUERSTEIN & T. W. KURTZ. 1993. Effects of ciglitazone on blood pressure and intracellular calcium metabolism. Hypertension **21:** 1020–1023.
61. ZHANG, F., J. R. SOWERS, J. L. RAM *et al.* 1994. Effects of pioglitazone on calcium channels in vascular smooth muscle. Hypertension **24:** 170–175.

Fatty Acid Regulation of Gene Expression

Its Role in Fuel Partitioning and Insulin Resistance[a]

STEVEN D. CLARKE,[b] REBECCA BAILLIE,[b]
DONALD B. JUMP,[c] AND MANABU T. NAKAMURA[b]

[b] Division of Nutritional Sciences
Department of Human Ecology
University of Texas
Austin, Texas 78712

[c] Departments of Physiology and Biochemistry
Michigan State University
East Lansing, Michigan 48824

INTRODUCTION

In the early stages of evolution, cellular growth and evolutionary success required that the developing life form respond to a myriad of environmental factors. In particular, the organism had to possess an ability to fulfill its nutrient needs and had to develop a sense for nutrient deficiency and excess in order to turn on pathways of synthesis or storage. Thus, nutrient control of gene expression probably evolved as one of the earliest environmental sensor mechanisms. This primitive humoral system likely operated to govern the expression and function of enzymes essential to cellular growth and survival. Nutrient regulation of gene expression remains today a fundamental player in growth and development, and is a key player in the development of nutritionally related pathophysiologies, such as diabetes, cancer, and heart disease. In this respect, dietary fatty acids have surfaced as significant players in the etiology of insulin resistance. The regulatory actions of fatty acids appear to involve two targets: (a) rapid and direct modification of gene transcription;[1] (b) modulation of signal transduction via manipulation of membrane fatty acid composition.[2] The role that membrane phospholipid fatty acid composition plays in hormonal signaling and receptor activity has been extensively pursued over the past 30 years,[3] but the notion that fatty acids rapidly and directly modify the transcription of specific genes is a

---

[a] This work was supported by grants from the USDA (No. NRICGP/USDA 92-3700-7465) and a Mead-Johnson Grant-in-Aid (to S. D. Clarke), by the National Institutes of Health (Grant No. DK 43220) (to D. B. Jump), and by the M. M. Love Chair for Nutritional, Cellular, and Molecular Sciences at the University of Texas (to S. D. Clarke).

rather recent discovery. Thus, our purpose is to focus upon the role that dietary fats play as modulators of gene transcription and to describe how changes in gene expression may alter fuel partitioning within a cell and among organs.

DIETARY FAT: IMPAIRED INSULIN BINDING AND GLUCOSE TRANSPORT

Historically, diabetes was viewed as a defect in glucose metabolism, and diabetic patients were treated with a low-carbohydrate, high-fat diet. However, over the past two decades, it has become increasingly clear that non-insulin-dependent diabetes mellitus (NIDDM) is a disorder of lipid metabolism as well as a disease of glucose intolerance. Moreover, several lines of evidence with humans and animal models indicate that dietary fat may play a causative role in the development of NIDDM.[4,5] In this regard, Anderson et al.[6] found that the oral glucose tolerance of normal-weight, obese, and mildly diabetic individuals decreased as the fat content of a liquid diet increased. This effect of dietary fat was not due to the removal of energy as carbohydrate from the diet because replacing comparable amounts of carbohydrate energy with calories derived from protein did not adversely affect glucose tolerance.[7]

The reduced metabolism of glucose that results from the ingestion of a high-fat diet appears to reflect an impairment in insulin signaling and receptor binding,[8-15] which in turn leads to (a) an increased level of hepatic glucose production and reduced hepatic glucose utilization,[12,13] (b) an impairment in adipose tissue glyceride-glycerol formation within the adipose tissue,[8,9] and (c) a reduction in glucose oxidation of the skeletal muscle.[12,13] This conclusion is supported by several lines of evidence from both human and animal model studies. As an example, hepatocyte and adipocyte membranes prepared from rats fed a high-lard, low-carbohydrate diet bound 50–75% less insulin than did rats fed a high-glucose diet.[8,9,11,14,15] This reduction in insulin binding was accompanied by a decrease in adipocyte glucose uptake and oxidation.[10,15] Similarly, euglycemic clamp measurements in rats fed a high-safflower-oil diet (versus high-glucose) revealed that the high-fat diet reduced whole body glucose disposal by 50% and that 29–61% of this suppressive effect of fat was attributable to impaired glucose oxidation in skeletal muscles.[11,12] This impairment in glucose utilization can partially be attributed to changes in two key metabolic events. First, dietary fat suppresses the fraction of catalytically active adipocyte and hepatocyte pyruvate dehydrogenase,[11,16] which in turn impairs the entry of pyruvate into the tricarboxylic acid cycle for oxidation or fatty acid biosynthesis. Second, prolonged ingestion of fat not only reduces the catalytic function of pyruvate dehydrogenase, but also reduces the amount of enzymatic protein via inhibition of pyruvate dehydrogenase gene transcription.[17] In addition, high-fat diets suppress the level of protein and mRNA for insulin-responsive glucose transporter-4.[18] Thus, it appears that the inhibition of gene expression by dietary fat may be the postinsulin receptor target that leads to an overall impairment in glucose utilization and to the development of insulin resistance.

FATTY ACIDS IN INSULIN SECRETION

Dietary fat not only precipitates insulin resistance in muscle, liver, and adipose tissue, but also appears to cause anomalies in glucose-stimulated insulin release from the β-cell.[5] In this regard, a key determinant in the glucose-dependent release of insulin is the synthesis of malonyl-CoA from glucose.[5] Thus, dietary factors that modulate the expression of pancreatic enzymes such as glucokinase, pyruvate kinase, and particularly acetyl-CoA carboxylase will in turn regulate the release of insulin from the pancreas. Recently, glucose was found to induce the transcription and mRNA level of pancreatic glucose transporter-2 and acetyl-CoA carboxylase.[19,20] In addition, glucose increased the amount of mRNA for islet pyruvate carboxylase and the Ela component of pyruvate dehydrogenase.[5] These changes in gene expression correlate nicely with the observation that exposing pancreatic islets for two days to elevated glucose in a partial pancreatectomy model resulted in a leftward shift in the dose dependence of glucose-induced insulin secretion.[5] On the other hand, prolonged exposure of islet cells to free fatty acids suppressed glucose-dependent insulin release[5] and significantly inhibited the transcription of acetyl-CoA carboxylase.[5,21] Presumably, the inhibition of acetyl-CoA carboxylase expression by fatty acids would lead to a reduction in β-cell content of malonyl-CoA, which in turn would impair glucose-stimulated insulin release.[5,21] It should be noted that the change in gene expression resulting from chronic exposure to high levels of free fatty acids is a long-term adaptation and that acute exposure of the β-cell to free fatty acids amplifies the glucose-dependent release of insulin.[5] In any case, suppression of pancreatic β-cell gene expression may explain how high-fat diets or obesity lead to impaired insulin secretion. In addition, since dietary fish oils rich in 20/22-carbon (n-3) fatty acids are more effective suppressors of glucokinase and acetyl-CoA carboxylase gene transcription,[1,22] this may also explain why dietary fish oils reduce insulin secretion.[23] The mechanisms by which fatty acids and glucose govern the expression of pancreatic glucokinase and acetyl-CoA carboxylase have not been elucidated, but it would not be surprising if these mechanisms were comparable to those governing gene expression in the liver (discussed later).

THE "FAT PARADOX": NOT ALL FATTY ACIDS IMPAIR GLUCOSE METABOLISM

Early studies addressing the impact of dietary fat on insulin resistance and glucose metabolism utilized diets that were exceedingly low in carbohydrate and generally employed lard as the primary, if not sole, source of fat. However, numerous studies over the past 25 years have demonstrated that saturated, monounsaturated, and polyunsaturated fatty acids do not regulate glucose and lipid metabolism in the same way or, more importantly, with the same outcomes.[1,2] For example, a diet with a high proportion of (n-6) polyenoic fatty acids was found to improve adipocyte insulin binding and insulin-stimulated glucose transport in both diabetic and nondiabetic rats when compared to a diet with a low (n-6)/saturated fat ratio.[11,15] In spite of demonstrating an improvement in glucose metabolism relative to saturated fat, a high (n-6) fat intake continues to precipitate an impairment in glucose metabolism.[12] However, a very interesting and significant observation is the following: if 20% of the (n-6) fatty acids are replaced with long-chain (n-3) fatty acids, whole body glucose metabolism is improved twofold and nearly reaches a level comparable to that observed with a high-carbohydrate diet.[13] The improvements in glucose disposal associated with (n-3) fatty acid ingestion may reflect the 50% increase in glucose transporter-4 content of skeletal muscle and the near-doubling of glucose transporter-4 and −1 of adipose tissue.[14,18] In contrast to the response of muscle and adipose tissues, hepatic utilization of glucose for fatty acid and triglyceride biosynthesis is markedly suppressed by dietary polyunsaturated fatty acids (PUFA),[1,24] particularly those fatty acids that have undergone delta-6 desaturation.[25] As an example, as little as 3% menhaden oil or 2.5% arachidonate in the diet suppressed hepatic fatty acid biosynthesis by 50% and 90%, respectively.[25] While dietary PUFA suppress hepatic lipogenesis, they concomitantly induce hepatic, and possibly muscle, fatty acid oxidation in peroxisomes and mitochondria, as well as increase hepatic ketone synthesis and raise hepatic glycogen levels.[13,24,26,27] Clearly, dietary fats vary markedly with respect to their impact on glucose metabolism and insulin resistance. Moreover, dietary polyunsaturated fats, notably those that have undergone delta-6 desaturation, appear to function as effective fuel partitioners (i.e., they modulate the organs and pathways of fuel metabolism). With these observations in mind, we have developed a hypothesis that proposes that dietary PUFA, particularly in the long-chain (n-3) fatty acids of fish oil, improve the anomalies in glucose and lipid metabolism associated with fat ingestion by coordinately modulating the transcription of genes encoding key enzymes of lipogenesis, fatty acid oxidation, and glucose transport. The outcome of such coordinate regulation is (a) a redirecting of hepatic glucose away from glycolysis and lipogenesis and toward glycogen production, combined with a concomitant increase in hepatic peroxisomal and mitochondrial fatty acid oxidation and a stimulation of hepatic ketone production; (b) a proliferation of peroxisomes in skeletal muscle and liver that leads to increased fatty acid flux through the peroxisomal fatty acid oxidative pathway, which in turn releases the negative feedback effect that fatty acids exert on skeletal muscle glucose metabolism; and (c) an improved muscle and adipose tissue glucose utilization due to a stimulation of glucose transporter activity.

GENE EXPRESSION AND POLYUNSATURATED FATTY ACIDS: A ROLE IN THE REGULATION OF FUEL PARTITIONING

Understanding how 20/22-carbon (n-3) fatty acids, and possibly 20:4 (n-6), improve insulin resistance via coordinate regulation of fatty acid oxidation and glucose utilization requires an understanding of the integrative changes that take place in the expression of key regulatory genes of liver, muscle, and adipose that are involved in lipid and glucose homeostasis. One factor associated with the development of insulin resistance is a high hepatic output of triglycerides, particularly those rich in saturated fatty acids.[2,5] In this respect, dietary polyunsaturated fats, notably those rich in 20:5 (n-3), markedly reduce hepatic triglyceride output by (a) suppressing hepatic *de novo* fatty acid biosynthesis,[1] (b) reducing hepatic triglyceride synthesis,[28] and (c) increasing hepatic fatty acid oxidation.[26]

Over three decades ago, Allmann and Gibson[29] discovered that adding 2% 18:2 (n-6) to a high-carbohydrate, fat-free diet of mice suppressed the rate of hepatic fatty acid biosynthesis and the activities of fatty acid synthase, malic enzyme, and glucose-6-phosphate dehydrogenase by nearly 70%. In contrast, supplementing the carbohydrate-rich diet with palmitate or oleate had no effect on hepatic lipogenesis or the activity of lipogenic enzymes. Over the past 25 years, several investigators have demonstrated that dietary polyenoic fatty acids of the (n-6) and (n-3) families suppress hepatic lipogenesis. These suppressive actions of PUFA occur at a level of intake that is fourfold to fivefold greater than that needed to fulfill the essential fatty acid requirement for optimal growth, and the effects are independent of carbohydrate intake or eicosanoid synthesis.[30]

The antilipogenic action of PUFA reflects the unique ability of these fatty acids to reduce the hepatic activities of enzymes in lipogenesis and glycolysis.[1,22] Such suppression of enzymatic activities does not represent a fatty acid–mediated impairment in enzyme catalytic efficiency (e.g., fatty acyl-CoA inhibition of acetyl-CoA carboxylase), but rather reflects a decrease in hepatic enzyme content caused by a PUFA-mediated suppression of enzyme synthesis.[1,25,30] A clearer understanding of how PUFA decrease enzyme synthesis came with the demonstration that PUFA decreased hepatic fatty acid synthase mRNA abundance and that this was the consequence of an inhibition of gene transcription.[22,31,32] The transcriptional response to dietary PUFA is very rapid. For example, replacing triolein in a high-carbohydrate diet with 5% menhaden oil resulted in a 60% suppression of transcription for the fatty acid synthase, pyruvate kinase, and S14 genes in less than 3 h.[22] Similarly, removing the PUFA from the diet induced gene transcription within 3 h.[32] Such a rapid response strongly suggests that PUFA directly modulate gene transcription rather than exert their influence by modifying membrane lipid fatty acid composition and altering hormone release or signaling. Moreover, the selective control of lipogenic gene expression can be re-created with hepatocytes in culture, which eliminates from consideration the idea that PUFA exert their effects by modifying insulin or glucagon release.[25]

In addition to suppressing rates of hepatic glycolysis and fatty acid biosynthesis by inhibiting the expression of genes encoding glycolytic and lipogenic enzymes, 20/22-carbon polyenoic (n-6) and (n-3) fatty acids are potent inhibitors of hepatic triglyceride synthesis.[28] Pathway flux studies indicate that 20:4 (n-6) and 20:5 (n-3) suppress the activity of diacylglycerol acyltransferase;[28] however, because this gene has not yet been cloned, it is impossible to determine if this effect is at the level of gene expression or simply altered catalytic function.

Polyunsaturated fatty acids not only inhibit the rates of hepatic synthesis of lipid, but they concomitantly increase the amount of mitochondrial and peroxisomal fatty acid oxidation, which is accompanied by an enhanced rate of ketogenesis.[24,26,30] A key factor for the enhanced hepatic oxidation of fatty acids is the induction of peroxisomal acyl-CoA oxidase.[26,27,33] Acyl-CoA oxidase catalyzes the initial rate-limiting step in peroxisomal fatty acid oxidation.[33] More importantly, diets rich in very long chain polyenoic fatty acids (e.g., fish oil) are associated with peroxisomal proliferation and greater acyl-CoA oxidase gene expression.[26,27] Thus, the hepatic capacity for peroxisomal oxidation is increased. In addition, the enhanced mitochondrial oxidation of fatty acids associated with fish oil ingestion may reflect an increased activity of carnitine palmitoyl transferase resulting from the release of its inhibition by malonyl-CoA because PUFA have suppressed acetyl-CoA carboxylase activity and thereby decreased malonyl-CoA synthesis.[5] Polyunsaturated fatty acids not only induce the expression of acyl-CoA oxidase, but they have recently been shown to stimulate the transcription of mitochondrial HMG-CoA synthase, a key regulatory enzyme in the synthesis of ketones.[34] This induction of oxidative and ketogenic enzymes associated with PUFA ingestion is paralleled by an increase in hepatic ketone output.[24] Clearly, PUFA function is to coordinately regulate the fate of hepatic fatty acids. Moreover, a defect in any of these adaptive responses would potentially increase hepatic triglyceride output and modulate peripheral tissue fuel metabolism.

The triglyceride content of skeletal muscle is highly correlated with insulin resistance and impaired glucose metabolism.[2,18] In this respect, Storlien et al.[2] have reported that triglyceride levels above 9 micromoles per gram lead to significant impairments in muscle glucose utilization as a result of Randle cycle action. One effect of dietary fats rich in 20/22-carbon fatty acids (e.g., fish oil) is to enhance hepatic fatty acid oxidation, particularly in the peroxisome.[26,35] Like the liver, skeletal muscle peroxisomal fatty acid oxidation may also increase in response to 20-carbon PUFA.[26] We have found that feeding rats a fat mixture of corn oil and fish oil (40%/20% of calories) induced an increase in the mRNA level of acyl-CoA oxidase in skeletal muscle by severalfold.[27] Dietary PUFA also induced peroxisomal acyl-CoA oxidase expression in kidney, small intestine, and to some extent adipose.[27] Peroxisomal fatty acid oxidation rates remain to be determined in order to establish that an increased expression of acyl-CoA oxidase in muscle is accompanied by greater peroxisomal fatty acid oxidation. If this is the case, enhanced peroxisomal fatty acid oxidation may explain how dietary (n-3) fatty acids improve skeletal muscle glucose utilization. Peroxisomal proliferation (i.e., induced acyl-CoA oxidase expression) in skeletal muscle would not only increase the muscle's capacity to oxidize fatty acids, but it would have a thermogenic effect as well because the ATP yield from peroxisomal fatty acid oxidation is 35-50% less than from mitochondrial oxidation.[33] In support of the concept that dietary fish oil may increase skeletal muscle fatty acid oxidation, Storlien's group[2] and Klimeš et al.[18] have reported that dietary fish oil reduced skeletal muscle triglyceride content and concomitantly improved glucose utilization. Thus, it appears that dietary 20-carbon (n-3) and (n-6) fatty acids improve glucose metabolism by coordinately regulating genes involved in fatty acid biosynthesis and fatty acid oxidation, which in turn is accompanied by an enhanced expression of the skeletal muscle glucose transporter-4 gene and an improved skeletal muscle glucose uptake.

The final coordinate action of dietary 20-carbon PUFA is at the level of adipose tissue glucose utilization.[36,37] While PUFA enhance glucose transport in the skeletal muscle,[13,18] they appear to suppress the expression of the insulin-stimulated glucose transporter in adipose tissue.[36,37] Tebbey et al.[36] have reported that exposing fully differentiated 3T3-L1 fat cells to 50 µM 20:4 (n-6) suppressed glucose transporter-4 mRNA abundance by 91%. Control of transporter activity by 20:4 (n-6) involved suppressed gene transcription and changes in the transcript half-life.[36] The decrease in glucose transporter-4 mRNA was paralleled by a comparable reduction in transporter-4 protein. Interestingly, the total cellular amount of glucose transporter-1 was unaffected by 20:4 (n-6), but the plasma membrane content of transporter-1 was greatly increased. Recent data from Pekala's group[37] indicate that 20:4 (n-6) mediates its control of glucose transporter-4 via a prostaglandin-dependent process.

In summary, it appears that 20-carbon (n-6) and (n-3) fatty acids modulate the expression of a collection of genes encoding proteins of fatty acid and glucose metabolism and that changes in the amount of these proteins govern the rates of pathway flux and thereby determine the mixture of macronutrient metabolism.

THE MOLECULAR MECHANISMS RESPONSIBLE FOR POLYUNSATURATED FATTY ACID REGULATION OF GENE EXPRESSION

Dietary fats appear to influence the onset and progression of various nutritionally related pathophysiologies such as insulin resistance by exerting an effect at two levels: (a) changes in membrane phospholipid composition and (b) direct control of the nuclear events that govern gene transcription. Thus, the beneficial as well as the detrimental effects of dietary fats may involve a combination of interactive regulatory mechanisms: (a) rapid changes in gene expression and (b) a long-term adaptive modulation of membrane composition leading to alteration in hormone signaling. In the latter scenario, dietary fats high in PUFA enrich plasma and microsomal membranes with long-chain polyenoic fatty acids. This enrichment subsequently alters hormone binding to cell-surface receptors, alters signal transduction proteins, and/or modifies the combination of lipid mediators (e.g., diacylglycerols).[2,3,5,11,15,23] Such changes are translated into modified signals that function to regulate gene expression via altered states of protein phosphorylation, acylation, etc. In addition to modifying the fatty acid composition of membrane phospholipids, fatty acids appear to have the ability to rapidly and directly regulate the transcription of several genes involved in lipid and glucose metabolism. Because of the rapidity with which dietary PUFA suppressed the transcription of lipogenic and glycolytic genes, we proposed early on that long-chain polyenoic fatty acids may be a ligand for a nuclear protein that functioned as a dominant modifier of gene expression. Our concept of a polyunsaturated fatty acid binding protein (PUFA-BP) was strengthened with the discovery of a family of transcription factors termed peroxisome proliferator activated receptors (PPARs).[38-41] PPARs are members of the steroidlike receptor superfamily and are found in nearly all tissues. Activation of PPARs appears to be a ligand-mediated event and the potential ligand activators include amphipathic carboxylates, herbicides, plasticizers, and most fatty acids.[38-41] Interestingly, peroxisomal proliferating agents, for example, clofibrate, reportedly suppress the expression of fatty acid synthase and inhibit hepatic lipogenesis, while concomitantly inducing peroxisomal genes including acyl-CoA oxidase.[42] These data have led many investigators to suggest that PPARs are the *trans*-factors responsible for mediating the fatty acid regulation of gene transcription. However, while it appears that PPARs may be responsible for the induction of acyl-CoA oxidase gene transcription, a number of recent *in vivo* and *in vitro* studies have failed to support the hypothesis that the PPARs mediate the PUFA inhibition of hepatic glycolytic and lipogenic genes.[27,43,44]

First, a polyunsaturated fatty acid response region (PUFA-RE) has been located in the proximal promoter of the pyruvate kinase and S14 genes.[45] However, the DNA sequence of the region does not correspond to a characteristic PPAR response element. Moreover, linker scanning mutation analysis of the PUFA-RE of the pyruvate kinase gene indicates that the site is an HNF-4 binding site and that PPARs will not bind to the PUFA-RE.[44,45] Interestingly, HNF-4, like PPAR, is a member of the steroid receptor superfamily, but there is no evidence at this time that fatty acids act as ligand activators for HNF-4. It should be noted that the PUFA-RE of the pyruvate kinase and S14 genes may not be common to all affected glycolytic and lipogenic genes. For example, a comparable PUFA-RE sequence is located in the proximal promoter and in the first intron of the fatty acid synthase gene, but functional analysis studies have failed to detect a PUFA-RE in this sequence (unpublished). Thus, it is possible that polyenoic fatty acids regulate gene transcription at multiple sites and by a variety of mechanisms.

The strongest evidence suggesting that fatty acids regulate gene transcription by modulating PPAR actions is derived from transcription *trans*-activation assays.[40,41] However, in these assays, saturated, monounsaturated, and polyunsaturated fatty acids all have equal potency as activators of PPARs and all stimulate reporter gene transcription equally well. Thus, the PPAR-like proteins do not appear to exhibit the selectivity required to explain the regulation of lipogenic and glycolytic gene transcription by (n-6) and (n-3) fatty acids.

The third and perhaps most-convincing piece of data indicating that PUFA do regulate lipogenic gene transcription through a PPAR-dependent process comes from our *in vivo* studies with the PPAR activator eicosatetraynoic acid (ETYA).[27] ETYA is an analogue of 20:4 (n-6) and is a potent activator of PPARs. When ETYA was fed to mice, it induced the expression of hepatic acyl-CoA oxidase as expected.[27] However, ETYA had no inhibitory effect on hepatic fatty acid synthase expression. In fact, ETYA prevented 18:2 (n-6) from inhibiting fatty acid synthase gene transcription.[27] Thus, fatty acid activation may be the mechanism by which 20-carbon polyenoic fatty acids induce peroxisome proliferation, but current data suggest that the suppression of lipogenic gene expression by (n-6) and (n-3) fatty acids involves a PPAR-independent process.

SUMMARY

Dietary polyenoic (n-6) and (n-3) fatty acids uniquely regulate fatty acid biosynthesis and fatty acid oxidation. They exercise this effect by modulating the expression of genes coding for key metabolic enzymes and, in doing this, PUFA govern the intracellular as well as the interorgan metabolism of glucose and fatty acids. During the past 20 years, we have gradually elucidated the cellular and molecular mechanism by which dietary PUFA regulate lipid metabolism. Central to this mechanism has been our ability to determine that dietary PUFA regulate the transcription of genes. We have only begun to elucidate the nuclear mechanisms by which PUFA govern gene expression, but one point is clear and that is that it is unlikely that one mechanism will explain the variety of genes governed by PUFA. The difficulty in providing a unifying hypothesis at this time stems from (a) the many metabolic routes taken by PUFA upon entering a cell and (b) the lack of identity of a specific PUFA-regulated *trans*-acting factor. Nevertheless, our studies have revealed that PUFA are not only utilized as fuel and structural components of cells, but also serve as important mediators of gene expression, and that in this way they influence the metabolic directions of fuels and they modulate the development of nutritionally related pathophysiologies such as diabetes.

REFERENCES

1. CLARKE, S. D. & D. B. JUMP. 1994. Annu. Rev. Nutr. **14**: 83–98.
2. STORLIEN, L. H., A. B. JENKINS, D. J. CHISHOLM, W. S. PASCOE, S. KHOURI & E. W. KRAEGEN. 1991. Diabetes **40**: 280–289.
3. SPECTOR, A. A. & M. A. YORK. 1985. J. Lipid Res. **26**: 1015–1035.
4. CLARKE, S. D. & D. R. ROMSOS. 1980. *In* Nutrition and the Adult. Vol. 3A, p. 141–158. Plenum. New York.
5. PRENTKI, M. & B. E. CORKEY. 1996. Diabetes **45**: 273–283.
6. ANDERSON, J. W., R. H. HERMAN & D. ZAKIM. 1973. Am. J. Clin. Nutr. **26**: 600–606.
7. ANDERSON, J. W. & R. H. HERMAN. 1975. Am. J. Clin. Nutr. **28**: 748–753.
8. KOLTERMAN, O. G., M. GREENFIELD, G. M. REAVEN, M. SAEKOW & J. M. OLEFSKY. 1979. Diabetes **28**: 731–736.
9. OLEFSKY, J. M. & M. SAEKOW. 1978. Endocrinology **103**: 2252–2263.
10. IP, C., H. M. TEPPERMAN, P. HOLOHAN & J. TEPPERMAN. 1976. J. Lipid Res. **17**: 588–599.
11. TEPPERMAN, H. M. & J. TEPPERMAN. 1985. Proc. Nutr. Soc. **44**: 211–220.
12. STORLIEN, L. H., D. E. JAMES, K. M. BURLEIGH, D. J. CHISHOLM & E. W. KRAEGEN. 1986. Am. J. Physiol. **250**: E576–E583.
13. STORLIEN, L. H., E. W. KRAEGEN, D. J. CHISHOLM, G. L. FORD, D. G. BRUCE & W. S. PASCOE. 1987. Science **237**: 885–888.
14. EZAKI, O., E. TSUJI, K. MOMOMURA, M. KASUGA & H. ITAKURA. 1992. Am. J. Physiol. **263**: E94–E101.
15. FIELD, C. J., E. A. RYAN, A. B. R. THOMSON & M. T. CLANDININ. 1989. J. Biol. Chem. **264**: 11143–11150.
16. BEGUM, N., H. M. TEPPERMAN & J. TEPPERMAN. 1982. Endocrinology **110**: 1914–1921.
17. DA SILVA, L. A., O. L. DE MARCUCCI & Z. R. KUHNLE. 1993. Biochim. Biophys. Acta **1169**: 126–134.
18. KLIMEŠ, I., E. ŠEBÖKOVÁ, A. MINCHENKO, T. MAASEN, R. MOSS, A. MITKOVÁ, M. WIERSMA & P. BOHOV. 1994. J. Nutr. Biochem. **5**: 389–396.
19. WAEBER, G., N. THOMPSON, N. A. HAEFLIGER & P. NICOD. 1994. J. Biol. Chem. **269**: 26912–26919.
20. BRUN, T., E. ROCHE, K. H. KIM & M. PRENTKI. 1993. J. Biol. Chem. **268**: 18905–18911.
21. BRUN, T., E. ROCHE, F. ASSIMACOPOULOS-JEANNET, B. E. CORKEY, K. H. KIM & M. PRENTKI. 1996. Diabetes **45**: 190–198.
22. JUMP, D. B., S. D. CLARKE, A. THELEN & M. LIIMATTA. 1994. J. Lipid Res. **35**: 1076–1084.
23. CHICCO, A., M. E. D'ALESSANDRO, L. KARABATAS, R. GUTMAN & Y. B. LOMBARDO. 1996. Ann. Nutr. Metab. **40**: 61–70.
24. WONG, S. H., P. J. NESTEL, R. P. TRIMBLE, G. B. STORER, R. J. ILLMAN & D. L. TOPPING. 1984. Biochim. Biophys. Acta **792**: 103–109.
25. CLARKE, S. D. & D. B. JUMP. 1993. *In* Nutrition and Gene Expression. CRC Press. Boca Raton, Florida.

26. TAKADA, R., M. SAITOH & T. MORI. 1994. J. Nutr. **124:** 469–474.
27. CLARKE, S. D., M. TURINI & D. JUMP. 1996. Prostaglandins Leukotrienes Essent. Fatty Acids. In press.
28. ODIN, R. S., B. A. ADKINS-FINKE, W. L. BLAKE, S. D. PHINNEY & S. D. CLARKE. 1987. Biochim. Biophys. Acta **921:** 378–391.
29. ALLMANN, E. W. & D. W. GIBSON. 1969. J. Lipid Res. **6:** 51–60.
30. CLARKE, S. D. & D. B. JUMP. 1993. Prog. Lipid Res. **32:** 139–149.
31. BLAKE, W. L. & S. D. CLARKE. 1990. J. Nutr. **120:** 1727–1729.
32. CLARKE, S. D., M. K. ARMSTRONG & D. B. JUMP. 1990. J. Nutr. **120:** 225–232.
33. REDDY, J. K. & G. P. MANNAERTS. 1994. Annu. Rev. Nutr. **14:** 343–370.
34. RODRIGUEZ, J. C., G. GIL-GOMEZ, F. G. HEGARDT & D. HARO. 1994. J. Biol. Chem. **269:** 18767–18772.
35. FLATMARK, T., A. NIILSSON, J. KRANNES, T. S. EIKHOM & M. H. FUKAMI. 1988. Biochim. Biophys. Acta **962:** 122–130.
36. TEBBEY, P. W., K. M. MCGOWAN, S. M. STEPHENS, T. M. BUTTKE & P. H. PEKALA. 1994. J. Biol. Chem. **269:** 639–644.
37. LONG, S. D. & P. H. PEKALA. 1996. J. Biol. Chem. **271:** 1138–1144.
38. GREEN, S. 1992. Biochem. Pharmacol. **43:** 393–401.
39. SCHMIDT, A., N. ENDO, S. J. RUTLEDGE, R. VOGEL, D. SHINAR & G. A. RODAN. 1992. Mol. Endocrinol. **6:** 1634–1641.
40. KELLER, H., C. DREYER, J. MEDIN, A. MAHFOUDI, K. OZATO & W. WAHLI. 1993. Proc. Natl. Acad. Sci. U.S.A. **90:** 2160–2164.
41. GOTTLICHER, M., E. WIDMARK, Q. LI & J. A. GUSTAFASSON. 1992. Proc. Natl. Acad. Sci. U.S.A. **89:** 4653–4657.
42. RUSTAN, A. C., E. A. CHRISTIANSEN & C. A. DREVON. 1992. Biochem. J. **283:** 333–339.
43. REN, B., A. THELEN & D. B. JUMP. 1996. J. Biol. Chem. **271:** 17167–17173.
44. JUMP, D. B., B. REN, S. D. CLARKE & A. THELEN. 1995. Prostaglandins Leukotrienes Essent. Fatty Acids **52:** 107–111.
45. LIIMATTA, M., H. C. TOWLE, S. D. CLARKE & D. B. JUMP. 1994. Mol. Endocrinol. **8:** 1147–1153.

Impact of Dietary Fatty Acid Composition on Insulin Action at the Nucleus

NANA A. GLETSU[a] AND M. THOMAS CLANDININ[a,b,c]

a Department of Agricultural, Food, and Nutritional Science
b Department of Medicine
Nutrition and Metabolism Research Group
University of Alberta
Edmonton, Alberta, Canada T6G 2P5

Membrane lipid composition can be altered by dietary fat or by disease states such as diabetes and obesity. Such alterations can affect the activities of several membrane-associated enzymes and receptors.[1] Increasing dietary polyunsaturated fatty acid increases membrane unsaturated fatty acid content and also alters gene expression. Feeding polyunsaturated fatty acids inhibits the expression of hepatic glycolytic and lipogenic enzymes including pyruvate kinase, fatty acid synthase, malic enzyme, and glucokinase.[2] The expression of many of these enzymes is also regulated by various hormones including insulin.[3-7] The mechanism by which polyunsaturated fatty acids and insulin influence gene expression in the cell nucleus is unknown. Several *cis*-acting elements that confer insulin or polyunsaturated fatty acid responsiveness to certain genes have been discovered in the promoter regions of these genes. DNA-binding transcription factors that recognize these elements have not yet been defined, nor have the intracellular mediators in the signaling cascade distal to insulin and polyunsaturated fatty acid cell-surface binding been delineated. Insulin receptors have been observed[8-12] in the nucleus of hepatocytes and adipocytes and it is speculated that these receptors may provide a mechanism by which insulin signaling reaches the nucleus. The insulin receptor tyrosine kinase may activate DNA-binding transcription factors by increasing their phosphorylated state. Dietary polyunsaturated fatty acids may affect insulin action at the nucleus by modifying the nuclear membrane environment of the insulin receptor. Alternatively, polyunsaturated fatty acids may influence the insulin control of transcription of certain genes by regulating this rate via an independent fashion. We have explored this question using feeding experiments in normal animals and using disease models characterized by abnormal insulin production and/or function such as diabetes and obesity. Our findings demonstrate that high polyunsaturated fatty acid feeding can improve insulin receptor binding to the nucleus in normal animals and in animals with obesity and diabetes. Obesity is a unique disease in which to study insulin function as insulin is overexpressed, yet the organism appears to be

---

[c] To whom all correspondence should be addressed.

relatively resistant to its effects. The enhanced rate of expression of insulin-responsive lipogenic genes found in animals with obesity was suppressed by high polyunsaturated fat feeding. Polyunsaturated fatty acids thus appear to have a dual action: altering insulin binding to the nuclear membrane and controlling the rate of lipogenic gene expression via a distinct mechanism. This paper will review some of our research suggesting that altered insulin action due to dietary fatty acid modification or disease will affect insulin regulation of gene expression.

DIETARY FAT ALTERS THE STRUCTURE OF THE NUCLEAR ENVELOPE

It has been well documented that plasma membrane lipid composition can be altered by changing the levels of specific fatty acid in the diet. Increasing the ratio of dietary polyunsaturated to saturated fat (P/S) increases the total polyunsaturated fatty acid content of adipocyte plasma membrane and, more specifically, the content of 18:2(ω-6) and 20:4(ω-6) in all membrane phospholipid classes.[13] It appears that the composition of the nuclear envelope is also responsive to dietary fatty acid. A similar relationship between dietary P/S ratio and lipid composition is seen in the hepatocyte nucleus. Rats and mice fed a high-P/S diet have higher levels of 18:2(ω-6) in phosphatidylcholine (PC), phosphatidylethanolamine (PE), and phosphatidylinositol (PI), and higher levels of 20:4(ω-6) in PC, compared to animals fed a low-P/S diet (TABLE 1).[1,14,15] The ratio of ω-6 to ω-3 polyunsaturated fatty acids (PUFA) in the nuclear membrane can also be manipulated by diet. Substituting increasing levels of 20:5(ω-3) and 22:6(ω-3) for 18:2(ω-6) in a low-P/S diet resulted in a displacement of ω-6 for ω-3 long-chain (LC) PUFA in the nuclear membrane.[16] Nuclear membrane fluidity and/or protein-lipid interactions and subsequently membrane protein function may be influenced by the increased incorporation of dietary polyunsaturated fatty acids in the nuclear membrane.

DIETARY FAT, DISEASE, AND THE NUCLEAR ENVELOPE

Obesity and diabetes are characterized by abnormalities in lipid metabolism. When compared to normal animals, obese and diabetic animals show marked changes in membrane fatty acid composition. For example, feeding similar diets to lean and obese animals results in lower levels of 20:4(ω-6) in membrane phospholipids in obese animals (TABLE 2).[17] Both streptozotocin-induced diabetes and genetically spontaneous diabetes have been associated with a decrease in the amount of LCPUFA, a decrease correctable by insulin therapy.[18-20] The decrease in membrane LCPUFA characteristic of both obesity and diabetes suggests that the process of desaturation-elongation of the essential fatty acids 18:2(ω-6) and 18:3(ω-3) is impaired. These conversions are catalyzed by Δ6-desaturase, responsible for the formation of 20:3(ω-6) from 18:2(ω-6), and Δ5-desaturase, responsible for the formation of 20:4(ω-6) from 20:3(ω-6). In genetically obese mice and rats, a lower Δ5-desaturase activity has

TABLE 1. Effect of High-P/S and Low-P/S Diets on Phospholipid Fatty Acid Composition of the Rat Liver Nuclear Envelope[a]

	Fatty Acids (% w/w)		
	High-P/S Diet	Low-P/S Diet	Chow Diet
Phosphatidylcholine			
16:0	18.5 ± 0.3[b]	17.0 ± 0.7[c]	24.3 ± 0.5[d]
18:0	20.8 ± 0.4[c]	26.4 ± 1.4[b]	18.4 ± 1.8[c]
18:2(6)	14.3 ± 1.4[b]	9.9 ± 1.0[c]	19.8 ± 1.6[d]
20:4(6)	29.0 ± 0.9[b]	25.2 ± 1.5[c]	15.6 ± 1.5[d]
Phosphatidylethanolamine			
16:0	13.4 ± 0.8[c]	14.0 ± 1.0[c]	21.1 ± 1.2[b]
18:0	28.3 ± 3.3	29.4 ± 3.2	25.5 ± 1.2
18:2(6)	6.8 ± 0.4[b]	3.9 ± 0.5[c]	9.2 ± 0.8[d]
20:4(6)	27.5 ± 2.3	28.1 ± 1.5[c]	22.1 ± 1.3[b]
Phosphatidylinositol			
16:0	9.8 ± 0.4[b,c]	6.5 ± 1.4[b]	15.0 ± 2.3[c]
18:0	44.0 ± 2.2[c]	51.8 ± 0.02[b]	44.0 ± 3.3[c]
18:2(6)	7.9 ± 0.4[b]	1.7 ± 0.1[c]	5.2 ± 0.3[d]
20:4(6)	36.8 ± 0.5	38.2 ± 1.5[c]	31.5 ± 0.5[b]
Phosphatidylserine			
16:0	18.6 ± 2.8	16.4 ± 3.8	13.7 ± 3.8
18:0	43.6 ± 0.8[c]	44.3 ± 5.5[c]	37.0 ± 3.5[b]
18:2(6)	6.8 ± 0.7	4.2 ± 0.9	5.8 ± 2.7
20:4(6)	20.0 ± 1.1	25.0 ± 6.7	18.2 ± 3.5

[a] Animals were fed the diets for 4 weeks. Fatty acids analyzed from the rat liver nuclear envelope included all saturated and unsaturated fatty acids of 14 to 24 carbons in chain length, but only four of these are presented. Data were statistically analyzed using analysis-of-variance procedures followed by a Duncan's multiple-range test. Values are the mean ± SD of three independent determinations and those without a common superscript are significantly different at $p < 0.05$ (8–10 animals/treatment).

been observed.[17,21,22] Streptozotocin-induced diabetes has been shown to be associated with decreased activity of microsomal Δ9- and Δ6-desaturase in the liver.[19] Evidence that abnormal insulin production, as seen in diabetes, and insulin resistance, characterized in obesity, induce impaired desaturation/elongation of membrane lipids indicates that these enzymes are insulin-dependent.[23]

Obesity and diabetes also induce changes in nuclear membrane composition. Nuclear membrane from a genetically spontaneous model of diabetes was associated with a decrease in 18:2(ω-6) and a decrease in 20:4(ω-6) in PC as diabetes progressed, suggesting a decrease in desaturase activity.[24] Decreased levels of 20:4(ω-6) observed in nuclear membrane phospholipid of streptozotocin-induced diabetic animals were found to be normalized by islet transplantation.[23] Streptozotocin-induced diabetes resulted in a lower content of 20:4(ω-6) in PE in the rat liver nuclear envelope.[14] Dietary fat was able to ameliorate the perturbations on the nuclear envelope fatty acid composition induced by diabetes and obesity. High-fat diets raised the levels of 20:4(ω-6) in nuclear membrane PC of animals with spontaneous diabetes (FIG. 1).[25]

TABLE 2. Fatty Acid Composition of Phosphatidylcholine from Brown Adipose Tissue Inner Mitochondrial Membrane from Lean (ob/−) and Obese (ob/ob) Mice Fed High-P/S or Low-P/S Diets[a]

Fatty Acid (mol%)	P/S = 0.25		P/S = 1.0	
	Lean	Obese	Lean	Obese
ΣSAT[b]	54.5 ± 1.4	46.8 ± 1.3[d]	49.5 ± 1.8[e]	45.0 ± 0.7[d]
ΣMONO[c]	16.8 ± 0.7	24.2 ± 1.8[d]	10.6 ± 0.9[e]	19.1 ± 0.7[d,e]
18:2(6)	16.0 ± 1.0	18.2 ± 1.1[d]	19.8 ± 0.9[e]	28.7 ± 1.4[d,e]
20:3(6)	0.3 ± 0.1	0.9 ± 0[d]	0.4 ± 0.2[e]	1.1 ± 1.1[d]
20:4(6)	5.3 ± 0.3	3.7 ± 0.2[d]	5.0 ± 0.2	3.5 ± 0.3[d]
Δ6-desaturase index [20:3(6)/18:2(6) × 100]	1.9	4.9[d]	1.9	3.8[d,e]
Δ5-desaturase index [20:4(6)/20:3(6) × 100]	16.6	4.1[d]	13.5	3.2[d,e]

[a] Values represent the mean ± SEM for $n = 6$ animals.
[b] Sum of saturated fatty acids.
[c] Sum of monounsaturated fatty acids.
[d] Significantly different from lean controls fed the same diet; $p < 0.05$.
[e] Significantly different from the same genotype fed the P/S diet; $p < 0.05$.

RELATIONSHIP BETWEEN DIETARY FAT, DISEASE, AND INSULIN BINDING IN THE NUCLEUS

Insulin receptor function is impaired in diabetes and obesity. The insulin receptor's α-subunit binds insulin at the cell surface and transduces a signal across the plasma membrane that activates the β-subunit, a tyrosine kinase, by autophosphorylation.[26] The insulin receptor tyrosine kinase then triggers the activation of a cascade of phosphorylated/dephosphorylated proteins that mediate the insulin signal within the cell. As the insulin receptor is a membrane-bound protein and the insulin receptor function is membrane-dependent, it may be influenced by factors that determine membrane constituents and the physical environment.[13] Changes in membrane lipid can be induced by dietary fat and disease states, and these changes have an impact on the function of the plasma membrane insulin receptor. A positive relationship was observed between membrane PUFA content upon high-fat, high-P/S feeding and insulin receptor ligand binding.[27] In rat adipocytes, a high-P/S diet induced an increase in insulin receptor function as assessed by insulin binding, which was coupled to increases in insulin-stimulated functions within the cell, including glucose transport, glucose oxidation, and lipid synthesis.

Insulin receptors in the nucleus can be quantitated by insulin binding assays using [125I]-labeled insulin and varying amounts of unlabeled insulin.[28] In the spontaneous diabetic B/B rat, insulin binding to nuclear receptors decreases as diabetes becomes more pronounced (FIG. 2).[24] Scatchard analysis reveals both a high- and a low-affinity binding site in the nuclei of control animals, but only a single class of binding sites in diabetic animals. Insulin binding levels are increased in both diabetic and control animals by a high-P/S diet compared to low-P/S or chow diets. However, high-PUFA

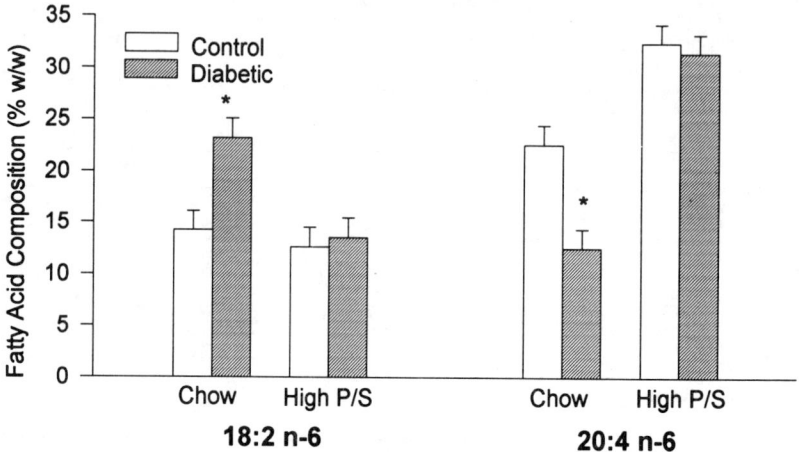

FIGURE 1. Effect of chow and high-P/S diets on the fatty acid profile of phosphatidylcholine of liver nuclei from 10 control animals and 10 diabetic animals. Values are the mean ± SE. Only major fatty acids are reported. (*) Significantly different from control.

feeding only partially restored the reduced levels of insulin binding found in diabetic animals. In diabetes induced by streptozotocin, nuclear insulin receptors had higher binding capacity compared to control animals.[23] Only one class of binding site was seen in diabetic animals, whereas high- and low-affinity binding was observed in control animals. Improved insulin binding in diabetes may be associated with the upregulation of insulin receptors induced by low circulating insulin. Insulin receptors may become more available or upregulated in insulin deficiency and conversely downregulated in response to states of insulin excess such as obesity. Consistent with previous reports, low-PUFA diets decreased and high-PUFA, high-ω-3 feeding increased insulin binding to nuclei in both control and diabetic animals. Obesity, a disease characterized by insulin resistance, was associated with reduced insulin binding to liver nuclei. In the genetically spontaneous obese mouse (ob/ob) model, obese mice had lower levels of insulin binding to hepatocyte nuclei compared to lean mice (FIG. 3).[10] High-PUFA feeding increased insulin binding in both lean and obese animals. Taken together, these results suggest that increasing the polyunsaturated fatty acid content of nuclear membrane phospholipids by high-PUFA feeding is associated with improved insulin binding capacity in the nucleus in both normal and disease states. Furthermore, not only is ligand binding to the insulin receptor increased by high-PUFA feeding, but so is receptor tyrosine kinase activity. It was reported that increased dietary intake of PUFA has resulted in an enhancement of the insulin receptor β-subunit tyrosine kinase activity in rat liver membranes[29] and that compounds such as lysophosphatidylcholine that increase membrane fluidity have a similar effect.[30]

It appears that dietary fat and disease states affect nuclear insulin receptor function probably by influencing its lipid environment. Understanding the mechanisms that regulate the levels of insulin receptor in the nucleus remains fundamental to the issue of how dietary fat and disease influence nuclear insulin receptor function. It has been

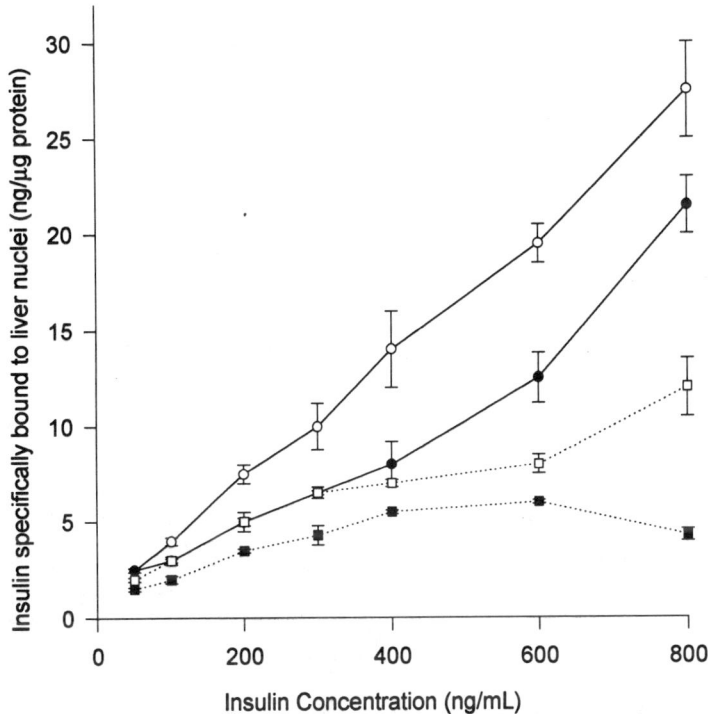

FIGURE 2. Total specific binding of insulin to liver nuclei of control and spontaneously diabetic B/B rats at 25°C. Total specific insulin bound to liver nuclei from control animals fed a high-P/S diet (O), control animals fed a low-P/S diet (●), diabetic B/B animals fed a high-P/S diet (□), and diabetic animals fed a low-P/S diet (■). Total specific insulin bound at each insulin concentration is a mean ± SE for 8 animals in each group at 800 ng/mL insulin ($p < 0.05$).

shown that insulin receptors are translocated to the nucleus from the plasma membrane via internalized endosomal pools upon insulin stimulation.[31,32] Recently, we have used both insulin binding and immunological methods to quantitate insulin receptors in the nucleus upon insulin stimulation *in vivo*. Insulin stimulation by oral glucose gavage to fasted animals resulted in an increase in insulin binding in isolated hepatocyte nuclei over a time course of 0 to 180 minutes.[56] Western blot analysis using an insulin receptor polyclonal antibody showed reduced receptor levels at fasting, but increases as early as 15 minutes following oral glucose gavage.[56] These results are consistent with observations in adipocytes in culture where addition of insulin caused an increase in nuclear insulin receptor levels in as early as 5 minutes.[9] Moreover, these receptors were found to have tyrosine kinase activity towards an exogenous substrate, poly-Glu_4:Tyr_1. Whether dietary lipids have an effect on nuclear insulin receptor tyrosine kinase activity has not yet been determined.

Insulin stimulation thus triggers internalization and intracellular trafficking of insulin receptors, resulting in the accumulation of receptors in the cell nucleus. Mem-

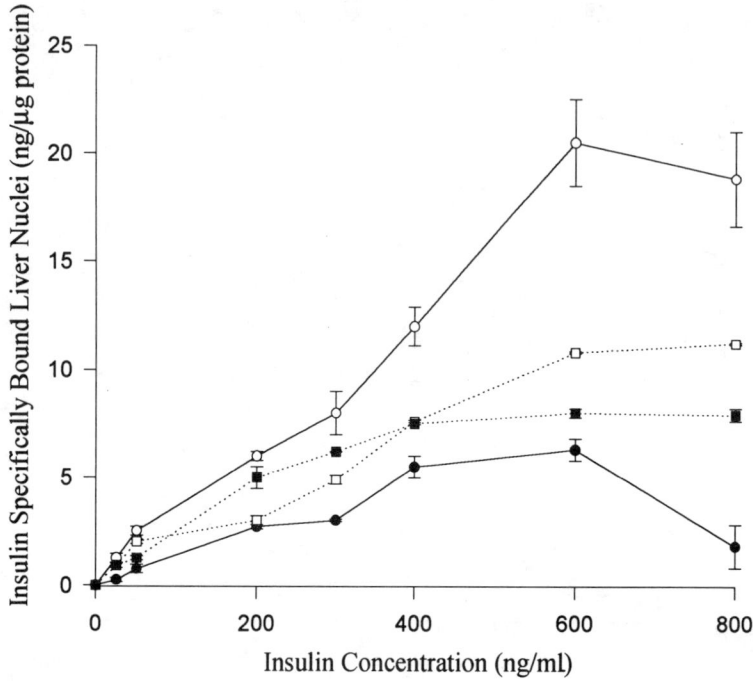

FIGURE 3. Total specific insulin bound at 25°C to liver nuclei. Total specific insulin bound at each insulin concentration is a group mean ± SE for 8 lean animals fed the high-P/S diet (○), 8 lean animals fed the low-P/S diet (●), 8 obese animals fed the high-P/S diet (□), and 8 obese animals fed the low-P/S diet (■).

brane environment may influence the insulin binding capacity and function of insulin receptors in the nucleus by interfering with the processes of insulin receptor intracellular trafficking and nuclear translocation. If the function of nuclear insulin receptors is important in insulin action, then the influence of membrane environment in nuclear insulin receptor function should be determined.

INFLUENCE OF DIETARY FAT ON INSULIN ACTION IN THE NUCLEUS: THE REGULATION OF EXPRESSION OF INSULIN-RESPONSIVE GENES

Recently, more attention is being focused upon the cell nucleus as an important site of action for insulin. Direct effects of insulin on the nucleus include (1) stimulation of mRNA efflux from the nucleus,[33] (2) stimulation of DNA and RNA synthesis,[34,35] and (3) regulation of transcriptional events.[36] Insulin affects the transcription of several genes, many of which are involved in glycolytic and lipogenic processes. Sites within the promoter regions of these genes that confer responsiveness to insulin have been identified in genes such as fatty acid synthase (FAS),[7] apo C-III,[37] and pyruvate kinase.[38] These regions are targets for yet-unidentified proteins (*trans*-acting factors), whose DNA-binding activity is regulated by insulin. The mechanism of transduction of the insulin signal from the cell surface to the transcription regulatory factors in the nucleus is not well understood. However, it is increasingly evident that nutritional and hormonal manipulations of insulin levels *in vivo* can influence the rate of transcription of insulin-responsive genes. For example, the expression of L-type pyruvate kinase, glucokinase, and the lipogenic genes in liver is increased by a fat-free carbohydrate/protein diet and decreased by fasting and the insulin-deficient diabetic state.[39-41] Obesity is associated with elevated insulin levels that normalize after weight reduction. Moreover, FAS and other lipogenic enzymes are overexpressed in obesity in comparison with lean controls.[42-44]

Several groups have reported that PUFA suppress the transcription of many insulin-responsive lipogenic genes, including FAS, malic enzyme (ME), stearoyl-CoA desaturase, and L-type pyruvate kinase.[3,45-47] Evidence suggests that the regulation of gene expression appears to be direct, acute, and predominantly due to an inhibition of gene transcription as opposed to an interference with a posttranscription process.[45,46] The potency of inhibition of a polyunsaturated fatty acid is determined by its degree of unsaturation and its chain length. The basic requirement for repression of gene expression is that the fatty acid should contain at least 18 carbons and possess at least two conjugated double bonds in the 9,12-positions.[48] For example, it has been found that saturated fatty acids and monounsaturated fatty acids do not suppress lipogenesis.[49-53] Within the family of PUFA, it has been found that ω-3 fatty acids are more potent inhibitors than ω-6 fatty acids and that the polyenoic products of $\Delta 5$- and $\Delta 6$-desaturase, 20:4(ω-6), 20:5(ω-3), and 22:6(ω-3), are more potent than their precursors, 18:2(ω-6) and 18:3(ω-3).[51,54]

How PUFA exert their effect on the nucleus is not yet known. As we have seen, dietary fatty acids travel from the plasma and become equilibrated in cellular lipids. Polyunsaturated fatty acids can enter the nucleus from cytosolic pools by incorporating directly into the nuclear membrane. It is speculated that PUFA may interact with a nuclear polyunsaturated fatty acid binding protein (PUFA-BP), with the latter becoming activated.[45] PUFA-BP then may recognize and bind to *cis*-acting elements within the promoter regions of PUFA-responsive genes or may activate some other yet-unidentified *trans*-acting factor. The *cis*-acting sequences within the promoter regions of the FAS gene have been identified.[45]

In the case of lipogenic gene expression, insulin seems to have a stimulatory effect, whereas PUFA seem to be inhibitory. The coordinate regulation of lipogenic genes has been studied *in vivo* using animal models of obesity and by feeding diets differing in dietary lipids. The question of whether PUFA affect lipogenic gene expression in the obese animal characterized by hyperinsulinemia was addressed in the following study. Eight-week-old lean and obese mice were fed diets of either a high- or low-P/S ratio for four weeks. Hepatic FAS and ME mRNA levels detected by Northern blotting were lower in lean animals compared to obese animals. A high-P/S diet lowered mRNA levels compared to a low-P/S diet in both lean and obese animals. Thus, it appears that PUFA are able to repress lipogenic gene overexpression in obese animals. The amount of mRNA found in obese animals fed a high-P/S diet was still significantly higher than that of lean animals fed either diet, which may be a manifestation of hyperinsulinemia in obese animals.[55] In a separate study using Wistar fatty and lean rats,[42] feeding lean animals a diet of corn oil containing high levels of PUFA following a two-day period of fasting was found to reduçe mRNA levels of hepatic lipogenic enzymes compared to feeding a hydrogenated fat diet. When obese animals were subjected to the same fasting and refeeding regimen using either a hydrogenated diet or a corn oil diet, no difference was found between the two diets in mRNA levels. The unresponsiveness of obese animals to PUFA suggests that there is a defect in polyunsaturated fatty acid suppression in obese animals. These apparently contradictory findings may have resulted from the different dietary regimens followed in the two studies. The fasting period preceding fat treatment in the latter study may have enhanced the differences seen due to the PUFA feeding in lean and obese animals. More research needs to be done to evaluate the efficacy of PUFA in suppressing lipogenic gene expression in the obese animal model.

DISCUSSION AND CONCLUSIONS

PUFA have long been recognized for their health benefits in the prevention of chronic human diseases such as obesity, non-insulin-dependent diabetes, and coronary heart disease. Insulin action is enhanced by PUFA feeding in adipocytes and hepatocytes due to an improvement in insulin receptor function. Insulin action in the nucleus is important for the regulation of the synthesis of glucose and lipid metabolic enzymes. Insulin action is impaired in obesity due to downregulation of receptor numbers and reduced receptor tyrosine kinase function associated with obesity-induced aberrations in lipid metabolism and hyperinsulinemia.

There appears to be two independent mechanisms by which PUFA can influence insulin action in the nucleus. A chronic effect of PUFA feeding occurs with incorporation of these fatty acids into the nuclear membrane. A higher membrane PUFA content improves insulin receptor kinetics, insulin binding affinity and capacity, and kinase activity. Enhancement of insulin receptor function may alleviate the insulin resistance associated with obesity, thereby reducing the stress upon the β-cell to produce more insulin. Another mechanism by which PUFA can influence insulin action in the nucleus has more acute effects. Dietary PUFA may rapidly incorporate into the cell nucleus and activate DNA-binding proteins. A high-fat, high-PUFA diet achieves inhibition of expression of lipogenic genes, which is a response opposite to that of insulin. If nuclear insulin receptor function in the nucleus is enhanced by PUFA, then why is there a repression of lipogenic gene expression after PUFA feeding? The answer to this question may lie in the assumption that there are at least two different mechanisms involved by which PUFA may affect gene expression and insulin action in the nucleus.

REFERENCES

1. CLANDININ, M. T., S. CHEEMA, C. J. FIELD, M. L. GARG, J. VENKATRAMAN & T. R. CLANDININ. 1991. Dietary fat: exogenous determination of membrane structure and cell function. FASEB J. **5:** 2761-2769.
2. JUMP, D. B., A. THELEN, S. CLARKE & M. LIIMATTA. 1994. Coordinate regulation of glycolytic and lipogenic gene expression by polyunsaturated fatty acids. J. Lipid Res. **35:** 1076-1084.
3. NTAMBI, J. M. 1995. The regulation of stearoyl-CoA desaturase (SCD). Prog. Lipid Res. **34(2):** 139-150.
4. GRANNER, D. K., K. SASAKI, T. ANDREONE & E. BEALE. 1986. Insulin regulates the expression of the phosphoenolpyruvate carboxykinase gene. Recent Prog. Horm. Res. **42:** 111-141.
5. MOLERO, C., M. BENITO & M. LORENZO. 1993. Regulation of malic enzyme gene expression by nutrients, hormones, and growth factors in fetal hepatocyte primary cultures. J. Cell. Physiol. **155:** 197-203.
6. LIIMATTA, M., H. C. TOWLE, S. CLARKE & D. B. JUMP. 1994. Dietary polyunsaturated fatty acids interfere with the insulin/glucose activation of L-type pyruvate kinase gene transcription. Mol. Endocrinol. **8:** 1147-1153.
7. MOUSTAID, N., R. S. BEYER & H. S. SUL. 1994. Identification of an insulin response element in the fatty acid synthase promoter. J. Biol. Chem. **269(8):** 5629-5634.
8. VIGNERI, R., I. D. GOLDFINE, K. Y. WONG, G. J. SMITH & V. PEZZINO. 1978. The nuclear envelope: the major site of insulin binding in rat liver nuclei. J. Biol. Chem. **253:** 2098-2103.
9. KIM, S. J. & C. R. KAHN. 1993. Insulin induces rapid accumulation of insulin receptors and increases tyrosine kinase activity in the nucleus of cultured adipocytes. J. Cell. Physiol. **157:** 217-228.
10. CHEEMA, S. K., J. VENKATRAMAN & M. T. CLANDININ. 1992. Insulin binding to liver nuclei from lean and obese mice is altered by dietary fat. Biochim. Biophys. Acta **1117:** 37-41.
11. RADULESCU, R. T. 1995. Insulin receptor α-subunit: a putative gene regulatory molecule. Med. Hypotheses **45:** 107-111.
12. PODLECKI, D. A., R. M. SMITH, M. KAO, P. TSAI, T. HUECKSTEADT, D. BRANDENBURG, R. LASHER, L. JARRET & J. M. OLEFSKY. 1987. Nuclear translocation of the insulin receptor, a possible mediator of insulin's long-term effects. J. Biol. Chem. **262(7):** 3362-3368.
13. FIELD, C. J., E. A. RYAN, A. B. R. THOMPSON & M. T. CLANDININ. 1990. Diet fat composition alters membrane phospholipid composition, insulin binding, and glucose metabolism in adipocytes from control and diabetic animals. J. Biol. Chem. **265:** 11143-11150.
14. KANG, J., J. T. VENKATRAMAN & M. T. CLANDININ. 1992. Influence of dietary lipids on the binding of adenosine to rat liver nuclear envelope. Diabetes Nutr. Metab. **5:** 167-174.
15. VENKATRAMAN, J. T., Y. A. LEFEBVRE & M. T. CLANDININ. 1986. Diet fat alters the structure and function of the nuclear envelope: modulation of membrane fatty acid composition, NTPase activity, and binding of triiodothyronine. Biochem. Biophys. Res. Commun. **135(2):** 655-661.
16. VENKATRAMAN, J. T., T. TOOHEY & M. T. CLANDININ. 1992. Does a threshold for the effect of dietary omega-3 fatty acid on the fatty acid composition of nuclear envelope phospholipid exist? Lipids **27:** 94-97.
17. CLANDININ, M. T., S. CHEEMA, D. PEHOWICH & C. J. FIELD. 1996. Effect of polyunsaturated fatty acids in obese mice. Lipids **31:** S13-S22.
18. MINNOUNI, V., M. NARCE & J. P. POISSON. 1992. Evidence for insulin-dependent hepatic microsomal gamma-linolenic acid chain elongation in spontaneously diabetic Wistar BB rats. Biochim. Biophys. Acta **1133:** 187-192.
19. FAAS, F. H. & W. J. CARTER. 1980. Altered fatty acid desaturation and microsomal fatty acid composition in the streptozotocin diabetic rat. Lipids **15:** 953-961.
20. POISSON, J. P. 1989. Essential fatty acid metabolism in diabetes. Nutrition **5:** 263-266.
21. CUNNANE, S. C., M. S. MANKER & D. F. HOROBIN. 1985. Abnormal essential fatty acid composition of tissue lipids in genetically diabetic mice is partially corrected by dietary linoleic and gamma-linolenic acids. Br. J. Nutr. **53:** 441-448.
22. GUESNET, P. H., J. M. BOUREE, M. GUERRE-MILLO, G. PASCAL & G. DURAND. 1990. Tissue phospholipid fatty acid composition in genetically lean (fa/-) or obese (fa/fa) Zucker female rats on the same diet. Lipids **25:** 517-522.

23. CHEEMA, S. K., R. V. RAJOTTE & M. T. CLANDININ. 1996. Effect of diet fat on liver nuclei membrane phospholipid fatty acid composition and insulin binding after islet transplantation. Biochim. Biophys. Acta. Submitted.
24. CHEEMA, S. K. & M. T. CLANDININ. 1996. Diet and diabetes induced change in insulin binding to the nuclear membrane in spontaneously diabetic rats in association with change in fatty acid composition of phosphatidylinositol. Biochim. Biophys. Acta. Submitted.
25. CHEEMA, S. K. & M. T. CLANDININ. 1993. Effect of dietary polyunsaturated fat on gene expression in liver. In Advances in Polyunsaturated Fatty Acid Research, p. 189–192. Elsevier. Amsterdam/New York.
26. KASUGA, M., F. A. KARLSSON & C. R. KAHN. 1982. Insulin stimulates the phosphorylation of the 95,000-dalton subunit of its own receptor. Science 215: 185–187.
27. FIELD, C. J., M. TOYOMIZU & M. T. CLANDININ. 1989. Relationship between diet fat, plasma membrane lipid composition, and insulin-stimulated functions in isolated adipocytes. In Biomembranes and Nutrition. Vol. 195, p. 319–332. Colloque INSERM.
28. GAMMELTOFT, S. 1984. Insulin receptors: binding kinetics and structure-function relationship of insulin. Physiol. Rev. **64**(4): 1321–1378.
29. FICKOVÁ, M., P. HUBERT, I. KLIMEŠ, C. STAEDEL, G. CRÉMEL, P. BOHOV & L. MACHO. 1994. Dietary fish oil and olive oil improve the liver insulin receptor tyrosine kinase activity in high sucrose fed rats. Endocr. Regul. **28**(4): 187–197.
30. MCCALLUM, C. D. & R. M. EPAND. 1995. Insulin receptor autophosphorylation and signaling is altered by modulation of membrane physical properties. Biochemistry **34**(6): 1815–1824.
31. CARPENTIER, J. L. 1992. The insulin receptor: what triggers and regulates its internalization. 1992. In Progress in Histo- and Cytochemistry: Histochemistry of Receptors. Vol. 26, p. 77–87. Fischer. Jena.
32. BURGESS, J. W., A. P. BEVAN, J. J. M. BERGERON & B. I. POSNER. 1992. Pharmacological doses of insulin equalize insulin receptor phosphotyrosine content, but not tyrosine kinase activity in plasmalemmal and endosomal membranes. Biochem. Cell Biol. **70**: 1151–1158.
33. SCHUMM, D. E. & T. C. WEBB. 1981. Insulin-modulated transport of RNA from isolated liver nuclei. Arch. Biochem. Biophys. **210**: 275–279.
34. RICHMAN, R. A., T. H. CLAUS, S. J. PILKIS & D. L. FRIEDMAN. 1976. Hormonal stimulation of DNA synthesis in primary cultures of adult rat hepatocytes. Proc. Natl. Acad. Sci. U.S.A. **73**: 3589–3593.
35. STEINER, D. F. 1966. Insulin and the regulation of hepatic biosynthetic activity. Vitam. Horm. **24**: 1–61.
36. CASTANO, J. G. 1991. Regulation of gene expression by insulin. Adv. Enzyme Regul. **31**: 185–194.
37. DAMMERMAN, M., L. A. SANDKUIYL, J. L. HALAAS, W. CHUNG & J. L. BRESLOW. 1993. An apolipoprotein C-III haplotype protective against hypertriglycemia is specified by promoter and 3′-untranslated region polymorphisms. Proc. Natl. Acad. Sci. U.S.A. **90**: 4562–4606.
38. YAMADA, K., T. NOGUCHI, J.-I. MATSUDA, M. TAKENAKA, K.-I. YAMAMURA & T. TANAKA. 1990. Tissue-specific expression of rat pyruvate kinase/chloramphenicol acetyltransferase fusion gene in transgenic mice and its regulation by diet and insulin. Biochem. Biophys. Res. Commun. **171**: 243–249.
39. PILKIS, S. J. & D. K. GRANNER. 1992. Molecular physiology of the regulation of hepatic gluconeogenesis and glycolysis. Annu. Rev. Physiol. **54**: 885–909.
40. CLARKE, S. D. & S. ABRAHAM. 1992. Gene expression: nutrient control of pre- and posttranscriptional events. FASEB J. **6**: 3146–3152.
41. IRITANI, N. 1992. A review: nutritional and hormonal regulation of lipogenic gene expression in rat liver. Eur. J. Biochem. **205**: 433–442.
42. IRITANI, N., H. HOSOMI, H. FUKUDA & H. IKEDA. 1995. Polyunsaturated fatty acid regulation of lipogenic enzyme gene expression in liver of genetically obese rats. Biochim. Biophys. Acta **1255**: 1–8.
43. DUGAIL, I., A. QUIGNARD-BOULANGE, X. LE LIEPVRE & B. ARDOUIN. 1992. Gene expression of lipid storage-related enzymes in adipose tissue of the genetically obese Zucker rat: coordinate increase in transcriptional activity and potentiation by hyperinsulinemia. Biochem. J. **281**(3): 607–611.
44. MAURY, J., T. ISSAD, D. PERDEREAU, B. GOUHOT, P. FERRE & J. GIRARD. 1993. Effect of acarbose

on glucose hemostasis, lipogenesis, and lipogenic enzyme gene expression in adipose tissue of weaned rats. Diabetologia **36**(6): 503–509.
45. CLARKE, S. D. & D. B. JUMP. 1993. Regulation of gene transcription by polyunsaturated fatty acids. Prog. Lipid Res. **32**(2): 139–149.
46. BLAKE, W. L. & S. D. CLARKE. 1990. Suppression of rat hepatic fatty acid synthase and S14 gene transcription by polyunsaturated fatty acids. J. Nutr. **120**: 1727–1729.
47. KATSURADA, A., N. IRITANI, H. FUKUDA, Y. MATSUMURA, N. NISHIMOTO, T. NOGUCHI & T. TANAKA. 1990. Effects of nutrients and hormones on transcriptional and post-transcriptional regulation of fatty acid synthase in the liver. Eur. J. Biochem. **190**: 435–441.
48. EMKEN, E. A., S. ABRAHAM & C. Y. LIN. 1987. Metabolism of *cis*-12-octadecanoic acid and *trans*-9, *trans*-octadecanoic acid and their influence on lipogenic enzyme activities in mouse liver. Biochim. Biophys. Acta **919**: 111–121.
49. CLARKE, S. D., M. K. ARMSTRONG & D. B. JUMP. 1990. Dietary polyunsaturated fats uniquely suppress rat liver fatty acid synthesis and S14 mRNA content. J. Nutr. **120**: 225–231.
50. MCDONOUGH, V. M., J. E. STUKEY & C. E. MARTIN. 1992. Specificity of unsaturated fatty acid-regulated expression of the *Saccharomyces cerevisiae* OLE 1 gene. J. Biol. Chem. **267**: 5931–5936.
51. NTAMBI, J. M. 1992. Dietary regulation of stearoyl-CoA desaturase 1 gene expression in mouse liver. J. Biol. Chem. **267**(15): 10925–10930.
52. SHILLABEER, G., J. HORNFORD, J. M. FORDEN, N. C. W. WONG & D. C. W. LAU. 1990. Hepatic and adipose tissue lipogenic enzyme mRNA levels are suppressed by high fat diets in the rat. J. Lipid Res. **31**: 623–631.
53. SONCINI, M., S-F. YET, Y. MOON, J-Y. CHUN & H. S. SUL. 1995. Hormonal and nutritional control of the fatty acid synthase promoter in transgenic mice. J. Biol. Chem. **270**(5): 30330–30343.
54. CLARKE, S. D. & D. B. JUMP. 1993. Fatty acid regulation of gene expression: a unique role for polyunsaturated fats. *In* Nutrition and Gene Expression, p. 227–246. CRC Press. Boca Raton, Florida.
55. CHEEMA, S. K. & M. T. CLANDININ. 1996. Diet fat alters expression of genes for enzymes of lipogenesis in lean and obese mice. Biochim. Biophys. Acta **1299**: 284–288.
56. GLETSU, N. A., W. T. DIXON, C. F. FIELD & M. T. CLANDININ. 1997. Translocation of active insulin receptor to the hepatocyte nucleus coincides with induction of insulin responsive gene expression. In preparation.

Molecular and Cellular Determinants of Triglyceride Availability[a]

E. ŠEBÖKOVÁ AND I. KLIMEŠ

*Diabetes and Nutrition Research Group
Institute of Experimental Endocrinology
Slovak Academy of Sciences
SK-833 06 Bratislava, Slovak Republic*

Dyslipidemia, including hypertriglyceridemia, low HDL concentrations, and alterations in LDL composition, is part of the insulin resistance (IR) syndrome and is regarded as an independent risk factor for coronary heart disease in subjects with impaired glucose tolerance or diabetes.[1-5] It has been shown in humans[6] and in rats[7-9] that both increased triglyceride (TG) levels in circulation and increased muscle TG stores are directly associated with IR.

While the genetic background is important, environmental factors play an essential role in the development of hypertriglyceridemia and IR.[2] Thus, recently, much attention has been paid to the composition of macronutrients in the diet, with special focus on fat composition. Increasing evidence exists demonstrating that dietary intake of polyunsaturated fatty acids (PUFA), particularly of the n-3 class, is associated with suppression of hypertriglyceridemia and improvement of insulin action.[8-17]

The plasma levels and tissue content of TG are determined by the balance between *de novo* synthesis of fatty acids (and consequently of TG), TG clearance in circulation (by lipoprotein lipase), and TG hydrolysis in tissues (by lipoprotein and/or hormone-sensitive lipase in adipose tissue, skeletal muscle, and the heart).[18-20] Thus, a defect at any step of this enzyme network may initiate a cascade of pathological events leading to hypertriglyceridemia.

REGULATION OF LIPOGENESIS

Understanding of the biochemistry and regulation of lipogenesis has resulted mostly from *in vitro* studies in rodents.[18] However, the biological importance, activity, and tissue distribution of the lipogenic pathways vary greatly among species. Whereas active lipogenesis occurs in liver and adipose tissue in the rat,[21] *de novo* lipogenesis is an inefficient process in humans.[22] It occurs predominantly in the liver and is of

---

[a] This work was supported by research grants of the Institute of Experimental Endocrinology, Slovak Academy of Sciences, Bratislava, provided by the Slovak Grant Agency for Sciences (GAV) (No. 2/544/92-95); the Natural Sciences and Engineering Research Council of Canada; the Canadian Diabetes Association; and a COST B5 grant.

minor significance in human adipose tissue, even under conditions of massive carbohydrate overfeeding.[23] The nutritional status of the organism, closely allied to changes in circulating insulin concentrations, appears to be the major determinant of the rate of lipogenesis. Lipogenesis is high in the fed state and following carbohydrate administration,[14-41] whereas it is suppressed by fasting,[27] high-fat diets,[29-31] and insulin deficiency.[38]

Insulin stimulates *de novo* lipogenesis in liver and adipose tissue by several mechanisms.[18,21] In adipose tissue, insulin stimulates glucose uptake, thereby increasing the supply of lipogenic substrate. Furthermore, several key enzymes in the lipogenic pathway, including fatty acid synthase (FAS), acetyl-CoA carboxylase (ACC), and malic enzyme (ME), are induced by insulin. In addition, the antilipolytic action of insulin reduces free fatty acid availability to the liver and thereby reduces the inhibitory effects of FA on hepatic lipogenesis. Similarly, the inhibition of lipolysis and the stimulatory effect of insulin on reesterification of FA in adipose tissue reduce the intra-adipocyte concentration of fatty acyl-CoA, an inhibitor of lipogenesis. Thus, the many actions of insulin serve to increase the biosynthesis of FA for subsequent storage in the form of TG. It should be emphasized, however, that the predominant source of FA substrate for TG synthesis in both the adipose tissue and liver comes from circulating free fatty acids derived either from adipose tissue lipolysis or from intravascular hydrolysis of lipoprotein TG by lipoprotein lipase.[21]

The stimulation of lipogenesis by insulin in response to nutritional changes involves a rapid activation of the key enzymes involved in FA synthesis. This phenomenon, however, is preceded and/or modulated by long-term transcriptional and translational events. It has been shown that food intake, in particular composition of the diet, may represent an important factor for the regulation of lipogenesis in liver and adipose tissue at the gene level.[19,24-28] Although techniques for evaluation of the molecular basis of lipogenesis are available, it is still rather complicated to carry out appropriate studies in humans, particularly due to ethical reasons. Therefore, various animal models of IR have been widely used to elucidate the molecular and cellular mechanisms that may be responsible for hypertriglyceridemia in human IR. The regulation of lipogenesis in these animal models will be discussed first.

REGULATION OF LIPOGENESIS IN ANIMAL MODELS OF INSULIN RESISTANCE

It is now generally accepted that feeding the animals a diet high in fat, particularly when the fat is saturated, is associated with development of IR and accumulation of TG in muscle tissues.[7,8] There was a variable effect on circulating TG levels, which may depend on the fat content and FA composition of the diet.[8] Despite an evident effect of the high-fat diet on muscle TG levels, the role of dietary fat content on *de novo* fat synthesis and storage is not so transparent. It has been demonstrated that the inclusion of large amounts of saturated fat (from 5% to 25%) has a minimal effect on the activity of lipogenic enzymes in liver or on the rate of *in vivo* fatty acid synthesis.[29] Other data, however, have demonstrated that the *de novo* fatty acid synthesis in liver and lipogenesis are suppressed when the animals are fed a high-fat diet containing at least 30% fat.[30,31] The divergence in these observations may be explained either by various fat content or by the fact that as much as 70% of the *de novo* fatty acid synthesis in the animals fed the high-fat diet takes place in the adipose tissue, as demonstrated earlier by Romsos and Leveille[32] and later by Nelson *et al.*[33] Supportive evidence for the inhibition of lipogenesis by a high-fat diet comes from recent data showing a decreased rate of fatty acid synthesis in the liver and in white adipose tissue of adult rats, which is paralleled with decreased expression of FAS and ACC.[34,35]

During the entire suckling period, the expression and activity of the major liver and white adipose tissue lipogenic enzymes directly involved in the lipogenic pathway — FAS, ACC, and ATP citrate lyase (ATP-CL) — were found to be negligible.[28] The transition from the high-fat diet of suckling to a high-complex-carbohydrate diet of laboratory chow led to increased TG in both tissues. The dramatic increase in liver FAS content after weaning to a high-carbohydrate/low-fat diet was associated with an acceleration of enzyme synthesis as measured by increased levels of tissue mRNA, but without a change in the rate of enzyme degradation. However, when the animals were weaned on a high-fat diet, the increase in lipogenesis was prevented by decreased expression and protein content of FAS, ACC, and ATP citrate lyase in adipose tissue and liver.[28]

Prolonged fasting (1–4 days) was found to be associated with a sharp (95%) decrease of FAS mRNA in white epididymal adipose tissue[27,36,37] and with a moderate decline in the liver.

In streptozotocin-induced diabetes associated with altered TG levels, gene expression for lipogenic enzymes in both liver and white adipose tissue (WAT) was markedly decreased.[38] On the contrary, the improvement of diabetes with vanadate treatment normalized mRNA levels only in liver, but not in WAT,[38] supporting the role of tissue specificity for the regulation of lipogenesis. Refeeding and insulin treatment of streptozotocin-treated (STZ) diabetic rats normalized the activity and gene expression of lipogenic enzymes in both the adipose tissue and liver.[38]

Neonatal streptozotocin rat model of type II diabetes: In spite of increased plasma and TG content in skeletal muscle and in the heart, we did not observe any change in liver lipogenesis when indexed by ACC activity[14] (TABLE 1).

In the hereditary hypertriglyceridemic (hHTG) rat, increased lipogenesis from tritiated water in the liver together with a mild rise in the secretion rate of TG from the liver (increase in post-Triton triglyceridemia) and the unchanged activity of the postheparin lipase indicate the hyperproductive nature of this form of hypertriglyceridemia.[39,40] In harmony with the data obtained *in vivo*, we also found increased expression of FAS and ME in the liver of these animals. On the contrary, the activity of the other lipogenic enzyme ACC and the activity of ME were not changed in the liver of hHTG rats when compared to controls (TABLES 2 and 3).

Feeding the animals high-fructose or high-sucrose diets leads to hypertriglyceridemia[41-43] and to the accumulation of TG in muscles.[9,14,15,44] The cascade of changes in lipid metabolism at high-sucrose or high-fructose intake is triggered in the liver

TABLE 1. Postprandial Serum and Skeletal Muscle Triglycerides (TG) and Activity of Acetyl-CoA Carboxylase (ACC) in Liver of STZ-induced Diabetic Rats (DM) and Control (NTG) Wistar Rats Fed Various Diets[a]

Strain	Diet	Plasma TG [mmol/L]	Muscle TG [nmol/mg]	ACC [nmol/mg]
NTG	B	3.2 ± 0.2^b	31.3 ± 0.1^b	1.27 ± 0.05^b
	FO	1.4 ± 0.2^c	29.7 ± 2.4^b	0.73 ± 0.07^c
	HS	5.0 ± 1.5^d	49.5 ± 2.4^c	2.10 ± 0.04^d
	HS+FO	2.0 ± 1.2^c	35.7 ± 3.0^b	0.92 ± 0.10^c
DM	B	3.0 ± 0.3^b	53.6 ± 2.8^c	1.26 ± 0.10^b
	FO	0.9 ± 0.1^e	34.2 ± 4.1^b	0.74 ± 0.06^c
	HS	8.0 ± 1.0^f	56.7 ± 5.8^c	1.85 ± 0.30^d
	HS+FO	$3.2 \pm 1.2^{b,c,e}$	61.1 ± 6.8^c	0.93 ± 0.10^c

[a] Data from reference 14. Mean values $\pm$ SEM ($n = 6$–8). Values without a common superscript (b–f) within a column are significantly different ($p < 0.05$). Terms: B = basal diet; FO = basal diet supplemented with fish oil; HS = high-sucrose diet; HS+FO = high-sucrose diet supplemented with fish oil.

by fructose, which is utilized to a much higher extent than glucose for the synthesis of fatty acids and TG.[41] They are then exported in the form of VLDL lipoproteins to the plasma compartment with a subsequent marked increase in their concentration.

The increased lipogenesis *in vivo* after high-sucrose feeding is documented also *in vitro* by measuring the activities of the main lipogenic enzymes. ACC activity is significantly increased after a high-sucrose diet in the liver of normal rats (TABLES 2 and 3), STZ diabetic rats[14] (TABLE 1), and hHTG rats (TABLES 2 and 3). Similarly, a high-sucrose diet is accompanied by an increase in activity and mRNA level for ME in both normotriglyceridemic rats and hHTG rats with already-activated lipogenesis.

TABLE 2. Postprandial Serum and Skeletal Muscle Triglycerides (TG) and Activity of Acetyl-CoA Carboxylase (ACC) and of Malic Enzyme (ME) in Liver of Hereditary Hypertriglyceridemic (hHTG) and Control (NTG) Wistar Rats Fed Various Diets[a]

Strain	Diet	Plasma TG [mmol/L]	Muscle TG [nmol/mg]	ME [ncat/mg]	ACC [nmol/mg]
NTG	B	1.7 ± 0.3^b	16.6 ± 4.3^b	3.20 ± 0.35^b	1.27 ± 0.05^b
	FO	0.9 ± 0.1^c	12.3 ± 1.6^b	1.55 ± 0.20^c	0.50 ± 0.04^c
	HS	3.1 ± 0.2^d	60.5 ± 9.8^c	7.50 ± 0.90^d	1.90 ± 0.20^d
	HS+FO	1.0 ± 0.1^c	38.3 ± 5.4^d	2.80 ± 0.50^b	1.00 ± 0.10^e
hHTG	B	3.2 ± 0.2^d	35.3 ± 3.0^d	1.10 ± 0.10^c	1.25 ± 0.10^b
	FO	1.0 ± 0.1^c	16.1 ± 2.5^b	2.30 ± 0.25^b	0.50 ± 0.05^c
	HS	5.2 ± 0.2^e	54.1 ± 9.2^c	4.20 ± 0.30^e	1.60 ± 0.06^d
	HS+FO	1.1 ± 0.1^c	29.5 ± 4.6^d	2.00 ± 0.05^b	0.70 ± 0.07^e

[a] Unpublished data of Šeböková and Klimeš. Data are expressed as the mean $\pm$ SEM ($n = 5$–8). Values without a common superscript (b–e) within a column are significantly different ($p < 0.05$). Terms: B = basal diet; FO = basal diet supplemented with fish oil; HS = high-sucrose diet; HS+FO = high-sucrose diet supplemented with fish oil.

TABLE 3. Relative Abundance of mRNA for Malic Enzyme (ME) and Fatty Acid Synthase (FAS) in Liver and of Lipoprotein Lipase (LPL) in Skeletal Muscle of Hereditary Hypertriglyceridemic (hHTG) and Control (NTG) Wistar Rats Fed Various Diets[a]

Strain	Diet	ME mRNA [AU]	FAS mRNA [AU]	LPL mRNA [AU]
NTG	B	0.47 ± 0.05[b]	1.23 ± 0.47[b]	6.6 ± 1.1[b]
	FO	0.81 ± 0.07[b]	1.85 ± 0.45[b]	24.2 ± 1.8[c]
	HS	7.70 ± 0.30[c]	14.1 ± 0.68[c]	2.6 ± 0.7[d]
	HS+FO	2.90 ± 0.20[d]	5.64 ± 0.34[d]	46.2 ± 2.5[e]
hHTG	B	1.66 ± 0.20[e]	2.76 ± 0.22[e]	7.9 ± 0.7[b]
	FO	2.48 ± 0.40[e]	1.06 ± 0.31[b]	26.2 ± 2.5[c]
	HS	10.4 ± 0.28[f]	5.80 ± 0.41[d]	7.3 ± 0.5[b]
	HS+FO	7.50 ± 0.80[g]	5.92 ± 0.43[d]	24.4 ± 1.5[c]

[a] Data from reference 37, except for the LPL data, which has not yet been published. Mean values ± SEM ($n = 4$). Values without a common superscript (b–g) within a column are significantly different ($p < 0.05$). Terms: B = basal diet; FO = basal diet supplemented with fish oil; HS = high-sucrose diet; HS+FO = high-sucrose diet supplemented with fish oil; AU = arbitrary units.

In animal models of IR associated with obesity, similar data on increased lipogenesis in liver have been reported. Thus, an overexpression of genes together with increased activities for FAS, ACC, and ME have been found in the liver of ob/ob mice.[31] In the genetically obese Zucker fa/fa rat, the increased lipogenic capacity of adipose tissue is due to the preferential channeling of nutrients into fatty acids and due to the activation of lipid storage-related enzymes. In particular, the expression of ME, glyceraldehyde-3-phosphate dehydrogenase (GAPDH), and FAS is stimulated in the adipose tissue of these rats. The overexpression of these genes is an early feature of the onset of obesity and is adipose tissue–specific since no differences are found in liver in comparison to normal rats.[45]

Taken together, the excessive dietary lipid supply associated with decreased activity and expression of lipogenic enzymes and/or increased endogenous lipid synthesis (after high-sucrose feeding or in other animal models of IR) may contribute to the development of hypertriglyceridemia. It should be noted here that the dietary manipulations lead to more pronounced changes in the activation of the lipogenic machinery than those that are hereditary-fixed.

REGULATION OF LIPOLYSIS

Alternatively, or in addition to an effect on the lipid biosynthetic pathway, the control of TG and fatty acid availability may be accomplished through changes in lipolytic enzymes involved in TG metabolism. The main role for lipases is to release fatty acids and make them available as an alternative fuel in order to meet the energy demands of tissues. In IR, preferential oxidation of fatty acids instead of glucose is demonstrated in muscle.[46,47] The potential sources for the lipid oxidation in muscle are (1) free fatty acids (FFA) from adipose tissue lipolysis through the action of the adipocyte protein hormone-sensitive lipase, (2) plasma lipoprotein TG fatty acids released via lipoprotein lipase (LPL), and (3) fatty acids derived from the intramuscular TG droplet likely provided through the action of lipoprotein and hormone-sensitive lipase, but within the muscle tissue.[48] Although it is assumed that most of the fatty acids oxidized by muscle come from adipose tissue in the form of FFA,[49] all three sources may contribute and the relative role of each needs to be delineated. Recently, a growing body of evidence is in favor of the assumption that modulation of the tissue LPL activity is a major mechanism controlling the destination of TG transport.[50]

Following an overnight fast (when blood insulin concentrations are low), fat is mobilized from adipose tissue stores by adipose tissue hormone-sensitive lipase to release FFA into plasma, where they circulate complexed with albumin and are available for uptake by tissues such as liver, skeletal muscle, and heart. Here, they can undergo oxidation or reesterification into TG.[21] After a mixed meal, the rise in blood insulin levels inhibits adipose tissue lipolysis and restrains FFA mobilization from endogenous fat stores. Exogenous fat from the meal is packaged by the intestine and released as TG-rich chylomicrons. Lipoprotein lipase (LPL), a secretory glycoprotein made in a number of tissues including adipose tissue and skeletal and cardiac muscle,[20] is transported to the glycocalyx of the capillary endothelium in the tissue of origin. Here, it carries out its hydrolytic function against the TG core of the TG-rich lipoproteins (chylomicrons and VLDL). Following the release of lipolysis products, the generated FFA are taken up by muscle and adipose tissue where they can be oxidized or stored after reesterification, respectively.[51,52] The highest activities of LPL are found in tissues that oxidize (heart, various types of skeletal muscle) and esterify (adipose tissue, lactating mammary gland) large quantities of fatty acids.[20]

Feeding, particularly of carbohydrate-rich meals,[53] is accompanied by elevated postprandial insulin levels, which results in activation of adipose tissue lipoprotein lipase (AT-LPL) and clearance of TG-rich lipoproteins. In contrast, the activity of LPL in skeletal muscle is decreased by insulin.[52,54] It is of interest that the magnitude of insulin-induced suppression of skeletal muscle LPL (delta SM-LPL) is inversely related to the insulin-mediated glucose uptake by the muscle.[54] As both carbohydrate and lipid oxidation are balanced in skeletal muscle, these data suggest that the regulation of LPL by insulin in skeletal muscle is appropriate for the metabolic needs of the tissue.

In insulin-resistant states, an alternative regulation of LPL exists. Numerous studies have described a decrease in the LPL catalytic activity in adipose tissue[55] and postheparin plasma[56] of NIDDM patients who are under poor glycemic control. Improved diabetes control reverses the defect in LPL activity—due to an increase in LPL synthesis— and reduces serum TG levels.[57] Yost and Eckel[58] have also demonstrated that the inhibition of AT-LPL by fat feeding usually seen in normal-weight subjects is no longer seen after isocaloric maintenance of the reduced obese state. Moreover, the response of AT-LPL to insulin is shifted to the right in obesity and is further diminished in NIDDM patients, suggesting that the greater degree of IR in NIDDM generates the lesser response.[48]

In skeletal muscle, insulin action also relates to the lipase response to insulin.

SM-LPL is increased to a greater degree by insulin in obese versus lean subjects. Moreover, when SM-LPL is plotted against the amount of glucose needed to maintain euglycemia during the hyperinsulinemic euglycemic clamp, an inverse correlation is seen. This suggests the following: when SM-LPL is further increased by insulin, then a greater amount of lipid and less glucose are oxidized.[48] Positive correlations between fasting LPL activity and the glucose infusion rate (amount of glucose needed to maintain euglycemia) have been demonstrated in skeletal muscle, but not in adipose tissue, when the data of control, nondiabetic-obese, and obese NIDDM have been analyzed together. However, the relative contribution of the intramuscular pool of TG fatty acids to meet the needs of skeletal muscle for lipid oxidation is not apparent.

Other studies aimed at the identification of possible factors contributing to the development of hypertriglyceridemia in subjects with obesity and/or IR have revealed that the ability of noradrenaline to stimulate lipolysis in fat cells in normal-weight subjects with obese first-degree relatives is markedly decreased.[59] The decreased lipolytic function of catecholamines in adipose tissue of these subjects was probably due, at least in part, to the decreased function of hormone-sensitive lipase (HSL) as demonstrated by suppressed levels of HSL in fat cells.[59] One may therefore speculate that impaired function of HSL is an early metabolic disturbance that may later lead to development of hyperlipidemia, obesity, and IR, especially under the conditions of increased energy intake.

A decreased ability of catecholamines to stimulate lipolysis was demonstrated also in fat cells from obese elderly men suffering from the metabolic syndrome.[60] Marked reduction of the maximum lipolytic response found in these patients was due to multiple defects in the lipolytic cascade. Both (i) reduced expression of β2 receptors leading to decreased binding of the hormone to the cell surface and (ii) decreased ability of cAMP to activate the hormone-sensitive lipase in adipocytes might contribute to the development of severe manifestations of the IR syndrome in these subjects.[60]

With the recent development of highly sophisticated methods of molecular biology, using high-quality antibodies and the cDNA probes, more detailed studies on the regulation of lipolysis have become possible. In harmony with this, it was demonstrated in both humans[61] and rodents[62] that changes in tissue LPL activity can be accompanied by changes at the transcriptional and translational level.

There are several situations in humans and in rodents as well where LPL activity is controlled at the level of gene expression.[63] The continuous modulation of adipose tissue LPL activity in response to feeding-fasting cycles is also associated with changes in mRNA levels, but these are often not of the same magnitude as the changes in LPL activity.[64] Furthermore, the changes in mRNA levels occur more slowly than the changes in LPL activity.[65] After a 16-hour fasting, LPL activity in guinea pig adipose tissue was found to be decreased by 80%, while LPL mass dropped only by 30%,[50] implying that posttranscriptional mechanisms are involved. In contrast, the increase of heart LPL activity during fasting occurred without any change in LPL mRNA nor in the rate of LPL synthesis, suggesting that in the heart other mechanisms are involved in the modulation of functional LPL activity.[49]

REGULATION OF LIPOLYSIS IN ANIMAL MODELS OF INSULIN RESISTANCE

Low LPL activity and decreased LPL mRNA levels are found in the adipose tissue of streptozotocin-diabetic rats.[66] Such defects in adipose tissue LPL are hypothesized to be the rate-limiting step in the development of hypertriglyceridemia in STZ-induced diabetes.[66] The results on heart LPL activity and mRNA levels, which were not changed in STZ-diabetic rats, further confirm the previous observation on the tissue specificity for LPL regulation at the gene level. However, it is more likely that the LPL expressed in adipose tissue is more important than that in the heart for the development of hypertriglyceridemia.[66,67] Other studies looking at the mechanisms of hypertriglyceridemia in STZ-diabetic rats have shown that the slow lipolysis of TG-rich particles with postheparin plasma lipase may be the basis of the increased TG levels in circulation of these animals in the fed state, but not during fasting.[68]

Acute administration of fructose *in vivo*, contrary to the administration of an equal amount of glucose, does not enhance the activity of LPL in adipose tissue of fasting rats,[69] despite the fact that *in vitro* both glucose and fructose enhance the LPL activity in incubated adipose tissue of fasting rats.[70] Feeding the high-fructose or high-sucrose diets to rats in chronic experiments has revealed conflicting data, showing an increase, decrease, or no change in the adipose tissue LPL (for review, see reference 71). As these dietary treatments differed in the amount of sugar and in the duration of feeding, it may be speculated that different kinetics of LPL activation may be responsible for the aforementioned fact. This explanation is in agreement with the data of Vrána,[71] who demonstrated a slight decrease in LPL activity in adipose tissue after fructose feeding to rats, which was dependent on the duration of feeding. However, when the data on LPL activity were compared to the data obtained in rats fed the basal laboratory chow, an increase in adipose tissue LPL activity was demonstrated.[71] In contrast, the activity of LPL in heart muscle and diaphragm of these animals was increased when compared to rats fed either glucose or standard laboratory chow, no matter the duration of the feeding. Since the activity of this lipolytic enzyme represents the acute activation under particular metabolic conditions of the organism, it seems more appropriate, especially in long-term dietary manipulations, to look at the expression of LPL in tissues.

Using a two-week dietary protocol, we have been able to show that the expression of LPL in skeletal muscle was decreased after high-sucrose feeding when compared to the data obtained in rats fed the basal diet (TABLE 3). When the hHTG rats (with increased TG availability in muscle) were fed the high-sucrose diet, LPL mRNA levels were not changed in skeletal muscle (TABLE 3). Moreover, compared with the data of Vrána,[71] there is an uncoupling between the LPL activity and the LPL mRNA levels in muscle after high-sucrose feeding. This is in contrast to the studies on LPL regulation during changes in nutritional status represented by fasting-feeding. Here, the activity of LPL in adipose tissue and in muscle was found to be related to the level of expression for LPL mRNA in those tissues.[72]

An alteration of LPL activity in adipose tissue and in the heart was demonstrated in rats with an intake of high-fat diets rich in saturated fatty acids.[34,73,74] The data of Benhizia et al.[34] on high-lard diet–induced inhibition of adipose tissue LPL activity, together with the striking opposite increase in plasma chylomicron concentrations, further support the important role of adipose tissue LPL in the control of TG levels. In other experiments, in spite of no changes in the fasting LPL activity in white and brown adipose tissue, it was shown that the LPL activity in both types of fat was significantly stimulated postprandially in response to high-carbohydrate and high-fat diets, respectively.[73] Tissue specificity of LPL regulation was further demonstrated in the heart and skeletal muscle, where increased fasting levels of LPL were found in high-fat-fed animals, but where LPL was suppressed postprandially in skeletal muscle, although not in the heart.[73]

In the hereditary hypertriglyceridemic (hHTG) rat, it seems that overproduction of VLDL TG due to the increased lipogenesis in the liver,[37,39] together with a defect in adipose tissue lipolysis (without affecting the postheparin lipoprotein lipase activity), could be the cause of hypertriglyceridemia.[75] In harmony with this, increased adipose tissue lipolysis was found in the hHTG line, leading to an increase in glycerol and in FFA concentrations in the circulation. The markedly enhanced FFA mobilization, accompanied by only a mild rise in serum FFA concentrations, indicates that an increased turnover of FA, when combined with hypertriglyceridemia, may lead to preferential utilization of lipids and to IR. However, the expression of LPL measured as mRNA levels in skeletal muscle of hHTG rats (TABLE 3) was not changed even after feeding the animals a high-sucrose diet.

In animal models of IR associated with obesity, increased lipolysis in adipose tissue has been demonstrated. In addition, increased activity,[76] together with increased gene expression,[45] for LPL was found in adipose tissue of genetically obese Zucker fa/fa rats. Since the overexpression of the LPL gene is already present in adipose tissue in suckling rats, which do not have such a high degree of obesity, these data indicate that a disharmony in LPL regulation could be an early feature contributing later to the development of massive hyperlipidemia and obesity of these animals.

Since only about 50% of the FA that are oxidized in muscle originate from the circulation, FA released from the TG stored in muscle lipid droplets after the hydrolysis are likely to be an important determinant of the preferential FA oxidation in the muscle in IR, especially in situations where increased TG storage is present. Evidence exists demonstrating that HSL, similar to that in adipose tissue, is present also in muscles and that this lipase could be responsible for intracellular lipolysis of TG.[77] Furthermore, correlations between intracellular lipolysis and LPL activity suggest a coordinated regulation of HSL and LPL in muscle to provide FA as the fuel for meeting the energy demands of the muscle. However, at the time when HSL is apparently hydrolyzing intracellular TG, the LPL is sent to the capillary beds in search of substrate.[77] Despite the availability of anti-HSL antibody and a specific cDNA, the regulation of this interesting enzyme with high potential for determining the muscle TG availability has not yet been extensively investigated at the molecular level.

DIETARY FATTY ACIDS AS DETERMINANTS OF TG AVAILABILITY

Focus on n-3 PUFA

In addition to the amount of fat in the diet, it is generally accepted that FA composition has a profound effect on the development of hypertriglyceridemia and IR.[7-9,12] Thus, much attention has been paid to PUFA, with special focus on n-3 FA. It has been known for more than two decades that dietary fish oil (FO) rich in highly unsaturated eicosapentaenoic acid (20:5 n-3) and docosahexaenoic acid (22:6 n-3) decreases plasma TG levels in healthy[16] as well as hypertriglyceridemic subjects.[16,78] Concerning its hypolipidemic potential, it was also shown that it is much more effective than equal amounts of polyunsaturated vegetable oils, rich in n-6 fatty acids.[7,9] In addition to the hypotriglyceridemic effect of FO in serum, which is not associated with accumulation of TG in the liver,[80] these long-chain n-3 FA have the potential to suppress the accumulation of TG in the heart and more importantly in skeletal muscle when included in high-sucrose or high-fat diets.[8,9,14,15] The TG-lowering effect of n-3 fatty acids may result from decreased FA synthesis, decreased TG synthesis or secretion, increased clearance, or some combination of these factors.

Studies using diets with n-3 fatty acids fed to normal rats clearly indicate that the activities of several lipogenic enzymes are markedly reduced[25,35,37] when compared with saturated FA diets and also when compared to diets rich in PUFA of the n-6 family. The reduced activities of lipogenic enzymes are a result of decreases in the quantity of enzyme present and, at least in the case of FAS and ACC, this is a result of a decline in the amount of mRNA for FAS and ACC due to decreased transcription of the respective genes.[81,82]

The composition of dietary FA seems to be of great importance also during the nutritional transition from the high-fat diet of suckling to the high-carbohydrate diet of weaning.[28] In these experiments, weaning to the high-carbohydrate diet with a high content of long-chain PUFA prevented the increase in FAS and ACC mRNA and/or enzyme activities that occurred normally in the liver.[28] Interestingly, there was no effect of dietary FA on lipogenic gene expression in white adipose tissue.

We have recently completed studies[37] in which saturated beef tallow or fish oil (Activepa TG) was fed to rats in diets containing a high amount (70 cal%) of sucrose (TABLES 2 and 3). We found that feeding diets high in n-3 PUFA led to a striking reduction of ME activity and of mRNA for ME and FAS in the liver of these high-sucrose diet–fed rats. In addition, although the inhibition of hepatic lipogenesis by PUFA was evident during the high-sucrose feeding, dietary FA composition had no effect on lipogenesis as indexed by ME gene expression in rat epididymal adipose tissue.[37]

Further supportive evidence for the possibility that the hypotriglyceridemic effect of fish oil in sucrose-induced hypertriglyceridemia is due to the inhibition of lipogenesis was obtained from correlation analyses of the circulating TG levels and the activity of lipogenic enzymes. These analyses revealed positive correlations between the circulating TG levels and the activity of ACC ($r = 0.65$, $n = 16$, $p < 0.01$) and ME ($r = 0.44$, $n = 16$, $p < 0.05$) in the liver (FIGURES 1A and 1B).

In another set of experiments, we have investigated whether altered activity and expression of lipogenic enzymes may account for the hypotriglyceridemic effect of FO under the condition of hereditary-fixed hypertriglyceridemia as seen in hHTG rats. Feeding the hHTG rat a diet high in n-3 PUFA suppressed the expression of FAS in liver. However, no effect on FAS gene expression was observed in hHTG rats when the hypertriglyceridemia was further exaggerated by high-sucrose feeding (TABLE 3), nor did the high content of n-3 FA in the diet suppress the ME mRNA level in liver of hHTG rats. In contrast, when the hHTG rats were fed a high-sucrose diet supplemented with fish oil, a striking reduction of ME mRNA and activity in liver of hHTG rats was demonstrated. Also, here, the results of the correlation analysis ($r = 0.76$, $n = 16$, $p < 0.001$) between ACC activity and plasma TG were consistent with the notion that the TG-lowering effect of FO may be mediated through the inhibition of lipogenic enzymes (data not shown).

A similar suppression of ACC in liver was also found when the n-3 FA were fed as a supplement to the basal or high-sucrose diet to rats with neonatal streptozotocin-induced type II diabetes[14] (TABLE 1). Moreover, the ACC activities correlated with the circulating TG levels ($r = 0.75$, $n = 16$, $p < 0.001$) in STZ-diabetic animals fed various FO-supplemented diets (data not shown).

The findings that diets enriched in n-3 FA reduce the activities and expression of enzymes required for FA synthesis are of particular interest because *de novo* synthesis of FA is a major determinant of hepatic TG secretion.[83] The effect of dietary n-3 FA on synthesis and secretion of TG has been studied *in vivo* in both humans[84] and animals,[85,86] as well as in isolated hepatocytes[87] and perfused livers.[88] Most of these studies have agreed that n-3 FA suppress TG synthesis and secretion. Based on the results of Rustan[89] and others, who found the inhibition of diacylglycerol acyltransferase by n-3 FA, it was suggested that dietary FO reduces plasma TG by combining inhibition of FA synthesis (by suppression of the enzymes for *de novo* synthesis), inhibition of TG synthesis (probably through the inhibition of diacylglycerol acyltransferase), and inhibition of TG secretion.

Another possible mechanism for lowering the TG level by dietary FO could be that the long-chain n-3 FA are preferentially oxidized and are thus unavailable for incorporation into the circulating or tissue TG pool.[19] If increased oxidation is to play the key role in decreasing the levels of TG, then the activity of LPL in muscle has to be changed to allow for increased delivery of FA to the tissues for oxidation. In support of this hypothesis are the results of Herzberg and Rogerson,[90] who have documented that LPL is increased in skeletal muscle and heart, but not in adipose tissue of rats fed a fructose-based diet containing FO. This information has been confirmed by Baltzell *et al.*,[91] who found that the activity of LPL was higher in the soleus muscle of rats fed menhaden oil compared with corn oil, although there was no effect on the hepatic lipase.

The role of dietary long-chain n-3 FA for modulation of the LPL activity has been further documented in our laboratory. We measured the expression of LPL in skeletal

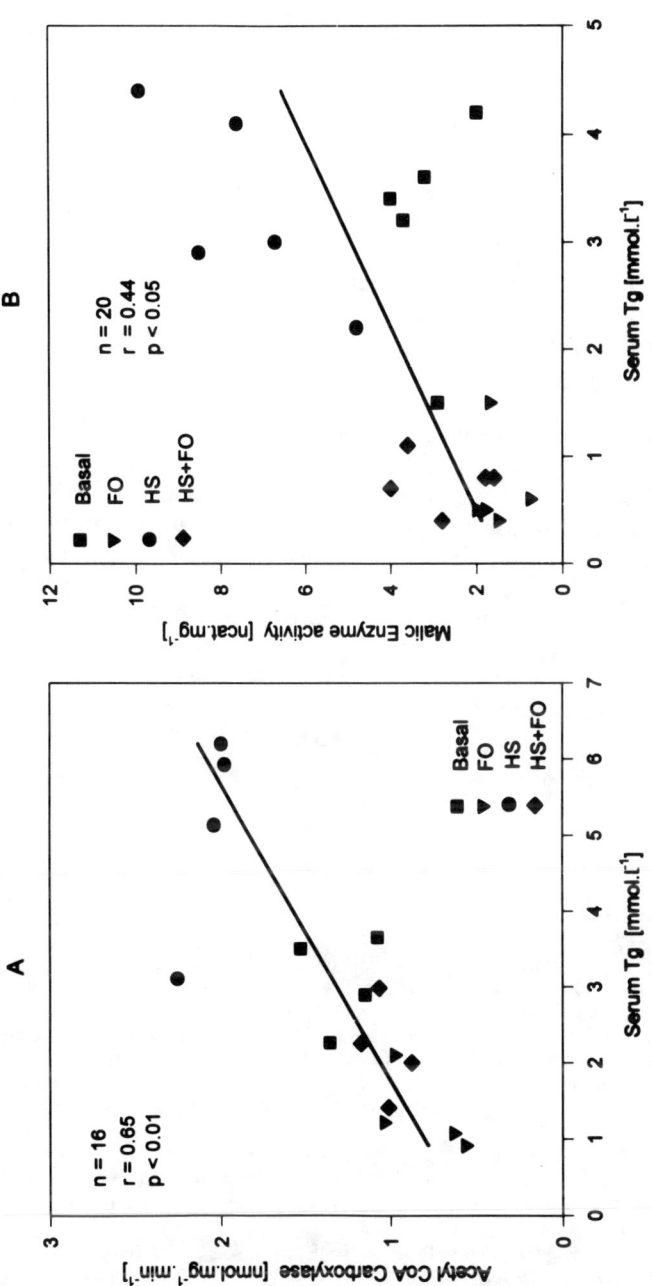

FIGURE 1. Correlation analysis between hepatic activity of acetyl-CoA carboxylase (A) or malic enzyme (B) and serum triglyceride (Tg) concentration in rats fed fish oil–supplemented diets. Terms: FO = basal diet supplemented with fish oil; HS = high-sucrose diet; HS+FO = high-sucrose diet supplemented with fish oil.

muscle of rats fed FO with or without a high-sucrose diet. In this experiment, the expression of LPL was increased in the soleus muscle when the n-3 FA were included in the basal or high-sucrose diet (TABLE 3). In addition, similar stimulation of LPL mRNA levels by n-3 FA was found in hereditary hypertriglyceridemic rats (TABLE 3).

The mechanisms by which dietary PUFA may alter the activity and expression of lipogenic and lipolytic enzymes have been difficult to discover. These processes may be regulated directly at the gene level,[25] affecting the nuclear mechanisms that change the expression of various genes encoding enzymes involved in lipid metabolism by controlling transcription, RNA processing, or mRNA stability.[24-26] Another interpretation may involve a long-term, adaptive modulation of membrane composition that leads to changes in hormone signaling.[92] Feeding dietary fats high in PUFA enriches cellular membranes with long-chain PUFA, which subsequently may alter the signaling mechanisms leading to the modification of key enzymes involved in lipid synthesis and metabolism.[92]

In summary, a good deal of evidence at the molecular and cellular level has now been accumulated to show that the FA composition of the diet is important and may be one of the critical regulators of TG availability. The key enzymes involved in TG synthesis and metabolism are part of the membrane compartments in the cell. Thus, it might be expected that specific changes in lipids (particularly to FA composition) could be involved in the regulation of their activity and expression in both physiological and pathological conditions. However, whether the accumulation of specific long-chain FA into tissue phospholipids will have a causal relationship to the regulation of TG availability remains to be established.

As dietary PUFA may regulate, at the gene level, the activity of lipogenic and lipolytic enzymes under particular dietary and genetic circumstances, it may be speculated that feeding PUFA in the early stages of development might reduce subsequent hypertriglyceridemia and/or IR.

REFERENCES

1. REAVEN, G. M. 1988. Diabetes 37: 1595–1607.
2. DEFRONZO, R. A., R. C. BONADONNA & E. FERRANNINI. 1992. Diabetes Care 15: 318–368.
3. FRAYN, K. N. 1993. Curr. Opin. Lipidol. 4: 197–204.
4. HOWARD, B. V. 1995. Diabetes Rev. 3: 423–432.
5. TASKINEN, M. 1995. Curr. Opin. Lipidol. 6: 153–154.
6. STANDL, E., N. LOTZ, T. DEXEL, H. JANKA & H. J. KOLB. 1980. Diabetologia 18: 463–469.
7. STORLIEN, L. H., A. B. JENKINS, D. J. CHISHOLM, W. S. PASCOE, S. KHOURI & E. W. KRAEGEN. 1991. Diabetes 40: 280–289.
8. STORLIEN, L. H., D. A. PAN, A. D. KRIKETOS & L. A. BAUR. 1993. Ann. N.Y. Acad. Sci. 683: 82–90.
9. KLIMEŠ, I., E. ŠEBÖKOVÁ, A. VRÁNA & L. KAZDOVÁ. 1993. Ann. N.Y. Acad. Sci. 683: 69–81.
10. KLIMEŠ, I., E. ŠEBÖKOVÁ, A. MINCHENKO, A. VRÁNA, M. FICKOVÁ, M. HROMADOVÁ, E. ŠVÁBOVÁ, P. BOHOV & L. KAZDOVÁ. 1991. In Lipoproteins and Atherosclerosis, p. 55–62. Fischer. Jena.
11. STORLIEN, L. H., E. W. KRAEGEN, D. J. CHISHOLM, G. L. FORD, D. G. BRUCE & W. S. PASCOE. 1987. Science 32: 885–888.
12. STORLIEN, L. H., L. A. BAUR, A. D. KRIKETOS, D. A. PAN, G. J. COONEY, A. B. JENKINS, G. D. GALVERT & L. V. CAMPBELL. 1996. Diabetologia 39: 621–631.

13. ŠEBÖKOVÁ, E., I. KLIMEŠ, M. HERMANN, M. HROMADOVÁ, P. BOHOV & M. HUETTINGER. 1992. Diabetes Nutr. Metab. **5**: 249-257.
14. ŠEBÖKOVÁ, E., I. KLIMEŠ, R. MOSS, P. ŠTOLBA, M. WIERSMA & A. MITKOVÁ. 1993. Ann. N.Y. Acad. Sci. **683**: 218-227.
15. ŠEBÖKOVÁ, E., I. KLIMEŠ, M. HERMANN, A. MINCHENKO, A. MITKOVÁ & M. HROMADOVÁ. 1993. Ann. N.Y. Acad. Sci. **683**: 183-191.
16. CONNOR, W. E., C. A. DEFRANCESCO & S. L. CONNOR. 1993. Ann. N.Y. Acad. Sci. **683**: 16-34.
17. SIMOPOULOS, A. P. 1994. Nutr. Today January/February: 12-16.
18. WAKIL, S. J., J. K. STOOPS & V. C. JOSHI. 1983. Annu. Rev. Biochem. **52**: 53.
19. HERZBERG, G. R. 1991. Physiol. Pharmacol. **69**: 1637-1647.
20. ECKEL, R. H. 1989. N. Engl. J. Med. **320**: 1060-1068.
21. FLAKOLL, P., M. G. CARLSON & A. CHERRINGTON. 1996. *In* Diabetes Mellitus, p. 121-131. Lippincott-Raven. Philadelphia/New York.
22. HELLERSTEIN, M. K., R. A. NEESE & J. M. SCHWARZ. 1993. Am. J. Physiol. **265**: E814-E819.
23. ACHESON, K. J., Y. SCHUTZ & T. BESSARD. 1988. Am. J. Clin. Nutr. **48**: 240-256.
24. CLARKE, S. D. & S. ABRAHAM. 1992. FASEB J. **6**: 3146-3152.
25. CLARKE, S. D. & D. B. JUMP. 1994. Annu. Rev. Nutr. **14**: 83-98.
26. KATSUDARA, A., N. IRITANI, H. FUKUDA, Y. MATSUMURA & N. NISHIMOTO. 1990. Eur. J. Biochem. **190**: 435-441.
27. IRITANI, N. 1992. Eur. J. Biochem. **205**: 433-442.
28. GIRARD, J., D. PERDEREAU, F. FOUFELLE, C. PRIP-BUUS & P. FERRE. 1994. FASEB J. **8**: 36-42.
29. HERZBERG, G. R. & N. JANMOHAMED. 1980. Br. J. Nutr. **43**: 571-579.
30. GODBOLE, V. & D. A. YORK. 1978. Diabetologia **14**: 191-197.
31. CLANDININ, M. T. & S. K. CHEEMA. 1996. Biochim. Biophys. Acta **1299**: 284-288.
32. ROMSOS, D. R. & G. A. LEVEILLE. 1974. Biochim. Biophys. Acta **360**: 1-11.
33. NELSON, G. J., D. S. KELLEY, P. C. SCHMIDT & C. M. SERRATO. 1987. Lipids **22**: 338-344.
34. BENHIZIA, F., I. HAINAULT, C. SEROUGNE, D. LAGRANCE, E. HAJDUCH, C. GUICHARD, M. I. MALEWIAK, A. QUIGNARD-BOULANGE, M. LAVAU & S. GRIGLIO. 1994. Am. J. Physiol. **267**: E975-E982.
35. HAINAULT, I., M. CARLOTTI, E. HAJDUCH, C. GUICHARD & M. LAVAU. 1993. Ann. N.Y. Acad. Sci. **683**: 98-101.
36. BECKER, D. J., L. N. ONGEMBA, V. BRICHARD, J. C. HENQUIN & S. M. BRICHARD. 1995. FEBS Lett. **371**: 324-328.
37. ŠEBÖKOVÁ, E., I. KLIMEŠ, D. GAŠPERÍKOVÁ, P. BOHOV, P. LANGER, M. LAVAU & M. T. CLANDININ. 1996. Biochim. Biophys. Acta **1303**: 56-62.
38. BRICHARD, S. M., L. N. ONGEMBA, J. GIRARD & J. C. HENQUIN. 1994. Diabetologia **37**: 1065-1072.
39. VRÁNA, A. & L. KAZDOVÁ. 1990. Transplant. Proc. **22**: 2579.
40. VRÁNA, A., L. KAZDOVÁ, Z. DOBEŠOVÁ, J. KUNEŠ, V. KŘEN, V. BÍLÁ, P. ŠTOLBA & I. KLIMEŠ. 1993. Ann. N.Y. Acad. Sci. **683**: 57-68.
41. VRÁNA, A. & L. KAZDOVÁ. 1986. Prog. Biochem. Pharmacol. **21**: 59-73.
42. VRÁNA, A. & P. FÁBRY. 1983. World Rev. Nutr. Diet. **42**: 56-101.
43. KLIMEŠ, I., M. FICKOVÁ, A. VRÁNA, P. BOHOV, E. ŠVÁBOVÁ, Š. ZÓRAD, L. ŠIKUROVÁ, L. KAZDOVÁ, V. BALÁŽ & L. MACHO. 1989. *In* Insulin and the Cell Membrane, p. 413-427. Harwood. Chur.
44. VRÁNA, A. & L. KAZDOVÁ. 1982. Nutr. Rep. Int. **26**: 743-749.
45. DUGAIL, I., A. QUIGNARD-BOULANGE, X. LE LIEPVRE, B. ARDOUIN & M. LAVAU. 1992. Biochem. J. **281**: 607-611.
46. RANDLE, P. J., P. B. GARLAND, C. N. HALES & E. A. NEWSHOLME. 1963. Lancet **1**: 785-789.
47. GROOP, L. C., C. SALORANTA, M. SHANK *et al.* 1991. J. Clin. Endocrinol. Metab. **72**: 96-107.
48. ECKEL, R. H., T. J. YOST & D. R. JENSEN. 1995. Int. J. Obes. **19**: S16-S21.
49. COPPACK, S. W., R. L. JUDD & J. M. MILES. 1993. Diabetes **42**: 38A.
50. OLIVECRONA, T., M. BERGO, M. HULTIN & G. OLIVECRONA. 1995. Can. J. Cardiol. **11**(suppl.): 73G-78G.
51. COPPACK, S. W., M. D. JENSEN & J. M. MILES. 1994. J. Lipid Res. **35**: 177-193.
52. FARESE, R. V., JR., T. J. YOST & R. H. ECKEL. 1991. Metabolism **40**: 214-216.
53. PYKALISTO, O. J., P. H. SMITH & J. D. BRUNZELL. 1975. J. Clin. Invest. **56**: 1108-1117.
54. KIENS, B., H. LITHELL, K. J. MIKINES & E. A. RICHTER. 1989. J. Clin. Invest. **84**: 1124-1129.

55. TASKINEN, M. & E. A. NIKKILA. 1979. Diabetologia 17: 351–356.
56. NIKKILA, E. A., J. K. HUTTUNEN & C. EHNHOLM. 1977. Diabetes 26: 11–21.
57. SIMSOLO, R. B., J. M. ONG, B. SAFFARI & P. A. KERN. 1992. J. Lipid Res. 33: 89–95.
58. YOST, T. J. & R. H. ECKEL. 1988. J. Clin. Endocrinol. Metab. 67: 259–264.
59. HELLSTRÖM, L., D. LANGIN, S. REYNISDOTTIR, M. DAUZATS & P. ARNER. 1996. Diabetologia 39: 921–928.
60. REYNISDOTTIR, S., K. ELLERFELDT, H. WAHRENBERG, H. LITHELL & P. ARNER. 1994. J. Clin. Invest. 93: 2590–2599.
61. KERN, P. A., J. M. ONG, B. SAFFARI & J. CARTY. 1990. N. Engl. J. Med. 322: 1053–1059.
62. BESSESEN, D. H., A. D. ROBERTSON & R. H. ECKEL. 1991. Am. J. Physiol. 261: E246–E251.
63. BRAUN, J. E. A. & D. L. SEVERSON. 1992. Biochem. J. 287: 337–347.
64. DOOLITTLE, M. H., O. BEN-ZEEV, J. ELOVSON, D. MARTIN & T. G. KIRCHGESSNER. 1990. J. Biol. Chem. 265: 4570–4577.
65. SEMB, H. & T. OLIVECRONA. 1989. Biochem. J. 262: 505–511.
66. TAVANGAR, K., Y. MURATA, M. E. PEDERSEN, J. F. GOERS, A. R. HOFFMAN & F. B. KRAEMER. 1992. J. Clin. Invest. 90: 1672–1678.
67. INADERA, H., J. TOSHIRO, Y. OKUBO, Y. ISHIKAWA, K. SHIRAI, Y. SAITO & S. YOSHIDA. 1992. Scand. J. Clin. Lab. Invest. 52: 797–802.
68. MAMO, J. C. L., T. HIRANO, A. SAINSBURY, A. K. FITZGERALD & T. G. REDGRAVE. 1992. Biochim. Biophys. Acta 1128: 132–138.
69. BAR-ON, H. & Y. STEIN. 1968. J. Nutr. 94: 95–105.
70. WEBB, W., P. J. NESTEL, C. FOXMAN & A. LYNCH. 1970. Nutr. Rep. Int. 1: 189–195.
71. VRÁNA, A., P. FÁBRY & L. KAZDOVÁ. 1974. Nutr. Metab. 17: 282–288.
72. LADU, M. J., H. KAPSAS & W. K. PALMER. 1991. Am. J. Physiol. 260: R953–R959.
73. BOIVIN, A., I. MONTPLAISIR & I. DESHAIES. 1994. Am. J. Physiol. 267: E620–E627.
74. DE GASQUET, P., S. GRIGLIO, E. PEQUIGNOT-PLANCHE & M. I. MALEWIAK. 1979. J. Nutr. 107: 199–212.
75. VRÁNA, A. & L. KAZDOVÁ. 1997. This volume.
76. BRAY, G. A., J. S. STER & T. W. CASTONGUAY. 1992. Am. J. Physiol. 262: E32–E39.
77. OSCAI, L. B., A. D. ESSIG & W. K. PALMER. 1990. J. Appl. Physiol. 69: 1571–1577.
78. GOODNIGHT, S. H., JR., W. S. HARRIS, W. E. CONNOR & D. ILLINGWORTH. 1982. Arteriosclerosis 2: 87–113.
79. HARRIS, W. S., W. E. CONNOR & M. P. MCMURRY. 1983. Metab. Clin. Exp. 32: 179–184.
80. GARG, M. L., A. B. R. THOMSON & M. T. CLANDININ. 1989. Biochim. Biophys. Acta 1006: 127–130.
81. BLAKE, W. L. & S. D. CLARKE. 1990. J. Nutr. 120: 1727–1729.
82. GOODRIDGE, A. G. 1987. Annu. Rev. Nutr. 7: 157–185.
83. TOPPING, D. L., R. P. TRIMBLE & G. B. STORER. 1987. Biochim. Biophys. Acta 927: 423–428.
84. ILLINGWORTH, D., W. S. HARRIS & W. E. CONNOR. 1984. Arteriosclerosis 4: 270–275.
85. LOMBARDO, Y. B., A. CHICCO, M. E. D'ALESSANDRO, M. MARTINELLI, A. SORIA & R. GUTMAN. 1996. Biochim. Biophys. Acta 1299: 175–182.
86. HERZBERG, G. R. & M. ROGERSON. 1988. J. Nutr. 118: 1061–1067.
87. WONG, S., M. REARDON & P. NESTEL. 1985. Metabolism 34: 900–905.
88. WONG, S. H., P. H. NESTEL, R. P. TRIMBLE, G. B. STORER, R. J. ILLMAN & D. L. TOPPING. 1984. Biochim. Biophys. Acta 792: 103–109.
89. RUSTAN, A. C., J. O. NOSSEN, E. N. CHRISTIANSEN & C. A. DREVON. 1988. J. Lipid Res. 29: 1417–1426.
90. HERZBERG, G. R. & M. ROGERSON. 1989. Lipids 24: 351–353.
91. BALTZELL, J. K., J. T. WOOTEN & D. A. OTTO. 1991. Lipids 26: 289–294.
92. CLANDININ, M. T., S. CHEEMA, C. J. FIELD, M. L. GARG, J. VENKATRAMAN & T. R. CLANDININ. 1991. FASEB J. 5: 2761–2769.

Dietary Fatty Acids, Insulin Resistance, and Diabetes

BARBARA V. HOWARD

Medlantic Research Institute
Washington, District of Columbia 20010

An important precursor of non-insulin-dependent diabetes mellitus (NIDDM) is the insulin resistance syndrome. This syndrome is defined as the presence of increased insulin concentrations in association with other disorders such as hyperglycemia, hypertension, dyslipidemia, central obesity, and sometimes hyperuricemia and renal dysfunction, an array of metabolic disorders accelerating the progression of atherosclerosis. Insulin resistance is specifically related to alterations in lipoprotein composition and metabolism that are associated with increased risk of cardiovascular disease [i.e., elevated very-low-density lipoprotein (VLDL), compositional changes in low-density lipoprotein (LDL), and decreased high-density lipoprotein (HDL) with concomitant compositional changes].[1] It is likely that this dyslipidemia originates with an increase in free fatty acid flux resulting from the central obesity that accompanies insulin resistance.

The prevalence of NIDDM, and concomitant insulin resistance, has escalated dramatically in this century in all industrialized nations.[2] This phenomenon is attributed in part to increases in dietary fat consumption that accompany Westernization and often result in obesity. Therefore, we briefly examine the epidemiologic and observational studies that may shed light on the long-term effect of dietary fat intake on insulin resistance.

DIETARY FAT AND NIDDM

The dramatic increase over the past century in the prevalence of NIDDM in every industrialized country links NIDDM with economic affluence and suggests that one or more aspects of an affluent lifestyle are risk factors for this condition. Dietary fat is generally implicated because cultural affluence is universally associated with an increase in the amount of fat in the diet.

Studies of populations that have undergone rapid changes in their way of life, such as Australian Aborigines,[3] Japanese migrant populations,[4] and American Indians,[5] provide indirect evidence that dietary fat induces a higher prevalence of diabetes. In these populations, the shift from traditional diets that were high in carbohydrates and fiber to diets containing large quantities of fat[6] is associated with a higher prevalence of diabetes. Data examined in prospective studies within populations strongly suggest that dietary fat intake increases the risk of developing diabetes. A prospective study analyzing the diet of Pima Indian women suggested an association between the incidence of diabetes and both fat and total calorie intake,[6] and a diet high in saturated fat and cholesterol was positively associated with fasting glucose in normoglycemic men in the Zutphen Study.[7] In a Dutch cohort of the Seven Countries Study, men with newly diagnosed diabetes had a higher fat intake 20 years prior compared to men with normal glucose tolerance.[8] During the 5-year follow-up of a large cohort study of male health professionals,[9] a strong positive association was found between obesity, body fat distribution, and weight gain and the risk of diabetes. In the San Luis Valley Diabetes Study,[10] the mean percentage of energy consumed as fat was associated with an increase in the risk of NIDDM, even after adjustment for fasting glucose and insulin. These conclusions support those of Fujimoto et al., who found that the percentage of dietary saturated fat was associated with increased incidence of NIDDM among second- and third-generation Japanese Americans.[11] Finally, indirect evidence for this hypothesis comes from a recent comparison of the Pima Indians of Arizona with their close relatives in Sonora, Mexico.[5] Mexican Pimas had much lower rates of diabetes and a diet intake averaging less than 23% of total energy from fat. In summary, these prospective studies support the implication of dietary fat as a potentiator of diabetes.

DIETARY FAT AND INSULIN RESISTANCE

The linkage of dietary fat with increased incidence of diabetes suggests that increased dietary fat intake could increase insulin resistance, a determinant of the incidence of NIDDM. Although studies in animal models provide overwhelming data on the effects of high-fat diets in inducing insulin resistance, the data in humans are equivocal. Several studies have compared high-fat/low-carbohydrate and low-fat/high-carbohydrate diets in a randomized fashion using direct measures of insulin action,[12-23] but they generally show no short-term effects on insulin action of modifications of dietary fat within the range generally acceptable in a human diet. Similarly, studies with fish oil supplementation were equivocal; two studies showed modest improvements[24,25] and two showed no change.[26,27]

In contrast to short-term intervention studies, there is now evidence from epidemiologic studies that appears to link higher fat intake, especially saturated fat, with insulin resistance. A study of insulin action in 45 lean and obese individuals with normal glucose tolerance revealed a significant relationship between a measure of insulin resistance and percentage of total dietary fat, although the relationship was not significant after adjustment for obesity.[28] In the Normative Aging Study, a significant independent correlation was found between total and saturated fat intake and fasting insulin concentrations, even after adjustment for adiposity;[29] this finding was echoed in the Kaiser-Permanente Women Twins Study, in which the percentage of total and saturated fat correlated with fasting insulin.[30] In women in the Zutphen Elderly Study, a percentage of saturated fat intake was positively related to glucose intolerance and degree of hyperinsulinemia, even after adjustment for adiposity.[31] Relationships of saturated fat intake and insulin resistance were obtained in the Veterans Administration Normative Aging Study,[32] which revealed associations between saturated fat intake, abdominal obesity, and hyperinsulinemia in a cohort of 878 men between the ages of 21 and 80. In the Zutphen[7] and San Luis Valley[10] studies, a higher percentage of monounsaturated fat was associated with increasing adiposity—a correlate of insulin resistance.

In sum, the evidence in humans suggests that consumption of a high-fat diet may be related to insulin resistance. Possible mechanisms of this association have been explored in recent studies. An investigation of tissues from normal patients and from those scheduled to undergo coronary artery bypass surgery revealed consistent relationships between the percentage of unsaturated lipids in muscle structural membrane lipids and insulin action.[33,34] Conversely, the more saturated the muscle membrane phospholipids, the more insulin resistance.[33] In a study of Pima Indians and of an Australian population, the ratio of n-6 to n-3 polyunsaturated fatty acid was correlated with fasting insulin and higher relative body weight.[35] A correlation was also found between saturated fat in muscle membrane and insulin resistance in a Swedish population.[36]

DIETARY FAT AND OBESITY

In addition to the action of dietary fat on insulin resistance, another way that high levels of dietary fat could increase the risk of diabetes is by inducing obesity. Animal studies suggest that diets with higher proportions of energy derived from dietary fat lead to body weight gain[37] and recent studies in humans now provide evidence to support this finding. In the analysis from the Women's Health Trial Feasibility Study, women placed on an ad-lib diet with 20% of the energy derived from fat lost weight compared to women on a diet in which fat was not limited.[38]

In a comparison of the effects of two diets containing 21% and 37% fat, the low-fat diet resulted in more weight loss,[39] and percentage of energy from fat has been shown to be a factor in predicting weight gain in a prospective study of men and women.[40] An observational study showed that obese individuals consume a diet with a higher percentage of fat than do their lean counterparts.[28] In an overfeeding study, diets containing 50% above the energy requirements of fat or carbohydrate were fed to 9 lean and 7 obese men.[41] Carbohydrate overfeeding resulted in 75-80% of the dietary energy being stored and fat overfeeding led to storage of 90-95% of the excess calories, which led the investigators to conclude that excess dietary fat leads to greater fat accumulation.

Additionally, a higher resting metabolic rate and increased sympathetic nervous system activity were observed in vegetarians who consumed a greater relative intake of carbohydrates and a lower intake of fat,[42] and insight into the mechanisms of the action of fat on calorie balance was provided by a study of the effect of lipid infusion on glucose metabolism in individuals with NIDDM.[43] In the latter study, prolonged postabsorbed hypoglycemia was induced by lipid infusion in NIDDM patients, suggesting that excess fat in individuals with NIDDM may have an even greater effect in stimulating weight gain. Finally, a recent analysis of the National Health and Nutrition Examination Survey (NHANES I) Epidemiologic Follow-up Study showed a positive correlation between fat intake and weight change only in men without any morbidity and a negative correlation between baseline percent fat intake and subsequent weight change over an 8- to 10-year period in women.[44]

CONCLUSIONS

The data in humans appear to indicate that a higher percentage of dietary fat leads to increased adiposity over time. Possible mechanisms include increased efficiency of conversion to stored fat relative to carbohydrate,[45] a failure of fat intake to drive fat oxidation,[46,47] and/or the relative failure of fat to provide appropriate satiation.[48] Animal studies suggest that the fatty acid profile of the diet is of primary importance[49] and the epidemiologic data in humans suggest that saturated fat intake is the likely culprit, although the data in humans are not conclusive.[44]

There also is evidence of the effect of dietary fat on insulin action, independent of changes in adiposity,[33] with the percentage of saturated fat being of major importance. These data are consistent with recent work in animals and humans showing that the dietary fatty acid profile affects the fatty acid profile of skeletal muscle, with insulin resistance being associated with a higher proportion of tissue saturated fats.

In general, this body of data supports the contention that the high dietary fat associated with a Westernized lifestyle increases the risk of diabetes. In most individuals with diabetes accompanied by obesity and insulin resistance, adherence to a low-fat diet that is high in complex carbohydrates is suggested. More human studies are needed, however, to further define dietary initiators of insulin resistance and to determine whether prolonged adherence to a low-fat diet can improve insulin sensitivity.

REFERENCES

1. HOWARD, B. V. 1993. Insulin, insulin resistance, and dyslipidemia. Ann. N.Y. Acad. Sci. **683**: 1–8.
2. HARRIS, M. L. 1985. Prevalence of non-insulin-dependent diabetes and impaired glucose tolerance. *In* Diabetes in America. NIH Publication No. 85-1468, p. VII–VI31. National Diabetes Data Group, United States Department of Health and Human Services. Washington, District of Columbia.
3. O'DEA, K. 1984. Marked improvement in carbohydrate and lipid metabolism in diabetic Australian Aborigines after temporary reversion to traditional lifestyle. Diabetes **33**(6): 596–603.
4. KAWATE, H., M. YARNAKIDO, Y. NISHIMOTO, P. H. BENNETT, R. F. HAMMAN & W. C. KNOWLER. 1979. Diabetes mellitus and its vascular complications in Japanese migrants on the island of Hawaii. Diabetes Care **3**: 161–170.
5. RAVUSSIN, E., M. E. VALENCIA, J. ESPARZA, P. H. BENNETT & L. O. SCHULZ. 1994. Effects of a traditional lifestyle on obesity in Pima Indians. Diabetes Care **17**(9): 1067–1074.
6. BENNETT, P. H., W. C. KNOWLER, H. R. BAIRD, W. J. BUTLER, D. J. PETTITT & J. M. REID. 1984. Diet and development of non-insulin-dependent diabetes mellitus: an epidemiological perspective. *In* Diet, Diabetes, and Atherosclerosis, p. 109–119. Raven Press. New York.
7. FESKENS, E. J. M. & D. KROMHOUT. 1990. Habitual dietary intake and glucose tolerance in euglycemic men: the Zutphen Study. Int. J. Epidemiol. **19**: 953–959.
8. FESKENS, E. J. M., S. M. VIRTANEN, L. RÄSÄNEN, J. TUOMILEHOTO, J. STENGÅRD, J. PEKKANEN, A. NISSINEN & D. KROMHOUT. 1995. Dietary factors determining diabetes and impaired glucose tolerance: a 20-year follow-up of the Finnish and Dutch cohorts of the Seven Countries Study. Diabetes Care **18**: 1104–1112.
9. CHAN, J. M.; E. B. RIMM, G. A. COLDITZ, M. J. STAMPFER & W. C. WILLETT. 1994. Obesity,

fat distribution, and weight gain as risk factors for clinical diabetes in men. Diabetes Care **17**(9): 961-969.
10. MARSHALL, J. A., S. HOAG, S. SHETTERLY & R. F. HAMMAN. 1994. Dietary fat predicts conversion from impaired glucose tolerance to NIDDM: the San Luis Valley Diabetes Study. Diabetes Care **17**(1): 50-56.
11. FUJIMOTO, W. Y., R. W. BERGSTRON, E. J. BOYKO, J. L. KINYOUN, D. L. LEONETTI, L. L. NEWELL-MORRIS, L. R. ROBINSON, W. P. SHUMAN, W. C. STOLOV, C. H. TSUNEHARA & P. W. WAHL. 1994. Diabetes and diabetes risk factors in second- and third-generation Japanese Americans in Seattle, Washington. Diabetes Res. Clin. Pract. **24**(suppl.): S43-S52.
12. HIMSWORTH, H. P. 1935. The dietetic factor determining the glucose tolerance and sensitivity to insulin of healthy men. Clin. Sci. **2**: 67-94.
13. BECK-NIELSEN, H., O. PEDERSEN & N. S. SORENSEN. 1978. Effects of diet on the cellular insulin binding and the insulin sensitivity in young healthy subjects. Diabetologia **15**: 289-296.
14. KOLTERMAN, O. G., M. GREENFIELD, G. M. REAVEN, M. SAEKOW & J. M. OLEFSKY. 1979. Effect of a high carbohydrate diet on insulin binding to adipocytes and on insulin action *in vivo* in man. Diabetes **28**: 731-736.
15. HJOLLUND, E., O. PEDERSEN, B. RICHELSEN, H. BECK-NIELSEN & N. S. SORENSEN. 1983. Increased insulin binding to adipocytes and monocytes and increased insulin sensitivity of glucose transport and metabolism in adipocytes from non-insulin-dependent diabetics after a low-fat/high-starch/high-fiber diet. Metabolism **32**: 1067-1075.
16. CHEN, M., R. N. BERGMAN & D. PORTE, JR. 1988. Insulin resistance and beta-cell dysfunction in aging: the importance of dietary carbohydrate. J. Clin. Endocrinol. Metab. **67**: 951-957.
17. FUKAGAWA, N. K., J. W. ANDERSON, G. HAGEMAN, V. R. YOUNG & K. MINAKER. 1990. High-carbohydrate, high-fiber diets increase peripheral insulin sensitivity in healthy young and old adults. Am. J. Clin. Nutr. **52**: 524-528.
18. BORKMAN, M., L. V. CAMPBELL, D. J. CHISHOLM & L. H. STORLIEN. 1991. Comparison of the effects on insulin sensitivity of high carbohydrate and high fat diets in normal subjects. J. Clin. Endocrinol. Metab. **72**(2): 432-437.
19. SWINBURN, B. A., V. L. BOYCE, R. N. BERGMAN, B. V. HOWARD & C. BOGARDUS. 1991. Deterioration in carbohydrate metabolism and lipoprotein changes induced by modern, high fat diet in Pima Indians and Caucasians. J. Clin. Endocrinol. Metab. **73**: 156-165.
20. GARG, A., S. M. GRUNDY & R. H. UNGER. 1992. Comparison of effects of high and low carbohydrate diets on plasma lipoproteins and insulin sensitivity in patients with mild NIDDM. Diabetes **41**: 1278-1285.
21. PARILLO, M., A. A. RIVELLESE, A. V. CIARDULLO, B. CAPALDO, A. GIACCO & S. GENOVESE. 1992. A high-monounsaturated-fat/low-carbohydrate diet improves peripheral insulin sensitivity in non-insulin-dependent diabetic patients. Metabolism **41**: 1373-1378.
22. RICCARDI, G. & M. PARILLO. 1993. Comparison of the metabolic effects of fat-modified *vs.* low fat diets. Ann. N.Y. Acad. Sci. **683**: 192-198.
23. HANNAH, J. S. & B. V. HOWARD. 1994. Dietary fats, insulin resistance, and diabetes. J. Cardiovasc. Risk **1**(1): 31-37.
24. CONNOR, W. E., M. J. PRINCE, D. ULLMANN, M. RIDDLE, L. HATCHER, F. E. SMITH & D. WILSON. 1993. The hypotriglyceridemic effect of fish oil in adult-onset diabetes without adverse glucose control. Ann. N.Y. Acad. Sci. **683**: 337-340.
25. MCVEIGH, G. E., G. M. BRENNAN, J. N. COHN, S. M. FINKELSTEIN, R. J. HAYES & G. D. JOHNSTON. 1994. Fish oil improves arterial compliance in non-insulin-dependent diabetes mellitus (NIDDM). Arterioscler. Thromb. **14**(9): 1425-1429.
26. MORGAN, W. A., P. RASKIN & J. ROSENSTOCK. 1995. A comparison of fish oil or corn oil supplements in hyperlipidemic subjects with NIDDM. Diabetes Care **18**(1): 83-86.
27. PUHAKAINEN, I., I. AHOLA & H. YKI-JÄRVINEN. 1995. Dietary supplementation with n-3 fatty acids increases gluconeogenesis from glycerol but not hepatic glucose production in patients with non-insulin-dependent diabetes mellitus. Am. J. Clin. Nutr. **61**: 121-126.
28. LOVEJOY, J. & M. DIGIROLAMO. 1992. Habitual dietary intake and insulin sensitivity in lean and obese adults. Am. J. Clin. Nutr. **55**: 1174-1179.
29. PARKER, D. R., S. T. WEISS, R. TROISI, P. A. CASSANO, P. S. VOKONAS & L. LANDSBERG. 1993. Relationship of dietary saturated fatty acids and body habitus to serum insulin concentration: the Normative Aging Study. Am. J. Clin. Nutr. **58**(2): 129-136.

30. MAYER, E. J., B. NEWMAN, C. P. QUESENBERRY, JR. & J. V. SELBY. 1993. Usual dietary fat intake and insulin concentrations in healthy women twins. Diabetes Care 16(11): 1459-1469.
31. FESKENS, E. J. M., J. G. LOEBER & D. KROMHOUT. 1994. Diet and physical activity as determinants of hyperinsulinemia: the Zutphen Elderly Study. Am. J. Epidemiol. 140: 350-360.
32. WARD, K. D., D. SPARROW, P. S. VOKONAS, W. C. WILLETT, L. LANDSBERG & S. T. WEISS. 1994. The relationships of abdominal obesity, hyperinsulinemia, and saturated fat intake to serum lipid levels: the Normative Aging Study. Int. J. Obes. 18: 137-144.
33. STORLIEN, L. H., B. V. HOWARD & A. P. SIMOPOULOS. 1996. Analysis of the relationship between dietary fat, obesity, and insulin action. Diabetologia. In press.
34. BORKMAN, M., L. H. STORLIEN, D. A. PAN, A. B. JENKINS, D. J. CHISHOLM & L. V. CAMPBELL. 1993. The relation between insulin sensitivity and the fatty acid composition of skeletal muscle phospholipids. N. Engl. J. Med. 328(4): 238-244.
35. STORLIEN, L. H., D. PAN, M. MILNER & S. LILLIOJA. 1993. Skeletal muscle membrane lipid composition is related to adiposity in man. Obes. Res. 1(suppl. 2): 77S.
36. VESSBY, B., S. TENGBLAD & H. LITHELL. 1994. Insulin sensitivity is related to the fatty acid composition of serum lipids and skeletal muscle phospholipids in 70-year-old men. Diabetologia 37: 1044-1050.
37. FLATT, J. P. 1993. Dietary fat, carbohydrate balance, and weight maintenance. Ann. N.Y. Acad. Sci. 683: 122-140.
38. SHEPPARD, L., A. R. KRISTAL & L. H. KUSHI. 1991. Weight loss in women participating in a randomized trial of low-fat diet. Am. J. Clin. Nutr. 54: 821-828.
39. PREWITT, T. E., D. SCHMEISSER, P. E. BOWEN, P. AYE, T. A. DOLECEK, P. LANGENBERG, T. COLE & L. BRACE. 1991. Changes in body weight, body composition, and energy intake in women fed high- and low-fat diets. Am. J. Clin. Nutr. 54: 304-310.
40. KLESGES, R. C., L. M. KLESGES, C. K. HADDOCK & L. H. ECK. 1992. A longitudinal analysis of the impact of dietary intake and physical activity on weight change in adults. Am. J. Clin. Nutr. 55: 818-822.
41. HORTON, T. J., H. DROUGAS, A. BRACHEY, G. W. REED, J. C. PETERS & J. O. HILL. 1995. Fat and carbohydrate overfeeding in humans: different effects on energy storage. Am. J. Clin. Nutr. 62: 19-29.
42. TOTH, M. J. & E. T. POEHLMAN. 1994. Sympathetic nervous system activity and resting metabolic rate in vegetarians. Metabolism 43(5): 621-625.
43. RIGALLEAU, V., C. GUILLOT, E. DETINGUY, A. IRON, G. DELERIS & H. GIN. 1994. Effect of lipid infusion on postabsorptive glucose metabolism in non-insulin-dependent diabetic patients. Metabolism 43(10): 1300-1304.
44. KANT, A. K., B. I. GRAUBARD, A. SCHATZKIN & R. BALLARD-BARBASH. 1995. Proportion of energy intake from fat and subsequent weight change in the NHANES I Epidemiologic Follow-up Study. Am. J. Clin. Nutr. 61(1): 11-17.
45. FLATT, J. P. 1978. The biochemistry of energy expenditure. In Obesity Research. Vol. II, p. 211-288. Newman. London.
46. SCHUTZ, Y., J. P. FLATT & E. JEQUIER. 1989. Failure of dietary fat intake to promote fat oxidation: a factor favoring the development of obesity. Am. J. Clin. Nutr. 50: 307-314.
47. ASTRUP, A., B. BUEMANN, N. J. CHRISTENSEN & S. TOUBRO. 1994. Failure to increase lipid oxidation in response to increasing dietary fat content in formerly obese women. Am. J. Physiol. 266(4): E592-E599.
48. BLUNDELL, J. E., S. GREEN & V. BURLEY. 1994. Carbohydrates and human appetite. Am. J. Clin. Nutr. 59: 728S-734S.
49. PAN, D. A., A. J. HULBERT & L. H. STORLIEN. 1994. Dietary fats, membrane phospholipids and obesity. J. Nutr. 124: 1555-1566.

Lipid Abnormalities in Muscle of Insulin-resistant Rodents

The Malonyl CoA Hypothesis[a]

N. B. RUDERMAN, A. K. SAHA, D. VAVVAS,
S. J. HEYDRICK, AND T. G. KUROWSKI

Diabetes Unit
Evans Department of Medicine
and
Department of Physiology
Boston University Medical Center
Boston, Massachusetts 02118

MUSCLE METABOLISM AFTER DENERVATION

Disuse or denervation has long been known to cause insulin resistance in skeletal muscle.[1-5] Thus, within 6-24 hours of denervation, the sensitivity and responsiveness to insulin of such processes as glucose transport, glycogen synthesis, and glycogen synthase activation are depressed in a variety of incubated muscles.[1,3,5] The mechanism for this insulin resistance is not well understood, although the observations that insulin binding to its receptor and activation of the receptor tyrosine kinase[2] and phosphatidylinositol 3-kinase[6] are not impaired indicate that distal events in the insulin signaling cascade are involved.

In searching for such altered signaling events, rat soleus muscles were studied 24 hours after sciatic nerve sectioning. Diacylglycerol (DAG) content and protein kinase C (PKC) activity were assayed because recent studies, albeit controversial, implicated PKC in the regulation of insulin action (reviewed in reference 7). As shown in TABLE 1, DAG mass was increased by 30% and membrane-associated PKC activity by 250% (using an assay for conventional PKC) in the 24-hour denervated soleus.[7] To assess whether an increase in *de novo* DAG synthesis contributed to the increase in DAG mass, the effect of insulin on glucose disposition was compared in control and denervated soleus muscles. As shown in TABLE 2, the incorporation of glucose into glycogen was depressed by over 90% after denervation, as reported by others;[1,3,5] however, glucose conversion to lactate and pyruvate was not depressed and its incorporation into lipids (totally into the glycerol moiety) was enhanced severalfold. TLC analysis indicated that these lipids included diacylglycerol, triglycerides (TG),

---

[a] This work was supported in part by grants from the Juvenile Diabetes Foundation and the NIH (Nos. DK 19514 and DK 49417).

TABLE 1. Effect of 24-h Denervation on Diacylglycerol (DAG) Mass and Protein Kinase C (PKC) Distribution in Rat Soleus Muscle[a]

		PKC Activity	
	DAG Mass (pmol/mg muscle)	100,000g Pellet (% of total)	Total (pmol phosphate/ min/mg muscle)
control	86 ± 6	15 ± 5	1.5 ± 0.5
denervated	121 ± 8[b]	38 ± 10[b]	1.6 ± 0.6

[a] Values are means ± SE for 9 (DAG mass) and 5 (PKC activity) muscles taken directly from rats. PKC was assayed on the basis of histone 3b phosphorylation in the presence and absence of Ca^{2+}, PS, and DAG. Reproduced from reference 7.
[b] $p < 0.05$ versus control.

and various phospholipids, all of which showed comparable increases (2-3-fold versus controls) in labeling.[7] Thus, there is a dichotomy of insulin action in denervated muscle: the stimulation of glycogen synthesis by insulin is nearly totally inhibited, but the stimulation of glucose incorporation into DAG and other glycerolipids is enhanced.

EFFECTS OF HYPERGLYCEMIA ON DAG MASS AND SYNTHESIS

In investigations that paralleled those with denervated muscle, we observed that adding insulin and glucose to the incubation medium increased the mass of DAG by as much as 30% and its *de novo* synthesis by 250-300% in the rat soleus muscle (FIGS. 1A and 1B).[8] In two studies (six pairs of muscles studied in each), differences in PKC activity similar to those between control and denervated muscle were observed when solei incubated for 20 minutes with no added insulin or glucose versus 30 mM glucose and 10 mU/mL insulin were compared.[9] In five other studies, however, no differences were observed. The reason for these discrepant results remains to be determined.

TABLE 2. Effect of 24-h Denervation on Glucose Disposition in Incubated Rat Soleus[a]

	Glucose Disposition (pmol glucose-carbon/mg muscle/15 min)			
Treatment	Glycogen	Lipids	Lactate + Pyruvate	CO_2
No insulin				
control	153 ± 27	71 ± 11	2060 ± 550	297 ± 59
denervated	94 ± 30	108 ± 20	2525 ± 362	192 ± 46
Insulin (10 mU/mL)				
control	2032 ± 272	276 ± 30	3448 ± 693	431 ± 80
denervated	195 ± 66	506 ± 50	3581 ± 701	382 ± 64

[a] Values are means ± SE of 8-12 experiments. Soleus muscles from sham-operated control and bilaterally denervated rats were incubated in the presence of 6 mM [U-^{14}C]-glucose. Reproduced from reference 7.

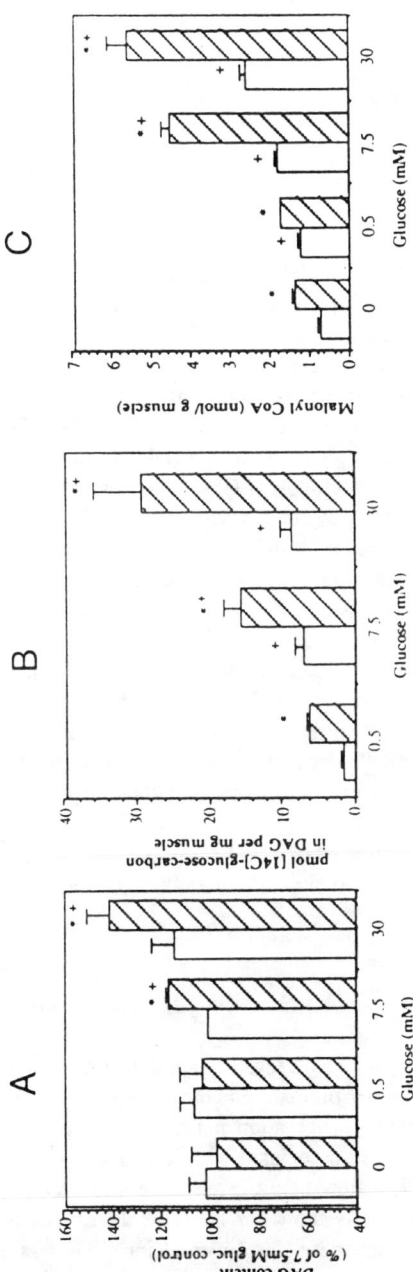

FIGURE 1. Results are means ± SE of at least 4–6 experiments. Muscles were incubated for 10 or 20 minutes ± insulin (10 mU/mL) and glucose at the indicated concentrations (□, control; ▨, 10 mU/mL insulin). DAG data are from Chen et al.[8] and malonyl CoA data from Saha et al.[16] The malonyl CoA concentration in muscles taken prior to incubation was over twice that of a muscle with 0.5 nM glucose in the absence of insulin (±2.0 versus ±0.8). (+) $p < 0.05$ versus muscles incubated with 0 or 0.5 mM glucose. (*) $p < 0.05$ versus no insulin.

TABLE 3. Malonyl CoA (nmol/g muscle)[a]

	Denervation	Electrical Stimulation
control	2.0 ± 0.3	2.2 ± 0.2
experimental	3.4 ± 0.5[b]	1.2 ± 0.1[b]

[a] Effects of denervation (24-h postsciatic nerve section) and intense contraction due to electrical stimulation (5/s for 5 minutes) of the sciatic nerve on malonyl CoA levels in rat soleus muscle. Results are means ± SE of 3-6 determinations. Adapted from reference 16.
[b] $p < 0.05$ versus control.

THE MALONYL CoA FUEL-SENSING MECHANISM AND INSULIN RESISTANCE

One potential modulator of the increased *de novo* synthesis of DAG and other glycerolipids in denervated muscle and muscle incubated with insulin and glucose could be malonyl CoA. Malonyl CoA is both an intermediate in the synthesis of fatty acids and an inhibitor of carnitine palmitoyl transferase 1 (CPT 1), the enzyme that catalyzes the transfer of long-chain fatty acyl CoA (LCFA CoA) into the mitochondria where they are oxidized. Thus, in liver and other tissues, malonyl CoA levels increase after eating and other events that increase plasma insulin and glucose, the oxidation of fatty acids is inhibited, and their incorporation into glycerolipids is accelerated.[10-13] Conversely, an opposite set of events occurs when malonyl CoA levels are decreased, as in starvation. As shown in TABLE 3, malonyl CoA levels are increased severalfold in rat soleus muscle after denervation and after incubation with insulin and glucose (FIG. 1C). In contrast, they decrease rapidly in muscle when it is fuel-deprived (FIG. 1C) or made to contract by electrical stimulation (TABLE 3). The latter finding is in agreement with an earlier report by Winder and coworkers,[14] who also observed a decrease in malonyl CoA after voluntary exercise.[15] Based on these observations, we have hypothesized that malonyl CoA and the enzyme that catalyzes its formation, acetyl-CoA carboxylase (ACC), are components of a fuel-sensing and signaling mechanism that responds to changes in both the fuel supply and energy expenditure of the muscle cell (FIG. 2).[16]

MALONYL CoA AND INSULIN RESISTANCE

In addition to lowering malonyl CoA levels, prior exercise and glucose deprivation enhance the ability of insulin to stimulate glucose transport and glycogen synthesis in skeletal muscle.[17,18] In contrast, denervation and sustained hyperglycemia,[19,20] which increase malonyl CoA levels, are associated with impaired insulin action. To study this interrelationship further, malonyl CoA levels were measured in muscle of other rodents with insulin resistance. As shown in TABLE 4, the concentration of malonyl CoA is increased in muscles of obese, insulin-resistant rodents such as KKA[y] (see reference 21) and Ob/Ob mice (unpublished). These mice are hyperinsulinemic; however, high concentrations of malonyl CoA were also found in muscle of the GK rat

FIGURE 2. The malonyl CoA fuel-sensing system. Our preliminary data suggest that changes in the fuel supply of the muscle cell increase or decrease the concentration of citrate (extramitochondrial) and that this in turn could affect malonyl CoA synthesis by increasing or decreasing the concentration of cytosolic acetyl CoA, the substrate for acetyl-CoA carboxylase, or by allosterically modifying the enzyme.[35] Presumably, malonyl CoA determines whether LCFA CoA in the muscle cell is oxidized or incorporated into triglycerides and other glycerolipids. Reproduced from reference 16.

(hyperglycemia only) (data not shown) and, as noted earlier, after denervation (no change in either plasma insulin or glucose). The high level of malonyl CoA in all of these insulin-resistant models was associated with an increase in the concentration of diacylglycerol and, where measured, of muscle triglycerides. In one rodent in which we observed an increase in the concentration of malonyl CoA, but not of DAG, namely, the Dahl salt-sensitive (Dahl-S) rat (normoglycemic, normoinsulinemic, but hypertriglyceridemic), insulin resistance was not present.[22] On the other hand, the Dahl-S rat became somewhat more insulin resistant and showed a greater increase in DAG content in muscle than did a control rat when fed a high-fat, high-sucrose diet. Interestingly, the Dahl rat also showed an increased predisposition to visceral obesity on this diet.

A link between high levels of malonyl CoA and insulin resistance was also suggested by studies with the insulin-sensitizing agent, pioglitazone. Numerous reports have described the ability of this and related drugs to diminish high plasma glucose and triglyceride concentrations in obese, hyperinsulinemic rodents (reviewed in reference 23). As reported elsewhere, pioglitazone also lowers the high levels of malonyl CoA both in these animals[21] and in the Dahl-S rat,[22] which is neither hyperinsulinemic nor hyperglycemic (FIG. 3). Although the precise mechanism of action of these agents remains to be determined, these findings raise the possibility that an effect on acetyl-CoA carboxylase or another determinant of malonyl CoA concentration is crucial to their action.

TABLE 4. Plasma Insulin and Glucose Levels and Lipid Metabolites in Soleus Muscle of Insulin-resistant Rodents[a]

Model	Plasma Insulin	Plasma Glucose	Muscle TG	Muscle DAG	Muscle Malonyl CoA
KKA[y] mouse	increased	increased	increased	increased	increased
Denervated rat	unchanged	unchanged	not determined	increased	increased
GK rat	decreased	increased	increased	increased	increased
Zucker rat	increased	unchanged	not determined	increased	not determined

[a] Based on data from our laboratory[16,21] and in the Zucker rat from Turinsky et al.[41]

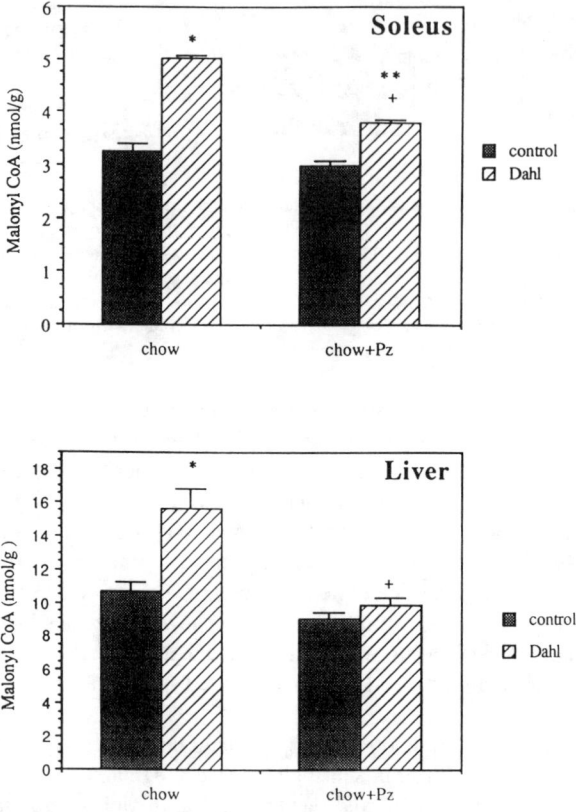

FIGURE 3. Effect of pioglitazone (PZ) on malonyl CoA concentrations in soleus muscle and liver of control and Dahl-S rats. Results are means ± SEM for 6–8 rats per group. Shaded bars = control rats; hatched bars = Dahl-S rats. Significantly different ($p < 0.001$) from (*) chow-fed control rats, (+) chow-fed Dahl-S rats, and (**) chow + pioglitazone–fed control rats. Adapted from reference 22.

MALONYL CoA AND INSULIN RESISTANCE: A HYPOTHESIS

A hypothetical schema depicting possible mechanisms for the association between malonyl CoA and insulin resistance is shown in FIGURE 4. According to the proposed model, sustained increases in the concentration of malonyl CoA, in the presence of high or normal levels of FFA, lead to insulin resistance in muscle by increasing the concentration of cytosolic LCFA CoA or a related metabolite such as DAG synthesized *de novo* or phosphatidic acid.[24] The increases in these lipids in turn could result in activation of one or more PKC isozymes[25,26] (see FIG. 4) and/or of an AMP-dependent protein kinase that can phosphorylate glycogen synthase[27] or other steps in the insulin-signaling cascade. In addition, fatty acids and LCFA CoA have been shown to cause alterations in protein acylation, membrane fluidity, and gene transcription. They can also directly inhibit glycogen synthase (reviewed in reference 21) and they may activate

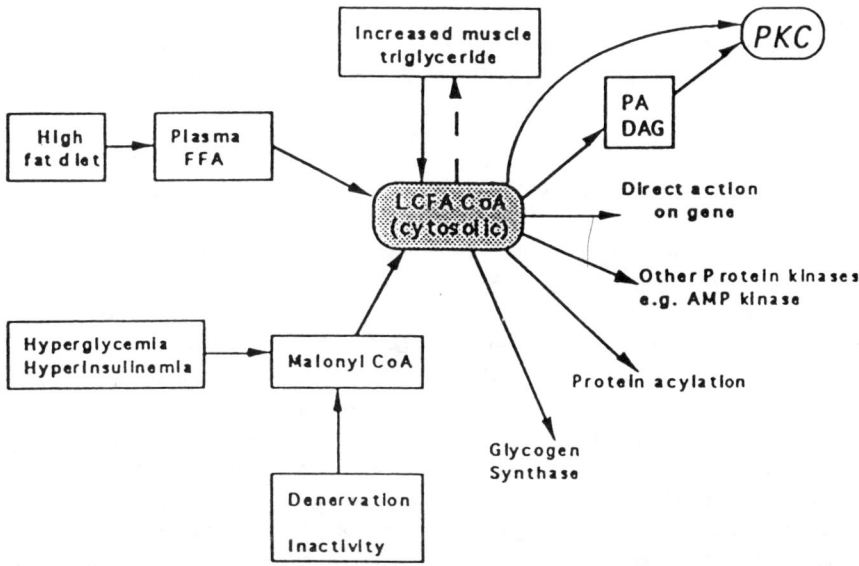

FIGURE 4. Hypothetical interrelations between the concentration of cytosolic long-chain fatty acyl CoA (LCFA CoA) and the development of insulin resistance in muscle. Factors that could increase the concentration of cytosolic LCFA CoA are indicated in open boxes. Possible mechanisms by which LCFA CoA or a related metabolite could affect the action of insulin are indicated on the right.

the hexosamine pathway,[28] which itself has been linked to insulin resistance in muscle.[29]

A key feature of the scheme in FIGURE 4 is that a common denominator in many insulin-resistant states may be an increase in intracellular levels of LCFA CoA, to which increases in the concentration of both malonyl CoA and free fatty acids (FFA) could contribute. The independent importance of FFA could explain why little, if any, insulin resistance was observed in the Dahl-S rat fed a standard chow diet, whereas it appeared to become more insulin-resistant than a control rat when fed a high-fat diet.

PROTEIN KINASE C

A potentially pivotal factor in the scheme outlined in FIGURE 4 and in theories of insulin resistance proposed by others is protein kinase C (PKC). Alterations in one or more PKC isozymes have been observed in insulin-resistant muscles of the Zucker rat[30] as well as in a variety of fat-fed[31] and fructose-fed rats.[32] Likewise, a redistribution of PKC activity has been observed in muscle made insulin-resistant by denervation (TABLE 1). Studies in adipose tissue[33] and cultured cells[34] suggest that the insulin resistance caused by hyperglycemia is due to activation of a PKC leading to phosphorylation of the insulin receptor and inhibition of its receptor kinase. In denervated muscle, no decrease in insulin binding to its receptor nor in activation of the receptor tyrosine kinase[11] has been observed. Thus, if the increase in membrane-associated PKC activity contributes to the impaired ability of insulin to stimulate glycogen synthesis in denervated muscle, another mechanism must be involved.

REGULATION OF MALONYL CoA LEVELS IN MUSCLE

The factors governing malonyl CoA formation have been studied in soleus muscles incubated with insulin and glucose for 20 minutes. Preliminary data indicate that the 2–5-fold increases in malonyl CoA observed in these muscles (FIG. 1C) are not due to increases in the intrinsic activity of acetyl-CoA carboxylase (ACC).[35] Instead, they appear to be related to increases in the cytosolic concentration of citrate, as reflected by changes in the concentrations of total cell citrate and of malate, an antiporter for citrate efflux from the mitochondria. Studies in cell-free systems indicate that citrate is both an allosteric activator of ACC and a precursor for its substrate, cytosolic acetyl CoA.[36] Whether it is acting in one or both capacities to regulate ACC activity *in vivo* remains to be determined. Whatever the explanation, these findings serve to link the malonyl CoA fuel-sensing mechanism to the glucose–fatty acid cycle of Randle,[37] in which increases in the concentration of citrate diminish glycolysis by allosterically inhibiting phosphofructokinase (see also reference 38). Presumably, these mechanisms complement each other in determining the relative use of fuels by skeletal muscle.

OTHER IMPLICATIONS OF THE MALONYL CoA FUEL-SENSING AND SIGNALING MECHANISM

A mechanism for regulating the concentration of malonyl CoA, similar to that described here, exists in the pancreatic beta-cell, where it plays a key role in the stimulation of insulin secretion by glucose,[11,39,40] and probably in the heart.[10] A logical prediction is that it will also be found in glucose-sensing cells in the central nervous system and in other cells in which the use of glucose as a fuel is determined by its availability. Whether a sustained and generalized increase in malonyl CoA concentration in these cells produces signaling abnormalities that lead to impaired insulin action, hyperinsulinemia, and other manifestations of the insulin resistance–obesity syndrome, as recently proposed,[16,21,39] requires further study.

ACKNOWLEDGMENTS

We wish to thank Barbara Corkey, Eleazar Shafrir, and Lee Witters for their advice during the course of these studies and David Tse for his assistance in preparing the manuscript.

REFERENCES

1. BURANT, C. F., S. K. LEMMON, M. K. TREUTELAAR & M. G. BUSE. 1984. Insulin resistance of denervated rat muscle: a model for impaired receptor-function coupling. Am. J. Physiol. **247**: E657–E666.
2. BURANT, C. F., M. K. TREUTELAAR & M. G. BUSE. 1986. In vitro and in vivo activation of the insulin receptor kinase in control and denervated skeletal muscle. J. Biol. Chem. **261**: 8985–8993.
3. FORSAYETH, J. R. & M. K. GOULD. 1982. Inhibition of insulin stimulated xylose uptake in denervated rat soleus muscle: a post-receptor event. Diabetologia **23**: 511–516.
4. NICHOLSON, W. F., P. A. WATSON & F. W. BOOTH. 1984. Glucose uptake and glycogen synthesis in muscle from immobilized limbs. J. Appl. Physiol. **56**: 431–435.
5. SMITH, R. L. & S. C. LAWRENCE. 1984. Insulin action in denervated rat hemidiaphragms: decreased hormonal stimulation of glycogen synthesis involves both glycogen synthase and glucose transport. J. Biol. Chem. **259**: 2201–2207.
6. CHEN, K. S., J. C. FRIEL & N. B. RUDERMAN. 1993. Regulation of phosphatidylinositol 3-kinase by insulin in rat skeletal muscle. Am. J. Physiol. **265**: E736–E742.
7. HEYDRICK, S. J., N. B. RUDERMAN, T. G. KUROWSKI, H. B. ADAMS & K. S. CHEN. 1991. Enhanced stimulation of diacylglycerol and lipid synthesis by insulin in denervated muscle: altered protein kinase C activity and possible link to insulin resistance. Diabetes **40**: 1707–1711.
8. CHEN, K. S., S. J. HEYDRICK, M. L. BROWN, J. C. FRIEL & N. B. RUDERMAN. 1994. Insulin increases a biochemically distinct pool of diacylglycerol in the rat soleus muscle. Am. J. Physiol. **266**: E479–E485.
9. CHEN, K. S., S. J. HEYDRICK, T. G. KUROWSKI & N. B. RUDERMAN. 1991. Diacylglycerol-protein kinase C signalling in rat skeletal muscle: a possible link to insulin resistance. Trans. Assoc. Am. Physicians **54**: 206–212.
10. AWAN, M. N. & E. D. SAGGERSON. 1993. Malonyl CoA metabolism in cardiac myocytes and its relevance to the control of fatty acid oxidation. Biochem. J. **295**: 61–65.
11. CHEN, S., A. OGAWA, M. OHNEDA, R. UNGER, D. W. FOSTER & J. D. MCGARRY. 1994. More direct evidence of a malonyl CoA-carnitine palmitoyltransferase I interaction as a key event in pancreatic β-cell signalling. Diabetes **43**: 878–883.
12. KUDO, N., A. J. BARR, R. L. BARR, S. DESAI & G. D. LOPASCHUK. 1995. High rates of fatty acid oxidation during reperfusion of ischemic hearts are associated with a decrease in malonyl CoA levels due to an increase in 5'-AMP-activated protein kinase inhibition of acetyl CoA carboxylase. J. Biol. Chem. **270**: 17513–17520.
13. CORKEY, B. E., M. C. GLENNON, K. S. CHEN, J. T. DEENEY, F. M. MATSCHINSKY & M. PRENTKI. 1989. A role for malonyl CoA in glucose-stimulated insulin secretion from clonal pancreatic β-cells. J. Biol. Chem. **264**: 21608–21612.
14. DUAN, C. & W. W. WINDER. 1992. Nerve stimulation decreases malonyl CoA in skeletal muscle. J. Appl. Physiol. **72**: 901–904.
15. WINDER, W. W., J. AROGYASAMI, I. M. ELAYAN & D. CARTMILL. 1990. Time course of exercise-induced decline in malonyl CoA in different muscle fiber types. Am. J. Physiol. **259**: E266–E271.
16. SAHA, A. K., T. G. KUROWSKI & N. B. RUDERMAN. 1995. A malonyl CoA fuel-sensing mechanism in muscle: effects of insulin, glucose, and denervation. Am. J. Physiol. **269**: E283–E289.
17. RICHTER, E. A., L. G. GARETTO, M. N. GOODMAN & N. B. RUDERMAN. 1982. Muscle glucose metabolism following exercise in the rat: increased sensitivity to insulin. J. Clin. Invest. **69**: 785–793.
18. GOODMAN, M. N. & N. B. RUDERMAN. 1979. Insulin sensitivity in rat skeletal muscle: effects of starvation and aging. Am. J. Physiol. **236**: 519–523.
19. RICHTER, E. A., B. F. HANSEN & S. A. HANSEN. 1988. Glucose-induced insulin resistance of skeletal-muscle glucose transport and uptake. Biochem. J. **252**: 733–737.
20. HANSEN, B. F., S. A. HANSEN, T. PLOUG, J. F. BAK & E. A. RICHTER. 1992. Effects of glucose and insulin on development of impaired insulin action in muscle. Am. J. Physiol. **262**: E440–E446.
21. SAHA, A. K., T. G. KUROWSKI, J. R. COLCA & N. B. RUDERMAN. 1994. Lipid abnormalities in tissue of the KKA^y mouse: effects of pioglitazone on malonyl CoA and diacylglycerol. Am. J. Physiol. **267**: E95–E101.

22. KUROWSKI, T. G., A. K. SAHA, B. A. CUNNINGHAM, R. I. HOLBERT, J. R. COLCA, B. E. CORKEY & N. B. RUDERMAN. 1995. Malonyl CoA and adiposity in the Dahl-salt sensitive rat: effect of pioglitazone. Metabolism **45:** 519–525.
23. COLCA, J. R. & D. R. MORTON. 1990. Antihyperglycemic thiazolidinediones: ciglitazone and its analogs. *In* New Anti-Diabetic Drugs, p. 255–261. Smith–Gordon. London.
24. LIMATOLA, C., D. SHAAP, W. H. MOOLEMAN & W. J. VAN BLITTERSWIJK. 1994. Phosphatidic acid activation of protein kinase C overexpressed in COS cells: comparison with other protein kinase C isotypes and other acidic lipids. Biochem. J. **304:** 1004–1008.
25. NISHIZUKA, Y. 1995. Protein kinase C and lipid signaling for sustained cellular responses. FASEB J. **9:** 484–496.
26. RASMUSSEN, H., C. M. ISALES, R. CALLE, D. THROCKMORTON, M. ANDERSON, J. GASALLA-HERRAIZ & R. MCCARTHY. 1995. Diacylglycerol production, Ca^{2+} influx, and protein kinase C activation in sustained cellular responses. Endocr. Rev. **16:** 649–681.
27. HARDIE, G. 1989. Regulation of fatty acid synthesis via phosphorylation of acetyl-CoA carboxylase. Lipid Res. **28:** 117–146.
28. HAWKINS, M., N. BARZILAI, R. LIU, W. CHEN & L. ROSETTI. 1996. Increased FFA and hyperglycemia result in a similar increase in carbon flux into the glucosamine (GlcN) pathway and peripheral insulin resistance. Diabetes **45:** 11A.
29. MCLAIN, D. A. & E. D. CROOK. 1996. Hexosamine and insulin resistance. Diabetes **45:** 1003–1009.
30. COOPER, D. G., J. E. WATSON & M. L. DAO. 1993. Decreased expression of protein kinase C α, β, ϵ in soleus muscle of Zucker obese (fa/fa) rats. Endocrinology **133:** 2241–2247.
31. SCHMITZ-PEIFFER, C., C. L. BROWNE, N. D. OAKES, A. WATKINSON, D. J. CHISHOLM, E. W. KRAEGEN & T. J. BIDEN. 1997. Alterations in the level and cellular locations of protein kinase C isozymes in skeletal muscle of the high-fat-fed rat model. Diabetes **46:** 169–178.
32. DONNELLY, R., M. J. REED, S. AZHAR & G. M. REAVEN. 1994. Expression of the major isozyme of protein kinase C in skeletal muscle, PKCθ, varies with muscle type and in response to fructose-induced insulin resistance. Endocrinology **135:** 2369–2374.
33. MULLER, H. K., M. KELLER, B. ERMEL, A. MUHLHOFFER, B. OBERMAIER-KUSSER, B. VOGT & H. U. HARING. 1991. Prevention by protein kinase C inhibitors of glucose-induced insulin receptor tyrosine kinase resistance in rat fat cells. Diabetes **40:** 1440–1448.
34. OLEFSKY, J. M., T. S. PILLAY & S. XIAO. 1996. Glucose-induced phosphorylation of the insulin receptor. J. Clin. Invest. **97:** 613–620.
35. SAHA, A. K., D. VAVVAS, T. G. KUROWSKI, A. APAZIDIS, L. A. WITTERS, E. SHAFRIR & N. B. RUDERMAN. 1997. Malonyl-CoA regulation in skeletal muscle: its link to cell citrate and the glucose-fatty acid cycle. Am. J. Physiol. **272:** E641–E648.
36. KIM, K-H., F. LOPEZ-CASILLAS, D. H. BAI, X. LUO & M. E. PAPE. 1989. Role of reversible phosphorylation of acetyl CoA carboxylase in long-chain fatty acid synthesis. FASEB J. **3:** 2250–2256.
37. RANDLE, P. J., P. B. GARLAND, C. N. HALES & E. A. NEWSHOLME. 1963. The glucose-fatty acid cycle: its role in insulin sensitivity and the metabolic disturbance of diabetes mellitus. Lancet **1:** 785–789.
38. MAIZELS, E. Z., N. B. RUDERMAN, M. N. GOODMAN & D. LAU. 1977. Effects of acetoacetate on glucose metabolism in the soleus and extensor digitorum longus muscles of the rat. Biochem. J. **162:** 557–568.
39. PRENTKI, M. & B. E. CORKEY. 1996. Are the β-cell signaling molecules malonyl CoA and cytosolic long-chain acyl CoA implicated in multiple tissue defects of obesity and NIDDM? Diabetes **45:** 273–283.
40. BRUNI, J., E. ROCHE, F. ASSIMACAPOLOUS-JEANNET, B. E. CORKEY, K. H. KIM & M. PRENTKI. 1996. Evidence for an anaplerotic malonyl CoA pathway in pancreatic β-cell signaling. Diabetes **45:** 190–198.
41. TURINSKY, J., B. P. BAYLY & D. M. O'SULLIVAN. 1990. 1,2-Diacylglycerol and ceramide levels in rat skeletal muscle and liver *in vivo*: studies with insulin, exercise, muscle denervation, and vasopressin. J. Biol. Chem. **265:** 7933–7938.

Pharmacological Strategies for Reduction of Lipid Availability

JAMES E. FOLEY, ROBERT C. ANDERSON,
PHILIP A. BELL, BRYAN F. BURKEY,
RHONDA O. DEEMS, CHRISTOPHER DE SOUZA,
AND BETH E. DUNNING

Department of Metabolic Diseases
Preclinical Research
Sandoz Research Institute
Sandoz Pharmaceutical Corporation
East Hanover, New Jersey 07936

FREE FATTY ACIDS AND INSULIN RESISTANCE

The chronic elevation of circulating free fatty acid (FFA) levels is associated with insulin resistance.[1] In normal individuals, excess energy is converted to FFAs by the liver and stored as triglyceride in adipocytes. The stored energy can be mobilized as needed via an increase in the rate of FFA release from the adipocytes. However, if the balance between storage and release is disturbed such that the FFAs become chronically elevated, insulin resistance is usually manifest.

Elevated FFA levels have been implicated in inducing metabolic derangements. For example, over 30 years ago, Randle demonstrated that FFAs have an impact on peripheral glucose utilization in muscle, principally via a FFA-induced suppression of pyruvate dehydrogenase (PDH).[2] Felber has recently shown that FFAs may be playing a role in glycogen storage[3] and Grill has suggested that they have an important role in regulating beta-cell function and thereby altering insulin secretion profiles.[4] Furthermore, recent studies by Bergman demonstrate that lowering of circulating FFAs is required for suppression of hepatic glucose production following a meal.[5]

The causative role of FFAs in the development or maintenance of insulin resistance has been difficult to assess. One reason for this could be that measurement of circulating FFA levels may be an inadequate measure of the role of FFAs since stored triglycerides and local fluxes may be major factors as well. In a normal individual, the stored triglyceride in tissues such as the liver, the muscle, and even the islet is relatively low. However, with chronic elevation of circulating FFA, as occurs in obesity, the triglyceride content of all of these tissues is predicted to be higher than normal, so local storage and release of FFAs may become important.

One may ask the following question: "If the circulating FFAs were reduced, would insulin sensitivity improve?" Acute experiments in humans suggest that the answer is no. However, such conclusions may be flawed because the locally stored triglycerides in these tissues could substitute for circulating FFAs as a source of FFAs and blunt any effect of a reduction in circulating levels.[6]

The basic premise of the present discussion is that altered regulation of FFA mobilization from adipocytes and abnormal storage and release of FFAs from other tissues contribute to the insulin-resistant state, at the level of the liver via suppression of hepatic glucose production, at the level of the muscle via decreased glucose utilization, and finally at the level of the islet via alterations in the insulin secretion profile.

FREE FATTY ACID MOBILIZATION FROM ADIPOCYTES

FFA mobilization from adipocytes is regulated by two processes. The first is lipolysis, the breakdown of triglyceride to FFAs and glycerol, and the second is glucose transport, the rate-limiting step in esterification of FFAs in adipocytes. *In vivo*, the data suggest that these two processes are about equally important under normal conditions.[7,8]

Previous studies have demonstrated a linear relationship between adipose cell size and the rates of basal and isoproterenol-stimulated lipolysis.[9] The rate of reesterification also tends to increase,[10] but with a smaller slope such that adipose cell size is limited by the more-rapid increase in lipolysis in the largest cells. Thus, increasing cell size in the face of increasing adiposity leads to higher circulating levels of FFAs and contributes to increasing insulin resistance. On the other hand, the smallest cells tend to store triglyceride more readily than larger cells due to their lower rate of lipolysis relative to their rate of reesterification. Increasing cell number in the face of increasing adiposity is predicted to limit further increases in insulin resistance due to a relatively reduced effect on circulating FFA levels.

Although it has been known for several decades that adipose cell size is an important determinant of the rates of lipolysis and reesterification, the direct link between adipocyte size and insulin resistance was lacking. A recent paper utilizing data from the Pima Indian Prospective Study indicates that enlargement of adipose cell size is a risk factor for the development of non-insulin-dependent diabetes mellitus (NIDDM).[11] This study also supports the hypothesis that, in this population, increased adipose cell size causes the increase in FFA levels that, in turn, at least partially explains the degree of insulin resistance.

Cross-sectional studies suggest that, as an individual progresses from being lean to moderately overweight, excess adiposity for the most part can be explained by increased adipose cell size. However, when an individual progresses from being moderately overweight to being severely obese, only a small increase in adipose cell size is observed.[12,13] Such data argue that increased adipose cell number must also play an important role in the etiology of obesity. The implication, however, is that increased adipose cell number becomes important only after adipocytes have become enlarged.

Given that increasing adipose cell size exacerbates insulin resistance, whereas increasing cell number does not, in the face of excess energy intake (overnutrition), what is the teleological advantage of increasing cell size rather than cell number? The answer may lie in the hypothesis that insulin resistance (accompanied by hyperinsulinemia) was a survival mechanism to enhance the ability of humans to store fat during periods of plenty in preparation for periods of famine. If this hypothesis were correct, then what would be the advantage of increasing adipose cell number before all fat cells have reached the apparent maximum cell size of about 3 µg lipid per cell?[14] One reasonable theory would be that increasing adipose cell number limits the overall degree of insulin resistance and thus counterbalances the insulin resistance survival mechanism.

The relative balance between increasing adipocyte size versus number may have implications for survival in populations such as Pima Indians, where food until 50 years ago was limited, and in Americans of European descent, where *ad libitum* feeding has been possible for large segments of the ancestral population for a millennium. In this regard, it is interesting to note that when lean Pima children were matched to lean children of European descent, abdominal adipocyte size was significantly larger in the Pima children.[15] Individual differences in the extent to which adiposity is associated with increased adipose cell size or number could explain why some individuals are more susceptible to the metabolic risks of obesity.

EFFECTS OF OVERFEEDING ON ADIPOCYTE FUNCTION, FREE FATTY ACIDS, AND INSULIN RESISTANCE

In an acute study, moderately overweight male subjects with a body mass index of 27 kg/m^2 and 25% fat were overfed 60% of their calories for two weeks. At the end of the overfeeding period, there was nearly a doubling of the relative number of small fat cells and the average adipose cell size tended to decrease from 0.63 to 0.56 µg lipid per cell, implying that increasing fat cell number was a significant consequence of this bout of overfeeding.[14]

There was a very large *decrease* in basal and maximal isoproterenol-stimulated *lipolysis* and a significant *increase* in basal and maximal insulin-stimulated *glucose transport* after the bout of overfeeding. The disproportionate changes of lipolysis and glucose transport could not be explained by the small tendency of average cell size to decrease during the two-week overfeeding period. Thus, it is likely that something other than adipocyte size led to the decrease in lipolysis and the increase in glucose transport. One probable contributing factor was the significant increase in fasting insulin levels (from 26 to 41 µU/mL).[14]

Consistent with the higher fasting insulin levels, with the decreased rate of lipolysis per cell, and with the increased reesterification per cell was a 39% decrease in FFA levels following the overfeeding ($p = 0.06$),[16] also consistent with decreased adipose cell size. In contrast to these effects of overfeeding in overweight individuals, in lean individuals FFA levels usually rise with increasing adiposity.[7] Taken together, the data suggest that increasing adipose cell size leads to increased FFA and that increased adipose cell number will prevent a rise in FFA levels and thus limit the degree of insulin resistance.

This hypothesis is supported by data from euglycemic clamps carried out before and after the bout of overfeeding in the studies in overweight individuals described above. Most glucose taken up in muscle during a euglycemic clamp is either stored as glycogen or oxidized to CO_2. In this study, there was a significant increase in glucose oxidation and a decrease in storage after overfeeding. Basically, the overall effect on total glucose utilization was a tendency for a 12% decrease, which is totally explained by the decrease in storage. Thus, the tendency for FFA levels to decrease was associated with an increase in glucose oxidation.[15]

MECHANISMS FOR CONTROLLING EXCESSIVE FATTY ACID RELEASE FROM ADIPOCYTES

The above discussion indicates that there is evidence that excessive FFA release from adipocytes contributes to the insulin-resistant state associated with obesity. Several mechanisms for restoring the appropriate balance between lipolysis and reesterification in adipocytes, and thereby normalizing FFA levels and insulin sensitivity, are immediately evident.

Increasing adipose cell number and thereby decreasing average adipose cell size is one such mechanism. A potential limitation of such an approach would be the apparent propensity for adipocytes to return to their former size, possibly via a hyperphagic signal to the brain,[17] and this is predicted to lead to weight gain.

Increasing the insulin sensitivity of antilipolysis and reesterification of FAs in adipocytes is another mechanism that has been considered.[8] Agonists to adipocyte receptors coupled to G_i-proteins such as the nicotinic acid and the adenosine A_1 receptors have been evaluated.[7] However, such an approach is predicted to be of limited value following a meal since insulin levels are usually high enough to maximally inhibit lipolysis and stimulate reesterification in this state. This approach, however, would be valuable in the fasted state when insulin levels are low. Unfortunately, difficulties with maintaining a sufficient duration of action with the known receptor agonists have reduced the viability of this approach as well.

Improving the insulin sensitivity of adipocytes would also be desirable early in a meal, before insulin has risen to prandial levels. During this time period, a reduction in circulating FFAs is predicted to lead to an increase in glucose oxidation in muscle, a decrease in Cori cycling, and a decrease in the utilization of FFAs as a source of fuel in liver. This would result in less glucose from the gut escaping into the peripheral circulation, a smaller glucose excursion, and smaller insulin excursions.[7,8,18] An improvement in the overall metabolic state is therefore predicted. Thus, nicotinic acid and adenosine A_1 agonists, or agents that stimulate early insulin release in response to a meal, would be predicted to improve the overall metabolic state if appropriately administered premeal.

PHARMACOLOGICAL APPROACHES

The thiazolidinediones represent a novel class of therapeutic agents that appear to improve insulin sensitivity *in vivo*. Compounds of this class, such as troglitazone, decrease blood glucose, insulin, FFA, and triglyceride levels in a variety of insulin-resistant animal models.[19] The mechanism of action of this class of agents has re-

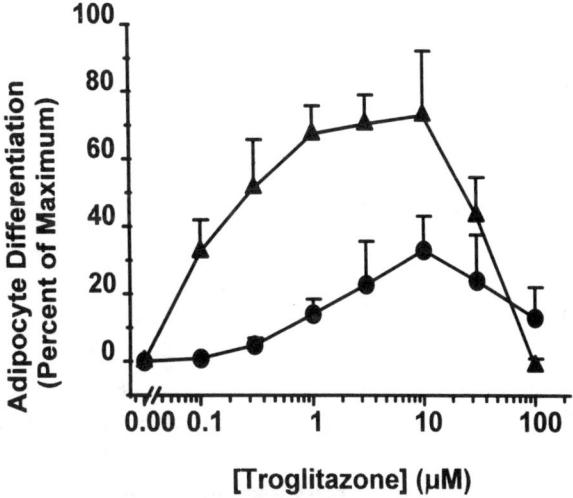

FIGURE 1. Effect of troglitazone on 3T3-L1 preadipocyte differentiation. Confluent 3T3-L1 preadipocytes were cultured in Dulbecco's Modified Eagle's Medium (DMEM) containing 10% fetal bovine serum and troglitazone at the concentrations indicated, in the presence (▲) or absence (●) of 160 nM insulin, with a change of medium every second day. After treatment for 7 days, cells were fixed in 10% formalin. Differentiation was evaluated by quantitation of lipid accumulation, determined by image analysis following staining with oil red O. Results are expressed relative to the extent of differentiation produced by a standard differentiation protocol,[28] utilizing treatment with insulin, dexamethasone, and isobutylmethylxanthine, and represent the mean ± SEM for 3 independent experiments.

mained obscure for a number of years, but recent studies suggest that the compounds are agonists for the gamma isoform of the peroxisome-proliferator-activated receptor, PPARγ, a member of the steroid hormone superfamily of ligand-activated nuclear transcription factors.[20] How this interaction leads to an improvement in insulin sensitivity is not fully established, but it is clear that these compounds can stimulate the conversion of adipocyte precursor cells into fully differentiated adipocytes.[19] For example, treatment of undifferentiated 3T3-L1 murine fibroblasts with troglitazone results in their differentiation into mature, lipid-containing adipocytes (FIG. 1). Stimulation of adipogenesis *in vivo* could lead, at least transiently, to a decrease in average fat cell size and to reduced FFA release and improved insulin sensitivity. In the longer term, however, increased adipogenesis leads to increased adiposity and weight gain in rodents[19,21] and possibly also in humans.[22] Reduced expression of the leptin (ob) gene as a consequence of thiazolidinedione treatment[21,23] may also have an impact on fat cell metabolism.

As mentioned earlier, the release of fatty acids from the adipocyte is dependent upon both the rate of lipolysis (determined by the activity of hormone-sensitive lipase) and the rate of reesterification of fatty acids (determined by the rate of glucose transport, the rate-limiting step for generation of α-glycerol phosphate). Hormone-sensitive lipase is activated via phosphorylation by cyclic AMP–dependent protein kinase; thus,

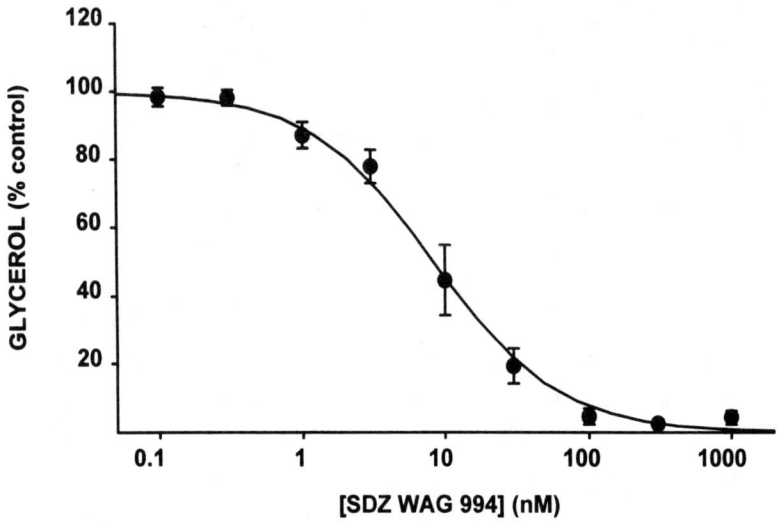

FIGURE 2. Effect of SDZ WAG 994 on lipolysis in rat adipocytes. Adipocytes, isolated from the epididymal fat pads of normal Sprague-Dawley rats by collagenase digestion, were incubated with adenosine deaminase (1 U/mL) and SDZ WAG 994 at the concentrations indicated for 60 min at 37°C. The medium was separated from the cells by centrifugation through dinonyl phthalate, and lipolysis was estimated by enzymatic determination of glycerol released into the infranatant. Results represent the mean ± SEM for 3–6 experiments.

agents that inhibit adenylate cyclase and thereby lower the cellular concentration of cyclic AMP will prevent the activation of hormone-sensitive lipase and will inhibit lipolysis. Among such agents is adenosine, acting via the adenosine A_1 receptor that is coupled negatively to adenylate cyclase through inhibitory G-proteins (G_i) in the adipocyte.[24,25] Adenosine A_1 receptor agonists also increase the sensitivity of adipocyte glucose transport to stimulation by insulin, by a process that appears to be independent of cyclic AMP,[26] and thus promote the reesterification of fatty acids as well as inhibiting their release.

SDZ WAG 994 (N-cyclohexyl-2'-O-methyl-adenosine) is a novel selective adenosine A_1 agonist and a potent antilipolytic agent both *in vitro* and *in vivo*.[27] In isolated rat adipocytes, removal of endogenous adenosine with adenosine deaminase results in stimulation of lipolysis to approximately 70% of the maximum isoproterenol-stimulated rate;[25] under such conditions, SDZ WAG 994 inhibits lipolysis with an IC_{50} of 8 nM (FIG. 2). Lipolysis inhibition by this agent is receptor-mediated and associated with inhibition of cyclic AMP production. Lipolysis stimulated by 1 mM isoproterenol could be overcome by a combination of SDZ WAG 994 and insulin, but not by either agent alone; this effect of the two agents in combination, under conditions where neither is effective alone, is anticipated to be a consequence of additive modes of action, that is, the inhibition of adenyl cyclase by SDZ WAG 994 and the stimulation of the low-K_m cAMP phosphodiesterase by insulin. SDZ WAG 994 also increased the sensitivity of glucose transport (determined by the incorporation of radioactivity

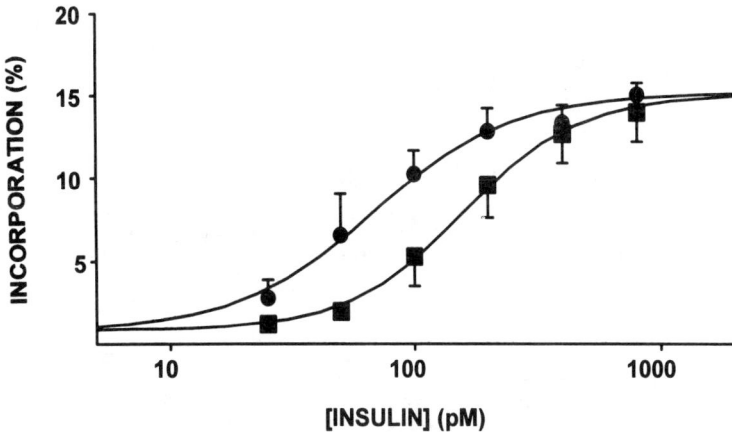

FIGURE 3. Effect of SDZ WAG 994 on the insulin sensitivity of glucose transport in rat adipocytes. Adipocytes, isolated from the epididymal fat pads of normal Sprague-Dawley rats by collagenase digestion, were incubated with adenosine deaminase (1 U/mL) and insulin at the concentrations indicated, in the presence or absence of 30 µM SDZ WAG 994, for 30 min at 37°C. [3-^{3}H]Glucose (final concentration, 50 µM, 0.5 µCi/mL) was then added and incubation was continued for a further 60 min. Incorporation of radioactivity from [3-^{3}H]glucose into cell lipids (a measure of glucose transport) was evaluated by extraction of the cell suspension (0.5 mL) with 5 mL of toluene-based scintillant, followed by liquid scintillation counting. Incorporation is expressed as % input radioactivity and represents the mean ± SEM for 4 experiments; ■, control; ●, SDZ WAG 994.

from [3-^{3}H]glucose into cell lipids) to insulin in the rat adipocyte, with an EC_{50} identical to that for inhibition of lipolysis. In the presence of a near-maximally effective concentration (30 nM) of the compound, the EC_{50} for insulin stimulation of [3-^{3}H]glucose incorporation into lipids was decreased from 160 to 69 pM (FIG. 3).

The effects of SDZ WAG 994 *in vivo* are consistent with *in vitro* actions described above. When given orally to 18-hour fasted normal rats, SDZ WAG 994 elicited a dose-related reduction in circulating levels of FFAs and of glucose (FIG. 4). This was accompanied by a reduction of fasting insulin (IRI) levels, even at a dose with no hypoglycemic effect (FIG. 4), suggesting that this antilipolytic agent enhances insulin sensitivity in the rat. To more directly address the possibility that SDZ WAG 994 increased insulin-sensitive glucose disposal, its effects were examined during hyperinsulinemic (~50 µU/mL) glucose clamps performed in normal fasted rats. Following the administration of SDZ WAG 994 or vehicle and insulin (1.5 mU/kg/min), blood glucose levels were clamped at the slightly hyperglycemic level of 100 mg/dL with a variable-rate glucose infusion. FIG. 5 illustrates the effects of SDZ WAG 994 to more than double the glucose infusion rate (GIR) necessary to maintain identical glucose levels.

The glucose-lowering effect of SDZ WAG 994 observed in the fasted state appeared to be dependent on its hypolipidemic action. Thus, when the FFA-lowering effect of SDZ WAG 994 was prevented by infusion of exogenous triglyceride (Intralipid) and heparin, the hypoglycemic effect of SDZ WAG 994 was virtually eliminated (FIG.

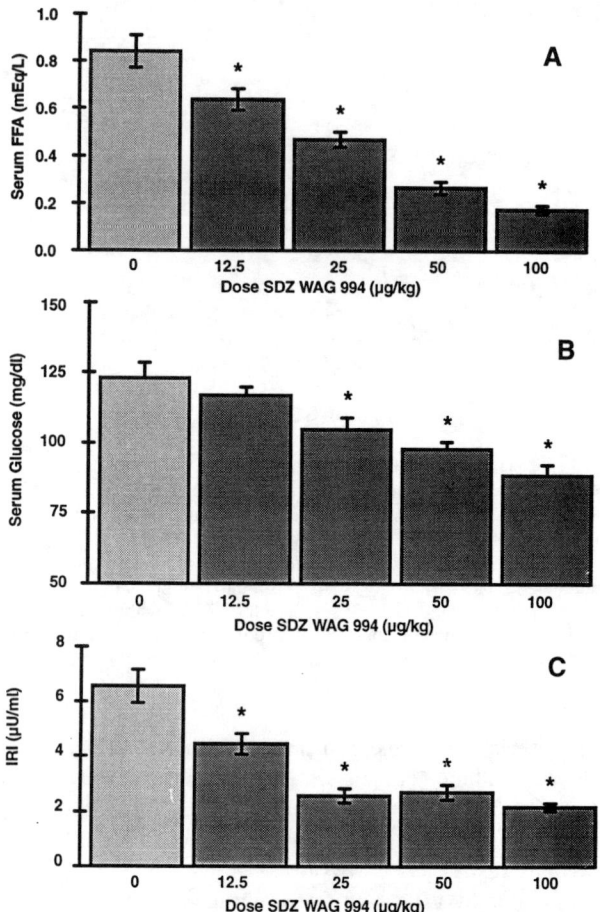

FIGURE 4. Serum levels of free fatty acids (FFA) (**panel A**), glucose (**panel B**), and immunoreactive insulin (IRI) (**panel C**) at 2 hours following oral administration of vehicle or increasing doses of SDZ WAG 994 to 18-hour fasted normal rats. Mean ± SEM; (∗) $p < 0.05$ or better; $n = 12$ rats/group. Normal adult male Sprague-Dawley rats provided with a standard chow diet and water *ad libitum* were fasted overnight and then received vehicle [0.5% carboxymethylcellulose (CMC), containing 0.2% Tween-80] or SDZ WAG 994 in vehicle, by oral gavage (10 mL/kg), at the doses indicated. Blood samples for measurement of FFA, glucose, and IRI were obtained by cardiac puncture under CO_2 narcosis. Serum FFAs were analyzed using an acyl-CoA synthetase and acyl-CoA oxidase method (Wako Pure Chemical Industries, Richmond, Virginia) and glucose levels were measured with a glucose oxidase technique on an automated analyzer (Yellow Springs Instruments, Yellow Springs, Ohio). Rat insulin was measured by radioimmunoassay using reagents from Linco, Research (St. Louis, Missouri).

6). Clearly, the adenosine A_1 agonist, SDZ WAG 994, is a potent lipid-lowering agent *in vivo* that can be used to improve the metabolic state as evidenced by significant lowering of insulin and glucose levels.

An even-simpler approach to reducing lipid availability and improving metabolic

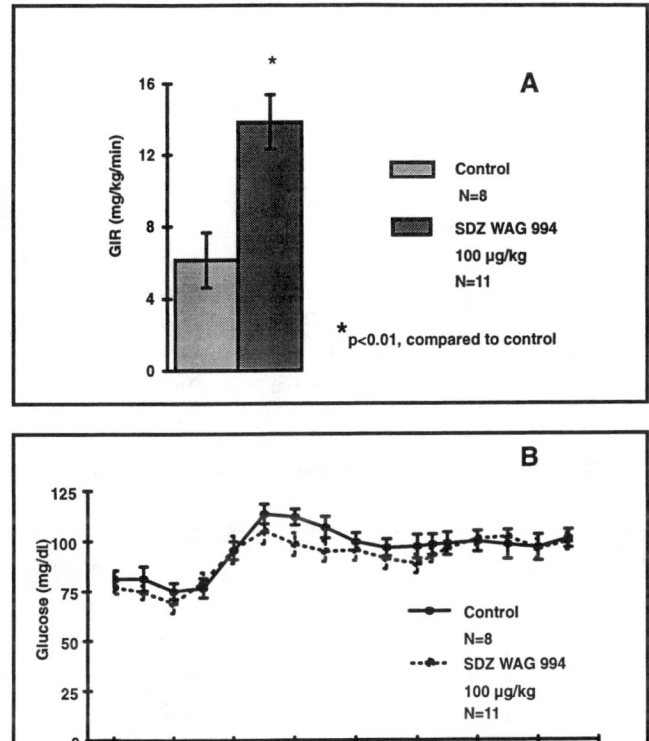

FIGURE 5. Glucose infusion rate during the final 50 minutes (GIR) (**panel A**) and blood glucose levels (**panel B**) throughout hyperinsulinemic clamps performed in chronically cannulated 18-hour fasted rats. Mean ± SEM. Two to 3 days prior to study, normal adult male Sprague-Dawley rats were implanted with jugular vein and carotid artery catheters under Nembutal anesthesia. Following recovery from surgery, rats were fasted overnight and then insulin was infused continuously at a rate of 1.5 mU/kg/min. Blood glucose levels were measured at 10-minute intervals and maintained constant with a variable-rate glucose infusion. The insulin infusion started at zero time and rats received either CMC vehicle or SDZ WAG 994 (100 μg/kg in CMC) by oral gavage at 30 minutes. Plasma insulin levels were no different in the two groups both prior to administration of compound (~10 μU/mL) and at 150 minutes during the clamp (~50 μU/mL).

control is to take advantage of the therapeutic properties of endogenously secreted insulin. However, since it is clear that chronic hyperinsulinemia is undesirable,[1] an optimal insulin secretagogue should be fast-acting and short-acting, that is, it should improve the kinetics of insulin secretion without producing chronic hyperinsulinemia. The compound A-4166 (trans-N-[[[1-methylethyl]cyclohexyl]carbonyl]-D-phenylalanine) represents such a new class of oral hypoglycemic agents.

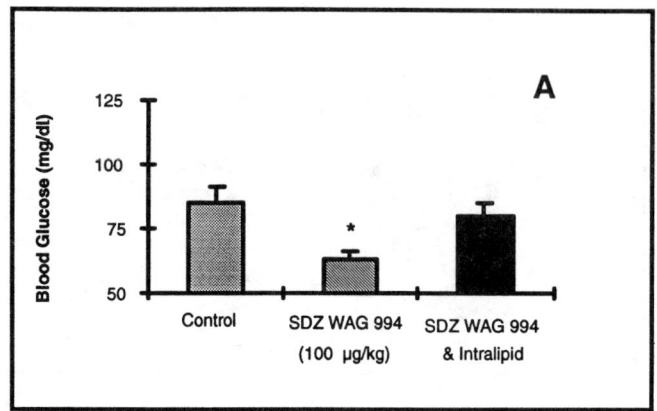

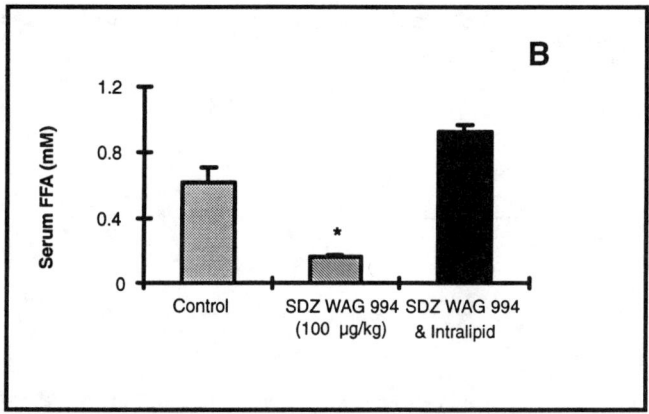

FIGURE 6. Blood glucose (**panel A**) and serum FFA levels (**panel B**) at 2 hours following oral administration of vehicle during saline infusion (control), SDZ WAG 994 during saline infusion, or SDZ WAG 994 during Intralipid/heparin infusion to 18-hour fasted normal rats. Mean ± SEM; (∗) $p < 0.05$ or better versus SDZ WAG 994 and Intralipid; $n = 3$–7 rats/group. Two to 3 days prior to study, normal adult male Sprague-Dawley rats were implanted with jugular vein and carotid artery catheters under Nembutal anesthesia. Following recovery from surgery, rats were fasted overnight and then received either vehicle or SDZ WAG 994 (100 µg/kg in vehicle) by oral gavage (10 mL/kg). Immediately after dosing, an intravenous infusion of either saline or Intralipid (0.3 g/kg/h) plus heparin (350 U/kg/h) was initiated. The Intralipid infusion rate was reduced to 0.2 g/kg/h (heparin reduced to 233 U/kg/h) after 30 minutes and was maintained at this rate for the remainder of the 2-hour study.

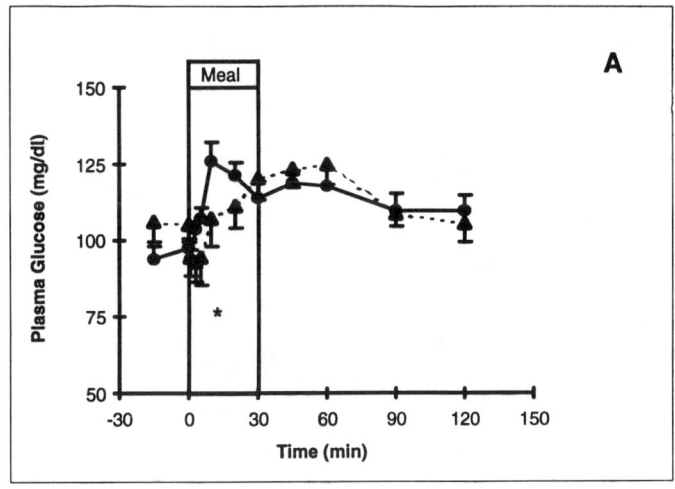
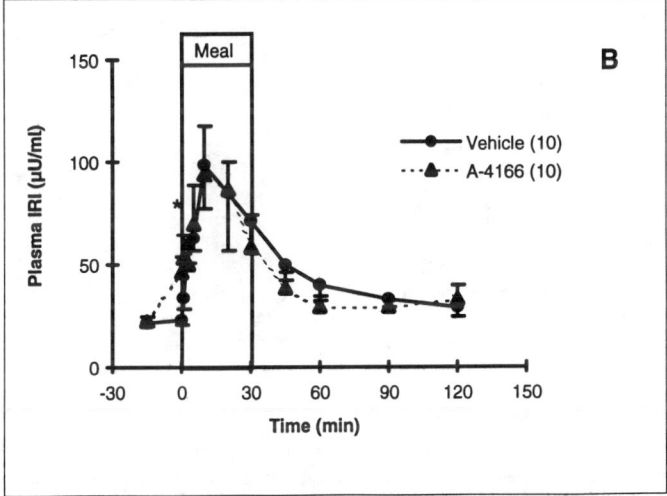

FIGURE 7. Plasma levels of glucose (**panel A**) and immunoreactive insulin (IRI) (**panel B**) during meal-tolerance tests in normal 13-hour fasted rats. Mean ± SEM; $n = 10$ rats/group; (∗) $p < 0.05$ versus vehicle-treated controls. Normal adult male Sprague-Dawley rats housed in a reverse-cycle room (lights off 6:00 A.M.–6:00 P.M.) were implanted with a jugular vein catheter at 2 weeks prior to study. Following recovery from surgery, rats were trained to consume their daily food ration in discrete 45-minute meals commencing at the beginning and the end of the dark cycle (7:00 A.M. and 5:00 P.M.). On the day of experiment, a baseline (−15 minutes relative to commencement of feeding) sample was obtained. At time −10 minutes, rats received either CMC (vehicle) (●) or A-4166 in CMC (▲) by oral gavage (2 mL/kg). A time zero sample was obtained and then animals were given access to powdered standard rat chow for 30 minutes. A-4166 did not affect the amount of food consumed during the meal. Samples were obtained at 1, 3, 5, 10, 20, and 30 minutes during the meal and at times 45, 60, 90, and 120 minutes. Plasma glucose, IRI, and FFA levels were measured as described above.

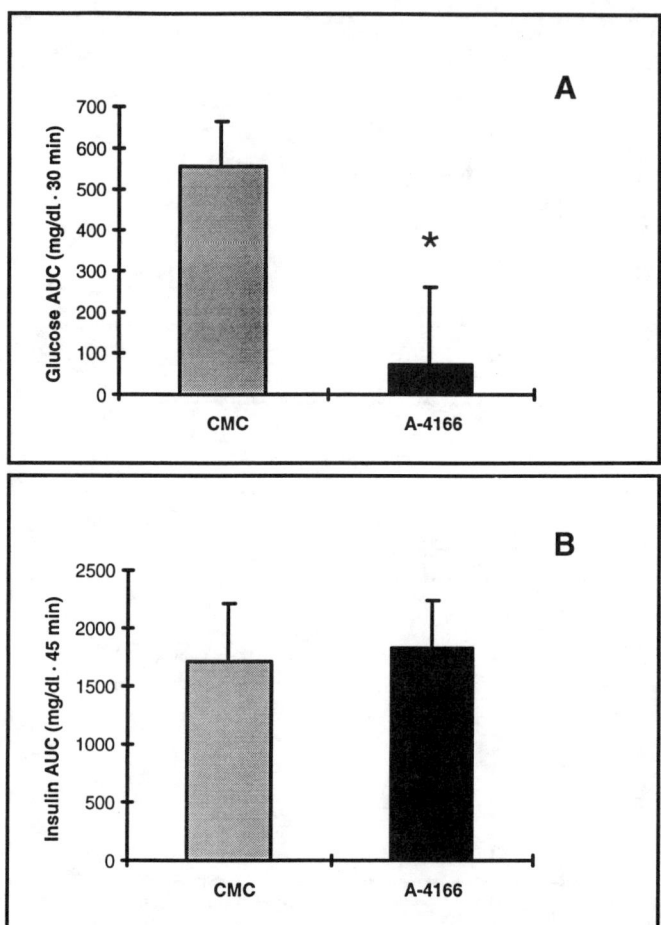

FIGURE 8. Area under the curve (AUC) for glucose from 0 to 30 minutes (**panel A**) and AUC for insulin from −15 to 30 minutes (**panel B**) during meal-tolerance tests in normal 13-hour fasted rats. Mean ± SEM; (∗) $p < 0.05$ versus vehicle-treated controls. Methods of the meal-tolerance test are described in FIGURE 7. AUC values over baseline were calculated using the trapezoidal rule.

A-4166, given orally just prior to a meal, can totally prevent prandial glucose excursions without increasing the overall exposure to insulin. The time course of the effects of A-4166 on plasma glucose and IRI levels during a 30-minute meal in normal cannulated rats is shown in FIG. 7. The only significant effect of A-4166 administered to normal rats (30 mg/kg, po, 10 minutes prior to initiation of a meal) on plasma insulin levels was an approximate doubling of IRI at time 0 (the last sample obtained prior to food presentation). However, glucose excursions were prevented. The apparent effects on prandial glucose excursions depicted in FIG. 7 may be minimized because the baseline glucose levels were higher in the drug-treated group than in the vehicle-treated

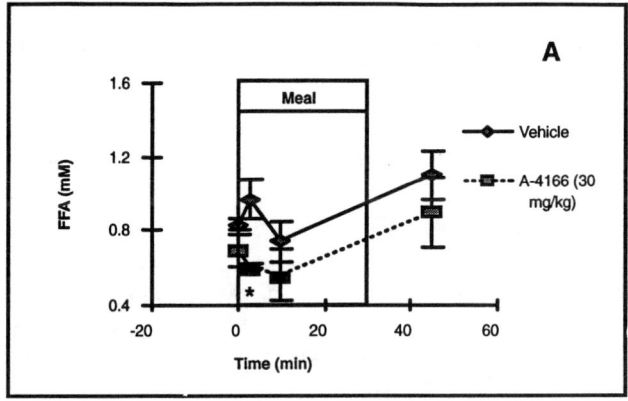

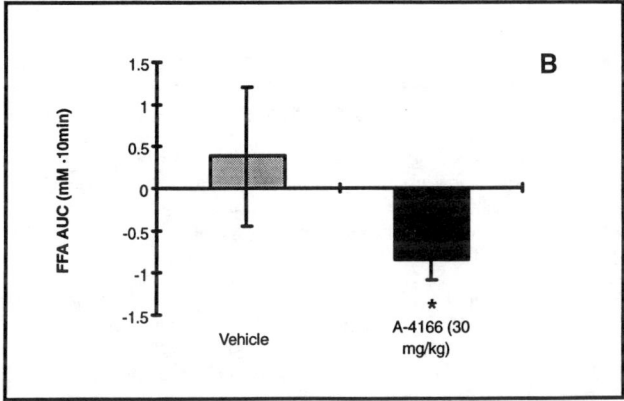

FIGURE 9. Plasma FFAs (**panel A**) and AUC for FFA from 0 to 10 minutes (**panel B**) during meal-tolerance tests in normal 13-hour fasted rats. Mean ± SEM; (∗) $p < 0.05$ versus vehicle-treated controls. Methods of the meal-tolerance test are described in FIGURE 7. AUC values over baseline were calculated using the trapezoidal rule.

group. The integrated glucose area above baseline during the 30-minute meal and the 45-minute integrated insulin area (from drug administration through the meal) are depicted in FIG. 8. It can be appreciated that meal-induced hyperglycemia was markedly reduced by A-4166. However, the brief stimulation of insulin release failed to significantly impact the total integrated IRI levels illustrated as the area under the curve (AUC) from times -15 to 30 minutes (relative to the onset of feeding).

As illustrated in FIG. 9, plasma FFA levels in vehicle-treated control rats were unaffected during the first 10 minutes of a meal. In contrast, FFA levels were significantly suppressed during the first 10 minutes of a meal in animals that received A-4166 at 10 minutes prior to the meal. As discussed earlier, it is likely that the timing of the transient suppression of FFAs that occurs during meals makes an important contribution to the overall metabolic response. Thus, a rapid-onset, short-duration insulin secretagogue such as A-4166 will be a safe and effective approach to improving prandial glucose and lipid metabolism.

CONCLUSIONS

There is a clear link between elevated lipids and insulin resistance. A principal culprit in this process appears to be the altered mobilization of FFAs from adipocytes that is associated with increased adipocyte size. Increasing adipocyte number, increasing the sensitivity of adipocytes to insulin at the start of a meal, and improving insulin secretion at the start of a meal represent potential pharmacological approaches to ameliorating this altered mobilization of FFA.

REFERENCES

1. REAVEN, G. M. 1988. Role of insulin resistance in human disease. Diabetes 37: 1595-1607.
2. RANDLE, P. J., C. N. HALES, P. B. GARLAND & E. A. NEWSHOLME. 1963. The glucose fatty-acid cycle: its role in insulin sensitivity and the metabolic disturbances of diabetes mellitus. Lancet I: 785-789.
3. FELBER, J. P., E. HAESLER & E. JEQUIER. 1993. Metabolic origin of insulin resistance in obesity with and without type 2 (non-insulin-dependent) diabetes mellitus. Diabetologia 36: 1221-1229.
4. ZHOU, Y. P. & V. E. GRILL. 1994. Long-term exposure of rat pancreatic islets to fatty acids inhibits glucose-induced insulin secretion and biosynthesis through a glucose fatty acid cycle. J. Clin. Invest. 93: 870-876.
5. REBERIN, K., G. M. STEIL, L. GETTY & R. N. BERGMAN. 1995. Free fatty acid as a link in the regulation of hepatic glucose output by peripheral insulin. Diabetes 44: 1038-1045.
6. BODEN, G., X. CHEN, J. RUIZ, J. V. WHITE & L. ROSSETTI. 1994. Mechanisms of fatty acid-induced inhibition of glucose uptake. J. Clin. Invest. 93: 2438-2446.
7. FOLEY, J. E. 1992. Rationale and application of fatty acid oxidation inhibitors in the treatment of diabetes mellitus. Diabetes Care 15: 773-784.
8. FOLEY, J. E. & R. C. ANDERSON. 1996. Fatty acid oxidation inhibitors. In Diabetes Mellitus: A Fundamental Clinical Text, p. 668-674. Lippincott-Raven. Philadelphia/New York.
9. ANDREWS, J., A. KASHIWAGI, M. A. VERSO, B. VASQUEZ, B. V. HOWARD & J. E. FOLEY. 1984. Effects of four-day fast on triglyceride mobilization in human adipocytes. Int. J. Obes. 8: 355-363.
10. BOGARDUS, C., S. LILLIOJA, D. MOTT, G. M. REAVEN, A. KASHIWAGI & J. E. FOLEY. 1984. Relationship between obesity and maximal insulin-stimulated glucose uptake *in vivo* and *in vitro* in Pima Indians. J. Clin. Invest. 73: 800-805.
11. PAOLISSO, G., P. A. TATARANNI, J. E. FOLEY, C. BOGARDUS, B. V. HOWARD & E. RAVUSSIN. 1995. A high concentration of fasting plasma non-esterified fatty acids is a risk factor for the development of NIDDM. Diabetologia 38: 1213-1217.
12. LILLIOJA, S., J. E. FOLEY, C. BOGARDUS, D. MOTT & B. V. HOWARD. 1986. Free fatty acid metabolism and obesity in man: *in vivo* and *in vitro* comparisons. Metabolism 35: 505-514.
13. HOWARD, B. V., C. BOGARDUS, E. RAVUSSIN, J. E. FOLEY, S. LILLIOJA, D. M. MOTT, P. H. BENNETT & W. C. KNOWLER. 1991. Studies of the etiology of obesity in Pima Indians. Am. J. Clin. Nutr. 53: 1577S-1585S.
14. KASHIWAGI, A., D. MOTT, C. BOGARDUS, S. LILLIOJA, G. M. REAVEN & J. E. FOLEY. 1985. The effects of short-term overfeeding on adipocyte metabolism in Pima Indians. Metabolism 34: 364-370.
15. ABBOTT, G. H. & J. E. FOLEY. 1987. Comparison of body composition, adipocyte size, and glucose and insulin concentrations in Pima Indian and Caucasian children. Metabolism 36: 576-579.

16. MOTT, D. M., S. LILLIOJA & C. BOGARDUS. 1986. Overnutrition induced decrease in insulin action for glucose storage *in vivo* and *in vitro* in man. Metabolism **35**: 160–165.
17. ZHANG, Y., R. PROENCA, M. MAFFEL, M. BARONE, L. LEOPOLD & J. M. FRIEDMAN. 1994. Positional cloning of the mouse obese gene and its human homologue. Nature **372**: 425–432.
18. KELLEY, D., M. MOKAN & T. VENEMAN. 1994. Impaired postprandial glucose utilization in non-insulin-dependent diabetes mellitus. Metabolism **43**: 1549–1557.
19. WHITCOMB, R. W. & A. R. SALTIEL. 1995. Thiazolidinediones. Exp. Opin. Invest. Drugs **4**: 1299–1309.
20. LEHMANN, J. M., L. B. MOORE, T. A. SMITH-OLIVER, W. O. WILKISON, T. M. WILLSON & S. A. KLIEWER. 1995. An antidiabetic thiazolidinedione is a high-affinity ligand for the nuclear peroxisome proliferator-activated receptor γ (PPARγ). J. Biol. Chem. **270**: 12953–12956.
21. ZHANG, B., M. P. GRAZIANO, T. W. DOEBBER, M. D. LEIBOWITZ, S. WHITE-CARRINGTON, D. M. SZALKOWSKI, P. J. HEY, M. WU, C. A. CULLINAN, P. BAILEY, B. LOLLMANN, R. FREDERICH, J. S. FLIER, C. D. STRADER & R. G. SMITH. 1996. Down-regulation of the expression of the *obese* gene by an antidiabetic thiazolidinedione in Zucker diabetic fatty rats and *db/db* mice. J. Biol. Chem. **271**: 9455–9459.
22. IWAMOTO, Y., K. KOSAKA, T. KUZUYA, Y. AKANUMA, Y. SHIGETA & T. KANEKO. 1996. Effects of troglitazone: a new hypoglycemic agent in patients with NIDDM poorly controlled by diet therapy. Diabetes Care **19**: 151–156.
23. KALLEN, C. B. & M. A. LAZAR. 1996. Antidiabetic thiazolidinediones inhibit leptin (*ob*) gene expression in 3T3-L1 adipocytes. Proc. Natl. Acad. Sci. U.S.A. **93**: 5793–5796.
24. STILES, G. L. 1992. Adenosine receptors. J. Biol. Chem. **267**: 6451–6454.
25. HONNOR, R. C., G. S. DHILLON & C. LONDOS. 1985. cAMP-dependent protein kinase and lipolysis in rat adipocytes. II. Definition of steady-state relationship with lipolytic and antilipolytic modulators. J. Biol. Chem. **260**: 15130–15138.
26. KURODA, M., R. C. HONNOR, S. W. CUSHMAN, C. LONDOS & I. W. SIMPSON. 1987. Regulation of insulin-stimulated glucose transport in the isolated rat adipocyte. J. Biol. Chem. **262**: 245–253.
27. WAGNER, H., M. MILAVEC-KRIZMAN, F. GADIENT, K. MENNINGER, P. SCHOEFFTER, C. TAPPARELLI, J. P. PFANNKUCHE & J. R. FOZARD. 1995. General pharmacological properties of SDZ WAG 994, a potent, selective, and orally active adenosine A_1 receptor agonist. Drug Dev. Res. **34**: 276–288.
28. FROST, S. C. & M. D. LANE. 1985. Evidence for the involvement of vicinal sulfhydryl groups in insulin-activated hexose transport by 3T3-L1 adipocytes. J. Biol. Chem. **260**: 2646–2652.

Effect of Oral Antidiabetics and Insulin on Lipids and Coronary Heart Disease in Non-Insulin-dependent Diabetes Mellitus

M. HANEFELD, T. TEMELKOVA-KURKTSCHIEV, AND C. KÖHLER

*Institute and Outpatient Clinic for Clinical Metabolic Research
Faculty of Medicine
Technical University of Dresden
D-01307 Dresden, Germany*

INTRODUCTION

Non-insulin-dependent diabetes mellitus (NIDDM) is prone to early and rapid development of atherosclerotic vascular disease. Coronary heart disease is the major cause of death in NIDDM.[1] Compared to age-matched nondiabetic populations, the relative risk for myocardial infarction in men is about 2 and in women about 4.[2] This, together with excessive incidence of stroke and occlusive arterial disease of the lower extremities, leads to a shortage of life expectancy of 5–15 years. Intensive epidemiological investigations have identified risk factors that contribute a large proportion of the excess risk for coronary heart disease: hypertension, dyslipidemia, smoking, and hyperfibrinogenemia. As shown in TABLE 1, the prevalence of the established risk factors is amplified in NIDDM.[3] Among these risk factors, the lipid triad – hypertriglyceridemia, low high-density lipoprotein (HDL), and hypercholesterolemia – plays a prominent role. Contrary to the nondiabetic population in NIDDM, hypertriglyceridemia has been confirmed to be an independent risk factor for myocardial infarction and all-cause mortality.[4,5] Low HDL-cholesterol is intricately related to syndrome X and was a significant risk factor for coronary heart disease in several, but not all, prospective studies with NIDDM.[1,5] Average cholesterol levels are not significantly different from nondiabetic populations. However, as shown in the MRFIT study, at a given cholesterol level diabetic men had a twofold higher incidence of coronary heart disease as compared to their nondiabetic counterparts.[6] These data suggest on the one hand the importance of the lipid triad and other established risk factors for the development of macroangiopathy in NIDDM, but they also indicate that a great part of the excess mortality in NIDDM may be due to diabetes-specific factors such as the quality of metabolic control, the impact of medical treatment, and insulin resistance/hyperinsulinemia. The question of whether antidiabetic therapy treatment

TABLE 1. Prevalence of Coronary Risk Factors (%) in Newly Detected NIDDM (DIS Cohort at Entry) and in Population[3]

Risk Factor	NIDDM (n = 1139)	Population (n = 1216)	Limits
HLP	17.6	7.6	triglyceride $\geq$ 2.8 mmol/L and/or cholesterol $\geq$ 7.8 mmol/L
Hypertriglyceridemia	11.3	3.4	
Hypercholesterolemia	3.5	3.7	
Mixed HLP	2.8	0.5	
Low HDL[a]			
M	14.3	–	<0.9 mmol/L
F	17.9	–	<1.1 mmol/L
Hypertension	53.0	17.3	blood pressure $\geq$ 160/95 mmHg and/or antihypertensive drugs
Smoking	34.0	30.3	tobacco $\geq$ 1 g/day
Obesity	49.0	8.2	IBWI:[b] M > 1.2 F > 1.3

[a] Five years after diabetes detection.
[b] IBWI = ideal body weight index.

has beneficial effects on the incidence of cardiovascular diseases in NIDDM has not yet been definitely answered.[7] Therefore, an analysis of the possible effects of oral antidiabetics and insulin on the coronary risk factor profile seems to be of importance to validate the different options of pharmacological treatment of NIDDM.

This paper will consider the following:

(1) lipoprotein anomalies, cardiovascular risk, and quality of metabolic control;
(2) does pharmacological treatment affect the lipid triad?;
(3) does lipid-lowering therapy reduce cardiovascular complications and mortality?;
(4) evidence for different effects of antidiabetic treatment on cardiovascular complications.

LIPOPROTEIN ANOMALIES, CARDIOVASCULAR RISK, AND GLUCOSE CONTROL

As already mentioned, hypertriglyceridemia and low HDL are the typical anomalies found in NIDDM. They are, however, intricately connected with other components of the metabolic syndrome (TABLE 2) or syndrome X, respectively.[8-10] These data are consistent with findings in the San Antonio Heart Study.[11] Thus, hypertriglyceridemia and low HDL in more than 80% of the cases coexist with multiple diseases in NIDDM. Hypertriglyceridemia leads to profound compositional changes in very-low-density lipoproteins (VLDL), low-density lipoproteins (LDL), and HDL: triglyceride enrichment and lower phospholipid and protein content.[12] This has at least

TABLE 2. Prevalence of Singular and Combined Risk Factors in Newly Detected Middle-aged NIDDM ($n = 1139$)[a]

Risk Factor	Prevalence (%)
none	9.2
obesity	12.7
hypertension	6.5
hypercholesterolemia (HCh)	1.4
hypertriglyceridemia (HTG)	1.0
smoking	5.4
obesity + hypertension	15.0
obesity + smoking	5.5
obesity + HCh	9.1
obesity + HTG	8.8
HCh + HTG	2.5
obesity + hypertension + HCh	4.4
obesity + hypertension + HTG	3.8
obesity + hypertension + smoking	5.5
obesity + HCh + HTG	7.7
obesity + hypertension + HCh + HTG	3.1
hypertension + HCh + HTG + smoking	5.0
obesity + hypertension + HCh + HTG + smoking	1.3

[a] The Diabetes Intervention Study: data at entry.

two functional consequences: the formation of small dense LDL that are more atherogenic and reduced reverse-cholesterol transport by abnormal HDL. As previously reported,[3] triglyceride concentrations are significantly correlated with serum insulin, whereas HDL-cholesterol is inversely correlated with insulin levels. In the Paris prospective study, the combination of hypertriglyceridemia and hyperinsulinemia potentiated the cardiovascular risk and mortality rate.[13] These results indicate that dyslipoproteinemia clusters with several risk factors that escalate the development of atherosclerosis in NIDDM, as recently reviewed by Genest and colleagues.[14] There are further modifications of lipoproteins that are strongly influenced by the quality of diabetes control that contribute to the excessive atherogenicity of LDL in NIDDM: glycosylation, oxidation, and increased malondialdehyde content.[15] These compositional and chemical anomalies of lipoproteins in NIDDM together with insulin resistance/hyperinsulinemia imply a higher lipid risk for cardiovascular diseases. There also is new evidence on the impact of the quality of glycemic control on lipoprotein anomalies. Glucose level is correlated with triglyceride concentrations and (inversely) with HDL-cholesterol.[16] HbA_{1C} as an indicator of the quality of metabolic control is inversely correlated with the oxidizability of LDL as measured by the lag time.[17] In poorly controlled NIDDM, up to 20% of the LDL particles are glycosylated.[16] Glyco-oxidation of LDL prolongs the removal of LDL by the LDL-receptor pathway and leads to increased uptake of modified LDL by macrophages. On the other hand, the HDL fractional catabolic rate is higher by glycosylation.[18]

Thus, perfect glycemic control without iatrogenic exogenous or endogenous hyper-

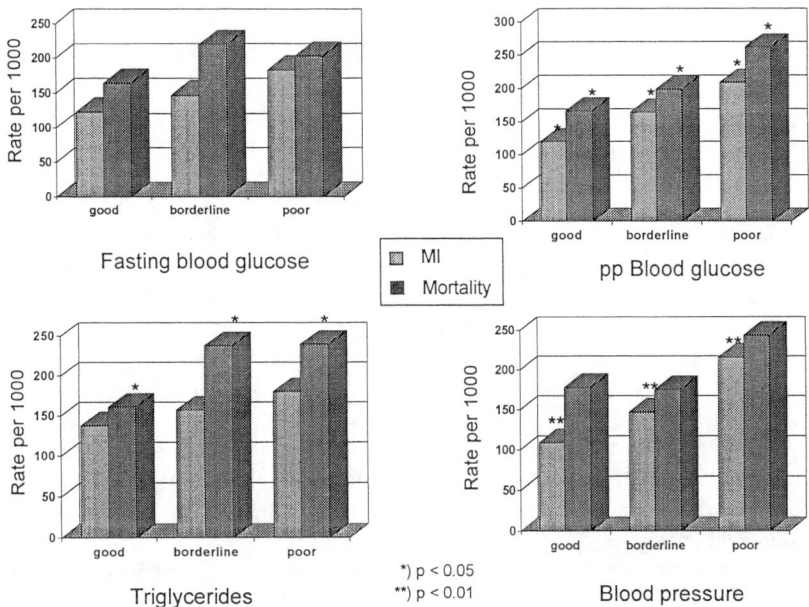

FIGURE 1. Incidence of myocardial infarction (MI) (n/1000) and all-cause mortality according to the categories of quality control of the NIDDM Policy Group: DIS, 11 years of follow-up, level at entry.[21]

insulinemia and insulin resistance is essential for correction of atherogenic lipoprotein anomalies in NIDDM. There are two recently published papers[19,20] that found the quality of HbA_1 control to be an independent risk factor for the incidence of myocardial infarction and mortality. In the Diabetes Intervention Study (DIS), the quality of control of postprandial hyperglycemia according to the categories of the NIDDM Policy Group[21] was of significant importance for the incidence of subsequent myocardial infarction and all-cause mortality (FIG. 1), whereas fasting blood glucose was not. Postprandial glucose peaks have been shown to strongly influence insulin sensitivity of hepatic gluconeogenesis, postprandial hyperlipidemia, and fibrinolysis.[22,23] According to the glucose-toxicity theory, perfect control of postprandial hyperglycemia may protect the B-cells of the pancreas. Impaired glucose tolerance is an independent cardiovascular risk factor in prospective studies.[24] This may explain why the control of postprandial hyperglycemia already in early stages of the disease is so important for the course of NIDDM and macroangiopathy.

The results of the DIS[21] and of other prospective trials[13,25,26] suggest that the metabolic syndrome associated with early diabetes may have a fatal impact on the risk of subsequent development of cardiovascular complications and survival. This could explain why intervention efforts in clinical NIDDM so far had no effect on macroangiopathy and excessive mortality.

EFFECT OF PHARMACEUTICAL ANTIDIABETIC TREATMENT ON LIPOPROTEIN METABOLISM

Lipid abnormalities are frequently caused or exaggerated by poor glucose control, overweight, or comedication, with adverse effects on lipoprotein metabolism. Therefore, the first approach to dyslipoproteinemia in NIDDM should be the optimization of diabetes control with diet, reduction of weight, and exercise of endurance type to improve the insulin sensitivity of the muscles. Body weight reduction to a desirable weight is particularly effective with respect to hypertriglyceridemia, low HDL, and hypertension.[27] Weight loss and exercise have favorable effects on insulin resistance and hyperinsulinemia that contribute to or cause the dyslipoproteinemia. A careful medical history and investigation should identify secondary hyperlipidemia due to lipid-affecting drugs such as β-blockers, thiazides, or retinoids. Renal diseases, hypothyroidism, etc., need to be considered as reasons for secondary dyslipidemia.

At the moment, there are three groups of oral antidiabetics—alpha-glucosidase inhibitors, biguanides, and sulfonylureas—and multiple insulin preparations on the market. Oral antidiabetics in advanced progress are thiazolidine derivatives[28] and amino-acid analogues.[29]

As recently demonstrated with the 6-year follow-up data of the UKPDS,[30] the results with standard drug therapies are far from ideal. Sulfonylureas, which primarily act on the ATP-sensitive potassium channels of the B-cell, lead to an inappropriate long-lasting hyperinsulinemia in obese NIDDM patients. This effect is associated with a weight gain of 3–5 kg as shown in the UKPDS and with 40% secondary failure.[31] Antihyperglycemic agents such as acarbose[32] and metformin[33] have the advantage of reducing postprandial hyperinsulinemia and of having no impact on body weight; rather, a small weight loss is induced. Only scarce information exists on the possible adjunct effects of oral antidiabetics on blood lipids. One reason that it is so difficult to validate the impact of antidiabetics on the lipoprotein profile is the fact that improved glycemic control per se leads to a substantial correction of most, but not all, lipoprotein abnormalities. Furthermore, many other diabetes-related factors that affect lipoprotein metabolism could bias the effect of antidiabetic drugs (TABLE 3). Therefore, in the following, we will only consider controlled studies or comparative investigations with a focus on the lipid triad.

Alpha-Glucosidase Inhibitors

Acarbose is the first representative of this new therapeutical principle that was introduced in therapeutical practice at the beginning of the 1990s.[34] By delaying the release of glucose from disaccharides, oligosaccharides, and polysaccharides, it initiates a cascade of events leading to lower VLDL production in the liver via reduced hyperinsulinemia (FIG. 2).[35] The effect on glycemic control and on the lipid triad is shown in TABLE 4. The major effect is on postprandial hypertriglyceridemia, with a reduction by about 20% versus placebo. HDL-cholesterol was not increased in placebo-controlled studies.[36] A small reduction in total cholesterol was reported in two

TABLE 3. Factors Affecting Lipoprotein Metabolism in Diabetes

Metabolic Control
  hyperglycemia, HbA$_{1C}$↑
  elevated FFA and lipolysis
  impaired FFA reesterification
  glycosylation
  increased oxidative stress
  weight changes

Hormonal Dysregulation
  insulin deficit (first peak, etc.)
  endogenous or exogenous hyperinsulinemia
  glucagon ↑ ↓
  stress hormones ↑

Antidiabetic Treatment

Comedication
  antihypertensives, diuretics, beta-blockers, alpha-blockers

Complications
  renal and liver diseases
  endothelial cell damage

---

studies.[37,38] We observed a significant decrease of total cholesterol in the highest tertile from 273 to 251 mg/dL versus an increase of 265 to 276 mg/dL in the placebo group ($p < 0.025$).[39] Nestel and coworkers[40] found in tracer kinetic studies with labeled glycerol a significant decrease in VLDL secretion after acarbose intake. In 14 patients with primary hyperlipidemia, Maruhama et al.[41] observed a significant drop in serum VLDL triglycerides and LDL-cholesterol after intake of 100 mg of acarbose tid. This was associated with changes in the fecal bacterial flora and intestinal bile

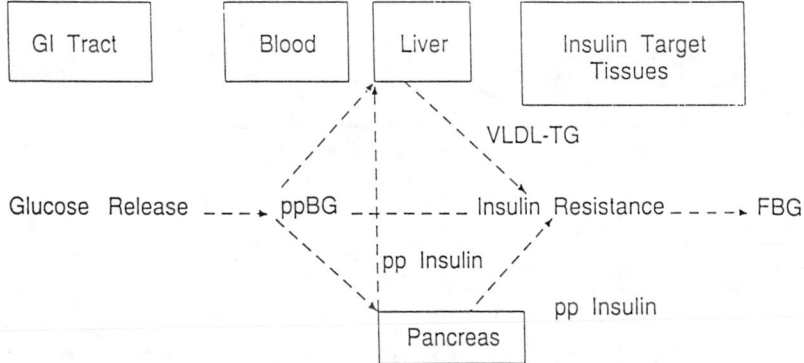

FIGURE 2. Paradigm of the mode of action of acarbose in NIDDM. Terms: BG = blood glucose, FBG = fasting blood glucose, GI = gastrointestinal, VLDL-TG = very-low-density lipoprotein triglycerides, (-->) reduced.

TABLE 4. Impact of Acarbose (A) Monotherapy on the Lipid Triad in Placebo (P)-controlled Studies of NIDDM Insufficiently Treated with Diet Alone (24 weeks, 100 mg tid)

Reference	Drug	n	BMI	HbA$_{1C}$ Change (Δ%)	TG Change (%)	Cholesterol Change (%)	HDL-Cholesterol Change (%)
110	A	91	33.8	(6.8)^a −0.1	(2.9)^a 11	(5.8)^a 3	(0.95)^a 3.2
	P	98	32.8	(6.6)^a 0.6	(2.3)^a 16	(5.7)^a 5	(1.1)^a 1.8
37	A	42	26	(10.0)^a −2.5	(2.1)^a −9.5	(6.2)^a −14.5	(1.1)^a 9.1
	P	44	26	(9.9)^a −1.1	(2.1)^a −4.8	(6.2)^a −4.8	(1.2)^a 8.3
44	A	38	28.8	(10.7)^a −0.7	no change	no change	no change
	P	39		(10.3)^a 0.3			
84	A	47	27.4	(9.3)^b −0.7	(2.4)^c −20.8	(5.7)^a 1.7	−
	P	46	27.7	(9.4)^b −0.1	(2.2)^c 0	(5.8)^a 3.4	
38	A	19	23.5	(11.1)^a −1.4	(1.5)^a −13.3	(5.5)^a 0	(1.2)^a 8.3
	P	18	22.9	(10.3)^a −0.4	(1.7)^a 5.9	(5.1)^a 1.9	(1.2)^a 8.3
42	A	28	26.5	(8.3)^a −1.0	(1.9)^a −30.2	(5.7)^a −10.4	(1.4)^a 6.3
	P	30	26.8	(8.3)^a 0.1	(1.6)^a −16.9	(5.7)^a 0	(1.4)^a 10.5
111	A	29		(9.1)^a −1.5	(3.5)^a −0.5	(6.3)^a 5.2	−
	P	30		(9.6)^a −0.9	(3.9)^a −7.3	(6.5)^a 2.9	

^{a} Prevalues in mmol/L.
^{b} HbA$_1$ prevalues in mmol/L.
^{c} One-hour postprandial values.

acid metabolism. This could contribute to increased cholesterol oxidation to bile acids. In placebo-controlled comparative investigations, acarbose was more effective in reducing triglycerides and cholesterol than glibenclamide and metformin, respectively (TABLE 5).[42,43] Acarbose, when used as an adjunct in NIDDM patients insufficiently

TABLE 5. Comparative Investigations on the Effect of Acarbose (A), Glibenclamide (G), or Metformin (M) versus Placebo (P) on the Lipid Triad in NIDDM Insufficiently Treated with Diet Alone

Reference	Drug	n	BMI	HbA$_{1C}$ Change (Δ%)	TG Change (%)	Cholesterol Change (%)	HDL-Cholesterol Change (%)
42	A	28	26.5	(8.3)^a −1.0	(1.9) −30.2	(5.7) −10.4	(1.4) 6.3
	P	30	26.8	(8.3) 0.1	(1.6) −16.9	(5.7) 0	(1.4) 10.5
	G	27	26.5	(8.3) −0.8	(2) −21.5	(5.6) −3.2	(1.6) 4.4
43	A	31	26.4	(9.6) −1.1	(1.7) −23.5	(6.3) −14.3	(1.4) 14.3
	P	32	26.3	(9.4) 0.3	(1.6) −12.5	(5.9) −1.7	(1.5) −6.7
	M	31	24	(9.7) −0.9	(1.4) −7.1	(5.8) 0	(1.6) 0

^{a} Numbers in parentheses: prevalues in mmol/L.

controlled with sulfonylurea, metformin, or insulin alone, had no effect on the lipid triad.[44] In a controlled study in NIDDM patients insufficiently controlled with glibenclamide alone, neither the addition of miglitol nor of metformin affected blood lipid levels.[45]

By extrapolation, acarbose has a moderate, but significant effect on postprandial hypertriglyceridemia and elevated cholesterol.

Biguanides

Metformin has been on the market for more than 30 years and is the only biguanide that is still accepted worldwide. Much has been learned about the mode of action of this drug in the last few years. It inhibits hepatic glucose production and enhances peripheral insulin sensitivity, resulting in a reduction of hyperinsulinemia.[46,47] Metformin also promotes a weight loss of about 1–3 kg in the first year after beginning of the treatment. Some smaller studies report a decrease in fasting triglycerides and cholesterol.[48,49] Schneider[50] has shown a correction of abnormal lipoprotein composition in NIDDM successfully treated with metformin that should be beneficial for the vascular risk. DeFronzo and colleagues[51] observed a significant decrease in LDL-cholesterol and triglycerides in a large multicenter study with NIDDM patients if compared with placebo and glibenclamide or if combined with glibenclamide. In other studies,[52,53] no effect on lipids was observed. The most extended data were derived from the UKPDS. After 3 years of follow-up, metformin, despite a highly significant reduction in HbA_{1C} of obese NIDDM, had no significant effect on the lipid triad in comparison to placebo (TABLE 6).[54] A further beneficial effect associated with the decrease in triglycerides was the reduction of PAI-1 with improvement of fibrinolysis.[52,55] In clinical practice, metformin is frequently used in combination with sulfonylureas if monotherapy fails to achieve sufficient metabolic control. This combination results in an average decrease of 0.5–1.0 HbA_{1C} percentage points as shown by DeFronzo[51] and others.[48,49] This is accompanied by a small decrease in total cholesterol and triglycerides. There exists, however, only one placebo-controlled study with regard to combined therapy that has shown a decrease in both lipid fractions

TABLE 6. Impact of Metformin versus Diet on the Lipid Triad in Newly Diagnosed Obese NIDDM[a]

	Changes by Treatment		
	Diet ($n = 177$)	Metformin ($n = 251$)	p
Glucose	+1.5	−0.5	<0.01
TG	+0.13	+0.06	n.s.
LDL-C	+0.11	−0.04	n.s.
HDL-C	+0.07	+0.07	n.s.

[a] The UKPDS experience, 3 years of follow-up;[54] n.s. = not significant.

together with a significant decrease in HbA_{1C}.[51] Thus, any weak improvement in lipid profile if metformin is used as an adjunct to sulfonylurea treatment may be mainly secondary due to improved glucose control.

Metformin was also tested with respect to lipid-lowering potency in some small studies with hypertensives[56-58] and hyperlipidemia type IIb without NIDDM. In hypertensives, triglycerides decreased by 22-27% and cholesterol by 4-18%, and HDL-cholesterol increased by 16-28%. In 24 subjects with hyperlipidemia, the only effect was a decrease in LDL-cholesterol by 12%.[59] Further investigations are needed to evaluate the potentials of metformin for the treatment of mild dyslipoproteinemia in diseases of the insulin resistance syndrome.

By extrapolation, metformin in some, but not all, studies has a moderate effect on triglycerides and cholesterol, but not on HDL-cholesterol. With respect to the risk of coronary heart disease, further therapeutical effects of metformin such as improved insulin sensitivity and fibrinolysis should be beneficial.

Sulfonylureas

This group of antidiabetics has been the most-frequently prescribed oral therapy for more than three decades. Contrary to biguanides and alpha-glucosidase inhibitors, there are a variety of compounds on the market. They remarkably differ in absorption, metabolism, elimination, and bioavailability. Nevertheless, they all primarily act by stimulating insulin secretion of the B-cell. This implies the hazard of hypoglycemia and weight gain by inappropriate hyperinsulinemia, in particular with long-acting agents such as glibenclamide.[31,60] The evidence for extrapancreatic metabolic and hormonal effects is scanty, at least for the first- and second-generation sulfonylureas. Glimepiride, a new sulfonylurea compound, is suggested to combine insulin secretion stimulatory effects with an improvement of peripheral insulin action.[61] First data show an increase in HDL-cholesterol with this drug.[62] Despite the long use of sulfonylureas in clinical practice, our knowledge about the effects on lipid metabolism is limited. Only very few controlled studies exist that prove the effect on the lipid triad.[63-68]

In some, but not all studies, reductions of triglycerides have been reported.[63,64] The UKPDS follow-up data show no significant effects on the lipid triad,[54] but a substantial weight gain of 3.5-4.8 kg after 6 years is observed.[69] Dunn et al.[65] and Taskinen et al.[64] found a small, but significant decrease in LDL-cholesterol, apoB, VLDL-cholesterol, and VLDL triglycerides. In an open crossover study with glibenclamide and metformin, each for three months in NIDDM insufficiently controlled with diet alone, Rains et al.[53] observed no effect on lipid fractions with sulfonylurea, despite a substantial improvement in glucose control. In this study, metformin reduced LDL-cholesterol by 0.34 mmol/L. With regard to HDL-cholesterol, comparative investigations with insulin revealed no difference[66,69] or lower levels[63,67,68] with sulfonylureas. Thus, Billingham et al.[63] concluded that abnormalities in lipoprotein concentrations were not corrected in patients treated with glibenclamide, but were near-normal in insulin-treated NIDDM. Improved glycemic control and long-lasting hyperinsulinemia may be of greater importance for changes in the lipoprotein profile than sulfonylureas per se (see TABLE 7).

TABLE 7. Comparative Effects of Diet, Sulfonylurea, and Insulin in Newly Diagnosed NIDDM[a]

	Changes by Treatment				
	Diet	Chlorpropamide	Glibenclamide	Insulin	p
Glucose	+1.1	−1.9	−1.0	−2.2	<0.001
TG	+0.14	−0.16	+0.16	=	<0.01
LDL-C	+0.13	−0.05	+0.14	+0.07	n.s.
HDL-C	+0.07	+0.06	+0.03	+0.06	n.s.

[a] The UKPDS experience: 3 years of follow-up; 1855 asymptomatic NIDDM subjects; mean age, 53 years; BMI, 27.5 kg/m^2; FBG, 8.7 mmol/L; lipids: TG, LDL-C, and HDL-C −1.78, 3.65, and 1.05 mmol/L; n.s. = not significant.

In summary, established sulfonylureas in short-term studies reduce triglycerides and have a weak, but inconsistent effect on LDL-cholesterol. There is no significant effect on the lipid triad in long-term studies. The reduction in triglycerides may be explained by improved glucose control and would be counterbalanced in the long run by harmful effects of hyperinsulinemia and weight gain induced by long-acting sulfonylureas such as glibenclamide.

Insulin

NIDDM is a progressive disease where deterioration in glycemic control occurs in the majority of cases, despite treatment with diet and oral antidiabetics; thus, 20–40% need insulin after 10 years. There are different options for the use of insulin in NIDDM. Since for a long time (even after failure of oral therapy to sufficiently control diabetes) the insulin secretion capacity for the second phase is still considerably high, combined treatment with sulfonylureas, acarbose, or metformin is frequently used as an interim regimen. This enables the patient to achieve good glycemic control with only one NPH-insulin injection at bedtime or in the morning.[70] The advantage of such a combined treatment with sulfonylureas has been reviewed recently in a meta-analysis.[71] In obese NIDDM with persistent high C-peptide secretion after glucagon injection, the combination of 50–100 mg of acarbose tid with a bedtime NPH-insulin injection was successful in achieving good-quality diabetes control.[72]

However, the therapeutic regimen often needs to be changed in the course of NIDDM and many patients eventually need intensified conventional insulin treatment with at least three to four insulin injections per day: regular insulin to the major meals and NPH-insulin for the night. The benefits of this intensified conventional treatment on microvascular complications have been convincingly demonstrated with IDDM in the DCCT trial.[73] Parallel to the reduction in HbA$_{1C}$, an improvement in the lipid profile was observed: LDL-cholesterol decreased by 34% in subjects with initial levels > 160 mg/dL, whereas body weight increased by 4.8 kg. The incidence of cardiovascular complications decreased by 41% in the 9 years of follow-up, which was not significant because of the small number of events in this rather young cohort of diabetics.

It is not our intention to review the multiple combinations with oral antidiabetics with insulin and the insulin regimen used in NIDDM with respect to their impact on the lipoprotein profile, not least because controlled studies that compare the efficacy and effects on associated risk factors are scanty or not done yet. Looking on the multiplicity of factors that affect the lipid-lowering action, including associated diseases of the insulin resistance syndrome, there is no perfect insulin regimen that can be generalized for all NIDDM patients. It is a consistent finding of the Diabetes Intervention Study and the UKPDS[32] that the results of conventional insulin treatment with two injections of combined insulin or high doses of basal insulin are disappointing with respect to diabetes. In these studies, insulin had no significant effect on the lipid triad.

In NIDDM patients with massive hypertriglyceridemia that reflects an absolute insulin deficiency, the lipoprotein profile was substantially improved by insulin treatment.[65] Correction of the hypertriglyceridemia due to a decrease in free fatty acid flow to the liver and improved VLDL clearance by stimulation of lipoprotein lipase activity, the key enzyme of VLDL removal, leads to a reduction of small dense LDL.[74] As with oral antidiabetics, improved glucose control by insulin therapy prevents or reverses glycosylation and oxidation of LDL and HDL. Insulin substitution diminishes the enhanced cholesterol synthesis and impaired LDL-receptor binding in poorly controlled NIDDM.[75] In cross-sectional investigations, insulin treatment reduces triglycerides and LDL-cholesterol[76,77] and increases HDL-cholesterol. However, in the UKPDS, no significant effect with insulin therapy on the lipid triad was detected after 6 years of follow-up. Subjects treated with insulin experienced a weight gain of 4.8 kg.[32]

Insulin treatment improved abnormalities in the lipoprotein composition,[10] but did not normalize it.[78] As compared to sulfonylureas, HDL-cholesterol was higher[67,68] or equal with insulin treatment.

It is also an open question whether an improved control of the lipid triad is possible with new insulin analogues such as Lys-Pro-insulin[79] that mimic the first insulin peak and allow a better control of postprandial hyperglycemia.

By extrapolation, insulin treatment leads to a substantial improvement in the atherogenic lipoprotein profile if given to NIDDM patients with severe insulin deficit. Even with perfect glucose control, no normalization in lipoprotein composition can be achieved. If used in early phases of NIDDM without absolute deficit, long-term studies reveal no benefit of diet versus oral antidiabetics. Beneficial effects on lipoprotein metabolism may be counterbalanced by peripheral hyperinsulinemia and weight gain due to inappropriate insulin therapy.

Differential Effects of Therapeutic Regimens on Risk Factor Profile

A summary of the possible differential effects of therapy variants on various components of the metabolic syndrome is given in TABLE 8. It has to be emphasized again that the net balance is strongly dependent on the stage of diabetes and on the relevant insulin secretion capacity. Obviously, antihyperglycemic agents that improve insulin

TABLE 8. Therapeutic Effects of Antidiabetic Therapy Variants on the Metabolic Syndrome in NIDDM

	Insulin	Sulfonylureas	Metformin	Acarbose
overweight	↑↑	↑	↓=	=↓
hyperglycemia	↓↓	↓	↓	↓
insulin resistance	↑=	↑=	↓	↓?
hyperinsulinemia	↑	↑	↓	↓
triglycerides	↓↑	=(↓)	=↓	↓=
cholesterol	=↓	=(↓)	=↓	=↓
HDL-C	↑	=	=↑	=
hypertension	=↑	=	=↓	=↓
atherosclerosis	↑?	↑?	↓?	↓?

sensitivity and reduce postprandial hyperinsulinemia, such as acarbose and metformin, are beneficial in the early stages of NIDDM and perhaps in IGT subjects. With the increasing failure of insulin secretion, sulfonylureas and/or insulin that can stimulate insulin secretion and substitute for it are superior with respect to the effects on the lipid triad and abnormalities of lipoprotein composition. Thus, the lesson from the studies with pharmacological treatment of NIDDM with respect to cardiovascular risk factors and dyslipoproteinemia in particular is that best results may be achieved with an individualized approach taking into account the pathophysiological constellation and the cluster of the metabolic syndrome.

RESULTS OF LIPID-LOWERING THERAPY IN CORONARY HEART DISEASE

Whereas the efficacy of lipid-lowering drugs to reduce the incidence of myocardial infarction and all-cause mortality could be recently confirmed in large-scale population-based primary[80] and secondary[81,82] prevention trials, only scarce information exists in NIDDM.

Fibric Acid Derivatives

Fibrates have been shown to reduce triglycerides by 30–50% and increase HDL-cholesterol by 8–15%. In addition, they decrease LDL-cholesterol by 10–20%.[83] The first study to test the preventive potentials of fibrates was the Diabetes Intervention Study.[84] In this double-blind placebo-controlled study with 1139 newly detected NIDDM patients without clinical coronary heart disease, clofibric acid was used in a 5-year follow-up. This drug significantly reduced triglycerides, but had only marginal effect on total cholesterol. The incidence of myocardial infarction and of all-ischemic heart disease was not different compared with the placebo group. Modern fibrates like bezafibrate, fenofibrate, and gemfibrozil have been shown to improve the lipid profile and the lipoprotein composition more efficiently. The Helsinki Heart Study using gemfibrozil produced a reduction of 10% in LDL-cholesterol and 43% in triglycerides and an increase in HDL-cholesterol of 10%, resulting in an overall reduction of 34% in the incidence of definite coronary endpoints compared to placebo.[85] This study included a subgroup of 135 diabetics with a similar lipid response to gemfibrozil. The diabetics in the placebo group exhibited a significantly higher incidence of coronary heart disease as the nondiabetics. The incidence of coronary heart disease in diabetics was reduced to the same frequency of events as in nondiabetics by gemfibrozil, although the reduction in diabetics did not reach statistical significance due to the small number of cases. In the Bezafibrate Coronary Atherosclerosis Intervention Trial (BECAIT) with nondiabetics, treatment with bezafibrate diminished the progression of atherosclerotic plaques. Interestingly, this beneficial effect was achieved with a correction of hypertriglyceridemia/low HDL in the absence of any significant decrease in LDL-cholesterol concentrations.[86]

There are two studies under way to test the preventive potentials of fibrates in diabetes. The Diabetes-Atherosclerosis Intervention Study (DAIS) is a randomized double-blind placebo-controlled follow-up regression study with NIDDM patients with prior coronary heart disease (PTCA, CABG, and/or lesions on baseline angiogram) using fenofibrate. In this multinational study, about 600 subjects will be included with a follow-up time of 3–5 years. The Fenofibrate Intervention and Event Lowering in Diabetes (FIELD) Study is a randomized trial of the effects of fenofibrate on coronary mortality and morbidity. Eight thousand diabetics aged 50–75 years will be considered, including cases with and without coronary heart disease at entry. Thus, a definite answer with regard to the potentials of fibrates for primary and secondary prevention of macroangiopathy in diabetes will not be given before the year 2000.

HMG-CoA Reductase Enzyme Inhibitors

Statins are most effective in reducing LDL-cholesterol by 25–40%, but they also reduce triglycerides by about 10% and increase HDL-cholesterol by about 5%. They were the first to convincingly demonstrate a significant decrease not only in coronary heart disease incidence, but also in all-cause mortality. The 4S secondary prevention trial was a landmark study to prove that aggressive lowering of LDL-cholesterol by simvastatin is associated with a significant decrease in all-cause mortality (by 30%) because of a highly significant reduction in coronary death rate (by 42%).[87] This study included 202 diabetics. The incidence of coronary heart disease in the simvastatin group was 24.9% versus 49.3% in the placebo group ($p < 0.01$). In analogy to the Helsinki Heart Study, subgroup analysis reveals that diabetics had the highest benefit among all high-risk groups. Similar beneficial effects have been seen in the recently published preliminary data of a secondary prevention trial in subjects with average cholesterol levels: the Cholesterol and Recurrent Events (CARE) Study.[82] In a primary intervention trial with pravastatin in the West of Scotland (WOS), Shepherd *et al*.[80] achieved a 26% reduction of LDL-cholesterol compared with placebo. Death from all cardiovascular causes was diminished by 32%. However, the small number of diabetics included and the low incidence rate do not allow one to draw conclusions on the potentials of HMG-CoA reductase inhibitors in primary prevention with regard to diabetes.

No data are available for the preventive use of bile acid sequestrants, probucol, or nicotinic acid in NIDDM. The latter cannot be recommended for treatment of dyslipoproteinemia in diabetics and IGT because of its adverse effect on glucose tolerance.

In the future, new drugs such as thiazolidinediones that address insulin resistance as well as dyslipoproteinemia and hypertension will be of interest. Troglitazone, the most-advanced compound, has been shown to significantly improve insulin sensitivity, increase HDL-cholesterol, and reduce triglycerides.[88,89]

In conclusion, the epidemiological data so far available indicate that lipid-lowering drugs are at least as effective in diabetics as in nondiabetics in the secondary prevention of coronary heart disease. With respect to primary prevention, a more-global approach that reduces the risk for conversion of IGT to diabetes and the deterioration of diabetes and macroangiopathy seems to be necessary.

PHARMACOLOGICAL TREATMENT OF NIDDM AND CORONARY HEART DISEASE

The effect of different antidiabetic agents on hard endpoints such as cardiovascular complications and life expectancy is the key question of any therapeutic option. Surprisingly enough, 75 years after introduction of insulin and 3 decades with the use of sulfonylureas and biguanides, this is still an open question in NIDDM for both insulin and oral antidiabetics.

Sulfonylureas

The first randomized large-scale intervention trial of diabetes, the University Group Diabetes Program (UGDP), resulted in an increase of cardiovascular complications — in particular, fatal myocardial events (TABLE 9)[90-93] — in the tolbutamide group. Tolbutamide is a first-generation short-acting sulfonylurea.

In the Diabetes Intervention Study (DIS) in striking analogy, the incidence of myocardial infarction was lowest in the groups treated with diet and insulin in contrast to sulfonylurea-taking patients. However, there was no difference in the incidence of ischemic ECG changes according to Minnesota criteria (TABLE 10). In DIS, glibenclamide was used. A retrospective study in Japan of over 2000 diabetics found no difference in cardiovascular mortality between those treated with diet, insulin, or sulfonylureas.[94] In a smaller controlled trial of secondary prevention of myocardial infarction, tolbutamide significantly improved survival in subjects with IGT.[95] Consistent with this finding in a controlled study of subjects with IGT, the tolbutamide group suffered from less coronary heart disease versus placebo.[96] The difference dis-

TABLE 9. Major Outcome of the UGDP[93]

	Treatment Group[a]			
	PLBO (n = 205)	TOLB (n = 204)	ISTD (n = 210)	IVAR (n = 204)
Death as of October 7, 1969				
All causes				
number	21	30	20	18
percentage	10.2	14.7	9.5	8.8
Cardiovascular cause				
number	10	26	13	12
percentage	4.9	12.7	6.2	5.9
Death as of March 7, 1973				
All causes				
number	41	45	33	32
percentage	20.0	22.1	15.7	15.7
Cardiovascular cause				
number	26	35	22	21
percentage	12.7	17.2	10.5	10.3

[a] PLBO = placebo, TOLB = tolbutamide, ISTD = insulin standard dosage, IVAR = insulin variable dosage.

appeared after 10 years of follow-up.[24] It looks as if sulfonylureas in IGT would have beneficial effect on macroangiopathy, but not in overt diabetes. This assumption is supported by interim reports from the UKPDS where, after 6 years of follow-up, no difference with respect to coronary events and survival between the diet, the insulin, the sulfonylurea, and the metformin groups could be observed.[97]

One explanation for this unexpected result should be the blockade of ATP-regulated K^+ channels for the cardiac cells by sulfonylureas.[98] This would have detrimental effects in the case of myocardial ischemia as typical for coronary atherosclerosis. On the other hand, high intracellular potassium content could protect against arrhythmias. In a study with NIDDM, glibenclamide significantly reduced the frequency of arrhythmias, while there was no change in the frequency of myocardial ischemic attacks.[99]

Recently, the impact of sulfonylurea treatment on cardiovascular disease has been extensively reviewed by Leibowitz and Cerasi.[100]

TABLE 10. Incidence (n/1000) of Coronary Events by Therapy Variants in NIDDM[a]

	Myocardial Infarction	Ischemic Heart Disease	All Coronary Events
insulin[b]	56	182	250
glibenclamide[b]	104	187	295
diet[b]	84	150	240

[a] Diagnosed by diet controlled at entry: DIS, 6 to 11 years of follow-up.
[b] At least since the fourth year of follow-up.

Biguanides

Only scarce information exists on the effect of biguanides on cardiovascular disease. In the UGDP, phenformin intake was stopped because of excess mortality.[101] Metformin appears to have potentially beneficial effects on cardiovascular risk in addition to its effects on lipid profiles. Increased fibrinolysis,[55] a small decrease in blood pressure,[56] and diminished platelet aggregation[102] have been reported in several, but not all studies. A placebo-controlled study in patients with peripheral arterial disease demonstrated improved blood flow during metformin treatment.[103] In the 10-year follow-up, metformin significantly reduced the frequency of intervention with vascular reconstruction surgery.[104] An answer on the possible preventive potentials of metformin for coronary heart disease will be presumably given by the UKPDS in 1998.

Alpha-Glucosidase Inhibitors

No data in humans are available so far for alpha-glucosidase inhibitors from clinical trials investigating cardiovascular effects. By extrapolation from the impact on blood lipids and hyperinsulinemia, beneficial effects on atherogenesis could be expected. The STOP-NIDDM trial with more than 1000 IGT subjects and other ongoing prospective trials with acarbose may give an answer to that question in 3 to 5 years from now.

Insulin

The Diabetes Control and Complication Trial (DCCT) has shown that the frequency of late complications in IDDM can be reduced by intensified insulin treatment.[73] There was also a tendency for a lower incidence of macroangiopathy. In the UGDP study, which was the first trial to test the relationship between insulin treatment and cardiovascular complications, neither a fixed dose regimen nor flexible application was superior to the diet group.[105] It is remarkable that, despite good glucose control (FIG. 3) with flexible insulin therapy to near-normal levels, the incidence of cardiovascular complications was identical to standard dosage results. In analogy, the UKPDS has not detected any difference in risk factor profile and coronary heart disease in the insulin-treated subgroup versus diet after 6 years of follow-up.[97] The same was the case in the DIS. These studies, however, have not used modern intensified conventional insulin regimens or pump treatment strategies. It is beyond the scope of this review to discuss the possible differences of insulin treatment variants that allow a lower dosage of insulin with less peripheral hyperinsulinemia with regard to cardiovascular risk. It may well be that inappropriately high dosage with a poor timing in obese NIDDM patients with persistent endogenous and exogenous hyperinsulinemia may be harmful, and not only by subsequent substantial weight gain. Thus, in two studies, a high insulin dosage was associated with a higher cardiovascular complication rate. In Pima Indians, a relationship between insulin use and fatal coronary heart

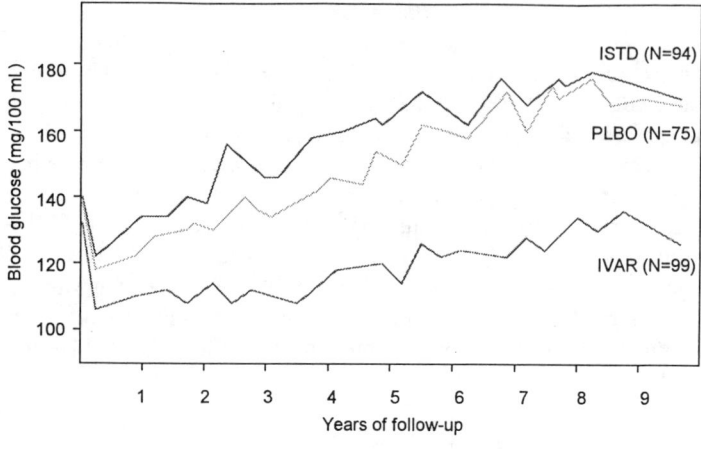

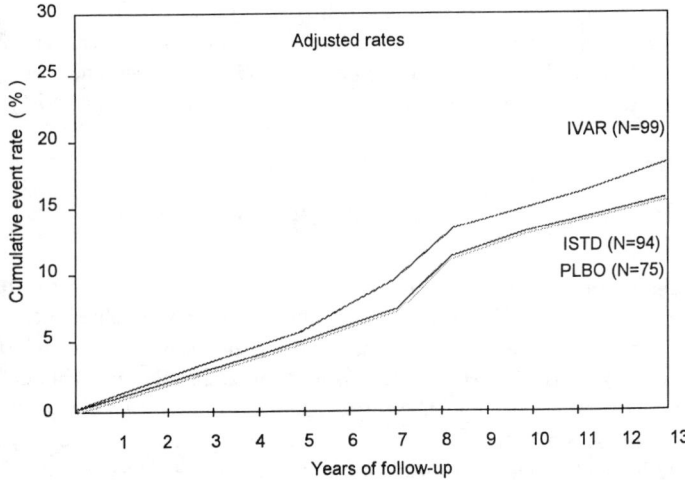

FIGURE 3. Effect of variable (IVAR) and standard dosage insulin (ISTD) treatment on the quality of blood glucose control and death from cardiovascular causes in NIDDM: UGDP, final results[105] (PLBO = placebo).

disease has been reported.[106] This population is known for a high prevalence of obesity and endogenous hyperinsulinemia.[107] In the Schwabing Study, insulin dosage was an independent risk factor for incidence of cardiovascular disease in multivariate analysis.[108]

Most of these studies may be biased by differences in the severity of diabetes, inappropriate insulin treatment regimens, or confounding variables. So far, the controlled prospective studies do not support the idea that insulin treatment is risky in NIDDM.[109] There is, however, a need for more physiological insulin treatment strategies in NIDDM that will better control postprandial hyperglycemia and lipoprotein abnormalities and avoid weight gain.

CONCLUSIONS

Abnormalities in lipoprotein concentrations and composition are common companions of NIDDM, with the lipid triad (hypertriglyceridemia, low HDL, and small dense LDL) as the major abnormality. The risk for cardiovascular disease and mortality from dyslipoproteinemia and hypercholesterolemia is amplified 2–4 times in NIDDM patients as compared to nondiabetics.

Besides secondary effects by improved glycemic control, acarbose and metformin have some beneficial effects on triglycerides and elevated cholesterol levels. Controversial results were reported on first- and second-generation sulfonylureas, with a decrease in triglycerides in some studies that is counterbalanced by weight gain in long-term trials. Insulin in late phases of NIDDM with absolute insulin deficiency reduces triglycerides and increases HDL-cholesterol. Lipoprotein composition is not normalized with insulin therapy. Application of insulin in early phases with hyperinsulinemia has no effect on the lipid triad.

Effective reduction in LDL-cholesterol by HMG-CoA reductase inhibitors significantly reduced the incidence of cardiovascular events in secondary prevention in subgroups of the 4S Study and CARE, indicating that diabetics should have at least the same benefit from aggressive LDL-cholesterol reduction as nondiabetics. A positive tendency was also seen by the use of fibrates in the Helsinki Heart Study and the BECAIT regression trial, with major effects on triglycerides and HDL-cholesterol.

Current pharmacological options (sulfonylureas, biguanides, insulin) have not been clearly proven to reduce or affect excessive cardiovascular morbidity and mortality in NIDDM. Studies with alpha-glucosidase inhibitors with respect to coronary heart disease have not been published yet. As far as the data allow, the best results, if any, have been obtained with IGT and early phases of diabetes. Carefully planned and designed long-term studies with modern quantitative measurement of atherosclerotic lesions and hard endpoints are needed that will test the benefits of pharmaceutical approaches in primary and secondary prevention trials for both diabetes and coronary heart disease. Obviously, a more-global strategy that attacks insulin resistance, hyperglycemia, and dyslipoproteinemia seems to be necessary to achieve a breakthrough in prevention of NIDDM and its complications.

REFERENCES

1. PYÖRÄLÄ, K., M. LAAKSO & M. UUSITUPA. 1987. Diabetes and atherosclerosis: an epidemiologic view. Diabetes Metab. Rev. 3(2): 463–524.
2. PANZRAM, G. 1987. Mortality and survival in type 2 (non-insulin-dependent) diabetes mellitus. Diabetologia 30(3): 123–131.

3. HANEFELD, M. & T. KURKTSCHIEV. 1995. Plasma lipids in diabetes. *In* New Horizons in Diabetes Mellitus and Cardiovascular Disease, pp. 89–96. Current Science. London.
4. HANEFELD, M., H. SCHMECHEL, U. JULIUS, S. FISCHER, J. SCHULZE, U. SCHWANEBECK, J. LINDNER, C. HORA & H. DUDE (DIS GROUP). 1991. Five-year incidence of coronary heart disease related to major risk factors and metabolic control in newly diagnosed non-insulin-dependent diabetes: the Diabetes Intervention Study (DIS). Nutr. Metab. Cardiovasc. Dis. **1:** 135–140.
5. RÖNNEMAA, T., M. LAAKSO, V. KALLIO, K. PYÖRÄLÄ, J. MARNIEMI & P. PUUKKA. 1989. Serum lipids, lipoproteins, and apolipoproteins and the excessive occurrence of coronary heart disease in non-insulin-dependent diabetic patients. Am. J. Epidemiol. **130**(4): 632–645.
6. STAMLER, J., O. VACCARO, J. NEATON & D. WENTWORTH (MULTIPLE RISK FACTOR INTERVENTION TRIAL RESEARCH GROUP). 1993. Diabetes, other risk factors, and 12-year cardiovascular mortality for men screened in the Multiple Risk Factor Intervention Trial. Diabetes Care **16:** 434–449.
7. GIUGLIANO, D. 1993. Does treatment of noninsulin-dependent diabetes mellitus reduce the risk of coronary heart disease? Curr. Opin. Lipidol. **4:** 227–240.
8. HANEFELD, M. & W. LEONHARDT. 1981. Das metabolische Syndrom. Dtsch. Gesundheitswes. **36:** 545–551.
9. HANEFELD, M. 1996. Das metabolische Syndrom: wurzel, mythen, und fakten. *In* Das Metabolische Syndrom, pp. 15–26. Fischer. Jena.
10. REAVEN, G. & M. GREENFELD. 1981. Diabetes hypertriglyceridemia: evidence for three clinical syndromes. Diabetes **30**(suppl. 2): 66–75.
11. FERRANNINI, E., S. M. HAFFNER, B. D. MITCHELL & M. P. STERN. 1991. Hyperinsulinaemia: the key feature of a cardiovascular and metabolic syndrome. Diabetologia **34**(6): 416–422.
12. AUSTIN, M. A. 1994. Small, dense low-density lipoprotein as a risk factor for coronary heart disease. Int. J. Clin. Lab. Res. **24**(4): 187–192.
13. FONTBONNE, A., E. ESCHWEGE, F. CAMBIEN, J. L. RICHARD, P. DUCIMETIERE, N. THIBAULT, J. M. WARNET, J. R. CLAUDE & G. E. ROSSELIN. 1989. Hypertriglyceridaemia as a risk factor of coronary heart disease mortality in subjects with impaired glucose tolerance or diabetes: results from the 11-year follow-up of the Paris Prospective Study. Diabetologia **32**(5): 300–304.
14. GENEST, J. J. & J. S. COHN. 1995. Clustering of cardiovascular risk factors: targeting high-risk individuals. Am. J. Cardiol. **76**(2): 8A–20A.
15. STEINBERG, D., S. PARTHASARATHY, T. E. CAREW, J. C. KHOO & J. L. WITZTUM. 1989. Beyond cholesterol: modifications of low-density lipoprotein that increase its atherogenicity. N. Engl. J. Med. **320**(14): 915–924.
16. AUSTIN, M. & K. L. EDWARD. 1996. Small, dense low-density lipoproteins, the insulin resistance syndrome, and non-insulin-dependent diabetes. Curr. Opin. Lipidol. **7:** 167–171.
17. CESTARO, B., R. GANDINI, P. VIANI, F. MARAFFI, G. CERVATO, C. MONTALTO, P. GATTI & R. MEGALI. 1994. Fluorescence-determined kinetics of plasma high oxidizability in diabetic patients. Biochem. Mol. Biol. Int. **32**(5): 983–994.
18. KESANIEMI, Y. A. 1985. Pathophysiology of low density lipoprotein and high density lipoprotein glucosylation. Monogr. Atheroscler. **13:** 63–73.
19. JANKA, H. U., B. BALLETSHOFER, A. BECKER *et al.* 1992. Das metabolische Syndrom als potenter kardiovaskulärer Risikofaktor für vorzeitigen Tod bei Typ-II-Diabetikern. Die Schwabinger Studie II – Untersuchung nach neun Jahren. Diabetes Stoffw. **1:** 2–7.
20. KUUSISTO, J., L. MYKKÄNEN, K. PYÖRÄLÄ & M. LAAKSO. 1994. NIDDM and its metabolic control predict coronary heart disease in elderly subjects. Diabetes **43**(8): 960–967.
21. HANEFELD, M., S. FISCHER, U. JULIUS *et al.* 1996. Risk factors for myocardial infarction and death in newly detected NIDDM: the Diabetes Intervention Study, 11-year follow-up. Diabetologia **39**. In press.
22. JUHAN-VAGUE, I., S. G. THOMPSON & J. JESPERSEN. 1993. Involvement of the hemostatic system in the insulin resistance syndrome: a study of 1500 patients with angina pectoris – the ECAT Angina Pectoris Study Group. Arterioscler. Thromb. **13**(12): 1865–1873.
23. CERIELLO, A., R. GIACOMELLO, G. STEL, E. MOTZ, C. TABOGA, L. TONUTTI, M. PIRISI, E. FALLETI & E. BARTOLI. 1995. Hyperglycemia-induced thrombin formation in diabetes: the possible role of oxidative stress. Diabetes **44**(8): 924–928.

24. JARRETT, R. J., P. MCCARTNEY & H. KEEN. 1982. The Bedford Survey: ten-year mortality rates in newly diagnosed diabetics, borderline diabetics, and normoglycaemic controls and risk indices for coronary heart disease in borderline diabetics. Diabetologia 22(2): 79-84.
25. ZIMMET, P., G. DOWSE, C. FINCH, S. SERJEANTSON & H. KING. 1990. The epidemiology and natural history of NIDDM—lessons from the South Pacific. Diabetes Metab. Rev. 6(2): 91-124.
26. HAFFNER, S. M., R. A. VALDEZ, H. P. HAZUDA, B. D. MITCHELL, P. A. MORALES & M. P. STERN. 1992. Prospective analysis of the insulin-resistance syndrome (syndrome X). Diabetes 41(6): 715-722.
27. HANEFELD, M. & M. WECK. 1989. Very low calorie diet therapy in obese non-insulin-dependent diabetes patients. Int. J. Obes. 13(suppl. 2): 33-37.
28. PETRIE, J. R. & R. DONNELLY. 1994. New pharmacological approaches to insulin and lipid metabolism. Drugs 47(5): 701-710.
29. SHINKAI, H. & Y. SATO. 1990. Hypoglycaemic action of phenylalanine derivatives. In New Antidiabetic Drugs, p. 249-254. Smith-Gordon. London.
30. UKPDS GROUP. 1995. Relative efficacy of randomly allocated diet, sulphonylurea, insulin, or metformin in patients with newly diagnosed non-insulin-dependent diabetes followed for three years. Br. Med. J. 310(6972): 83-88.
31. GERICH, J. E. 1989. Oral hypoglycemic agents. N. Engl. J. Med. 321(18): 1231-1245.
32. HANEFELD, M., S. FISCHER, J. SCHULZE, M. SPENGLER, M. WARGENAU, K. SCHOLLBERG & K. FÜCKER. 1991. Therapeutic potentials of acarbose as a first-line drug in NIDDM insufficiently treated with diet alone. Diabetes Care 14(8): 732-737.
33. BAILEY, C. J. 1992. Biguanides and NIDDM. Diabetes Care 15(6): 755-772.
34. PULS, W., U. KEUP, H. P. KRAUSE, G. THOMAS & F. HOFFMEISTER. 1977. Glucosidase inhibition: a new approach to the treatment of diabetes, obesity, and hyperlipoproteinaemia. Naturwissenschaften 64(10): 536-537.
35. HANEFELD, M. 1993. Acarbose efficacy review. Pract. Diabetes 6: 21-27.
36. LEONHARDT, W., M. HANEFELD, S. FISCHER & J. SCHULZE. 1994. Efficacy of alpha-glucosidase inhibitors on lipids in NIDDM subjects with moderate hyperlipidaemia. Eur. J. Clin. Invest. 24(suppl. 3): 45-49.
37. BRAUN, D. & H. J. MITZKAT. 1993. 24-Wöchige klinische Prüfung zur Wirksamkeit von Acarbose im Vergleich zu Placebo bei Typ II-Diabetikern mit unzureichendem Stoffwechselergebnis unter alleiniger Diättherapie. Diabetes Stoffw. 2: 116-117.
38. HOTTA, N., H. KAKUTA, T. SANO, H. MATSUMAE, H. YAMADA, S. KITAZAWA & N. SAKAMOTO. 1993. Long-term effect of acarbose on glycaemic control in non-insulin-dependent diabetes mellitus: a placebo-controlled double-blind study. Diabetic Med. 10: 134-138.
39. LEONHARDT, W., M. HANEFELD, S. FISCHER, J. SCHULZE & M. SPENGLER. 1991. Positive Wirkung einer Acarbose-Behandlung auf die Serumlipide nicht-insulinbedürftiger Diabetiker. Arzneim. Forch. Drug Res. 41: 735-738.
40. NESTEL, P. J. 1988. Lower triglyceride production with carbohydrate-rich diet during treatment with acarbose. In Acarbose for the Treatment of Diabetes Mellitus, p. 68. Springer-Verlag. Berlin/New York.
41. MARUHAMA, Y., A. NAGASAKI, Y. KANAZAWA, H. HIRAKAWA, Y. GOTO, H. NISHIYAMA, Y. KISHIMOTO & T. SHIMOYAMA. 1980. Effects of a glucoside-hydrolase inhibitor (Bay g 5421) on serum lipids, lipoproteins, bile acids, fecal fat and bacterial flora, and intestinal gas production in hyperlipidemic patients. Tohoku J. Exp. Med. 132(4): 453-462.
42. HOFFMANN, J. & M. SPENGLER. 1994. Efficacy of 24-week monotherapy with acarbose, glibenclamide, or placebo in NIDDM patients: the Essen Study. Diabetes Care 17(6): 561-566.
43. HOFFMANN, J. & M. SPENGLER. 1996. Diabetes Care. Submitted.
44. CHIASSON, J. L., R. G. JOSSE, J. A. HUNT, C. PALMASON, N. W. RODGER, S. A. ROSS, E. A. RYAN, M. H. TAN & T. M. WOLEVER. 1994. The efficacy of acarbose in the treatment of patients with non-insulin-dependent diabetes mellitus: a multicenter controlled clinical trial. Ann. Intern. Med. 121(12): 928-935.
45. HANEFELD, M., D. PETZINNA, S. RAPTIS, Y. RYBKA, G. SCHERNTHANER & E. STANDL. 1995. Wirkungsvergleich von Miglitol und Metformin bei auf Glibenclamid eingestellten Typ-II-Diabetikern: eine kontrollierte multizentrische Studie. Diabetes Stoffw. 4: 153.
46. BAILEY, C. J. & R. C. TURNER. 1995. Metformin. N. Engl. J. Med. 334(9): 574-579.

47. SCHERNTHANER, G. 1992. Kritische Analyse der antidiabetischen Therapie mit Metformin: Stoffwechselwirkungen, antiatherogene Effekte, und Kontraindikationen. Akt. Endokr. Stoffw. **13:** 44-50.
48. HERMANN, L. S., B. SCHERSTEN, P. O. BITZEN, T. KJELLSTRÖM, F. LINDGARDE & A. MELANDER. 1994. Therapeutic comparison of metformin and sulfonylurea, alone and in various combinations: a double-blind controlled study. Diabetes Care **17**(10): 1100-1109.
49. BAILEY, C. J. 1993. Metformin—an update. Gen. Pharmacol. **24**(6): 1299-1309.
50. SCHNEIDER, J. 1991. Effects of metformin on dyslipoproteinemia in non-insulin-dependent diabetes mellitus. Diabete Metab. **17:** 185-190.
51. DEFRONZO, R. A. & A. M. GOODMAN. 1995. Efficacy of metformin in patients with non-insulin-dependent diabetes mellitus: the Multicenter Metformin Study Group. N. Engl. J. Med. **333**(9): 541-549.
52. GRANT, P. J., N. A. STICKLAND, N. A. BOOTH & C. R. M. PRENTICE. 1991. Metformin causes a reduction in basal and post-venous occlusion plasminogen activator inhibitor-1 in type 2 diabetic patients. Diabetic Med. **8:** 361-365.
53. RAINS, S. G., G. A. WILSON, W. RICHMOND & R. S. ELKELES. 1988. The effect of glibenclamide and metformin on serum lipoproteins in type 2 diabetes. Diabetic Med. **5**(7): 653-658.
54. CULL, C. A., R. D. CARTER, S. E. MANLEY, R. HOLMAN & R. C. TURNER. 1994. Hyperlipidemia in type 2 diabetes: comparative effects of diet, sulfonylurea, insulin, and metformin therapies. Diabetic Med. **11**(suppl. 2): 21.
55. VAGUE, P., I. JUHAN-VAGUE, M. C. ALESSI, C. BADIER & J. VALADIER. 1987. Metformin decreases the high plasminogen activator inhibition capacity, plasma insulin, and triglyceride levels in non-diabetic obese subjects. Thromb. Haemostasis **57**(3): 326-328.
56. GIUGLIANO, D., N. DEROSA, G. DI-MARO, R. MARFELLA, R. ACAMPORA, R. BUONINCONTI & F. D'ONOFRIO. 1993. Metformin improves glucose, lipid metabolism, and reduces blood pressure in hypertensive, obese women. Diabetes Care **16**(10): 1387-1390.
57. LANDIN, K., L. TENGBORN & U. SMITH. 1994. Metformin and metoprolol CR treatment in non-obese men. J. Intern. Med. **235**(4): 335-341.
58. MENDOZA, S. G., A. FAIETA, H. CARRASCO et al. 1994. Metformin lowers blood pressure, insulin resistance, triglycerides, and endogenous estradiol in hypertensive men. Clin. Res. **42:** 338A.
59. PENTIKAINEN, P. J., E. VOUTILAINEN, A. ARO, M. UUSITUPA, I. PENTTILA & H. VAPAATALO. 1990. Cholesterol-lowering effect of metformin in combined hyperlipidemia: placebo-controlled double-blind trial. Ann. Med. **22**(5): 307-312.
60. GROOP, L. C. 1992. Sulfonylureas in NIDDM. Diabetes Care **15**(6): 737-754.
61. DRAEGER, E. 1995. Clinical profile of glimepiride. Diabetes Res. Clin. Pract. **28**(suppl.): S139-S146.
62. IYAKU, R. et al. 1993. Effect of glimepiride on the lipid triad in NIDDM. J. Clin. Ther. Med. **9:** 1107-1129.
63. BILLINGHAM, M. S., J. J. MILLES, C. J. BAILEY & R. A. HALL. 1989. Lipoprotein subfraction composition in non-obese newly diagnosed non-insulin-dependent diabetes after treatment with diet and glibenclamide. Diabetes Res. **11:** 13-20.
64. TASKINEN, M. R., W. F. BELTZ, I. HARPER, R. M. FIELDS, G. SCHONFELD, S. M. GRUNDY & B. V. HOWARD. 1986. Effects of NIDDM on very-low-density lipoprotein triglyceride and apolipoprotein B metabolism: studies before and after sulfonylurea therapy. Diabetes **35**(11): 1268-1277.
65. DUNN, F. L., P. RASKIN, D. W. BILHEIMER & S. M. GRUNDY. 1984. The effect of diabetic control on very-low-density lipoprotein-triglyceride metabolism in patients with type II diabetes mellitus and marked hypertriglyceridemia. Metabolism **33:** 117-123.
66. HUUPPONEN, R. K., J. S. VIIKARI & H. SAARIMAA. 1984. Correlation of serum lipids with diabetes control in sulfonylurea-treated diabetic patients. Diabetes Care **7**(6): 575-578.
67. LISCH, H. J. & S. SAILER. 1981. Lipoprotein patterns in diet, sulfonylurea, and insulin-treated diabetics. Diabetologia **20:** 118-122.
68. CALVERT, G. D., J. J. GRAHAM, T. MANNIK, P. H. WISE & R. A. YEATES. 1978. Effects of therapy on plasma high-density-lipoprotein-cholesterol concentration in diabetes mellitus. Lancet **2:** 66-68.
69. TURNER, R. C. & R. R. HOLMAN. 1995. Lessons from UK Prospective Diabetes Study. Diabetes Res. Clin. Pract. **28**(suppl.): S151-S157.
70. GROOP, L. C., E. WIDEN, A. EKSTRAND, C. SALORANTA, A. FRANSSILA-KALLUNKI, C. SCHALIN-

JANTTI & J. G. ERIKSSON. 1992. Morning or bedtime NPH insulin combined with sulfonylurea in treatment of NIDDM. Diabetes Care 15(7): 831–834.
71. PUGH, J. A., M. L. WAGNER, J. SAWYER, G. RAMIREZ, M. TULEY & S. J. FRIEDBERG. 1992. Is combination sulfonylurea and insulin therapy useful in NIDDM patients? Diabetes Care 15(8): 953–959.
72. HANEFELD, M. et al. 1996. Unpublished results.
73. DCCT RESEARCH GROUP. 1993. The effect of intensive treatment of diabetes on the development and progression of long-term complications in insulin-dependent diabetes mellitus. N. Engl. J. Med. 329: 977–986.
74. BRUNZELL, J. D. & A. CHAIT. 1990. Lipoprotein pathophysiology and treatment. In Ellenberg and Rifkin's Diabetes Mellitus: Theory and Practice, p. 756–767. Elsevier. Amsterdam/New York.
75. SCOPPOLA, A., G. TESTA, S. FRONTONI, E. MADDALONI, S. GAMBRADELLA, G. MENZINGER & A. LALA. 1995. Effects of insulin on cholesterol synthesis in type II diabetes mellitus. Diabetes Care 18: 1362–1369.
76. TURNER, R. C. & R. R. HOLMAN. 1990. Insulin use in NIDDM: rationale based on pathophysiology of disease. Diabetes Care 13: 1011–1020.
77. GENUTH, S. 1990. Insulin use in NIDDM. Diabetes Care 13: 1240–1264.
78. BAGDADE, J. D., W. E. BUCHANAN, T. KUUSI & M. R. TASKINEN. 1990. Persistent abnormalities in lipoprotein composition in noninsulin-dependent diabetes after intensive insulin therapy. Arteriosclerosis 10(2): 232–239.
79. HOWEY, D. C., R. R. BOWSHER, R. L. BRUNELLE & J. R. WOODWORTH. 1994. [Lys(B28), Pro(B29)]-human insulin: a rapidly absorbed analogue of human insulin. Diabetes 43: 396–402.
80. SHEPHERD, J., S. COBBE, I. FORD et al. 1995. Prevention of coronary heart disease with pravastatin in men with hypercholesterolemia. N. Engl. J. Med. 333(20): 1301–1307.
81. SCANDINAVIAN SIMVASTATIN SURVIVAL STUDY GROUP. 1994. Randomised trial of cholesterol lowering in 4444 patients with coronary heart disease: the Scandinavian Simvastatin Survival Study (4S). Lancet 344(8934): 1383–1389.
82. CHOLESTEROL AND RECURRENT EVENTS STUDY. 1996. Presented at the American College of Cardiology Meeting, Orlando.
83. LARSEN, M. & D. R. ILLINGWORTH. 1994. Drug treatment of dyslipoproteinemia. Med. Clin. North Am. 78: 225–244.
84. HANEFELD, M., S. FISCHER, H. SCHMECHEL, G. ROTHE, J. SCHULZE, H. DUDE, U. SCHWANEBECK & U. JULIUS. 1991. Diabetes Intervention Study: multi-intervention trial in newly diagnosed NIDDM. Diabetes Care 14(4): 308–317.
85. MANNINEN, V., M. O. ELO, M. H. FRICK, K. HAAPA, O. P. HEINONEN, P. HEINSALMI, P. HELO, J. K. HUTTUNEN, P. KAITANIEMI, P. KOSKINEN et al. 1988. Lipid alterations and decline in the incidence of coronary heart disease in the Helsinki Heart Study. JAMA 260(5): 641–651.
86. ERICSSON, C. G., A. HAMSTEN, J. NILSSON, L. GRIP, B. SVANE & U. DE FAIRE. 1996. Angiographic assessment of effects of bezafibrate on progression of coronary artery disease in young male postinfarction patients. Lancet 347(9005): 849–853.
87. KJEKSHUS, J. & T. R. PEDERSEN. 1995. Reducing the risk of coronary events: evidence from the Scandinavian Simvastatin Survival Study (4S). Am. J. Cardiol. 76(9): 64C–68C.
88. NOLAN, J. J., B. LUDVIK, P. BEERDSEN, M. JOYCE & J. OLEFSKY. 1994. Improvement in glucose tolerance and insulin resistance in obese subjects treated with troglitazone. N. Engl. J. Med. 331(18): 1188–1193.
89. SUTER, S. L., J. J. NOLAN, P. WALLACE, B. GUMBINER & J. M. OLEFSKY. 1992. Metabolic effects of new oral hypoglycemic agent CS-045 in NIDDM subjects. Diabetes Care 15(2): 193–203.
90. KOLATA, G. 1979. Controversy over study of diabetes drugs continues for nearly a decade. Science 203: 986–990.
91. SCHOR, S. 1971. The University Group Diabetes Program: a statistician looks at the mortality results. J. Am. Med. Assoc. 217: 1673–1675.
92. SELTZER, H. 1972. A summary of criticisms of the findings and conclusions of the University Group Diabetes Program (UGDP). Diabetes 21: 976–979.
93. PROUT, T. E. 1976. A progress report on the University Group Diabetes Program. Int. J. Clin. Pharmacol. Biopharm. 12(1-2): 244–251.
94. OHNEDA, A., Y. MARUHAMA, H. ITABASHI, S. I. OIKAWA, T. KOBAYASHI, K. HORIGOME, M.

CHIBA, Y. KAI, T. SAKAI, F. OKUGUCHI & J. NIHEI. 1978. Vascular complications and longterm administration of oral hypoglycemic agents in patients with diabetes mellitus. Tohoku J. Exp. Med. **124**(3): 205-222.
95. PAASIKIVI, J. & F. WAHLBERG. 1971. Preventive tolbutamide treatment and arterial disease in mild hyperglycaemia. Diabetologia **7**(5): 323-327.
96. KEEN, H., R. J. JARRETT, C. CHLOUVERAKIS & D. R. BOYNS. 1968. The effect of treatment of moderate hyperglycaemia on the incidence of arterial disease. Postgrad. Med. J. Suppl., p. 960-965.
97. UK PROSPECTIVE DIABETES STUDY GROUP. 1995. Overview of 6 years therapy of type-II-diabetes: a progressive disease. Diabetes **44**: 1249-1258.
98. PANTEN, U., M. SCHWANSTECHER & C. SCHWANSTECHER. 1992. Pancreatic and extrapancreatic sulfonylurea receptors. Horm. Metab. Res. **24**(12): 549-554.
99. CACCIAPUOTI, F., R. SPIEZIA, U. BIANCHI, D. LAMA, M. D'AVINO & M. VARRICCHIO. 1991. Effectiveness of glibenclamide on myocardial ischemic ventricular arrhythmias in non-insulindependent diabetes mellitus. Am. J. Cardiol. **67**(9): 843-847.
100. LEIBOWITZ, G. & E. CERASI. 1996. Sulfonylurea treatment of NIDDM patients with cardiovascular disease: a mixed blessing? Diabetologia **39**: 503-514.
101. KLIMT, C. R., G. L. KNATTERUD, C. L. MEINERT & T. E. PROUT. 1970. The University Group Diabetes Program: a study of the effect of hypoglycemic agents on vascular complications in patients with adult-onset diabetes. I. Design, methods, and baseline characteristics. II. Mortality results. Diabetes **19**: 747-830.
102. COLLIER, A., H. H. WATSON, A. W. PATRICK, C. A. LUDLAM & B. F. CLARKE. 1989. Effect of glycaemic control, metformin and gliclazide on platelet density and aggregability in recently diagnosed type 2 (non-insulin-dependent) diabetic patients. Diabete Metab. **15**(6): 420-425.
103. SIRTORI, C. & C. PASIK. 1994. Re-evaluation of a biguanide, metformin: mechanism of action and tolerability. Pharmacol. Res. **30**: 187-228.
104. SIRTORI, C., G. FRANCESCHINI, G. GIANFRANCESCHI, M. SIRTORI, G. MONTANARI, E. BOSISIO, E. MANTERO & A. BONDIOLI. 1984. Metformin improves peripheral vascular flow in nonhyperlipidemic patients with arterial disease. J. Cardiovasc. Pharmacol. **6**(5): 914-923.
105. KNATTERUD, G., C. KLIMT, M. GOLGNER et al. 1982. Effects of hypoglycemic agents on vascular complications in patients with adult-onset diabetes. VIII. Evaluation of insulin therapy: final report. Diabetes **31**(suppl. 5): 1-26.
106. NELSON, R. G., M. L. SIEVERS, W. C. KNOWLER, B. A. SWINBURN, D. J. PETTITT, M. F. SAAD, I. M. LIEBOW, B. V. HOWARD & P. H. BENNETT. 1990. Low incidence of fatal coronary heart disease in Pima Indians despite high prevalence of non-insulin-dependent diabetes. Circulation **81**(3): 987-995.
107. SAAD, M. F., S. LILLIOJA, B. L. NYOMBA, C. CASTILLO, R. FERRARO, M. DEGREGORIO, E. RAVUSSIN, W. C. KNOWLER, P. H. BENNETT, B. V. HOWARD et al. 1991. Racial differences in the relation between blood pressure and insulin resistance. N. Engl. J. Med. **324**(11): 733-739.
108. JANKA, H. U., A. G. ZIEGLER, E. STANDL & H. MEHNERT. 1987. Daily insulin dose as a predictor of macrovascular disease in insulin-treated non-insulin-dependent diabetics. Diabete Metab. **13**(3)(part 2): 359-364.
109. WINGARD, D. L. & E. L. BARRET-CONNOR. 1995. Is insulin really a heart disease risk factor? Diabetes Care **18**: 1299-1304.
110. CONIFF, R. F., J. A. SHAPIRO & T. B. SEATON. 1995. A double-blind placebo-controlled trial evaluating the safety and efficacy of acarbose for the treatment of patients with insulin-requiring type II diabetes. Diabetes Care **18**(7): 928-932.
111. RYBKA, J., A. GREGOROVA, A. ZYMDLENA & P. JARON. 1990. Klinische Studie mit Acarbose. Drug Invest. **2**(4): 264-267.

Relationships between Fatty Acid Composition and Insulin-induced Oxidizability of Low-Density Lipoproteins in Healthy Men[a]

T. PELIKÁNOVÁ, E. TVRZICKÁ, L. KAZDOVÁ, AND A. ŽÁK

Postgraduate Medical School
Institute for Clinical and Experimental Medicine
and
First Medical Faculty
Charles University
140 00 Prague 4, Czech Republic

The central role of oxidative modification of low-density lipoproteins (LDL) in the pathogenesis of atherosclerosis has been suggested by several *in vitro* and *in vivo* studies.[1-6] It is clear now that oxidized LDL may itself elicit a number of biological responses that could contribute to atherogenesis. For example, oxidized LDL directly deliver various lipid oxides and hydroperoxides to target cells, acting as cytotoxins, monocyte chemoattractants, stimulators of cholesterol ester accumulation by macrophages, and inhibitors of macrophage movement.[5,6]

Detailed understanding of the mechanisms involved in the regulation of LDL oxidizability could have direct implications in the treatment and prevention of atherosclerosis. Numerous naturally occurring dietary components, that is, vitamins E and C, carotenes, selenium, xanthines, and flavonoids, have been recognized to possess antioxidant properties.[7] Although insulin resistance and the accompanying hyperinsulinemia are considered by some authors to be one of the risk factors of atherosclerosis,[8,9] little attention has been paid to the role of insulin in the regulation of LDL oxidizability. To date, only Rifici *et al.*[10] have reported that insulin and IGF-I at supraphysiological concentrations stimulate the mononuclear cell–mediated oxidation of LDL *in vitro*, but they did not see a lasting effect of clamp-induced hyperinsulinemia on LDL oxidizability that could be measured *ex vivo*. This study, however, was not controlled and only 3 diabetics and 2 healthy subjects were included.

Another confounding issue is the role of fatty acids (FA) in lipid peroxidation and atherogenesis. Oxidation of LDL is a free-radical process in which the polyunsaturated fatty acids contained in the LDL are degraded by a lipid peroxidation process and the susceptibility of LDL to oxidation depends upon the amount and type of FA

[a] This study was supported by the Health Ministry of the Czech Republic (Project No. 3550-3).

present in the LDL particles. Compared to monoenes, n-6 polyunsaturated FA enhance LDL oxidizability and similar results have been obtained in most studies with n-3 polyunsaturates.[11-14] Regardless of the effects on LDL oxidizability, both types of polyunsaturated FA have shown a beneficial effect on the development of atherosclerosis.[15,16] The seeming inconsistencies between the findings of these two distinct research areas underline the complexity of the pathogenetic mechanisms of atherosclerosis.

The aim of our study was (a) to evaluate, in healthy volunteers, the effect of acute hyperinsulinemia on oxidative modification of LDL and FA composition of LDL particles (phosphatidylcholine and LDL cholesterol esters) and (b) to test the hypothesis that the insulin-induced changes (if any) in the susceptibility of LDL to oxidative modification are related to the FA pattern of LDL. Euglycemic hyperinsulinemic clamp (75 µU/mL) lasting 10 hours was used for the *in vivo* induction of hyperinsulinemia and compared to a time-controlled study with saline infusion.

MATERIALS AND METHODS

Subjects

The study group consisted of 11 normal-weight healthy men. All subjects had normal glucose tolerance (confirmed by the oral glucose tolerance test) and none had a family history of diabetes mellitus. They were not taking any drugs nor were they on any special diet; their average dietary intake, assessed by dietary records covering three consecutive days before examination, contained 14.5% proteins, 41.5% fat, and 44% carbohydrates. The clinical characteristics of the study group are summarized in TABLE 1. Informed consent was obtained from all the individuals after the purpose, nature, and potential risks of the study had been explained.

TABLE 1. Characteristics of the Study Group ($n = 11$)[a]

Age (years)	33.8 ± 6.6
BMI (kg/m^2)	23.7 ± 2.5
Fasting PG (mmol/L)	4.64 ± 0.34
PG$_{120}$ (mmol/L)	4.44 ± 0.63
S-TG (mmol/L)	1.15 ± 0.65
NEFA (µmol/L)	345 ± 87
Cholesterol (mmol/L)	4.05 ± 0.64
HDL cholesterol (mmol/L)	1.16 ± 0.28
LDL cholesterol (mmol/L)	2.36 ± 0.62
α-tocopherol (mmol/L)	23.0 ± 4.13
β+γ-tocopherol (mmol/L)	2.39 ± 1.19

[a] Abbreviations: BMI, body mass index; PG, plasma glucose; PG$_{120}$, plasma glucose at 120 min of the oral glucose tolerance test; S-TG, serum triglycerides; NEFA, nonesterified fatty acids. Values are the mean ± SD.

Procedures

The subjects were examined on an outpatient basis, after overnight fasting with only tap water ad libitum. They were instructed to adhere to their ordinary lifestyle and to avoid changes in food intake, alcohol consumption, and exercise. Each subject underwent two examinations performed in random order at a 2-week interval.
The euglycemic hyperinsulinemic clamp study lasted 10 hours and was conducted as previously described.[17] Briefly, Teflon cannula (Venflon; Viggo, Helsingborg, Sweden) was inserted into an antecubital vein for the infusion of all test substances. A second cannula was inserted retrogradely into a wrist vein for blood sampling, and the hand was placed in a heated (65°C) box in order to achieve venous blood arterialization. A primed-continuous insulin infusion (1 mU/kg·min of Actrapid HM; Novo Nordisk, Copenhagen, Denmark) was administered. Plasma glucose concentrations during the clamp were maintained at the fasting levels by continuous infusion of 15% glucose. Plasma immunoreactive insulin (IRI) concentrations were measured before and at every 60 min of each study. Blood samples for the measurement of serum lipids [LDL isolation, LDL oxidizability, FA composition, cholesterol, HDL cholesterol, triglycerides (TG)] and tocopherols were taken before (0 min) and at 600 min. The time-controlled studies with saline infusion (4 mmol/h) and tap water drinking (150 mL/h) were performed using an identical protocol of blood sampling.

Laboratory Methods

Plasma glucose concentrations were measured using the glucose oxidase method on a Beckman Analyzer (Beckman Instruments, Fullerton, California). FIRI was determined by radioimmunoassay using an OPiDI RIA-INS kit (Institute of Atomic Energy, Swierk, Poland). Serum nonesterified fatty acids (NEFA) were measured using the colorimetric method,[18] and serum cholesterol, HDL cholesterol, and serum TG were measured by standard enzymatic methods using Peridochrom-Triglyceride CHOD-PAP and Cholesterin CHOD-PAP monotests (Boehringer Mannheim, Germany). LDL particles were isolated using rate-gradient ultracentrifugation.[19,20] Total lipids of LDL were extracted using chloroform/methanol,[21] and cholesterol esters and phosphatidylcholine fractions were separated by thin-layer chromatography. Methylesters of FA were analyzed using capillary gas-liquid chromatography as previously described.[22] The susceptibility of LDL to *in vitro* Cu^{2+}-stimulated oxidation (at 37°C) was determined by three different methods. First, continuous monitoring of the conjugated diene (CD) formation:[23] The initial amount of dienes present in the LDL preparation was calculated using the initial 234-nm absorbance and the molar absorptivity of CD (29,500 L/mol·cm). The lag time (the interval between the addition of $CuSO_4$ and the intercept of the tangent of the slope of the absorbance curve during the propagation phase), the maximal rate of diene production (calculated from the slope of the absorbance curve during the propagation phase), and the total amount of conjugated diene production were evaluated. Second, formation of thiobarbituric acid–reactive substances (TBARS) was measured.[24] Third, measurement of the oxidative modification of the

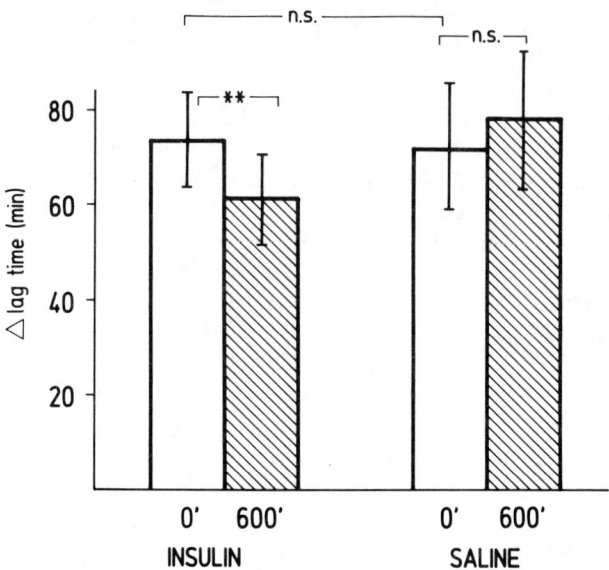

FIGURE 1. Lag time of diene production at 0 min and 600 min of the hyperinsulinemic clamp study (insulin) versus saline infusion in healthy control subjects ($n = 11$). Statistical significance: (**) $p < 0.01$; n.s. = not significant.

protein moiety by the emission fluorescence spectra at 430 nm (excitation, 355 nm) was employed.[25] Serum α-, β+γ-tocopherol concentrations were assayed by high-performance liquid chromatography.[26]

Statistical Analysis

All data are expressed as means ± SD. The data were analyzed by analysis of variance (ANOVA) with repeated measures using BMDP386 statistical software program 2V (University of California Press, Los Angeles, California).

RESULTS

The mean plasma glucose concentrations were comparable during the clamp and control studies (4.8 ± 0.3 mmol/L with coefficients of variation of up to 4% versus 4.7 ± 0.3 mmol/L). The fasting IRI level was 7.2 ± 2.5 mU/L. It rose to 74 ± 13 mU/L during the clamp study and remained unchanged during the control study (6.2 ± 2.8 mU/L).

Ten-hour hyperinsulinemia led to a significant decrease in the lag time of diene production ($p < 0.01$) (FIG. 1), despite no significant changes in the maximal rate of

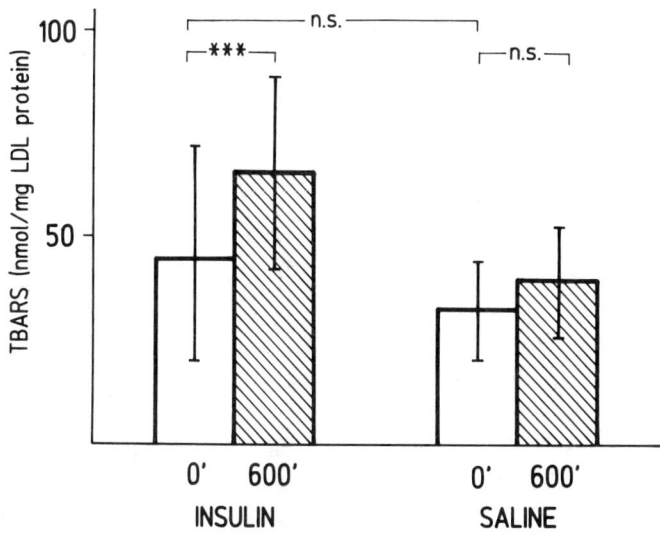

FIGURE 2. Thiobarbituric acid–reactive substances (TBARS) formation at 0 min and 600 min of the hyperinsulinemic clamp study (insulin) versus saline infusion in healthy control subjects ($n = 11$). Statistical significance: (∗∗∗) $p < 0.01$; n.s. = not significant.

diene production (clamp: 13.38 ± 3.63 versus 14.66 ± 3.02 nmol/min·g of LDL protein; saline: 12.57 ± 3.26 versus 12.47 ± 3.19 nmol/min·g of LDL protein) or in the total amount of conjugated diene production (clamp: 512 ± 126 versus 551 ± 119 nmol/mg of LDL protein; saline: 443 ± 86 versus 450 ± 71 nmol/mg of LDL protein). Furthermore, insulin compared to saline led to a significant increase in TBARS formation (FIG. 2) ($p < 0.01$) and fluorescence intensity ($p < 0.05$) (clamp: 281 ± 115 versus 424 ± 158 [$p < 0.05$]; saline: 321 ± 133 versus 370 ± 101 [not significant]).

TABLES 2 and 3 show the FA composition of LDL phosphatidylcholine and LDL cholesterol esters. No significant differences were found between the FA pattern at 0 min and after 10-hour insulin infusion. The insulin-induced decrease in the lag time of diene production was significantly associated with the content of monoenoic acids (FIG. 3) and n-3 FA (FIG. 4) in LDL phosphatidylcholine. Correlations with the content of saturated and n-6 family FA were not statistically significant; similarly, no statistical significance has been found between LDL oxidizability and FA composition of LDL cholesterol esters.

The concentrations of serum lipids and α-, $\beta + \gamma$-tocopherols are given in TABLE 1. There were no significant changes in serum cholesterol, HDL cholesterol, LDL cholesterol, and apo B during insulin infusion. Serum TG and vitamin concentrations comparably decreased during insulin and saline studies. There were no significant correlations between the parameters of LDL oxidizability and the levels of serum lipids or tocopherols.

TABLE 2. The Fatty Acid Composition of LDL Phosphatidylcholine at 0 min and after 10-h (600-min) Insulin or Saline Infusion in Healthy Control Subjects ($n = 11$)[a]

Fatty Acid (%)	Insulin		Saline	
	0 min	600 min	0 min	600 min
saturated	44.95 ± 2.01	44.56 ± 1.71	44.80 ± 2.13	44.06 ± 2.84
14:0	0.30 ± 0.13	0.30 ± 0.08	0.39 ± 0.35	0.25 ± 0.12
16:0	29.23 ± 2.19	28.64 ± 0.77	29.32 ± 2.87	29.02 ± 3.43
18:0	15.44 ± 1.06	15.61 ± 0.80	15.08 ± 1.02	14.79 ± 1.18
monounsaturated	15.84 ± 0.47	16.36 ± 0.51	15.98 ± 0.92	16.09 ± 1.81
16:1	0.84 ± 0.10	0.87 ± 0.09	0.90 ± 0.23	0.85 ± 0.13
18:1 n-7	2.58 ± 0.37	2.71 ± 0.43	2.63 ± 0.51	2.76 ± 0.56
18:1 n-9	12.42 ± 0.46	12.78 ± 0.42	12.45 ± 0.63	12.50 ± 0.85
n-6 family	34.60 ± 1.99	34.14 ± 2.12	34.12 ± 2.30	34.08 ± 2.31
18:2	24.25 ± 1.51	22.87 ± 1.70	23.84 ± 1.94	22.43 ± 1.94
18:3	0.13 ± 0.02	0.15 ± 0.02	0.15 ± 0.10	0.14 ± 0.03
20:3	2.44 ± 0.43	2.78 ± 0.57	2.32 ± 0.37	2.72 ± 0.53
20:4	7.53 ± 0.90	8.02 ± 0.95	7.54 ± 1.10	8.49 ± 1.16
22:4	0.25 ± 0.04	0.30 ± 0.11	0.28 ± 0.14	0.30 ± 0.08
n-3 family	4.60 ± 0.86	4.94 ± 0.99	5.09 ± 0.73	5.76 ± 0.63
18:3	0.24 ± 0.04	0.20 ± 0.05	0.29 ± 0.10	0.22 ± 0.05
20:5	1.14 ± 1.23	1.23 ± 0.47	1.17 ± 0.34	1.27 ± 0.33
22:5	0.76 ± 0.07	0.86 ± 0.11	0.78 ± 0.11	0.93 ± 0.11
22:6	2.45 ± 0.42	2.66 ± 0.54	2.85 ± 0.46	3.34 ± 0.41

[a] Values show the relative percentages of each FA in the total FA (mean ± SD). The differences between 0 min and 600 min and between insulin and saline are not statistically significant.

DISCUSSION

The mechanisms responsible for the association between hyperinsulinemia and atherogenesis remain unresolved. We have shown in our study that even relatively short, 10-hour, hyperinsulinemia, induced *in vivo* by the euglycemic clamp technique, enhances LDL oxidizability, which is thought to be the proatherogenic factor. The increase in LDL oxidizability has been confirmed by three different methods: (a) a decrease of the lag time of the diene production, even though the total amount of dienes did not change, (b) an increase in TBARS production, and (c) an increase in fluorescence intensity, which is related to the extent of oxidative modification of apolipoprotein B. The results are in agreement with those of a previous *in vitro* study reporting that incubation with insulin at supraphysiological concentrations stimulates the mononuclear cell–mediated oxidation of LDL,[10] as measured by TBARS production, fluorescence intensity, trinitrobenzene sulfonic acid reactivity, and electrophoretic mobility of LDL. Unlike us, the above investigators were not able to confirm the lasting effect of *in vivo* 5-hour hyperinsulinemia (150 mU/L) induced by euglycemic clamp on LDL oxidizability.[10] This study, however, was not controlled and only 3 diabetics and 2 healthy subjects were included.

It has been shown that the susceptibility of LDL to oxidative modification is influ-

TABLE 3. The Fatty Acid Composition of LDL Cholesterol Esters at 0 min and after 10-h (600-min) Insulin or Saline Infusion in Healthy Control Subjects ($n = 11$)[a]

Fatty Acid (%)	Insulin		Saline	
	0 min	600 min	0 min	600 min
saturated	18.86 ± 2.28	19.04 ± 1.95	17.94 ± 2.04	15.93 ± 2.39
14:0	0.86 ± 0.32	0.89 ± 0.61	0.78 ± 0.44	0.46 ± 0.34
16:0	13.70 ± 1.53	13.19 ± 1.37	12.92 ± 1.17	11.04 ± 2.04
18:0	4.30 ± 1.15	4.95 ± 1.32	4.25 ± 1.53	5.67 ± 3.96
monounsaturated	25.71 ± 2.06	26.19 ± 2.74	24.71 ± 3.48	25.69 ± 3.14
16:1	3.14 ± 0.48	3.02 ± 0.54	3.26 ± 0.96	2.40 ± 0.68
18:1 n-7	1.70 ± 0.13	1.78 ± 0.23	1.65 ± 0.24	1.78 ± 0.22
18:1 n-9	20.88 ± 1.88	21.39 ± 2.70	19.80 ± 2.89	21.51 ± 3.28
n-6 family	53.9 ± 2.14	53.02 ± 3.90	55.43 ± 4.36	56.30 ± 3.33
18:2	48.29 ± 2.17	47.16 ± 3.81	49.29 ± 4.47	49.45 ± 3.33
18:3	0.58 ± 0.08	0.64 ± 0.33	0.61 ± 0.19	0.66 ± 0.12
20:3	0.59 ± 0.08	0.63 ± 0.17	0.61 ± 0.17	0.66 ± 0.17
20:4	4.25 ± 0.75	4.21 ± 0.87	4.87 ± 0.83	5.37 ± 1.04
22:4	0.20 ± 0.26	0.38 ± 0.55	0.11 ± 0.36	0.16 ± 0.22
n-3 family	1.53 ± 0.37	1.76 ± 0.69	1.91 ± 0.49	2.08 ± 0.50
18:3	0.49 ± 0.07	0.45 ± 0.11	0.56 ± 0.21	0.54 ± 0.14
20:5	0.78 ± 0.26	1.01 ± 0.50	0.92 ± 0.35	1.11 ± 0.47
22:5	0.06 ± 0.02	0.06 ± 0.02	0.06 ± 0.05	0.06 ± 0.02
22:6	0.19 ± 0.09	0.23 ± 0.17	0.36 ± 0.14	0.36 ± 0.08

[a] Values show the relative percentages of each FA in the total FA (mean ± SD). The differences between 0 min and 600 min and between insulin and saline are not statistically significant.

enced by dietary fats.[11-14] In our study, we have found a relationship even between an insulin-induced increase in LDL oxidizability and the FA pattern of LDL phosphatidylcholine. The magnitude of the drop in the lag time of diene production (Δ lag time), indicating an increase of LDL oxidizability after insulin infusion, correlated negatively with the amount of monounsaturated FA and positively with the content of n-3 family FA in LDL phosphatidylcholine. The results concerning the content of monounsaturated FA are in good agreement with previous studies stressing the protective role of monounsaturates against LDL oxidation and atherogenesis.[11,14,27-29] Conflicting data have been reported on the effect of n-3 FA, showing no effect[30,31] or, like our data, increased oxidizability of LDL.[12,13,32-34] We have not found any relation between the insulin-induced increase in LDL oxidizability and the content of saturated and n-6 family FA in LDL, although previous results consistently reported that n-6 FA-enriched LDL are oxidized faster than LDL enriched with other FA.[11,32,35]

The insulin infusion in the present study did not alter the FA composition of phosphatidylcholine and cholesterol esters in LDL particles. This finding is in accordance with our previous study, where we did not see any changes in serum phospholipids and cholesterol esters after 5-hour clamp-induced supraphysiological hyperinsulinemia.[36] Therefore, the insulin-induced increase in LDL oxidizability could not be mediated by changes in the FA pattern and thus other mechanisms should be involved.

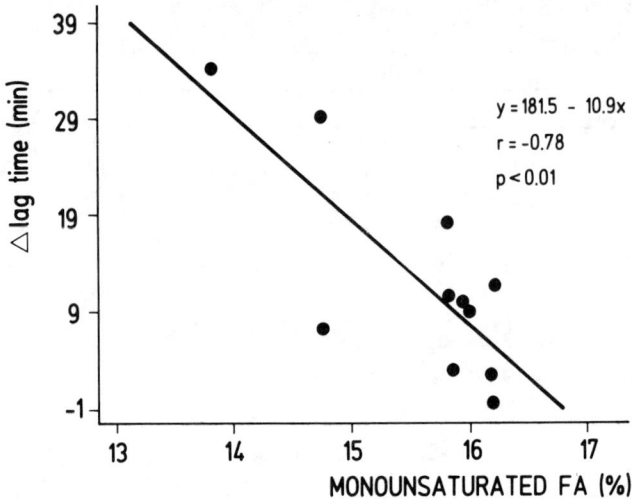

FIGURE 3. Relationship between insulin-induced Δ lag time of diene production (lag time at 0 min minus lag time at 600 min) and the content of monounsaturated FA in LDL phosphatidylcholine.

Nevertheless, insulin has been shown to stimulate FA desaturation and *de novo* synthesis of FA,[37,38] and the long-term effect of hyperinsulinemia on FA composition and subsequently on LDL oxidation cannot be excluded.

It has been suggested that the susceptibility of LDL to oxidative stress depends

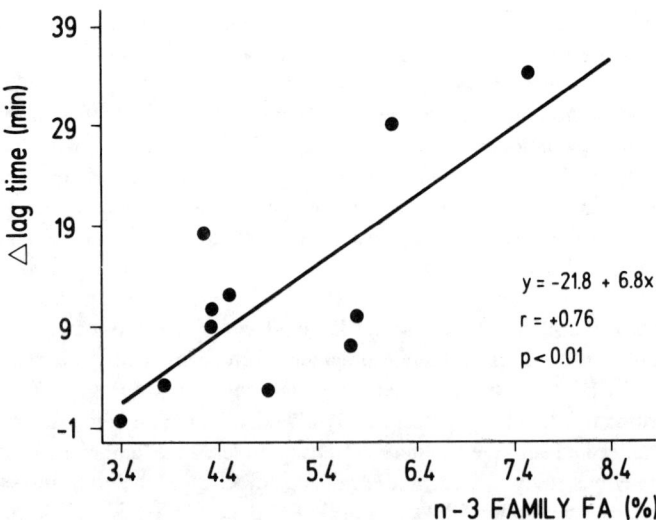

FIGURE 4. Relationship between insulin-induced Δ lag time of diene production (lag time at 0 min minus lag time at 600 min) and the content of n-3 family FA in LDL phosphatidylcholine.

on the balance between the LDL content of polyunsaturated fats and antioxidants, tocopherol in particular.[7,39] We did not find any significant relationship between the insulin-induced increase in LDL oxidizability and serum tocopherol levels. Nevertheless, the variable levels of antioxidants could be the reason for the seeming inconsistency between the pro-oxidative role of polyunsaturated FA *in vitro* and their *in vivo* antiatherogenic effect.[15,16] Another explanation could be the age of the LDL particles. Diets rich in polyunsaturated FA make LDL particles disappear from the blood more quickly, through an enhanced uptake via LDL receptors,[40] and older plasma lipoproteins have been shown to be more susceptible to oxidation.[41] The presence of older LDL particles or increased amounts of small dense LDL particles may be the reason for the increased oxidizability of LDL associated with elevated levels of serum triglycerides or LDL cholesterol.[42,43]

Because we did not find any significant changes in the concentrations of serum tocopherols and lipids, the increase in LDL oxidizability seen after 10-hour insulin infusion cannot be explained by changes in the above parameters; thus, other mechanisms should be responsible. It was shown that the insulin effect on cell-mediated LDL oxidation is dependent on the presence of glucose in the media.[10] It is possible that insulin stimulates LDL oxidizability via an increase in glucose utilization, resulting in increased O_2^- production[44] or in increased thiol formation and the secondary generation of O_2^- in the media by the reoxidation of thiols.[45]

We conclude that acute hyperinsulinemia increases LDL susceptibility to oxidative modification in healthy control subjects. Despite the modulatory role of FA composition in LDL susceptibility to oxidation, the insulin-induced increase in LDL oxidizability is not mediated by changes in the FA pattern of phosphatidylcholine or cholesterol esters of LDL particles.

ACKNOWLEDGMENTS

The skillful technical assistance of D. Lapešová, the kind help of R. Prahl with the manuscript preparation, and the help of V. Lánská with the statistical evaluation of the data are gratefully acknowledged.

REFERENCES

1. WITZUM, J. L. & D. STEINBERG. 1991. J. Clin. Invest. **88:** 1785–1792.
2. YLA-HERTTUALA, S., W. PALINSKI, M. E. ROSENFELD, S. PARTHASARATHY, T. E. CAREW, J. WITZUM & D. STEINBERG. 1989. J. Clin. Invest. **84:** 1086–1095.
3. SALONEN, T. J., S. YLA-HERTTUALA, R. YAMAMOTO, S. BUTLER, H. KORPELA, R. SALONEN, K. NYSSONEN, W. PALINSKI & J. WITZUM. 1992. Lancet **339:** 883–887.
4. REGNSTROM, J., J. NILSSON, P. TORNVALL, C. LANDOU & A. HAMSTEN. 1992. Lancet **339:** 1183–1186.
5. PENN, M. S. & G. M. CHISOLM. 1994. Atherosclerosis **108**(suppl.): S21–S29.
6. STEINBERG, D., S. PARTHASARATHY, T. E. CAREW, J. C. KHOO & J. L. WITZUM. 1989. N. Engl. J. Med. **320:** 915–924.
7. DECKER, E. A. 1995. Nutr. Rev. **53:** 49–58.
8. STOUT, R. W. 1990. Diabetes Care **13:** 631–654.

9. STERN, M. P. 1995. Diabetes **44**: 369–374.
10. RIFICI, V. A., S. H. SCHNEIDER & A. K. KHACHADURIAN. 1994. Atherosclerosis **107**: 99–108.
11. REAVEN, P. D., B. J. GRASSE & D. L. TRIBBLE. 1994. Arterioscler. Thromb. **14**: 557–566.
12. OOSTENBRUG, G. S., R. P. MENSINK & G. HORNSTRA. 1994. Eur. J. Clin. Nutr. **48**: 895–898.
13. SUZUKAWA, M., M. ABBEY, P. R. C. HOWE & P. J. NESTEL. 1995. J. Lipid Res. **36**: 473–484.
14. NENSETER, M. S. & C. A. DREVON. 1996. Curr. Opin. Lipidol. **7**: 8–13.
15. HORROBIN, D. F. & Y. S. HUANG. 1987. Int. J. Cardiol. **17**: 241–255.
16. KROMHOUT, D., E. B. BOSSCHIETER & C. COULANDER. 1985. N. Engl. J. Med. **312**: 1205–1209.
17. DEFRONZO, R. A., J. D. TOBIN & R. ANDRES. 1979. Am. J. Physiol. **237**: E214–E227.
18. NOVAK, M. 1965. J. Lipid Res. **6**: 431–433.
19. SCHUMACKER, D. L. & D. L. PUPPIONE. 1986. *In* Methods in Enzymology. Vol. 128, p. 155–170. Academic Press. Orlando.
20. NARUSZEWITZ, M., E. SELINGER & J. DAVIGNON. 1992. Metabolism **41**: 1215–1224.
21. FOLCH, L., M. LEES & C. H. SLOANE-STANLEY. 1957. J. Biol. Chem. **226**: 497–509.
22. TVRZICKÁ, E., E. CVRČKOVÁ, B. MÁCHA & M. JIRÁSKOVÁ. 1994. J. Chromatogr. **656**: 51–57.
23. KLEINVELD, H. A., H. L. M. HAK-LEMMERS, A. F. H. STALENHOEF & P. N. M. DEMACKER. 1992. Clin. Chem. **38**: 2066–2072.
24. THOMAS, C. E. & D. J. REED. 1988. J. Pharmacol. Exp. Ther. **245**: 492–500.
25. ESTERBAUER, H., G. JURGENS, O. QUEHENBERGER & E. KOLLER. 1987. J. Lipid Res. **28**: 495–509.
26. CATIGNANI, G. L. 1986. Methods Enzymol. **123**: 215–219.
27. PARTHASARATHY, S., J. C. KHOO, E. MILLER, J. BARNETT, J. L. WITZUM & D. STEINBERG. 1990. Proc. Natl. Acad. Sci. U.S.A. **87**: 3894–3898.
28. BERRY, E. M., S. EISENBERG, D. HARATZ, Y. FRIEDLANDER, Z. NORMAN, N. A. KAUFMAN & Y. STEIN. 1991. Am. J. Clin. Nutr. **53**: 899–907.
29. BONANOME, A., A. PAGNAN, S. BIFFANTI, O. OPPORTUNO, F. SORGATO, M. DORELLA, M. MAIORINO & F. URSINI. 1992. Arterioscler. Thromb. **12**: 529–533.
30. SAITO, H., K. J. CHANG, Y. TAMURA & S. YOSHIDA. 1991. Biochem. Biophys. Res. Commun. **175**: 61–67.
31. FRANKEL, E. N., E. J. PARKS, R. XU, B. O. SCHNEEMAN, P. A. DAVIS & J. B. GERMAN. 1994. Lipids **29**: 233–236.
32. THOMAS, M. J., T. THORNBURG, J. MANNING, K. HOOPER & L. L. RUDEL. 1994. Biochemistry **33**: 1828–1834.
33. WHITMAN, S. C., J. R. FISH, M. L. RAND & K. A. ROGERS. 1994. Arterioscler. Thromb. **14**: 1170–1176.
34. HARATZ, D., Y. DABACH, G. HOLLANDER, M. BEN-NAIM, R. SCHWARTZ, E. M. BERRY, O. STEIN & Y. STEIN. 1991. Atherosclerosis **90**: 127–139.
35. ABBEY, M., G. B. BELLING, M. NOAKES, F. HIRATA & P. J. NESTEL. 1993. Am. J. Clin. Nutr. **57**: 391–398.
36. PELIKÁNOVÁ, T., M. KOHOUT, J. BAŠE, Z. ŠTEFKA, J. KOVÁŘ, L. KAZDOVÁ & J. VÁLEK. 1991. Clin. Chim. Acta **203**: 329–337.
37. HUANG, Y. S., D. F. HORROBIN, M. S. MANKU, J. MITCHELL & M. A. RYAN. 1984. Lipids **19**: 367–370.
38. HUANG, Y. S., K. FUJII, R. TAKAHASHI, J. MITCHELL & D. F. HORROBIN. 1985. IRCS Med. Sci. **13**: 1145–1146.
39. ESTERBAUER, H., H. PUHL, M. DIEBER-ROTHENEDER, G. WAEG & H. RABL. 1991. Ann. Med. **23**: 573–581.
40. GOODNIGHT, S. H., W. S. HARRIS, W. E. CONNOR & D. R. ILLINGWORTH. 1982. Arteriosclerosis **2**: 87–93.
41. WALZEM, R. L., S. WATKINS, E. N. FRANKEL, R. J. HANSEN & J. B. GERMAN. 1995. Proc. Natl. Acad. Sci. U.S.A. **92**: 7460–7464.
42. HIRAMATSU, K. & S. ARIMORI. 1988. Diabetes **37**: 832–837.
43. DEJAGER, S., E. BRUCKERT & M. J. CHAPMAN. 1993. J. Lipid Res. **34**: 295–308.
44. HENECKE, J. W., H. ROSEN, L. A. SUZUKI & A. CHAIT. 1987. J. Biol. Chem. **262**: 10098–10103.
45. SPARROW, C. P. & J. OLSZEWSKI. 1993. J. Lipid Res. **34**: 1219–1228.

Small Dense Low-Density Lipoprotein (LDL) in Non-Insulin-dependent Diabetes Mellitus (NIDDM)

Impact of Hypertriglyceridemia

T. TEMELKOVA-KURKTSCHIEV,
M. HANEFELD, AND W. LEONHARDT

Institute and Outpatient Clinic for Clinical Metabolic Research
Faculty of Medicine
Technical University of Dresden
D-01307 Dresden, Germany

INTRODUCTION

An excess mortality rate, particularly due to cardiovascular complications, has been well established in non-insulin-dependent diabetes mellitus (NIDDM).[1] Macrovascular diseases are the most important overall reason of morbidity and mortality in NIDDM patients and these are substantially more common among diabetics than in the nondiabetic population.[1,2] Similar to nondiabetics, hypercholesterolemia plays a significant role in atherogenesis in NIDDM as well.[3] However, NIDDM patients display a threefold-higher CHD death rate in comparison to nondiabetics for the same cholesterol levels.[3] Therefore, other factors associated with diabetes must be involved. Along with hyperglycemia itself,[4,5] a cluster of risk factors, such as dyslipidemia,[6] hypertension,[7,8] obesity,[8] insulin resistance,[9] hyperinsulinemia,[10] and hemostatic disturbances, was suggested to be responsible for the increased prevalence of macrovascular complications in NIDDM.

Contrary to nondiabetics, hypertriglyceridemia (HTG) in NIDDM was found to be an independent risk factor for atherosclerosis.[11] Furthermore, the prevalence of HTG in NIDDM is two to three times higher than in nondiabetics.[12] Although there is some evidence for the direct role of triglyceride-rich lipoproteins in the pathogenesis of atherosclerosis, HTG may also promote risk indirectly.[13] Thus, HTG was shown to be associated with alterations of lipoprotein size[14] and composition,[15] such as triglyceride enrichment of LDL and HDL, which were found to predict strongly increased cardiovascular mortality in NIDDM.[16] It was also demonstrated that plasma triglyceride levels correlate inversely with LDL size.[17] Hypertriglyceridemic subjects are characterized by the preponderance of the small dense LDL, which were recently shown to be particularly atherogenic.[14,18-21] Numerous recent studies demonstrate convincingly their apparent association with the occurrence of CHD.[18-22] The increased atherogenicity of small dense LDL seems to be due to a reduced binding of these particles to the LDL receptor,[23] an increased susceptibility to oxidation,[24,25] and an increased ability to penetrate into the arterial wall.[26] Recent observations showed that the small dense LDL are an integral feature of the insulin resistance syndrome.[27,28] The LDL phenotype B was found to be independently associated with each aspect of this syndrome.[27] LDL size was shown to be significantly related to specific insulin, proinsulin, and the fasting proinsulin/insulin ratio in nondiabetic subjects.[29] A twofold- to threefold-higher frequency of LDL pattern B was found in normolipemic NIDDM patients.[30,31] Thus, small dense LDL are associated with both HTG and NIDDM. Therefore, it would be of interest to examine the LDL subfraction pattern in NIDDM patients with respect to different triglyceride levels.

The aim of our study was to evaluate the relationship of mild, moderate, and severe hypertriglyceridemia and LDL subfraction distribution and composition in NIDDM patients. A comparison was also made with normolipemic diabetic and nondiabetic subjects, and correlations were assessed between small dense LDL and triglycerides, glycemic control, insulin, BMI, and other metabolic parameters.

MATERIALS AND METHODS

Forty NIDDM patients with HbA1c levels between 6.5% and 10.5%, stable body weight, and duration of diabetes for more than 2 years were compared to 20 healthy controls, matched for sex and age. The diabetic patients either were treated with a diet only ($n = 8$) or were receiving oral antidiabetic drugs, such as glibenclamide ($n = 20$), acarbose ($n = 7$), and metformin ($n = 5$). Insulin treatment was an exclusion criterion. According to their plasma triglyceride levels, the diabetic patients were divided into four groups: (1) normotriglyceridemic ($n = 12$), (2) with mild HTG (1.7 < TG < 2.3 mmol/L) ($n = 11$), (3) with moderate HTG (2.3 < TG < 4.6 mmol/L) ($n = 11$), and (4) with severe HTG (TG > 4.6 mmol/L) ($n = 6$). As shown in TABLE 1, the different NIDDM groups were well matched for glycemic control, as assessed by fasting plasma glucose and hemoglobin A1c. The healthy controls and normotriglyceridemic diabetics did not differ significantly for BMI and plasma lipids. The controls ($n = 20$) were examined for blood count, liver and kidney function, and metabolic parameters (lipids, HbA1c, blood glucose) and were included only if the results were within the normal range. The characteristics of the examined subjects are presented in TABLE 1.

LDL subclasses (large buoyant LDL1, density 1.025 to 1.034 g/mL; middle LDL2, density 1.034 to 1.044 g/mL; and dense LDL3, density 1.044 to 1.060 g/mL) were isolated directly from plasma by nonequilibrium density-gradient ultracentrifugation as described by Griffin *et al.*[32] Fresh plasma was overlayed with a six-step curvilinear salt gradient in the density (d) range from 1.060 to 1.019 kg/L in Ultra-clear tubes using a peristaltic pump. Ultracentrifugation was carried out at 40,000 rpm for 24 hours in a Beckman centrifuge (Model L8-80M) using an SW 40 rotor, after which the LDL subfractions were eluted by upward displacement and detected by continuous monitoring of absorbance at 280 nm with a UV spectrophotometer (DU 640, Beckman Instruments). The LDL subfraction pattern comprises overlapping populations of par-

TABLE 1. Clinical and Biochemical Characteristics of Subjects

Variable[a]	Controls (n = 20)	NTG-NIDDM (n = 12)	Mild HTG (n = 11)	Moderate HTG (n = 11)	Severe HTG (n = 6)
Age (years)	47.5 ± 11.5	58.5 ± 7.6	58.9 ± 6.8	48.8 ± 9.3	53.7 ± 8.5
Sex (M/F)	10/10	7/5	5/6	7/4	5/1
BMI (kg/m^2)	24.6 ± 2.2	26.8 ± 2.1	27.0 ± 2.2	28.6 ± 2.2	28.1 ± 1.9
FPG (mM)	4.58 ± 0.5	11.3 ± 2.9^b	11.8 ± 2.5^b	11.5 ± 2.6^b	11.7 ± 2.3^b
HbA1c (%)	5.04 ± 0.5	8.36 ± 1.4^b	8.42 ± 1.2^b	8.36 ± 1.3^b	8.12 ± 1.3^b
TG (mM)	1.06 ± 0.5	1.11 ± 0.3	1.92 ± 0.3^{b,c}	2.92 ± 0.5^{b,c}	7.14 ± 2.2^{b,c}
CH (mM)	5.24 ± 0.7	5.22 ± 1.1	5.96 ± 0.8^{b,c}	6.48 ± 0.8^{b,c}	7.98 ± 1.0^{b,c}
HDL-C (mM)	1.33 ± 0.3	1.13 ± 0.3	0.96 ± 0.2^b	0.94 ± 0.1^b	0.80 ± 0.1^{b,c}

[a] Terms: BMI = body mass index, FPG = fasting plasma glucose, TG = triglycerides, CH = cholesterol, HDL-C = high-density lipoprotein cholesterol.
[b] $p < 0.05$ versus healthy controls.
[c] $p < 0.05$ versus normotriglyceridemic NIDDM.

ticles from which three distinct subfractions could be resolved, corresponding in size and density to LDL1, LDL2, and LDL3 as originally defined by Krauss and Blanche.[33] The individual subfraction areas were quantified (Peakfit program, Jandel) and expressed as a percentage of the total LDL mass.

Subfractionation into large, triglyceride-rich, very-low-density lipoproteins or VLDL1 (Sf 60–400), small VLDL or VLDL2 (Sf 20–60), IDL (Sf 12–20), and LDL (Sf 0–12) was carried out by a modification of the cumulative flotation ultracentrifugation described by Lindgren et al.[34] Fresh plasma was overlayed with a six-step discontinuous salt gradient ranging from $d = 1.0988$ to 1.0588 kg/L in an Ultra-clear tube using a peristaltic pump. Four consecutive runs were carried out in a Beckman ultracentrifuge (Model L8-80M), after which each subfraction was isolated by aspiration from the top using a Pasteur pipette: (1) at 39,000 rpm for 1 h, 38 min for VLDL1, (2) at 18,500 rpm for 15 h, 41 min for VLDL2, (3) at 39,000 rpm for 2 h, 35 min for IDL, and (4) at 30,000 rpm for 21 h, 10 min for LDL. Compositional analysis was performed for each subfraction by established methods: the protein content was determined by the modified method of Lowry after Markwell et al.;[35] total and free cholesterol, triglycerides, and phospholipids were measured enzymatically on a Ciba Corning Express 500 analyzer, using commercially available test kits (Boehringer Mannheim, Germany). Esterified cholesterol was calculated by the following difference: esterified cholesterol = (total cholesterol − free cholesterol) × 1.68.

HDL cholesterol was examined by precipitation with polyethylene glycol (Quantolip HDL, Immuno AG, Heidelberg, Germany). Plasma glucose was measured by the glucose oxidase method. Hemoglobin A1c was examined by high-performance liquid chromatography on a Diamat analyzer (Bio Rad Laboratories, Germany). Insulin and C-peptide were determined by RIA (Webster Texas kits).

Statistical analysis of data was conducted with the SPSS/PC+ program. The Mann-Whitney U test was used for group comparisons. Correlations between relevant parameters were assessed using the Spearman correlation coefficient.

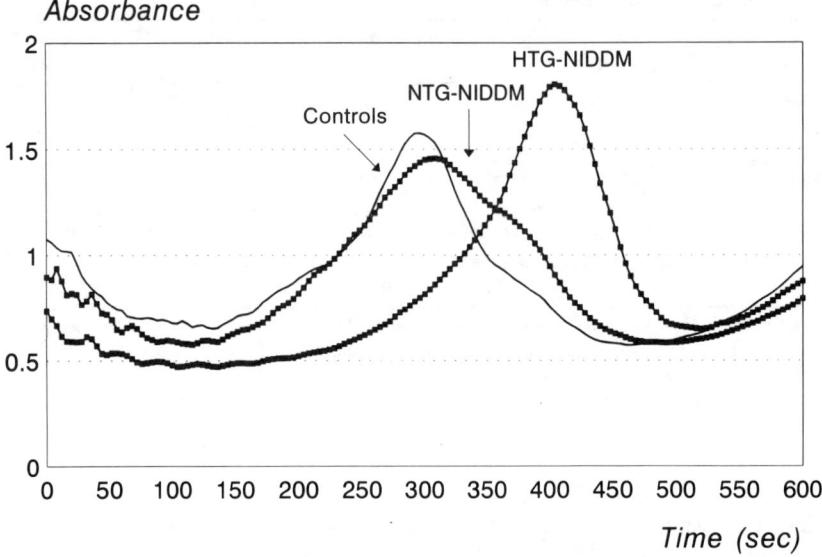

FIGURE 1. LDL subclass pattern in NIDDM.

RESULTS

The healthy controls and the normotriglyceridemic diabetics showed a predominance of the middle dense LDL2 subclass (FIG. 1). The NIDDM patients with HTG, whether mild, moderate, or severe, displayed a similar LDL pattern, characterized by a significant shift in the distribution of apolipoprotein B along the LDL density range towards the higher hydrated density region, with a prevalence of the dense LDL3 subfraction (LDL B pattern) (FIGS. 1 and 2). The percentage of the LDL3 subfraction from the total LDL mass was somewhat higher in the normotriglyceridemic diabetics than in the controls (FIG. 2). The effect of HTG on the LDL subfraction distribution was substantial: the higher the plasma triglycerides, the higher the percentage of LDL3 and the lower the percentage of LDL2 and LDL1 within the total LDL mass.

The percentage of individuals with the LDL B pattern was almost six times higher among the diabetics in comparison with the control group (FIG. 3). It was found that the number of individuals with the LDL B pattern among the normolipemic diabetics was only slightly higher than among the controls. The prevalence of the atherogenic LDL B pattern was strongly positively related to triglyceride levels (FIG. 3).

As shown in FIG. 4, the normotriglyceridemic diabetics do not differ from the controls for subfraction distribution. HTG is mostly associated with an increase of the large, triglyceride-rich, VLDL1 subfraction. In addition, significant compositional alterations of lipoprotein subfractions were established even in the normotriglyceridemic NIDDM patients, as seen by triglyceride enrichment of all subfractions studied,

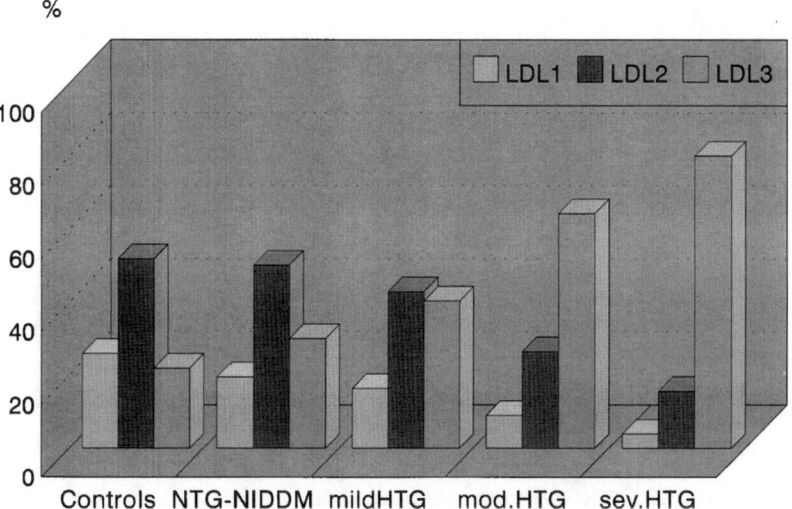

FIGURE 2. LDL subfraction pattern in NIDDM.

protein depletion of VLDL and IDL, and free cholesterol enrichment of VLDL1 and VLDL2 (TABLE 2). These anomalies were sustained by a concomitant HTG (TABLE 3).

No correlations were found between glycemic control, as assessed by hemoglobin A1c, or BMI and any of the metabolic parameters studied. There was a significant positive correlation between plasma triglycerides and the concentration of VLDL1

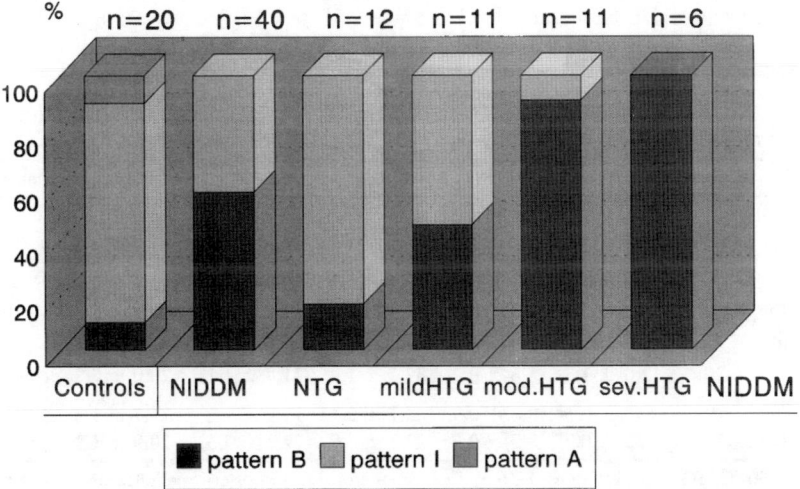

FIGURE 3. Prevalence of LDL B pattern among NIDDM patients with different TG plasma levels.

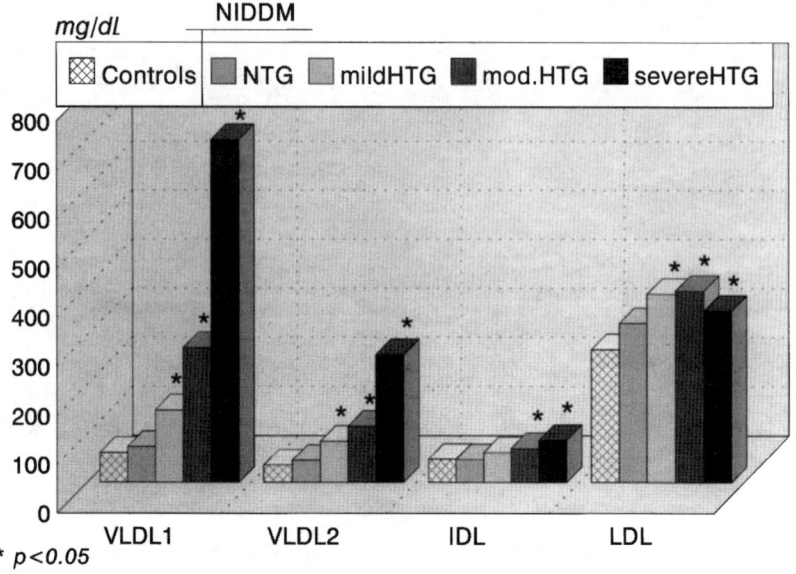

FIGURE 4. Concentration of apo B–containing subfractions in NIDDM patients in relation to triglyceride levels.

($r = 0.99$, $p < 0.001$), VLDL2 ($r = 0.71$, $p < 0.001$), IRI ($r = 0.55$, $p < 0.01$), and C-peptide ($r = 0.53$, $p < 0.01$). The LDL3 subfraction was significantly positively correlated with plasma triglycerides ($r = 0.75$, $p < 0.001$), VLDL1 concentration ($r = 0.86$, $p < 0.001$), and serum insulin levels ($r = 0.55$, $p < 0.01$) and inversely correlated with HDL cholesterol levels ($r = -0.51$, $p < 0.01$).

TABLE 2. Composition of Apo B–containing Lipoprotein Subfractions in Normolipemic NIDDM[a]

	TG	FCH	ECH	PR	PL	Subjects
VLDL1	62.6 ± 2.9[b]	5.0 ± 1.3[b]	9.33 ± 2.5	9.0 ± 1.40[b]	14.0 ± 2.0	NIDDM
(Sf 60–400)	59.2 ± 1.9	2.9 ± 1.0	11.8 ± 1.7	12.1 ± 1.7	14.0 ± 1.6	controls
VLDL2	38.0 ± 3.9[b]	7.7 ± 2.0[b]	19.0 ± 6.9	15.5 ± 2.4[b]	19.8 ± 5.4	NIDDM
(Sf 20–60)	35.7 ± 1.7	6.4 ± 0.7	22.1 ± 1.3	16.4 ± 2.2	19.4 ± 1.2	controls
IDL	14.9 ± 3.5[b]	8.9 ± 0.8	35.2 ± 3.6	20.2 ± 1.1[b]	20.8 ± 1.5	NIDDM
(Sf 12–20)	12.0 ± 1.3	8.6 ± 0.7	37.7 ± 1.9	21.8 ± 1.1	20.0 ± 0.9	controls
LDL	4.46 ± 0.5[b]	8.4 ± 0.9	42.7 ± 2.4	22.8 ± 2.2	21.7 ± 1.8	NIDDM
(Sf 0–12)	3.52 ± 0.7	8.9 ± 0.9	42.3 ± 3.0	24.3 ± 3.3	20.9 ± 1.2	controls

[a] Terms: TG = triglycerides, FCH = free cholesterol, ECH = esterified cholesterol, PR = protein, PL = phospholipids.
[b] $p < 0.05$.

TABLE 3. Triglyceride Percentage in Apo B-containing Lipoprotein Subfractions in NIDDM

Subjects[a]	VLDL1	VLDL2	IDL	LDL
HTG-NIDDM	63.8 ± 2.5	41.4 ± 5.2[b]	17.3 ± 3.0[b]	6.06 ± 1.6[b]
ATG-NIDDM	64.1 ± 3.6	37.4 ± 3.1	14.3 ± 2.5	5.07 ± 1.7
NTG-NIDDM	62.6 ± 2.9	38.0 ± 3.9	14.9 ± 3.5	4.46 ± 0.5

[a] Terms: NTG = normotriglyceridemic, ATG = with acceptable triglycerides (<2.3 mmol/L), HTG = hypertriglyceridemic.
[b] $p < 0.05$ versus NTG-NIDDM.

DISCUSSION

Our patients are representative of moderately obese NIDDM. Their quality of metabolic control with diet alone or with oral antidiabetics was acceptable and did not differ between the categories of triglyceride level. However, even in the normolipemic NIDDM subjects, slight alterations of the lipoprotein subfraction distribution and composition could be observed. This confirms other published observations indicating that diabetics with acceptable control and normal triglycerides exhibit atherogenic abnormalities in their lipoprotein profile,[31] which could contribute to the excessive mortality and prevalence of CHD among diabetics. Furthermore, we observed a close relationship between triglyceride level and changes in the concentration of VLDL and IDL subfractions, which was particularly impressive for VLDL1 and VLDL2. Interestingly, the LDL concentration reached its maximum already in mild HTG, with a decrease in severe HTG, as is well known from the chylomicronemia syndrome. A detailed analysis of the LDL subfraction pattern, however, reveals that with increasing TG levels there is a shift within the LDL towards LDL3, that is, to the small dense LDL that are known to be particularly atherogenic, as already worked out in the INTRODUCTION. The preponderance of the small dense LDL even in the diabetic patients with a mild TG elevation indicates the possibility for multiplying the effect of both factors—insulin resistance and hypertriglyceridemia. As previously reported, small dense LDL were significantly correlated to VLDL1[14] and to serum insulin.[29] Furthermore, we found an inverse correlation between LDL3 and HDL cholesterol levels, which is a classical feature of the insulin resistance syndrome. This indicates that plasma triglycerides, VLDL1, small dense LDL, and hyperinsulinemia are linked in the metabolic syndrome, or syndrome X. Thus, our findings suggest a central role of insulin resistance/hyperinsulinemia together with HTG for the pathophysiology of atherogenic lipoprotein anomalies in NIDDM.

REFERENCES

1. DE GRAUW, W. J., E. H. VAN DE LISDONK, H. J. VAN DEN HOOGEN & C. VAN WEEL. 1995. Diabetes Med. **12**(2): 117–122.
2. MANSON, J. E., G. A. COLDITZ, M. J. STAMLER et al. 1991. Arch. Intern. Med. **151**: 1141–1147.

3. STAMLER, J., O. VACCARO, J. NEATON & D. WENTWORTH (MULTIPLE RISK FACTOR INTERVENTION TRIAL RESEARCH GROUP). 1993. Diabetes Care **16:** 434–449.
4. BROWNLEE, M. 1992. Diabetes Care **15:** 1835–1843.
5. MYKKANEN, L., M. LAAKSO & K. PYORALA. 1992. Diabetes Care **15(8):** 1020–1030.
6. LAAKSO, M., S. LEHTO, I. PENTTILA & K. PYORALA. 1993. Circulation **88(4):** 1421–1430.
7. HYPERTENSION IN DIABETES STUDY (HDS). 1993. J. Hypertens. **11(3):** 319–325.
8. KAPLAN, N. M. 1989. Arch. Intern. Med. **149:** 1514–1520.
9. REAVEN, G. M. 1988. Diabetes **37:** 1595–1607.
10. FERRANNINI, E., S. M. HAFFNER, B. D. MITCHELL & M. P. STERN. 1991. Diabetologia **34:** 416–422.
11. FONTBONNE, A., E. ESCHWEGE, F. CAMBIEN, J. L. RICHARD, P. DUCIMETIERE, N. THIBULT, J. M. WARNET, J. R. CLAUDE & G. E. ROSSELIN. 1989. Diabetologia **32:** 300–304.
12. FISCHER, S., M. HANEFELD, U. SCHWANEBECK, J. SCHULZE & U. JULIUS. 1995. Diabetes Stoffwechsel. **4:** 349–356.
13. SHEPHERD, J. & C. J. PACKARD. 1991. Metabolic consequences of hypertriglyceridemia. *In* Atherosclerosis Reviews. Vol. 22, p. 1–8. Raven Press. New York.
14. GRIFFIN, B. A., D. J. FREEMAN, G. W. TAIT, J. THOMSON, M. J. CASLAKE, C. J. PACKARD & J. SHEPHERD. 1994. Atherosclerosis **106(2):** 241–253.
15. DECKELBAUM, R. J., E. GRANOT, Y. OSCHRY, L. ROSE & S. EISENBERG. 1984. Arteriosclerosis **4:** 225–231.
16. UUSITUPA, M. I. J., L. K. NISKANEN, O. SIITONEN, E. VOUTILAINEN & K. PYORALA. 1993. Diabetologia **36:** 1175–1184.
17. WILLIAMS, P. T., K. M. VRANIZEN & R. M. KRAUSS. 1992. J. Lipid Res. **33:** 765–774.
18. ROHEIM, P. S. & B. F. ASZTALOS. 1995. Clin. Chem. **41:** 147–152.
19. CORESH, J., P. O. KWITEROVICH, H. H. SMITH & P. S. BACHORIK. 1993. J. Lipid Res. **34(10):** 1687–1697.
20. KRAUSS, R. M. 1995. Am. J. Cardiol. **75:** 53–57.
21. RAJMAN, I., S. MAXWELL, R. CRAMB & M. KENDALL. 1994. Q. J. Med. **87:** 709–720.
22. AUSTIN, M. A. 1994. Int. J. Clin. Lab. Res. **24:** 187–192.
23. DE GRAF, J., J. C. HENDRIKS, D. W. SWINKELS, P. N. DEMACKER & A. F. STALENHOEF. 1993. Artery **20(4):** 201–230.
24. SEVANIAN, A., J. HWANG, H. HODIS, G. GAZZOLATO, P. AVOGARO & G. BITTOLOBON. 1996. Arterioscler. Thromb. Vasc. Biol. **16:** 784–793.
25. CHAIT, A., R. L. BRAZG, D. L. TRIBBLE & R. M. KRAUSS. 1993. Am. J. Med. **94(4):** 350–356.
26. BJORNHEDEN, T., A. BABYI, G. BONDJERS & O. WIKLUND. 1996. Atherosclerosis **123:** 43–56.
27. SELBY, J. V., M. A. AUSTIN, B. NEWMAN, D. ZHANG, C. P. QUESENBERY, E. J. MAYER & R. M. KRAUSS. 1993. Circulation **88:** 381–387.
28. AUSTIN, M. A. & K. L. EDWARDS. 1996. Curr. Opin. Lipidol. **7:** 167–171.
29. HAFFNER, S. M., L. MYKKANEN, D. ROBBINS, R. VALDEZ, H. MIETTINEN, B. V. HOWARD, M. P. STERN & R. BOWSHER. 1995. Diabetologia **38:** 1328–1336.
30. ABATE, N., G. L. VEGA, A. GARG & S. M. GRUNDY. 1995. Atherosclerosis **118:** 111–122.
31. FEINGOLD, K. R., C. GRUNFELD, M. PANG, W. DOERRLER & R. M. KRAUSS. 1992. Arterioscler. Thromb. **12:** 1496–1502.
32. GRIFFIN, B. A., M. J. CASLAKE, B. YIP, G. W. TAIT, C. J. PACKARD & J. SHEPHERD. 1990. Atherosclerosis **83:** 59–67.
33. KRAUSS, R. M. & P. J. BLANCHE. 1992. Curr. Opin. Lipidol. **3:** 377–383.
34. LINDGREN, F. G., L. C. JENSEN & F. T. HATCH. 1972. The isolation and quantitative analysis of serum lipoproteins. *In* Blood Lipids and Lipoproteins: Quantitation, Composition, and Metabolism, p. 181–274. Wiley. New York.
35. MARKWELL, M., S. M. HAAS, L. L. BIEBER & N. E. TOLBERT. 1978. Anal. Biochem. **87:** 206–210.

Does Dietary Fat Influence Insulin Action?[a]

L. H. STORLIEN,[b] A. D. KRIKETOS,[b,c]
A. B. JENKINS,[b] L. A. BAUR,[c] D. A. PAN,[b,c]
L. C. TAPSELL,[b] AND G. D. CALVERT[b,d]

[b] Metabolic Research Center
Department of Biomedical Science
University of Wollongong
Wollongong NSW 2522, Australia

[c] Departments of Endocrinology and of Pediatrics and Child Health
University of Sydney
Sydney NSW 2006, Australia

[d] Medical Research Unit
Illawarra Regional Hospital
and
University of Wollongong
Wollongong NSW 2522, Australia

BACKGROUND

Defects in insulin action ("insulin resistance") and associated hyperinsulinemia are strongly linked to the development of a cluster of prevalent diseases including non-insulin-dependent diabetes mellitus (NIDDM), obesity, hyperlipidemias, hypertension, and heart disease—the "Syndromes of Insulin Resistance" or the "Metabolic Syndrome." A common opening in analysis of NIDDM is that "dietary intervention is the keystone to therapy for NIDDM." Despite this and the plethora of dietary guidelines for NIDDM, there are surprisingly little hard data on the role of diet in modulating insulin action.

Early work from Himsworth,[1,2] results from short-term studies employing liquid formula diets providing a remarkably high percentage (75–85%) of calories from carbohydrate,[3-5] and the driving force from dietary guidelines for cardiovascular disease have conspired over the past 30 years to create guidelines for NIDDM that emphasize very low fat intake and very high carbohydrate intake. This is despite the fact that the characteristic problem of diabetes is an inability to appropriately metabolize carbohydrates.

---

[a] This work was supported by the National Health and Medical Research Council (Australia), the Diabetes Australia Research Trust, the Egg Industry Research and Development Council, the Hoechst Diabetes Research Fund, the Meadow Lea Foods and Grains Research and Development Corporation, and the Rebecca L. Cooper Medical Research Foundation.

With the relatively recent development of techniques to assess insulin action *in vivo* and their adaptation to experimental animals, the true relationship between dietary fat intake and insulin action is now becoming clearer. For so many aspects of the overall problem area, dietary fat subtypes have been revealed as the critical elements and we are now coming to a real understanding of the "Oils Ain't Oils" message. Thus, this short review will focus particularly on fat subtypes and how they might differentially influence insulin action.

DIETARY FAT AND INSULIN ACTION IN EXPERIMENTAL ANIMALS

The work in experimental animals has largely been limited to rodents. Early studies were confined to *in vitro* investigations in adipocytes and isolated skeletal muscles.[6-8] The results were consistent. Ad libitum feeding of high-fat diets led to impairment of insulin-stimulated glucose uptake. There was very little information about mechanisms, but observations about this time showing that high-fat feeding resulted in increased accumulation of stored triglyceride in muscle[9,10] were certainly consistent with the reciprocal relationship between lipid and carbohydrate metabolism first highlighted by Randle and coworkers.[11,12] Impairments at the transport step might also include downregulation of GLUT-4,[13,14] although that may be only with an extremely high percentage of dietary fat[15] or where major changes in glucose transporter numbers can be achieved. Thus, recent data from transgenic mice demonstrate that increased GLUT-4 levels can prevent the impairment in glycemic control induced by a high safflower fat oil diet.[16] Possible intracellular defects include reduced insulin activation of tyrosine kinase[17] and/or reduced proportion of glycogen synthase in the I (or active) form in skeletal muscle.[18]

A major advance was made when the hyperinsulinemic, euglycemic clamp technique was developed in unanesthetized, chronically cannulated rats and successfully combined with 2-deoxyglucose to allow assessment of diet effects on whole-body insulin action and in individual tissues.[19,20] The early studies showed that high-fat diets fed even equicalorically over short periods of time led to profound whole-body, and tissue-specific, insulin resistance.[21,22] The insulin resistance was widespread among tissues including the liver and skeletal muscle, the latter being the most important tissue for insulin-stimulated glucose uptake. Similarly, when insulin-resistant suckling rats were weaned onto high-carbohydrate diets they became insulin-sensitive, but remained insulin-resistant when weaned onto a high-fat diet.[23]

As noted above, a leading possibility for the mechanism underlying high-fat feeding insulin resistance is lipid oversupply and its known and postulated effects on glucose metabolism. However, lipid subclasses have very different effects on lipogenesis and both fasting and prandial lipemic excursions.[24] In particular, the n-3 (or omega-3) fats suppress hepatic lipogenesis and reduce circulating triglyceride levels.[24-26] Interestingly, when n-3 fatty acids were introduced into high-fat diets in rats, insulin resistance could be virtually eliminated.[27,28]

Two aspects of these beneficial effects were important. First, both the storage and nonstorage components of insulin-stimulated glucose metabolism in skeletal muscle could be assessed by combining the hyperinsulinemic, euglycemic clamp with tracer doses of both labeled 2-deoxyglucose to track total glucose uptake and labeled glucose to assess glycogen storage. The results showed that there was an improvement in both oxidative and storage components of insulin-stimulated glucose disposal in skeletal muscle following equicaloric substitution of n-3 fatty acids into a high-fat diet.[27] This observation suggested either multiple defects or, more parsimoniously, a single defect at an early common point in the glucose metabolic pathway, that is, the plasma membrane level. Second, and congruent with this possibility, was the finding that insulin action was positively linked with the incorporation of highly unsaturated fatty acids into muscle phospholipid, the major structural lipid component of membranes. The role of membrane lipid fatty acid composition is potentially important and is discussed at greater length in the section below entitled MUSCLE INSULIN RESISTANCE AND MEMBRANE LIPID SUBTYPES.

Finally, in the rodent work, there is a consistent theme across a range of our studies indicating that manipulations that result in increases in lipid availability, either circulating or stored in muscle, result in impaired insulin action.[29] Conversely, dietary or pharmaceutical interventions that decrease lipid availability improve insulin action. In the case of dietary fats it was, of course, the known effects of n-3 fatty acids on suppressing hepatic lipogenesis that prompted our original studies and the results are certainly consistent with improvement in insulin action by reduction in lipid availability. In this regard, a fascinating new area of metabolic control must be mentioned. It is now clear from the pioneering work of Clarke and coworkers that polyunsaturated fatty acids (PUFAs) in particular can act as potent regulators of gene expression of enzymes such as fatty acid synthase[30,31] (see also Clarke et al. in this volume). Control of endogenous lipogenesis obviously will modulate lipid supply, but, equally, endogenous lipogenesis is a source of saturated fatty acids. As well as their much publicized role in elevating cholesterol levels, saturated fatty acids are poorly oxidized for energy compared to unsaturated fatty acids[32,33] and their increased presence in muscle membrane phospholipids is associated with insulin resistance in both rodents and humans.[28,34–36] This area of gene regulation by particular fatty acid subclasses offers an intriguing new way to view lipids and metabolism.

In summary, the work from rodents demonstrates that high-fat diets can induce insulin resistance, but the type of lipid appears more important than the amount. The mechanisms underlying these observations are still not entirely clear. However, alterations in the physical nature of membranes, modulation of lipid supply, and control of gene expression offer productive areas of investigation.

DIETARY FAT AND INSULIN ACTION IN HUMANS

It is clearly recognized that insulin resistance is a major metabolic abnormality in the etiology of NIDDM, even if actual disease expression eventually requires a relative failure of the pancreatic beta-cells. Dietary guidelines have emphasized reduction in fat intake with the implicit assumption of an improvement in insulin action either directly or indirectly following weight loss. Surprisingly, the evidence for these assumptions is not strong. While there is a literature on diet and glucose tolerance, glucose tolerance is a complex interaction between insulin action and insulin secretion. In the present context, it is likely that dietary fat subtypes, via their incorporation into beta-cell membranes and/or modulation of the production of eicosanoids and other PUFA metabolic products, are potent modulators of insulin secretion.[37-39] Thus, glucose tolerance will reflect effects on both insulin secretion and action that are unable to be separated.

There are relatively few studies in humans where dietary intake has been manipulated and where insulin action has been measured by currently accepted techniques. Even among the studies that could be identified where dietary lipid amount[40-46] and type[47-51] had been altered, there is little consensus about outcome. Some of the positive results were in studies with extraordinarily high levels of carbohydrate (69-75% of calories[40,45]), while with more realistic levels the results were generally negative. An important aspect of all of these studies, however, is that they were of short duration (i.e., at most a few weeks). We have now completed a one-year intervention where a fat-modified diet that emphasized the monounsaturated fat, oleic, and n-3 fatty acids was compared to an attempt at the currently recommended high-carbohydrate diet.[52] The fat modification was readily achieved, but there were only modest increases in carbohydrate intake on the recommended diet. The effects on insulin action were assessed by the hyperinsulinemic, euglycemic clamp technique. Overall, there was an impressive improvement in insulin action (and a number of lipid variables) on the fat-modified diet in contrast to a general deterioration in metabolic control on the control diet.[52]

Another approach to intervention studies is to examine the evidence from epidemiological (used in the broad sense of cross-sectional, prospective, and retrospective population) studies linking dietary measures with clinical outcomes. The problem here is that there are again relatively few studies that have measured insulin action directly. However, fasting insulin levels in the normoglycemic population are a reasonable index of insulin action and can be usefully employed. A second difficulty in these studies is the major inaccuracies involved in self-reported food intake. This problem has been truly exposed by the use of the doubly labeled water technique for measuring energy expenditure.[53,54] Again, though, these inaccuracies may be more related to the quantitative assessment of overall intake than to the relative intakes of the major macronutrients and their subclasses. However, even in the face of this methodological adversity, a reasonable pattern of outcomes can be discerned.

We could find only a single published cross-sectional study[55] where insulin action was directly assessed. A food-frequency questionnaire was used to measure habitual intake in a group of glucose-tolerant individuals with a range of body weights. The frequently sampled intravenous glucose tolerance test (FSIGT) was used to calculate an insulin sensitivity index (S_i). A significant, negative relationship was found between log S_i and the percentage of energy consumed as fat. However, when S_i was adjusted for body mass index (BMI), the percent of energy consumed as fat was no longer a significant, independent predictor of insulin sensitivity, indicating the possibility that the effect of dietary fat was mediated via changes in whole-body adiposity.

Other studies investigating dietary variables have relied on fasting insulin or C-peptide to index insulin action. Here, there is some consistency. Both high total fat and saturated fat intake have been positively related to fasting insulin or C-peptide levels in a number of different populations.[56-58] However, there were a number of other interesting aspects to the results of these studies. In contrast to the consistent deleterious effects of saturated fats, Feskens and coworkers found a beneficial effect of polyunsaturated fats.[58] Second, the effects of dietary fats in the studies of Mayer et al.[57] were more pronounced in sedentary individuals, whereas increased levels of physical activity attenuated the relationship between dietary fat intake and fasting insulin. This result is in line with the more recent analysis by Flatt[59] that accords both dietary fat and physical activity preeminent place in the energy balance equation; moreover, it is a recognition that, while percent intake as fat has been steady or even falling in most Western societies over the past decade or two, obesity and diabetes continue to rise dramatically.

Since it is not unreasonable to assume that insulin resistance plays a major role in progress to glucose intolerance and NIDDM, then it is also instructive to look at outcome measures in light of habitual dietary intake. In the Zutphen Elderly Study, Feskens and coworkers initially reported that the percentage of energy consumed as saturated fat was positively related to the relative likelihood of developing glucose intolerance.[60] Similarly, results from the San Luis Valley study[61] showed that dietary fat intake as a percent of total energy was higher in glucose-intolerant and NIDDM individuals compared to BMI-matched controls. Further, when glucose-intolerant individuals from this population are followed for periods of 1 to 3 years, poor outcomes (remaining glucose-intolerant or progressing to NIDDM) are more likely to be seen in individuals with relatively high intakes of total, saturated, and monounsaturated fats (but not polyunsaturated fat), even when adjustment is made for relevant variables such as age, sex, and ethnicity.[62]

In an interesting variation, the fatty acid profile of serum cholesterol esters (reflecting dietary fatty acid profile over relatively long periods) was related to the risk of developing NIDDM at a 10-year follow-up.[63] Increased saturated fatty acids and decreased linoleic acid (n-6 PUFA) were found in the serum cholesterol esters of those who developed NIDDM, compared to those who had not. The authors suggest that the lower level of linoleic acid may reflect lower vegetable fat intake. In this regard, the study by Colditz et al.[64] also showed a modest, but significant, relationship between decreased vegetable fat intake (by dietary analysis) and development of NIDDM in a 6-year follow-up of over 80,000 women.

Finally, Feskens and coworkers[65] have now completed a 20-year follow-up of the Finnish and Dutch cohorts of the Seven Countries Study. High-fat, and particularly high-saturated-fat, intake was positively associated with subsequent development of glucose intolerance and NIDDM. However, fish intake (high in n-3 PUFAs) was protective.

In summary, while intervention studies have provided conflicting results, there is reasonably good consensus from epidemiological studies. High-fat, particularly saturated fat, intake is associated with indices of insulin resistance and a greater likelihood of developing glucose intolerance and NIDDM. PUFAs may be protective, although the evidence is not strong. Monounsaturated (oleic) fat intake has been shown to be beneficial in short-term trials when compared to a high-carbohydrate diet.[66,67] However, epidemiological studies show a deleterious effect of monounsaturated fat intake and both measures of insulin resistance[57,68] and diabetes outcome.[62] This may reflect the fact that these studies were done in areas where much of the monounsaturated fat comes in tandem with the saturated fats of meat and dairy products. This issue needs further work, particularly in regions such as the Mediterranean where saturated and monounsaturated fat intakes are not necessarily linked.

DIETARY FAT AND INSULIN RESISTANCE: IS THE RELATIONSHIP VIA OBESITY?

A number of the experimental animal and human studies also speak to the issue of whether the effect of fat intake on insulin action is mostly via its effects on increasing total body adiposity, a variable closely linked to insulin resistance.[56,69,70]

Certainly, there are a number of ways in which an increased level of dietary fats might lead to obesity: increased efficiency of storage;[71] a failure to provoke fat oxidation,[72] especially in some subgroups[73] (although see Surina et al.[74]); and/or a relatively deficient effect on satiating processes (see Blundell and Macdiarmid in this volume).

There is a very large experimental animal literature on dietary fat intake and development of obesity. While there is considerable genetic heterogeneity in proneness to obesity, overall the balance of results suggest that an increased percentage of dietary fat leads to an increase in weight gain and adiposity. However, the type of fat again seems important[75-77] (see reference 78 for a recent review) and the more unsaturated the fatty acid profile of the diet, the more resistant the animals are to obesity. This would be consistent with the relative ease with which the n-3 PUFAs and monounsaturated fats in particular are used for energy compared to saturated fats.[32]

Previous human studies would also seem to be roughly consistent with a positive relationship between dietary fat intake (as a percentage of macronutrients) and weight gain.[55,57,79-82] However, the percent of energy intake consumed as fat accounts for only a very small part of the variance in weight gain and a recent analysis of the NHANES I Epidemiologic Follow-up Study shows remarkably little correlation between baseline percent energy intake as fat and subsequent weight change in the next 8 to 10 years.[83] Of course, here we must contend with the problems noted earlier in obtaining accurate dietary intake data. As with the experimental animal data, again there is some evidence that the type of fat is important. In at least two studies, body fatness has been significantly related to the percentage of saturated and monounsaturated fat, but not to polyunsaturated fat.[79,81] Finally, increased adiposity itself may not be a single variable. Central adiposity appears to be the major culprit in a range of metabolic complications.[84] Recent work from our laboratory has shown, in a cross-sectional study on Pima Indians, that both muscle triglyceride accumulation and waist/thigh ratio, as a measure of central adiposity, independently relate to insulin action (Pan *et al.*, *Diabetes*, in press). Further, when those variables are taken into account, total adiposity no longer significantly relates to insulin action.

That brings us to the question of whether the insulin resistance associated with some, but not all, types of dietary fat intake can be explained solely by the dietary fat effect on adiposity. Here, the issue is not clear. In the cross-sectional study where insulin sensitivity was directly assessed,[55] the relationship between dietary fat intake and insulin sensitivity was no longer significant when the relationship was adjusted for BMI. This suggested that the relationship was mediated by a dietary fat–induced increase in adiposity. In contrast, there are now a number of studies where insulin action has been indexed indirectly (fasting insulin or C-peptide levels) and related to both adiposity and dietary fat intake. In these studies, dietary fat intake and in particular saturated fat intake were still related to insulin action even after adjustment for adiposity.[56,57,68,85] Importantly, in the study of Feskens and coworkers,[68] the percentage of polyunsaturated fat was inversely related to their measure of insulin action both before and after correction for adiposity. This again highlights the different effects of different dietary fats.

In summary, the balance of evidence suggests that an increased proportion of fat in the diet does lead to increased adiposity. However, this conclusion is only particularly true if that fat is saturated. PUFAs, particularly of the n-3 variety, may be neutral or even protective against weight gain. In addition, there is reasonable evidence demonstrating that total, and saturated, fat intake is associated with insulin resistance independent of any effects on adiposity. Finally, there is a great range of genetic susceptibility to dietary fat and there may be an interaction with a predisposition to high or low levels of physical activity.

MUSCLE INSULIN RESISTANCE AND MEMBRANE LIPID SUBTYPES

As discussed above, defects in both the storage and nonstorage components of insulin-stimulated glucose metabolism in skeletal muscle were observed following high-fat feeding.[27] This was suggestive of a defect early in the metabolism of glucose. Subsequent work showed a tight relationship between insulin action and elevation of the percentage of highly unsaturated phospholipids in skeletal muscle membranes.[28] Phospholipids are structural lipids essentially confined to membranes. They form a major lipid component of the membrane bilayer: from proportionally the largest component of plasma membrane to almost all the lipid in mitochondrial membrane.

The studies in rats were followed up by a number of studies in humans. First, our laboratory demonstrated a clear relationship between the percentage of more unsaturated lipids in the muscle structural membrane lipids and insulin action[34] in normal adult Australian, largely Caucasian, subjects: that is, the more unsaturated the fatty acids in muscle phospholipid, the better the insulin action; conversely, the more saturated the muscle membrane phospholipids, the more insulin resistant the individual. Similar relationships have now been shown in the diabetes- and obesity-prone Pima

Indians of Arizona in the United States[36] and in an adult population in Sweden.[35] Interestingly, while in the Australian and Pima subjects the relationships were positive and tightest between the polyunsaturated fatty acids and insulin action, in the Swedish study the highest correlations were negative and occurred with saturated fatty acids. The levels of n-3 fatty acids are particularly high in the Swedish population, probably reflecting a high dietary intake. In contrast, the level of n-3 fatty acids in the Pima Indian muscle membranes was remarkably low, something like one-fifth of the Swedish levels and less than half of the Australian values. While these observations presumably reflect, at least in part, a difference in the dietary intake of n-3 fatty acids, there is also very likely to be a genetic predisposition to incorporation of specific fatty acids into membranes. In this regard, it is interesting that an intestinal fatty acid binding protein on chromosome 4q has been linked with insulin resistance in the Pima Indian population.[86]

While much of the focus has been on n-3 PUFAs, the overall results and relationships between individual PUFAs and insulin action suggest that n-6 PUFAs are also beneficial. This is consistent with the apparent protective effect of linoleic acid against subsequent development of NIDDM.[63,64] However, when we plot the muscle membrane phospholipid n-6/n-3 ratio against fasting insulin (as a measure of insulin resistance) across the Australian and Pima Indian subjects that we have studied, the higher the n-6/n-3 ratio, the higher the fasting insulin and the higher the relative body weight.[87] This is consistent with work from Raheja et al.[88] showing a beneficial effect on glycemic control in NIDDM subjects as a result of reducing the dietary n-6/n-3 ratio. In the past 30–40 years, driven from the perspective of cardiovascular disease, there has been a major increase in n-6 PUFA intake and, equally, a major rise in the n-6/n-3 ratio. We now need well designed studies to explore the possible negative effect of this on the tendency to develop NIDDM and obesity.

Finally, there are dietary sources of the long-chain (20+ carbon backbone) highly unsaturated fatty acids that, when incorporated into membrane structural lipids, are associated with good insulin action. Seafood is perhaps the best known source, particularly of the n-3 PUFAs. However, the overall unsaturation index (UI, the number of double bonds per fatty acyl group times 100—a measure of overall lipid unsaturation) of the average "Western" diet is much less ($\approx$ 80) than the UI of skeletal muscle phospholipid, which averaged 160 to 170 in the studies described above.[34-36,89] This means that even the major dietary PUFAs, linoleic n-6 and α-linolenic n-3 fatty acids, must be both elongated and desaturated to be transformed into the fatty acids that our results have shown to be associated with insulin sensitivity. This is accomplished by the fatty acid desaturase and elongase enzymes. Activity of these enzymes can be roughly indexed from product/precursor ratios. Thus, the ratio of 20:4 n-6 to 20:3 n-6 gives a crude measure of the activity of the Δ5-desaturase enzyme. While not an optimal method for measuring enzyme activity, our results in humans[34,36] have shown good relationships between reduced Δ5-desaturase and elongase activities and both insulin resistance and obesity. Finally, new data in infants suggest that there is a range of endogenous desaturase and elongase activities across individuals and that this may play a role in early signs of glucose dysregulation.[90] In this regard, agents that influence the activity of desaturase enzymes may play a central role in development of disease. It is interesting here to note that it has been suggested that a high intake of *trans* fatty acids may lead to the development of obesity and insulin resistance.[91] *Trans* fatty acids have been shown to inhibit the activity of desaturase enzymes (see reference 92).

Insulin action must involve the insertion and movement of peptides (e.g., insulin receptors and glucose transporters) in the plasma membrane. Changing the characteristics of the membrane, by changing the phospholipid fatty acid composition, could then have potent effects on the ability of insulin to effect translocation/insertion of glucose transporters and/or their intrinsic activity[93] when inserted into the plasma membrane. Equally, changing the membrane lipid profile will change the "leakiness" of membranes to ions and thus the energy requirements to maintain ionic homeostasis.[94,95] These may provide mechanistic explanations for the observed correlational relationships between muscle membrane phospholipid, insulin action, and obesity. Those possibilities are the subject of current work.

INTERACTIONS BETWEEN DIETARY FAT, INSULIN ACTION, AND THE STRESS AXIS

Increased stress responsivity is often seen as a major player in the etiology of the Metabolic Syndrome and elevated cortisol is linked particularly to increased central adiposity (see reference 96 for example). In the Zucker fa/fa "fatty" genetic rodent model of obesity and insulin resistance, a pronounced increase in the cortisol response to restraint stress has been demonstrated.[97] In addition, an increased stress responsivity has been shown to be predictive of high weight gain on a high-fat/high-sucrose diet.[98] In turn, fat feeding itself results in an increased responsivity to physiological stressors.[99]

There is now evidence that the levels of storage triglyceride in skeletal muscle are inversely related to insulin action in rats and humans.[28,100] These results can then be coupled with the observation that, for a given stimulus, the rate of lipolysis in muscle is proportional to the amount in storage[101] and the demonstrated inhibition of glucose utilization by increased lipid availability.[12,102,103] Increased storage lipid in muscle might then result in inhibition of insulin-stimulated glucose metabolism via a heightened nervous system–generated lipolytic response to insulin[104] during the prandial period, countering insulin's normal role in inhibition of lipolysis. The factors that control the accumulation of storage triglyceride in muscle, and the amount and timing of their availability, are matters of importance for future research.

SUMMARY AND CONCLUSIONS

What is clear from the research thus far is that dietary fat intake does influence insulin action. However, whether the effect is good, bad, or indifferent is strongly related to the fatty acid profile of that dietary fat. The evidence has taken many forms, including *in vitro* evidence of differences in insulin binding and glucose transport in cells grown with different types of fat in the incubation medium,[105,106] *in vivo* results in animals fed different fats,[27,28] relationships demonstrated between the membrane structural lipid fatty acid profile and insulin resistance in humans,[34,35,89] and finally epidemiological evidence linking particularly high saturated fat intake[56,60–62,68,85] with hyperinsulinemia and increased risk of diabetes. This contrasts with the lack of relationship, or even possible protective effect, of polyunsaturated fats. In particular, habitual increased n-3 polyunsaturated dietary fat intake (as fish fats)[58,107,108] would appear to be protective against the development of glucose intolerance. It is reassuring that the patterns of dietary fatty acids that appear beneficial for insulin action and energy balance are also the patterns that would seem appropriate in the fight against thrombosis and cardiovascular disease.[109] Mechanisms, though, still need to be defined. However, there are strong indicators that defining the ways in which changes in the fatty acid profile of membrane structural lipids are achieved, and in turn influence relevant transport events, plus understanding the processes that control accumulation and availability of storage lipid in muscle may be fruitful avenues for future research.

One of the problems of moving the knowledge gained from research at the cellular level through to the individual and on to populations is the need for more accommodating research designs. *In vitro* studies may provide in-depth insights into intricate mechanisms, but they do not give the "big picture" for practical recommendations. On the other hand, correlational studies tend to be fairly blunt instruments, requiring large numbers that are very often not feasible if a greater depth of understanding of the biological processes is to be incorporated. There may be benefit in turning to the clinical case study as a framework for a more comprehensive analysis of the links between dietary fats and insulin action. The real challenge is to keep the depth of analysis rigorous enough to be able to explain and accommodate individual variation (i.e., the diversity of both environmental and genetic backgrounds) while at the same time satisfying the cultural need to provide appropriate overall dietary guidelines.

Finally, David Kritchevsky brought to our attention a delightful quote from Mark Twain: "There is something fascinating about science. One gets such a wholesale return of conjecture for such a trifling investment of fact." In the field of dietary fats and the Metabolic Syndrome, this quotation is, unfortunately, apt. Much more research is necessary to define how dietary fats really work to affect insulin action. Well designed, long-term studies in "free range" humans must be undertaken if dietary guidelines for the Metabolic Syndrome are to be based on anything more than a "trifling" amount of "fact."

REFERENCES

1. HIMSWORTH, H. P. 1935. The dietetic factor determining the glucose tolerance and sensitivity to insulin of healthy men. Clin. Sci. **2**: 67–94.
2. HIMSWORTH, H. P. & R. B. KERR. 1939. Insulin-sensitive and insulin-insensitive types of diabetes mellitus. Clin. Sci. **41**: 119–152.
3. BRUNZELL, J. D., R. L. LERNER, W. R. HAZZARD, D. PORTE & E. L. BIERMAN. 1971. Improved glucose tolerance with high carbohydrate feeding in mild diabetes. N. Engl. J. Med. **284**: 521–524.

4. ANDERSON, J. W., R. H. HERMAN & D. ZAKIM. 1973. Effect of high glucose and high sucrose diets on glucose tolerance of normal men. Am. J. Clin. Nutr. **26:** 600–607.
5. ANDERSON, J. W. 1977. Effect of carbohydrate restriction and high carbohydrate diets on men with chemical diabetes. Am. J. Clin. Nutr. **30:** 402–408.
6. LAVAU, M. & C. SUSINI. 1975. [U-14C]Glucose metabolism *in vivo* in rats rendered obese by a high fat diet. J. Lipid Res. **16:** 134–142.
7. SUSINI, C. & M. LAVAU. 1978. *In-vitro* and *in-vivo* responsiveness of muscle and adipose tissue to insulin in rats rendered obese by a high-fat diet. Diabetes **27:** 114–120.
8. GRUNDLEGER, M. L. & S. W. THENEN. 1982. Decreased insulin binding, glucose transport, and glucose metabolism in soleus muscle of rats fed a high fat diet. Diabetes **31:** 232–237.
9. BRINGOLF, M., N. ZARAGOZA, D. RIVIER & J. P. FELBER. 1972. Studies on the metabolic effects induced in the rat by a high-fat diet: inhibition of pyruvate metabolism in diaphragm *in vitro* and its relation to the oxidation of fatty acids. Eur. J. Biochem. **26:** 360–367.
10. SCHINDLER, C. & J. P. FELBER. 1986. Study on the effect of high fat diet on diaphragm and liver glycogen and glycerides in the rat. Horm. Metab. Res. **18:** 91–93.
11. RANDLE, P. J., P. B. GARLAND, C. N. HALES & E. A. NEWSHOLME. 1963. The glucose fatty-acid cycle, its role in insulin sensitivity, and the metabolic disturbances of diabetes mellitus. Lancet **1:** 785–789.
12. RANDLE, P. J., A. L. KERBEY & J. ESPINAL. 1988. Mechanisms decreasing glucose oxidation in diabetes and starvation: role of lipid fuels and hormones. Diabetes Metab. Rev. **4:** 623–638.
13. LETURQUE, A., C. POSTIC, P. FERRÉ & J. GIRARD. 1991. Nutritional regulation of glucose transporter in muscle and adipose tissue of weaned rats. Am. J. Physiol. **260:** E588–E593.
14. KAHN, B. B. & O. PEDERSEN. 1993. Suppression of GLUT4 expression in skeletal muscle of rats that are obese from high fat feeding, but not from high carbohydrate feeding or genetic obesity. Endocrinology **132:** 13–22.
15. WAKE, S., J. SOWDEN, L. STORLIEN, D. JAMES & P. CLARK. 1991. Effects of exercise training and dietary manipulation on insulin-regulatable glucose-transporter mRNA in rat muscle. Diabetes **40:** 275–279.
16. IKEMOTO, S., K. S. THOMPSON, M. TAKAHASHI, H. ITAKURA, M. D. LANE & O. EZAKI. 1995. High fat diet–induced hyperglycemia: prevention by low level expression of a glucose transporter (GLUT4) minigene in transgenic mice. Proc. Natl. Acad. Sci. U.S.A. **92:** 3096–3099.
17. NAGY, L. & G. GRUNBERGER. 1990. High-fat feeding induces tissue-specific alteration in a proportion of activated insulin receptors in rats. Acta Endocrinol. Copenh. **122:** 361–368.
18. HEDESKOV, C. J., K. CAPITO, H. ISLIN, S. E. HANSEN & P. THAMS. 1992. Long-term fat-feeding-induced insulin resistance in normal NMRI mice: postreceptor changes of liver, muscle, and adipose tissue metabolism resembling those of type 2 diabetes. Acta Diabetol. **29:** 14–19.
19. KRAEGEN, E. W., D. E. JAMES, A. B. JENKINS & D. J. CHISHOLM. 1985. Dose-response curves for *in vivo* insulin sensitivity in individual tissues in rats. Am. J. Physiol. **248:** E353–E362.
20. JENKINS, A. B., S. M. FURLER & E. W. KRAEGEN. 1986. 2-Deoxy-glucose metabolism in individual tissues of the rat *in vivo*. Int. J. Biochem. **18:** 311–318.
21. STORLIEN, L. H., D. E. JAMES, K. M. BURLEIGH, D. J. CHISHOLM & E. W. KRAEGEN. 1986. Fat feeding causes widespread *in vivo* insulin resistance, decreased energy expenditure, and obesity in rats. Am. J. Physiol. **251:** E576–E583.
22. KRAEGEN, E. W., D. E. JAMES, L. H. STORLIEN, K. M. BURLEIGH & D. J. CHISHOLM. 1986. *In vivo* insulin resistance in individual peripheral tissues of the high fat fed rat: assessment by euglycaemic clamp plus deoxyglucose administration. Diabetologia **29:** 192–198.
23. ISSAD, T., C. COUPE, M. PASTOR-ANGLADA, P. FERRÉ & J. GIRARD. 1988. Development of insulin-sensitivity at weaning in the rat: role of the nutritional transition. Biochem. J. **251:** 685–690.
24. WEINTRAUB, M., R. ZECHNER, A. BROWN, S. EISENBERG & J. BRESLOW. 1988. Dietary polyunsaturated fats of the w-6 and w-3 series reduced postprandial lipoprotein levels. J. Clin. Invest. **82:** 1884–1893.
25. WONG, S. H., P. J. NESTEL, R. P. TRIMBLE, G. B. STORER, R. J. ILLMAN & D. L. TOPPING. 1984. The adaptive effects of dietary fish and safflower oil on lipid and lipoprotein metabolism in perfused rat liver. Biochim. Biophys. Acta **792:** 103–109.
26. PHILLIPSON, B. E., D. W. ROTHROCK, W. E. CONNOR, W. S. HARRIS & D. R. ILLINGWORTH. 1985. Reduction of plasma lipids, lipoproteins, and apoproteins by dietary fish oils in patients with hypertriglyceridemia. N. Engl. J. Med. **312:** 1210–1216.

27. STORLIEN, L. H., E. W. KRAEGEN, D. J. CHISHOLM, G. L. FORD, D. G. BRUCE & W. S. PASCOE. 1987. Fish oil prevents insulin resistance induced by high-fat feeding in rats. Science **237**: 885-888.
28. STORLIEN, L. H., A. B. JENKINS, D. J. CHISHOLM, W. S. PASCOE, S. KHOURI & E. W. KRAEGEN. 1991. Influence of dietary fat composition on development of insulin resistance in rats: relationship to muscle triglyceride and w-3 fatty acids in muscle phospholipids. Diabetes **40**: 280-289.
29. STORLIEN, L. H., M. BORKMAN, A. B. JENKINS, D. A. PAN & L. V. CAMPBELL. 1991. Dietary modification of insulin action *in vivo*. *In* Diabetes 1991, p. 25-28. Elsevier. Amsterdam/New York.
30. BLAKE, W. L. & S. D. CLARKE. 1990. Suppression of rat hepatic fatty acid synthase and S 14 gene transcription by dietary polyunsaturated fat. J. Nutr. **120**: 1727-1729.
31. CLARKE, S. D. & D. B. JUMP. 1993. Regulation of gene transcription by polyunsaturated fatty acids. Prog. Lipid Res. **32**: 139-149.
32. LEYTON, J., P. J. DRURY & M. A. CRAWFORD. 1987. Differential oxidation of saturated and unsaturated fatty acids *in vivo* in the rat. Br. J. Nutr. **57**: 383-393.
33. PAN, D. A. & L. H. STORLIEN. 1993. Dietary lipid profile is a determinant of tissue phospholipid fatty acid composition and rate of weight gain in rats. J. Nutr. **123**: 512-519.
34. BORKMAN, M., L. H. STORLIEN, D. A. PAN, A. B. JENKINS, D. J. CHISHOLM & L. V. CAMPBELL. 1993. The relationship between insulin sensitivity and the fatty acid composition of phospholipids of skeletal muscle. N. Engl. J. Med. **328**: 238-244.
35. VESSBY, B., S. TENGBLAD & H. LITHELL. 1994. Insulin sensitivity is related to the fatty acid composition of serum lipids and skeletal muscle phospholipids in 70-year-old men. Diabetologia **37**: 1044-1050.
36. PAN, D. A., S. LILLIOJA, M. R. MILNER, A. D. KRIKETOS, L. A. BAUR, C. BOGARDUS & L. H. STORLIEN. 1995. Skeletal muscle membrane lipid composition is related to adiposity and insulin action. J. Clin. Invest. **96**: 2802-2808.
37. ROBERTSON, R. P. 1986. Arachidonic acid metabolite regulation of insulin secretion. Diabetes Metab. Rev. **2**: 261-296.
38. ROBERTSON, R. P. 1988. Eicosanoids as pluripotential modulators of pancreatic islet function. Diabetes **37**: 367-370.
39. OPARA, E. C., M. GARFINKEL, V. S. HUBBARD, W. M. BURCH & O. E. AKWARI. 1994. Effect of fatty acids on insulin release: role of chain length and degree of unsaturation. Am. J. Physiol. **266**: E635-E639.
40. KOLTERMAN, O. G., M. GREENFIELD, G. M. REAVEN, M. SAEKOW & J. M. OLEFSKY. 1979. Effect of a high carbohydrate diet on insulin binding to adipocytes and on insulin action in man. Diabetes **28**: 731-736.
41. BECK-NEILSEN, H., O. PEDERSEN & N. SCHWARTZ-SORENSEN. 1978. Effects of diet on the cellular insulin binding and insulin sensitivity in young healthy subjects. Diabetologia **15**: 289-296.
42. BORKMAN, M., L. V. CAMPBELL, D. J. CHISHOLM & L. H. STORLIEN. 1991. High-carbohydrate low-fat diets do not enhance insulin sensitivity in normal subjects. J. Clin. Endocrinol. Metab. **72**: 432-437.
43. HJØLLUND, E., O. PEDERSEN, B. RICHELSEN, H. BECK-NEILSEN & N. SCHWARTZ-SORENSEN. 1983. Increased insulin binding to adipocytes and monocytes and insulin sensitivity of glucose transport and metabolism in adipocytes from non-insulin-dependent diabetics after a low-fat/high-fiber diet. Metabolism **32**: 1067-1075.
44. CHEN, M., R. N. BERGMAN & D. PORTE. 1988. Insulin resistance and B-cell dysfunction in aging: the importance of dietary carbohydrate. J. Clin. Endocrinol. Metab. **67**: 951-957.
45. SWINBURN, B. A., V. L. BOYCE, R. N. BERGMAN, B. V. HOWARD & C. BOGARDUS. 1991. Deterioration in carbohydrate metabolism and lipoprotein changes induced by modern, high fat diet in Pima Indians and Caucasians. J. Clin. Endocrinol. Metab. **73**: 156-165.
46. PARILLO, M., A. A. RIVELLESE, A. V. CIARDULLO, B. CAPALDO, A. GIACCO, S. GENOVESE & G. RICCARDI. 1992. A high-monounsaturated-fat/low-carbohydrate diet improves peripheral insulin sensitivity in non-insulin-dependent diabetic patients. Metabolism **41**: 1373-1378.
47. POPP-SNIJDERS, C., J. A. SCHOUTEN, R. J. HEINE, J. VAN DER MEER & E. A. VAN DER VEEN. 1987. Dietary supplementation of omega-3 polyunsaturated fatty acids improves insulin sensitivity in non-insulin-dependent diabetes. Diabetes Res. **4**: 141-147.

48. FASCHING, P., K. RATHEISER, W. WALDHAUSL, M. ROHAC, W. OSTERRODE, P. NOWWTNY & H. VIEHAPPER. 1991. Metabolic effects of fish-oil supplementation in patients with impaired glucose tolerance. Diabetes 40: 583-589.
49. GLAUBER, H. S., P. WALLACE, K. GRIVER & G. BRECHTEL. 1988. Adverse metabolic effects of omega-3 fatty acids in non-insulin-dependent diabetes mellitus. Ann. Intern. Med. 108: 663-668.
50. BORKMAN, M., D. CHISHOLM, S. FURLER, L. STORLIEN, E. KRAEGEN, L. SIMONS & C. CHESTERMAN. 1989. Effects of fish oil supplementation on glucose and lipid metabolism in NIDDM. Diabetes 38: 1314-1319.
51. HEINE, R. J., C. MULDER, C. POPP-SNIJDERS, J. VAN DER MEER & E. A. VAN DER VEEN. 1989. Linoleic-acid-enriched diet: long-term effects on serum lipoprotein and apolipoprotein concentrations and insulin sensitivity in noninsulin-dependent diabetic patients. Am. J. Clin. Nutr. 49: 448-456.
52. FANAIAN, M., J. SZILASI, L. STORLIEN & G. D. CALVERT. 1996. The effect of modified fat diet on insulin resistance and metabolic parameters in type II diabetes. Diabetologia 39(suppl. 1): A7.
53. LIVINGSTONE, M. B. E., A. M. PRENTICE, W. A. COWARD, S. M. CEESAY, J. L. STRAIN, P. G. MCKENNA, G. B. NEVIN, M. E. BARKER & R. J. HICKEY. 1990. Simultaneous measurement of free-living energy expenditure by the doubly labeled water method and heart-rate monitoring. Am. J. Clin. Nutr. 52: 59-65.
54. LIGHTMAN, S. W., K. PISARSKA, E. R. BERMAN, M. PESTONE, H. DOWLING, E. OFFENBACHER, H. WEISEL, S. HESHKA, D. E. MATTHEWS & S. B. HEYMSFIELD. 1992. Discrepancy between self-reported and actual caloric intake and exercise in obese subjects. N. Engl. J. Med. 327: 1893-1898.
55. LOVEJOY, J. & M. DIGIROLAMO. 1992. Habitual dietary intake and insulin sensitivity in lean and obese adults. Am. J. Clin. Nutr. 55: 1174-1179.
56. PARKER, D. R., S. T. WEISS, R. TROISI, P. A. CASSANO, P. S. VOKONAS & L. LANDSBERG. 1993. Relationship of dietary saturated fatty acids and body habitus to serum insulin concentration: the Normative Aging Study. Am. J. Clin. Nutr. 58: 129-136.
57. MAYER, E. J., B. NEWMAN, C. P. QUESENBURY, JR. & J. V. SELBY. 1993. Usual dietary fat intake and insulin concentrations in healthy women twins. Diabetes Care 16: 1459-1469.
58. FESKENS, E., C. BOWLES & D. KROMHOUT. 1991. Inverse association between fish intake and risk of glucose intolerance in normoglycemic elderly men and women. Diabetes Care 14: 935-941.
59. FLATT, J. P. 1995. McCollum Award Lecture, 1995: Diet, lifestyle, and weight maintenance. Am. J. Clin. Nutr. 62: 820-836.
60. FESKENS, E. J. M. & D. KROMHOUT. 1990. Habitual dietary intake and glucose tolerance in euglycemic men: the Zutphen Study. Int. J. Epidemiol. 19: 953-959.
61. MARSHALL, J. A., R. F. HAMMAN & J. BAXTER. 1991. High-fat, low-carbohydrate diet and the etiology of non-insulin-dependent diabetes mellitus: the San Luis Valley Diabetes Study. Am. J. Epidemiol. 134: 590-603.
62. MARSHALL, J. A., S. HOAG, S. SHETTERLY & R. F. HAMMAN. 1994. Dietary fat predicts conversion from impaired glucose tolerance to NIDDM. Diabetes Care 17: 50-56.
63. VESSBY, B., A. ARO, E. SKARFORS, L. BERGLUND, I. SALMINEN & H. LITHELL. 1994. The risk to develop NIDDM is related to the fatty acid composition of the serum cholesterol esters. Diabetes 43: 1353-1357.
64. COLDITZ, G. A., J. E. MANSON, M. J. STAMPFER, B. ROSNER, W. C. WILLETT & F. E. SPEIZER. 1992. Diet and risk of clinical diabetes in women. Am. J. Clin. Nutr. 55: 1018-1023.
65. FESKENS, E. J. M., S. M. VIRTANEN, L. RÄSÄNEN, J. TUOMILEHTO, J. STENGÅRD, J. PEKKANEN, A. NISSINEN & D. KROMHOUT. 1995. Dietary factors determining diabetes and impaired glucose tolerance: a 20-year follow-up of the Finnish and Dutch cohorts of the Seven Countries Study. Diabetes Care 18: 1104-1112.
66. CAMPBELL, L. V., P. E. MARMOT, J. A. DYER, M. BORKMAN & L. H. STORLIEN. 1994. The high-monounsaturated fat diet as a practical alternative for NIDDM. Diabetes Care 17: 177-182.
67. GARG, A., J. P. BANTLE, R. R. HENRY, A. M. COULSTON, K. A. GRIVER, S. K. RAATZ, L. BRINKLEY, Y. CHEN, S. M. GRUNDY, B. A. HUET & G. M. REAVEN. 1994. Effects of varying

carbohydrate content of diet in patients with non-insulin-dependent diabetes mellitus. JAMA **271:** 1421-1428.
68. FESKENS, E. J. M., J. G. LOEBER & D. KROMHOUT. 1994. Diet and physical activity as determinants of hyperinsulinemia: the Zutphen Elderly Study. Am. J. Epidemiol. **140:** 350-360.
69. COON, P. J., E. M. ROGUS, D. DRINKWATER, D. C. MULLER & A. P. GOLDBERG. 1992. Role of body fat distribution in the decline in insulin sensitivity and glucose tolerance with age. J. Clin. Endocrinol. Metab. **75:** 1125-1132.
70. KOHRT, W. M., J. P. KIRWAN, M. A. STATEN, R. E. BOUREY, D. S. KING & J. O. HOLLOSZY. 1993. Insulin resistance in aging is related to abdominal obesity. Diabetes **42:** 273-281.
71. FLATT, J. P. 1978. The biochemistry of energy expenditure. *In* Obesity Research II, p. 211-288. Newman. London.
72. SCHUTZ, Y., J. P. FLATT & E. JÉQUIER. 1989. Failure of dietary fat intake to promote fat oxidation: a factor favoring the development of obesity. Am. J. Clin. Nutr. **50:** 307-314.
73. ASTRUP, A., B. BUEMANN, N. J. CHRISTENSEN & S. TOUBRO. 1994. Failure to increase lipid oxidation in response to increasing dietary fat content in formerly obese women. Am. J. Physiol. **266:** E592-E599.
74. SURINA, D. M., W. LANGHANS, R. PAULI & C. WENK. 1993. Meal composition affects postprandial fatty acid oxidation. Am. J. Physiol. **264:** R1065-R1070.
75. PAN, D. A. & L. H. STORLIEN. 1993. Effect of dietary lipid profile on the metabolism of w-3 fatty acids: implications for obesity prevention. *In* Omega-3 Fatty Acids: Metabolism and Biological Effects, p. 97-106. Birkhäuser. Basel.
76. SHIMOMURA, Y., T. TAMURA & M. SUZUKI. 1990. Less body fat accumulation in rats fed a safflower oil diet than in rats fed a beef tallow diet. J. Nutr. **120:** 1291-1296.
77. HAINAULT, I., M. CARLOTTI, E. HAJDUCH, C. GUICHARD & M. LAVAU. 1993. Fish oil in a high lard diet prevents obesity, hyperlipemia, and adipocyte insulin resistance in rats. Ann. N.Y. Acad. Sci. **683:** 98-101.
78. PAN, D. A., A. J. HULBERT & L. H. STORLIEN. 1994. Dietary fats, membrane phospholipids, and obesity. J. Nutr. **124:** 1555-1566.
79. DREON, D. M., B. FREY-HEWITT, N. ELLSWORTH, P. T. WILLIAMS, R. B. TERRY & P. D. WOOD. 1988. Dietary fat:carbohydrate ratio and obesity in middle-aged men. Am. J. Clin. Nutr. **47:** 995-1000.
80. MILLER, W. C., A. K. LINDEMAN, J. WALLACE & M. NIEDERPRUEM. 1990. Diet composition, energy intake, and exercise in relation to body fat in men and women. Am. J. Clin. Nutr. **52:** 426-430.
81. ROMIEU, I., W. C. WILLETT, M. J. STAMPFER, G. A. COLDITZ, L. SAMPSON, B. ROSNER, C. H. HENNEKENS & F. E. SPEIZER. 1988. Energy intake and other determinants of relative weight. Am. J. Clin. Nutr. **47:** 406-412.
82. TUCKER, L. A. & M. J. KANO. 1992. Dietary fat and body fat: a multivariate study of 205 females. Am. J. Clin. Nutr. **56:** 616-622.
83. KANT, A. K., B. I. GRAUBARD, A. SCHATZKIN & R. BALLARD-BARBASH. 1995. Proportion of energy intake from fat and subsequent weight change in the NHANES I Epidemiologic Follow-up Study. Am. J. Clin. Nutr. **61:** 11-17.
84. DESPRÉS, J. P. 1994. Dyslipidaemia and obesity. Bailliere's Clin. Endocrinol. Metab. **8:** 629-660.
85. MARON, D. J., J. M. FAIR & W. L. HASKELL. 1991. Saturated fat intake and insulin resistance in men with coronary artery disease: The Stanford Risk Intervention Project Investigators and Staff. Circulation **84:** 2020-2027.
86. PROCHAZKA, M., S. LILLIOJA, J. F. TAIT, W. C. KNOWLER, D. M. MOTT, M. SPRAUL, P. H. BENNETT & C. BOGARDUS. 1993. Linkage of chromosomal markers on 4q with a putative gene determining maximal insulin action in Pima Indians. Diabetes **42:** 514-519.
87. STORLIEN, L. H., D. A. PAN, A. D. KRIKETOS, J. O'CONNOR, I. D. CATERSON, G. J. COONEY, A. B. JENKINS & L. A. BAUR. 1995. Skeletal muscle membrane lipids and insulin resistance. Lipids **31:** S261-S265.
88. RAHEJA, B. S., S. M. SADIKOT, R. B. PHATAK & M. B. RAO. 1993. Significance of the n-6/n-3 ratio for insulin action in diabetes. Ann. N.Y. Acad. Sci. **683:** 258-271.
89. STORLIEN, L. H., D. A. PAN, M. MILNER & S. LILLIOJA. 1993. Skeletal muscle membrane lipid composition is related to adiposity in man. Obes. Res. **1**(suppl. 2): 77S.

90. BAUR, L. A., J. O'CONNOR, D. A. PAN & L. H. STORLIEN. 1995. Determinants of skeletal muscle membrane lipid composition in young children. J. Paediatr. Child Health **31:** A10.
91. SIMOPOULOS, A. P. 1994. Is insulin resistance influenced by dietary linoleic acid and *trans* fatty acids? Free Radical Biol. Med. **17:** 367-372.
92. BEYERS, E. C. & E. A. EMKEN. 1991. Metabolites of *cis*, *trans* and *trans*, *cis* isomers of linoleic acid in mice and incorporation into tissue lipids. Biochim. Biophys. Acta **1082:** 275-284.
93. ROSHOLT, M. N., P. A. KING & E. S. HORTON. 1994. High-fat diet reduces glucose transporter responses to both insulin and exercise. Am. J. Physiol. **266:** R95-R101.
94. ELSE, P. L. & A. J. HULBERT. 1987. Evolution of mammalian endothermic metabolism: "leaky" membranes as a source of heat. Am. J. Physiol. **253:** R1-R7.
95. BRAND, M. D. 1990. The contribution of the leak of protons across the mitochondrial inner membrane to standard metabolic rate. J. Theor. Biol. **145:** 267-286.
96. BJÖRNTORP, P. 1991. Visceral fat accumulation: the missing link between psychosocial factors and cardiovascular disease? J. Intern. Med. **230:** 195-201.
97. GUILLAUME-GENTIL, C., F. ROHNER-JEANRENAUD, F. ABRAMO, G. E. BESTETTI, G. L. ROSSI & B. JEANRENAUD. 1990. Abnormal regulation of the hypothalamo-pituitary-adrenal axis in the genetically obese fa/fa rat. Endocrinology **126:** 1873-1879.
98. LEVIN, B. E. 1992. Intracarotid glucose-induced norepinephrine response and the development of diet-induced obesity. Int. J. Obes. **16:** 451-457.
99. PASCOE, W., G. SMYTHE & L. STORLIEN. 1991. Enhanced responses to stress induced by fat-feeding in rats: relationship between hypothalamic noradrenaline and blood glucose. Brain Res. **550:** 192-196.
100. PAN, D. A., S. LILLIOJA & L. H. STORLIEN. 1995. Muscle lipid composition is related to body fatness and insulin action in man. Int. J. Obes. **19:** 213 (abstract).
101. STANDL, E., N. LOTZ, T. DEXEL, H. JANKA & H. J. KOLB. 1980. Muscle triglycerides in diabetic subjects. Diabetologia **18:** 463-469.
102. FELLEY, C. P., E. M. FELLEY, G. D. VAN MELLE, P. FRASCAROLO, E. JÉQUIER & J. FELBER. 1989. Impairment of glucose disposal by infusion of triglycerides in humans: role of glycemia. Am. J. Physiol. **256:** E747-E752.
103. GROOP, L. C., R. C. BONADONNA, D. C. SIMONSON, A. S. PETRIDES, M. SHANK & R. A. DEFRONZO. 1992. Effect of insulin on oxidative and nonoxidative pathways of free fatty acid metabolism in human obesity. Am. J. Physiol. **263:** E79-E84.
104. ANDERSON, E. A., R. P. HOFFMAN, T. W. BALON, C. A. SINKEY & A. L. MARK. 1991. Hyperinsulinemia produces both sympathetic neural activation and vasodilation in normal humans. J. Clin. Invest. **87:** 2246-2252.
105. GRUNFELD, C., K. BAIRD & C. R. KAHN. 1981. Maintenance of 3T3-L1 cells in culture media containing saturated fatty acids decreases insulin binding and insulin action. Biochem. Biophys. Res. Commun. **103:** 219-226.
106. YOREK, M., E. LEENEY, J. DUNLAP & B. GINSBERG. 1989. Effect of fatty acid composition on insulin and IGF-1 binding in retinoblastoma cells. Invest. Ophthalmol. Visual Sci. **30:** 2087-2092.
107. DYERBERG, J. & H. O. BAN. 1979. Haemostatic function and platelet polyunsaturated fatty acids in Eskimos. Lancet **ii:** 433-435.
108. ADLER, A. I., E. J. BOYKO, C. D. SCHRAER & N. J. MURPHY. 1994. Lower prevalence of impaired glucose tolerance and diabetes associated with daily seal oil or salmon consumption among Alaska natives. Diabetes Care **17:** 1498-1501.
109. STORLIEN, L. H., A. D. KRIKETOS, G. D. CALVERT, L. A. BAUR & A. B. JENKINS. 1996. Fatty acids, triglycerides, and syndromes of insulin resistance. Prostaglandins Leukotrienes Essent. Fatty Acids. In press.

Monounsaturated and Marine ω-3 Fatty Acids in NIDDM Patients[a]

A. A. RIVELLESE

Department of Clinical and Experimental Medicine
Federico II University Medical School
80131 Naples, Italy

METABOLIC EFFECTS OF HIGH-MUFA DIETS IN COMPARISON WITH HIGH-CHO DIETS

In the last decade, dietary recommendations for diabetic patients, especially those from the American Diabetes Association, have somewhat changed. Reduction of saturated fat remains the cornerstone of these recommendations. However, instead of the low-fat/high-carbohydrate diet suggested in 1986, the current advice is that the percentage of calories coming from carbohydrate (CHO) and fat—only unsaturated fat, of course—be varied and individualized on the basis of each patient's habits and metabolic features.[1]

This new set of advice is based on the results of some studies that have evaluated, in NIDDM patients, the effects of diets equally reduced in saturated fat, but rich in either unsaturated fat or CHO. With respect to the type of unsaturated fat, at least in NIDDM and in the last few years, monounsaturated fatty acids (MUFA) have been generally used more than polyunsaturated fatty acids (PUFA). Taken together, these studies show that high-MUFA diets have some metabolic advantages compared to high-CHO diets. In fact, with the high-MUFA diet, blood glucose control, especially in the postprandial period, improves, postprandial insulin levels decrease, insulin sensitivity improves, triglyceride levels decrease, and HDL cholesterol concentrations either remain unchanged or increase.[2-4]

Moreover, it has recently been shown that the advantages of a high-MUFA diet can be particularly evident in NIDDM patients with a more-severe form of metabolic impairment, such as patients treated with oral drugs, compared to those treated by diet alone (therefore with a milder metabolic impairment).[5] In fact, only in the group of patients treated with glibenclamide was there a significant increase in postprandial blood glucose with the high-CHO diet compared to the high-MUFA diet (TABLE 1). On the other hand, in the diet group, there was no significant change in postprandial blood glucose, probably because, in this milder form of glucose intolerance, the higher CHO intake produces a compensatory increase in insulin secretion, as shown by the significant increase in postprandial insulin concentrations, that allows plasma glucose

---

[a] The experiments presented in this paper were supported by the Italian National Research Council (Grant Nos. 04663PST75 and 91.00226.PF41).

TABLE 1. Metabolic Effects of the High-MUFA and the High-CHO Diet in Two Groups of NIDDM Patients (mean ± SEM)

	Diet Group ($n = 9$)		Glibenclamide Group ($n = 9$)	
	High-MUFA	High-CHO	High-MUFA	High-CHO
Postprandial blood glucose (mmol/L)	8.9 ± 0.6	9.7 ± 0.9	11.0 ± 1.8^a	13.6 ± 1.4
Postprandial insulin (pmol/L)	192 ± 28^b	248 ± 32	202 ± 24	226 ± 19
Plasma cholesterol (mmol/L)	5.1 ± 0.3	5.1 ± 0.3	4.4 ± 0.2	4.4 ± 0.3
Plasma triglyceride (mmol/L)	1.12 ± 0.2^c	1.36 ± 0.2	1.14 ± 0.1^c	1.40 ± 0.3
HDL cholesterol (mmol/L)	1.0 ± 0.05	1.0 ± 0.05	1.1 ± 0.05	1.1 ± 0.1

^{a} $p < 0.02$.
^{b} $p < 0.01$.
^{c} $p < 0.05$.

levels to remain unchanged (TABLE 1). The advantages of the high-MUFA diet in terms of lipid metabolism are equal in both groups of patients: the high-CHO diet induced a similar significant increase in triglycerides, without changes in total cholesterol and HDL cholesterol (TABLE 1). The increase in triglycerides reported in all the studies comparing high-MUFA diets with high-CHO diets is evident also in patients with very low triglyceride levels, such as those participating in the study shown in TABLE 1.

Considering the importance that is currently being placed on postprandial triglyceride levels as a possible independent cardiovascular risk factor, it is important to evaluate the effects of MUFA and CHO also on postprandial lipid levels. On this point, the data in the literature are scanty and almost controversial. Most studies report that the higher triglyceride levels on high-CHO diets at fasting are maintained also in the postprandial phase,[4,6] even if the increase in the postprandial period is quite comparable in both diets.[7] However, in some of these studies, there were differences between the two diets in terms of saturated fat, which could influence the results. In one of our experiments aimed at comparing MUFA and CHO, we looked at the postprandial levels of total and VLDL triglycerides. Of course, with both diets, there was a significant increase in postprandial triglycerides compared to fasting levels; however, even if the total and VLDL triglycerides were significantly higher at fasting with the high-CHO diet, in the postprandial phase there were no significant differences and, if anything, six hours after the meal both total and VLDL triglycerides tended to be lower with the high-CHO diet compared to the high-MUFA diet (TABLE 2).

These data suggest that the increase in triglycerides induced by high-CHO intake could be a phenomenon more relevant only in the postabsorptive state; the fact that the effect could be limited only to a part of the day could help explain why, from an epidemiological point of view, high-CHO diets are associated with lower rather than higher incidences of cardiovascular mortality.

TABLE 2. Effects of the High-MUFA and the High-CHO Diet on Fasting and Postprandial Total and VLDL Triglycerides in NIDDM Patients ($n = 8$) (mean ± SD)

	High-MUFA			High-CHO		
	0 h	3 h	6 h	0 h	3 h	6 h
Total TG (mmol/L)	1.03 ± 0.38[a]	2.06 ± 0.84	1.99 ± 0.89	1.32 ± 0.58	2.17 ± 0.89	1.81 ± 0.58
VLDL TG (mmol/L)	0.58 ± 0.35[a]	1.65 ± 0.80	1.54 ± 0.86	0.83 ± 0.50	1.72 ± 0.81	1.34 ± 0.55

[a] $p < 0.05$ versus high-CHO diet.

In relation to the CHO-induced hypertriglyceridemia, although the phenomenon has been known for years, the mechanisms by which this occurs are still not perfectly clear. In most studies performed in the 1970s and 1980s, there were differences in saturated fat that could directly affect lipoprotein metabolism. However, it seems that the principal event in the CHO-induced hypertriglyceridemia is represented by the increase in the synthesis and secretion of VLDL (it is not perfectly clear whether the increase is limited only to triglycerides or whether it involves the whole particle).[8,9] As to the VLDL clearance, the data in the literature are few and contrasting, but overall it seems that a reduction in the clearance, when occurring after a high-CHO diet, is of secondary and limited importance. Of course, one of the main steps in the catabolic pathway of VLDL is represented by the activity of lipolytic enzymes, mainly lipoprotein lipase (LPL), and one of the most recent studies on this topic shows that there is no difference in fasting postheparin LPL and HL activity (measured at different times after heparin) in NIDDM patients treated with a high-MUFA or a high-CHO diet.[7] This result reinforces the concept that the catabolic pathway of VLDL is not influenced very much by differences in the intake of CHO or MUFA.

METABOLIC EFFECTS OF HIGH-MUFA DIETS IN COMPARISON WITH HIGH-CHO/HIGH-FIBER DIETS

CHO is a very heterogeneous family of nutrients with different metabolic effects. Therefore, the approach with a high-MUFA diet must be compared not only with a diet rich in CHO, represented mainly by foods rich in starch, but also with diets rich in both starch and fiber, especially soluble fiber, which has important metabolic effects, even if their use still meets some criticism and skepticism.[10-12]

When we have compared, in NIDDM patients, the metabolic effect of these three dietary approaches—high-MUFA, high-starch, and high-starch plus high-fiber (≈ 50 g/day) diets—with a similar level of saturated fat, the advantages of the high-MUFA diet compared to the high-starch diet in terms of postprandial blood glucose were lost when the effect of the same diet was compared to that of the diet rich in both starch and fiber. Indeed, with this diet, there was a significant improvement in postprandial blood glucose. The same happened for the effect on insulin-stimulated glucose disposal, measured by the euglycemic hyperinsulinemic clamp. The slight, but significant improvement obtained with the high-MUFA diet compared to the high-CHO diet was no longer evident when compared to the high-starch/high-fiber diet (if anything, with the latter diet, glucose utilization tends to improve) (FIG. 1).

Therefore, from these different comparisons and also taking into account the well-known effects of fiber per se on lipid metabolism, it is reasonable to conclude that the best dietary approach for diabetic patients remains the high-CHO/high-fiber diet. However, since some people are not able to comply with this kind of approach, in these patients a high-MUFA diet could be particularly useful instead of a high-starch diet. This dietary approach could also be particularly appropriate in patients who are unable to comply with a high-fiber diet and, moreover, who have a more-severe metabolic impairment or are hypertriglyceridemic.

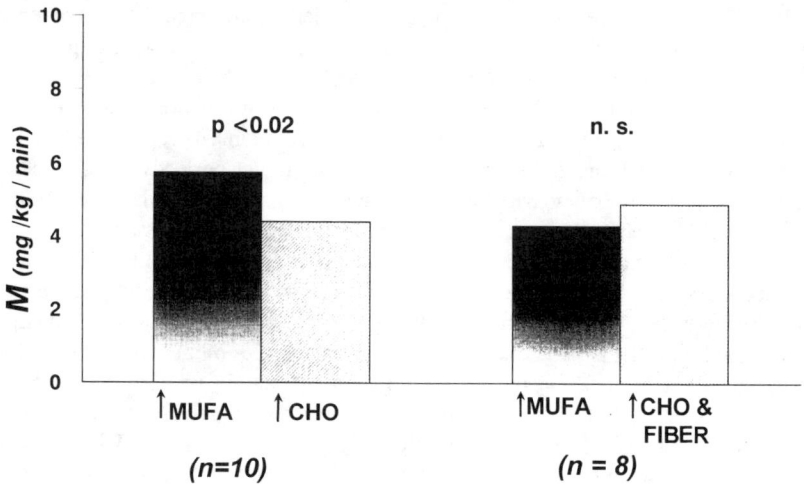

FIGURE 1. Insulin-stimulated peripheral glucose disposal in NIDDM patients after high-MUFA, high-CHO, and high-CHO + high-fiber diets.

Although it is necessary in experimental conditions to consider separately each of these two approaches, in everyday life it is right to ask whether the two diets really need to be alternatives. An important answer to this question comes directly from hyperlipidemic patients who participated in a long-term dietary experiment. These patients were randomly allocated a high-unsaturated-fat diet and a high-CHO/high-fiber diet after a run-in period on a control diet, relatively rich in saturated fat. The metabolic results of this experiment showed the two diets to have similar beneficial effects on lipids (both cholesterol and triglycerides) with no change in fasting and postprandial blood glucose and serum insulin with both diets.[13] With respect to the compliance to the two dietary approaches, patients succeeded in following the dietary recommendations quite well. Concerning dietary fiber, they increased substantially the intake of fiber, but were unable to reach the amount recommended (40 g/day). In fact, the average intake was about 30 g/day; however, this was still able to have beneficial effects on plasma cholesterol and to counteract the untoward effects of the high-starch diet on triglycerides and blood glucose concentrations. Although patients were unable to reach the amount of fiber recommended, the high-fiber diet was more acceptable than the high-MUFA diet. In fact, with the latter diet there were as many as 13 dropouts, while with the high-fiber diet only 6 patients refused to continue. The main reason for dropping out during the high-MUFA diet was that the diet was not satisfying because of the limited range of foods. Moreover, large amounts of olive oil—the main source of MUFA in our diet—are difficult to consume without proper amounts of food; this is also true in Mediterranean countries, where the habitual consumption of olive oil is high.

Therefore, it seems that the best compromise in achieving optimal compliance and optimal metabolic control would be to recommend a moderate increase in fiber-rich foods combined with the use of monounsaturated fat. This kind of diet is very similar to the type of diet consumed in Mediterranean regions, characterized by a low incidence of cardiovascular diseases.

POSSIBLE UNTOWARD EFFECTS OF HIGH-MUFA DIETS

A high-MUFA diet is a diet rich in fat; from a clinical point of view, it is therefore important to consider one of the possible untoward effects of this kind of diet, namely, weight gain. This aspect is relevant in NIDDM patients, who are characterized by a large prevalence of overweightness and obesity. At present, it is not possible to give a definite answer to this question as long-term studies are still needed. In the meantime, it is important to keep in mind that (1) population studies generally show that high-fat diets are associated with obesity,[14] (2) high-fat diets are more likely to be associated with overeating than high-CHO diets,[14] and (3) excess energy as fat is stored more efficiently than excess energy as CHO, which, in the long run, could cause weight gain.[15] Therefore, on the basis of these indications, when a high-MUFA diet is prescribed to NIDDM patients, it is important to keep body weight strictly under control.

METABOLIC EFFECTS OF MARINE ω-3 FATTY ACIDS IN NIDDM PATIENTS

The metabolic effects of ω-3 fatty acids have been extensively studied in NIDDM patients, with very controversial results, especially in relation to blood glucose control.[16-18] In the last few years, new attention has been given to the possible effects of this type of fatty acid on insulin resistance. In fact, it has been shown that ω-3 fatty acids are able to significantly improve insulin resistance in rats fed a high-saturated-fat diet.[19] Moreover, it has also been shown that insulin resistance in animals is significantly and inversely correlated with the amount of ω-3 fatty acids contained in the phospholipids of muscle membranes.[19]

On the basis of these results, Storlien et al. have hypothesized that ω-3 fatty acids are able to influence insulin action through changes in the composition of cell membranes, particularly in the muscles. If this hypothesis is true, the effects of ω-3 fatty acids on insulin action in humans could become evident only after an adequate period of time. For this reason, we have performed a long-term experiment (six months) aimed at evaluating the effects of a moderate supplementation of fish oil (2.7 g/day for the first two months, 1.7 g/day for the remaining four months) on blood glucose control, insulin resistance, and lipid metabolism in NIDDM patients with hypertriglyceridemia. The study was performed according to a randomized double-blind placebo-controlled design with a parallel group sequence. Some of the most-important results of the study are shown in TABLE 3. Fish oil had a significant hypotriglyceridemic effect in the long run, without any untoward effects on blood glucose control. However, after six months, fish oil was unable to change peripheral glucose utilization (measured by a euglycemic hyperinsulinemic clamp).[20]

TABLE 3. Effects of Fish Oil on Plasma Lipids, LDL Size, Blood Glucose Control, and Peripheral Glucose Utilization (M) in NIDDM Patients with Hypertriglyceridemia (mean ± SEM)

	Fish Oil ($n = 8$)		Placebo ($n = 8$)	
	Baseline	Six Months	Baseline	Six Months
Plasma TG (mmol/L)	3.8 ± 0.3	2.9 ± 0.2[b]	3.3 ± 0.4	3.1 ± 0.3
Plasma Cholesterol (mmol/L)	6.3 ± 0.6	5.7 ± 0.3	5.8 ± 0.3	6.2 ± 0.4
LDL size (Å)	254.2 ± 1.7	255.3 ± 2.2	253.7 ± 2.0	252.7 ± 1.7
Fasting BG (mmol/L)	10.2 ± 1.8	10.9 ± 0.5	9.2 ± 0.6	10.3 ± 1.0
PP BG (mmol/L)[a]	12.9 ± 1.8	12.1 ± 1.7	11.5 ± 1.2	12.9 ± 1.6
HbA1c (%)	7.3 ± 0.4	8.3 ± 0.5	6.9 ± 0.6	7.7 ± 0.5
M (mg/kg/min)	4.0 ± 0.8	3.9 ± 0.5	3.5 ± 0.6	4.1 ± 0.5

[a] PP BG = postprandial blood glucose.
[b] $p < 0.01$ versus baseline.

Another aspect that seems to be emerging in the field of cardiovascular diseases is the pattern of LDL subfractions. As a matter of fact, it has been shown that the prevalence of smaller LDL subfractions (the so-called LDL pattern B) is associated with a higher prevalence of cardiovascular diseases.[21] Moreover, it is also clear that one of the main determinants of this pattern is represented by the levels of plasma triglycerides since the higher the triglyceride levels, the higher the amount of small LDL.[22] Since the main and most-important metabolic effect of fish oil, in the long run, remains the hypotriglyceridemic effect, it is right to postulate that this may act also on LDL size. Unfortunately, despite the significant reduction in triglycerides, there was no change in the major peak of LDL size (TABLE 3), as assessed by non-denatured polyacrylamide gradient gel electrophoresis.[23]

In conclusion, moderate amounts of fish oil in NIDDM patients reduce plasma triglycerides and do not deteriorate blood glucose control, do not improve insulin resistance, and do not modify LDL size. Other effects of fish oil (mainly on hemostatic variables, vascular reactivity, and so on) should be of major importance in accounting for its possible protection against cardiovascular diseases.[24]

ACKNOWLEDGMENT

We thank Rosanna Scala for her linguistic expertise.

REFERENCES

1. ADA POSITION STATEMENT. 1994. Nutrition recommendations and principles for people with diabetes mellitus. Diabetes Care 17: 519–522.
2. PARILLO, M., A. A. RIVELLESE, A. V. CIARDULLO, B. CAPALDO, A. GIACCO, S. GENOVESE & G. RICCARDI. 1992. A high-monounsaturated-fat/low-carbohydrate diet improves peripheral insulin sensitivity in non-insulin-dependent diabetic patients. Metabolism 41: 1373–1378.
3. GARG, A., S. M. GRUNDY & R. H. UNGER. 1992. Comparison of effects of high and low car-

bohydrate diets on plasma lipoproteins and insulin sensitivity in patients with mild NIDDM. Diabetes **41**: 1278–1285.
4. GARG, A., A. BONANOME, S. M. GRUNDY, Z-J. ZHANG & R. H. UNGER. 1988. Comparison of a high carbohydrate diet with a high monounsaturated fat diet in patients with non-insulin-dependent diabetes mellitus. N. Engl. J. Med. **319**: 829–834.
5. PARILLO, M., A. GIACCO, A. V. CIARDULLO, A. A. RIVELLESE & G. RICCARDI. 1996. Does a high carbohydrate diet have different effects in NIDDM patients treated with diet alone or hypoglycemic drugs? Diabetes Care **19**: 498–500.
6. CHEN, I. Y.-D., S. SWAMI, R. SKOWRONSKI, A. M. COULSTON & G. REAVEN. 1993. Effects of variations in dietary fat and carbohydrate intake on postprandial lipemia in patients with non-insulin-dependent diabetes mellitus. J. Clin. Endocrinol. Metab. **76**: 347–351.
7. BLADES, B. & A. GARG. 1995. Mechanisms of increase in plasma triacylglycerol concentrations as a result of high carbohydrate intakes in patients with non-insulin-dependent diabetes mellitus. Am. J. Clin. Nutr. **62**: 996–1001.
8. BOOGAERTS, J. R., M. MCNEAL, J. ARCHAMBAULT-SCHEXNAYDER & R. A. DAVIS. 1984. Dietary carbohydrate induces lipogenesis and very low density lipoprotein synthesis. Am. J. Physiol. **246**: E77–E83.
9. CIANFLONE, K., S. DAHAN, J. C. MONGE & A. D. SNIDERMAN. 1992. Pathogenesis of carbohydrate-induced hypertriglyceridemia using HepG2 cells as a model system. Arterioscler. Thromb. **12**: 271–277.
10. RICCARDI, G. & A. A. RIVELLESE. 1991. Effects of dietary fiber and carbohydrate on glucose and lipoprotein metabolism in diabetic patients. Diabetes Care **14**: 1115–1125.
11. JENKINS, D. J. A., T. M. S. WOLEVER, A. VAN KETESHEWER-ROOK et al. 1993. Effects on blood lipids of very high intakes of fiber in diets low in saturated fat and cholesterol. N. Engl. J. Med. **329**: 21–26.
12. NUTTAL, F. Q. 1993. Dietary fiber in the management of diabetes. Diabetes **42**: 503–508.
13. RIVELLESE, A. A., P. AULETTA, G. MAROTTA, G. SALDALAMACCHIA, A. GIACCO, V. MASTRILLI, O. VACCARO & G. RICCARDI. 1994. Long term metabolic effects of two dietary methods for treating hyperlipidemia. Br. Med. J. **308**: 227–231.
14. DONFORTH, E. 1985. Diet and obesity. Am. J. Clin. Nutr. **41**: 1132–1145.
15. HORTON, T. J., H. DROUGAS, A. BRACHEY, G. W. REED, J. C. PETERS & J. O. HILL. 1995. Fat and carbohydrate overfeeding in humans: different effects on energy storage. Am. J. Clin. Nutr. **62**: 19–29.
16. ANNUZZI, G., A. RIVELLESE, B. CAPALDO, L. DI MARINO, C. IOVINE, G. MAROTTA & G. RICCARDI. 1991. A controlled study on the effects of ω-3 fatty acids on lipid and glucose metabolism in non-insulin-dependent diabetic patients. Atherosclerosis **87**: 65–73.
17. AXELROD, L. 1989. Omega-3 fatty acids in diabetes mellitus: gift from the sea? Diabetes **38**: 539–543.
18. SIMOPOULOS, A. P. 1991. Omega-3 fatty acids in health and disease and in growth and development. Am. J. Clin. Nutr. **54**: 438–463.
19. STORLIEN, L. H., L. A. BAUR, A. D. KRIKETOS, A. PAN, G. J. COONEY, A. B. JENKINS, G. D. CALVERT & L. V. CAMPBELL. 1996. Dietary fats and insulin action. Diabetologia **39**: 621–631.
20. RIVELLESE, A. A., A. MAFFETTONE, C. IOVINE, L. DI MARINO, G. ANNUZZI, M. MANCINI & G. RICCARDI. 1996. Fish oil and insulin resistance in hypertriglyceridemic NIDDM. Diabetes Care. In press.
21. AUSTIN, M. A., M. C. KING, K. M. VRANISAN & R. M. KRAUSS. 1990. Atherogenic lipoprotein phenotype: a proposed genetic marker for coronary heart disease risk. Circulation **82**: 495–506.
22. GRIFFIN, R. A., D. J. FREEMAN & G. W. TAIT. 1994. Role of plasma triglyceride in the regulation of plasma low density lipoprotein (LDL) subfractions; relative contribution of small, dense LDL to coronary heart risk. Atherosclerosis **106**: 241–253.
23. KRAUSS, R. M. & D. J. BURKE. 1982. Identification of multiple subclasses of plasma low density lipoproteins in normal humans. J. Lipid Res. **23**: 97–104.
24. NORDOY, A. 1996. Fish consumption and cardiovascular disease: a reappraisal. Nutr. Metab. Cardiovasc. Dis. **6**: 103–109.

Omega-3 and Omega-6 Fatty Acids in the Insulin Resistance Syndrome

Lipid and Lipoprotein Metabolism and Atherosclerosis

A. C. RUSTAN,[a] M. S. NENSETER,[b] AND C. A. DREVON[b]

[a]Department of Pharmacology
School of Pharmacy
[b]Institute for Nutrition Research
University of Oslo
N-0316 Oslo, Norway

INTRODUCTION

Patients with the insulin resistance syndrome (IRS) are characterized by hypertriglyceridemia, decreased plasma high-density lipoprotein (HDL) cholesterol, small dense low-density lipoprotein (LDL) particles, hypertension, central obesity, glucose intolerance, non-insulin-dependent diabetes mellitus (NIDDM), increased uric acid, and abnormalities of the fibrinolytic system.[1,2] Patients with this cluster of metabolic alterations are at increased risk for coronary heart disease (CHD).[1,2]

This paper will focus on the role of polyunsaturated omega-3 and omega-6 fatty acids (PUFA) in relation to IRS. Both types of PUFA may influence several of the abnormalities of this syndrome, such as dyslipidemia, the reactivity of platelets and leukocytes, and vascular changes. In particular, the effects of PUFA on lipid and lipoprotein metabolism, on oxidation of LDL, and on development of atherosclerosis will be highlighted.

POLYUNSATURATED FATTY ACIDS (PUFA)

Lipids are important nutrients providing 25–45% of the dietary energy in most affluent societies. Fatty acids in the diet are very important for storage and transport of energy, insulation, and mechanical protection. In addition, PUFA play numerous essential biological roles[3] and recently much attention has been given to the effects of certain PUFA as both a prophylactic and a therapeutic measure.

In Western diets, PUFA are mainly omega-6 (ω-6) fatty acids, predominantly linoleic acid (18:2, ω-6) and to a lesser extent arachidonic acid (20:4, ω-6). The intake of α-linolenic acid (18:3, ω-3) varies between 1 and 2 grams per day in most populations, whereas the intake of very-long-chain omega-3 fatty acids like eicosapentaenoic acid (EPA, 20:5, ω-3), docosapentaenoic acid (DPA, 22:5, ω-3), and docosahexaenoic acid (DHA, 22:6, ω-3) is between 0 and 10 grams per day.[4,5] Most of the human intervention studies with very-long-chain omega-3 fatty acid supplementation have used 2-10 grams/day.[6] The suggested optimum intake of PUFA per day is 3-4 grams for omega-3 and 5-12 grams for omega-6 fatty acids.[5]

CHEMISTRY AND ORIGIN

Omega-3 and omega-6 fatty acids are essential fatty acids and cannot be interconverted *in vivo*. Omega-6 fatty acids are derived from vegetables, vegetable oils, and meat. Some vegetable oils contain α-linolenic acid, but they do not contain any very-long-chain (>18 carbon atoms) omega-3 fatty acids. Omega-3 fatty acids with >18 carbon atoms are mostly made by phytoplankton and are transferred via the nutrition chain to higher animals. The most-common very-long-chain omega-3 fatty acids (EPA and DHA) are found in marine animals containing the highest amount of fat.[7] The concentrations of EPA and DHA in fish and fish products are highly variable. To obtain concentrations high enough to achieve therapeutic effects, it is usually necessary to give dietary supplements of concentrated products.

MODES OF ACTION

Eicosanoids

Dietary PUFA are incorporated into cell membranes where they affect membrane characteristics and give rise to biologically active compounds.[8,9] α-Linolenic acid can be converted into EPA and DHA, and linoleic acid into dihomo-γ-linolenic acid and arachidonic acid. EPA, dihomo-γ-linolenic acid, and arachidonic acid are all precursors of bioactive eicosanoids,[8] including prostaglandins, thromboxanes, prostacyclins, and leukotrienes, which are important for several cellular functions like platelet aggregability, endothelial cell motility, and growth of smooth muscle cells, white blood cells, and endothelial cells. Leukotrienes are also important chemotactic factors. In addition, it is important to recognize that omega-3 fatty acids are precursors of eicosanoids that in general are less potent than eicosanoids derived from the omega-6 fatty acid family.[10] Differences in biological actions of the eicosanoids are likely to be important for their role in health and disease. It has been suggested that excessive production of omega-6 eicosanoids from arachidonic acid may give rise to pathophysiological signaling.[9] Dietary omega-3 fatty acids may act in two different ways to diminish the signaling by omega-6 fatty acids: first, by competition for enzymes involved in elongation and desaturation and by incorporation into tissue esters and conversion into eicosanoids to reduce formation of omega-6-derived eicosanoids; second, by forming weak omega-3 eicosanoids that compete at cellular receptor sites to reduce the signaling by omega-6 eicosanoids.[9] The ratio between omega-3 and omega-6 fatty acids in the diet is important. Currently, this ratio might be too low and the optimal ratio is an important continuing research question.

Altered Substrate Specificity

Omega-3 fatty acids may execute their action on certain biological systems by having a different substrate specificity as compared to other fatty acids. For instance, EPA is a much-poorer substrate than all other examined fatty acids for esterification to cholesterol and diacylglycerol, in cultured cells as well as in isolated microsomes.[11,12] Certain omega-3 fatty acids are preferred substrates for certain desaturases.[13] It is also likely that the preferential incorporation of omega-3 fatty acids into certain classes of phospholipids depends upon being preferred substrates for the enzymes responsible for esterification to phosphoacylglycerols.

Membrane Fluidity

When high amounts of very-long-chain omega-3 fatty acids are ingested, there is an increased incorporation of EPA and DHA into membrane phospholipids, promoting some alterations in the physical characteristics of the membrane.[14] The activities of some membrane proteins are changed as a result of altered fluidity of the membrane or different hydrophobic interactions.[15] By changing the fatty acyl composition of phospholipids surrounding the insulin receptor in muscle, omega-3 fatty acids may improve insulin action in rats.[16,17] It has been shown that the flexibility of membranes from blood cells in animals fed fish oil is markedly increased and this might be of importance for the microcirculation. Moreover, whole blood viscosity is decreased during fish oil feeding.[18] Increased incorporation of very-long-chain omega-3 fatty acids into plasma lipoproteins also changes the physical properties of LDL, promoting a decreased melting point of core cholesterol esters that may influence the viscosity of plasma.[19]

Acylation of Proteins

Some proteins are acylated with stearic, palmitic, or myristic acid. This acylation of proteins is important for the anchoring of proteins to the membrane and is crucial for the function of these proteins.[20] Although it is generally the saturated fatty acids that have been covalently linked to specific proteins, it has been shown that proteins in platelets are acylated with omega-3 and omega-6 fatty acids.[21,22]

Nuclear Receptor Interactions

As regulators of gene expression, PUFA (or their metabolites) are thought to affect the activity of transcription factors, which in turn target key *cis*-linked elements associated with specific genes. Whether this targeting involves DNA-protein interaction or the interaction of PUFA-regulated factors is unclear.[23] The first example of a fatty acid-activated nuclear factor is the peroxisome-proliferator-activated receptor (PPAR).[24] It is possible that PPAR or similar proteins bind certain fatty acids and they may be natural ligands for PPAR.[25] It has been observed that PUFA inhibit the transcription of several hepatic lipogenic and glycolytic genes independent of PPAR.[26] Interestingly, arachidonic acid itself decreases the transcription of glucose transporters in adipocytes.[27] The nuclear actions of PUFA may clarify the mechanisms behind their metabolic effects and may lead to a better understanding of the role of PUFA in disease processes such as IRS.

Lipid Peroxidation

One of the concerns with the intake of omega-3 and omega-6 fatty acids has been their high degree of unsaturation and thereby the possibility of promoting increased oxidation of LDL and foam cell formation as part of the development of atherosclerosis (see below). On the other hand, lipid peroxidation may be beneficial under some circumstances. Omega-3 fatty acids suppress mammary tumor growth by forming lipid oxidation products that possess cytotoxic effects.[28]

PUFA IN RELATION TO CARDIOVASCULAR DISEASES

The progression and possibly initiation of CHD and inflammatory diseases appear to be influenced by the type of fat ingested (TABLE 1).[10,29] Saturated fat and cholesterol have the strongest predictive value for incidence of CHD, whereas very-long-chain omega-3 fatty acids have been shown to exert a number of beneficial effects upon the cardiovascular system.[10] Briefly, omega-3 fatty acids decrease platelet[30] and leukocyte reactivity,[31,32] inhibit lymphocyte proliferation,[33] and may slightly decrease blood pressure.[34,35] Omega-3 fatty acids may also beneficially influence vessel wall characteristics[36] and blood rheology.[8,10,18,37] Furthermore, these compounds have been shown to prevent ventricular tachyarrhythmias in animals.[38,39] The role of omega-3 fatty acids with respect to coagulation and fibrinolysis still remains controversial.[10,40,41] Omega-3 fatty acids also exert favorable metabolic changes with respect to CHD by decreasing fasting and postprandial triglycerides (see below)[37,42] and may improve insulin sensitivity, as shown in animal studies.[16,17]

It should be noted that fish (and thereby very-long-chain omega-3 fatty acids) has

TABLE 1. Dietary Fatty Acids May Influence Metabolic and Cardiovascular Events[a]

	Negative Influence	Positive Influence
Coronary artery disease	saturated	omega-3 and monoenes
Stroke	saturated	?
Blood pressure	saturated	omega-3
Insulin resistance	saturated	omega-3 (?)
Blood clotting and fibrinolysis	?	omega-3 (?) and omega-6 (?)
Function of platelets	?	omega-3 and omega-6 (?)
Hyperlipidemia	saturated	omega-3, omega-6, and monoenes
Oxidation of LDL	omega-6 (?)	monoenes
Atherogenesis (leukocyte reactivity, immunological functions)	saturated and monoenes (?)	omega-3 and omega-6
Endothelial dysfunction	?	omega-3 (?)
Cardiac arrhythmias	saturated	omega-3 and omega-6

[a] Omega-3, very-long-chain omega-3 fatty acids (EPA and DHA); omega-6, mainly linoleic acid (18:2, ω-6); monoenes, oleic acid (*cis* 18:1, ω-9); saturated fatty acids, mainly myristic and palmitic acid (14:0 and 16:0).

been an important part of the human diet for most coastal populations for thousands of years. This gives some confidence that intake of omega-3 fatty acids from fish is reasonably safe. In addition, in some populations with very high omega-3 fatty acid intake, for example, Greenland Eskimos[43] and some coastal Japanese,[44] the mortality from ischemic heart disease is low. Epidemiological studies also indicate that fish consumption is protective against CHD.[45,46] In contrast, there are epidemiological studies showing no connection between fish intake and CHD.[47,48] α-Linolenic acid may also exert some beneficial effect with respect to coronary events, although this is controversial.[49,50]

Omega-6 fatty acids (mainly linoleic acid) have also been shown to have beneficial effects with respect to cardiovascular diseases. Omega-6 fatty acids lower plasma cholesterol, prevent development of atherosclerosis,[51] and possess antiarrhythmic properties in animal models.[38] γ-Linolenic acid (GLA, 18:3, ω-6), which is produced by Δ6-desaturation of linoleic acid, has also been suggested to be of importance for prevention of CHD.[52,53] Supplementation with GLA in combination with omega-3 fatty acids has been shown to improve abnormal lipid and thromboxane metabolism in NIDDM patients.[54] With respect to arachidonic acid, it has been suggested that elevated arachidonic acid levels, when associated with increased levels of omega-3 fatty acids, may be beneficial.[55,56]

PLASMA LIPIDS AND LIPOPROTEINS

Effects of Omega-6 Fatty Acids

Replacement of saturated fat with polyunsaturated fat decreases the plasma concentrations of total and LDL cholesterol.[57,58] A meta-analysis of omega-6 PUFA versus monounsaturated fat (MUFA) gives evidence that there is no significant difference in LDL or HDL cholesterol levels when oils high in either MUFA or PUFA are exchanged in the diet.[59] Moreover, triglyceride (TG) levels were modestly lower on diets high in PUFA.[59] However, Howard et al.[60] recently observed that substitution of PUFA for MUFA resulted in a progressive decline in total and LDL cholesterol and in less TG elevation, without effect on HDL cholesterol. Omega-6 fatty acids may also alter chylomicron metabolism, but they have less impact than omega-3 fatty acids.[61,62]

Effects of Omega-3 Fatty Acids

The impact of fish oil-derived omega-3 fatty acids on blood lipoprotein levels has been examined in many studies in both animals and humans over the last 15 years. Studies in humans have demonstrated the potent, sustainable TG-lowering effect of these fatty acids.[6,42,63]

Metabolic experiments in humans and rats have shown that omega-3 fatty acids promote the reduced production of VLDL and chylomicrons by the liver and intestine, respectively.[6,64–69] By using cultured cells and animals, we have shown that EPA decreases the hepatic synthesis of TG by interfering with the esterification sequence[11,12,70,71] and by increasing peroxisomal and mitochondrial fatty acid oxidation.[71,72] We further observed that the postprandial plasma concentration of free fatty acids (FFA) was markedly decreased by feeding fish oil to rats, which may promote the reduced availability of FFA for hepatic TG synthesis (FIG. 1).[71] Singer et al.[73] also observed that the intake of fish oil in humans decreased the plasma concentration of FFA. In another study, we have demonstrated that the replacement of saturated fat with EPA and DHA in rats also decreased plasma FFA and glycerol.[67] Recent evidence suggests that the mobilization of fatty acids from adipose tissue is reduced and that the animals fed omega-3 fatty acids preferentially oxidize carbohydrates (improved insulin sensitivity), as evaluated in whole animal experiments.[67,74] Furthermore, the reduction of plasma TG and FFA after omega-3 fatty acid supplementation may limit hypertrophy of certain adipose tissues (FIG. 1).[67,74,75]

Chronic supplementation with fish oil in humans lowers fasting as well as postprandial TG levels.[61,64,76] The mechanism by which omega-3 fatty acids reduce postprandial lipemia remains obscure (FIG. 1). It has been shown that fish oil reduces postprandial TG concentrations without accelerating lipid-emulsion removal rates, suggesting decreased chylomicron production or secretion.[64] Moreover, chronic exposure of EPA to human intestinal cells (CaCo-2) reduces the rate of chylomicron

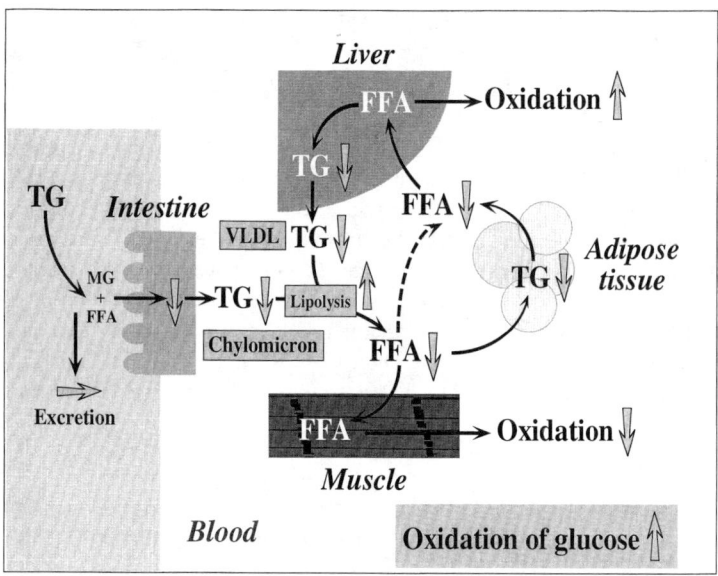

FIGURE 1. Fish oil omega-3 fatty acids—mechanisms of action for lowering of postprandial plasma triglycerides and free fatty acids (see text for details).[67,71,77] Terms: TG, triglycerides; MG, monoacylglycerol; FFA, free (unesterified) fatty acids; VLDL, very-low-density lipoprotein.

secretion.[68] In contrast, it has recently been observed that omega-3 fatty acids enhance chylomicron lipolysis in the rat without changing chylomicron production.[77]

The amount of cholesterol ester in nascent VLDL secreted from cultured rat hepatocytes is also markedly reduced in the presence of EPA.[11] This might lead to the decrease in plasma LDL concentration observed in some reports.[6] Reevaluation of several studies, however, has brought up the possibility that the reported fall in LDL concentration after fish oil supplementation could be due to the concomitant reduction in dietary intake of saturated fat.[6] It would appear that the intake of high amounts of long-chain omega-3 fatty acids in patients with type IV and V hyperlipoproteinemia may cause a concomitant fall in plasma concentration of VLDL and an increase in LDL.[6,78] The reason for this response is not known, but it is frequently observed that LDL concentration increases when VLDL concentration is reduced in subjects by drugs or diets. The possibility that this apparently adverse effect may diminish with time has been suggested in long-term trials.[79-81] The safety of fish oil supplementation was also supported in these studies.

There is some evidence that VLDL secreted in the presence of fish oil omega-3 fatty acids are smaller in size,[82] although we did not observe any size alteration.[19] The size of lipoprotein particles could be of metabolic importance since Packard et al.[83] have shown that small VLDL are more prone to end up as LDL particles as compared to large VLDL.

In most studies on lipoprotein metabolism in humans ingesting gram quantities of omega-3 fatty acids, there is a small, but significant increase in plasma concen-

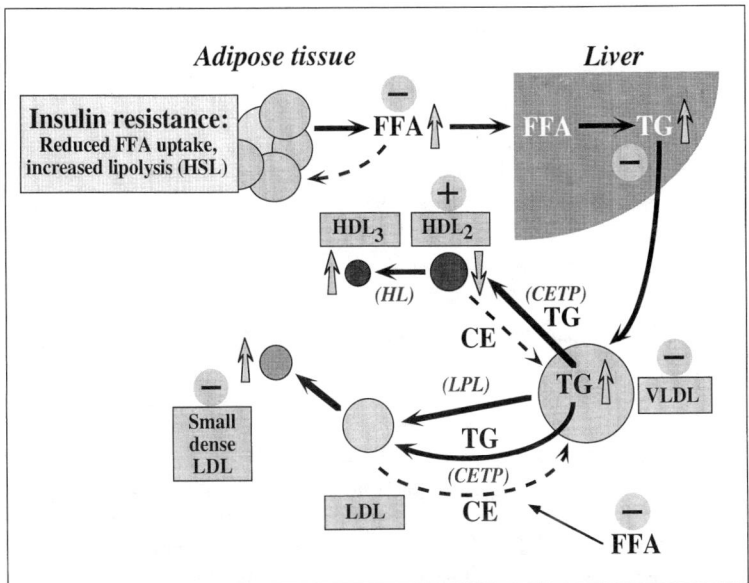

FIGURE 2. Omega-3 fatty acids and the dyslipidemia in IRS: mechanisms by which elevated concentrations of free fatty acids could lead to derangements in the plasma profiles of all lipoprotein classes and possible preventive effects of omega-3 fatty acids (see text). Terms: FFA, free (unesterified) fatty acids; TG, triglycerides; CE, cholesteryl ester; LPL, lipoprotein lipase; VLDL, very-low-density lipoprotein; LDL, low-density lipoprotein; HDL, high-density lipoprotein; CETP, cholesteryl ester transfer protein; HL, hepatic lipase; HSL, hormone-sensitive lipase; +, increased by omega-3 fatty acids; −, decreased by omega-3 fatty acids.

trations of HDL (HDL_2).[6,10,81] This increase in HDL is often taking place simultaneously with a fall in VLDL concentration. The increased concentration of HDL cholesterol could be explained by the reduced concentration of FFA in plasma,[73,78] thus causing a reduced flux of cholesterol ester from HDL to LDL and VLDL via the cholesteryl ester transfer protein (CETP) (FIG. 2).[83,84]

It has recently been observed that EPA is primarily responsible for the hypotriglyceridemic effect of fish oil in humans and rats[72,85,86] and that DHA, in contrast to EPA, does not inhibit hepatic TG production.[87] The possibility that EPA and DHA possess different metabolic properties with respect to lipoprotein metabolism should be considered.

The effects of α-linolenic acid on serum lipids and lipoproteins seem to be different from those of EPA and DHA and comparable to those of linoleic acid.[88]

Omega-3 Fatty Acids in Relation to NIDDM

Fish oil intake is associated with a low incidence of diabetes mellitus.[89] In patients with NIDDM, dietary supplementation with very-long-chain omega-3 fatty acids consistently decreased plasma triacylglycerol, whereas LDL and HDL cholesterol concentrations remained unchanged in most studies.[10,89] Although these alterations in plasma lipids may be beneficial in patients with NIDDM, earlier studies have shown that omega-3 fatty acids impair glycemic control and therefore should be used with caution. However, most of the recent trials show no adverse effects of fish oil omega-3 fatty acids, suggesting that they may be used as adjunctive therapy in hypertriglyceridemic NIDDM patients.[90] Moreover, two recent long-term trials with omega-3 fatty acids in patients with coronary artery disease and essential hypertension have shown no alteration in glucose metabolism.[35,79]

Omega-3 Fatty Acids in Relation to Dyslipidemia in IRS

IRS is associated with hypertriglyceridemia, decreased plasma HDL, the occurrence of small dense LDL particles, extended postprandial lipemia, insulin resistance, and elevated plasma free fatty acids (FFA). A single-pathway hypothesis that may give rise to dyslipidemia in IRS suggests that the elevation of plasma FFA is caused by decreased uptake in adipose tissue and increased lipolysis due to insulin resistance (FIG. 2).[91] This could lead to derangements in the plasma profiles of all lipoprotein classes: increased VLDL, decreased HDL, hyperapolipoprotein B, the formation of small dense LDL particles, and the absence of normal HDL (HDL_2). Omega-3 fatty acids may improve peripheral and hepatic insulin sensitivity as shown in animal studies[16,17,89,92] and may have the potential to beneficially affect several steps in the pathway outlined in FIG. 2.

LIPID PEROXIDATION, OXIDATION OF LDL, AND ATHEROSCLEROSIS

The safety of dietary supplementation with PUFA on a long-term basis needs to be considered. The effects of PUFA on LDL oxidation have been thoroughly studied since oxidized LDL has been suggested to play an important role in the development of atherosclerosis.[93-95] Thus, the "oxidation hypothesis" proposes that oxidative modification of LDL and possibly other lipoproteins[96] is a pivotal event in the initiation and progression of atherosclerosis (FIG. 3). Small dense LDL particles found in IRS are strongly associated with an increased risk for CHD[97] and it should be noted that these particles are more susceptible to oxidation and less protected by vitamin E supplementation.[98]

Effects of Omega-6 Fatty Acids

Oxidative modification of LDL involves the peroxidation of polyunsaturated fatty acids (FIG. 3), and the effects of dietary PUFA on LDL oxidation have recently been reviewed in reference 99. LDL particles rich in omega-6 fatty acids (linoleic acid) have been shown to be more susceptible to *in vitro* oxidative modification as compared

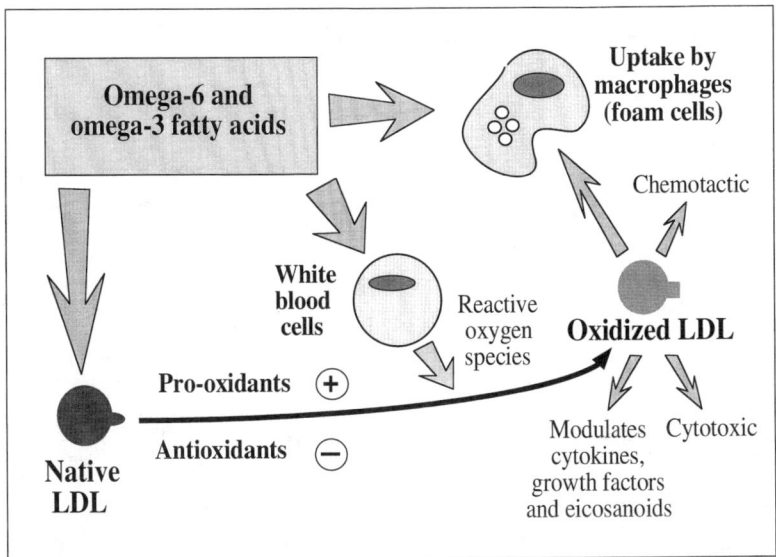

FIGURE 3. Oxidation of low-density lipoprotein (LDL) and polyunsaturated fatty acids (PUFA). Oxidation of LDL is influenced by the dietary supply of PUFA. The concentration of PUFA in LDL, white blood cells, and macrophages, as well as the different pro-oxidant and antioxidant systems, may alter the oxidation of LDL and the subsequent uptake into macrophages. Some of the biological effects of oxidized LDL are also shown.

to LDL particles enriched with monounsaturated fatty acids (oleic acid) in both clinical trials and animal experiments (TABLE 2). It has been suggested that the concentration of PUFA in LDL may be a more-important factor affecting the rate of oxidation rather than the degree of unsaturation.[100] Supplementation with α-tocopherol (vitamin E) may prevent the oxidation of omega-6 fatty acid-enriched LDL particles.[99,100] In contrast, diets highly enriched in linoleic acid can overwhelm the antioxidant effects of high levels of α-tocopherol.[101] Regardless of a diet-induced increase in the susceptibility of LDL to lipid peroxidation, an increased polyunsaturated/saturated ratio in the diet results in less atherosclerotic lesion formation in both human and animal models,[102-104] suggesting that the increase in LDL oxidation potential is compensated for by beneficial effects. Interestingly, it has recently been observed that atherogenic diets enriched with omega-6 fatty acids protect African green monkeys from coronary artery atherosclerosis when compared with dietary monounsaturated and saturated fat, by preventing cholesteryl oleate accumulation in LDL and in the coronary arteries in these primates.[51]

Effects of Omega-3 Fatty Acids

Conflicting reports have appeared on the effect of very-long-chain omega-3 fatty acids on the susceptibility of LDL to oxidative modification.[99] In some studies, it was

TABLE 2. Oxidation of Low-Density Lipoprotein (LDL): Summary of Clinical Trials and Animal Studies with PUFA Supplementation[a]

	LDL Oxidation	Design
Omega-6 fatty acids		
clinical trials	↑	ω-6 vs. mono and ctr. (7)
animal studies	↑	ω-6 vs. mono (1); ω-6 vs. saturated, mono, and ω-3 (1)
Omega-3 fatty acids		
clinical trials	↑	ω-3 vs. mono (1); ± ω-3 (2); ω-3 vs. ω-6 (1); ω-3 vs. ctr. (1)
	↔	ω-3 vs. ω-6 (2); ω-3 vs. mono (1); ± ω-3 (2)
animal studies	↑	ω-3 vs. ctr. (1)
	↓	± ω-3 (1)

[a] ω-6, mainly linoleic acid–based vegetable oils; mono, oleic acid–enriched vegetable oils; ω-3, fish oil or EPA and DHA; ctr., control/baseline diet; ±, before and after supplementation; number of studies in parentheses.[99,105,118]

observed that ingestion of fish oil in both normolipidemic and hypertriglyceridemic subjects enhanced the oxidation of LDL as compared to baseline values, whereas other reports showed no effects on lipid peroxidation in humans supplemented with omega-3 fatty acids as compared to corn oil, olive oil, or baseline values (TABLE 2).[99] In addition, it has recently been shown that omega-3 fatty acids did not enhance LDL susceptibility to oxidation in hypertriglyceridemic hemodialyzed subjects.[105] In animal studies, reduced susceptibility of rabbit LDL to *in vitro* oxidation has been reported after intake of EPA.[106]

Despite the apparently conflicting *in vitro* data, the *in vivo* effects of fish oil-enriched LDL are more consistent. Thus, no difference was observed in the minipigs between fish oil- and control oil-supplied groups with respect to atherosclerotic lesion development.[107] These data support morphological data from two previous reports, showing that supplementation with EPA and DHA to cholesterol-fed rabbits prevented or delayed the development of fatty streaks.[108,109] Omega-3 fatty acids also reduced the progression of the development of coronary atherosclerosis in pigs.[110] One explanation for these findings might be the induction of antioxidant mechanisms *in vivo* that attenuated the increased potential of omega-3 fatty acid-enriched LDL to undergo oxidative modification. It has been reported that hepatic antioxidant enzyme activities increased in mice fed EPA.[111] It should be noted that worsening of atherosclerosis by omega-3 fatty acid supplementation has been shown in some animal studies.[10]

In addition to incorporation of omega-3 fatty acids into lipoproteins, dietary supplementation with omega-3 fatty acids results in their incorporation into cellular membranes, with potential effects upon several physiological systems. Fisher *et al.*[112,113] demonstrated that omega-3 fatty acid supplementation reduced free radical production in stimulated human monocytes and polymorphonuclear leukocytes. Sirtori *et al.*[114] reported reduced superoxide generation by monocytes after dietary intake of omega-3 fatty acids. Vossen *et al.*[115] observed that EPA- and DHA-enriched endothelial cells exhibited lower susceptibility to lipid peroxidation than expected from their unsaturation index. Changes in cellular free radical production[113,114,116] as well as changes in membrane oxidizability[115] are likely to influence cell-mediated oxidative modification of LDL *in vivo*. Thus, one mechanism by which omega-3 fatty acids may exert a possible protection against lipid peroxidation might be through their tight packing in complex membrane lipids,[117] which could make the double bonds less available for oxygen or free radical interaction.

SUMMARY

Dietary fatty acids appear to be of significant importance for several of the most-common diseases in modern societies. To obtain more knowledge about the health consequences of dietary fatty acids, we depend upon a better understanding of the mechanisms of action of these fatty acids *in vivo*. With regard to the IRS, omega-3 PUFA may exert beneficial effects upon many of the associated pathophysiological metabolic changes. Omega-3 PUFA reduce fasting and postprandial TG, may improve insulin sensitivity (as shown in animal experiments), decrease platelet and leukocyte reactivity, alter immunological functions, and may slightly decrease blood pressure. Omega-3 PUFA may also beneficially influence vessel wall characteristics and blood rheology. Furthermore, both types of PUFA (omega-3 and omega-6) have been shown to inhibit cardiac arrhythmias in animals. The role of omega-3 PUFA in blood clotting and fibrinolysis still remains controversial, whereas omega-6 fatty acids may lead to increased oxidation of lipoproteins. Regardless of the effects on LDL oxidizability, both types of PUFA have shown beneficial effects on the development of atherosclerosis. As yet, little is known about the effect of specific omega-6 fatty acids with respect to the IRS. Potential adverse effects of dietary PUFA must not be neglected, but should be viewed in light of the beneficial effects of these agents.

REFERENCES

1. REAVEN, G. M. 1992. The role of insulin resistance and hyperinsulinemia in coronary heart disease. Metabolism 41: 16-19.
2. REAVEN, G. M. 1995. Pathophysiology of insulin resistance in human disease. Physiol. Rev. 75: 473-486.
3. BJERVE, K. 1989. n-3 fatty acid deficiency in man: pathogenetic mechanisms and dietary requirements. Omega-3 News 4: 1-4.
4. DYERBERG, J. 1989. Coronary heart disease in Greenland Inuit: a paradox. Implications for Western diet patterns. Arct. Med. Res. 48: 47-54.
5. DREVON, C. A. 1990. n-6 and n-3 fatty acids—how much and which balance? Scand. J. Clin. Nutr. 34: 56-61.
6. HARRIS, W. S. 1989. Fish oils and plasma lipid and lipoprotein metabolism in humans: a critical review. J. Lipid Res. 30: 785-807.
7. SIMOPOULOS, A. P., R. R. KIFER & R. E. MARTIN. 1986. *In* Health Effects of Polyunsaturated Fatty Acids in Sea Foods, p. 453-455. Academic Press. New York.
8. LEAF, A. & P. C. WEBER. 1988. Cardiovascular effects of n-3 fatty acids. N. Engl. J. Med. **318:** 549-557.
9. LANDS, W. E. M. 1993. Eicosanoids and health. Ann. N.Y. Acad. Sci. **676:** 46-59.

10. SCHMIDT, E. B. & J. DYERBERG. 1994. Omega-3 fatty acids: current status in cardiovascular medicine. Drugs **47**: 405-424.
11. RUSTAN, A. C., J. Ø. NOSSEN, H. OSMUNDSEN & C. A. DREVON. 1988. Eicosapentaenoic acid inhibits cholesterol esterification in cultured parenchymal cells and isolated microsomes from rat liver. J. Biol. Chem. **263**: 8126-8132.
12. RUSTAN, A. C., J. Ø. NOSSEN, E. N. CHRISTIANSEN & C. A. DREVON. 1988. Eicosapentaenoic acid reduces hepatic synthesis and secretion of triacylglycerol by decreasing the activity of acyl-coenzyme A:1,2-diacylglycerol acyltransferase. J. Lipid Res. **29**: 1417-1426.
13. BRENNER, R. R. 1989. Factors influencing fatty acid chain elongation and desaturation. In The Role of Fats in Human Nutrition. Second edition, p. 45-80. Academic Press. New York/London.
14. SALEM, N. 1989. N-3FA: molecular and biochemical aspects. In Current Topics in Nutrition and Disease, p. 109-228. Liss. New York.
15. SLATER, S. J., M. KELLY, M. D. YEAGER, J. LARKIN, C. HO & C. D. STUBBS. 1996. Polyunsaturation in cell membranes and lipid bilayers and its effects on membrane proteins. Lipids **31**: S189-S192.
16. STORLIEN, L. H., A. B. JENKINS, D. J. CHISHOLM, W. S. PASCOE, S. KHOURI & E. W. KRAEGEN. 1991. Influence of dietary fat composition on development of insulin resistance in rats: relationship to muscle triglyceride and omega-3 fatty acids in muscle phospholipid. Diabetes **40**: 280-289.
17. LIU, S., V. E. BARACOS, H. A. QUINNEY & M. T. CLANDININ. 1994. Dietary omega-3 and polyunsaturated fatty acids modify fatty acyl composition and insulin binding in skeletal-muscle sarcolemma. Biochem. J. **299**: 831-837.
18. ERNST, E. 1989. Effects of n-3 fatty acids on blood rheology. J. Intern. Med. Suppl. **225**: 129-132.
19. NENSETER, M. S., A. C. RUSTAN, S. LUND-KATZ, E. SØYLAND, G. MAELANDSMO, M. C. PHILLIPS & C. A. DREVON. 1992. Effect of dietary supplementation with n-3 polyunsaturated fatty acids on physical properties and metabolism of low density lipoprotein in humans. Arterioscler. Thromb. **12**: 369-379.
20. MCILHINNEY, R. A. 1990. The fats of life: the importance and function of protein acylation. Trends Biochem. Sci. **15**: 387-391.
21. MUSZBEK, L. & M. LAPOSATA. 1993. Covalent modification of proteins by arachidonate and eicosapentaenoate in platelets. J. Biol. Chem. **268**: 18243-18248.
22. HALLAK, H., L. MUSZBEK, M. LAPOSATA, E. BELMONTE, L. F. BRASS & D. R. MANNING. 1994. Covalent binding of arachidonate to G protein alpha subunits of human platelets. J. Biol. Chem. **269**: 4713-4716.
23. CLARKE, S. D. & D. B. JUMP. 1994. Dietary polyunsaturated fatty acid regulation of gene transcription. Annu. Rev. Nutr. **14**: 83-98.
24. ISSEMANN, I. & S. GREEN. 1990. Activation of a member of the steroid hormone receptor superfamily by peroxisome proliferators. Nature **347**: 645-650.
25. GÖTTLICHER, M., E. WIDMARK, Q. LI & J. Å. GUSTAFSSON. 1992. Fatty acids activate a chimera of the clofibric acid-activated receptor and the glucocorticoid receptor. Proc. Natl. Acad. Sci. U.S.A. **89**: 4653-4657.
26. CLARKE, S. D. & D. B. JUMP. 1996. Polyunsaturated fatty acid regulation of hepatic gene transcription. Lipids **31**: S7-S11.
27. LONG, S. D. & P. H. PEKALA. 1996. Regulation of GLUT4 gene expression by arachidonic acid: evidence for multiple pathways, one of which requires oxidation to prostaglandin E_2. J. Biol. Chem. **271**: 1138-1144.
28. GONZALEZ, M. J. 1995. Fish oil, lipid peroxidation, and mammary tumor growth. J. Am. Coll. Nutr. **14**: 325-335.
29. DREVON, C. A. 1992. Marine oils and their effects. Nutr. Rev. **50**: 38-45.
30. MUTANEN, M. & R. FREESE. 1996. Polyunsaturated fatty acids and platelet aggregation. Curr. Opin. Lipidol. **7**: 14-19.
31. KIM, D. N., A. EASTMAN, J. E. BAKER, A. MASTRANGELO, S. SETHI, J. S. ROSS, J. SCHMEE & W. A. THOMAS. 1995. Fish oil, atherogenesis, and thrombogenesis. Ann. N.Y. Acad. Sci. **748**: 474-480.
32. ENDRES, S., R. DE CATERINA, E. B. SCHMIDT & S. D. KRISTENSEN. 1995. n-3 polyunsaturated fatty acids: update 1995. Eur. J. Clin. Invest. **25**: 629-638.
33. SØYLAND, E., M. S. NENSETER, L. BRAATHEN & C. A. DREVON. 1993. Very long chain n-3

and n-6 polyunsaturated fatty acids inhibit proliferation of human T-lymphocytes *in vitro*. Eur. J. Clin. Invest. **23**: 112–121.
34. KNAPP, H. R. 1996. n-3 fatty acids and human hypertension. Curr. Opin. Lipidol. **7**: 30–33.
35. TOFT, I., K. H. BØNAA, O. C. INGEBRETSEN, A. NORDØY & T. JENSSEN. 1995. Effects of n-3 polyunsaturated fatty acids on glucose homeostasis and blood pressure in essential hypertension: a randomized, controlled trial. Ann. Intern. Med. **123**: 911–918.
36. WILLIAMS, M. A., R. W. ZINGHEIM, I. B. KING & A. M. ZEBELMAN. 1995. Omega-3 fatty acids in maternal erythrocytes and risk of preeclampsia. Epidemiology **6**: 232–237.
37. PSCHIERER, V., W. O. RICHTER & P. SCHWANDT. 1995. Primary chylomicronemia in patients with severe familial hypertriglyceridemia responds to long-term treatment with (n-3) fatty acids. J. Nutr. **125**: 1490–1494.
38. MCLENNAN, P. L. 1993. Relative effects of dietary saturated, monounsaturated, and polyunsaturated fatty acids on cardiac arrhythmias in rats. Am. J. Clin. Nutr. **57**: 207–212.
39. PEPE, S. & P. L. MCLENNAN. 1996. Dietary fish oil confers direct antiarrhythmic properties on the myocardium of rats. J. Nutr. **126**: 34–42.
40. SANDERS, T. A. B. 1996. Effects of unsaturated fatty acid on blood clotting and fibrinolysis. Curr. Opin. Lipidol. **7**: 20–23.
41. TREMOLI, E., S. ELIGINI, S. COLLI, P. MADERNA, P. RISE, F. PAZZUCCONI, F. MARANGONI, C. R. SIRTORI & C. GALLI. 1994. n-3 fatty acid ethyl ester administration to healthy subjects and to hypertriglyceridemic patients reduces tissue factor activity in adherent monocytes. Arterioscler. Thromb. **14**: 1600–1608.
42. HARRIS, W. S. 1996. n-3 fatty acids and lipoproteins: comparison of results from human and animal studies. Lipids **31**: 243–252.
43. KROMANN, N. & A. GREEN. 1980. Epidemiological studies in the Upernavik district, Greenland: incidence of some chronic diseases 1950–1974. Acta Med. Scand. **208**: 401–406.
44. HIRAI, A., T. HAMAZAKI, T. TERANO, T. NISHIKAWA, Y. TAMURA, A. KAMUGAI & J. JAJIKI. 1980. Eicosapentaenoic acid and platelet function in Japanese. Lancet **2**: 1132–1133.
45. BURR, M. L., A. M. FEHILY, J. F. GILBERT, S. ROGERS, R. M. HOLLIDAY, P. M. SWEETNAM, P. C. ELWOOD & N. M. DEADMAN. 1989. Effects of changes in fat, fish, and fibre intakes on death and myocardial reinfarction: diet and reinfarction trial (DART). Lancet **2**: 757–761.
46. KATAN, M. B. 1995. Fish and heart disease: what is the real story? Nutr. Rev. **53**: 228–230.
47. MORRIS, M. C., J. E. MANSON, B. ROSNER, J. E. BURING, W. C. WILLETT & C. H. HENNEKENS. 1995. Fish consumption and cardiovascular disease in the physicians' health study: a prospective study. Am. J. Epidemiol. **142**: 166–175.
48. ASCHERIO, A., E. B. RIMM, M. J. STAMPFER, E. L. GIOVANNUCCI & W. C. WILLETT. 1995. Dietary intake of marine n-3 fatty acids, fish intake, and the risk of coronary disease among men. N. Engl. J. Med. **332**: 977–982.
49. DE LORGERIL, M., S. RENAUD, N. MAMELLE, P. SALEN, J. L. MARTIN, I. MONJAUD, J. GUIDOLLET, P. TOUBOUL & J. DELAYE. 1994. Mediterranean alpha-linolenic acid-rich diet in secondary prevention of coronary heart disease. Lancet **343**: 1454–1459.
50. MCLENNAN, P. L. & J. A. DALLIMORE. 1995. Dietary canola oil modifies myocardial fatty acids and inhibits cardiac arrhythmias in rats. J. Nutr. **125**: 1003–1009.
51. RUDEL, L. L., J. S. PARKS & J. K. SAWYER. 1995. Compared with dietary monounsaturated and saturated fat, polyunsaturated fat protects African green monkeys from coronary artery atherosclerosis. Arterioscler. Thromb. Vasc. Biol. **15**: 2101–2110.
52. HORROBIN, D. F. 1993. Fatty acid metabolism in health and disease: the role of delta-6-desaturase. Am. J. Clin. Nutr. **57**: 736S–737S.
53. KARLSTAD, M. D., S. J. DEMICHELE, W. D. LEATHEM & M. B. PETERSON. 1993. Effect of intravenous lipid emulsions enriched with gamma-linolenic acid on plasma n-6 fatty acids and prostaglandin biosynthesis after burn and endotoxin injury in rats. Crit. Care Med. **21**: 1740–1749.
54. TAKAHASHI, R., J. INOUE, H. ITO & H. HIBINO. 1993. Evening primrose oil and fish oil in non-insulin-dependent diabetes. Prostaglandins Leukotrienes Essent. Fatty Acids **49**: 569–571.
55. COLLIER, G. R. & A. J. SINCLAIR. 1993. Role of N-6 and N-3 fatty acids in the dietary treatment of metabolic disorders. Ann. N.Y. Acad. Sci. **683**: 322–330.
56. SINCLAIR, A. J. & N. J. MANN. 1996. Short-term diets rich in arachidonic acid influence plasma phospholipid polyunsaturated fatty acid levels and prostacyclin and thromboxane production in humans. J. Nutr. **126**: 1110S–1114S.

57. MENSINK, R. P. & M. B. KATAN. 1992. Effect of dietary fatty acids on serum lipids and lipoproteins: a meta-analysis of 27 trials. Arterioscler. Thromb. 12: 911–919.
58. HEGSTED, D. M., L. M. AUSMAN, J. A. JOHNSON & G. E. DALLAL. 1993. Dietary fat and serum lipids: an evaluation of the experimental data. Am. J. Clin. Nutr. 57: 875–883.
59. GARDNER, C. D. & H. C. KRAEMER. 1995. Monounsaturated versus polyunsaturated dietary fat and serum lipids: a meta-analysis. Arterioscler. Thromb. Vasc. Biol. 15: 1917–1927.
60. HOWARD, B. V., J. S. HANNAH, C. C. HEISER, K. A. JABLONSKI, M. C. PAIDI, L. ALARIF, D. C. ROBBINS & W. J. HOWARD. 1995. Polyunsaturated fatty acids result in greater cholesterol lowering and less triacylglycerol elevation than do monounsaturated fatty acids in a dose-response comparison in a multiracial study group. Am. J. Clin. Nutr. 62: 392–402.
61. HARRIS, W. S. 1994. Chylomicron metabolism and ω3 and ω6 fatty acids. World Rev. Nutr. Diet. 76: 23–25.
62. BERGERON, N. & R. J. HAVEL. 1995. Influence of diets rich in saturated and omega-6 polyunsaturated fatty acids on the postprandial responses of apolipoproteins B-48, B-100, E, and lipids in triglyceride-rich lipoproteins. Arterioscler. Thromb. Vasc. Biol. 15: 2111–2121.
63. HARRIS, W. S. 1996. Dietary fish oil and blood lipids. Curr. Opin. Lipidol. 7: 3–7.
64. HARRIS, W. S. & F. MUZIO. 1993. Fish oil reduces postprandial triglyceride concentrations without accelerating lipid-emulsion removal rates. Am. J. Clin. Nutr. 58: 68–74.
65. NESTEL, P. J., W. E. CONNOR, M. F. REARDON, S. CONNOR, S. WONG & R. BOSTON. 1984. Suppression by diets rich in fish oil of very low density lipoprotein production in man. J. Clin. Invest. 74: 82–89.
66. MURTHY, S., E. ALBRIGHT, S. N. MATHUR & F. J. FIELD. 1990. Effect of eicosapentaenoic acid on triacylglycerol transport in CaCo-2 cells. Biochim. Biophys. Acta 1045: 147–155.
67. RUSTAN, A. C., B.-E. HUSTVEDT & C. A. DREVON. 1993. Dietary supplementation of very long-chain n-3 fatty acids decreases whole body lipid utilization in the rat. J. Lipid Res. 34: 1299–1309.
68. RANHEIM, T., A. GEDDE-DAHL, A. C. RUSTAN & C. A. DREVON. 1994. Effect of chronic incubation of CaCo-2 cells with eicosapentaenoic acid (20:5, n-3) and oleic acid (18:1, n-9) on triacylglycerol production. Biochem. J. 303: 155–161.
69. OTTO, D. A., J. K. BALTZELL & J. T. WOOTEN. 1992. Reduction in triacylglycerol levels by fish oil correlates with free fatty acid levels in ad libitum fed rats. Lipids 27: 1013–1017.
70. NOSSEN, J. Ø., A. C. RUSTAN, S. H. GLOPPESTAD, S. MÅLBAKKEN & C. A. DREVON. 1986. Eicosapentaenoic acid inhibits synthesis and secretion of triacylglycerols by cultured rat hepatocytes. Biochim. Biophys. Acta 879: 56–65.
71. RUSTAN, A. C., E. N. CHRISTIANSEN & C. A. DREVON. 1992. Serum lipids, hepatic glycerolipid metabolism, and peroxisomal fatty acid oxidation in rats fed omega-3 and omega-6 fatty acids. Biochem. J. 283: 333–339.
72. WILLUMSEN, N., J. SKORVE, S. HEXEBERG, A. C. RUSTAN & R. K. BERGE. 1993. The hypotriglyceridemic effect of eicosapentaenoic acid in rats is reflected in increased mitochondrial fatty acid oxidation followed by diminished lipogenesis. Lipids 28: 683–690.
73. SINGER, P., M. WIRTH & I. BERGER. 1990. A possible contribution of decrease in free fatty acids to low serum triglyceride levels after diets supplemented with n-6 and n-3 polyunsaturated fatty acids. Atherosclerosis 83: 167–175.
74. RUSTAN, A. C., B.-E. HUSTVEDT & C. A. DREVON. 1996. Postprandial decrease in plasma unesterified fatty acids during omega-3 fatty acid feeding is not caused by accumulation of fatty acids in adipose tissue. Submitted.
75. PARRISH, C. C., D. A. PATHY, J. G. PARKES & A. ANGEL. 1991. Dietary fish oils modify adipocyte structure and function. J. Cell. Physiol. 148: 493–502.
76. WEINTRAUB, M. S., R. ZECHNER, A. BROWN, S. EISENBERG & J. L. BRESLOW. 1988. Dietary polyunsaturated fats of the w-6 and w-3 series reduce postprandial lipoprotein levels: chronic and acute effects of fat saturation on postprandial lipoprotein metabolism. J. Clin. Invest. 82: 1884–1893.
77. HARRIS, W. S., B.-E. HUSTVEDT, E. HAGEN, M. H. GREEN, G. P. LU & C. A. DREVON. 1997. N-3 fatty acids and chylomicron metabolism in the rat. J. Lipid Res. 38: 503–515.
78. SULLIVAN, D. R., T. A. SANDERS, I. M. TRAYNER & G. R. THOMPSON. 1986. Paradoxical elevation of LDL apoprotein B levels in hypertriglyceridemic patients and normal subjects ingesting fish oil. Atherosclerosis 61: 129–134.

79. ERITSLAND, J., H. ARNESEN, I. SELJEFLOT & A. T. HØSTMARK. 1995. Long-term metabolic effects of n-3 polyunsaturated fatty acids in patients with coronary artery disease. Am. J. Clin. Nutr. **61:** 831–836.
80. LEAF, A., M. B. JORGENSEN, A. K. JACOBS, G. COTE, D. A. SCHOENFELD, J. SCHEER, B. H. WEINER, J. D. SLACK, M. A. KELLETT, A. E. RAIZNER *et al.* 1994. Do fish oils prevent restenosis after coronary angioplasty? Circulation **90:** 2248–2257.
81. SAYNOR, R. & T. GILLOTT. 1992. Changes in blood lipids and fibrinogen with a note on safety in a long-term study on the effects of n-3 fatty acids in subjects receiving fish oil supplements and followed for seven years. Lipids **27:** 533–538.
82. HORNSTRA, G. 1989. The significance of fish and fish oil-enriched food for prevention and therapy of ischemic cardiovascular disease. *In* The Role of Fats in Human Nutrition. Second edition, p. 151–236. Academic Press. New York/London.
83. PACKARD, C. J., A. MUNRO, A. R. LORIMER, A. M. GOTTO & J. SHEPHERD. 1984. Metabolism of apolipoprotein B in large triglyceride-rich very low density lipoproteins of normal and hypertriglyceridemic subjects. J. Clin. Invest. **74:** 2178–2192.
84. BARTER, P. J. 1990. Enzymes involved in lipid and lipoprotein metabolism. Curr. Opin. Lipidol. **1:** 518–523.
85. RAMBJØR, G. S., A. I. WÅLEN, S. L. WINDSOR & W. S. HARRIS. 1996. Eicosapentaenoic acid is primarily responsible for hypotriglyceridemic effect of fish oil in humans. Lipids **31:** S45–S49.
86. FRØYLAND, L., H. VAAGENES, D. K. ASIEDU, A. GARRAS, O. LIE & R. K. BERGE. 1996. Chronic administration of eicosapentaenoic acid and docosahexaenoic acid as ethyl esters reduced plasma cholesterol and changed the fatty acid composition in rat blood and organs. Lipids **31:** 169–178.
87. WILLUMSEN, N., S. HEXEBERG, J. SKORVE, M. LUNDQUIST & R. K. BERGE. 1993. Docosahexaenoic acid shows no triglyceride-lowering effects, but increases the peroxisomal fatty acid oxidation in liver of rats. J. Lipid Res. **34:** 13–22.
88. VALSTA, L. M., M. JAUHIAINEN, A. ARO, I. SALMINEN & M. MUTANEN. 1995. The effect on serum lipoprotein levels of two monounsaturated fat-rich diets differing in their linoleic and α-linolenic acid contents. Nutr. Metab. Cardiovasc. Dis. **5:** 129–140.
89. MALASANOS, T. H. & P. W. STACPOOLE. 1991. Biological effects of omega-3 fatty acids in diabetes mellitus. Diabetes Care **14:** 1160–1179.
90. HARRIS, W. S. 1996. Do ω3 fatty acids worsen glycemic control in NIDDM? Int. Soc. Study Fatty Acids Lipids (ISSFAL) Newsl. **2:** 6–9.
91. LAWS, A. 1996. Free fatty acids, insulin resistance, and lipoprotein metabolism. Curr. Opin. Lipidol. **7:** 172–177.
92. CHICCO, A., M. E. DALESSANDRO, L. KARABATAS, R. GUTMAN & Y. B. LOMBARDO. 1996. Effect of moderate levels of dietary fish oil on insulin secretion and sensitivity, and pancreas insulin content in normal rats. Ann. Nutr. Metab. **40:** 61–70.
93. STEINBERG, D., S. PARTHASARATHY, T. E. CAREW, J. C. KHOO & J. L. WITZTUM. 1989. Beyond cholesterol: modifications of low-density lipoprotein that increase its atherogenicity. N. Engl. J. Med. **320:** 915–924.
94. WITZTUM, J. L. 1993. Role of oxidised low density lipoprotein in atherogenesis. Br. Heart J. **69:** S12–S18.
95. PALINSKI, W., V. A. ORD, A. S. PLUMP, J. L. BRESLOW, D. STEINBERG & J. L. WITZTUM. 1994. ApoE-deficient mice are a model of lipoprotein oxidation in atherogenesis: demonstration of oxidation-specific epitopes in lesions and high titers of autoantibodies to malondialdehyde-lysine in serum. Arterioscler. Thromb. **14:** 605–616.
96. MABILE, L., R. SALVAYRE, M. J. BONNAFE & A. NEGRE SALVAYRE. 1995. Oxidizability and subsequent cytotoxicity of chylomicrons to monocytic U937 and endothelial cells are dependent on dietary fatty acid composition. Free Radical Biol. Med. **19:** 599–607.
97. AUSTIN, M. A. 1994. Small, dense low-density lipoprotein as a risk factor for coronary heart disease. Int. J. Clin. Lab. Res. **24:** 187–192.
98. REAVEN, P. D., D. A. HEROLD, J. BARNETT & S. EDELMAN. 1995. Effects of vitamin E on susceptibility of low-density lipoprotein and low-density lipoprotein subfractions to oxidation and on protein glycation in NIDDM. Diabetes Care **18:** 807–816.
99. NENSETER, M. S. & C. A. DREVON. 1996. Dietary polyunsaturates and peroxidation of low density lipoprotein. Curr. Opin. Lipidol. **7:** 8–13.

100. THOMAS, M. J., T. THORNBURG, J. MANNING, K. HOOPER & L. L. RUDEL. 1994. Fatty acid composition of low-density lipoprotein influences its susceptibility to autoxidation. Biochemistry 33: 1828-1834.
101. REAVEN, P. D., B. J. GRASSE & D. L. TRIBBLE. 1994. Effects of linoleate-enriched and oleate-enriched diets in combination with alpha-tocopherol on the susceptibility of LDL and LDL subfractions to oxidative modification in humans. Arterioscler. Thromb. 14: 557-566.
102. WOLFE, M. S., J. K. SAWYER, T. M. MORGAN, B. C. BULLOCK & L. L. RUDEL. 1994. Dietary polyunsaturated fat decreases coronary artery atherosclerosis in a pediatric-aged population of African green monkeys. Arterioscler. Thromb. 14: 587-597.
103. GOODNIGHT, S., JR., W. S. HARRIS, W. E. CONNOR & D. R. ILLINGWORTH. 1982. Polyunsaturated fatty acids, hyperlipidemia, and thrombosis. Arteriosclerosis 2: 87-113.
104. KIM, D. N., J. SCHMEE, K. T. LEE & W. A. THOMAS. 1984. Hypo-atherogenic effect of dietary corn oil exceeds hypo-cholesterolemic effect in swine. Atherosclerosis 52: 101-113.
105. BONANOME, A., F. BIASIA, M. DE LUCA, G. MUNARETTO, S. BIFFANTI, M. PRADELLA & A. PAGNAN. 1996. n-3 fatty acids do not enhance LDL susceptibility to oxidation in hypertriacylglycerolemic hemodialyzed subjects. Am. J. Clin. Nutr. 63: 261-266.
106. SAITO, H., K. J. CHANG, Y. TAMURA & S. YOSHIDA. 1991. Ingestion of eicosapentaenoic acid–ethyl ester renders rabbit LDL less susceptible to Cu^{2+}-catalyzed oxidative modification. Biochem. Biophys. Res. Commun. 175: 61-67.
107. WHITMAN, S. C., J. R. FISH, M. L. RAND & K. A. ROGERS. 1994. n-3 fatty acid incorporation into LDL particles renders them more susceptible to oxidation *in vitro*, but not necessarily more atherogenic *in vivo*. Arterioscler. Thromb. 14: 1170-1176.
108. PIATTI, G., F. BORELLA, M. A. BERTI & P. C. BRAGA. 1993. Delaying effects of dietary eicosapentaenoic-docosahexaenoic acids on development of "fatty streaks" in hypercholesterolemic rabbits: a morphological study by scanning electron microscopy. Drugs Exp. Clin. Res. 19: 175-181.
109. GUSAIN, D., I. SHARMA & V. P. DIXIT. 1994. Antiatherosclerotic effects of fish extract feeding in rabbits. J. Ethnopharmacol. 41: 59-63.
110. SASSEN, L. M., J. M. LAMERS, W. SLUITER, J. M. HARTOG, D. H. DEKKERS, A. HOGENDOORN & P. D. VERDOUW. 1993. Development and regression of atherosclerosis in pigs: effects of n-3 fatty acids, their incorporation into plasma and aortic plaque lipids, and granulocyte function. Arterioscler. Thromb. 13: 651-660.
111. DEMOZ, A., N. WILLUMSEN & R. K. BERGE. 1992. Eicosapentaenoic acid at hypotriglyceridemic dose enhances the hepatic antioxidant defense in mice. Lipids 27: 968-971.
112. FISHER, M., K. S. UPCHURCH, P. H. LEVINE, M. H. JOHNSON, C. H. VAUDREUIL, A. NATALE & J. J. HOOGASIAN. 1986. Effects of dietary fish oil supplementation on polymorphonuclear leukocyte inflammatory potential. Inflammation 10: 387-392.
113. FISHER, M., P. H. LEVINE, B. H. WEINER, M. H. JOHNSON, E. M. DOYLE, P. A. ELLIS & J. J. HOOGASIAN. 1990. Dietary n-3 fatty acid supplementation reduces superoxide production and chemiluminescence in a monocyte-enriched preparation of leukocytes. Am. J. Clin. Nutr. 51: 804-808.
114. SIRTORI, C. R., E. GATTI, E. TREMOLI, C. GALLI, G. GIANFRANCESCHI, G. FRANCESCHINI, S. COLLI, P. MADERNA, F. MARANGONI, P. PEREGO et al. 1992. Olive oil, corn oil, and n-3 fatty acids differently affect lipids, lipoproteins, platelets, and superoxide formation in type II hypercholesterolemia. Am. J. Clin. Nutr. 56: 113-122.
115. VOSSEN, R. C., M. C. VAN DAM-MIERAS, G. HORNSTRA & R. F. ZWAAL. 1995. Differential effects of endothelial cell fatty acid modification on the sensitivity of their membrane phospholipids to peroxidation. Prostaglandins Leukotrienes Essent. Fatty Acids 52: 341-347.
116. FISHER, M. & P. H. LEVINE. 1991. Effects of dietary omega 3 fatty acid supplementation on leukocyte free radical production. World Rev. Nutr. Diet. 66: 245-249.
117. APPLEGATE, K. R. & J. A. GLOMSET. 1986. Computer-based modeling of the conformation and packing properties of docosahexaenoic acid. J. Lipid Res. 27: 658-680.
118. BITTOLO-BON, G., G. CAZZOLATO, P. ALESSANDRINI, S. SOLDAN, G. CASALINO & P. AVOGARO. 1993. Effects of concentrated DHA and EPA supplementation on LDL peroxidation and vitamin E status in type II B hyperlipidemic patients. *In* Omega-3 Fatty Acids: Metabolism and Biological Effects, p. 51-58. Birkhäuser. Basel.

Omega-6/Omega-3 Fatty Acid Ratio and *Trans* Fatty Acids in Non-Insulin-dependent Diabetes Mellitus

ARTEMIS P. SIMOPOULOS

Center for Genetics, Nutrition, and Health
Washington, District of Columbia 20009

INTRODUCTION

Over the last 20 years, many studies and clinical investigations have been carried out on the metabolism of polyunsaturated fatty acids (PUFA) in general and omega-3 fatty acids in particular. Today, we know that omega-3 fatty acids are essential for normal growth and development and may play an important role in the prevention and treatment of coronary artery disease, hypertension, diabetes, arthritis, other inflammatory and autoimmune disorders, and cancer.[1-7] Research has been carried out in animal models, tissue cultures, and human beings. The original observational studies have given way to controlled clinical trials. Great progress has taken place in our knowledge on the physiological and molecular mechanisms of the various fatty acids in health and disease. This paper discusses the evolutionary aspects of diet in terms of the balance of the omega-6 and omega-3 fatty acid intake; the increase in *trans* fatty acids as a result of industrialization, agribusiness, and food processing; and their effects on non-insulin-dependent diabetes mellitus (NIDDM).

EVOLUTIONARY ASPECTS OF DIET, OMEGA-6/OMEGA-3 FATTY ACID BALANCE, AND *TRANS* FATTY ACIDS

On the basis of estimates from studies in Paleolithic nutrition and modern-day hunter-gatherer populations, it appears that human beings evolved on a diet that was much lower in saturated fatty acids than is today's diet.[8] Furthermore, the diet contained small, but roughly equal amounts of omega-6 and omega-3 PUFAs (1-2/1) and much less *trans* fatty acids than today's diet (FIG. 1).[9] The current Western diet is very high in omega-6 fatty acids ($\omega 6/\omega 3$ = 20-30/1) because of the indiscriminate recommendation to substitute omega-6 fatty acids for saturated fats in order to lower serum cholesterol levels.[10] The intake of omega-3 fatty acids is much lower today

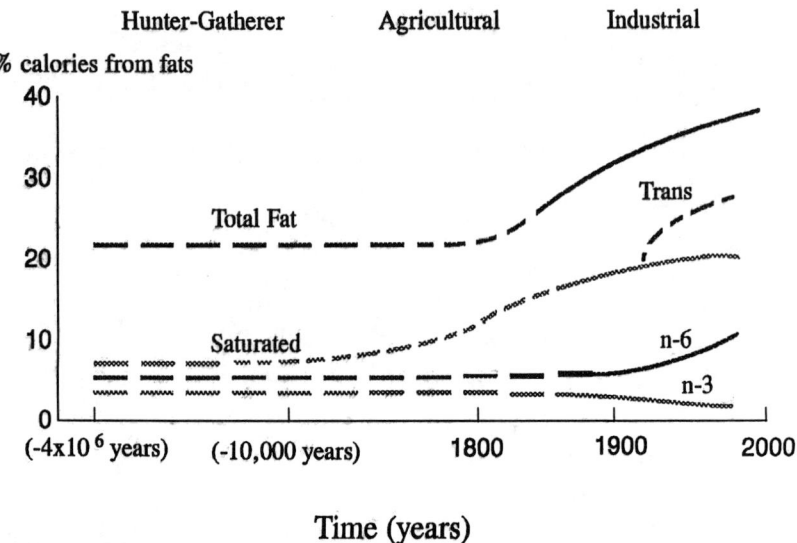

FIGURE 1. Hypothetical scheme of the relative percentages of fat and different fatty acid families in human nutrition as extrapolated from cross-sectional analyses of contemporary hunter-gatherer populations and from longitudinal observations and their putative changes during the preceding 100 years in relation to the recent increase in the frequency of coronary artery disease (CAD). *Trans* fatty acids, the result of the hydrogenation process, have only been part of the food supply during this century.

because of the decrease in fish consumption and the industrial production of animal feeds rich in grains containing omega-6 fatty acids, leading to production of meat rich in omega-6 and poor in omega-3 fatty acids.[11] The same is true for farmed fish in aquaculture[12] and eggs.[13] Even cultivated vegetables contain less omega-3 fatty acids than plants in the wild.[14,15] In summary, modern agriculture, with its emphasis on production, has decreased the omega-3 fatty acid content in many foods: green leafy vegetables, animal meats, eggs, and even fish.

Similarly, the hydrogenation of vegetable oils rich in omega-6 fatty acids to form margarine has led to increased amounts of *trans* fatty acids in the food supply.[16] *Trans* fatty acids are formed during hydrogenation, a process that solidifies liquid vegetable oils by adding hydrogen atoms to the double bonds of the unsaturated fatty acids in the *trans* instead of the *cis* position on the fatty acid molecule. The degree of hydrogenation dictates how much *trans* fatty acid is in the product. The concentration of *trans* fatty acids varies with the extent and type of processing of oil. Salad oils contain 8% to 17% *trans*, shortenings 14% to 60%, and margarines 16% to 70%. In food composition tables, the *trans* fatty acids are listed as monounsaturated fatty acids and this has led to confusion about the amount of saturated fat in the United States diet. In the United States, *trans* fatty acids contribute 3% to 7% of the fat consumed (TABLE 1).[16,17] *Trans* fatty acids occur rarely or in small amounts in nature.

TABLE 1. Partial List of Foods Containing *Trans* Fatty Acids in the American Diet[16,17]

Margarines
Shortening
Cooking oils
Salad dressings
Mayonnaise
Foods prepared with fats and oils high in *trans* fatty acids
  generally consumed by Americans
    bread/rolls
    cakes
    cheese-corn snacks/corn snacks
    cookies
    crackers
    doughnuts
    french fried potatoes
    fried chicken
    fried fish
    potato chips
    snacks (miscellaneous)

BIOLOGICAL EFFECTS OF OMEGA-6 AND OMEGA-3 FATTY ACIDS

Linoleic acid (LA) and alpha-linolenic acid (LNA) and their long-chain derivatives are important components of animal and plant cell membranes. When humans ingest fish or fish oil, the eicosapentaenoic acid (EPA) and docosahexaenoic acid (DHA) from the diet partially replace the omega-6 fatty acids, especially arachidonic acid (AA), in the membranes of probably all cells, but especially in the membranes of platelets, erythrocytes, neutrophils, monocytes, and liver cells (reviewed in reference 1).

Because of the increased amounts of omega-6 fatty acids in the Western diet, the eicosanoid metabolic products from AA, specifically prostaglandins, thromboxanes, leukotrienes, hydroxy fatty acids, and lipoxins, are formed in larger quantities than those formed from omega-3 fatty acids, specifically EPA. The eicosanoids from AA are biologically active in very small quantities and, if they are formed in large amounts, they contribute to the formation of thrombus and atheromas; to allergic and inflammatory disorders, particularly in susceptible people; and to proliferation of cells. Thus, a diet rich in omega-6 fatty acids shifts the physiological state to one that is prothrombotic and proaggregatory, with increases in blood viscosity, vasospasm, and vasoconstriction and decreases in bleeding time. Bleeding time is decreased in groups of patients with hypercholesterolemia,[18] hyperlipoproteinemia,[19] myocardial infarction, other forms of atherosclerotic disease, and diabetes (obesity and hypertriglyceridemia). Atherosclerosis is a major complication in NIDDM patients. Therefore, in this paper, the antiatheromatous and antithrombotic aspects of omega-3 fatty acids are reviewed. Bleeding time is longer in women than in men and longer in young than in old people.

TABLE 2. Ethnic Differences in Fatty Acid Concentrations in Thrombocyte Phospholipids and Percent of All Deaths from Cardiovascular Disease[a]

	Europe, United States	Japan	Greenland Eskimos
Arachidonic acid, C20:4 ω6 (%)	26	21	8.3
Eicosapentaenoic acid, C20:5 ω3 (%)	0.5	1.6	8.0
ω6:ω3	52	13	1
Cardiovascular mortality (%)	45	12	7

[a] Modified from reference 1.

There are ethnic differences in bleeding time that appear to be related to diet. TABLE 2 shows that the higher the ratio of omega-6/omega-3 fatty acids in platelet phospholipids, the higher the death rate from cardiovascular disease.[1]

There is substantial agreement that ingestion of fish or fish oils has the following effects: platelet aggregation to epinephrine and collagen is inhibited, thromboxane A_2 production is decreased, whole blood viscosity is reduced, and erythrocyte membrane fluidity is increased.[2,3,20-23] Fish oil ingestion increased the concentration of plasminogen activator and decreased the concentration of plasminogen activator inhibitor 1 (PAI-1).[24] *In vitro* studies have demonstrated that PAI-1 is synthesized and secreted in hepatic cells in response to insulin, and population studies indicate a strong correlation between insulinemia and PAI-1 levels. In patients with types IIb and IV hyperlipoproteinemia and in another double-blind clinical trial involving 64 men aged 35 to 40 years, ingestion of omega-3 fatty acids decreased the fibrinogen concentration.[25] Two other studies did not show a decrease in fibrinogen, but in one a small dose of cod-liver oil was used[26] and in the other the study consisted of normal volunteers and was of short duration. A recent study noted that fish and fish oil increase fibrinolytic activity, indicating that 200 g/day of lean fish or 2 g of omega-3 EPA and DHA improves certain hematologic parameters implicated in the etiology of cardiovascular disease.[27]

In summary, the antithrombotic effects of EPA and DHA supplementation suggest that their ingestion leads to a return to a more-physiologic state and away from a prothrombotic and atherogenic state in both animal models and human beings.

Ingestion of omega-3 fatty acids not only increases the production of PGI_3, but also of PGI_2 in tissue fragments from the atrium, aorta, and saphenous vein obtained at surgery in patients who received fish oil two weeks prior to surgery.[28] Omega-3 fatty acids inhibit the production of platelet-derived growth factor (PDGF) in bovine endothelial cells.[29] PDGF is a chemoattractant for smooth muscle cells and a powerful mitogen. Thus, the reduction in its production by endothelial cells, monocytes/macrophages, and platelets could inhibit both the migration and proliferation of smooth muscle cells, monocytes/macrophages, and fibroblasts in the arterial wall. Insulin increases the growth of smooth muscle cells, leading to increased risk for the development of atherosclerosis. Omega-3 fatty acids increase endothelium-derived relaxing factor (EDRF).[30] EDRF, presumably nitric oxide, facilitates relaxation in large arteries and vessels. In the presence of EPA, endothelial cells in culture increase the release of relaxing factors, indicating a direct effect of omega-3 fatty acids on the cells.

In animal experiments, rats fed diets high in omega-6 fatty acids developed insulin resistance. Partially substituting fish oil restored normal insulin action.[31] LNA produced similar effects, except when the diet had high amounts of 18:2 ω6 and some AA, in which case it was necessary to use EPA and DHA. When diets deficient in omega-3 fatty acids were used, all fats induced insulin resistance. These findings led to the conclusion that a deficiency of EPA and DHA or a high omega-6/omega-3 ratio in dietary fats could contribute to insulin resistance. Considering that the current Western diet has an absolute and relative deficiency of omega-3 fatty acids, it is obvious that genetic predisposition in the present dietary environment is conducive to the development of NIDDM. Furthermore, in designing clinical interventions, it is evident that consideration should be given to saturated fat intake, omega-6/omega-3 fatty acid ratio, and *trans* fatty acid intake. The ratio of omega-6/omega-3 fatty acids should be similar to the Paleolithic diet.[8] Although precise information is lacking, it is considered to be 1–2/1 (ω6/ω3) rather than 20–30/1, which is the ratio in most Western diets.

Studies carried out in India indicate that the higher 18:2 ω6/18:3 ω3 = 20/1 in their food supply led to increases in the prevalence of NIDDM in the population, whereas a diet with a ratio of 6:1 led to decreases.[32] A recent prospective study in the United States showed that 18:3 ω3 intake is negatively related to the development of coronary artery disease,[33] and 18:3 ω3 added to a Mediterranean type of diet in patients with one episode of myocardial infarction decreased mortality by 70%.[34]

TRANS FATTY ACIDS AND THEIR EFFECTS ON ESSENTIAL FATTY ACID METABOLISM, LIPID LEVELS, PLATELET AGGREGATION, AND OTHER ASPECTS OF ATHEROSCLEROSIS

The physical properties of *trans* fatty acids are like those of saturated fatty acids. *Trans* fatty acids have adverse effects on various enzyme systems involved in lipid metabolism of rats.[35-38]

Trans fatty acids (*cis*, *trans* and *trans*, *trans* linoleates, 18:2 ω6) are devoid of essential fatty acid (EFA) activity and, as the sole dietary source of fat, they retard growth to a greater degree than an EFA-deficient diet in rats.[37] Mahfouz *et al.*[39] observed inhibition of delta-9 and delta-6 desaturases by various isomers of *trans* 18:1 ω9. *Trans*, *trans* 18:2 ω6 decreased the conversion of linoleic to gamma-linolenic acid in rat liver microsomal preparations.[40] *Trans*, *cis* 18:2 ω6 also inhibited the desaturation of 18:2 ω6. Various studies involving rodents both *in vivo*[41-45] and *in vitro*[39,40,46-50] and human fibroblasts *in vitro*[51] have adequately demonstrated that *trans* fatty acids impair the microsomal desaturation and chain elongation of both LA and LNA to their long-chain metabolites in the 20- and 22-carbon atoms, specifically AA and DHA.

As early as 1969, Spritz and Mishkel suggested that many of the effects of unsat-

TABLE 3. Adverse Effects of *Trans* Fatty Acids[a]

Increase
  Low-density lipoprotein (LDL)
  Platelet aggregation
  Lipoprotein (a) [Lp(a)]
  Body weight
  Cholesterol transfer protein (CTP)
  Abnormal morphology of sperm (in male rats)
Decrease or Inhibit
  Decrease or inhibit incorporation of other fatty acids into cell membranes
  Decrease high-density lipoprotein (HDL)
  Inhibit delta-6 desaturase (interfere with elongation and desaturation of essential fatty acids)
  Decrease serum testosterone (in male rats)
  Cross the placenta and decrease birth weight (in humans)

[a] From reference 57.

urated or polyunsaturated fatty acids were due to their cross-sectional area.[52] They further suggested and showed that the *trans* fatty acids should have the same effect as saturated fats in raising serum cholesterol concentrations. It was the first demonstration that changes in the structure of PUFA changed their biologic effects. *Trans* fatty acids raise triglycerides in comparison to butterfat[53] and raise LDL cholesterol in comparison to oleic acid; also, they lower HDL, whereas oleic acid does not.[54] Nestel et al.[55] also showed that a diet consisting of elaidic acid at 7% energy intake (about two times as high as the Australian diet, but within the range of the United States diet) increased LDL cholesterol and significantly elevated Lp(a) as compared to the other three diets: (1) enriched with butterfat (lauric, myristic, palmitic); (2) oleic acid-rich; and (3) palmitic-rich. *Trans* fatty acids lower HDL and increase triglycerides, Lp(a), and platelet aggregation.[56] Furthermore, these effects occurred even when *trans* fatty acids comprised 7% of the energy, which is within the range of *trans* fatty acid intake in the American diet.[16] TABLE 3 shows the various adverse effects of *trans* fatty acids.[57] Of significance is the fact that *trans* fatty acids increase fat cell size in animals. Large fat cells were predictive of NIDDM in Pima Indians.[58]

FATTY ACIDS AND DIABETES

NIDDM is a multigenic and multifactorial disorder. A number of environmental factors contribute to insulin resistance (TABLE 4). The importance of genetic predisposition and its interaction with diet and exercise in the development of NIDDM has been known since the time of Hippocrates, but progress on the genetics of NIDDM has been slow until recently.[58-60] The observation that fatty acids can regulate gene expression in adipose cells may introduce a link between the composition of diets and the hyperplastic/hypertrophic response of white adipose tissue; also, the characterization of fatty acid-responsive genes may produce some clues about the development of the insulin-resistant state and cell hypertrophy.[61] NIDDM is characterized

TABLE 4. Environmental Factors Influencing Insulin Action, Increasing Insulin Resistance, or Decreasing Sensitivity to Insulin and Secondary Hyperinsulinemia[a]

(1) Insulin action declines with increases in body weight (although, in some patients, insulin resistance precedes the obese state)
(2) Physical activity decreases insulin resistance
(3) Weight loss decreases insulin resistance
(4) Saturated fat intake increases insulin resistance
(5) Linoleic acid (18:2 ω6) in muscle phospholipid is positively related to hyperinsulinemia
(6) Long-chain polyunsaturated fatty acids (C20-C22) in muscle phospholipids are inversely related to hyperinsulinemia and positively related to insulin sensitivity

[a] Modified from reference 60.

by hyperglycemia in the presence of insulin resistance, hypertriglyceridemia, and the development of vascular complications. Males with NIDDM have a threefold and females have a fivefold increase in cardiovascular mortality compared to the nondiabetic population, and this increased risk is also carried by nondiabetic first-degree relatives of NIDDM subjects. Clustering of atherogenic (dyslipidemia, hypertension) and thrombotic (PAI-1, factor VII, fibrinogen) risk factors in association with insulin resistance may explain the increased risk. EPA and DHA decrease PAI-1, factor VII, and serum fibrinogen levels (reviewed in reference 62).

Clustering of hypertension, lipid abnormalities, obesity, and diabetes in the same subject may indicate a shared genetic background.[63,64] Apolipoprotein (Apo) E 4/3 and 4/4 phenotypes in patients with central obesity were shown to be associated with high insulin concentration and a most-abnormal insulin-glucose ratio.[65] Therefore, in selecting patients for clinical intervention, Apo E polymorphism needs to be taken into consideration since women with Apo E 3/2 phenotype are less susceptible and those with Apo E 4/3 and 4/4 are more susceptible to hyperinsulinemia and related abnormalities characteristic of insulin resistance, including hypertension.[65]

In 1993, Borkman et al.[66] showed that the hyperinsulinemia/insulin resistance is inversely associated with the amount of C20-C22 fatty acids in muscle cell membrane phospholipids in patients with coronary heart disease and normal volunteers. This decrease in C20-C22 could occur (1) by a lower dietary intake since current Western diets are characterized by a decrease of EPA and DHA relative to both LA (18:2 ω6) and AA; (2) by increased dietary intake of *trans* fatty acids interfering with desaturation and elongation of LA (18:2 ω6) and LNA (18:3 ω3), leading to lower AA, EPA, and DHA; (3) by genetic defects of delta-6 and delta-5 desaturases or other genetic defects interfering with the transport or binding of C20-C22; or (4) by the very high LA (18:2 ω6) dietary intake leading to decreased production of AA as well as interfering with the desaturation and elongation of LNA (18:3 ω3) to EPA and DHA (FIG. 2, TABLE 5).[60,67]

An increase in C20-C22 long-chain PUFA—AA, EPA, and DHA—leads to (1) increases in membrane fluidity, (2) increases in the number of insulin receptors, and (3) increases in insulin action. In humans, the ratio of omega-6/saturated fatty acids in serum phospholipids correlates with insulin sensitivity. *Trans* fatty acids are incorporated into cell membrane phospholipids, leading to decreased fluidity of mem-

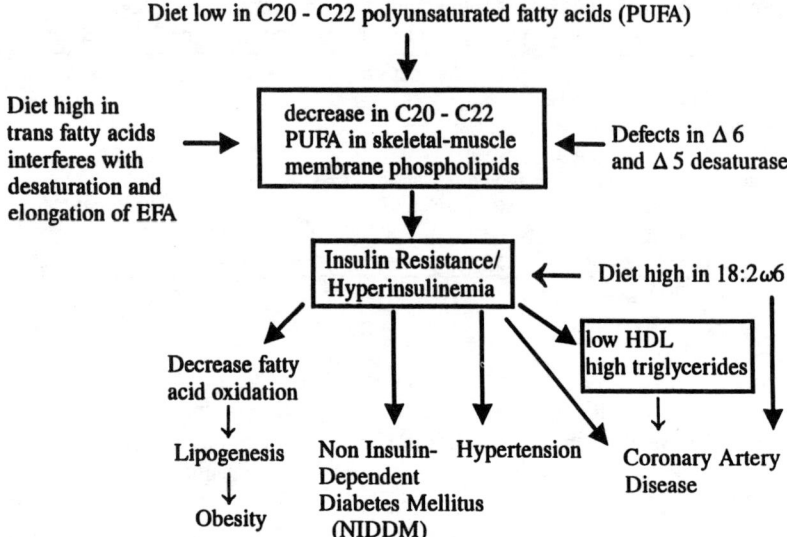

FIGURE 2. A hypothetical scheme of the effects of dietary C20-C22 PUFA on the composition of the C20-C22 PUFA in skeletal-muscle membrane phospholipids and their relationship to insulin resistance/hyperinsulinemia and chronic disease (obesity, NIDDM, hypertension, coronary artery disease).

branes and decreased insulin binding to its receptor, followed by impaired insulin action, insulin resistance, and hyperinsulinemia. In animal experiments, isomeric *trans* fatty acids increase 18:2 ω6, while lowering AA in tissue phospholipids, indicating inhibition of delta-6 desaturase.

There have been about 23 studies on the effects of omega-3 fatty acids in patients with NIDDM (TABLE 6).[68] In most studies, fish oils decreased serum triglyceride levels significantly; however, in some studies, there was an increase in plasma glucose levels. In many of these studies, though, the number of subjects was small, the dose of omega-3 fatty acids was higher than 3 g/day, and controls were lacking. A dose of 3 g/day decreased triglyceride levels significantly. At the level of 3 g of omega-3 fatty acids per day, only one study showed an increase in blood glucose level.

TABLE 5. Factors Decreasing C20-C22 Long-Chain Polyunsaturated Fatty Acids (PUFA) in Muscle Phospholipids[a]

Decreased intake of C20-C22
Increased linoleic acid (18:2 ω6) intake
Factors interfering with desaturation and elongation of 18:2 ω6 and 18:3 ω3 to C20-C22, such as increased *trans* fatty acid intake
Genetic defects in delta-6 and delta-5 desaturases
Genetic defects in intestinal fatty acid binding protein (IFABP)
Increased catabolism of arachidonic acid (AA)

[a] Modified from reference 60.

TABLE 6. Published Effects of ω3 FA on Fasting Glucose and Triglyceride Levels in NIDDM Patients[a]

Effect on Glucose	Effect on Triglycerides	Number of Data Sets
→	→	7
→	↓	11
↑	↓	3
→	not reported	1
↑	not reported	1

[a] Modified from reference 68.

In a randomized double-blind placebo-controlled crossover trial, Connor et al.[69] used 6 g of omega-3 fatty acids for six months, in addition to their usual oral therapy. Fasting serum glucose levels increased by 11% during the omega-3 fatty acid phase and by 8% during the placebo phase (olive oil), a net increase of 3%, which is not statistically significant. Similarly, there was no significant change in glycosylated hemoglobin concentrations, while there was a 43% decrease in fasting triglyceride levels, which is highly significant. This study is the largest- and longest-reported placebo-controlled trial. It has convincingly demonstrated that omega-3 fatty acids along with oral therapy for diabetes can lower triglycerides without an adverse effect on glycemic control.

In conclusion, EPA and DHA are beneficial in decreasing cardiovascular risk factors in patients with NIDDM without adversely affecting glycemic control. There is a need to carry out clinical trials with determination of cardiovascular disease endpoints in addition to measuring risk factors, such as changes in lipid levels, and abnormalities in carbohydrate metabolism. Because diabetes is a multigenic disease, stratification of patients and controls by genetic markers such as Apo E or IFABP is necessary. In developing protocols, it is essential to consider a ratio of omega-6/omega-3 of about 1–2/1, while decreasing saturated fat intake and keeping *trans* fatty acids to 2% of the total energy intake.

REFERENCES

1. SIMOPOULOS, A. P. 1991. Omega-3 fatty acids in health and disease and in growth and development. Am. J. Clin. Nutr. **54:** 438–463.
2. SIMOPOULOS, A. P., R. R. KIFER & R. E. MARTIN, Eds. 1986. Health Effects of Polyunsaturated Fatty Acids in Seafoods. Academic Press. New York.
3. GALLI, C. & A. P. SIMOPOULOS. 1989. Dietary W3 and W6 Fatty Acids: Biological Effects and Nutritional Essentiality. Plenum. New York.
4. SIMOPOULOS, A. P., R. R. KIFER, R. E. MARTIN & S. M. BARLOW, Eds. 1991. Health Effects of W3 Polyunsaturated Fatty Acids in Seafoods. Vol. 66. World Rev. Nutr. Diet. Karger. Basel.
5. GALLI, C., A. P. SIMOPOULOS & E. TREMOLI, Eds. 1994. Fatty Acids and Lipids: Biological Aspects. Vol. 75. World Rev. Nutr. Diet. Karger. Basel.
6. GALLI, C., A. P. SIMOPOULOS & E. TREMOLI, Eds. 1994. Effects of Fatty Acids and Lipids in Health and Disease. Vol. 76. World Rev. Nutr. Diet. Karger. Basel.
7. SALEM, N., JR., A. P. SIMOPOULOS, C. GALLI, M. LAGARDE & H. R. KNAPP, Eds. 1996. Fatty Acids and Lipids from Cell Biology to Human Disease. Lipids Volume 31 (suppl.), p. S1–S326.

8. EATON, S. B. & M. KONNER. 1985. Paleolithic nutrition: a consideration of its nature and current implications. N. Engl. J. Med. **312:** 283-289.
9. SIMOPOULOS, A. P. 1995. The Mediterranean food guide: Greek column rather than an Egyptian pyramid. Nutr. Today **30**(2): 54-61.
10. REPORT OF THE NATIONAL CHOLESTEROL EDUCATION PROGRAM EXPERT PANEL ON DETECTION, EVALUATION, AND TREATMENT OF HIGH BLOOD CHOLESTEROL IN ADULTS. 1988. Arch. Intern. Med. **148:** 36-69.
11. CRAWFORD, M. A. 1968. Fatty acids in free-living and domestic animals. Lancet **1:** 1329-1333.
12. VAN VLIET, T. & M. B. KATAN. 1990. Lower ratio of n-3 to n-6 fatty acids in cultured than in wild fish. Am. J. Clin. Nutr. **51:** 1-2.
13. SIMOPOULOS, A. P. & N. SALEM, JR. 1989. n-3 fatty acids in eggs from range-fed Greek chickens. N. Engl. J. Med. **321:** 1412.
14. SIMOPOULOS, A. P. & N. SALEM, JR. 1986. Purslane: a terrestrial source of omega-3 fatty acids. N. Engl. J. Med. **315:** 833.
15. SIMOPOULOS, A. P., H. A. NORMAN & J. E. GILLASPY. 1995. Purslane in human nutrition and its potential for world agriculture. *In* Plants in Human Nutrition. Vol. 77, p. 47-74. World Rev. Nutr. Diet. Karger. Basel.
16. DUPONT, J., P. J. WHITE & E. B. FELDMAN. 1991. Saturated and hydrogenated fats in food in relation to health. J. Am. Coll. Nutr. **10:** 577.
17. ENIG, M. G., S. ATAL, M. KEENEY & J. SAMPUGNA. 1990. Isomeric *trans* fatty acids in the U.S. diet. J. Am. Coll. Nutr. **9:** 471.
18. BROX, J. H., J. E. KILLIE, B. OSTERUD, S. HOLME & A. NORDOY. 1983. Effects of cod liver oil on platelets and coagulation in familial hypercholesterolemia (type IIa). Acta Med. Scand. **213:** 137.
19. JOIST, J. H., R. K. BAKER & G. SCHONFELD. 1979. Increased *in vivo* and *in vitro* platelet function in type II- and type IV-hyperlipoproteinemia. Thromb. Res. **15:** 95.
20. WEBER, P. C. & A. LEAF. 1991. Cardiovascular effects of w3 fatty acids: atherosclerosis risk factor modification by w3 fatty acids. World Rev. Nutr. Diet. **66:** 218-232.
21. BOTTIGER, L. E., J. DYERBERG & A. NORDOY. 1989. n-3 fish oils in clinical medicine. J. Intern. Med. **225**(suppl. 1): 1.
22. LEWIS, R. A., T. H. LEE & K. F. AUSTEN. 1986. Effects of omega-3 fatty acids on the generation of products of the 5-lipoxygenase pathway. *In* Health Effects of Polyunsaturated Fatty Acids in Seafoods. Academic Press. New York.
23. CARTWRIGHT, I. J., A. G. POCKLEY, J. H. GALLOWAY *et al.* 1985. The effects of dietary w-3 polyunsaturated fatty acids on erythrocyte membrane phospholipids, erythrocyte deformability, and blood viscosity in healthy volunteers. Atherosclerosis **55:** 267.
24. BARCELLI, U. O., P. GLASS-GREENWALT & V. E. POLLAK. 1985. Enhancing effect of dietary supplementation with omega-3 fatty acids on plasma fibrinolysis in normal subjects. Thromb. Res. **39:** 307.
25. RADACK, K., C. DECK & G. HUSTER. 1989. Dietary supplementation with low-dose fish oils lowers fibrinogen levels: a randomized, double-blind controlled study. Ann. Intern. Med. **111:** 757.
26. SANDERS, T. A. B., M. VICKERS & A. P. HAINES. 1981. Effect on blood lipids and hemostasis of a supplement of cod-liver oil, rich in eicosapentaenoic and docosahexaenoic acids, in healthy young men. Clin. Sci. **61:** 317.
27. BROWN, A. J. & D. C. K. ROBERTS. 1991. Fish and fish oil intake: effect on hematological variables related to cardiovascular disease. Thromb. Res. **64:** 169.
28. DE CATERINA, R., D. GIANNESSI, A. MAZZONE *et al.* 1990. Vascular prostacyclin is increased in patients ingesting n-3 polyunsaturated fatty acids prior to coronary artery bypass surgery. Circulation **82:** 428.
29. FOX, P. L. & P. E. DICORLETO. 1988. Fish oils inhibit endothelial cell production of a platelet-derived growth factor-like protein. Science **241:** 453.
30. SHIMOKAWA, H. & P. M. VANHOUTTE. 1988. Dietary cod-liver oil improves endothelium dependent responses in hypercholesterolemic and atherosclerotic porcine coronary arteries. Circulation **78:** 1421.
31. STORLIEN, L. H., A. B. JENKINS, D. J. CHISHOLM, W. S. PASCOE, S. KHOURI & E. W. KRAEGEN. 1991. Influence of dietary fat composition on development of insulin resistance. Diabetes **40:** 280-289.
32. RAHEJA, B. S., S. M. SADIKOT, R. B. PHATAK & M. B. RAO. 1993. Significance of the n-6/n-3 ratio for insulin action in diabetes. Ann. N.Y. Acad. Sci. **683:** 258-271.

33. ASCHERIO, A., E. B. RIMM, E. L. GIOVANNUCCI, D. SPIEGELMAN, M. STAMPFER & W. C. WILLETT. 1996. Dietary fat and risk of coronary heart disease in men: cohort follow-up study in the United States. Br. Med. J. **313**: 84–90.
34. DE LORGERIL, M., S. RENAUD, N. MAMELLE, P. SALEN, J.-L. MARTIN, I. MONJAUD, J. GUIDOLLET, P. TOUBOUL & J. DELAYE. 1994. Mediterranean alpha-linolenic acid–rich diet in secondary prevention of coronary heart disease. Lancet **343**: 1454–1459.
35. EMKEN, E. F. 1984. Nutrition and biochemistry of *trans* and positional fatty acid isomers in hydrogenated oils. Annu. Rev. Nutr. **4**: 339–376.
36. HWANG, D. H. & J. E. KINSELLA. 1978. The effects of *trans* linoleic acid on the concentration of serum prostaglandins $F_{2\alpha}$ and platelet aggregation. Prostaglandins Med. **1**: 121–130.
37. PRIVETT, O. S., F. PHILLIPS, H. SHIMASAKI, T. NOZAWA & E. C. NICKELL. 1977. Studies of effects of *trans* fatty acids in the diet on lipid metabolism in essential fatty acid deficient rats. Am. J. Clin. Nutr. **30**: 1009–1017.
38. HOY, C.-E. & G. HOLMER. 1990. Influence of dietary linoleic acid and *trans* fatty acids on the fatty acid profile of cardiolipins in rats. Lipids **25**: 455–459.
39. MAHFOUZ, M. M., S. JOHNSON & R. T. HOLMAN. 1980. The effect of isomeric *trans* 18:1 acids on the desaturation of palmitic, linoleic, and eicosa-8,11,14-trienoic acids by rat liver microsomes. Lipids **15**: 100–107.
40. BRENNER, R. R. & R. O. PELUFFO. 1969. Regulation of unsaturated fatty acid biosynthesis: effect of unsaturated fatty acids of 18 carbons on the microsomal desaturation of linoleic acid into gamma-linolenic acid. Biochim. Biophys. Acta **176**: 471–479.
41. ANDERSON, R. L., C. S. FULLMER & E. J. HOLLENBACH. 1975. Effects of the *trans* isomers of linoleic acid on the metabolism of linoleic acid in rats. J. Nutr. **105**: 393–400.
42. KINSELLA, J. E., D. H. HWANG, P. YU, J. MAI & J. SHIMP. 1979. Prostaglandins and their precursors in tissues from rats fed on *trans*, *trans*-linoleate. Biochem. J. **184**: 701–704.
43. HWANG, D. H., P. CHANMUGAM & R. ANDING. 1982. Effects of dietary 9-*trans*, 12-*trans* linoleate on arachidonic acid metabolism in rat platelets. Lipids **17**: 307–313.
44. LAWSON, L. D., E. G. HILL & R. T. HOLMAN. 1983. Suppression of arachidonic acid in lipids of rat tissues by dietary mixed isomeric *cis* and *trans* octadecenoates. J. Nutr. **113**: 1827–1835.
45. BRUCKNER, G., S. GOSWAMI & J. E. KINSELLA. 1984. Dietary trilinolelaidate: effects on organ fatty acid composition, prostanoid biosynthesis, and platelet function in rats. J. Nutr. **114**: 58–67.
46. LAWSON, L. D. & F. A. KUMMEROW. 1979. Beta-oxidation of the coenzyme A esters of elaidic, oleic, and stearic acids and their full-cycle intermediates by rat heart mitochondria. Biochim. Biophys. Acta **573**: 245–254.
47. MAHFOUZ, M. M., T. L. SMITH & F. A. KUMMEROW. 1984. Effect of dietary fats on desaturase activities and the biosynthesis of fatty acids in rat-liver microsomes. Lipids **19**: 214–222.
48. DE SCHRIJVER, R. & O. S. PRIVETT. 1982. Interrelationship between dietary *trans* fatty acids and the 6- and 9-desaturases in the rat. Lipids **17**: 27–34.
49. SHIMP, J. L., G. BRUCKNER & J. E. KINSELLA. 1982. The effects of dietary trilinolelaidin on fatty acid and acyl desaturases in rat liver. J. Nutr. **112**: 722–735.
50. CHERN, J. & J. E. KINSELLA. 1983. The effects of unsaturated fatty acids on the synthesis of arachidonic acid in rat kidney cells. Biochim. Biophys. Acta **750**: 465–471.
51. ROSENTHAL, M. D. & M. A. DOLORESCO. 1984. The effects of *trans* fatty acids on fatty acyl delta-5 desaturation by human skin fibroblasts. Lipids **19**: 869–874.
52. SPRITZ, N. & M. A. MISHKEL. 1969. Effects of dietary fats on plasma lipids and lipoproteins: an hypothesis for the lipid-lowering effect of unsaturated fatty acids. J. Clin. Invest. **48**: 78–86.
53. ANDERSON, J. T., R. GRANDE & A. KEYS. 1961. Hydrogenated fats in the diet and lipids in the serum of man. J. Nutr. **75**: 388–394.
54. MENSINK, R. P. & M. B. KATAN. 1990. Effect of dietary *trans* fatty acids on high-density and low-density lipoprotein cholesterol levels in healthy subjects. N. Engl. J. Med. **323**: 439–445.
55. NESTEL, P., M. NOAKES, B. BELLING, R. MCARTHER, P. CLIFTON, E. JANUS & M. ABBEY. 1992. Plasma lipoprotein lipid and Lp(a) changes with substitution of elaidic acid for oleic acid in the diet. J. Lipid Res. **33**: 1029–1036.
56. GAUTHERON, P. & S. RENAUD. 1972. Hyperlipidemia-induced hypercoagulable state in rat: role of an increased activity of platelet phosphatidylserine in response to certain dietary fatty acids. Thromb. Res. **1**: 353–370.
57. SIMOPOULOS, A. P. 1995. Evolutionary aspects of diet: obesity and reference standards. *In* Obesity: New Directions in Assessment and Management. Charles Press. Philadelphia.
58. TATARANNI, P. A., L. J. BAIER, G. PAOLISSO, B. V. HOWARD & E. RAVUSSIN. 1996. Role of lipids

in development of noninsulin-dependent diabetes mellitus: lessons learned from the Pima Indians. Lipids 31: S267–S270.
59. SIMOPOULOS, A. P., V. HERBERT & B. JACOBSON. 1995. The Healing Diet. Macmillan Co. New York.
60. SIMOPOULOS, A. P. 1994. Is insulin resistance influenced by dietary linoleic acid and *trans* fatty acids? Free Radical Biol. Med. **17**(4): 367–372.
61. AMRI, A. 1991. Fatty acids are inducers of the aP_2 gene expression. J. Lipid Res. **32**: 1449–1456.
62. SIMOPOULOS, A. P. 1996. Part 1: Metabolic effects of omega-3 fatty acids and essentiality. *In* Handbook of Lipids in Human Nutrition. CRC Press. Boca Raton, Florida.
63. DESPRES, J.-P., S. MOORJANI, P. J. LUPIEN, A. TREMBLAY, A. NADEAU & C. BOUCHARD. 1992. Genetic aspects of susceptibility to obesity and related dyslipidemias. Mol. Cell. Biochem. **113**: 151–169.
64. CARMELLI, D., L. R. CARDON & R. FABSITZ. 1994. Clustering of hypertension, diabetes, and obesity in adult male twins: same genes or same environments. Am. J. Hum. Genet. **55**: 566–573.
65. UUSITUPA, M. I. J., L. KARHUNEN, A. RISSANEN, A. FRANSSILA-KALLUNKI, L. NISKANEN, K. KERVINEN & Y. ANTERO KESANIEMI. 1996. Apolipoprotein E phenotype modifies metabolic and hemodynamic abnormalities related to central obesity in women. Am. J. Clin. Nutr. **64**: 131–136.
66. BORKMAN, M., L. H. STORLIEN, D. A. PAN, A. B. JENKINS, D. J. CHISHOLM & L. V. CAMPBELL. 1993. The relation between insulin sensitivity and the fatty-acid composition of skeletal-muscle phospholipids. N. Engl. J. Med. **328**: 238–244.
67. SIMOPOULOS, A. P. 1994. Fatty acid composition of skeletal muscle membrane phospholipids, insulin resistance, and obesity. Nutr. Today **29**(1): 12–16.
68. HARRIS, W. S. 1996. Do w3 fatty acids worsen glycemic control in NIDDM? ISSFAL Newsl. **3**(2): 6–9.
69. CONNOR, W. E., M. J. PRINCE, D. ULLMANN *et al.* 1993. The hypotriglyceridemic effect of fish oil in adult-onset diabetes without adverse glucose control. Ann. N.Y. Acad. Sci. **683**: 337–340.

Dietary Fats and Hypertension
Focus on Fish Oil

PETER R. C. HOWE[a]

CSIRO Division of Human Nutrition
Adelaide SA 5000, Australia

HYPERTENSION – NONPHARMACOLOGICAL APPROACHES

Hypertension is not only a major risk factor for cardiovascular disease, but it is probably the most prevalent medically treatable chronic condition in Western societies, affecting more than 20% of the adult population. Improvements in its detection and management are an important factor in the progressive decline of cardiovascular mortality.[1,2] Yet, despite the availability of antihypertensive medication, an unacceptably high proportion of hypertension is still poorly controlled.[2] Moreover, an increasing incidence of hypertension may pose a new health threat to developing countries as they become more affluent and adopt Western lifestyles.

Thus, there is a need to identify and promote diet and lifestyle modifications that can help to prevent or treat hypertension. Such nonpharmacological therapies must be as thoroughly substantiated as antihypertensive drugs; otherwise, they are likely to be viewed with skepticism by the medical profession and ignored by patients. Unfortunately, compared with the vast expenditure on antihypertensive drug development, there is little financial incentive to evaluate natural therapies. Moreover, their effects on blood pressure tend to be more subtle and harder to demonstrate. Nevertheless, some consensus has been reached in recent years on dietary strategies that can help to control blood pressure, particularly at the labile stage before drug treatment becomes necessary, that is, in borderline or mild hypertension.[3] The best established of these are weight reduction, salt restriction, and alcohol moderation. However, alternative options to supplement the diet, for example, with potassium, calcium, or polyunsaturated fatty acids, may prove more acceptable and may achieve greater compliance than dietary restrictions.

No doubt, the main health issue impacting on our eating habits is the link between saturated fat and coronary heart disease.[4] Reducing saturated fat intake is the primary intervention strategy for hyperlipidemia, which, together with obesity, insulin resistance, and hypertension, is recognized as one of a cluster of associated risk factors for cardiovascular disease.[5] However, hypertension must also be acknowledged as an independent risk factor, not necessarily associated with metabolic abnormalities, but attributable to a wide range of pathogenic mechanisms. Without knowing the un-

---

[a] Present address: Department of Biomedical Science, University of Wollongong, Wollongong NSW 2522, Australia.

derlying causes, teasing out the effects of dietary change on specific risk factors can be difficult. Determining the influence of dietary fat on blood pressure typifies this difficulty. If adoption of a low fat diet is accompanied by weight loss, it can also improve at least two related risk factors, viz. obesity and hypertension. In fact, blood pressure is estimated to fall by 1.6/1.3 mmHg per kg of weight reduction in overweight subjects.[6] While there have been many attempts to demonstrate direct effects of lipids on blood pressure, only a few intervention trials were strictly controlled for other dietary effects. In these cases, large changes in the total[7] or saturated fat[8,9] content of the diet had no effect on blood pressure. It now appears that, without weight loss, reduction of fat intake per se is unlikely to lower blood pressure.[10-12]

ANTIHYPERTENSIVE EFFECTS OF ω-6 POLYUNSATURATED FATTY ACIDS

On the other hand, it appears that polyunsaturated fatty acids can exert specific effects on blood pressure. Hence, the *type* of fat may be more important than the quantity. In studies to assess the effects of saturated fat on blood pressure, the ratio of polyunsaturated to saturated fat in the diet is usually raised by adding vegetable oils rich in the ω-6 polyunsaturated fatty acid, linoleic acid (LA), and it has been claimed that LA supplementation per se can lower blood pressure by facilitating the production of vasodilatory and natriuretic eicosanoids.[13-15] Moreover, diets rich in γ-linolenic acid (GLA), an intermediate in the conversion of LA to arachidonic acid (the immediate precursor of eicosanoid synthesis), have been shown to counteract hypertension in experimental animals.[16-18] Nevertheless, this issue remains controversial; the observed effects of ω-6 fatty acids vary between different animal models of hypertension[19] and are influenced by the stage of development of the hypertension[20] and by the basal dietary intakes of essential fatty acids and sodium chloride.[17,21]

Theoretically, LA supplementation in salt-sensitive or borderline spontaneously hypertensive rats fed a high-salt, essential fatty acid–deficient diet might be expected to enhance natriuresis by increasing renal prostaglandin E_2 (PGE_2) production, thus counteracting the hypertensive effect of the sodium load. In support of this concept, Dahl salt-sensitive rats have abnormally low renal PGE_2, which can be increased by feeding LA;[22] inhibiting prostaglandin synthesis with indomethacin can prevent the hypotensive effect of an LA-rich diet;[23] and direct administration of arachidonic acid has been shown to lower blood pressure.[24] On the other hand, the extent to which arachidonic acid can be increased by LA supplementation is limited[25,26] and other studies have failed to show an increase in PGE_2 production.[15,20] More importantly, the conditions under which antihypertensive effects of LA have been obtained in animal models must be considered of doubtful relevance to the management of clinical hypertension. Indeed, clinical trials in which LA intake has been selectively increased show no antihypertensive benefit.[8,9,26-28]

The limiting step in the synthesis of arachidonic acid is the conversion of LA to GLA by δ-6 desaturase.[25,26] Hence, GLA-rich oils show greater antihypertensive potential than LA supplements in animal studies.[16-18] However, clinical confirmation is still lacking. The final intermediate between GLA and arachidonic acid is dihomo-γ-linolenic acid (DGLA), which is a direct precursor of the vasodilatory prostaglandin PGE_1. To test the hypothesis that GLA may lower blood pressure by increasing PGE_1 production, we conducted a clinical trial with a purified DGLA supplement and, despite doubling plasma DGLA levels after 6 weeks, blood pressure was unaffected.[29]

ANTIHYPERTENSIVE EFFECTS OF ω-3 POLYUNSATURATED FATTY ACIDS

In contrast to the difficulty in establishing clinical antihypertensive benefits of dietary ω-6 fatty acid supplementation, evidence that increased consumption of the long-chain ω-3 fatty acids in fish and fish oil can lower blood pressure is growing rapidly.[30] The most-likely mechanism is a shift in eicosanoid production away from the 2-series prostaglandins derived from arachidonic acid to the 3-series prostaglandins derived from eicosapentaenoic acid (EPA), the 20-carbon ω-3 equivalent of arachidonic acid, together with competitive inhibition of arachidonic acid metabolism by both EPA and the 22-carbon ω-3 fatty acid docosahexaenoic acid (DHA).[25,26,31] This could lead to reduced synthesis of a variety of cyclooxygenase, lipoxygenase, and cytochrome P-450–dependent monooxygenase products of arachidonic acid, including potent vasoconstrictors such as thromboxane A_2 (TxA_2), $PGF_{2\alpha}$, and 12-, 19-, and 20-hydroxyeicosatetraenoic acid (HETE).[32-35] The contribution of these eicosanoids to hypertension is demonstrated by reductions of vascular tone and blood pressure when their formation is selectively inhibited in hypertensive rats.[34-38]

Although arachidonic acid produces both vasoconstrictor and vasodilator eicosanoids, the net effect of endogenous eicosanoids tends to be vasodilatory and antihypertensive.[30] The ability of ω-6 supplementation to enhance this effect by increasing arachidonic acid levels is limited. Increased consumption of long-chain ω-3 fatty acids, on the other hand, has the potential to suppress the production of the 2-series eicosanoids, vasoconstrictor as well as dilator, while producing 3-series eicosanoids, which are either inactive or vasodilatory.[31,39] Whether or not this causes a fall in blood pressure will depend on the net contribution of eicosanoid mechanisms in different vascular beds.

If hypertension is induced by renal impairment or exacerbated by salt loading, suppression of renal PGE_2 production by ω-3 fatty acids might be expected to facilitate sodium retention and oppose any antihypertensive effect of fish oil.[17,40,41] This may account for the limited effects of dietary fish oil observed in some experimental models of hypertension. In the spontaneously hypertensive rat (SHR) and its stroke-prone substrain, SHRSP, it appears that the vasoconstrictor eicosanoids may predominate.[32-35] Thus, ω-6 supplementation is less likely to lower blood pressure in these strains, whereas ω-3 supplementation, by substituting inactive 3-series eicosanoids for the vasoconstrictor eicosanoids, may have greater antihypertensive potential than in normotensive strains. Our studies in SHR and SHRSP support this hypothesis: we found that feeding an LA-rich diet had no effect on blood pressure compared with diets containing monounsaturated and/or saturated fat,[42] whereas feeding fish oil attenuated the development of hypertension and had little effect on blood pressure in normotensive Wistar/Kyoto (WKY) rats.[41-44] These observations are consistent with the early finding by Schoene and Fiore that blood pressure levels were lower in SHR and SHRSP, but not WKY, fed a menhaden oil diet compared with rats fed a corn oil diet.[32]

A wide range of mechanisms may account for the vascular effects of fish oil (see review by Chin and Dart[45]). Our observations in ganglion-blocked rats suggest that feeding fish oil can lower basal (nonsympathetically mediated) vascular tone.[41] Enhancement of endothelial (nitric oxide–mediated) vasodilation and inhibition of intrinsic thromboxane-like vasoconstriction are possible mechanisms. Clinical studies of forearm blood flow have confirmed the importance of endothelial mechanisms in the vasorelaxant effect of fish oil.[45] Fish oil supplementation can also attenuate vasoconstrictor responses to the sympathetic neurotransmitter, noradrenaline, and, in particular, to angiotensin II (AII), which is strongly implicated in the pathogenesis of spontaneous hypertension.[46] It is of interest that TxA_2 has been postulated to mediate the constrictor effects of AII.[47] Alternatively, ω-3 fatty acids may attenuate vasoconstrictor responsiveness directly by altering calcium transport mechanisms in vascular smooth muscle cells.[45] This hypothesis is supported by observations on the direct application of ω-3 fatty acids to isolated smooth muscle preparations in the presence of cyclooxygenase inhibitors.[48]

Apart from influencing acute hemodynamic mechanisms, it is tempting to speculate that the ω-3 fatty acids can retard the development of hypertension by counteracting vascular hypertrophy, which, in SHR, appears to be at least partly attributable to a trophic influence of intravascular AII.[46] Fish oil may also exert an antihypertrophic influence by inhibiting production of 12-HETE and platelet-derived growth factor.[49] Consistent with this hypothesis, if rats are supplemented with fish oil during the early developmental stages of hypertension, when hypertrophic influences are critical, the rise in blood pressure is usually inhibited.[32,41-44] In contrast, we have found it difficult to demonstrate any blood pressure reduction by ω-3 fatty acids in rats with established hypertension.[43]

CLINICAL TRIALS WITH ω-3 SUPPLEMENTS

Despite the uncertainty about antihypertensive effects and underlying mechanisms in animal models, there is now a reasonable consensus from clinical trials that blood pressure can be lowered by increasing the dietary intake of ω-3 fatty acids.[30] The question is which ω-3 fatty acids and how much. Recent meta-analyses of clinical trials to assess the antihypertensive potential of fish oil supplementation indicate that the extent of blood pressure reduction depends on the initial level of blood pressure and on the dose of ω-3 fatty acids so that clinically significant effects (estimated to be as large as 5.5/3.5 mmHg) may be expected in hypertensive (but not normotensive) subjects, given an average dose of approximately 3 g of ω-3 fatty acids/day.[50,51] This dose relates to the sum of EPA and DHA present in the supplement; concentrations of other ω-3 fatty acids in fish oil preparations are usually considered negligible.

The shorter-chain ω-3 fatty acid, α-linolenic (ALA), is present in high levels (~50%) in linseed (flaxseed) oil and, to a lesser extent (~10%), in canola, and it is thought that these vegetable oils may also have the potential to lower blood pressure.[52] However, incorporation of ALA in tissue phospholipids and its desaturation and elongation to form EPA is limited and critically dependent on the relative intake of LA.[26] Blood pressure reductions have been reported in some, although not all,[27] studies with subjects consuming flaxseed oil[26] or eggs from hens fed flaxseed oil,[53] but the quantities required are either not feasible or uneconomical compared with fish oil. It is unlikely that canola supplementation could yield sufficient ALA to affect blood pressure. In a long-term study of ALA supplementation following myocardial infarction, cardiovascular morbidity was reduced by consumption of margarines containing rapeseed oil, but there was no change in blood pressure.[54] Nevertheless, increased consumption of ALA-rich oils such as canola can favorably modify the ω-6/ω-3 balance. In Australia, the ratio of ω-6/ω-3 fatty acids derived from edible oils has decreased from 26 to 9 in the last decade (A. Green, personal communication). However, this trend is likely to minimize any demonstrable effects of further dietary ω-3 supplementation, in the same way as a high intake of fish in the basal diet was found to negate the antihypertensive effect of fish oil in the Tromsö study.[55]

EPA-rich fish oil preparations have been used in most clinical trials, with the assumption that EPA is the principal mediator of antihypertensive eicosanoid mechanisms. In early studies, large doses (up to 50 g/day) of triglyceride preparations such as MaxEPA, containing 18% EPA and 12% DHA, were required to demonstrate significant reductions of blood pressure.[28] More recent availability of highly concentrated fish oil extracts such as Omacor, containing >85% EPA and DHA as methyl esters, has enabled antihypertensive effects to be obtained with more acceptable levels of supplementation (as little as 4 g/day).[55-58] However, experimental animal studies with purified ω-3 fatty acids now indicate that DHA, which also inhibits eicosanoid synthesis, may have greater efficacy than EPA in a range of cardiovascular functions, including blood pressure reduction.[59] This hypothesis is supported by clinical trials in which VLDL triglycerides, fibrinogen and TxA_2 production, bleeding time, and blood pressure were lowered by supplementing the diet with DHA-rich fish oil[60,61] or by consuming moderate quantities of fish,[61,62] as the tissue concentration of DHA in fish usually exceeds that of EPA. As further studies are undertaken to confirm the antihypertensive efficacy of DHA, there is renewed interest in DHA-rich oils such as tuna oil, with DHA/EPA ratios up to 5:1, and microalgal sources of DHA such as DHASCO (40% DHA with negligible EPA).[63] Once the optimal DHA/EPA ratio has been established, appropriate levels of supplementation should be determined in long-term trials of antihypertensive efficacy. So far, no valid trial has exceeded 6 months of duration and attempts to determine a dose-response relationship have been limited. However, as more-refined and potent ω-3 fatty acid preparations are developed with a view to pharmaceutical registration for specific health indications, more extensive evaluation is likely.

The modest antihypertensive effect of fish oil may be enhanced by combination with other nonpharmacological modalities, such as sodium restriction. In a study with elderly normotensives, we found that fish oil combined with a low-sodium diet caused blood pressure to fall by 7/5 mmHg, considerably more than with either treatment alone.[64] The rationale for evaluating this particular combination was based on the above-mentioned experimental animal studies suggesting that fish oil might exacerbate the hypertensive effect of a high dietary sodium intake by suppressing renal synthesis of prostaglandin E_2, an important mediator of natriuresis.[40] In contrast, we found that sodium restriction could potentiate the antihypertensive effect of fish oil in the animal model.[43] Although we were unable to recruit untreated hypertensives for our clinical trial, we recognized that the elderly are an important risk group for hypertension — more than 40% develop hypertension by 65 years of age — with management often complicated by polypharmacy for other chronic disorders. Moreover, the sensitivity of blood pressure to sodium intake increases with advancing age.[65] Thus, a low-sodium diet with ω-3 supplementation represents a worthwhile nonpharmacological strategy for the management of blood pressure in the elderly. It may also help to prevent the development of hypertension in predisposed subjects, although there is little likelihood that this could be evaluated.

FISH OIL AS AN ADJUNCT THERAPY

Nonpharmacological treatments tend to be limited to those with uncomplicated mild hypertension. At higher levels of blood pressure or when other risk factors are present, antihypertensive drugs are prescribed and the dietary or lifestyle changes are often ignored, even though maintaining these strategies as an adjunct to drug therapy could be very beneficial, not only in terms of blood pressure management, but also with respect to other associated risk factors.[66] It is important to recognize, however, that dietary treatments, like drug treatments, are not necessarily additive, so the efficacy of specific combinations needs to be tested on an individual basis. Singer *et al.* found that, when fish oil was coadministered with the β-blocking drug propranolol, blood pressure reduction was significantly greater than with the antihypertensive drug alone.[67] Yet, in other studies where fish oil supplementation was given to drug-treated hypertensives, there was no further reduction of blood pressure.[68-70] In these studies, the numbers taking each class of antihypertensive drug were too small to draw conclusions about the lack of efficacy of specific treatment combinations.

We are now conducting a series of dietary intervention trials to evaluate ω-3 supplements in patients treated with different classes of antihypertensive drug. We have confirmed and extended Singer's observations in which fish oil and propranolol were given to previously untreated hypertensives[67] by showing that fish oil supplementation could further reduce blood pressure in hypertensive patients recruited through general practices whose blood pressure had already been normalized by long-term treatment with diuretics and/or β-blockers.[71,72] However, our experience with ACE inhibitors[73] indicates that the antihypertensive effect of fish oil will not necessarily be additive with other types of antihypertensive drug.

We tested the effects of sodium restriction and fish oil supplementation in patients managed on long-term ACE inhibitor monotherapy, once again in a general practice setting.[73] Although sodium restriction caused a further fall in systolic pressure, there was no change of blood pressure with the fish oil, either alone or in combination with the sodium restriction. It is unlikely that the dose of fish oil was inadequate;

there were substantial changes in other parameters (triglycerides, platelet TxA_2 production). Moreover, the dose of ω-3 fatty acids given (3.3 g/day) was equivalent to that which lowered blood pressure in our study of hypertensives treated with β-blockers/diuretics. It appears that the lack of synergism may be attributable to the specific nature of the interaction with ACE inhibitors. If, as suggested earlier, ω-3 fatty acids are able to inhibit the vasoconstrictor and/or hypertrophic actions of AII, then inhibition of AII production by ACE inhibitors may render this mechanism redundant. However, considering that ACE inhibitors can facilitate eicosanoid synthesis through bradykinin-dependent mechanisms,[74] the explanation for their lack of interaction with fish oil is likely to be more complex.

Despite the inability of fish oil to further reduce blood pressure in hypertensives treated with ACE inhibitors, it may nevertheless help to reduce their drug requirement. In an extension of our study, general practitioners, although blinded to the nature of the supplements, were encouraged to titrate the dose of ACE inhibitor to maintain the desired level of blood pressure in their patients; there was a trend for greater dose reductions in patients taking fish oil than in those taking the placebo supplement (unpublished observations).

We are currently evaluating an ω-3 supplement as adjunct therapy in hypertensives treated with calcium blockers, the other relatively new major class of antihypertensive drug. One study with this combination has been published.[75] In that study, the calcium blocker was administered to previously untreated hypertensives only after they had failed to respond to fish oil supplementation, and there was no difference between the fish oil and placebo groups in the antihypertensive effect of the drug.

REDUCTION OF ASSOCIATED RISK FACTORS

Regardless of whether ω-3 fatty acid supplementation can further reduce blood pressure in patients being treated with a particular class of antihypertensive drug, it has considerable potential to offset other associated cardiovascular risk factors and possibly afford protection against hypertension-induced target organ damage. These "fringe benefits" may prove to be more valuable than any additional antihypertensive effect of ω-3 supplementation. A major factor in the transition from early, but effective antihypertensive drugs such as β-blockers and diuretics to the newer and far-more-expensive ACE inhibitors and calcium blockers was the recognition that the former classes of drug may have adverse effects on lipid metabolism, raising VLDL cholesterol and triglycerides and decreasing HDL cholesterol, thus compromising the overall cardiovascular benefit for many patients.[76] In our study of hypertensives treated with β-blockers and diuretics, fish oil treatment not only improved blood pressure, but at the same time caused a 21% reduction of plasma triglycerides and a small, but significant, rise in HDL_2 cholesterol, thus counteracting the main adverse effect of these drugs and presumably restoring their full potential to reduce cardiovascular morbidity.[71] The improvement of lipids is of particular significance for non-insulin-dependent diabetics with hypertension, especially since concerns about adverse hyperglycemic effects have been allayed.[77] We have recently shown that a moderate dose of fish oil can lower blood pressure and triglycerides in diabetics, without elevating blood glucose.[78] However, as LDL is more susceptible to oxidation in diabetics,[79] the oxidative potential of fish oil[80,81] may raise additional concerns in diabetes.

Other cardiovascular risk factors prominent in hypertension may also be improved by ω-3 fatty acid supplementation.[82] Although some concern had been expressed about the risk of hemorrhage in hypertensives taking large doses of fish oil, it is now acknowledged that the risk is less than with low-dose aspirin, which inhibits platelet TxA_2 production to a greater extent without providing the fibrinolytic and endothelial benefits ascribed to fish oil.[81-83] For a large proportion of hypertensives with significant vascular disease, the potential of a fish oil supplement to reduce the risk of coronary or cerebral thrombosis is likely to be a more-important consideration. Moreover, there is increasing evidence that ω-3 supplementation can help to prevent hypertensive damage by acting directly on mechanisms within target organs and, in some cases, this may be independent of its effects on systemic blood pressure.[84]

A series of animal experiments by McLennan and colleagues have defined an antiarrhythmic function for ω-3 fatty acids[85] that could account for the reported protection against sudden death following myocardial infarction in patients eating moderate amounts of fish.[86] This benefit may be attributable to the relatively high content of DHA in fish, as DHA was more effective than EPA in preventing the experimental arrhythmias.[59] Of perhaps greater significance for hypertensives is the recent observation that ω-3 supplementation can reduce the oxygen demand of the working heart, thus improving cardiac performance under ischemic conditions (P. McLennan, personal communication). We are currently evaluating the ability of dietary fish oil to prevent hypertension-induced cardiac failure. Our initial experiments in salt-loaded SHRSP, a useful model of malignant hypertension, have shown reductions of cardiac hypertrophy and myocardial fibrosis with fish oil supplementation (unpublished observations). However, these protective effects were elicited during the development of hypertension and may be attributable simply to the antihypertensive effect of the fish oil. It will be important to establish whether fish oil supplementation can help to sustain cardiac function once the hypertension and cardiac hypertrophy are established.

Another example of major significance is the rapidly growing evidence that fish oil supplementation can help to preserve renal function in a variety of pathological conditions, especially those of an inflammatory or immune nature.[87-90] De Caterina et al. have summarized the evidence from different animal models of renal disease in which dietary fish oil confers protection[89] and have postulated a key role for inflammatory cytokines, cell adhesion molecules, and the inhibition of their expression by ω-3 fatty acids.[90] We have now shown that ω-3 fatty acid supplementation can also prevent renal damage due to hypertension.[91] Feeding diets enriched with pure EPA or DHA to salt-loaded SHRSP from one to four months of age retarded the early development of proteinuria, a classical marker of renal disease, whereas a GLA-enriched diet was ineffective. In older animals, with established hypertension, only DHA supplementation was able to inhibit the proteinuria. The antiproteinuric effects were not closely related to the reduction of systemic blood pressure, suggesting that the ω-3 fatty acids were acting directly on intrarenal mechanisms to confer protection. Renal production of TxA_2 and 12-HETE were markedly suppressed by both EPA and DHA supplementation, suggesting a role for eicosanoids in either the initiation or progression of renal damage. Others have found that the lipoxygenase inhibitor, phenidone, which also suppresses renal 12-HETE production, can prevent the development of proteinuria in salt-loaded SHRSP.[92]

In subsequent studies, we have shown that supplementing the diet with fish oil, but not canola oil, inhibits the development of proteinuria in salt-loaded SHRSP.[93,94] The effect of fish oil is dose-dependent (with a threshold concentration in the diet between 2.5% and 5%) and is associated with a reduction in the ratio of ω-3/ω-6 fatty acids in the kidney and the prevention of histopathological changes in glomeruli and tubules.[94] So far, we have been unable to completely dissociate the antihypertensive and antiproteinuric effects of the fish oil. However, in clinical trials with normotensive subjects, fish oil supplementation has been shown to benefit other forms of renal disease without changes in blood pressure.[87,88] Moreover, the unique ability of ACE inhibitors to inhibit proteinuria and renal disease, independent of hypertension or blood pressure reduction,[95] suggests that, even in established hypertension, maintenance of intrarenal microvascular function may be a critical factor in preventing the progression of renal failure. Hypertension is second to diabetes as the main cause of end-stage renal disease, the prevalence of which is still increasing.[96] Regardless of whether or not the renal protective effects of fish oil in salt-loaded SHRSP are attributable to blood pressure reduction, they could be of major therapeutic importance. However, they should also be tested under conditions of normal or reduced salt intake before proceeding to clinical evaluation.

CONCLUSIONS

Reduction of dietary fat, particularly saturated fat, is a key strategy for preventing cardiovascular disease, but it is unlikely to lower blood pressure unless accompanied by weight loss. However, modifying the type of polyunsaturated fatty acid in the diet can influence blood pressure by altering vascular eicosanoid synthesis. In particular, supplementing the diet with fish or fish oil can lower blood pressure in hypertensives. The extent of reduction appears dependent on the degree of hypertension, the level of sodium intake, and the dose of long-chain ω-3 fatty acids, particularly DHA. As an adjunct to drug therapy, ω-3 supplementation can improve the overall cardiovascular risk profile and further decrease blood pressure (depending on the type of concomitant medication). Independent of the extent of blood pressure reduction, it may also help to prevent hypertensive damage by acting directly on target organs. Thus, ω-3 fatty acid supplementation is a promising strategy for the prevention and management of hypertension-related cardiovascular disease and it warrants further long-term evaluation.

REFERENCES

1. THE FIFTH REPORT OF THE JOINT NATIONAL COMMITTEE ON DETECTION, EVALUATION, AND TREATMENT OF HIGH BLOOD PRESSURE. 1993. Arch. Intern. Med. **153**: 154–183.

2. SHEA, S. 1994. Hypertension control. Am. J. Public Health **84:** 1725-1727.
3. KAPLAN, N. M. 1991. Long-term effectiveness of nonpharmacological treatment of hypertension. Hypertension **18**(suppl. I): I153-I160.
4. WHO TECHNICAL REPORT SERIES. 1990. Diet, Nutrition, and the Prevention of Chronic Diseases. No. 797, p. 203. WHO. Geneva.
5. REAVEN, G. M. 1988. Role of insulin resistance in human disease. Diabetes **37:** 1595-1607.
6. STAESSEN, J., R. FAGARD, P. LIJNEN & A. AMERY. 1989. Body weight, sodium intake, and blood pressure. J. Hypertens. **7**(suppl. I): S19-S23.
7. MENSINK, R. P., M.-C. JANSSEN & M. B. KATAN. 1988. Effect on blood pressure of two diets differing in total fat, but not in saturated and polyunsaturated fatty acids in healthy volunteers. Am. J. Clin. Nutr. **47:** 976-980.
8. SACKS, F. M., I. L. ROUSE, M. J. STAMPFER, L. M. BISHOP, C. F. LENHERR & R. J. WALTHER. 1987. Effect of dietary fats and carbohydrate on blood pressure of mildly hypertensive patients. Hypertension **10:** 452-460.
9. MARGETTS, B. M., L. J. BEILIN, B. K. ARMSTRONG, I. L. ROUSE, R. VANDONGEN, K. D. CROFT & E. J. MCMURCHIE. 1985. Blood pressure and dietary polyunsaturated and saturated fats: a controlled trial. Clin. Sci. **69:** 165-175.
10. BEILIN, L. J. 1987. Diet and hypertension: critical concepts and controversies. J. Hypertens. **5**(suppl. 5): S447-S457.
11. SACKS, F. M. 1989. Dietary fats and blood pressure: a critical review of the evidence. Nutr. Rev. **47:** 291-300.
12. PIETINEN, P. & A. ARO. 1990. The role of nutrition in the prevention and treatment of hypertension. Adv. Nutr. Res. **8:** 35-78.
13. COMBERG, H. U., S. HEYDEN & C. G. HAMES. 1978. Hypotensive effect of dietary prostaglandin precursor in hypertensive man. Prostaglandins **15:** 193-197.
14. HOFFMANN, P., C. H. TAUBE, K. PONICKE, U. ZEHL, J. BEITZ, W. FORSTER, L. SOMOVA, V. ORBETZOVA & F. DAVIDOVA. 1982. Alterations in renal and aortic prostaglandin E and F formation correlate with blood pressure increase in salt-loaded rats after dietary linoleate deficiency. Arch. Int. Pharmacodyn. Ther. **259:** 40-58.
15. IZUMI, Y., T. W. WEINER, R. FRANCO-SAENZ & P. J. MULROW. 1986. Effects of dietary linoleic acid on blood pressure and renal function in subtotally nephrectomized rats. Proc. Soc. Exp. Biol. Med. **183:** 193-198.
16. ENGLER, M. M. 1993. Comparative study of diets enriched with evening primrose, black currant, borage, or fungal oils on blood pressure and pressor responses in spontaneously hypertensive rats. Prostaglandins Leukotrienes Essent. Fatty Acids **49:** 809-814.
17. MILLS, D. E., R. P. WARD, M. MAH & L. DEVETTE. 1989. Dietary n-6 and n-3 fatty acids and salt-induced hypertension in the borderline hypertensive rat. Lipids **24:** 17-24.
18. SINGER, P., V. MORITZ, M. WIRTH, I. BERGER & D. FORSTER. 1990. Blood pressure and serum lipids from SHR after diets supplemented with evening primrose, sunflower seed, or fish oil. Prostaglandins Leukotrienes Essent. Fatty Acids **40:** 17-20.
19. SREENIVAS, R. R. & T. A. KOTCHEN. 1996. Attenuation of experimental hypertension by dietary linoleic acid is model dependent. J. Am. Coll. Nutr. **15:** 92-96.
20. HOFFMANN, P., C. TAUBE, L. HEINROTH-HOFFMANN, A. FAHR, J. BEITZ, W. FORSTER, W. S. POLESHUK & C. M. MARKOV. 1985. Antihypertensive action of dietary polyunsaturated fatty acids in spontaneously hypertensive rats. Arch. Int. Pharmacodyn. Ther. **276:** 222-235.
21. WATANABE, Y., Y.-S. HUANG, V. A. SIMMONS & D. F. HORROBIN. 1989. The effect of dietary n-6 and n-3 polyunsaturated fatty acids on blood pressure and tissue fatty acid composition in spontaneously hypertensive rats. Lipids **24:** 638-644.
22. TOBIAN, L., M. GANGULI, A. GOTO *et al.* 1982. Influence of renal prostaglandins and dietary linoleate on hypertension in Dahl S rats. Hypertension **4**(suppl. II): II149-II153.
23. MATHIAS, M. & J. DUPONT. 1985. Quantitative relationships between dietary linoleate and prostaglandin biosynthesis. Lipids **20:** 791-796.
24. OATES, J. A., H. W. SEYBERTH, O. OLEZ *et al.* 1976. Prostaglandins in the etiology and treatment of cardiovascular disease. *In* Proceedings of the Sixth International Congress of Pharmacology. Volume 5, p. 139-150.
25. KINSELLA, J. E., B. LOKESH & R. A. STONE. 1990. Dietary n-3 polyunsaturated fatty acids and amelioration of cardiovascular disease: possible mechanisms. Am. J. Clin. Nutr. **52:** 1-28.

26. SINGER, P., I. BERGER, M. WIRTH, W. GODICKE, W. JAEGER & S. VOIGT. 1986. Slow desaturation and elongation of linoleic and a-linolenic acids as a rationale of eicosapentaenoic acid-rich diet to lower blood pressure and serum lipids in normal, hypertensive, and hyperlipemic subjects. Prostaglandins Leukotrienes Med. 24: 173-193.
27. KESTIN, M., P. CLIFTON, G. B. BELLING & P. J. NESTEL. 1990. n-3 fatty acids of marine origin lower systolic blood pressure and triglycerides, but raise LDL cholesterol compared with n-3 and n-6 fatty acids from plants. Am. J. Clin. Nutr. 51: 1028-1034.
28. KNAPP, H. R. & G. A. FITZGERALD. 1989. The antihypertensive effects of fish oil: a controlled study of polyunsaturated fatty acid supplements in essential hypertension. N. Engl. J. Med. 320: 1037-1043.
29. NESTEL, P. J., P. M. CLIFTON, M. NOAKES, R. MCARTHUR & P. R. C. HOWE. 1993. Enhanced blood pressure response to dietary salt in elderly women, especially those with small waist:hip measurements. J. Hypertens. 11: 1387-1394.
30. HOWE, P. R. C. 1995. Can we recommend fish oil for hypertension? Clin. Exp. Pharmacol. Physiol. 22: 199-203.
31. GIBSON, R. A. 1988. The effects of diets containing fish and fish oils on disease risk factors in humans. Aust. N.Z. J. Med. 18: 713-722.
32. SCHOENE, N. W. & D. FIORE. 1981. Effect of a diet containing fish oil on blood pressure in spontaneously hypertensive rats. Prog. Lipid Res. 20: 569-570.
33. MASFERRER, J. & K. M. MULLANE. 1988. Modulation of vascular tone by 12(R)-, but not 12(S)-hydroxyeicosatetraenoic acid. Eur. J. Pharmacol. 151: 487-490.
34. FU-XIANG, D., J. SKOPEC, A. DIEDRICH & D. DIEDRICH. 1992. Prostaglandin H2 and thromboxane A2 are contractile factors in intrarenal arteries of spontaneously hypertensive rats. Hypertension 19: 795-798.
35. MCGIFF, J. C. 1991. Cytochrome P-450 metabolism of arachidonic acid. Annu. Rev. Pharmacol. Toxicol. 31: 339-369.
36. UDERMAN, H. D., E. K. JACKSON, D. PUETT & R. J. WORKMAN. 1984. Thromboxane synthetase inhibitor UK38,485 lowers blood pressure in the adult spontaneously hypertensive rat. J. Cardiovasc. Pharmacol. 6: 969-972.
37. STERN, N., K. NOZAWA, M. GOLUB, P. EGGENA, E. KNOLL & M. L. TUCK. 1993. The lipoxygenase inhibitor phenidone is a potent hypotensive agent in the spontaneously hypertensive rat. Am. J. Hypertens. 6: 52-58.
38. OSANI, T., T. KIKUCHI, Y. YOKONO, H. MATSUMURA, O. MINAMI, R. AKIBA et al. 1991. Role of renomedullary thromboxane A_2 in development of DOCA-salt hypertension. Clin. Exp. Hypertens. A13: 159-188.
39. DREVON, C. A. 1992. Marine oils and their effects. Nutr. Rev. 50: 38-45.
40. CODDE, J. P., L. J. BEILIN, K. D. CROFT & R. VANDONGEN. 1987. The effect of dietary fish oil and salt on blood pressure and eicosanoid metabolism of spontaneously hypertensive rats. J. Hypertens. 5: 137-142.
41. HOWE, P. R. C., Y. LUNGERSHAUSEN, P. F. ROGERS, J. F. GERKENS, R. J. HEAD & R. M. SMITH. 1991. Effects of dietary sodium and fish oil on blood pressure development in stroke-prone spontaneously hypertensive rats. J. Hypertens. 9: 639-644.
42. HEAD, R. J., M. MANO, S. BEXIS, P. R. C. HOWE & R. M. SMITH. 1991. Dietary fish oil administration retards the development of hypertension and influences vascular neuroeffector function in the stroke prone spontaneously hypertensive rat (SHRSP). Prostaglandins Leukotrienes Essent. Fatty Acids 44: 119-122.
43. HOWE, P. R. C., P. F. ROGERS & Y. LUNGERSHAUSEN. 1991. Blood pressure reduction by fish oil in adult rats with established hypertension—dependence on sodium intake. Prostaglandins Leukotrienes Essent. Fatty Acids 44: 113-117.
44. BEXIS, S., Y. K. LUNGERSHAUSEN, M. T. MANO, P. R. C. HOWE, J. KONG, D. L. BIRKLE, D. A. TAYLOR & R. J. HEAD. 1994. Dietary fish oil administration retards blood pressure development and influences vascular properties in the spontaneously hypertensive rat, but not in the stroke-prone spontaneously hypertensive rat. Blood Pressure 3: 120-126.
45. CHIN, J. P. F. & A. M. DART. 1995. How do fish oils affect vascular function? Clin. Exp. Pharmacol. Physiol. 22: 71-81.
46. LI, P. & E. K. JACKSON. 1987. A possible explanation of genetic hypertension in the spontaneously hypertensive rat. Life Sci. 41: 1903-1908.

47. WILCOX, C. S. & W. J. WELCH. 1990. Thromboxane mediation of the pressor response to infused angiotensin II. Am. J. Hypertens. **3:** 242-249.
48. ENGLER, M. B., M. M. ENGLER & P. C. URSELL. 1994. Vasorelaxant properties of n-3 polyunsaturated fatty acids in aortas from spontaneously hypertensive and normotensive rats. J. Cardiovasc. Risk **1:** 75-80.
49. FOX, P. L. & P. E. DICORLETO. 1988. Fish oils inhibit endothelial cell production of platelet-derived growth factor-like protein. Science **241:** 453-456.
50. MORRIS, M. C., F. SACKS & B. ROSNER. 1993. Does fish oil lower blood pressure? A meta-analysis of controlled trials. Circulation **88:** 523-533.
51. APPEL, L. J., E. R. MILLER, A. J. SEIDLER & P. K. WHELTON. 1993. Does supplementation of diet with fish oil reduce blood pressure? Arch. Int. Med. **153:** 1429-1438.
52. BERRY, E. M. & J. HIRSCH. 1986. Does dietary linolenic acid influence blood pressure? Am. J. Clin. Nutr. **44:** 336-340.
53. OH, S. Y., J. RYUE, C. H. HSIEH & D. E. BELL. 1991. Eggs enriched in omega-3 fatty acids and alterations in lipid concentrations in plasma and lipoproteins and in blood pressure. Am. J. Clin. Nutr. **54:** 689-695.
54. DELORGERIL, M., S. RENAUD, N. MAMELLE *et al.* 1994. Mediterranean alpha-linolenic acid-rich diet in secondary prevention of coronary heart disease. Lancet **343:** 1454-1459.
55. BONAA, K., K. S. BJERVE, B. STRAUME, I. T. GRAM & D. THELLE. 1990. Effect of eicosapentaenoic and docosahexaenoic acids on blood pressure in hypertension. N. Engl. J. Med. **322:** 795-801.
56. TOFT, I., K. BONAA, O. C. INGEBRETSEN, A. NORDOY & T. JENSSEN. 1995. Effects of n-3 polyunsaturated fatty acids on glucose homeostasis and blood pressure in essential hypertension. Ann. Intern. Med. **123:** 911-918.
57. MACKNESS, M. I., D. BHATNAGAR, P. N. DURRINGTON, H. PRAIS, B. HAYNES, J. MORGAN & L. BORTHWICK. 1994. Effects of a new fish oil concentrate on plasma lipids and lipoproteins in patients with hypertriglyceridaemia. Eur. J. Clin. Nutr. **48:** 859-865.
58. GRUNDT, H., D. W. T. NILSEN, O. HETLAND, T. AARSLAND, I. BAKSAAS, T. GRANDE & L. WOIE. 1995. Improvement of serum lipids and blood pressure during intervention with n-3 fatty acids was not associated with changes in insulin levels in subjects with combined hyperlipidaemia. J. Int. Med. **237:** 249-259.
59. MCLENNAN, P., P. HOWE, M. ABEYWARDENA, R. MUGGLI, D. RAEDERSTORFF, M. MANO, T. RAYNER & R. HEAD. 1996. The cardiovascular protective role of docosahexaenoic acid. Eur. J. Pharmacol. **300:** 83-89.
60. SANDERS, T. A. B. & A. HINDS. 1992. The influence of a fish oil high in docosahexaenoic acid on plasma lipoprotein and vitamin E concentrations and haemostatic function in healthy male volunteers. Br. J. Nutr. **68:** 163-173.
61. COBIAC, L., P. M. CLIFTON, M. ABBEY, G. B. BELLING & P. J. NESTEL. 1991. Lipid, lipoprotein, and hemostatic effects of fish vs. fish oil n-3 fatty acids in mildly hyperlipidemic males. Am. J. Clin. Nutr. **53:** 1210-1216.
62. SINGER, P., M. WIRTH, S. VOIGT *et al.* 1985. Blood pressure- and lipid-lowering effect of mackerel and herring diet in patients with mild essential hypertension. Atherosclerosis **56:** 223-235.
63. KYLE, D. J., K. D. B. BOSWELL, R. M. GLADUE & S. E. REEB. 1992. Designer oils from microalgae as nutritional supplements. *In* Biotechnology and Nutrition, p. 451-468. Butterworths-Heineman. London.
64. COBIAC, L., P. J. NESTEL, L. M. H. WING & P. R. C. HOWE. 1992. A low-sodium diet supplemented with fish oil lowers blood pressure in the elderly. J. Hypertens. **10:** 87-92.
65. GROBBEE, D. E. & A. HOFMAN. 1986. Does sodium restriction lower blood pressure? Br. Med. J. **293:** 27-29.
66. THE WORLD HYPERTENSION LEAGUE. 1993. Nonpharmacological intervention as an adjunct to the pharmacological treatment of hypertension: a statement by WHL. J. Hum. Hypertens. **7:** 159-164.
67. SINGER, P., S. MELZER, M. GOSCHEL *et al.* 1990. Fish oil amplifies the effect of propranolol in mild essential hypertension. Hypertension **16:** 682-691.
68. WING, L. M. H., P. J. NESTEL, J. P. CHALMERS, I. ROUSE, M. J. WEST, A. J. BUNE, A. L.

TONKIN & A. E. RUSSELL. 1990. Lack of effect of fish oil supplementation on blood pressure in treated hypertensives. J. Hypertens. **8:** 339–343.
69. MARGOLIN, G., G. HUSTER, C. J. GLUECK *et al.* 1991. Blood pressure lowering in elderly subjects: a double-blind crossover study of n-3 and n-6 fatty acids. Am. J. Clin. Nutr. **53:** 562–572.
70. GRAY, D. G., C. G. GOZZIP, J. H. EASTBAM & M. L. KASHYAP. 1996. Fish oil as an adjuvant in the treatment of hypertension. Pharmacotherapy **16:** 295–300.
71. LUNGERSHAUSEN, Y. K., M. ABBEY, P. J. NESTEL & P. R. C. HOWE. 1994. Reduction of blood pressure and plasma triglycerides by omega-3 fatty acids in treated hypertensives. J. Hypertens. **12:** 1041–1045.
72. LUNGERSHAUSEN, Y. K. & P. R. C. HOWE. 1994. Improved detection of a blood pressure response to dietary intervention with 24-hour ambulatory monitoring. Am. J. Hypertens. **7:** 1115–1117.
73. HOWE, P. R. C., Y. K. LUNGERSHAUSEN, L. COBAIC, G. DANDY & P. J. NESTEL. 1994. Effect of sodium restriction and fish oil supplementation on BP and thrombotic risk factors in patients treated with ACE inhibitors. J. Hum. Hypertens. **8:** 43–49.
74. ZUSMAN, R. M. 1984. Renin- and non-renin-mediated antihypertensive actions of converting enzyme inhibitors. Kidney Int. **25:** 969–983.
75. LANDMARK, K., E. THAULOW, J. HYSING, H. H. MUNDAL, J. ERITSLAND & I. HJERMANN. 1993. Effects of fish oil, nifedipine, and their combination on blood pressure and lipids in primary hypertension. J. Hum. Hypertens. **7:** 25–32.
76. AMES, R. P. 1988. Antihypertensive drugs and lipid profiles. Am. J. Hypertens. **1:** 421–427.
77. HARRIS, W. S. 1996. Do omega-3 fatty acids worsen glycemic control in NIDDM? ISSFAL Newsl. **3:** 6–9.
78. LUNGERSHAUSEN, Y. K., P. R. C. HOWE, P. M. CLIFTON *et al.* 1997. Evaluation of an omega-3 fatty acid supplement in diabetics with microalbuminuria. This volume.
79. BABIY, A. V., J. M. GEBICKI, D. R. SULLIVAN & K. WILLEY. 1992. Increased oxidizability of plasma lipoproteins in diabetic patients can be decreased by probucol therapy and is not due to glycation. Biochem. Pharmacol. **43:** 995–1000.
80. SUZUKAWA, M., M. ABBEY, P. R. C. HOWE & P. J. NESTEL. 1995. Effects of fish oil fatty acids on low density lipoprotein size, oxidizability, and uptake by macrophages. J. Lipid Res. **36:** 473–484.
81. LEAF, A. 1992. Health claims: omega-3 fatty acids and cardiovascular disease. Nutr. Rev. **50:** 150–154.
82. SIMOPOULOS, A. P. 1991. Omega-3 fatty acids in health and disease and in growth and development. Am. J. Clin. Nutr. **54:** 438–463.
83. MALLE, E. & G. M. KOSTNER. 1993. Effects of fish oils on lipid variables and platelet function indices. Prostaglandins Leukotrienes Essent. Fatty Acids **49:** 645–663.
84. SHIMAMURA, T. & A. C. WILSON. 1991. Influence of dietary fish oil on the aortic, myocardial, and renal lesions of SHR. J. Nutr. Sci. Vitaminol. **37:** 581–590.
85. PEPE, S. & P. L. MCLENNAN. 1996. Dietary fish oil confers direct antiarrhythmic properties on the myocardium of rats. J. Nutr. **126:** 34–42.
86. BURR, M. L., A. M. FEHILY, J. F. GILBERT *et al.* 1989. Effects of changes in fat, fish, and fiber intakes on death and myocardial reinfarction: diet and reinfarction trial (DART). Lancet ii: 757–761.
87. DONADIO, J. V., JR. 1991. Omega-3 polyunsaturated fatty acids: a potential new treatment of immune renal disease. Mayo Clin. Proc. **66:** 1018–1028.
88. DE CATERINA, R., R. CAPRIOLI, D. GIANNESSI, R. SICARI, C. GALLI, G. LAZZERINI, W. BERNINI, L. CARR & P. RINDI. 1993. n-3 fatty acids reduce proteinuria in patients with chronic glomerular disease. Kidney Int. **44:** 843–850.
89. DE CATERINA, R., S. ENDRES, S. D. KRISTENSEN & E. B. SCHMIDT. 1994. n-3 fatty acids and renal diseases. Am. J. Kidney Dis. **24:** 397–415.
90. DE CATERINA, R., M. I. CYBULSKY, S. K. CLINTON, M. A. GIMBONE, JR. & P. LIBBY. 1994. The omega-3 fatty acid docosahexaenoate reduces cytokine-induced expression of proatherogenic and proinflammatory proteins in human endothelial cells. Arterioscler. Thromb. **14:** 1829–1836.

91. RAYNER, T. E. & P. R. C. HOWE. 1995. Purified omega-3 fatty acids retard the development of proteinuria in salt-loaded hypertensive rats. J. Hypertens. **13:** 771–780.
92. MUNSIFF, A. V., P. N. CHANDER, S. LEVINE & C. T. STIER. 1992. The lipoxygenase inhibitor phenidone protects against proteinuria and stroke in stroke-prone spontaneously hypertensive rats. Am. J. Hypertens. **5:** 56–63.
93. HOBBS, L. M., T. E. RAYNER & P. R. C. HOWE. 1996. Dietary fish oil prevents the development of renal damage in salt-loaded stroke-prone spontaneously hypertensive rats. Clin. Exp. Pharmacol. Physiol. **23:** 508–513.
94. WATSON, C., L. JABLONSKIS, T. RAYNER & P. HOWE. 1995. Dose-dependent antihypertensive and renal effects of dietary omega-3 fatty acids in salt-loaded stroke-prone spontaneously hypertensive rats (SHRSP). Proc. Aust. Soc. Clin. Exp. Pharmacol. Toxicol. **2:** 88.
95. LEWIS, E. J., L. G. HUNSICKER, R. P. BAIN & R. D. ROHDE. 1993. The effect of angiotensin-converting-enzyme inhibition on diabetic nephropathy. N. Engl. J. Med. **329:** 1456–1462.
96. EGGERS, P. W. 1988. Effect of transplantation on the Medicare End-Stage Renal Disease Program. N. Engl. J. Med. **318:** 223–229.

Postprandial Triglyceride High Response and the Metabolic Syndrome[a]

J. SCHREZENMEIR,[b] S. FENSELAU,[b] I. KEPPLER,[b]
J. ABEL,[b] B. ORTH,[b] CH. LAUE,[b]
W. STÜRMER,[c] U. FAUTH,[d] M. HALMAGYI,[d]
AND W. MÄRZ[e]

[b] *Institute of Physiology and Biochemistry of Nutrition*
Federal Research Center
D-24103 Kiel, Germany

[c] *Medizinische Klinik*
University of Würzburg
D-87080 Würzburg, Germany

[d] *Clinic for Anesthesiology*
Gutenberg University
D-55101 Mainz, Germany

[e] *Medizinische Klinik*
University of Freiburg
D-79106 Freiburg, Germany

INTRODUCTION

A remarkable interindividual variability in postprandial (pp) triglyceride (TG) responses has been emphasized by several authors on the basis of investigations in small groups of volunteers.[1–9] To clarify whether excessive triglyceride responses in some individuals were due to variance in meal composition or reflected a distinct phenomenon,

(1) a standardized manufactured liquid formula was designed[10,11] containing components that were shown to challenge pp TG metabolism: saturated fat, sucrose, and ethanol;[12–16]
(2) this oral metabolic tolerance test (oMTT) was applied to a homogeneous group of 113 healthy male volunteers of similar age (25.0 ± 0.3 years) and body mass index (BMI = 22.4 ± 0.4 kg/m^2)[11] to avoid any variance in TG response by these variables.[5,17–19]

---

[a] This work was supported by the Stifterverband für die Deutsche Wissenschaft by the award of an H.- and L.-Schilling-Professorship to J. Schrezenmeir, and by the Institut Danone für Ernährung.

The bimodal frequency distribution of TG peak values following this standardized load in this homogeneous cohort indicated the existence of two distinct clusters: one group displaying pp TG peak values below 260 mg/dL (so-called normal responders = NR) and another exceeding 260 mg/dL (so-called high responders = HR).[11]

The phenomenon of TG high response accounted for 15% of the total population investigated. Fasting TG also displayed a bimodal frequency distribution with a discriminative value between both groups at 150 mg/dL. Although fasting and pp TG maxima were highly correlated, the normal fasting group and the NR, as well as the high fasting group and the HR, were not identical: 53% of the HR had fasting TG < 150 mg/dL.[11]

Interestingly, the HR had higher fasting insulin levels than the NR, but the low fasting TG group did not differ in insulin from the high fasting TG group.[11] This suggested that pp rather than fasting TG levels are related to the metabolic syndrome, which is characterized by insulin resistance, hyperinsulinemia, and dyslipoproteinemia.

In agreement with this assumption, we found not only higher TG and insulin levels in HR, but also higher pp glucose levels, indicating insulin resistance,[20] and a higher proinsulin/insulin ratio, indicating impaired proinsulin processing,[20,21] which is also well regarded as a typical feature of the metabolic syndrome.

In this article, further evidence is furnished of the link between the phenomenon of TG high response and the insulin resistance syndrome.

METHODS AND SUBJECTS

Experimental Protocol of the Oral Metabolic Tolerance Test (oMTT)

All participants were asked to abstain from food, nicotine, and drink (except for water) for 12 hours before the oMTT. A fasting blood sample was taken at 8:00 A.M. Participants then consumed within 15 min a 500-mL standardized manufactured drink that supplied 4221 kJ energy, 30 g protein (11.9 E%), 75 g carbohydrates (29.6 E%), 10 g ethanol (6.9 E%), 58 g fat (51.6 E%), 600 mg cholesterol, and 30,000 IU vitamin A as retinylpalmitate. Carbohydrates were sucrose and lactose (93% and 7% by weight, respectively). Fatty acids were 65% saturated and 35% monounsaturated and polyunsaturated. Blood was sampled repetitively from the lower cubital vein during the following 9 h. No food or strong physical activity was allowed during this time.

Experimental Protocol of the Oral Glucose Tolerance Test (oGTT)

Participants were instructed to consume a diet high in carbohydrates for 3 days before the test. After an overnight fasting period, a 100-g oral glucose load (Dextro®-OGT, Boehringer Mannheim, Germany) was administered at 8:00 A.M. Blood was sampled before and 15, 30, 45, 60, 120, 150, and 180 min after ingestion to determine glucose and insulin levels.

Analysis

Total serum triglycerides, glucose, and free fatty acids (FFA) were all determined enzymatically: triglycerides according to a modification of Bucolo and David[22] using a kit of Merck, Darmstadt, Germany; glucose by using a test kit–UNI KIT Roche, Glukose HK–of Hoffmann-La Roche, Basel, Switzerland; and FFA by using a colorimetric test kit of Boehringer Mannheim, Germany. Insulin was determined by radioimmunoassay (Serono, Freiburg, Germany) or it was determined in the course of specific determination of "true" insulin, proinsulin, and 32,33 split proinsulin by a two-site immunoradiometric assay.[23] The measurement of abdominal fat was done by ultrasound according to Armellini et al.[24] Energy expenditure was assessed by indirect calorimetry with a ventilated hood system (Deltatrac, Datex). Apolipoprotein E polymorphism was analyzed by immunoblotting as described by März et al.[25] Enzyme activities of LPL and HTGL were determined in fasting samples of subjects after injection of 60 IU/kg body weight of heparin iv according to Zechner[26] and Schotz et al.[27]

STUDY DESIGN

Reproducibility Study

In a former experiment, 113 healthy young men [mean age, 25.0 ± 0.3 years; body mass index (BMI), 22.7 ± 0.2 kg/m^2] with fasting TG levels < 200 mg/dL had ingested an oMTT as described earlier.[11] Blood samples were drawn before the test meal and 4, 5, 6, and 9 h after the load. TG, glucose, insulin, and FFA were determined. The subjects were not taking any medication. For reinvestigation, out of the 86 NR and 14 HR classified in the first experiment, 10 participants per group were selected at random for the second experiment to study the reproducibility of the phenomenon of pp TG high response. Blood was sampled at 0.5 h pp and hourly until 9 h pp. In this study, only TG levels were measured. Mean age and BMI of the participants were not different from those in the first experiment.

"True" Insulin, Proinsulin, and 32,33 Split Proinsulin

Twenty-three NR and 11 HR ingested an oMTT. Fasting and pp blood samples were taken during a 9-h observation period. TG, glucose, true insulin, proinsulin, and 32,33 split proinsulin values were determined by using a two-site immunoradiometric assay. The mean age of the subjects was as follows–HR: 24.9 ± 0.7 years; NR: 25.4 ± 0.4 years. BMI was 25.4 ± 1.3 kg/m^2 in HR and 23.1 ± 0.5 kg/m^2 in NR. Mean age and BMI of the participants in the two groups were not different. They were not taking any medication.

Assessment of Abdominal Fat and Postprandial Energy Expenditure

The energy expenditure after an oMTT was assessed by indirect calorimetry with a ventilated hood system (Deltatrac I, Datex, Germany) at 30 min before and during 30-min periods until 9 h pp. Subcutaneous and intra-abdominal fat mass was determined by ultrasound according to Armellini et al.[24] Twelve NR and 6 HR participated in this study. Their mean age was as follows—HR: 26.0 ± 0.7 years; NR: 25.4 ± 0.6 years. BMI was 23.7 ± 0.5 kg/m^2 in HR and 22.1 ± 1.1 kg/m^2 in NR. Mean age and BMI of the participants in the two groups did not differ statistically. They were not taking any medication.

Apolipoprotein E Phenotypes

The apolipoprotein E phenotype of 93 NR, 14 HR, and 21 NIDDM patients was analyzed in a fasting blood sample by immunoblotting as described.[25] NR and HR did not differ with respect to age or BMI (25.0 ± 0.3 versus 24.6 ± 0.7 years and 22.3 ± 0.3 versus 22.9 ± 0.6 kg/m^2, respectively). They were not taking any medication. The mean age of the NIDDM patients was 60.3 ± 11.4 years and their mean BMI was 29.1 ± 5.0 kg/m^2. Fourteen patients were treated with sulfonylurea and 7 with insulin.

Study in Sons of Parents Suffering from the Insulin Resistance Syndrome

An oMTT was performed in 27 male subjects whose mother or father suffered from the insulin resistance syndrome comprising android obesity, dyslipoproteinemia, hypertension, and NIDDM. Their mean age was 31.2 years; their mean BMI was 27.9 ± 0.7 kg/m^2. Twenty-seven healthy male subjects without any family history matched in pairs for age (± 5 years) and body weight (± 1 kg/m^2) served as controls. Their mean age was 31.3 years and their mean BMI was 28.0 ± 0.7 kg/m^2. After ingestion of an oMTT, TG, glucose, and insulin were determined before and 0.5, 1, 2, 3, 4, 5, 6, 7, 8, and 9 hours after ingestion. In this study, an insulin sensitivity index was calculated from glucose and insulin levels by forming the following quotient: 100/[blood glucose (mmol/L) × serum insulin (mU/l)].

oGTT versus oMTT

In this study, oral glucose tolerance and oral metabolic tolerance were compared. Fifteen NR and 10 HR were subjected to (1) an oGTT and (2) an oMTT in random order. TG, glucose, and insulin were determined. NR and HR did not differ with respect to age or BMI (24.8 ± 0.6 versus 25.6 ± 0.6 years and 24.3 ± 0.7 versus 23.4 ± 0.7 kg/m^2, respectively). They were not taking any medication.

LPL and HTGL

Enzyme activities were assayed in the fasting state in 10 NR and 8 HR according to Zechner.[26] NR and HR did not differ with respect to age or BMI (24.2 ± 0.4 versus 23.6 ± 1.4 years and 21.8 ± 0.7 versus 24.0 ± 1.3 kg/m^2, respectively). They were not taking any medication.

Calculations and Statistical Analysis

Statistical evaluation was done using the SPSS/PC 4.01 package. Arithmetic means ± SEM are given throughout. Differences between NR and HR were tested by a nonpaired t test. Those data settings having $p < 0.05$ were reported as significant. Correlations between parameters were calculated by linear regression. Apo E frequencies were compared between groups by chi-square testing.

RESULTS

Reproducibility Study

Triglycerides

In an earlier study,[11] we found a bimodal frequency distribution of TG response to a standardized oMTT in 113 healthy subjects. The participants were classified as either HR if postprandial TG maxima were ⩾260 mg/dL or NR if postprandial TG maxima were <260 mg/dL. In this study, 15% were classified as HR. In a consecutive study, 10 subjects of each group were given an oMTT again. Nine out of 10 NR and 5 out of 10 HR showed a response of the same type. In the subjects formerly classified as NR, the mean TG maximum was 177.0 ± 18.6 mg/dL; in the HR, the mean TG maximum was 253.0 ± 41.0 mg/dL. The correlation coefficient between TG in the first and in the second test was $r = 0.48$ ($p < 0.05$) (FIG. 1).

Insulin, Glucose, and FFA

In both experiments, insulin levels of HR were higher than those of NR, in both the fasting state and pp, with a significant difference between NR and HR in the fasting state and 2 h pp ($p < 0.05$). Also, mean glucose levels of HR were higher than those of NR; this difference was significant at 3 and 4 h pp ($p < 0.05$). Fasting FFA were lower in HR in both experiments, but pp FFA were higher in HR (FIG. 2).

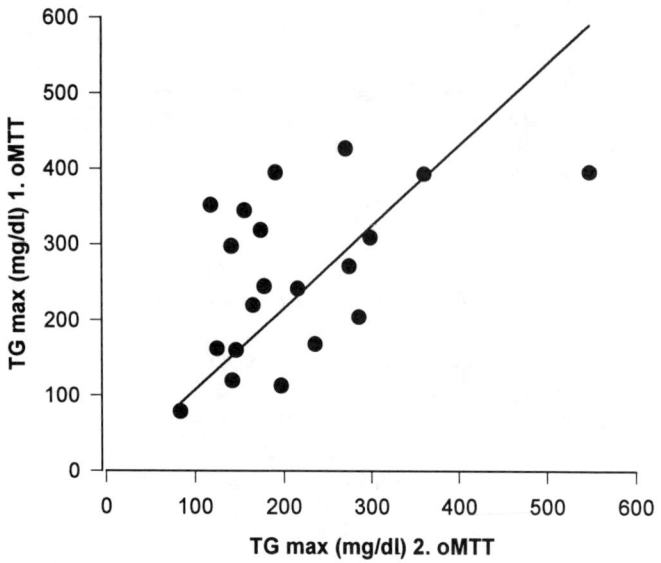

FIGURE 1. Reproducibility of triglyceride high response. In a first experiment (1. oMTT), 86 subjects were classified as normal responders (with TG max < 260 mg/dL) and 14 were classified as high responders (with TG ⩾ 260 mg/dL). Ten participants per group were selected at random for the second oMTT (2. oMTT) to study the reproducibility of the phenomenon of pp TG high response. The pp TG maxima of 10 NR and 10 HR after the first and second oMTT were correlated ($r = 0.48$, $p < 0.05$).

"True" Insulin, Proinsulin, and 32,33 Split Proinsulin

In this study, TG mounted again to higher values in HR (329.0 ± 45.0 mg/dL) than in NR (156.0 ± 17.0 mg/dL, $p < 0.001$). Glucose levels were slightly higher in HR than in NR, with a significant difference at 4 h pp ($p < 0.05$). Glucose maxima were 103.0 ± 9.4 mg/dL in HR and 98.7 ± 4.9 mg/dL in NR. HR showed higher AUC (area under the curve) values for "true" insulin, proinsulin, and 32,33 split proinsulin. The difference between the groups was most pronounced with respect to proinsulin. AUC values at 0–4 h pp were 1.32-fold higher for insulin and 2.93-fold higher for proinsulin in HR versus NR. The proinsulin/insulin ratio (P/I) and the split proinsulin/insulin ratio (S/I) expressed as AUC values after the oMTT tended to be higher in HR than in NR (FIG. 3). The proinsulin/insulin quotient increased with increasing glucose levels in HR, but not in NR.

Assessment of Abdominal Fat and Postprandial Energy Expenditure

The assessment of adipose tissue showed higher subcutaneous fat tissue in HR (2.38 ± 0.3 cm) versus NR (1.52 ± 0.2 cm) ($p = 0.02$) and higher intra-abdominal

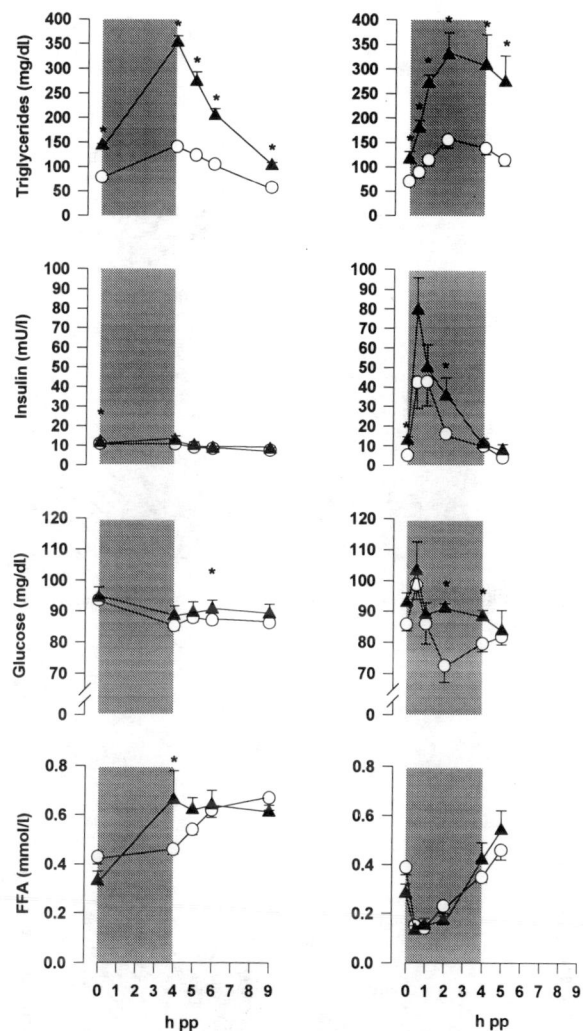

FIGURE 2. TG, insulin, glucose, and FFA of the first oMTT (left) in 86 NR (○) and 14 HR (▲) and of the second oMTT (right) in those subjects who showed reproducible response [9 NR (○) and 5 HR (▲)] (mean ± SEM; *, $p < 0.05$).

fat mass in HR (4.2 ± 0.35 cm) than in NR (3.1 ± 0.2 cm) ($p = 0.04$), despite similar BMI in both groups (FIG. 4).[28]

The resting energy expenditure was similar in both groups (HR: 73.2 ± 4.0 kcal/h; NR: 72.9 ± 3.2 kcal/h) (FIG. 5). However, 6.5 to 9 h after ingestion of the oMTT, the dietary-induced thermogenesis diverged between HR (Net AUC, 9.2 ± 7.3 kcal/h) and NR (0.73 ± 9.85 kcal/h) ($p = 0.03$).

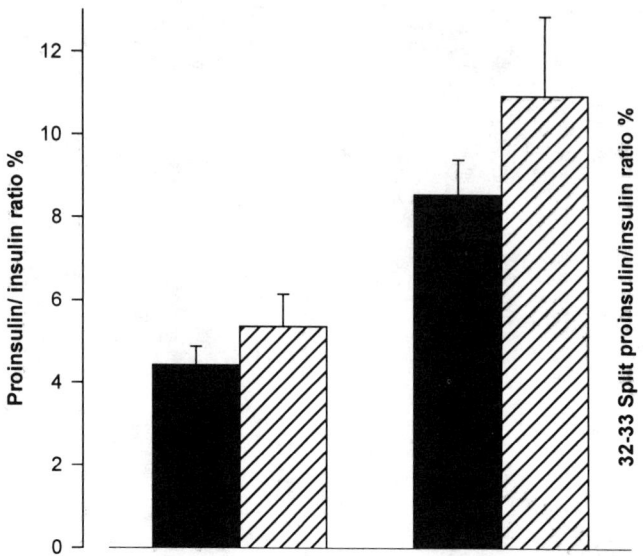

FIGURE 3. The proinsulin/insulin (P/I) and the 32,33 split proinsulin/insulin (S/I) ratio in 23 TG NR (■) and 11 TG HR (▨) after oMTT expressed as area under the curve (AUC) (mean ± SEM).

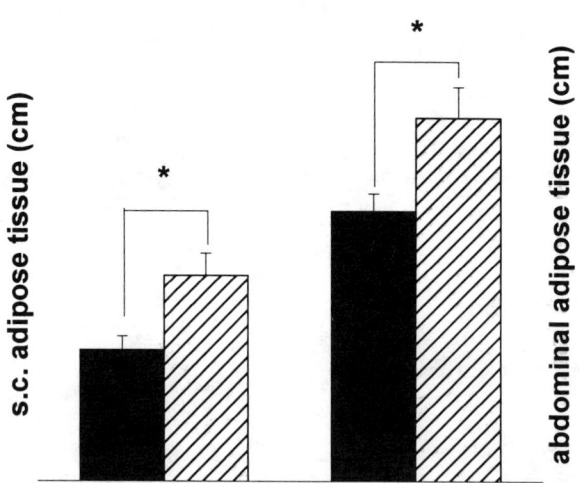

FIGURE 4. Subcutaneous (s.c.) and abdominal adipose tissue in 10 normal responders (■) (BMI, 22.3 ± 0.4 kg/m^2) and 6 high responders (▨) (BMI, 23.4 ± 0.6 kg/m^2) assessed by sonography (mean ± SEM; *, $p < 0.05$) — s.c. tissue: distance between skin and linea alba, 1 cm above umbilicus; abdominal tissue: distance between peritoneum and aorta, 1 cm above umbilicus.

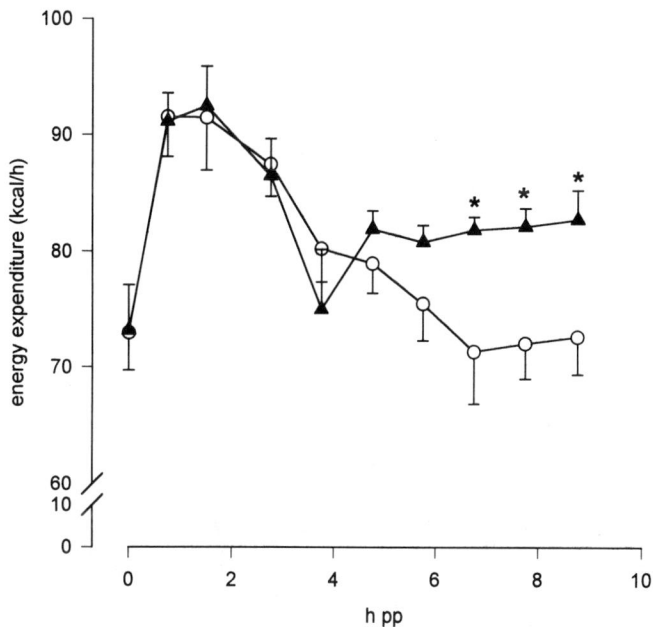

FIGURE 5. Resting and pp energy expenditure in 12 NR (○) and 6 HR (▲) (*, $p < 0.05$).

Apolipoprotein E Phenotypes

The apo E 3/3 phenotype was most frequent in all of the groups studied. The NR revealed the highest frequency of the apo E 3/4 phenotype, followed by the NIDDM patients and the HR. The apo E phenotype 2/3 was observed in 4% of the NR, in 17.5% of the HR, and in 19% of the NIDDM patients. Consideration of the allele ε2 frequency showed a significant overrepresentation in HR ($p = 0.05$) and in NIDDM ($p = 0.02$) compared to NR, as well as comparable prevalence in HR and NIDDM.[29]

oGTT versus oMTT

TG increased to significantly ($p = 0.001$) higher maxima in the 10 HR (300.0 ± 13.0 mg/dL) compared to the NR (176.0 ± 7.5 mg/dL) after ingestion of the oMTT. Insulin values were higher and glucose values increased to significantly higher values in HR (118.0 ± 6.7 mg/dL) compared to NR (102.0 ± 3.5 mg/dL) ($p < 0.05$). In contrast to this, the oGTT did not differ between both groups: glucose values increased to 152.0 ± 3.5 in HR versus 151.0 ± 5.6 mg/dL in NR. Insulin values showed no difference after oGTT too (FIG. 6).

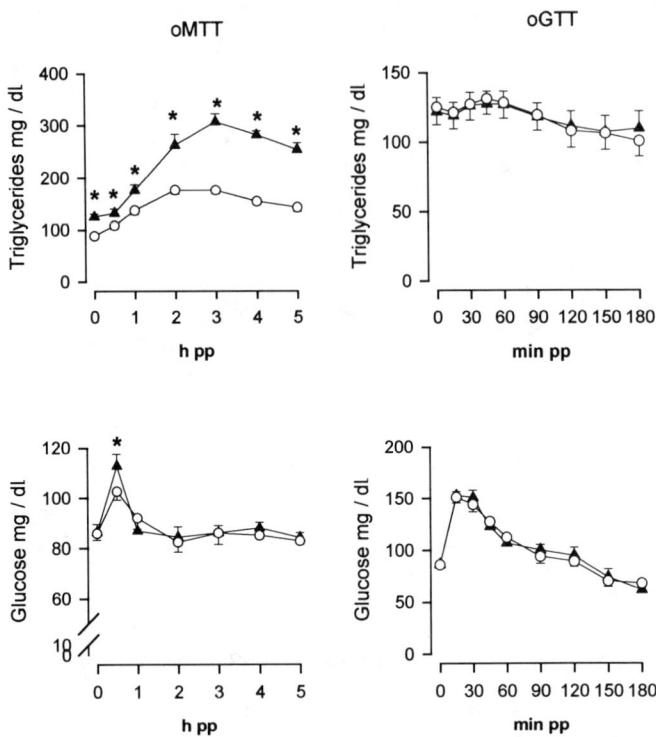

FIGURE 6. TG serum levels and venous blood glucose levels after oMTT (left) and after oGTT (right) in 15 NR (○) and 10 HR (▲) (mean ± SEM; $*$, $p < 0.05$).

Study in Sons of Parents Suffering from the Metabolic Syndrome

As expected, the sons of affected index subjects showed significantly higher fasting and pp TG levels (FIG. 7A). Eighteen out of 27 sons (66%) from parents (mother or father) suffering from the insulin resistance syndrome showed a high TG response after ingestion of the oMTT (maximal pp > 260 mg/dL), whereas only 6 out of 27 controls (22%) did so. When the HR sons and the NR sons of the affected parents were separately compared with NR controls, the HR sons showed a diminished calculated insulin sensitivity index in the 2–8-hour pp period ($p = 0.043$), whereas the NR sons had a sensitivity similar to the controls (FIGS. 7B and 7C).

LPL and HTGL

Lipoprotein lipase (LPL) activity was 6.35 ± 2.05 mmol/L/h in HR and 5.16 ± 1.93 mmol/L/h in NR ($p > 0.05$, not significant). The activity of hepatic triglyceride lipase (HTGL) was significantly different in HR and NR (8.41 ± 3.7 mmol/L/h versus 5.24 ± 2.4 mmol/L/h; $p < 0.04$).

DISCUSSION

Following the administration of a standardized manufactured formula containing saturated fat, sucrose, and ethanol as a challenge to the metabolism of TG, a population of 25-year-old, normal-weight men fell into two clusters based on a bimodal frequency distribution of pp TG peaks. About 15% of the subjects presented with excessive TG responses of ⩾260 mg/dL.[11]

This phenomenon of TG HR was proven to be reproducible (FIG. 1). The correlation coefficient of 0.48, however, indicated that reproduction was not attained in every case.

In the second oMTT performed, again the fasting and pp insulin levels and the pp glucose levels were higher in those classified as HR by the first oMTT than in those classified as NR (FIG. 2). This confirmed the formerly documented findings of "hyperinsulinemia"[11,20] and reduced insulin sensitivity in HR,[20] which stimulated the suggestion of a link between TG HR and the insulin resistance syndrome.[10,11]

Indeed, this was supported by the increased proinsulin/insulin ratio and 32,33 split proinsulin/insulin ratio (FIG. 3), which are typical and early signs of the insulin resistance syndrome.[30] They indicate an impaired processing of proinsulin to insulin at a very early stage of the insulin resistance syndrome, where only lipid tolerance, but not glucose tolerance, is impaired (FIG. 6). The fact that P/I increased with increasing glucose levels in HR,[20] but remained constant in NR independently of glucose levels, clearly addresses a qualitative difference between HR and NR. Evidently, NR are able to secrete insulin in adequate proportions to proinsulin over the physiological range of glucose levels, whereas HR are not.

In agreement with the constitutional features of the insulin resistance syndrome, abdominal adipose tissue was higher in HR than in NR (FIG. 4),[28] although both groups were selected for normal weight and statistically did not differ in BMI. The likewise increased subcutaneous adipose tissue in HR was also assessed in the abdominal region and evidently has to be discriminated from subcutaneous fat in other regions of the body, too.[31]

Abdominal adipose tissue is known to have a higher metabolic rate and is more sensitive to sympathetic stimuli.[32] Therefore, it was not unexpected that energy expenditure was higher in HR than in NR (FIG. 5). As demonstrated by van Gaal et al.,[33] in obese subjects, the resting energy expenditure was correlated with BMI and pp thermogenesis with abdominal adipose tissue. Thus, HR and NR with similar (and normal) BMI did not differ in resting energy expenditure, but presented with significantly higher pp thermogenesis (FIG. 5). This was accompanied by a higher catecholamine release and an exaggerated adrenal reactivity.[28]

Apolipoprotein E polymorphism has a well-acknowledged impact on pp TG[5,34] and some authors have reported an increased apo E2 prevalence in type-2 diabetes.[35]

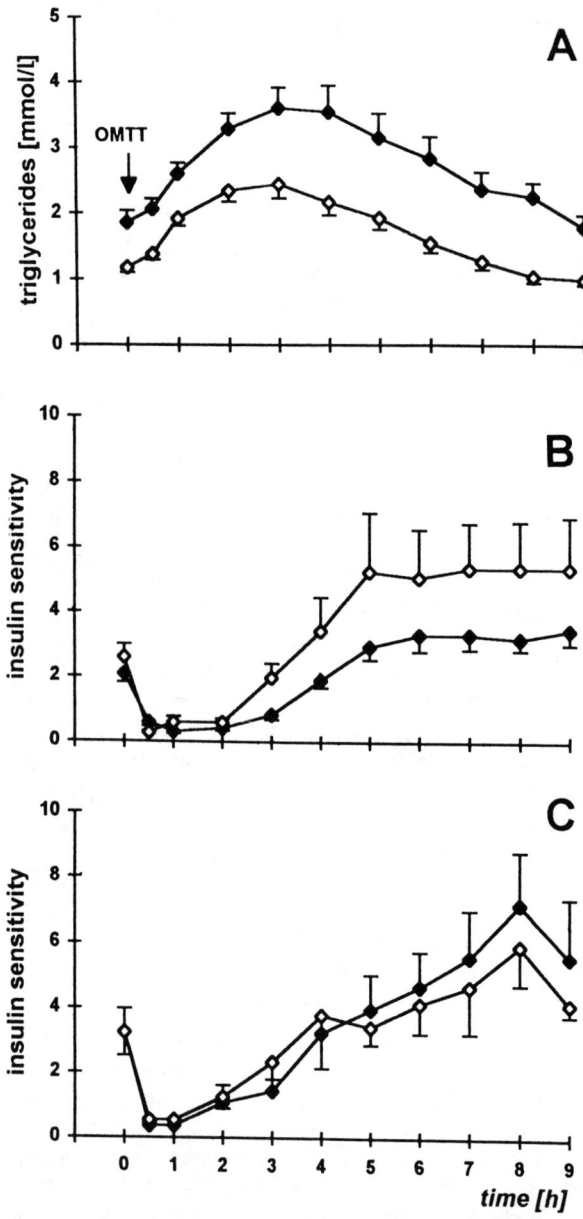

FIGURE 7. (A) TG levels after oMTT in 27 sons (♦) of a father or mother suffering from the insulin resistance syndrome (type-2 diabetes + hypertension + obesity + dyslipoproteinemia) and in 27 controls (◊) without any family history matched for BMI and age (mean ± SEM). (B) Insulin sensitivity calculated as [100/insulin (mU) × glucose (mmol)] after oMTT in 16 HR sons with family history (♦) and 16 NR controls (◊). (C) Insulin sensitivity index after oMTT in 8 NR sons with family history (♦) and 8 NR controls (◊).

This increased prevalence in type-2 diabetes was confirmed by our study.[29] The very similar prevalence that we found in diabetics and HR, differing significantly from that in NR, suggests a common genetic background of HR and diabetics. Since the apo E2 allele accounts for no more than approximately 18%, it certainly does not explain the phenomenon of HR or even type-2 diabetes. Based on its TG-increasing effect or by any other reason, it may, however, contribute to their manifestation.

Hyperinsulinemia, reduced insulin sensitivity, impaired proinsulin processing, higher abdominal adipose tissue and increased alimentary thermogenesis, and increased apo E2 prevalence were biochemical and clinical indicators for a link of TG HR to the insulin resistance syndrome, which is characterized by these features. The results in the offspring of subjects suffering from this syndrome, however, are ongoing evidence for such a link. In fact, the sons of affected index subjects had significantly higher pp TG levels after the oMTT (FIG. 7A). After all, two-thirds presented with TG high response. Interestingly, those offspring presenting with HR showed a diminished insulin sensitivity, whereas those with normal TG response did not (FIGS. 7B and 7C). This holds true when only those subjects were considered, whose fasting TG values were normal (<150 mg/dL). It may be suggested that the affected offspring could be discriminated from unaffected offspring with the aid of the oMTT. A fasting TG value ≥ 150 mg/dL as well may be used in this respect. A fasting level < 150 mg/dL, however, does not seem to be valuable as an exclusion criterion.

Therefore, the pp TG response to a lipid-containing load seems to imply a crucial event. According to this, the above-mentioned metabolic alterations seen in HR after the oMTT—particularly higher glucose and insulin levels—were not seen after an oGTT (FIG. 6). This may be due to the higher pp TG and FFA levels in HR derived from the dietary fat in the oMTT. In contrast to the fasting state, chylomicrons postprandially appear in the bloodstream; from them, chylomicron remnants and FFA are released with the aid of lipoprotein lipase. In fact, the high fasting TG levels in HR mainly rely on VLDL elevations, whereas the higher pp TG response essentially is caused by chylomicrons.[10] From these increased TG-rich lipoproteins, more remnants and FFA postprandially are released in HR.[10] Since FFA are known to impair insulin sensitivity,[36,37] this may explain why the administration of sucrose combined with fat did unveil the metabolic disturbance in HR and resulted in insulin resistance, whereas merely giving glucose in the oGTT did not.

The increased release of FFA from the TG-rich lipoproteins was further promoted by the increased LPL and hepatic lipoprotein lipase activity in HR. This increased lipase activity may be explained by the higher levels of insulin, which is known to stimulate the enzyme.[38] Since, recently, an association of LPL activity and the risk of atherosclerosis was shown,[39] this may contribute to atherogenesis like the hyperinsulinemia and high TG, FFA, and catecholamines, which we have found in HR.

In conclusion, HR were characterized by metabolic disturbances in the sense of a "premetabolic syndrome" before fasting dyslipoproteinemia, obesity, hypertension, and diabetes were present. These alterations may be detected with the presented oMTT before oGTT is impaired. Furthermore, the metabolic data (high glucose levels in spite of high insulin; however, low fasting FFA and high pp FFA accompanied by activation of lipoprotein lipase) indicated a dissociation of insulin sensitivity between the glucose and the lipid branch of insulin action.[20,40]

ACKNOWLEDGMENTS

We acknowledge the kind support by B. Clark and C. N. Hales (Department of Biochemistry, University of Cambridge, United Kingdom) in determining "true" insulin, proinsulin, and split proinsulin.

REFERENCES

1. GROOT, P. H. E., W. A. H. J. VAN STIPHOUT, X. H. KRAUSS, H. JANSEN, A. VAN TOL, E. VAN RAMSHORST, S. CHIN-ON, A. HOFMAN, S. R. CROSSWELL & L. HAVEKES. 1991. Postprandial lipoprotein metabolism in normolipidemic men with and without coronary artery disease. Atheroscler. Thromb. **11:** 653–662.
2. BAGGIO, G., R. FELLIN, M. R. BAIOCCHI, S. MARTINI, G. BALDO, E. MANZATO & G. CREPALDI. 1980. Relationship between triglyceride-rich lipoproteins (chylomicrons and VLDL) and HDL2 and HDL3 in the postprandial phase in humans. Atherosclerosis **87:** 271–276.
3. PATSCH, J. R., S. PRASAD, A. M. GOTTO & G. BENGTSSON-OLIVECRONA. 1984. Postprandial lipemia: a key for the conversion of HDL2 into HDL3 by hepatic lipase. J. Clin. Invest. **74:** 2017–2023.
4. PATSCH, J. R., S. PRASAD, A. M. GOTTO & W. PATSCH. 1987. HDL2: relationship of the plasma level of this lipoprotein species to its composition, to the magnitude of postprandial lipemia, and to the activities of lipoprotein lipase and hepatic lipase. J. Clin. Invest. **80:** 341–347.
5. WEINTRAUB, S., S. EISENBERG & J. L. BRESLOW. 1987. Different patterns of postprandial lipoprotein metabolism in normal, type IIa, type III, and type IV hyperlipoproteinemic individuals. J. Clin. Invest. **79:** 1110–1119.
6. REAVEN, G. M., R. L. LEARNER, M. P. STERN & J. W. FARQUHAR. 1967. Role of insulin in endogenous hypertriglyceridemia. J. Clin. Invest. **46:** 1756–1767.
7. SCHMAHL, F. W., P. PRICKLER, W. PÖTTER, B. TSCHIRCKEWAHN & F. H. HECKERS. 1983. Are fasting triglycerides sufficient for determining the risk factor status of hypertriglyceridemia of the atherosclerotic vascular disease? VASA **12:** 303–356.
8. BARR, S. I., B. A. KOTTE & T. S. MAO. 1985. Postprandial distribution of apolipoprotein C-II and C-III in normal subjects and patients with mild hypertriglyceridemia: comparison of meals containing corn oil and medium-chain triglyceride oil. Metabolism **34:** 983–992.
9. WEBER, P., J. SCHREZENMEIR, S. FENSELAU, S. AUSIEKER, R. PROBST, H. D. ZUCHHOLD, W. PRELLWITZ & J. BEYER. 1993. Prolonged postprandial increment in triglycerides and decreased postprandial response of very low density lipoproteins in type 2 diabetics following an oral lipid load. Ann. N.Y. Acad. Sci. **683:** 315–321.
10. SCHREZENMEIR, J., P. WEBER, R. PROBST, H. K. BIESALSKI, C. LULEY, W. PRELLWITZ, U. KRAUSE & J. BEYER. 1992. Postprandial pattern of triglyceride-rich lipoproteins in normal-weight humans after an oral lipid load: exaggerated triglycerides and altered insulin response in some subjects. Ann. Nutr. Metab. **36:** 186–196.
11. SCHREZENMEIR, J., I. KEPPLER, S. FENSELAU, P. WEBER, H. K. BIESALSKI, R. PROBST, CH. LAUE, H. D. ZUCHHOLD, W. PRELLWITZ & J. BEYER. 1993. The phenomenon of a high triglyceride response to an oral lipid load in healthy subjects and its link to the metabolic syndrome. Ann. N.Y. Acad. Sci. **683:** 302–314.
12. BERGERON, N. & R. J. HAVEL. 1995. Influence of diets rich in saturated and omega-6 polyunsaturated fatty acids on the postprandial responses of apolipoprotein B-48, B-100, E, and lipids in triglyceride-rich lipoproteins. Arterioscler. Thromb. Vasc. Biol. **15:** 2111–2121.
13. GRANT, K. I., M. P. MARAIS & M. A. DHANSAY. 1994. Sucrose in a lipid-rich meal amplifies the postprandial excursion of serum and lipoprotein triglyceride and cholesterol concentrations by decreasing triglyceride clearance. Am. J. Clin. Nutr. **59:** 853–860.
14. SUPERKO, H. R. 1992. Effects of acute and chronic alcohol consumption on postprandial lipemia in healthy normotriglyceridemic men. Am. J. Cardiol. **69:** 701–704.

15. BREWSTER, A. C., H. G. LANKFORD, M. G. SCHWARTZ & J. F. SULLIVAN. 1966. Ethanol and alimentary lipemia. Am. J. Clin. Nutr. **19:** 255-259.
16. GINSBERG, H., J. OLEFSKY, J. W. FARQUHAR & G. REAVEN. 1974. Moderate ethanol ingestion and triglyceride levels. Ann. Intern. Med. **80:** 143-149.
17. NESTEL, P. J. 1964. Relationship between plasma triglycerides and removal of chylomicrons. J. Clin. Invest. **43:** 943-949.
18. BAGGIO, G., C. GABELLI, R. FELLIN, M. R. BAIOCCHI, S. MARTINI, G. BALDO, E. MANZATO & G. CREPALDI. 1983. Metabolism of lipoproteins in the postprandial phase. Prog. Biochem. Pharmacol. **19:** 129-140.
19. LEWIS, G. F., N. M. O'MEARA, P. A. SOLTYS, J. D. BLACKMAN, P. H. IVERIUS, A. F. DRUETZLER, G. S. GETZ & K. S. POLONSKY. 1990. Postprandial lipoprotein metabolism in normal and obese subjects: comparison after the vitamin A fat loading test. J. Clin. Endocrinol. Metab. **71:** 1041-1050.
20. SCHREZENMEIR, J. 1996. Hyperinsulinemia, hyperproinsulinemia, and insulin resistance in the metabolic syndrome. Experientia **5:** 426-432.
21. SCHREZENMEIR, J., B. CLARK, S. FENSELAU, J. ABEL, J. BEYER & C. N. HALES. 1994. Incomplete proinsulin processing in subjects with postprandial high triglyceride response (premetabolic syndrome). Exp. Clin. Endocrinol. **102**(suppl. 2): 175.
22. BUCOLO, G. & H. DAVID. 1973. Quantitative determination of serum triglycerides by the use of enzymes. Clin. Chem. **19:** 476-482.
23. SOBEY, W. J., S. F. BEER, C. A. CARRINGTON, P. M. S. CLARK, B. H. FAANK, I. P. GRAY, S. D. LUZIO, D. P. OWENS, A. E. SCHNEIDER, K. SIDDLE, R. C. TEMPLE & C. N. HALES. 1989. Sensitive and specific two-site immunoradiometric assays for human insulin, proinsulin, 65-66 split, and 32-33 split proinsulin. Biochem. J. **260:** 535-541.
24. ARMELLINI, F., M. ZAMBONI, L. RIGO, I. BERGAMO-ANDREIS, R. ROBBO, M. DE MARCHI & O. BOSELLO. 1991. Sonography detection of small intra-abdominal fat variations. Int. J. Obes. **15:** 847-852.
25. MÄRZ, W., S. CEZANNE & W. GROSS. 1991. Phenotyping of lipoprotein E by immunoblotting in immobilized pH gradients. Electrophoresis **12:** 59-63.
26. ZECHNER, R. 1990. Rapid and simple isolation procedure for lipoprotein lipase from human milk. Biochim. Biophys. Acta **1044:** 20-25.
27. SCHOTZ, M. C., A. S. GARFINKEL, R. J. HEUBOTTER & J. E. STEWART. 1970. A rapid assay for lipoprotein lipase. J. Lipid Res. **11:** 68-69.
28. SCHREZENMEIR, J., S. FENSELAU, B. STRAUSBERGER & L. SCHANDT. 1995. Association of increased adrenal reactivity with abdominal adipose tissue, high postprandial triglyceride response, and increased dietary-induced thermogenesis. Exp. Clin. Endocrinol. **103**(suppl. 1): 82.
29. FENSELAU, S., A. DIBA, I. KEPPLER, P. WEBER, W. MÄRZ, J. BEYER & J. SCHREZENMEIR. 1994. Apolipoprotein phenotypes in NIDDM, high, and normal triglyceride responders. Diabetologia **37**(suppl. 1): A86.
30. WARD, W. K., E. C. LA CAVA, T. L. PAQUETTE, J. C. BLARD, B. J. WALLUM & D. J. R. PORTE. 1987. Disproportionate elevation of immunoreactive proinsulin in type 2 (non-insulin-dependent) diabetes mellitus and in experimental insulin resistance. Diabetologia **30:** 698-702.
31. VAN GAAL, L. 1996. Personal communication.
32. WAHRENBERG, H., F. LÖNNQUIST & P. ARNER. 1989. Mechanisms underlying regional differences in lipolysis in human adipose tissue. J. Clin. Invest. **84:** 458-467.
33. VAN GAAL, L. F., J. L. VANUYTSEL, G. A. VANSANT & I. M. DE LEEUW. 1994. Sex hormones, body fat distribution, resting metabolic rate, and glucose-induced thermogenesis in premenopausal obese women. Int. J. Obes. **18:** 333-338.
34. ASSMANN, G., G. SCHMITZ, H. J. MENZEL & H. SCHULTE. 1984. Apolipoprotein E polymorphism and hyperlipidemia. Clin. Chem. **30:** 641-643.
35. VOGELBERG, K. H. & E. MANCY. 1988. Apo E2 phenotypes in type 2 diabetics with and without insulin therapy. Klin. Wochenschr. **66:** 690-693.
36. FERRANNINI, E., E. J. BARRET, S. BEVILACQUA & R. A. DEFRONZO. 1983. Effect of fatty acids in glucose production and utilization in man. J. Clin. Invest. **72:** 1737-1747.

37. BODEN, G., X. CHEN, J. RUIZ, J. V. WHITE & L. ROSSETTI. 1994. Mechanism of fatty acid–induced inhibition of glucose uptake. J. Clin. Invest. **93:** 2438–2446.
38. ECKEL, R. H. 1989. Lipoprotein lipase: a multifunctional enzyme relevant to common metabolic diseases. N. Engl. J. Med. **320:** 1060–1068.
39. GOLDBERG, I. J. 1996. Lipoprotein lipase and lipolysis: central roles in lipoprotein metabolism and atherosclerosis. J. Lipid Res. **37:** 693–707.
40. SCHREZENMEIR, J., I. KEPPLER, J. ABEL, S. FENSELAU, P. WEBER & J. BEYER. 1993. Dissociation of insulin sensitivity towards glucose and fat metabolism at an early stage of the metabolic syndrome. Int. J. Obes. **17**(suppl. 2): 97.

Evaluation of an Omega-3 Fatty Acid Supplement in Diabetics with Microalbuminuria

YVONNE K. LUNGERSHAUSEN,[a]
PETER R. C. HOWE,[a,b] PETER M. CLIFTON,[a]
CHRISTOPHER R. T. HUGHES,[c] PAT PHILLIPS,[d]
JOHN J. GRAHAM,[e] AND DAVID W. THOMAS[f]

[a] CSIRO Division of Human Nutrition
Adelaide SA 5000, Australia

[c] Diabetes Clinic
Modbury Hospital
Modbury SA 5092, Australia

[d] Endocrine and Diabetes Service
The Queen Elizabeth Hospital
Woodville South SA 5011, Australia

[e] Ashford Specialist Center
Ashford SA 5035, Australia

[f] Department of Chemical Pathology
Women's and Children's Hospital
North Adelaide SA 5006, Australia

INTRODUCTION

While other causes of cardiovascular mortality are declining, the incidence of end-stage renal disease is still increasing, with diabetes, particularly non-insulin-dependent diabetes mellitus (NIDDM), being the single most-important cause, accounting for about 30% of all cases.[1] The development of diabetic nephropathy is characterized by abnormal urinary albumin excretion, progressing from microalbuminuria (excretion rates of 20–200 µg/min) to macroalbuminuria (rates exceeding 200 µg/min) or overt proteinuria.[2] Excretion rates less than 20 µg/min are regarded as normal. Microalbuminuria is a strong predictor of cardiovascular mortality in both insulin-dependent (IDDM) and NIDDM diabetics.[3] Early screening and intervention in diabetics with microalbuminuria can prevent or at least delay the progression of nephropathy, with important economic benefits.[4]

Blood pressure control and improved glycemic management have been identified as important factors in preventing or treating diabetic nephropathy.[2] Angiotensin-

[b] To whom all correspondence should be addressed. Present address: Department of Biomedical Science, University of Wollongong, Wollongong NSW 2522, Australia.

converting enzyme (ACE) inhibitors are the antihypertensive treatment of choice, having been shown to be effective in reducing proteinuria in both experimental[5] and clinical diabetes.[6-9] The protective effect of the ACE inhibitors appears to be independent of reductions in systemic blood pressure and has been demonstrated even in normotensive IDDM[6-8] and NIDDM[9] patients with microalbuminuria. Thus, apart from counteracting the hypertension that can result from declining renal function, ACE inhibitors may have a primary role in preventing renal dysfunction, possibly by lowering intraglomerular pressure through inhibition of intrarenal angiotensin II.[5]

Other treatments that have been shown to reduce albumin excretion in diabetics are those that may influence renal hemodynamics by acting on eicosanoid mechanisms. They include thromboxane synthetase inhibitors,[10,11] prostacyclin analogues,[12] and ω-3 fatty acids.[13] The ω-3 fatty acids in fish oil produce trienoic eicosanoids that can alter inflammatory mediators and vascular reactivity, both of which are important in glomerular injury. Indeed, fish oil supplementation has been shown to be efficacious in numerous experimental models of inflammatory and immune-related renal disease.[14] Moreover, its ability to reduce proteinuria in patients with chronic glomerular disease[15] and retard IgA nephropathy[16] appears to be independent of blood pressure reduction.

Although there have been many studies in diabetic patients examining the effects of fish oil supplements on lipids and hemostatic parameters, very few have looked at albumin excretion. Two studies in patients with IDDM and established proteinuria showed no effect of fish oil on proteinuria or GFR,[17,18] although one showed a reduction in the elevated transcapillary escape rate for albumin, which was not related to blood pressure reduction.[18] Both studies used cod-liver oil, which is a relatively crude ω-3 supplement. In contrast, microalbuminuria was reduced in a group of predominantly NIDDM patients when pure eicosapentaenoic acid (EPA) was given.[13] Thus, the dose of EPA may have been limiting in the former studies. Alternatively, the nephropathy may have passed a critical stage beyond which ω-3 supplementation is less effective.

The lack of recent studies to further examine the possible benefits of fish oil in diabetic nephropathy probably reflects concerns arising from a number of early studies, particularly in NIDDM patients, which showed a worsening of fasting blood glucose levels and increases in glycated hemoglobin levels with fish oil supplementation.[19,20] However, it is now recognized that the use of relatively large doses of fairly impure fish oil extracts (<30% ω-3 fatty acids) may have contributed to these observations.[21] Hence, the purpose of the present study was to see whether a moderate dose of a highly refined fish oil extract (>85% ω-3 fatty acids) could ameliorate microalbuminuria in diabetics without adversely affecting glycemic management.

METHODS

Location and Ethical Approval

The study was conducted in Adelaide at the Nutrition Research Clinic, CSIRO Division of Human Nutrition, and at the Diabetes Clinic, Modbury Hospital, in accordance with National Health and Medical Research Council Guidelines for Good Clinical Research Practice. The use of these two centers for patient visits helped to accommodate the transport needs of the subjects. Patients were also recruited through diabetes clinics at The Queen Elizabeth Hospital and at Ashford Hospital in Adelaide. Hence, independent ethics approvals for the study were obtained from the institutional review boards of the CSIRO Division of Human Nutrition and each of the collaborating hospitals.

Subject Recruitment

Potential subjects were identified by reviewing patient case notes in diabetes clinics at Modbury Hospital, The Queen Elizabeth Hospital, and Ashford Hospital. They were contacted about the study if their case notes indicated that they might fulfill the following inclusion/exclusion criteria:

Inclusion criteria — confirmed IDDM or NIDDM
    — urinary albumin excretion rate of 20–200 µg/min (30–300 mg/24 h)
Exclusion criteria — evidence of proliferative retinopathy
    — use of ACE inhibitors or nonsteroidal anti-inflammatory drugs (NSAID)
    — $HbA_{1c} > 10\%$
    — excessive alcohol intake (average > 40 g daily)
    — smoking more than 20 cigarettes per day.

Those contacted were informed of the nature of the study and received detailed written information about the aims and protocol. After obtaining informed consent, we asked volunteers to complete a health and lifestyle questionnaire giving details of their past medical history, current medications, dietary intakes, and smoking and drinking habits. This information was used to confirm the suitability of each volunteer as assessed from the hospital case notes. Those who appeared suitable for inclusion were provided with urine collection bottles and were asked to attend a prestudy screening appointment.

All volunteers were asked to fast for 12 hours before attending the screening appointment (2 IDDM patients who were unable to fast had a very light breakfast; their data were excluded from blood glucose analyses) and to provide a timed overnight urine collection from the preceding day for determination of albumin and creatinine excretion rates. They were asked to have an empty bladder prior to blood pressure measurement. After sitting quietly for 5 minutes, blood pressure readings were taken at 1-minute intervals using a portable automated sphygmomanometer (Dinamap Model 845XT, Critikon, Tampa, Florida) with the appropriate-size cuff on the right arm. Blood pressure was averaged from three readings after discarding the first reading. Height and weight were recorded and body mass index (BMI) was calculated as weight (kg)/height (m^2). A blood sample was taken to determine blood glucose levels, glycated hemoglobin levels (HbA_{1c}), plasma cholesterol, and triglycerides. Provided that the values for urinary albumin and HbA_{1c} were within the predetermined range stated in the entry criteria, subjects were enrolled in the intervention trial. Their medical

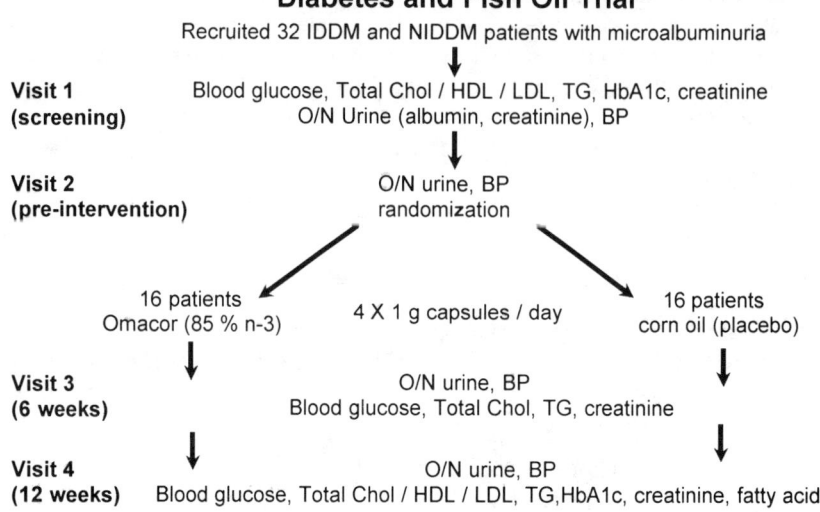

FIGURE 1. Study protocol (see METHODS for details and abbreviations).

practitioners were advised of their involvement in the study and asked to notify any proposed changes in medication, for example, prescription of an ACE inhibitor or NSAID or changes to prescriptions of antihypertensive or lipid-lowering drugs, so that their continuing participation in the study could be reviewed.

Study Design

The study was planned as a randomized, double-blind, parallel comparison of fish oil or placebo treatment for 12 weeks in diabetic subjects with microalbuminuria (FIG. 1). Subjects were stratified into two treatment groups matched for sex and diabetic classification (IDDM or NIDDM).

The 12-week intervention trial commenced with a pretreatment visit at 1–2 weeks after the screening appointment. A second timed overnight urine collection was obtained and the blood pressure measurements were repeated. Weight, medication, and state of health were recorded; then, each subject was issued with encoded containers of capsules, either Omacor fish oil or placebo (corn oil), and instructed to take four 1-g capsules daily for 12 weeks. The daily dose of ω-3 fatty acids obtained from Omacor was 3.4 g, that is, 2 g eicosapentaenoic acid (EPA) and 1.4 g docosahexaenoic acid (DHA). They were advised to take the capsules with food at regular intervals throughout the day to avoid possible gastrointestinal side effects.

Subjects returned to the clinic at 6-week intervals. They attended at the same time of day for each visit, bringing with them a timed overnight urine collection for albumin and creatinine determination, and were asked to fast for 12 hours beforehand. Their weight, medication, and state of health were monitored; their blood pressure was recorded; and a blood sample was taken.

Fasting blood samples taken at the screening appointment and after 6 and 12 weeks of intervention were analyzed for plasma glucose and creatinine and for serum total cholesterol and triglycerides. Additional determinations of serum HDL cholesterol, LDL cholesterol, and plasma HbA_{1c} were made only at the initial and final appointments. Plasma fatty acids were determined only at the final appointment.

Compliance of the subjects with their treatment was monitored by counting the unused capsules returned at each visit. Subjects were also contacted by telephone every 2–3 weeks to check on compliance with the supplements and to provide verbal encouragement. Compliance with the urine collection was assessed by checking the procedure with each subject and by monitoring creatinine excretion. Compliance with the oil capsules was also assessed by comparing the fasting plasma fatty acid levels.

Laboratory Determinations

Urines were analyzed by the Department of Chemical Pathology at the Women's and Children's Hospital, Adelaide. A modified rate reaction (Jaffe reagent, Beckman Instruments) was used for the measurement of creatinine and an immunoturbidimetric method (Atlantic Antibodies) was used for the measurement of albumin. Plasma glucose, HbA_{1c}, and creatinine and serum total cholesterol, triglycerides, HDL cholesterol, and LDL cholesterol determinations were performed by the State Pathology Laboratory, Gribbles Pathology Services.

Plasma fatty acids were analyzed by the CSIRO Division of Human Nutrition, as follows. Lipids were extracted from plasma with chloroform/methanol/0.1 M HCl (4:2:1) containing 0.005% butylated hydroxytoluene as an antioxidant, methylated with dry HCl in methanol (5% w/v), and extracted with petroleum spirit. Contaminating free fatty acids were removed by chromatography on 15-mm Biosil A columns. The eluate was dried under nitrogen, samples were taken up in isooctane, and an aliquot was injected onto a vitreous silica (30 M × 0.53 mm i.d.) cross-linked free fatty acid phase gas-liquid chromatography column for separation of fatty acids by using a Hewlett Packard 5711A gas chromatograph (Hewlett Packard, Avondale, Pennsylvania) with hydrogen as the carrier gas.

Statistical Analyses

Within-subject changes in the above-mentioned parameters were analyzed using a split-plot ANOVA for repeated measurements. Differences between treatment and placebo groups were considered significant if $p < 0.05$. It was estimated that a minimum of 30 subjects (15 per group) would give 80% power to obtain a 25% difference with treatment in the main outcome measure, that is, the change in urinary albumin excretion.

TABLE 1. Characteristics of Participants in the Treatment Groups[a]

Supplement	Corn Oil (Placebo)	Fish Oil (Omacor)
IDDM	6	5
NIDDM	10	11
Sex	12m, 4f	12m, 4f
Age (years)	58 ± 3	51 ± 3
BMI (kg/m^2)	31 ± 1	30 ± 2

[a] Age and BMI values are mean ± SEM.

RESULTS

Participant Profile

A total of 52 volunteers were screened for the trial, 32 of whom met the entry criteria and were enrolled. Most of those excluded had overnight albumin excretion rates below 20 µg/min. Two subjects with HbA$_{1c}$ levels of 11%, but who were otherwise suitable, were also enrolled. Another 2 were subsequently withdrawn, 1 due to an unrelated illness and the other due to gastrointestinal side effects resulting from consumption of the oil supplement provided. Thus, 32 participants (24 males, 8 females) ranging in age from 29 to 74 years (mean age = 55 years) actually completed the study. Eleven of these had been diagnosed with IDDM and 21 with NIDDM. TABLE

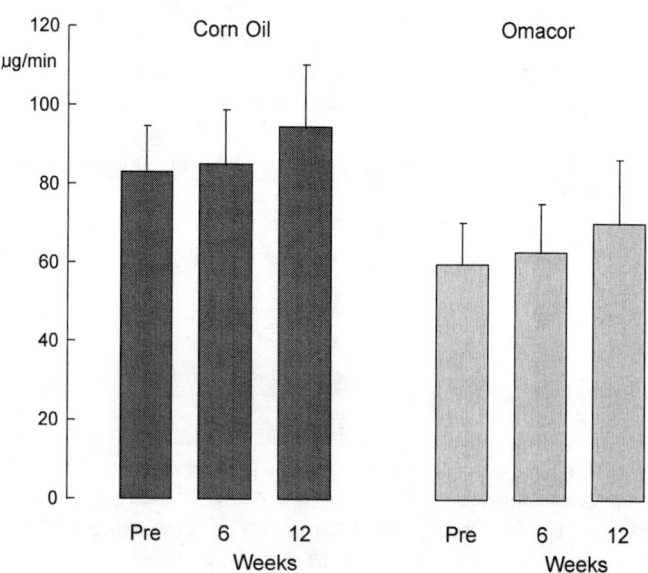

FIGURE 2. Urinary albumin excretion rate (mean ± SEM) in subjects taking fish oil (Omacor) or a corn oil placebo (Pre: averaged data from screening and preintervention visits).

TABLE 2. Creatinine Clearance (mL/min)[a]

Supplement	Corn Oil (Placebo)	Fish Oil (Omacor)
screening visit	112 ± 13	120 ± 8
6 weeks	105 ± 12	113 ± 9
12 weeks	108 ± 12	118 ± 9
Δ(0-12 weeks)	−3.7 ± 5.5	−2.1 ± 7.0

[a] Values are mean ± SEM.

1 shows their division into two treatment groups, stratified according to age, sex, and type of diabetes. Participants in the two groups were equally overweight. Apart from their diabetic medication, 8 subjects were taking antihypertensive drugs and 7 were taking lipid-lowering drugs. Their case notes indicated that preexisting levels of medication remained constant throughout the study.

Urinary Albumin and Creatinine

The final analysis of albumin excretion rates included only those subjects whose average baseline values (at the screening and preintervention visits) were within the range for microalbuminuria. Thus, 1 subject whose screening-visit value was acceptable had to be omitted from analysis after obtaining an exceedingly high value at the subsequent preintervention visit. There was no treatment effect on albumin excretion, which continued to rise (nonsignificantly) over 12 weeks from 60 to 70 µg/min and from 83 to 95 µg/min in the Omacor and corn oil groups, respectively (FIG. 2). Similarly, there was no change in creatinine clearance, which was used as a crude index of glomerular filtration, in either treatment group (TABLE 2). Average creatinine clearance rates were normal in both cases.

Fatty Acids and Lipids

The plasma fatty acid profiles determined in all subjects at the end of the intervention period indicated a high level of compliance with the supplements. In the group taking Omacor, the relative proportion of ω-3 plasma fatty acids was three times higher than in the group taking corn oil ($p < 0.05$, unpaired t test). The greatest change was seen with EPA (fivefold increase); DHA increased to a lesser extent (see FIG. 3).

The measurements of serum lipids are shown in FIG. 4. Total cholesterol, HDL cholesterol, and LDL cholesterol levels were similar in the two groups and were unaffected by the intervention. Plasma triglycerides were slightly higher initially in the placebo group (2.7 ± 0.4 mM) and continued to rise (to 3.1 ± 0.5 mM). However, in the Omacor group, they fell from 1.9 ± 0.2 mM to 1.5 ± 0.2 mM during the intervention. Thus, the within-subject decrease in triglycerides over 12 weeks with Omacor treatment (−0.43 ± 0.17 mM) was significantly different ($p = 0.013$, ANOVA) from the increase (0.44 ± 0.28 mM) seen in the placebo group.

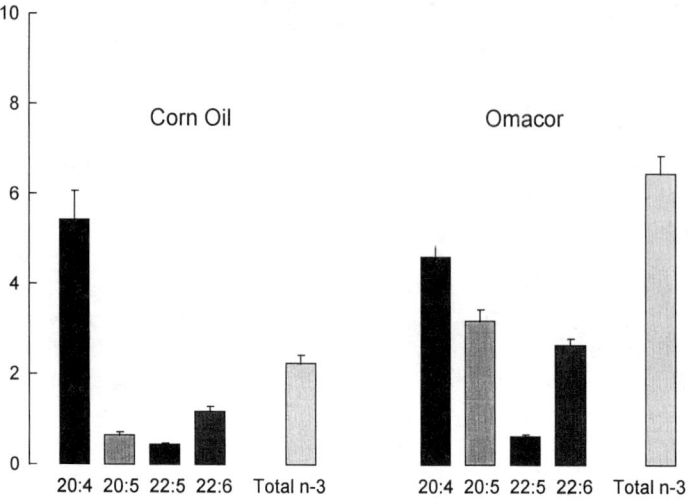

FIGURE 3. Selected long-chain fatty acids (mean ± SEM of % of total) in plasma of subjects taking corn oil or fish oil (Omacor) for 12 weeks (20:4 arachidonic acid; 20:5 eicosapentaenoic acid; 22:5 docosapentaenoic acid; 22:6 docosahexaenoic acid; total n-3 = 20:5 + 22:5 + 22:6).

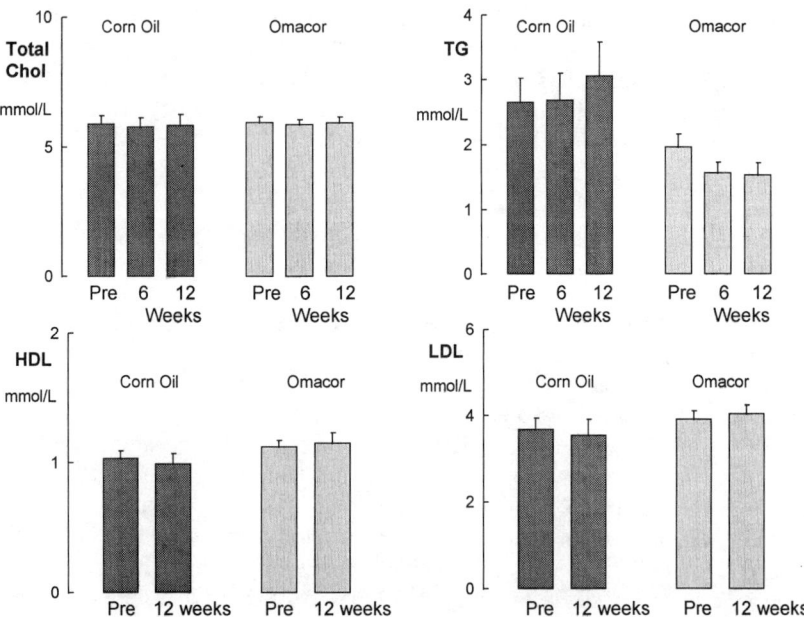

FIGURE 4. Serum lipid concentrations (mean ± SEM) in subjects taking corn oil or fish oil (Omacor) for 12 weeks (Chol: cholesterol; HDL: high-density lipoprotein cholesterol; LDL: low-density lipoprotein cholesterol; TG: triglycerides; Pre: screening-visit measurement).

TABLE 3. Blood Pressure (mmHg)[a]

Supplement	Corn Oil (Placebo)		Fish Oil (Omacor)	
	Systolic	Diastolic	Systolic	Diastolic
average baseline	140 ± 4	75 ± 2	139 ± 5	81 ± 3
6 weeks	138 ± 4	75 ± 3	134 ± 4	77 ± 3
12 weeks	138 ± 4	73 ± 2	133 ± 4	80 ± 3

[a] Values are mean ± SEM.

Blood Pressure

TABLE 3 shows the blood pressures recorded during intervention. Values for the screening and preintervention visits have been averaged. Even though preintervention averages for systolic and diastolic pressures were just within the normal range, the fish oil treatment caused systolic pressure to fall over the 12-week period (see FIG. 5). The average within-subject reduction (-5.9 ± 2.4 mmHg) was significantly different ($p = 0.039$; ANOVA using initial blood pressure as a covariate) from the change in systolic pressure obtained with the corn oil placebo (-0.4 ± 1.5 mmHg). There was no significant change in diastolic blood pressure.

Metabolic Indices

TABLE 4 shows average measurements of weight, fasting plasma glucose, and HbA_{1c} taken during the intervention. There was no significant change of weight in

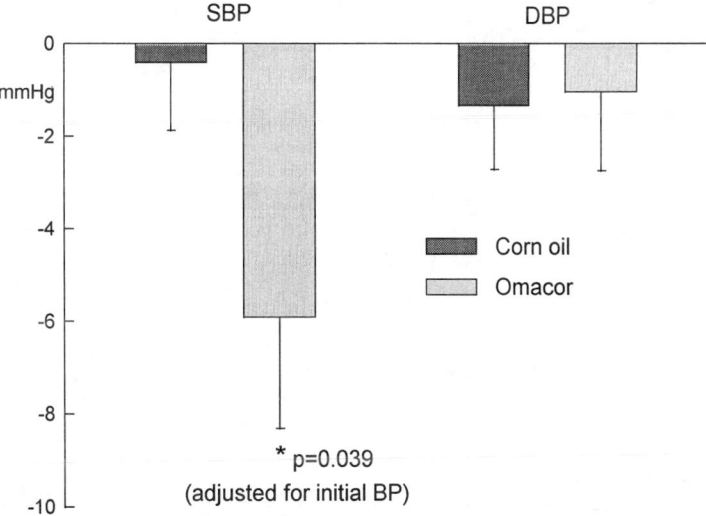

FIGURE 5. Within-subject changes in blood pressure (mean ± SEM) after 12 weeks of corn oil or fish oil (Omacor) supplementation (SBP: systolic blood pressure; DBP: diastolic blood pressure).

TABLE 4. Metabolic Indices[a]

	BMI (kg/m^2)		HbA$_{1c}$ (%)		Fasting Glucose (mM)	
	Corn Oil	Omacor	Corn Oil	Omacor	Corn Oil	Omacor
screening	30.8 ± 1.3	29.7 ± 1.6	8.5 ± 0.3	8.5 ± 0.3	11.2 ± 1.1	11.6 ± 0.9
6 weeks	31.1 ± 1.4	29.8 ± 1.6	—	—	11.2 ± 1.0	12.4 ± 1.1
12 weeks	31.1 ± 1.3	29.8 ± 1.6	9.2 ± 0.4	9.4 ± 0.4	11.0 ± 1.1	12.7 ± 1.1
Δ(0–12 weeks)			0.69 ± 0.22[b]	0.88 ± 0.24[b]	−0.2 ± 1.1	1.1 ± 0.8

[a] Values are mean ± SEM.
[b] $p < 0.01$.

either group. Fasting glucose levels were unaffected by corn oil supplementation, but rose slightly with the fish oil supplementation. However, the rise was not significant. There was a slight rise in plasma HbA_{1c} concentration over the 12-week period in both treatment groups. The rise was not significantly different between the two groups.

DISCUSSION

The main finding in this study was a failure of fish oil supplementation to reduce urinary albumin excretion in diabetics with microalbuminuria. Others have found that fish oil treatment is unable to regress overt proteinuria in diabetics.[17,18] However, if the ω-3 fatty acids in fish oil can act directly on intrarenal mechanisms to prevent damage,[14] one might expect intervention to be more successful at an early stage in the pathogenic process, that is, during microalbuminuria. Indeed, Hamazaki et al.[13] reported a significant reduction of albumin excretion in a similar mix of patients with IDDM and NIDDM. They administered a smaller amount of EPA (1.8 g/day) than that given in our study, but for twice as long a period. However, an ACE inhibitor has been shown to reduce microalbuminuria in IDDM patients from a moderate level (about 70 µg/min) to less than 20 µg/min within 6 weeks, without affecting blood pressure,[7] so it is unlikely that we needed a longer intervention period in order to see a reduction in albumin excretion. The study by Hamazaki et al. was small (16 patients), with assessment limited to spot urine analyses, and warranted further evaluation. However, our observations taken together with the other reported trials[17,18] suggest that a direct benefit of fish oil in diabetic microalbuminuria is unlikely. The reported antiproteinuric effects of drugs acting on renal eicosanoid mechanisms offer a possible alternative therapeutic approach. Meanwhile, the ACE inhibitors still seem to be the best option for early intervention in diabetic nephropathy.

Despite the lack of effect on urinary albumin excretion, Omacor supplementation had beneficial effects on other cardiovascular risk factors in our patients, viz. reductions of blood pressure and plasma triglycerides. Fish oils have been shown to lower blood pressure in treated and untreated hypertensives and, under certain circumstances, in normotensive subjects.[22] In diabetics, however, blood pressure-lowering effects have been variable, possibly because the doses used were marginal.[23] Blood pressure was unaffected in the study by Hamazaki et al. using purified EPA.[13] This may be attributed to the small dose or, alternatively, to the lack of DHA, as it appears that DHA may be more efficacious than EPA in modifying cardiovascular risk factors.[24] Using Omacor, we were able to provide a relatively high dose of EPA and DHA in a small quantity of oil. There have been other recent reports showing that the same modest dose of Omacor (4 g/day) is able to reduce blood pressure in normotensive and mildly hypertensive subjects.[25,26] Since elevated blood pressure can exacerbate the progression of diabetic nephropathy,[27] the modest reduction observed in this trial in predominantly normotensive diabetics may provide long-term renal benefit, despite the lack of a direct effect on albumin excretion. However, it should be noted that, when ACE inhibitors are used to treat diabetic nephropathy, any antihypertensive effect of fish oil is likely to be redundant.[28]

The well-established hypotriglyceridemic effect of fish oil also offers an important potential benefit to diabetics, particularly NIDDM patients who are at greater risk of cardiovascular complications.[1,2] There are few treatments for hypertriglyceridemia as simple and effective as fish oil supplementation that can be used safely in combination with cholesterol-lowering drugs. Unfortunately, its use has been limited by the reported adverse effect on glycemic control in NIDDM patients.[19,20] This could be linked to its triglyceride-lowering effect as the ω-3 fatty acids, by impairing hepatic triglyceride synthesis, may divert substrates from lipogenesis to gluconeogenesis. In addition, the ω-3 fatty acids may reduce glucose-stimulated insulin secretion.[20] However, Harris, in a recent review of 16 trials with fish oil in NIDDM, notes that a beneficial reduction of triglycerides was seen in most cases without detrimental effects on glycemic control and suggests that, even if fasting glucose levels are raised in some patients, the overall therapeutic benefits of fish oil supplementation could be expected to outweigh the cost of increased hypoglycemic medication.[21]

Since poor glycemic control predisposes to diabetic nephropathy,[2] it is not surprising that our microalbuminuric patients had relatively high levels of fasting glucose and HbA$_{1c}$. It is encouraging, therefore, that we found no evidence of adverse effects on glycemic control that could be attributed to the ω-3 supplementation. There was a slight worsening of HbA$_{1c}$ during the trial, but this occurred with both Omacor and placebo supplements. This outcome is consistent with other studies that have shown beneficial effects using the same dose of Omacor (4 g/day) without changes in fasting glucose or insulin responses[25,26] and supports Harris' view that a dose of 3–5 g/day of ω-3 fatty acids can be used safely as *adjunctive* therapy to lower triglycerides in NIDDM.[21]

ACKNOWLEDGMENTS

We wish to thank Novo Nordisk Australia for supporting this study and Pronova Biocare for providing the Omacor and placebo capsules.

REFERENCES

1. EGGERS, P. W. 1988. Effect of transplantation on the Medicare End-Stage Renal Disease Program. N. Engl. J. Med. **318:** 223–229.
2. JERUMS, G., M. COOPER, R. GILBERT, R. O'BRIEN & J. TAFT. 1994. Microalbuminuria in diabetes. Med. J. Aust. **161:** 265–268.
3. DAMSGAARD, E. M., A. FROLAND, O. D. JORGENSEN & C. E. MOGENSEN. 1990. Microalbuminuria as predictor of increased mortality in elderly people. Br. Med. J. **300:** 297–300.
4. BORCH-JOHNSEN, K., H. WENZEL, G. C. VIBERTI & C. E. MOGENSEN. 1993. Is screening and intervention for microalbuminuria worthwhile in patients with insulin-dependent diabetes? Br. Med. J. **306:** 1722–1725.
5. ANDERSON, S., H. G. RENNKE, D. L. GARCIA & B. M. BRENNER. 1989. Short- and long-term effects of antihypertensive therapy in the diabetic rat. Kidney Int. **36:** 526–536.
6. LEWIS, E. J., L. G. HUNSICKER, R. P. BAIN & R. D. ROHDE. 1993. The effect of angiotensin-converting-enzyme inhibition on diabetic nephropathy. N. Engl. J. Med. **329:** 1456–1462.
7. MARRE, M., M. HALLAB, A. BILLIARD, J. J. LE JEUNE, F. BLED, A. GIRAULT & P. FRESSINAUD.

1991. Small doses of ramipril to reduce microalbuminuria in diabetic patients with incipient nephropathy independently of blood pressure changes. J. Cardiovasc. Pharmacol. **18:** S165–S168.
8. MATHIESON, E. R., E. HOMMEL, J. GIESE & H-H. PARVING. 1991. Efficacy of captopril in postponing nephropathy in normotensive insulin-dependent diabetic patients with microalbuminuria. Br. Med. J. **303:** 81–87.
9. RAVID, M., H. SAVIN, I. JUTRIN, T. BENTAL, B. KATZ & M. LISHNER. 1993. Long-term stabilizing effect of angiotensin-converting enzyme inhibition on plasma creatinine and on proteinuria in normotensive type II diabetic patients. Ann. Intern. Med. **118:** 577–581.
10. BARNETT, A. H., B. A. LEATHERDALE, A. POLAK, M. TOOP, K. WAKELIN, J. R. BRITTON, J. BENNETT, D. ROWE & K. DALLINGER. 1984. Specific thromboxane synthetase inhibition and albumin excretion rate in insulin-dependent diabetes. Lancet **i:** 1322–1324.
11. KONTESSIS, P. S., S. L. JONES, S. E. BARROW, P. D. STRATTON, P. ALESSANDRINI, S. DE COSMO, J. M. RITTER & G. C. VIBERTI. 1993. Effect of selective inhibition of thromboxane synthesis on renal function in diabetic nephropathy. J. Lab. Clin. Med. **121:** 415–423.
12. SHINDO, H., M. TAWATA, N. YOKOMORI, Y. HOSAKA, M. OHTAKA & T. ONAYA. 1993. Iloprost decreases urinary albumin excretion rate in patients with diabetic nephropathy. Diabetes Res. Clin. Pract. **21:** 115–122.
13. HAMAZAKI, T., E. TAKAZAKURA, K. OSAWA, M. URAKAZE & S. YANO. 1990. Reduction in microalbuminuria in diabetics by eicosapentaenoic acid ethyl ester. Lipids **25:** 541–545.
14. DE CATERINA, R., S. ENDRES, S. D. KRISTENSEN & E. B. SCHMIDT. 1994. n-3 fatty acids and renal diseases. Am. J. Kidney Dis. **24:** 397–415.
15. DE CATERINA, R., R. CAPRIOLI, D. GIANNESSI, R. SICARI, C. GALLI, G. LAZZERINI, W. BERNINI, L. CARR & P. RINDI. 1993. n-3 fatty acids reduce proteinuria in patients with chronic glomerular disease. Kidney Int. **44:** 843–850.
16. DONADIO, J. V., JR., E. J. BERGSTRALH, K. P. OFFORD, D. C. SPENCER & K. E. HOLLEY. 1994. A controlled trial of fish oil in IgA nephropathy. N. Engl. J. Med. **331:** 1194–1199.
17. ROSSING, P., B. V. HANSEN, F. S. NIELSEN, B. MYRUP & H.-H. PARVING. 1995. Fish oil in diabetic nephropathy (abstract). Eur. J. Endocrinol. **132**(suppl. 1): 23.
18. JENSEN, T., S. STENDER, K. GOLDSTEIN, G. HOLMER & T. DECKERT. 1989. Partial normalization by dietary cod-liver oil of increased microvascular albumin leakage in patients with insulin-dependent diabetes and albuminuria. N. Engl. J. Med. **321:** 1572–1577.
19. GLAUBER, H., P. WALLACE, K. GRIVER & G. BRECHTEL. 1988. Adverse metabolic effect of omega-3 fatty acids in non-insulin-dependent diabetes mellitus. Ann. Intern. Med. **108:** 663–668.
20. KASIM, S. E. 1993. Dietary marine fish oils and insulin action in type 2 diabetes. Ann. N.Y. Acad. Sci. **683:** 250–257.
21. HARRIS, W. S. 1996. Do omega-3 fatty acids worsen glycemic control in NIDDM? ISSFAL Newsl. **3:** 6–9.
22. HOWE, P. R. C. 1995. Can we recommend fish oil for hypertension? Clin. Exp. Pharmacol. Physiol. **22:** 199–203.
23. MALASANOS, T. H. & P. W. STACPOOLE. 1991. Biological effects of omega-3 fatty acids in diabetes mellitus. Diabetes Care **14:** 1160–1179.
24. MCLENNAN, P., P. HOWE, M. ABEYWARDENA, R. MUGGLI, D. RAEDERSTORFF, M. MANO, T. RAYNER & R. HEAD. 1996. The cardiovascular protective role of docosahexaenoic acid. Eur. J. Pharmacol. **300:** 83–89.
25. MACKNESS, M. I., D. BHATNAGAR, P. N. DURRINGTON, H. PRAIS, B. HAYNES, J. MORGAN & L. BORTHWICK. 1994. Effects of a new fish oil concentrate on plasma lipids and lipoproteins in patients with hypertriglyceridaemia. Eur. J. Clin. Nutr. **48:** 859–865.
26. TOFT, I., K. H. BONAA, O. C. INGEBRETSEN, A. NORDOY & T. JENSSEN. 1995. Effects of n-3 polyunsaturated fatty acids on glucose homeostasis and blood pressure in essential hypertension. Ann. Intern. Med. **123:** 911–918.
27. BAKRIS, G. L. 1993. Hypertension in diabetic patients – an overview of interventional studies to preserve renal function. Am. J. Hypertens. **6:** 140S–147S.
28. HOWE, P. R. C., Y. K. LUNGERSHAUSEN, L. COBIAC, G. DANDY & P. J. NESTEL. 1994. Effect of sodium restriction and fish oil supplementation on BP and thrombotic risk factors in patients treated with ACE inhibitors. J. Hum. Hypertens. **8:** 43–49.

Paleodiet and Its Relation to Atherosclerosis

J. LIETAVA,[a] M. THURZO,[b] AND A. DUKÁT[a]

[a] Second Department of Internal Medicine
Comenius University
Bratislava, Slovak Republic

[b] Department of Anthropology
Slovak National Museum
Bratislava, Slovak Republic

The study of paleodiets may provide important insight into the understanding of civilization diseases and especially of atherosclerosis, a leading cause of mortality in Western countries. Dietary factors contribute to the development of atherosclerosis.[1] Although nutrition is not the only risk factor in atherogenesis, this paper will consider only the relationship of diet and atherosclerosis.

It is speculated that prehistoric nutritional habits are genetically coded.[2] This inherited message could be projected into "civilization diseases" caused by profound dietetic changes. However, it is necessary to recognize which diet is genetically determined or which diet was typical for our human ancestors.

Analysis of the paleodiet is hindered by a lack of relevant information and may be misleading because of misinterpretation of dietary evolution due to the limited number of early hominid finds[3] and the limited validity of analytical methods.[4] Reconstruction of the paleodiet is often based on the ancient method of Posidonius of Apomea: "from the life of primitive or barbarian societies."[5] Transposition of dietetic habits of recent natural (primitive) societies to prehistoric cultures may contain a high degree of false extrapolation. It is also questionable to which degree recent hunter-gatherer societies (exploiting limited subsistence resources and inconvenient geoclimatic territories) correspond to the socioeconomic conditions of Paleolithic hunters, which used the huge natural resources of big game.

The great French philosopher, Jean-Jacques Rousseau, introduced into science an image of the "noble savage," innocent beings living in natural state, free of civilization's decay.[6] The idea of the noble savage reflects the ancient Sumerian and Greek myths of the golden age of mankind, the era without hunger, disease, or death,[7] an era of nostalgic appeal, but which never occurred. To these myths belongs also the concept of an ideal diet — mainly vegetarian. Although a purely vegetarian diet may correspond to the earliest hominid dietary habits, it might be deceiving to apply this model to prehistoric populations.

Because of the normal postmortem decomposition of soft tissue, we currently have no older evidence of atherosclerosis than that acquired from Egyptian mummies. How-

TABLE 1. Diet and Human Evolution[a]

Time (Years Ago)	Type	Paleodiet	Method
10,000	H. sapiens sapiens	Herb. > Carn. (frequent malnutrition or periodic famine)	Isotope biochemistry
35,000	H. sapiens sapiens	Carn. > Herb.	Isotope biochemistry Standard anthropological methods
40,000–250,000	H. sapiens neanderthalensis	Carn. > Herb. Carn. >> Herb.?	Microscopic, microwear, and particle analysis
1,000,000	H. erectus	Fire—food processing?	Comparative anthropology > isotope analysis Same
2,000,000	H. erectus	Herb. >> Carn. Raw diet Hunting (?) or butchering of large animals	
4,000,000	H. habilis Australopithecines	Herb. Predation and scavenging	Same
6,000,000	Proconsul, Sivapithecus, Dryopithecus, etc.	Herb. (98%) Browsers and grazers	

[a] Paleodiet during several stages of hominoid and hominid evolution reconstructed according to paleoanthropological, isotopic, and microwear analysis. Abbreviations: Herb., herbivorous-like diet; Carn., carnivorous-like diet.

TABLE 2. Changes in Estimated Body Mass and Supposed Energy Expenditure in Recent Hominoids and Prehistoric Hominids during Evolution[a]

Species	Body Mass [kg] M	F	Energy [kcal]
Extant apes			
Pongo pygmaeus	69	37	
Pan troglodytes	49	41	
Pan paniscus	39	31	
Gorilla gorilla	140	70	
Gracile early hominids			
Australopithecus afarensis	45	29	1500–2000
A. africanus	41	30	
Robust early hominids			
A. robustus	40	32	
A. boisei	49	34	4000–5000 [M] 2000–4000 [F]
Homo habilis	52	32	
Homo erectus	50–70	40–60	
Homo sapiens neanderthalensis	65–70	50–58	
Homo sapiens sapiens	65	54	2800

[a] Data compiled from several sources.[3,9,10] Abbreviations: M, male; F, female.

ever, the close relationship between modern diet and atherosclerosis allows us some reasonable considerations of the role of prehistoric diet in the development of the disease.

Progression of atherosclerotic disease and its major complications (ischemic heart disease, stroke, and renal failure) may be schematically attributed to three main factors:

(1) the diet inducing atherosclerosis;
(2) exposure time (life longevity during which diet may cause the disease);
(3) energy expenditure (which may balance an increased energy intake).

DIET

During hominid evolution, except in modern times, there was a gradual shift from almost totally vegetarian habits to a diet with increasing intake of meat (see TABLE 1). These changes were closely connected with an improvement of stone tool production and of scavenging and hunting potential of early man.[8] Increased protein intake was associated with a gradual increase of estimated body mass of our direct human ancestors during evolution (see TABLE 2).

Considering the rather heterogenic group of "ramapiths" as a common human and ape ancestor, the early evolution of hominoids and hominids originated in southern and eastern Africa. The study and subsequent reconstruction of the local natural resources[11] may serve as a model starting point for dietary analysis. Early hominids consumed prevailingly vegetarian and strictly raw diet. Demonstration of slight differences between grazers and browsers is irrelevant considering the development of atherosclerosis due to the low risk potential of the vegetarian diet, but is useful for demonstration of the ability to prove the dietary differences in extremely old fossils.

Using carbon isotope analysis in two African fossilized baboon species dated to 1.7 to 1.9 million years ago (mya), Lee-Thorp et al.[12] found a preference for graminivorous diet (primarily seeds-oriented) (C4) in one group and an omnivorous diet (C3) in the other. Ciochon et al.[13] analyzed phytoliths (species-specific stony particles formed from dissolved silica in plants) in the *Gigantopithecus blacki*, which lived in southeastern Asia between 0.3 to 1.5 mya. Microscopic analysis revealed that this individual ate mainly bamboo, with minor additions of grasses and fruits.

Slight differences in diet were demonstrated also between the robust *Paranthropus* and more gracile *Australopithecus* species (dated to ca. 2.0 mya).[14] A fast Fourier transform of dental microwear indicated that the masticatory-potent *Paranthropus* preferred hard food items (date palm seeds, palm nuts, and bark), while *Australopithecus* ate a more-soft frugivorous diet (fruits, leaves). Recent observations of some primates (chimpanzees, baboons) confirm a dominant vegetarian diet with very little (<5%) intake of insects and meat.[3,15,16]

Dietary diversity of the species mentioned above could be based on anatomical and functional characteristics, climatic differences between regions, or food-acquisition techniques. In any case, the analysis demonstrates an almost strictly vegetarian uncooked diet during the first several million years of human evolution—a diet that is antisclerotic in nature.

Homo erectus was the first hominid who, through cultural innovations, managed to expand beyond the African homeland approximately 2 mya. The most significant of these innovations may have been the controlled use of fire and the construction of dwellings.[8]

Archeological evidence also proved the hunting skill of *Homo erectus*: several finds of stone and/or bone tools associated with dismembered elephants or hippos are evidence for hunting/scavenging activities.[3,17] Controlled use of fire is documented from East and South Africa (1.5 mya), France, and Spain (ca. 0.7 to 1.0 mya). Baked bones may also indicate the role of fire in food processing.[3,8]

Archeological evidence, paleoecological records, and recent isotope bone analysis indicate that the subsistence of *Homo sapiens neanderthalensis* was basically determined by hunting and scavenging, leading to a prevalence of meat in the diet.[18,19] Some scientists even support the hypothesis that the cold paleoclimatic geography in northern parts of Europe practically eliminated available vegetable resources, so the Neanderthal diet might have resembled that of precontact Eskimos. In any case, *Homo sapiens neanderthalensis* in the middle Paleolithic period was the first-known hominid that exploited more animal food resources than vegetable ones for survival.

The late Paleolithic diet was excellently reconstructed by Eaton and Konner,[20] who in their analysis assumed a dietary ratio of 35% of meat and 65% of vegetable foods (TABLE 3). Such a paleodiet is characterized by high cholesterol, high fiber, and low fat contents. Despite the high cholesterol level, the calculated ratio between unsat-

TABLE 3. Comparison of Calculated Paleodiet with the Current American Diet and Recommended Diet[a]

	Late Paleolithic Diet	Current American Diet	Recommended Diet
Total energy (%)			
protein	34	12	12
carbohydrate	45	46	58
fat	21	42	30
P:S ratio	1.41	0.44	1.00
Cholesterol (mg)	591	600	300
Fiber (g)	45.7	19.7	30-60
Na (mg)	690	2300-6900	1100-3300
Ca (mg)	1580	740	800-1200
Ascorbic acid (mg)	392.3	87.7	45

[a] Assuming the paleodiet contains 35% meat and 65% vegetable. From reference 20. Abbreviations: P, polyunsaturated fats; S, saturated fats. Recommended diet—United States Senate Select Committee Recommendations.

urated and saturated fats was favorable for the prevention of atherosclerosis. Moreover, wild animal fat contains a very high percentage of omega-3 fatty acids (about 4%), which possess antisclerotic properties and rheologically mitigate thrombogenesis, a substantial factor triggering the fatal clinical complications of atherosclerosis.[21] The antisclerotic nature of the diet is supported by a low sodium and a high calcium and ascorbic acid content. High sodium intake is directly related to hypertension, another very potent contributor to atherosclerosis, and to the high cardiovascular mortality in Western countries.[22]

The Neolithic transition (ca. 10,000 to 3000 B.C.), that is, the change from hunting/gathering food sources in the Paleolithic period to agricultural production, occurred very quickly if one is considering the total human evolution. However, agriculture, including domestication of animals and pastoralism, was an adaptive strategy lasting for several thousand years with scattered geographical distribution. Introduction of an agricultural pattern was determined by climatic conditions and by availability of traditional food supplies. Moreover, even centuries of cultivation of some cereal crops may not have greatly influenced the general longitudinal dietary pattern. Thus, van der Merwe and Vogel[23] demonstrated a sudden increase of maize consumption in the Illinois Valley only between 1000 and 1200 C.E., despite the evidence of maize cultivation from the Middle Woodland period (100-200 C.E.). Such a dietary shift to this highly nutritious, but biologically disadvantaged, maize diet (e.g., with increased occurrence of porotic hyperostosis—a skeletal marker of anemia or chronic pathogenic stress)[24,25] was accompanied by an increase in morbidity,[24] although atherosclerosis cannot be considered in this case because of missing paleopathological evidence.

On the other side, typically nomadic or pastoral people who seemingly subsisted on meat, like Huns or Avars in early medieval times, were exploiting agriculture in food production[26] and consumed a partially vegetarian diet.[27]

TABLE 4. Life Expectancy and Evolution[a]

	% of Living Population at Age of:		Life Expectancy
	5 Years	20 Years	(Years)
Lower Paleolithic	44.0	28.0	14.5
Upper Paleolithic	55.5	42.8	21.4
Neolithic	65.9	61.8	22.1
Roman Empire	74.0	60.1	30.4
Early Medieval	67.5	48.6	26.7
Modern Slovakia			67.8 (men)

[a] Life expectancy of population living from lower Paleolithic to modern times. After reference 28.

EXPOSURE TIME

Progression of atherosclerosis may last decades until clinical symptoms are evident. This means, in the normal population, that the major clinical manifestations of atherosclerosis endanger life in middle and late adulthood. A similar pattern might be expected also in our human ancestors.

Life-expectancy tables of early hominids are based on anthropological estimation of the ages at death (TABLE 4). Analyzing the life expectancy, it is possible to conclude that all the ancestors living before *Homo erectus* had lived for too short a time to develop an atherosclerotic disease. For example, the fossilized remains of *Homo erectus pekinensis* from Zhoukoudian belong prevailingly to young individuals (40% of the specimens were under the age of 14 years), with only a very few individuals (2.6%) dying older than 50 years of age.[3]

Of the entire early *Homo sapiens* sample (*Homo sapiens neanderthalensis*) living between 35,000 and 250,000 years ago, fewer than 10% of the individuals died older than 35 years of age;[18] in fact, among important specimens from Iraq and France, it is difficult to find an individual older than 40 years at death.[29]

The modern forms of *Homo sapiens* appeared before 35,000 years ago. In the Paleolithic period, the average life span did not exceed 20 years because of high neonatal and infant mortality. On the other hand, several individuals reached their forties—the age when they theoretically could develop atherosclerotic disease.

During the Neolithic and Bronze Age, the life expectancy was gradually increasing, with some deviations from average depending on economic or climatic determinants in a given culture.

More detailed information is available from Hellenistic and Roman eras. The average length of life was still very short (about 30 years) due to the above-mentioned high infant mortality, but analyses of the curriculum vitae of known Greek and Roman personalities show that they lived to about 60 years, close to the current life span. Pliny Secundus,[30] in his *Naturalis Historia*, described the clinical features typical for clinical manifestation of atherosclerosis. The disease was limited mainly to upper classes, a numerically negligible part of the population. However, there are no regular reports on the health status of lower classes.

In the Medieval Period, the life span increased to the theoretical atherosclerotic limit; however, the overpopulation crises increased the level of general malnutrition.[31] Moreover, the infectious pandemias regularly decimated the population and provide a broad field for consideration of the inverse relationship between inflammatory and civilization diseases.[32]

As to the break point for the recent pandemia of atherosclerotic disease, it seems that life span must be considered together with other known risk factors. The steep increase in agriculture production and the profound social/economic changes in the 18th and 19th centuries led to an increase of caloric intake in the broad population. Every region had its own break point of intensive agriculture. For example, in England, it could be traced to 1845 and 1846. In 1845, the parliament approved a progressive tax reduction on imported sugar, with a subsequent increase of consumption from 5–10 kg/year to 45 kg/year by the end of the century. The Corn Law from 1846 and the introduction of porcelain floor mills led to mass consumption of white bread, which is nutritionally deprived in comparison with the whole-grain type. Analogous introductions of nutritionally deprived, but energetically exaggerated diets can be observed in the middle and eastern European regions in the 20th century.[33]

ENERGY EXPENDITURE

Body size and energy requirement (TABLE 2) correlate in a weak linear relationship. To analyze the prehistoric condition, we need to omit several unpredictable variables like local food niche, predation preference, climatic conditions, and social organization of hunter-gatherer societies. The body size significantly influences a behavioral pattern and vice versa, so anthropological characteristics are closely related to such parameters as locomotion, bipedalism, and sexual dimorphism. The heavier creatures were less arboreally mobile and were forced to higher hunter-gatherer activity in savannah-like territory.[34] Also, dietary intake had a profound influence on stature—early European Paleolithic hunters with their high protein intake were about 12 cm taller than Neolithic farmers living in the same territory on a prevailingly unbalanced vegetable diet.[34]

Some authors hypothesized that prehistoric male hominids had periodically high energy expenditure (and thus also high need) due to the high demand during hunting activities.[11] According to recent studies, manually working lumbermen or miners have a daily energy expenditure of about 4500 kcal, while harvesting farmers may reach even 7000 kcal, calculated on an 8-hour working day.[35] The high expenditure may be the reason for fueling the body fat and also an explanation for the consumption of bone marrow—fatty and calorically very rich food. However, recent observations of primitive hunter-gatherer societies (living in hostile and infertile climatic conditions) do not support such a hard-fight-for-food theory. Observation in the !Kung society living in the Kalahari Desert (southwest Africa) documented that they, "with a modest subsistence effort of 2.4 workdays per week, produced an adequate diet and surplus."[3] Assuming the size of a Paleolithic hunting band as 25–30 persons or a tribe as 100 persons, one mammoth weighing several tons would provide caloric surplus for a few weeks (based on the analysis of Eaton and Konner).[20] Several Late Paleolithic Venuses indicate a high storage of body fat.[36] Applying recent knowledge[37] on waist circumference or hip-to-waist ratio on these artistic presentations, we can surely obtain high-risk ratios for the development of atherosclerosis.

Actually, Zimmerman[38] proved atherosclerosis in several precontact Eskimo frozen bodies, even in women in their late thirties and at fertile age, which is recently judged as protective against atherosclerosis.

DISCUSSION

We are facing a peculiar logistic paradox: the Western diet is judged to be of lower quality than the Paleolithic one, but the life span has almost tripled up to the present time. Although there is only anthropological, but not genetic, evidence on the presumed linkage between *Homo erectus* and *Homo sapiens*, it is assumed that *erectus'* basically raw vegetarian diet may be encoded in our present genome.

However, the prehistoric diet, especially during the last 35,000 years (the verified existence of *Homo sapiens sapiens*), exhibits a wide variability of dietetic composition due to various subsistence strategies and geoclimatic conditions of Eurasia.[39] Similar variability in meat-to-vegetable ratios, in maritime to terrestrial food sources, or in caloric intake is observed in recent European people, with conspicuous differences between nations and with profound intrapopulation changes within very short time frames.[33] Therefore, from a genetic aspect, it may be dubious to suppose either vegetarian or meat prehistoric diet as typical or as genetically encoded. However, such dietary diversity may enhance a level of genome polymorphism.

Despite the traditional belief in sudden dietary changes during the Neolithic transition, these changes could be better explained by an increase of caloric intake (mainly an increase of carbohydrates) than by qualitative changes.[40]

Theories of "thrifty genes" hypothesize a role for genetic adaptation to periodic long-lasting famines for development of obesity and diabetes mellitus in American Indians.[41,42]

Similar processes could also be attributed to atherosclerosis. It is reasonable to expect that a steep increase in food production and consumption induced by technicalization of agriculture could promote excessive body fat formation and atherogenesis in some predisposed populations. Recently, the highest prevalence of clinical atherosclerotic events is present in nations that were nutritionally deprived in the near-past (Ireland, eastern Europe, eastern Finland). European "thrifty gene" populations overcame the subsistence problem by increasing the production of supplementary vegetative resources, mainly grains with a high caloric content of starch.[2,22,33] The historically short transition time with sudden caloric excess is connected with increased morbidity.

Therefore, the contribution of modern diet to atherogenesis may be determined more probably by inadequate adaptation of "thrifty genes" to caloric excess than maladaptation of the prehistoric hominid genome to recent qualitative dietary change.

REFERENCES

1. Ross, R. 1992. The pathogenesis of atherosclerosis. *In* Heart Disease, p. 1106–1124. Saunders. Philadelphia.
2. Lutz, W. 1994. Die Kohlenhydrat-Theorie. Wien. Med. Wochenschr. **114:** 387–392.
3. Nelson, H. & R. Jurmain. 1988. Introduction to Physical Anthropology. Fourth edition. West Pub. St. Paul, Minnesota.
4. Lambert, J. B. & G. Grupe, Eds. 1993. Prehistoric Human Bone–Archaeology at the Molecular Level. Springer-Verlag. Berlin/New York.
5. Phillips, E. D. 1964. The Greek vision of prehistory. Antiquity **38:** 171–178.
6. Stumpf, S. E. 1993. Socrates to Sartre: A History of Philosophy. Fifth edition. McGraw-Hill New York.
7. Brentjes, B. 1968. Von Shanidar bis Akkad. Urania-Verlag. Leipzig/Jena/Berlin.
8. Feder, K. L. & M. A. Park. 1989. Human Antiquity: An Introduction to Physical Anthropology and Archaeology. Mayfield Pub. Mountain View, California.
9. Kay, R. F. 1985. Dental evidence for the diet of *Australopithecus*. Annu. Rev. Anthropol. **14:** 315–341.
10. McHenry, H. M. 1992. How big were early hominids? Evol. Anthropol. **1**(1): 15–20.
11. Peters, Ch. R. 1979. Toward an ecological model of African Plio-Pleistocene adaptation. Am. Anthropol. **81:** 261–278.
12. Lee-Thorp, J. J., N. J. van der Merwe & C. K. Brain. 1989. Isotopic evidence for dietary differences between two extinct baboon species from Swartkrans. J. Hum. Evol. **18:** 183–190.
13. Ciochon, R. L., D. R. Piperno & R. G. Thompson. 1990. Opal phytoliths found on the teeth of the extinct ape *Gigantopithecus blacki*: implications for paleodietary studies. Proc. Natl. Acad. Sci. U.S.A. **87:** 8120–8124.
14. Grine, F. E. & R. F. Kay. 1988. Early hominid diets from quantitative image analysis of dental microwear. Nature **333:** 765–768.
15. Peters, Ch. R. & E. M. O'Brien. 1981. Early hominid plant-food niche: insights from an analysis of plant exploitation by *Homo*, *Pan*, and *Papio* in eastern and southern Africa. Curr. Anthropol. **22:** 127–140.
16. Luchterhand, K., W. C. McGrew, M. J. Sharman, P. J. Baldwin & E. G. Tutin. 1982. On early hominid plant-food niches. Curr. Anthropol. **23:** 211–217.
17. Zihlman, A. & N. Tanner. 1978. Gathering and hominid adaptation. *In* Female Hierarchies, p. 163–194. Beresford. Chicago.
18. Stringer, C. & C. Gamble. 1993. In Search of the Neanderthals: Solving the Puzzle of Human Origin. Thames & Hudson. London.
19. Bocherens, H., M. Fizet, A. Mariotti, B. Lange-Badre, B. Vandermeerch, J. P. Borel & G. Bellon. 1991. Isotopic biogeochemistry (^{13}C, ^{15}N) of fossil vertebrate collagen: application to the study of a past food web including Neanderthal man. J. Hum. Evol. **20:** 481–492.
20. Eaton, S. B. & M. Konner. 1985. Paleolithic nutrition: a consideration of its nature and current implications. N. Engl. J. Med. **312:** 283–289.
21. Stein, B. & V. Fuster. 1992. Clinical pharmacology of platelet inhibitors. *In* Thrombosis in Cardiovascular Disorders. Saunders. Philadelphia.
22. James, W. P. T. 1988. Healthy nutrition: preventing nutrition-related diseases in Europe. WHO Reg. Publ. Eur. Ser. No. 24.
23. van der Merwe, N. & J. C. Vogel. 1978. ^{13}C content of human collagen as a measure of prehistoric diet in Woodland North America. Nature **276:** 815–816.
24. Cohen, M. N. & G. J. Armalagos. 1984. Paleopathology at the Origin of Agriculture. Academic Press. New York.
25. Stuart-Macadam, P. 1992. Porotic hyperostosis: a new perspective. Am. J. Phys. Anthropol. **87:** 39–47.
26. Beranová, M. 1986. Die Archäologie über die Pflanzenproduktion bei den Hunnen, Awaren, und Protobulgaren. Památky Archeologické **76:** 81–103.
27. Thurzo, M. & J. Lietava. 1990. Intrapopulation distribution of strontium level at Košice Šebastovce, Avar period (7th–8th C.A.D.) cemetery in Slovakia. *In* Proceedings of Abstracts of the Third Anthropological Congress Dedicated to Aleš Hrdlička. ČSAS pri ČSAV. Prague.
28. Acsádi, G. & J. Nemeskéri. 1970. History of Human Life Span and Mortality. Akad. Kiadó. Budapest.

29. TRINKAUS, E. & D. D. THOMPSON. 1987. Femoral diaphyseal histomorphometric age determinations for the Shanidar 3, 4, 5, and 6 Neanderthals and Neanderthal longevity. Am. J. Phys. Anthropol. **72:** 123-129.
30. PLINIUS STARŠÍ (PLINIUS SECUNDUS). 1962. Kapitoly o přírode (*Naturalis Historia*). Antická Knihovna. Prague.
31. RUSSEL, W. M. S. 1983. The paleodemographic view. *In* Disease in Ancient Man, p. 217-253. Clarke Irwin. Toronto.
32. BROTHWELL, D. R. & A. T. SANDISON, Eds. 1967. Diseases in Antiquity. Thomas. Springfield, Illinois.
33. WHO. 1995. Food and Health Indicators from Europe and Selected Countries around the World. Version 3. Danish Catering Center (WHO Collaborating Center). Herlev, Denmark.
34. ANGEL, J. L. 1975. Paleoecology, paleodemography, and health. *In* Population, Ecology, and Social Evolution, p. 167-190. Mouton. The Hague.
35. KOLESÁR, J. & Z. MIKEŠ. 1981. Ergometria v klinickej praxi (Ergometry in Clinical Praxis). Martin. Osveta.
36. PONTIUS, A. A. 1986. Stone-age art "Venuses" as heuristic clues for types of obesity: contribution to "iconodiagnosis." Percept. Mot. Skills **63**(part 1): 544-546.
37. DUNCAN, B. B., L. E. CHAMBLESS, M. I. SCHMIDT, M. SZKLO, A. R. FOLSOM, M. A. CARPENTER & J. R. CROUSE III. 1995. Correlates of body fat distribution: variation across categories of race, sex, and body mass in the Atherosclerosis Risk in Communities Study. Ann. Epidemiol. **5:** 192-200.
38. ZIMMERMAN, M. R. 1985. Paleopathology in Alaskan mummies. Am. Sci. **73:** 20-25.
39. ROSS, E. B. 1987. An overview of trends in dietary variations from hunter-gatherer to modern capitalist societies. *In* Food and Evolution: Toward a Theory of Human Food Habits, p. 7-55. Temple University Press. Philadelphia.
40. GRUPE, G. 1995. On Stone Age human diet. Hum. Evol. **10**(3): 245-249.
41. NEEL, J. 1962. Diabetes mellitus: a "thrifty" genotype rendered detrimental by "progress"? Am. J. Hum. Genet. **14:** 353-362.
42. WENDORF, M. 1989. Diabetes, the ice-free corridor, and the Paleo-Indian settlement of North America. Am. J. Phys. Anthropol. **79:** 503-520.

Passive Overconsumption
Fat Intake and Short-Term Energy Balance[a]

JOHN E. BLUNDELL AND JENNIE I. MACDIARMID

BioPsychology Group
Department of Psychology
University of Leeds
Leeds LS2 9JT, United Kingdom

REAL WORLD AND LABORATORY RESEARCH

In considering the relationship between fat intake and energy balance, two sets of issues come into prominence. In the "real world"—here meaning natural circumstances surrounding living, working, and relaxing—what happens when people are enveloped in a food supply containing an abundance of high-energy-dense foods (usually high in fat), which are extremely palatable and actively promoted, and which form part of modern culture? Under these circumstances, are people able to maintain an energy intake equivalent to (but no greater than) their energy expenditure (which may be quite low)? Can people eat freely without fear of overconsumption? In addition, what happens when people try to reduce their fat intake? Is it possible to undereat without inducing compensatory eating responses?

It is accepted that "real world" events are often very difficult to monitor and measure. This is exemplified by the problems involved in achieving good estimates of habitual energy intakes. Indeed, it is widely recognized that reported values of food intake derived from various techniques represent substantial underestimates of the actual energy and nutrients consumed.[1] Moreover, considering the alarming and continuing increase in the frequency of obesity (which can be accurately measured), it can be surmised that people experience great difficulty in preventing their energy intakes from rising above their energy expenditure. It is relevant to inquire what features of human beings—biological, psychological, or both—allow them to drift with apparent ease into a positive energy balance with the inevitable gain in weight.[2]

In this domain of science, one of the objectives of laboratory research should be to model the happenings in the real world. In practice, what usually happens is that researchers attempt to demonstrate how the physiological system responds when subjected to various nutritional challenges. What happens when people under experimental protocols are forced to consume particular amounts of fat or allowed to eat

---

[a] The research reported in this article was supported by grants from the Biotechnology and Biological Sciences Research Council (BBSRC) and the United Kingdom Ministry of Agriculture, Fisheries, and Foods (MAFF), AGRO-FOOD Link scheme.

unlimited amounts? To what extent can these types of studies inform us about habitual "real world" capabilities?

FAT-PLUS AND FAT-MINUS

One requirement of laboratory research should be to disclose the consequences of eating a surfeit of dietary fat and the effects of reducing fat intake so as to create a deficit (compared with a normal habitual level of consumption). These constitute the fat-plus and fat-minus manipulations.[3] Does the physiological system respond symmetrically and equally to these two types of challenges?

SATIETY SIGNALS RELATED TO FAT INTAKE

To provide a biological perspective for understanding the effect of dietary fat on the control of appetite (food intake) and short-term energy balance, it is useful initially to consider the potential inhibitory responses (satiety signals) that could be generated by fat ingestion.

Much interest in peripheral sites of action for the control of appetite has focused on peptidergic inhibition of food intake. Many peripherally administered peptides lead to an anorectic response, and good experimental evidence for natural roles exists for cholecystokinin (CCK), pancreatic glucagon, bombesin, and somatostatin.[4] Recent research has supported the status of CCK as a hormone that mediates satiation and possibly early-phase satiety.[5] The consumption of protein or fat stimulates the release of CCK, which activates CCK-A-type receptors in the pyloric region of the stomach. This signal is transmitted via vagal afferents to the nucleus of the tractus solitarius, where it is relayed to the medial zones of the hypothalamus, including the paraventricular nucleus and the ventromedial hypothalamus. The anorectic effect of systemically administered CCK can be blocked by vagotomy[6] and by the selective CCK-A receptor antagonist, devazepide (MK-329).[7] Significantly, many reports now demonstrate that the CCK-A-type antagonist administered alone leads to an increase in food intake in experimental animals.[8] Interestingly, trypsin inhibitors that block the inactivation of CCK produce a suppression of food intake in animals and in humans.[9]

A serotonin link in the mediation of CCK satiety has been proposed[10] and this probably depends on the 5-HT$_{2C}$ subtype.[11] Indeed, metergoline attenuates CCK-induced anorexia, whereas devazepide antagonizes 5-HT-induced inhibition of feeding.[12] These mechanisms indicate ways in which the ingestion of dietary fat could trigger neurochemical responses that mediate satiety.

Another route by which dietary fat could induce a behavioral response is via enzyme systems responsible for the digestion of fat. In particular, pancreatic procolipase is a cofactor for lipase that is necessary for optimal fat digestion in the intestine during a meal. In rats, the 100-amino-acid procolipase is cleaved by trypsin to colipase and a pentapeptide, Val-Pro-Asp-Pro-Arg. This peptide, VPDPR or enterostatin, decreases food intake in rats.[13] Moreover, enterostatin appears to selectively reduce intake of a high-fat diet.[14] Enterostatin, with the suggested structure Val-Pro-Gly-Pro-Arg, also is increased after high-fat feeding and after administration of CCK-8.[15] These data suggest ways in which peripheral satiety signals could be generated by fat consumption.

Another class of satiety signals is believed to arise during the absorptive or postabsorptive phase. The products of food digestion may be metabolized in peripheral tissues or organs, or they may enter the brain directly. It has been argued that the degree of oxidative metabolism of glucose and free fatty acids in the liver constitutes a significant source of information useful for the control of appetite.[16] If oxidative metabolism is a satiety signal for fat, it is important to recognize that, according to this hypothesis, the oxidative signal could arise from the metabolism of ingested fat or from fuels derived from internal adipose stores.[17] The oxidation of fat may constitute a signal for the suppression of eating. Experimental evidence indicates that the inhibition of fat oxidation by methyl palmoxirate[18] or 2-mercaptoacetate[19] causes an increase in feeding. It follows that any suppression of appetite arising from this mechanism need not be tightly synchronized with the ingestion of fat.

In principle, many products of digestion and humoral peptides that activate the enzymes that metabolize those products could bind to specific receptors or alter some aspect of neuronal metabolism. In each case, the brain is informed about some aspect of the metabolic state resulting from food consumption. One interesting possibility concerns the glycoprotein apolipoprotein A-IV, which is produced exclusively by the human small intestine.[20] The output of A-IV into intestinal lymph increases as a result of feeding fat (e.g., see reference 21). The intravenous administration of apolipoprotein A-IV decreases meal size in rats[22] and A-IV may constitute a physiological signal for satiation after the consumption of fat.[23] Although apolipoprotein A-IV levels do respond to dietary fat in humans, the hypothesis that A-IV is a metabolic signal is weakened by the observation that concentrations are subject to a rapidly acting autoregulatory mechanism.[24]

FAT AS A RISK FACTOR FOR OVERCONSUMPTION

Although the body appears to contain potent physiological responses that are triggered by fat ingestion, many studies have demonstrated that people who consume high-fat foods (either through personal choice or in experimental situations) tend to overconsume energy. The opposite occurs with high-carbohydrate foods. Interestingly, however, in some studies in which the proportions of fat and carbohydrate were varied, the results showed the two to have equivalent satiating power.[25] However, because of the potent action of protein on satiety, it is clearly necessary when comparing fat and carbohydrate to keep protein constant. In one study in which fat displayed a satiety action similar to carbohydrate, the fat foods contained much higher amounts of protein.[26] Consequently, the apparently strong satiating action of fat could have been due to the presence of the protein. However, the satiating power of dietary fat may be altered according to its association in a food with either protein or carbohydrate.[27] Such an effect could explain instances where fat caused a potent inhibition of subsequent eating (e.g., see reference 28).

There is another aspect, however, that may be more important when measuring effects of fat ingestion—the difference between satiation and satiety. Satiation is the process in operation while foods are being eaten; satiety is the state engendered as a consequence of consumption. Satiation controls meal size; satiety measures the capacity of food to control subsequent hunger and eating, commonly called satiating efficiency.[29] In considering dietary fat as a risk factor in overconsumption, the effect on satiation is likely to be much more important than that on postingestive satiety.[30]

PASSIVE OVERCONSUMPTION

Interestingly, most experiments on fat and appetite have used some variation of the preload or fixed-meal presentation in which subjects are required to consume an obligatory amount of fat (or fat mixture). The consequences of this mandatory consumption (determined by the researcher) are then measured. This procedure offers a measure of satiety (not satiation). It is likely that this procedure provides only limited information about the effect of fat on appetite and may preclude complete understanding of the effect that dietary fat has on energy intake. An alternative procedure, sometimes called concurrent evaluation,[31] can be used to assess the effect that the fat content of foods has on the amount willingly consumed (satiation) when subjects are provided with a range of foods and allowed to eat freely to comfortable fullness. Use of this procedure has demonstrated that subjects consume much greater quantities of energy from a range of high-fat foods than from foods high in general carbohydrates[32] or sucrose.[33] This effect, particularly strong in obese subjects,[34] has been termed high-fat hyperphagia[35] or passive overconsumption.[36] The effect is almost certainly due, in large part, to the high-energy density of the high-fat foods; hence, it can be regarded as a passive form of high consumption rather than eating as actively driven. However, the use of the term passive means only that there is no deliberate intention on the part of the eater to overconsume. People are not normally endeavoring to eat as much as they possibly can. Indeed, a number of mechanisms contribute to the occurrence of passive overconsumption. For example, ingested fat can stimulate food intake[36] and the oral sensory effects of fat appear to stimulate intake in rats.[37] The combination of fat and alcohol can stimulate overeating independently of energy density.[38]

Although high-fat foods induce remarkably high levels of energy intake (mainly fat energy) in meals or snacks, the experimental design still permits a measure of postingestive effects on satiety. Interestingly, these very high levels of fat energy consumed do not induce any noticeable intensification of satiety when compared with much lower energy intakes of high-carbohydrate foods. Therefore, high-fat foods can readily stimulate high intakes of fat energy with no proportionate increase in satiating power. Consequently, a single meal (or snack) of high-fat foods normally leads to a significant increase in that day's food consumption. These effects provide an explanation for the relationship of the eating pattern to the long-term overconsumption of energy on a high-fat (low food quotient) diet.[39]

The methodology outlined above is important in that it demonstrates overconsumption in connection with the intake of high-fat foods, as seems to happen in real life.

This indicates a genuine point of contact between "real world" events and laboratory research. The methodology emphasizes that it is necessary to measure the effects of fatty foods while they are being consumed (satiation) as well as after consumption (satiety). Indeed, it certainly appears as if dietary fat's most-important action on appetite is while it is being consumed rather than afterwards.

THE FAT PARADOX

In considering the relationship between fat and satiety, a paradox becomes apparent. On the one hand, fat in the intestine appears to generate potent satiety signals.[40] On the other hand, exposure to high-fat foods leads to a form of passive overconsumption that suggests that fat has a weak effect on satiety.[33] The paradox can be expressed as the puzzle of fat-induced satiety and high-fat hyperphagia.

As noted above, it has been demonstrated that fat induces physiological responses that should inhibit food intake. However, when rats are placed on high-fat diets or given fat supplements, they take in excessive amounts of energy and rapidly gain weight. Moreover, human subjects given a range of high-fat foods also increase their energy intake and gain weight compared with subjects eating a medium- or low-fat diet (see above). In addition, high-fat foods markedly increase meal size (measured in terms of energy), an effect particularly marked in obese subjects.[34] What is the explanation for the apparent contradiction between fat-induced satiety signals and the easy overconsumption of high-fat foods?

Although emulsified fat delivered to the intestine (duodenum or jejunum) produces prompt satiety signals, fat consumed orally takes some time to reach the intestine in similar form, and its action is likely to be diluted by other nutrients. Hence, consumed fat may engender more slowly arising satiety signals. Two features of fat favor the rapid consumption of energy: fat produces potent oral stimulation, which facilitates intake; and high-fat foods normally have a high-energy density, which means that a large amount of fat energy can be consumed before fat-induced satiety signals become operative. The signals are apparently too delayed to prevent the intake of large amounts of this food. One of the consequences of a food supply containing readily available, very palatable high-fat foods is that the natural fat-induced satiety mechanism becomes overwhelmed.

Indeed, in a series of experiments using a research design that measures both satiation and satiety, we have observed striking differences with subjects allowed to consume from ranges of high-CHO or high-fat foods on separate occasions.[41] With the high-fat foods (high-energy density), subjects invariably consume a significantly greater amount of energy (sometimes double) when eating to a point of comfortable fullness. However, the actual weight of food consumed (total mass) is significantly less with the high-fat foods. The main function of a satiety signal should be to limit the amount of food being put into the mouth (an inhibition of eating behavior). Therefore, the fat paradox can be resolved.

The dietary override of physiological satiety signals has a number of implications. First, it is clear that the effect of fat per se should be separated from the effects of high-fat foods. References to the effect of dietary fat and satiety should recognize that the effects of fat delivered in controlled amounts to the intestine may demonstrate effects that are still present, but eclipsed when subjects eat large quantities of high-fat foods. Second, the recognition of this paradox suggests nutritional and pharmaceutical strategies for the reduction of fat consumption. Any technique that would advance or intensify fat-induced satiety signals or that would prevent the effects of high-fat foods on satiation would prevent the passive overconsumption of fat energy that hinders good control of appetite.

One further implication of the uncoupling of eating behavior and energy intake is that, through passive overconsumption, fatty foods have a strong tendency to generate an immediate positive energy balance. This is apparent under periods of rest, normal activity, and even high levels of exercise (cycling or running).[42] Since this positive energy (and fat) balance does not generate any strong compensatory responses, the short-term positive energy balance will probably be incorporated into body energy stores.

FAT INTAKE AND FAT CONSUMERS

One useful procedure to investigate the effect of fat on short-term energy balance is to use laboratory experiments to model natural eating patterns. In this way, the effect of foods of varying types upon satiation and satiety (precisely monitored) during free consumption can be measured. Another procedure is to characterize natural high- and low-fat consumers identified through large-scale surveys or epidemiological studies.

The Leeds High Fat Study[43] identified such individuals in the Leeds (United Kingdom) community. In addition, the database arising from the Dietary and Nutritional Survey of British Adults (DNSBA)[44] was reanalyzed to define for that sample the relationship between dietary fat, food consumption, and body mass index (BMI).

High- and low-fat consumers were defined as consuming >45% or ≤35% of their food energy as fat, respectively. When nutrient intake is defined as the percentage of total food energy consumed, it can be seen that carbohydrate intake falls as fat intake increases (FIG. 1).

What foods contribute to the overall intake of dietary fat? The DNSBA data indicate that the greatest contributor to fat in the diet is meat and meat products, which provide 25% of the total intake of fat. Both men and women in the high-fat groups derived the greatest percentage of their fat intake from meat and meat products, whereas cereal and cereal products (including cakes, pastries, biscuits, and bread) contributed the most fat to the diet in the low-fat groups. Comparing high- and low-fat consumers, the male high-fat consumers ate significantly more high-fat meat products (bacon, lamb, meat pies, sausages, burgers, fried whitefish), butter, whole milk, cheese, eggs, savory snacks (i.e., crisps), desserts, and alcohol. They also ingested less breakfast cereal (including high-fiber cereals), skim milk, low-fat spreads, chicken, yogurt, potatoes (not fried), fruit, and table sugar. The women's dietary patterns were similar to the men's, but there were fewer differences between high- and low-fat groups in terms of consumption of high-fat meat products. Bacon, meat pies, and sausages were the only meat products that differentiated the female high- and low-fat groups. These

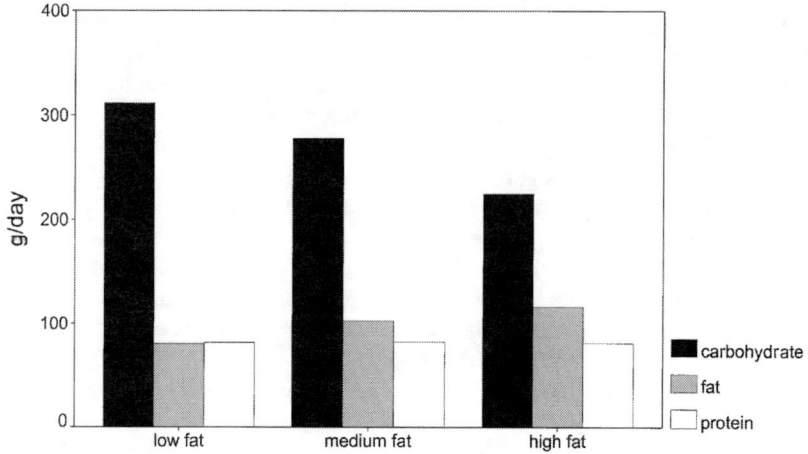

FIGURE 1. Relationships among macronutrient intake in three groups of individuals consuming high, medium, and low amounts of fat. As the percentage of fat intake increases across the groups, the carbohydrate intake falls.

differences were consistent with variations in food intake found between high- and low-fat consumers in the Leeds study. These data draw attention to some of the types of foods likely to be most responsible for generating passive overconsumption and contributing to the short-term positive energy balance.

THE PUZZLE OF SUGAR AND FAT IN THE DIET

In recent years, there has emerged a growing confusion over the role of sugar and fat in energy balance and their influence over body weight. On the one hand, it has long been contested that foods high in sugar and fat are specifically targeted by food cravers[45] and therefore represent a basis for overeating. This idea gets partial support from the observation that obese women rank sweet fat foods highly in their listing of preferred foods.[46] More recently, the combination of sweet high-fat foods in the diet has been regarded as a potential source of weight gain[47] and this has been accompanied by a recommendation to reduce both fat and sugar in the diet as a means of curtailing the rise in obesity.

On the other hand, a good deal of publicity has been given to the notion of the "sugar-fat seesaw."[48] This concept is based on the following observation: when the energy consumed as sugar or fat is separately expressed as a percentage of the total energy consumed, then a reciprocal relationship is seen. Because of this inverse association, it is argued by some that it is therefore impossible to reduce intakes of both fat and sugar since as one is reduced the other will automatically rise.[49]

Recent analyses of some large databases appear to support the sugar-fat seesaw.

For example, combining data from the MONICA study and the Scottish Heart Study, the dietary components of fat and sugar were calculated as quintiles and then related to BMI.[50] First, ascending quintiles with the highest fat consumption were inversely related to the quintiles of sugar intake. Second, a high BMI was associated with the highest quintile of fat intake (and therefore the lowest of sugar). Accordingly, BMI was positively related to the fat:sugar ratio. This pattern is also reflected in the analysis of the United Kingdom database forming the Diet and Nutrition Survey of British Adults (DNSBA).[51] Here, a high intake of sugar (as a percentage of total energy) was associated with the lowest BMI. Overall, these studies appear to confirm that the intake of fat is clearly related to obesity, while the intake of sugar is associated with leanness. It appears to follow that sugar and fat are indeed inversely related in the diet and therefore to eat less of one of them will inevitably mean eating more of the other.

However, the issue is not completely resolved by these analyses of large databases. It has been noted earlier, in the Leeds High Fat Study, that fat and carbohydrate intakes are inversely related when these nutrients are expressed as percentages of energy in the total diet consumed. However, as shown in FIG. 2, when expressed in absolute terms as grams of nutrient consumed, there is a positive relationship. When a similar analysis for fat and sugar is carried out on a large database such as the DNSBA (OPCS) survey, these two dietary components are inversely related when expressed as percentages (the sugar-fat seesaw), but are positively related when expressed as grams.[52] Consequently, different outcomes are obtained when exactly the same data are represented in two different ways. Does this mean that the sugar-fat seesaw is an artifact of the way in which the data are analyzed? Does it mean that a simultaneous reduction of fat and sugar intakes can or cannot be achieved?

At the present time, it is not clear whether the observed relationships between fat and sugar reflect processes going on in physiology, nutrition, or food choice (eating behavior). It is clear, though, that when people are allowed to eat from ranges of high-fat or high-sucrose foods, then passive overconsumption only occurs with fat.[53] It follows that fat will readily promote overconsumption, while sucrose will probably prevent it. As long as there is a clear separation of foods containing either fat or sucrose, processes of food choice will probably ensure that high intakes of sucrose (in people with a sweet tooth) will be inversely related to fat intake (people who are fat preferrers). Of course, there will be some individuals who have a preference for sweet items and fatty foods. Moreover, what happens when fat and sugar are combined in a single food (biscuits, cakes, chocolate, pastries)? Are these eaten by the "sweet-tooth" or "fat-tooth" people, or by both? Does the presence of sweetness (sucrose) in biscuits, pastries, etc., raise their palatability and therefore make them more likely to be consumed, thereby inevitably raising fat intake? The expression of nutrient intakes in absolute weights demonstrates that large amounts of sucrose and fat can coexist in the diets of particular individuals.

One interpretation is that low-fat diets are not likely to stimulate passive overconsumption or to result in a short-term positive energy balance. Low-fat consumers will certainly be eating large amounts of carbohydrate, some of which will be sugars. However, this sugar does not appear to constitute a threat of obesity. In contrast, high-fat foods have the potential to cause overconsumption and sweet fat foods are particularly

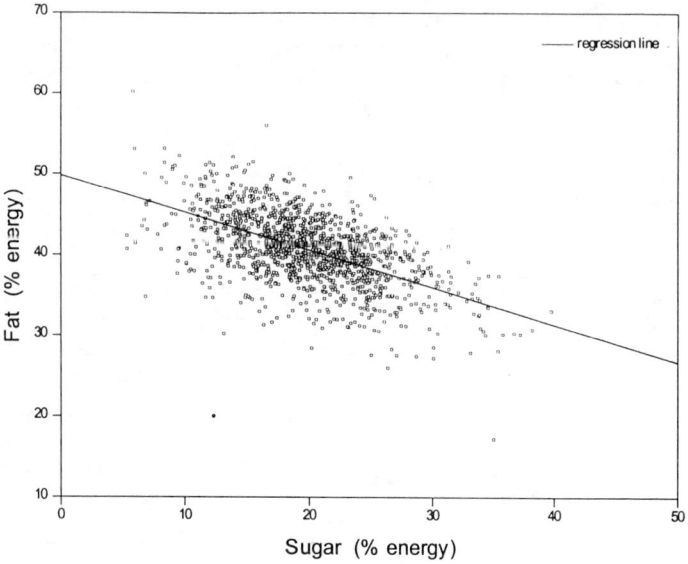

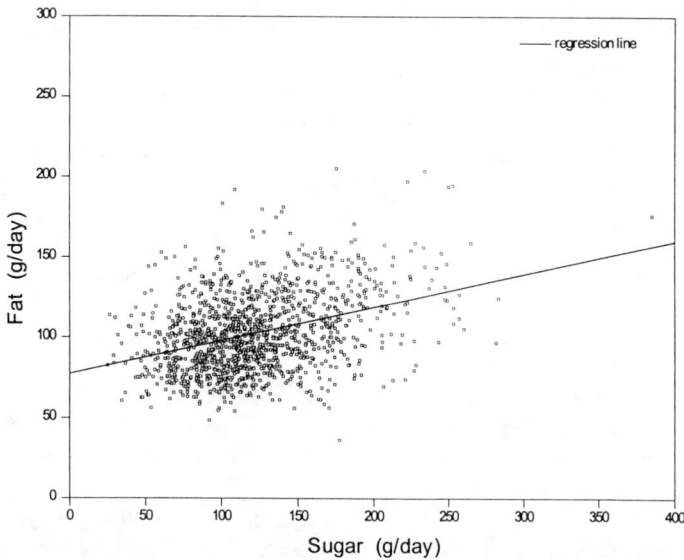

FIGURE 2. Scatterplots of intakes of total fat and total sugar in which the components are expressed as percentages of food energy (upper panel) or as absolute intakes in grams (lower panel).

potent under some conditions.[54] In some individuals, the sweet characteristics of fatty foods will facilitate consumption (through processes influencing food choice); therefore, sweetness will here be associated with a high-fat intake and will contribute to a positive energy balance and weight gain. Sweetness, in the form of sucrose or high-intensity sweeteners, can contribute to a positive energy balance in those high-fat consumers who also have a sweet tooth.

HIGH-FAT CONSUMPTION AND BMI

Of course, one consequence of a positive energy balance induced by exposure to high-fat foods should be an increase in body weight. Evidence linking dietary fat and BMI already exists. It follows that natural high- and low-fat consumers should show differences in average BMI and the frequency of obesity. Computations have been carried out with the national database (DNSBA). The sample was defined according to the percentage and the absolute amount of fat consumed. To ensure the inclusion of valid data, subjects who reported being ill or who were dieting during the study period were excluded from the analysis. In addition, it is argued that a ratio of energy intake (EI) to basal metabolic rate (BMR) of less than 1.2 is not compatible with habitual intake and normally signifies underreporting of energy intake.[1] Therefore, all subjects with an EI:BMR of less than 1.2 were excluded. Consequently, the analysis was carried out on those subjects most reliably identified as habitual high- and low-fat consumers.

The most-meaningful relationship is the distribution of BMI for the high- and low-fat groups, classified by fat grams consumed per day (FIG. 3). There is a greater positive skew in the distribution of BMI of high-fat consumers. Defining obesity as a BMI of more than 30, there were 19 times more obese subjects in the high-fat group than there were in the low-fat group. Indeed, among the sample of low-fat consumers, only one person was obese. Thus, a low-fat diet offers a good deal of protection against the development of obesity.

However, not everyone in the sample identified as eating a high-fat diet was obese. Indeed, there were many normal-weight and even some underweight people among the high-fat consumers. This suggests that some people are able to resist the weight-increasing properties of high-fat diets. This resistance could be behavioral or physiological. For example, some individuals may expend extremely high levels of energy in physical activity or they may have high BMRs or high rates of fat oxidation. Although a high-fat diet facilitates weight gain and promotes obesity, it does not appear to constitute a biological imperative that inevitably leads to obesity. Of course, in cross-sectional data such as these, it is not possible to know the length of time for which people have been consuming a particular type of diet. This would obviously affect their body weight.

DAY-TO-DAY FLUCTUATIONS IN FAT INTAKE

If it is accepted that the passive overconsumption that can occur when people are exposed to high-fat foods can in turn generate a positive energy balance that is not

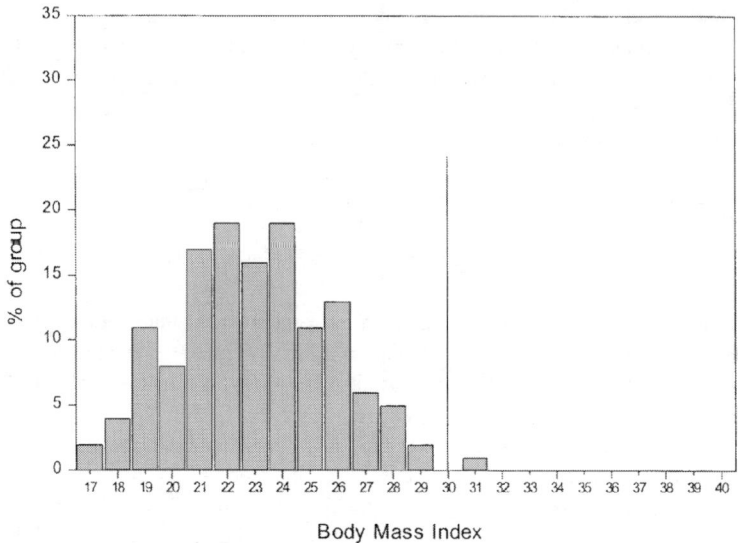

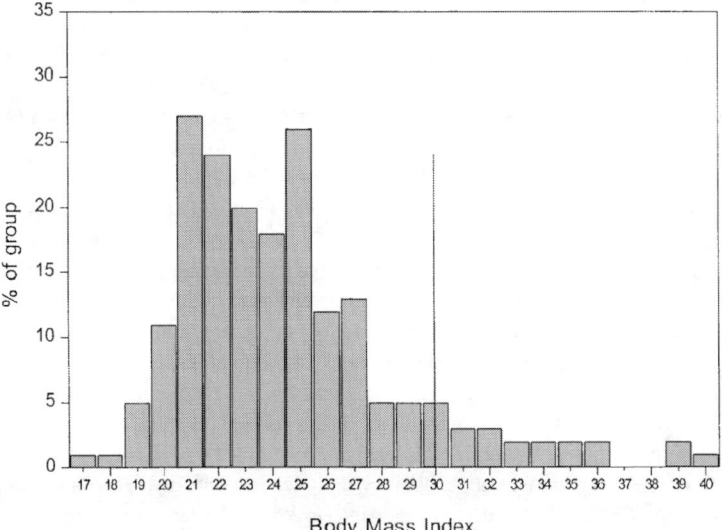

FIGURE 3. Frequency distributions of body mass index (BMI) among low-fat (upper panel) and high-fat (lower panel) consumers. Groups are defined by the amount (weight in grams) of fat consumed. The ordinate shows the percentage of each sample.

offset by later compensation, then fluctuations in fat intake become particularly important. This issue could have an effect on the relationship between high- and low-fat consumers and BMI. For example, individuals with a low mean intake of dietary fat (<35% food energy) could occasionally consume a surfeit of fat at a particular meal or during a particular day. This high consumption would probably not endure long enough to induce a rise in fat oxidation and, therefore, the positive energy balance would be left to be incorporated into stores. Note that it has been demonstrated that an increase in fat oxidation only occurs after three to seven days of consistent high-fat feeding.[55,56] With shorter periods of high-fat eating, the body would be left to deal with accumulating positive fat balances. It is therefore important to examine day-to-day fluctuations in fat intake.

In the Leeds High Fat Study, a small cohort of high- and low-fat consumers taken from the large sample have been investigated in detail using food-frequency questionnaires, seven-day weighed-food diaries, and intensive interviewing. From the diary records, daily profiles of nutrient and energy intake have been plotted. Examples of a low-fat and a high-fat consumer are shown in FIGURE 4.

The profiles indicate that for these two individuals there is considerable variation around the mean daily fat intake. This is true for most of the individuals in the cohort. For the low-fat consumer, the daily mean fat intake is 63.7 g (the United Kingdom average for women is 78.0 g). However, on two of the seven days of the food record, intake markedly exceeds this value, reaching 140 g on one day. For the high-fat consumer, the mean daily intake of 154 g fails to reveal that daily fluctuations vary between 80 and 210 g.

The effect of daily fluctuations such as these on biological endpoints such as obesity (or other disease states) is not known, but the effects may be different from those expected from laboratory studies in which constant high- or low-fat manipulations are administered to subjects. For a low-fat consumer, it can be surmised that occasional days of high-fat (and high-energy) intake would lead to a positive fat (and energy) balance; if this were not compensated by later behavioral or metabolic adjustment, then a small increase in body energy stores would occur. With repeated occurrences, this would be expected to have an effect on BMI or on the distribution of body fat (reflected in the waist-hip ratio).

A SCENARIO FOR A HUMAN PREDICAMENT

It has been estimated that the average daily fat intake for men in the United Kingdom is 108 g.[57] However, experimental studies and data collected from free-living individuals indicate that people can consume well over 150 g of fat in a single meal and upwards of 300 g in a full day. It appears to be relatively easy to consume large amounts of fat. Since people do not consciously set out with the intention of eating as much fat as they can, we have called this passive overconsumption. The mechanisms that cause this to happen include the high-energy density of fat, the very high palatability of high-fat foods, culturally approved high-fat food habits, an aggressive marketing

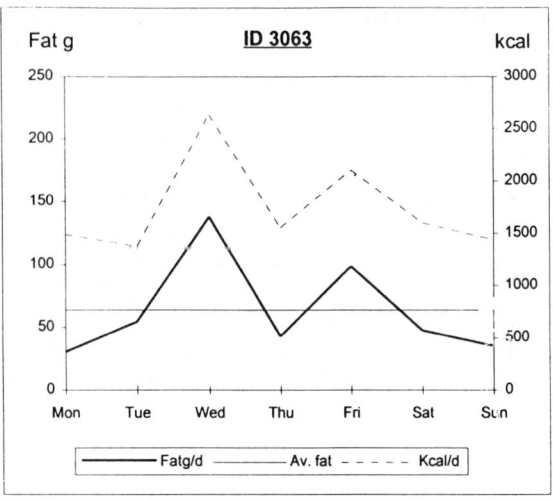

Age: 43   BMI: 22.7   female
FAT: 63.7g (33.6%)   Kcal: 1741   EI:BMR 1.31

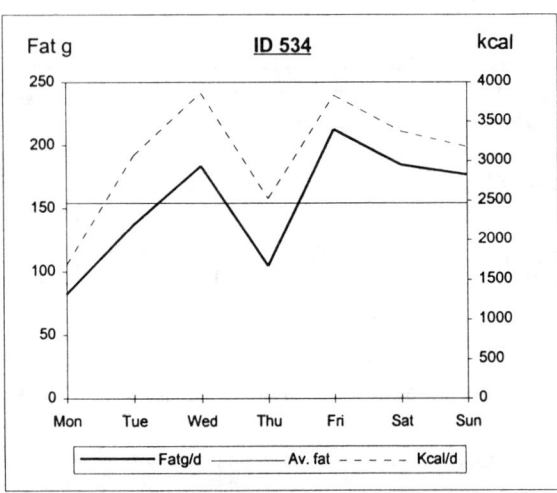

Age: 52   BMI: 29.0   male
FAT: 154.3g (46.0%)   Kcal: 3075   EI:BMR 1.63

FIGURE 4. Profiles of a low-fat consumer (upper graph) and a high-fat consumer (lower graph) indicating the large day-to-day variation in fat intake and the occasional very-high-fat intakes that can be reached even by a low-fat eater with a lower average daily intake.

by segments of the food industry, and human preferences for foods with a fatty texture and the flavors associated with fat. Taken together, these facilitatory factors overwhelm the first line of inhibitory processes, namely, the preabsorptive fat-induced physiological satiety signals. The major effect of high-fat foods on the short-term positive energy balance occurs during consumption, while the food is in contact with the orosensory receptors. This means that the processes controlling satiation are too weak or too slow to prevent the intake of an amount of energy. The subsequent disproportionately weak effect of fat on postingestive satiety works against the development of energetic compensation (a reduction of later intake to compensate for the high-fat load).[58]

Postabsorptive metabolic adjustments that follow food consumption (fuel utilization) do not appear to provide a strong feedback for the control of food intake. At least these processes do not appear to be strong enough to modulate habitual patterns of food intake held in place by long-term conditioning, physiological rhythms, and environmental contingencies. In addition, the induction of fat oxidation to a level commensurate with fat intake does not appear to occur until after several days of uniform high-fat feeding. Any satiety signal associated with this fat oxidation is obviously too late to prevent the consumption of the (already) eaten fat and does not appear strong enough to disturb habitual eating patterns.

Consequently, although the physiological system does contain inhibitory processes that influence the termination of eating episodes and the duration of the intervals between episodes, these processes can be overwhelmed by the superpotent fat-containing food products that can generate a positive energy balance (particularly when energy expenditure is low). The human psychobiological predilection to derive pleasure from eating allows the environment to overcome physiology.

REFERENCES

1. GOLDBERG, G. R., A. E. BLACK, S. A. JEBB, T. J. COLE, P. R. MURGATROYD, W. A. COWARD & A. M. PRENTICE. 1991. Critical evaluation of energy intake data using fundamental principles of energy physiology. 1: Derivation of cut-off values to identify underrecording. Eur. J. Clin. Nutr. 45: 569–581.
2. BLUNDELL, J. E. & N. A. KING. 1996. Overconsumption as a cause of weight gain: behavioural-physiological interactions in the control of food intake (appetite). In The Origins and Consequences of Obesity, p. 138–158. Ciba Foundation. London.
3. BLUNDELL, J. E., V. J. BURLEY, J. R. COTTON & C. L. LAWTON. 1993. Dietary fat and the control of energy intake: evaluating the effects of fat on meal size and post-meal satiety. Am. J. Clin. Nutr. 57(suppl.): 772S–778S.
4. SMITH, G. P. 1988. Humoral mechanisms in the control of body weight. In Perspectives in Behavioral Medicine – Eating Regulation and Dyscontrol, p. 59–65. Erlbaum. Hillsdale, New Jersey.
5. GREENOUGH, A., G. COLE & J. E. BLUNDELL. 1996. Relationship between hunger, anxiety, nausea, and energy intake during intravenous cholecystokinin (CCK-8) infusion. Int. J. Obes. 20(4): 71.
6. SMITH, G. P., C. JEROME & R. NORGREN. 1985. Afferent axons in abdominal vagus mediate satiety effect of cholecystokinin in rats. Am. J. Physiol. 249: R638–R641.
7. DOURISH, C. T., J. COUGHLAN, D. HAWLEY, M. CLARK & S. D. IVERSEN. 1988. Blockade of CCK-induced hypophagia and prevention of morphine tolerance by the CCK antagonist L-364,718. In CCK Antagonists, p. 307–325. Alan R. Liss. New York.
8. HEWSON, G., G. E. LEIGHTON, R. G. HILL & J. HUGHES. 1988. The cholecystokinin receptor antagonist L364,718 increases food intake in the rat by attenuation of the action of endogenous cholecystokinin. Br. J. Pharmacol. 93: 79–84.

9. HILL, A. J., S. R. PEIKIN, C. A. RYAN & J. E. BLUNDELL. 1990. Oral administration of proteinase inhibitor II from potatoes reduces energy intake in man. Physiol. Behav. **48:** 241-246.
10. COOPER, S. J., C. T. DOURISH & D. J. BARBER. 1990. Reversal of the anorectic effect of (+)fenfluramine in the rat by the selective cholecystokinin antagonist MK-329. Br. J. Pharmacol. **99:** 65-70.
11. POESCHLA, B. D., J. GIBBS, K. J. SIMANSKY, D. GREENBERG & G. P. SMITH. 1992. Cholecystokinin-induced satiety depends upon activation of 5-HT$_{1C}$ receptors. Am. J. Physiol. **264:** R62-R64.
12. GRIGNASCHI, G., B. MANTELLI, C. FRACASSO, M. ANELLI, S. CACCIA & R. SAMANIN. 1993. Reciprocal interaction of 5-hydroxy-tryptamine and cholecystokinin in the control of feeding patterns in rats. Br. J. Pharmacol **109:** 491-494.
13. ERLANSON-ALBERTSSON, C. & A. LARSSON. 1988. The activation peptide of pancreatic procolipase decreases food intake in rats. Regul. Pept. **22:** 325-331.
14. ERLANSON-ALBERTSSON, C., J. MEI, S. OKADA, D. YORK & G. A. BRAY. 1991. Pancreatic procolipase propeptide, enterostatin, specifically inhibits fat intake. Physiol. Behav. **49:** 1191-1194.
15. MEI, J., R. C. BOWYER, A. M. T. JEHANLI, G. PATEL & C. ERLANSON-ALBERTSSON. 1993. Identification of enterostatin, the pancreatic procolipase activation peptide in the intestine of rat: effect of CCK-8 and high fat feeding. Pancreas **8:** 488-493.
16. FRIEDMAN, M. I., M. G. TORDOFF & I. RAMIREZ. 1986. Integrated metabolic control of food intake. Brain Res. Bull. **17:** 855-859.
17. FRIEDMAN, M. I. 1991. Metabolic control of calorie intake. *In* Chemical Senses. Volume 4, p. 19-38. Dekker. New York.
18. FRIEDMAN, M. I. & M. G. TORDOFF. 1986. Fatty acid oxidation and glucose utilization interact to control food intake in rats. Am. J. Physiol. **251:** R840-R847.
19. LANGHANS, W. & E. SCHARRER. 1987. Evidence for a vagally mediated satiety signal derived from hepatic fatty acid oxidation. J. Auton. Nerv. Syst. **18:** 13-21.
20. SHERMAN, J. R. & R. B. WEINBERG. 1988. Serum apolipoprotein A-IV and lipoprotein cholesterol in patients undergoing total parenteral nutrition. Gastroenterology **95:** 394-401.
21. HAYASHI, H., D. F. NUTTING, K. FUJIMOTO, J. A. CARDELLI, D. BLACK & P. TSO. 1990. Transport of lipid and apolipoproteins A-I and A-IV in intestinal lymph of the rat. J. Lipid Res. **31:** 1613-1625.
22. FUJIMOTO, K., H. MACHIDORI, R. IWAKIRI, K. YAMAMOTO, J. FUJISAKI *et al.* 1993. Effect of intravenous administration of apolipoprotein A-IV on patterns of feeding, drinking, and ambulatory activity of rats. Brain Res. **608:** 233-237.
23. FUJIMOTO, K., J. A. CARDELLI & P. TSO. 1992. Increased apolipoprotein A-IV in rat mesenteric lymph after lipid meal acts as a physiological signal for satiation. Am. J. Physiol. **262:** G1002-G1006.
24. WEINBERG, R. B., C. DANTZKER & C. S. PATTON. 1990. Sensitivity of serum apolipoprotein A-IV levels to changes in dietary fat content. Gastroenterology **98:** 17-24.
25. ROLLS, B. J., S. KIM, A. L. MCNELIS, M. W. FISCHMAN, R. W. FOLTIN & T. H. MORAN. 1991. Time course of effects of preloads high in fat or carbohydrate on food intake and hunger ratings in humans. Am. J. Physiol. **260:** R756-R763.
26. FOLTIN, R. W., M. W. FISCHMAN, T. H. MORAN, B. J. ROLLS & T. H. KELLY. 1990. Caloric compensation for lunches varying in fat and carbohydrate content by humans in a residential laboratory. Am. J. Clin. Nutr. **52:** 969-980.
27. COTTON, J. R., V. J. BURLEY & J. E. BLUNDELL. 1993. Fat and satiety: effect of fat in combination with either protein or carbohydrate. Int. J. Obes. **17(2):** 63.
28. SEPPLE, C. P. & N. W. READ. 1990. The effect of pre-feeding lipid on energy intake from a meal. Gut **31:** 158-161.
29. KISSILEFF, H. R., L. P. GRUSS, J. THORNTON & H. A. JORDAN. 1984. The satiating efficiency of foods. Physiol. Behav. **32:** 319-332.
30. GOLAY, A. & E. BOBBIONI. 1997. The role of dietary fat in obesity. Int. J. Obes. **21(suppl. 3):** S2-S11.
31. KISSILEFF, H. R. 1984. Satiating efficiency and a strategy for conducting food loading experiments. Neurosci. Biobehav. Rev. **8:** 129-135.
32. BLUNDELL, J. E., S. GREEN & V. J. BURLEY. 1994. Carbohydrates and human appetite. Am. J. Clin. Nutr. **59(suppl.):** 728S-734S.
33. GREEN, S. M., V. J. BURLEY & J. E. BLUNDELL. 1994. Effect of fat- and sucrose-containing foods on the size of eating episodes and energy intake in lean males: potential for causing overconsumption. Eur. J. Clin. Nutr. **48:** 547-555.

34. LAWTON, C. L., V. J. BURLEY, J. K. WALES & J. E. BLUNDELL. 1993. Dietary fat and appetite control in obese subjects: weak effects on satiation and satiety. Int. J. Obes. **17:** 409–416.
35. BLUNDELL, J. E. & A. TREMBLAY. 1995. Appetite control and energy (fuel) balance. Nutr. Res. Rev. **8:** 225–242.
36. BLUNDELL, J. E. & V. J. BURLEY. 1990. Evaluation of the satiating power of dietary fat in man. In Progress in Obesity Research 1990, p. 453–457. Libbey. London.
37. TORDOFF, M. G. & D. R. REED. 1991. Sham-feeding sucrose or corn oil stimulates food intake in rats. Appetite **17:** 97–103.
38. TREMBLAY, A. & S. ST.-PIERRE. 1996. The hyperphagic effect of a high-fat diet and alcohol intake persists after control for energy density. Am. J. Clin. Nutr. **63:** 479–482.
39. LISSNER, L., D. A. LEVITSKY, B. J. STRUPP, H. KALKWARF & D. A. ROE. 1987. Dietary fat and the regulation of energy intake in human subjects. Am. J. Clin. Nutr. **46:** 886–892.
40. WELCH, I., K. SAUNDERS & N. W. READ. 1985. Effect of ileal and intravenous infusions of fat emulsions on feeding and satiety in human volunteers. Gastroenterology **89:** 1293–1297.
41. BLUNDELL, J. E., J. R. COTTON, H. DELARGY, S. GREEN, N. A. KING & C. L. LAWTON. 1995. The fat paradox: fat-induced satiety signals versus high fat overconsumption. Int. J. Obes. **19:** 832–835.
42. KING, N. A. & J. E. BLUNDELL. 1995. High fat foods overcome the energy expenditure induced by high intensity cycling or running. Eur. J. Clin. Nutr. **49:** 114–123.
43. MACDIARMID, J., V. HAMILTON, J. CADE & J. E. BLUNDELL. 1996. Leeds High Fat Study (LHFS): factors associated with high fat consumption. Int. J. Obes. **20**(4): 48.
44. MACDIARMID, J. I., J. E. CADE & J. E. BLUNDELL. 1996. High and low fat consumers, their macronutrient intake, and body mass index: further analysis of the National Diet and Nutrition Survey of British Adults. Eur. J. Clin. Nutr. **50:** 505–512.
45. DREWNOWSKI, A. 1987. Changes in mood after carbohydrate consumption. Am. J. Clin. Nutr. **46:** 703.
46. DREWNOWSKI, A., C. KURTH, J. HOLDEN-WILTSE & V. SAARI. 1992. Food preferences in human obesity: carbohydrates versus fats. Appetite **18:** 207–221.
47. EMMETT, P. M. & K. W. HEATON. 1995. Is extrinsic sugar a vehicle for dietary fat? Lancet **345:** 1537–1540.
48. MCCOLL, K. A. 1988. The sugar-fat seesaw. Nutr. Bull. **13:** 114–119.
49. GIBNEY, M. J. 1990. Dietary guidelines: a critical appraisal. J. Hum. Nutr. Diet. **3:** 245–254.
50. BOLTON-SMITH, C. & M. WOODWARD. 1994. Dietary composition and fat to sugar ratios in relation to obesity. Int. J. Obes. **18:** 820–828.
51. GIBSON, S. A. 1996. Are high-fat, high-sugar foods and diets conducive to obesity? Int. J. Food Sci. Nutr. **47:** 405–415.
52. MACDIARMID, J. I., J. E. CADE & J. E. BLUNDELL. 1995. Extrinsic sugar as vehicle for dietary fat. Lancet **346:** 696–697.
53. GREEN, S. M. & J. E. BLUNDELL. 1996. Effect of fat- and sucrose-containing foods on the size of eating episodes and energy intake in lean dietary restrained and unrestrained females: potential for causing overconsumption. Eur. J. Clin. Nutr. **50:** 625–635.
54. BLUNDELL, J. E. & S. M. GREEN. 1996. Effect of sucrose and sweeteners on appetite and energy intake. Int. J. Obes. **20**(2): S12–S17.
55. SCHRAUEN, P., W. D. VAN MARKEN LICHTENBELT, W. H. M. SARIS & K. K. WESTERTERP. 1996. Adaptation of fat oxidation to a high fat diet. Int. J. Obes. **20**(4): 81.
56. JEBB, S. A., A. M. PRENTICE, G. R. GOLDBERG, P. R. MURGATROYD, A. E. BLACK & W. A. COWARD. 1996. Changes in macronutrient balance during over- and under-feeding assessed by 12-d continuous whole-body calorimetry. Am. J. Clin. Nutr. **64:** 259–266.
57. GREGORY, J., K. FOSTER, H. TYLER & M. WISEMAN. 1990. The Dietary and Nutritional Survey of British Adults, p. 393. H. M. Stationery Office. London.
58. STUBBS, R. J., P. R. MURGATROYD, G. R. GOLDBERG & A. M. PRENTICE. 1993. Carbohydrate balance and the regulation of day to day food intake in humans. Am. J. Clin. Nutr. **57:** 897–903.

Dietary Fats and Thermogenesis

A. TREMBLAY AND C. BOUCHARD

Physical Activity Sciences Laboratory
PEPS
Laval University
Québec, Canada G1K 7P4

INTRODUCTION

The demonstration of a significant relationship between energy and macronutrient balance, both in animals[1] and in humans,[2] has been one of the most-significant scientific contributions in the study of energy balance over the last decade. It implies that energy balance is likely to occur under free-living conditions when macronutrient balance is also reached. In addition, this concept has drawn attention to the fact that the precision with which substrate balance is maintained is not the same for alcohol, protein, carbohydrate, and lipid.

Numerous factors contribute to the stability of a given macronutrient balance over a period of time. These include the ability of a substrate to inhibit subsequent food intake and to promote its utilization. The stability of a given substrate balance also depends on the metabolic pathways that contribute to the disposal of an excess substrate intake. As described in this paper, fat balance seems to be the most-vulnerable component of an excess intake. Moreover, this vulnerability appears to be particularly pronounced in individuals predisposed to obesity. However, why obesity-prone individuals accumulate more fat before restoring body weight stability has never been clearly established. This paper focuses on this issue and considers environmental and genetic factors that can influence energy and fat balance and the predisposition to obesity.

DIET COMPOSITION AND ENERGY BALANCE

Numerous human studies have shown over the last decade that an increase in the fat content of the diet is associated with an increase in daily energy intake and body fatness.[3-6] This hyperphagic effect of fat is related to its reduced ability to inhibit subsequent food intake and to promote satiety in comparison to protein and carbohydrate.[7-10]

The impact of alcohol on energy intake has also been investigated in epidemiological and experimental studies. In general, it has been demonstrated that alcohol does not inhibit the intake of macronutrients to the extent that it would fully compensate for its energy content.[11-14] Recent data obtained in our laboratory revealed that the hyperphagic effect of alcohol is additive to that resulting from a high-fat diet[15] (FIG. 1). Furthermore, we also recently reported that the overfeeding associated with the

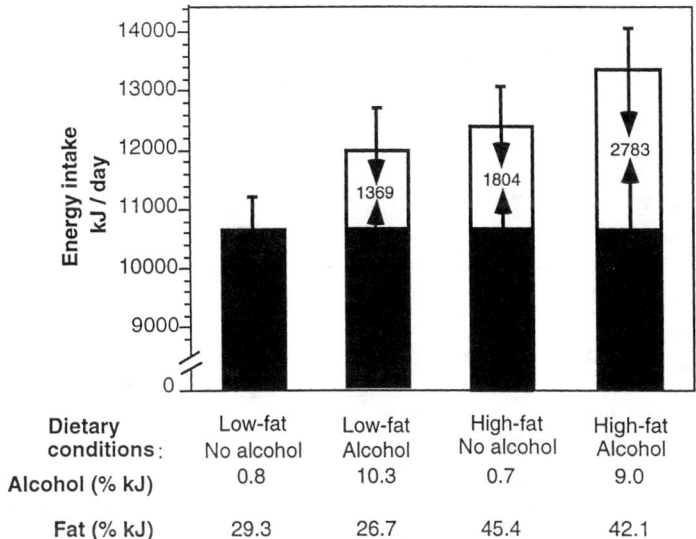

FIGURE 1. Mean daily increase in energy intake (open bars) above the value (black bars) observed in the low-fat, no-alcohol condition. Values are presented in relation to the mean percentage of energy as alcohol and fat in each condition. Reprinted from reference 15 with permission.

combination of alcohol and high-fat foods is not fully explained by their high-energy density.[16] These observations suggest that, under nutritional conditions characterized by high-alcohol and high-fat intake, satiety cannot be reached without overfeeding. They also suggest that some afferent signals eliciting satiety are less influenced by fat and alcohol than by carbohydrate or protein.

SUBSTRATE OXIDATION AND ENERGY BALANCE

An increase in the fat content of a meal does not promote a significant increase in postprandial fat oxidation,[17] which contrasts with the stronger potential of carbohydrate to favor its own utilization. In addition, both carbohydrate[2] and alcohol[18] are known to inhibit fat oxidation. Experimental evidence also suggests that the composition of ingested fat affects postprandial fat utilization. For instance, it has been reported[19-21] that polyunsaturated fat has a greater potential to increase lipid oxidation and energy expenditure compared to saturated fat. Other data suggest that this effect of polyunsaturated fat is associated with a stimulation of sympathetic nervous system activity.[22] These observations indicate that, in the short term, fat oxidation is not increased in proportion to fat intake, particularly saturated fat, and is even decreased by the intake of other energy substrates.

The inability of fat intake to promote its utilization is, however, not unlimited.

After several days of high-fat intake, fat oxidation seems to increase in proportion to the amount of ingested fat, but this effect is not seen in all individuals. This was demonstrated by Thomas et al.,[23] who found a significant positive association between fat oxidation and intake after several days of high-fat feeding in lean subjects. On the other hand, essentially no increase in fat oxidation was observed in response to the high-fat diet in obese individuals.

Other experimental data also tend to confirm that the risk for excess body fat storage is related to a reduced capacity to oxidize fat. The comparison of postobese and lean individuals is an approach that has been frequently used to determine whether individuals prone to obesity are characterized by a reduced potential to oxidize fat. In general, fat oxidation was found to contribute a lower fraction of daily energy needs in postobese subjects than in lean subjects.[24-27] Recent evidence also suggests that this difference is more likely to be observed in the postprandial state.[28] Since fat gain is associated with an increase in fat oxidation,[29,30] an increase in body fat storage in people predisposed to obesity likely represents a necessary adaptation to restore energy and fat balance when exposed to an environment promoting positive energy balance. The association between fat oxidation and body weight stability has also been investigated in prospective designs. For instance, Zurlo et al.[31] demonstrated that an increased RQ, reflecting a reduced relative fat oxidation, is a risk factor for body weight gain over a follow-up period of a few years. This finding was subsequently confirmed by other investigators.[32]

The possibility that the reduced potential for *in vivo* fat utilization in obesity-prone individuals is related to a low skeletal muscle metabolism has been recently examined. These investigations have shown that a low skeletal muscle oxidative potential is associated with an increase in total body fat[33,34] and visceral fat deposition.[35] This is concordant with the observation that a high level of skeletal muscle lipoprotein lipase activity predicts a high potential to oxidize fat.[36]

It has not been possible up to now to clearly establish whether the relative ability to use fat calories is attributable to environmental or genetic factors. Among environmental factors, physical activity is probably the one that exerts the strongest impact on fat utilization. Indeed, prolonged vigorous physical activity is known to increase the postexercise metabolic rate and fat oxidation.[37-39] In trained individuals, the regular practice of aerobic exercise is associated with an increase in sympathetic nervous system activity,[38,39] which is likely mediated by a specific increase in skeletal muscle beta-adrenergic receptor stimulation.[40] Since overweight individuals are known to be less active under free-living conditions[41,42] or in a confined environment such as a respiratory chamber,[43] it is likely that a decrease in physical activity participation predisposes to an increase in body fat stores.

Genetic factors also influence energy expenditure and substrate oxidation. Several studies have shown that a significant fraction of variation in the resting metabolic rate,[44,45] energy cost of standardized activities,[45] and resting and exercise RQ[45] was explained by genetic heritability. Finally, recent results obtained in our laboratory suggest that cycles of profound body weight loss and regain can substantially reduce daily energy expenditure.[46]

In summary, the evidence described in this section supports the concept that fat gain is a necessary adaptation in some individuals to restore energy and fat balance

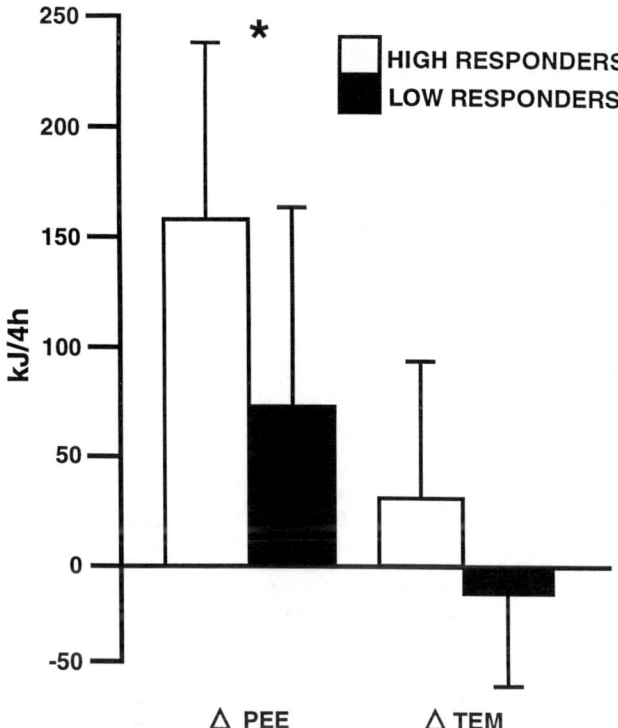

FIGURE 2. Changes in postprandial energy expenditure (PEE) and thermic effect of a meal (TEM; PEE above resting metabolic rate) in subjects displaying a high or a low response in postprandial plasma insulin to overfeeding. $\bar{X} \pm$ SD; $*$, $P < 0.05$. Reprinted from reference 53 with permission.

in a context promoting excess energy intake. This is likely explained by both environmental and genetic factors modifying the regulation of energy balance.

GENOTYPE-ENVIRONMENT INTERACTION AND ENERGY BALANCE

A paradigm incorporating interactions between environmental and genetic factors probably offers the most-realistic explanation for the increase in the prevalence of obesity that has been observed in industrialized countries in this century. This concept suggests that some individuals are at greater risk to develop obesity in an environment promoting positive fat and energy balance.

Our laboratory has been engaged in the study of genotype-environment interaction effects on energy balance for more than a decade. In a series of studies, we have subjected monozygotic twins to standardized overfeeding or exercise-training protocols to determine whether individuals sharing the same genetic background were more alike in their response compared to subjects having no genes in common by descent. These studies have generally demonstrated a more-homogenous response within than

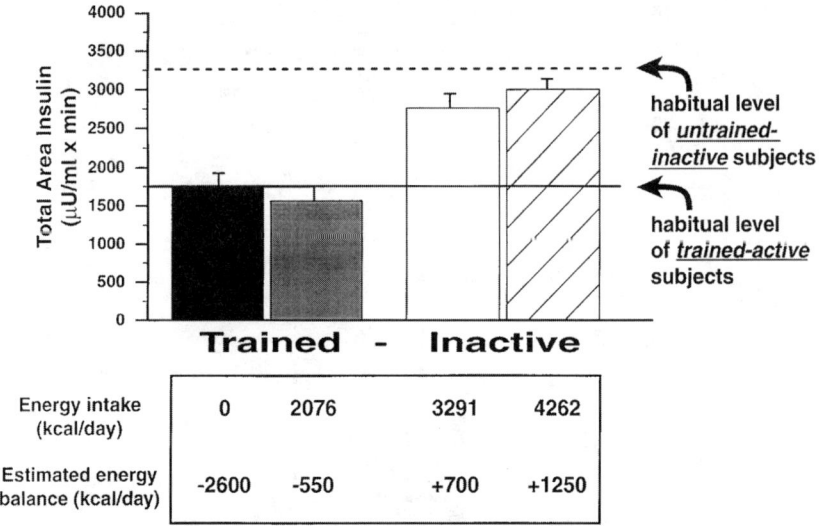

FIGURE 3. Effect of short-term detraining on the insulin response to intravenous glucose in trained individuals subjected to different levels of energy intake. The insulin response is compared to the habitual level measured in untrained-inactive and trained-active subjects. Adapted from reference 55.

between twin pairs for resting and exercise energy expenditure and body composition as a result of both overfeeding[47,48] and exercise-training.[49,50] This supports the notion that one's genotype is partly responsible for the physiological response to environmental factors altering energy balance. This observation is important, but it also emphasizes the need for this research to be pursued at the molecular level. Thus, it would be relevant to consider genes coding for molecules involved in regulatory mechanisms linking peripheral metabolism and neurosystems that influence energy balance. Insulin plays a role in this context and the discussion of its potential effects on energy balance is particularly relevant in this symposium.

INSULIN AND REGULATION OF ENERGY BALANCE

The peripheral effects of insulin on substrate disposal and anabolic processes are well established, but the search for mechanisms still remains the object of many investigations. Insulin also participates in the regulation of energy balance via central effects that influence both energy intake and expenditure. Hyperinsulinemia with euglycemia increases sympathetic nervous system activity, as reflected by an increase in plasma norepinephrine[51] and muscle sympathetic nerve activity.[52,53] As illustrated in FIGURE 2, we have shown that, under conditions of experimental overfeeding, a high increase in postprandial insulin is associated with a greater increase in postprandial energy expenditure.[54] This agrees with a prospective study in Pima Indians that revealed that hyperinsulinemia is associated with a reduced long-term body weight gain.[55] These observations are also concordant with results presented in FIGURE 3, which demonstrate that a considerable increase in plasma insulin occurs in trained individuals after several days of detraining, provided that this interruption of exercise is accompanied by positive energy balance.[56]

Recent studies have also identified potential neuropeptides that could mediate the effects of insulin on energy balance. This is the case for neuropeptide Y (NPY), whose gene expression, level in the arcuate nucleus, and probably orexigenic effects are inhibited by insulin.[57] Another potential mediator of the effects of insulin is corticotropin-releasing hormone (CRH), whose supplementation decreases insulinemia and interrupts weight gain in obese rats.[58] Since NPY[59] and CRH[60] also influence sympathetic activity, they and their receptors represent prime candidate genes for the study of molecular mechanisms affecting individual differences in energy balance.

CONCLUSIONS

A lifestyle characterized by high-fat and alcohol intake as well as sedentariness favors positive fat and energy balance. These factors also impose an increased demand on regulatory mechanisms responsible for the maintenance of body weight stability. Insulin seems to be involved in these mechanisms since hyperinsulinemia occurs when energy intake exceeds expenditure. Hyperinsulinemia can induce a substantial increase in sympathetic nervous system activity and is related to an increase in thermogenesis under conditions of positive energy balance. Hyperinsulinemia is also associated with a reduced risk for long-term body weight gain. An increased risk to develop the metabolic syndrome potentially represents a detrimental side effect of a mechanism that promotes energy balance and body weight stability.

REFERENCES

1. FLATT, J. P. 1987. Dietary fat, carbohydrate balance, and weight maintenance: effects of exercise. Am. J. Clin. Nutr. **45:** 296–306.
2. JÉQUIER, E. 1994. Carbohydrates as a source of energy. Am. J. Clin. Nutr. 59(suppl.): 682S–685S.
3. LISSNER, L., D. A. LEVITSKY, B. J. STRUPP, H. J. KALKWARF & D. A. ROE. 1987. Dietary fat and regulation of energy intake in human subjects. Am. J. Clin. Nutr. **46:** 886–892.
4. DREON, D. M., B. FREY-HEWITT, N. ELLSWORTH, P. T. WILLIAMS, R. B. TERRY & P. D. WOOD. 1988. Dietary fat: carbohydrate ratio and obesity in middle-aged men. Am. J. Clin. Nutr. **47:** 995–1000.
5. ROMIEU, I., W. C. WILLETT, M. J. STAMPFER, G. A. COLDITZ, L. SAMPSON, B. ROSNER, C. H. HENNEKENS & F. E. SPEIZER. 1988. Energy intake and other determinants of relative weight. Am. J. Clin. Nutr. **47:** 406–412.
6. TREMBLAY, A., G. PLOURDE, J. P. DESPRÉS & C. BOUCHARD. 1989. Impact of dietary fat content and fat oxidation on energy intake in humans. Am. J. Clin. Nutr. **49:** 799–805.
7. WALLS, E. K. & H. S. KOOPMANS. 1989. Effect of intravenous nutrient infusions on food intake in rats. Physiol. Behav. **45:** 1223–1226.
8. WALLS, E. K. & H. S. KOOPMANS. 1992. Differential effects of intravenous glucose, amino acids, and lipid on daily food intake in rats. Am. J. Physiol. **262:** R225–R234.
9. LAWTON, C. L., V. J. BURLEY, J. K. WALES & J. E. BLUNDELL. 1993. Dietary fat and appetite control in obese subjects: weak effects on satiety. Int. J. Obes. **17:** 409–416.
10. ROLLS, B. J., S. KIM-HARRIS, M. W. FISCHMAN, R. W. FOLTIN, T. H. MORAN & S. A. STONER.

1994. Satiety after preloads with different amounts of fat and carbohydrate: implications for obesity. Am. J. Clin. Nutr. **60:** 476-487.
11. DE CASTRO, J. M. & S. OROZCO. 1990. The effects of moderate alcohol intake on the spontaneous eating patterns of humans: evidence of unregulated supplementation. Am. J. Clin. Nutr. **52:** 246-253.
12. BEBB, H. T., H. B. HOUSER, J. C. WITSCHI, A. S. LITELL & R. K. FULLER. 1971. Calorie and nutrient contribution of alcoholic beverages to the usual diets of 155 adults. Am. J. Clin. Nutr. **24:** 1042-1052.
13. VEENSTRA, J., J. A. A. SCHENKEL, A. M. J. VAN ERP-BAART, H. A. M. BRANTS, K. F. A. M. HULSHOF, C. KISTEMAKER, G. SCHAAFSMA & T. OCKHUIZEN. 1993. Alcohol consumption in relation to food intake and smoking habits in the Dutch National Food Consumption Survey. Eur. J. Clin. Nutr. **47:** 482-489.
14. GRUCHOW, H. W., K. A. SOBOCINSKI, J. J. BARBORIAK & J. G. SCHELLER. 1985. Alcohol consumption, nutrient intake, and relative body weight among U. S. adults. Am. J. Clin. Nutr. **42:** 289-295.
15. TREMBLAY, A., E. WOUTERS, M. WENKER, S. ST-PIERRE, C. BOUCHARD & J. P. DESPRÉS. 1995. Alcohol and high-fat diet: a combination favoring overfeeding. Am. J. Clin. Nutr. **62:** 639-644.
16. TREMBLAY, A. & S. ST-PIERRE. 1996. The hyperphagic effect of high-fat and alcohol persists after control for energy density. Am. J. Clin. Nutr. **63:** 479-482.
17. FLATT, J. P., E. RAVUSSIN, K. J. ACHESON & E. JÉQUIER. 1985. Effects of dietary fat on postprandial substrate oxidation and on carbohydrate and fat balances. J. Clin. Invest. **76:** 1119-1124.
18. SUTER, P. M., Y. SCHUTZ & E. JÉQUIER. 1992. The effect of ethanol on fat storage in healthy subjects. N. Engl. J. Med. **326:** 983-987.
19. JONES, P. J. H., P. B. PENCHARZ & M. T. CLANDININ. 1985. Whole body oxidation of dietary fatty acids: implications for energy utilization. Am. J. Clin. Nutr. **42:** 769-777.
20. JONES, P. J. H. & D. A. SCHOELLER. 1988. Polyunsaturated:saturated ratio of diet fat influences energy substrate utilization in the human. Metabolism **37:** 145-151.
21. CLANDININ, M. T., L. C. H. WANG, R. V. RAJOTTE, M. A. FRENCH, Y. GOH & E. S. KIELO. 1995. Increasing the dietary polyunsaturated fat content alters whole-body utilization of 16:0 and 10:0. Am. J. Clin. Nutr. **61:** 1052-1057.
22. MATSUO, T., Y. SHIMOMURA, S. SAITOH, K. TOKUYAMA, H. TAKEUCHI & M. SUZUKI. 1995. Sympathetic activity is lower in rats fed a beef tallow diet than in rats fed a safflower oil diet. Metabolism **44:** 934-939.
23. THOMAS, C. D., J. C. PETERS, G. W. REED, N. N. ABUMRAD, M. SUN & J. O. HILL. 1992. Nutrient balance and energy expenditure during ad libitum feeding of high-fat and high-carbohydrate diets in humans. Am. J. Clin. Nutr. **55:** 934-942.
24. LEAN, M. E. J. & W. P. T. JAMES. 1988. Metabolic effects of isoenergetic nutrient exchange over 24 hours in relation to obesity in women. Int. J. Obes. **8:** 641-648.
25. BUEMANN, B., A. ASTRUP, N. J. CHRISTENSEN & J. MADSEN. 1992. Effect of moderate cold exposure on 24-h energy expenditure: similar response in postobese and nonobese women. Am. J. Physiol. **263:** E1040-E1045.
26. BUEMANN, B., A. ASTRUP, J. MADSEN & N. J. CHRISTENSEN. 1992. A 24-hr energy expenditure study on reduced-obese and non-obese women: effect of β-blockade. Am. J. Clin. Nutr. **56:** 662-670.
27. LARSON, D. E., R. T. FERRARO, D. S. ROBERTSON & E. RAVUSSIN. 1995. Energy metabolism in weight-stable postobese individuals. Am. J. Clin. Nutr. **62:** 735-739.
28. RABEN, A., H. B. ANDERSON, N. J. CHRISTENSEN, J. MADSEN, J. J. HOLST & A. ASTRUP. 1994. Evidence for an abnormal postprandial response to a high fat meal in women predisposed to obesity. Am. J. Physiol. **267:** E549-E559.
29. SCHUTZ, Y., A. TREMBLAY, R. L. WEINSIER & K. M. NELSON. 1992. Role of fat oxidation in the long-term stabilization of body weight in obese women. Am. J. Clin. Nutr. **55:** 670-674.
30. ASTRUP, A., B. BUEMANN, P. WESTERN, S. TOUBRO, A. RABEN & N. J. CHRISTENSEN. 1994. Obesity as an adaptation to a high-fat diet: evidence from a cross-sectional study. Am. J. Clin. Nutr. **59:** 350-355.
31. ZURLO, F., S. LILLIOJA, D. P. A. ESPOSITO, B. L. NYOMBA, I. RAZ, M. F. SAAD, B. A. SWINBURN, W. C. KNOWLER, C. BOGARDUS & E. RAVUSSIN. 1990. Low ratio of fat to carbohydrate oxidation as predictor of weight gain: study of 24-h RQ. Am. J. Physiol. **259:** E650-E657.

32. SEIDELL, J. C., D. C. MULLER, J. D. SORKIN & R. ANDRES. 1992. Fasting respiratory exchange ratio and resting metabolic rate as predictors of weight gain: the Baltimore Longitudinal Study on Aging. Int. J. Obes. **16:** 667–674.
33. SIMONEAU, J.-A. & C. BOUCHARD. 1995. Skeletal muscle metabolism and body fat content in men and women. Obes. Res. **3:** 23–29.
34. KRIKETOS, A. D., D. A. PAN, S. LILLIOJA, G. J. COONEY, L. A. BAUR, M. R. MILNER, J. R. SUTTON, A. B. JENKINS, C. BOGARDUS & L. H. STORLIEN. 1996. Interrelationships between muscle morphology, insulin action, and adiposity. Am. J. Physiol. **270:** R1332–R1339.
35. COLBERG, S. R., J.-A. SIMONEAU, F. LELAND THAETE & D. E. KELLEY. 1995. Skeletal muscle utilization of free fatty acids in women with visceral obesity. J. Clin. Invest. **95:** 1846–1853.
36. FERRARO, R. T., R. H. ECKEL, D. E. LARSON, A. M. FONTVIEILLE, R. RISING, D. R. JENSEN & E. RAVUSSIN. 1993. Relationship between skeletal muscle lipoprotein lipase activity and 24-hour macronutrient oxidation. J. Clin. Invest. **92(1):** 441–445.
37. BAHR, R., I. INGNES, O. VAAGE, O. M. SEJERSTED & E. A. NEWSHOLME. 1987. Effect of duration of exercise on excess postexercise O_2 consumption. J. Appl. Physiol. **62:** 485–490.
38. TREMBLAY, A., J. P. COVENEY, J. P. DESPRÉS, A. NADEAU & D. PRUD'HOMME. 1992. Increased resting metabolic rate and lipid oxidation in exercise-trained individuals: evidence for a role of beta-adrenergic stimulation. Can. J. Physiol. Pharmacol. **70:** 1342–1347.
39. POEHLMAN, E. T., A. W. GARDNER, P. J. ARCIERO, M. I. GORAN & J. CALLES-ESCANDON. 1994. Effects of endurance training on total fat oxidation in elderly persons. J. Appl. Physiol. **76:** 2281–2287.
40. PLOURDE, G., S. ROUSSEAU MIGNERON & A. NADEAU. 1993. Effect of endurance training on beta-adrenergic system in three different skeletal muscle. J. Appl. Physiol. **74:** 1641–1646.
41. JOHNSON, M. L., B. S. BURKE & J. MAYER. 1956. Relative importance of inactivity and overeating in the energy balance of obese high school girls. Am. J. Clin. Nutr. **4(1):** 37–43.
42. BULLEN, B. A., R. B. REED & J. MAYER. 1964. Physical activity of obese and non-obese adolescent girls appraised by motion picture sampling. Am. J. Clin. Nutr. **14:** 211–223.
43. RISING, R., I. T. HARPER, A. M. FONTVIEILLE, R. T. FERRARO, M. SPRAUL & E. RAVUSSIN. 1994. Determinants of total daily energy expenditure: variability in physical activity. Am. J. Clin. Nutr. **59:** 800–804.
44. FONTAINE, E., R. SAVARD, A. TREMBLAY, J. P. DESPRÉS, E. POEHLMAN & C. BOUCHARD. 1985. Resting metabolic rate in monozygotic and dizygotic twins. Acta Genet. Med. Gemellol. **34:** 41–47.
45. BOUCHARD, C., A. TREMBLAY, A. NADEAU, J. P. DESPRÉS, G. THÉRIAULT, M. R. BOULAY, G. LORTIE, C. LEBLANC & G. FOURNIER. 1989. Genetic effect in resting and exercise metabolic rates. Metabolism **38:** 364–370.
46. ST-PIERRE, S., B. ROY & A. TREMBLAY. 1996. A case study on energy balance during an expedition through Greenland. Int. J. Obes. **20:** 493–495.
47. POEHLMAN, E. T., A. TREMBLAY, E. FONTAINE, J. P. DESPRÉS, A. NADEAU & C. BOUCHARD. 1986. Genotype dependency of dietary-induced thermogenesis: its relation with hormonal changes following overfeeding. Metabolism **35:** 30–36.
48. TREMBLAY, A., E. T. POEHLMAN, A. NADEAU, J. DUSSAULT & C. BOUCHARD. 1987. Heredity and overfeeding-induced changes in submaximal exercise VO2. J. Appl. Physiol. **62(2):** 539–544.
49. POEHLMAN, E. T., A. TREMBLAY, A. NADEAU, J. DUSSAULT, G. THÉRIAULT & C. BOUCHARD. 1986. Heredity and changes in hormones and metabolic rates with short-term training. Am. J. Physiol. **250:** E711–E717.
50. BOUCHARD, A., A. TREMBLAY, J. P. DESPRÉS, G. THÉRIAULT, A. NADEAU, P. J. LUPIEN, S. MOORJANI, D. PRUD'HOMME & G. FOURNIER. 1994. The response to exercise with constant energy intake in identical twins. Obes. Res. **2(5):** 400–410.
51. ROWE, J. W., J. B. YOUNG, K. L. MINAKER, A. L. STEVEN, J. PALLOTTA & L. LANSBERG. 1981. Effect of insulin and glucose infusions on sympathetic nervous system activity in normal man. Diabetes **30:** 219–225.
52. BERNE, C., J. FAGIUS, T. POLLARE & P. HEMJDAHL. 1992. The sympathetic response to euglycaemic hyperinsulinemia. Diabetologia **35:** 873–879.
53. VOLLENWEIDER, P., D. RANDIN, L. TAPPY, E. JÉQUIER, P. NICOD & U. SCHERRER. 1994. Impaired insulin-induced sympathetic neural activation and vasodilation in skeletal muscle in obese humans. J. Clin. Invest. **93:** 2365–2371.

54. TREMBLAY, A., A. NADEAU, J. P. DESPRÉS & C. BOUCHARD. 1995. Hyperinsulinemia and regulation of energy balance. Am. J. Clin. Nutr. **61:** 827–830.
55. SCHWARTZ, M. W., E. J. BOYKO, S. E. KAHN, E. RAVUSSIN & C. BOGARDUS. 1995. Reduced insulin secretion: an independent predictor of body weight gain. J. Clin. Endocrinol. Metab. **80:** 1571–1576.
56. TREMBLAY, A., A. NADEAU, D. RICHARD & J. LEBLANC. 1983. The influence of exercise and nutrition on plasma glucose and insulin of human subjects. *In* Health Risk Estimation, Risk Reduction, and Health Promotion, p. 87–93. Canadian Public Health Association. Ottawa.
57. SCHWARTZ, M. J., J. MARKS, A. J. SIPOLS, D. G. BASKIN, S. C. WOOD, S. E. KAHN & D. PORTE. 1991. Central insulin administration reduces neuropeptide Y mRNA expression in the arcuate nucleus of food-deprived lean (Fa/Fa), but not obese (fa/fa) Zucker rats. Endocrinology **128:** 2645–2647.
58. JEANRENAUD, B. 1990. Neuroendocrinology and evolutionary aspects of experimental obesity. *In* Progress in Obesity Research 1990, p. 409–421. Libbey. London.
59. EGAWA, M., H. YOSHIMATSU & G. BRAY. 1991. Neuropeptide Y suppresses sympathetic activity in interscapular brown adipose tissue in rats. Am. J. Physiol. **260:** R328–R334.
60. BROWN, M. R. & L. A. FISHER. 1985. Corticotropin-releasing factor: effects on the autonomic nervous system and visceral systems. Fed. Proc. **44:** 243–248.

Fat Metabolism in the Predisposition to Obesity[a]

ARNE ASTRUP, ANNE RABEN,
BENJAMIN BUEMANN, AND SØREN TOUBRO

*Research Department of Human Nutrition
and
KVL Center for Food Research
The Royal Veterinary and Agricultural University
1958 Frederiksberg, Copenhagen, Denmark*

DIETARY FAT AND OBESITY

Overweight and obesity have become highly prevalent conditions, in both industrialized and many developing countries. The prevalence of obesity has increased markedly and so have its complications, which are among the leading causes of premature mortality, such as coronary heart disease, stroke, type-2 diabetes, and certain cancers.[1] Obesity develops as an interaction between a genetic predisposition and certain environmental factors, such as energy-dense fat-rich diets and low physical fitness caused by a sedentary lifestyle.

Numerous physiological and psychological phenomena have been recognized that influence energy intake, but their relative importance with respect to body weight regulation has been very complicated to elucidate under free-living conditions. An important reason appears to be the considerable redundancy in the systems controlling food intake. Although temporary disruptions may occur when one control site is activated or inactivated, body weight tends to become reestablished at some plateau.[2] There is evidence, however, to suggest that only subjects with the genetic predisposition to obesity gain weight when exposed to a high-fat diet over extended periods of time. In cross-sectional studies, the univariate correlations between body fat and dietary fat content and level of physical activity may be surprisingly high (FIG. 1). In longitudinal studies, weight gain is only weakly associated with dietary fat content at baseline, but the association between dietary fat content and weight gain is considerably stronger in overweight and obese subjects and particularly in those at familial risk. It is not clear whether the reason for the less-effective negative feedback on appetite caused by energy derived from fat rather than from carbohydrate and protein is a result of the higher energy density of fats or due to unrelated fat-specific properties. These properties are difficult to dissociate because foods rich in fat have

---

[a] This study was supported by the Danish Medical Research Council, the P. Carl Petersen Foundation, and the Danish Research and Development Program for Food Technology.

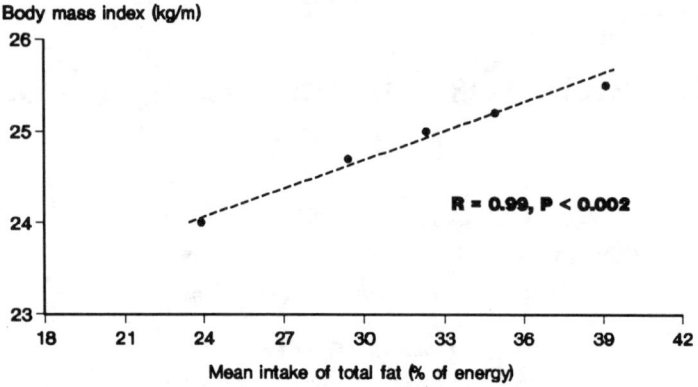

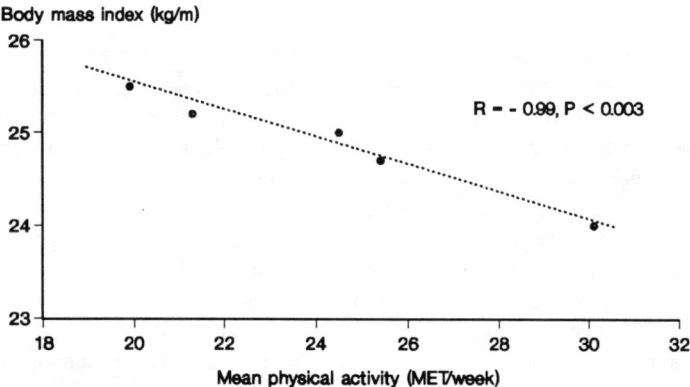

FIGURE 1. Simple correlations between quintiles of body fat and (top) dietary fat and (bottom) physical activity in 43,757 health professionals in the United States. (Calculated from Ascherio et al. 1996. Br. Med. J. **313**: 84–90.)

a high-energy density and vice versa. There is, however, evidence in support of the fat oxidation capacity having a metabolically driven, genetically determined component, which may play an important role for overconsumption of energy in predisposed individuals facing a high-fat diet. Therefore, it is pertinent to achieve an in-depth understanding of the physiological regulation of fat oxidation by identifying the factors determining its rate and to elucidate the extent to which fat oxidation rate is influenced by genes. Moreover, whether the genes express their effect indirectly through food choices or by regulation of the activities of enzymes and levels and sensitivity of hormones controlling fat transport and β-oxidation is not known.

TABLE 1. Macronutrient and Energy Balance[a]

Day 0		Day 1
Macronutrient Balance		*Energy Balance*
↑ 1 MJ CHO balance	⇒	↓ 0.4 MJ (6%)
↑ 1 MJ protein balance	⇒	↓ 0.9 MJ (2%)
↑ 1 MJ fat balance	⇒	↑ 0.4 MJ (20%)

[a] The relationship between one day's macronutrient balances and the subsequent day's energy intake (balance) based on an experiment of 7 days of ad libitum intake of diets of constant composition, during which energy intake and expenditure were monitored.[28] For every megajoule of increased protein balance on one day, a change in the subsequent day's energy balance of −0.4 MJ, etc., was produced.

REGULATION OF FAT BALANCE

The traditional concept of the energy balance equation, which describes weight gain as an excessively positive energy imbalance, can be usefully replaced by a series of macronutrient balance equations in which gains in body fat stores are viewed specifically as an imbalance of fat. The background for this replacement is that calories are not equal in their metabolic effects, interconversion between the four macronutrients is negligible, and an oxidative hierarchy operates in inverse proportion to the size of available stores for each macronutrient. Changes in amino acid, glucose, and alcohol oxidation occur readily in adjusting to protein, carbohydrate, and alcohol intakes. When considered over a few days, regulation appears to be geared primarily toward maintenance of appropriate glycogen reserves.

Alcohol is most readily oxidized because it cannot be stored. Oxidation of carbohydrate and protein is also under tight autoregulatory feedback control; oxidation increases in direct response to intake. By contrast, there is virtually no acute feedback between fat intake and fat oxidation. Fat oxidation is primarily a function of the gap between total energy expenditure and the amounts of alcohol, protein, and carbohydrate energy consumed,[3] which results in a much less accurately maintained fat balance. Although regulatory responses serving to achieve fat balance exist, their effectiveness to inhibit the expansion of the fat mass seems to be limited (TABLE 1). Leptin, a hormone secreted from adipose tissue at an increased rate when the fat mass is expanded, is supposed to inhibit food intake and increase sympathetic nervous system activity through a central action.[4] The finding that obese subjects remain obese in spite of 10-fold-higher circulating leptin concentrations suggests that its penetration through the blood-brain barrier is insufficient, that receptor action is impaired, or that lipostatic mechanisms are insufficient to restrict energy intake in this category of individuals. However, long-term fat and energy balances tend to remain close to zero over prolonged periods once a weight maintenance plateau has been reached (i.e., less than 2–3% error relative to energy turnover when considered over a year).[5]

In conclusion, protein, carbohydrate, and ethanol oxidation rates relate well to the respective intake of these nutrients. Daily fat oxidation, however, relates poorly with daily variations in fat consumption.[6] This is a likely explanation as to why obesity develops particularly among individuals with a genetically determined low fat oxidation capacity when diets are high in fat and when physical activity is limited.[7]

FACTORS DETERMINING THE RATE OF FAT OXIDATION

The fat oxidation rate and the composition of the fuel mix that is oxidized are mainly determined by *energy requirements*, *energy balance*, and the *dietary macronutrient composition*. Overall, substrate oxidation is dictated by the body's need to regenerate the ATP used in performing its metabolic functions, in maintaining body temperature, and in moving, at rates depending upon an individual's size and physical activity. The absolute rates of total oxidation equal the energy expenditure and energy requirements, which in turn are determined by the size of fat-free mass and physical activity and, to a smaller extent, by the size of fat mass, plasma concentrations of T_3 and androstenedione, and sympathetic nervous system activity.[8] Thus, on a standardized diet with fixed macronutrient composition, a subject with a large fat-free mass and a high level of physical activity will oxidize greater amounts of all macronutrients expressed in g/day (or kJ/day) than a person with a lower fat-free mass or lower activity levels. When the oxidation rates are expressed as a proportion of total oxidation (i.e., fraction or percent) or if the oxidation rates are adjusted for the metabolically active fat-free mass, differences in an individual's energy requirements are accounted for. These proportions of rates are referred to as *the composition of the fuel mix*.

The composition of the fuel mix oxidized to drive oxidative phosphorylation varies considerably during the day and from one day to another due to the influence of mainly energy balance and due to the composition of the meals.

Energy Balance and Dietary Fat Content

The body's energy balance is changing all the time and never reaches a steady state, except when viewed over a certain period (e.g., one week). This is due to the fact that energy is expended continuously, whereas energy intake takes place intermittently at meals. A positive energy balance increases RQ due to an increase in carbohydrate oxidation and a decrease in fat oxidation. Conversely, a negative energy balance causes a decrease in RQ due to suppression of glucose oxidation and stimulation of fatty acid oxidation. Meal ingestion may be regarded as an acutely invoked positive energy balance, and even meals with high-fat and very low carbohydrate and protein content increase glucose oxidation and hence RQ. In the resting postabsorptive phase, energy balance becomes negative and RQ decreases due to inhibition of glucose oxidation and increased fatty acid oxidation partly as a result of the stimulation of lipolysis. The same positive relationship between energy balance and RQ applies when viewed over 24 hours, which can be demonstrated when measurements are carried out in a respiratory chamber.[9]

If the macronutrient composition of the diet is fixed and fed to subjects to exactly meet energy requirements, the body will inevitably oxidize macronutrients in the same proportions as in the diet after some days. Calorimeter studies suggest that, under these conditions, it takes 3–4 days to reach macronutrient balance (RQ equals FQ) in normal individuals, providing the initial difference in composition is <10–15% (FQ = food quotient, i.e., the ratio of CO_2 produced to O_2 consumed during the biological oxidation of a representative sample of the diet).[10] However, the highly variable conditions in real life make the situation more complex. Diet composition and energy balance vary considerably from meal to meal and from day to day, and the metabolic demand for substrate influences appetite. Differences in physical fitness and constitutional differences in muscle LPL activity and oxidative pathways cause fundamental changes in substrate use, which will influence energy intake and macronutrient preferences. In the experimental setup, though, it is crucial to acknowledge that the antecedent diet exerts an important impact on the subsequent day's RQ.

Because fat oxidation (%) and glucose oxidation (%) are inversely related, it is of interest to know how the former influences the latter and vice versa. In the classical view, fat regulates glucose metabolism. This concept, however, has recently been challenged[11] and it has been proposed that the opposite is the case. At present, evidence for the existence of the so-called glucose–fatty acid cycle *in vivo* is equivocal.[12] In addition there are indications that carbohydrate may regulate fat metabolism, possibly through malonyl-CoA, which is involved in the regulation of FFA transportation into the mitochondria. Carbohydrate ingestion increases plasma glucose and decreases fat utilization by inhibiting both plasma FFA oxidation and intramuscular triglyceride utilization (FIG. 2). This effect is induced by insulin and by a muscle effect, probably via malonyl-CoA. Consequently, the carbohydrate content of the diet and the energy balance are crucial for the rate of fat oxidation and hence fat balance. These normal physiological regulatory mechanisms do not exclude the fact that fat oxidation also is influenced by muscle LPL activity, β-hydroxy-acyl-CoA dehydrogenase (BOAC),[13,14] and sympathetic nervous system activity.[15]

Genetic Influence on Fat Oxidation

Direct evidence for a genetic influence on RQ was delivered by Dériaz et al.,[16] who studied the relationship between DNA variation at the genes coding for the Na,K-ATPase peptides, RQ, and body fat. Postabsorptive RQ was found to be associated with the $α_2$-gene and linked with the β-gene of the Na,K-ATPase, which suggests that these, or neighboring genes, influence RQ. Twin studies also support the heritability of RQ. Bouchard et al. found greater similarity in RQ during exercise among monozygotic twins than among dizygotic twins.[17] Moreover, in the Quebec Family Study involving 300 individuals from 75 nuclear families, the genetic heritability of RQ was 20%.[17] It cannot be ruled out, however, that the genetic effect is indirect, that is, mediated through food preferences, which in turn influence RQ. Against this possibility are other studies where diet composition and energy intake have been rigorously controlled, such as in a 100-day controlled overfeeding study where significant within-pair resemblance was found for changes in RQ, in both the postabsorptive and postprandial state.[17] Also, 24-h RQ studies suggest that the oxidation pattern is under a

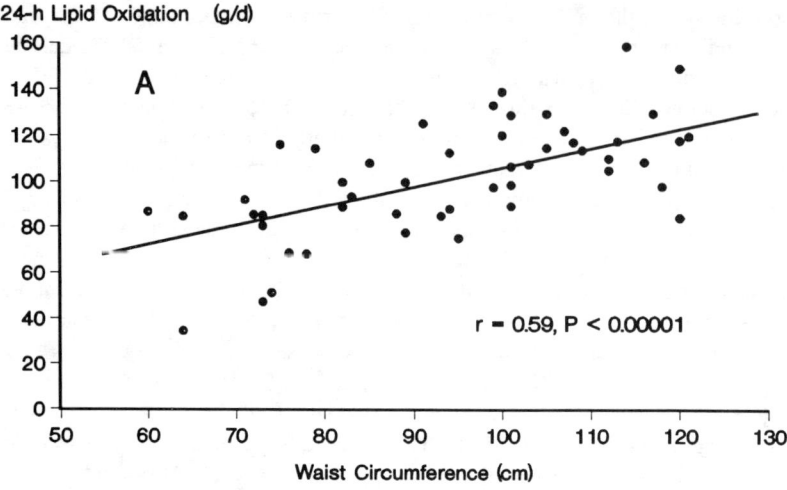
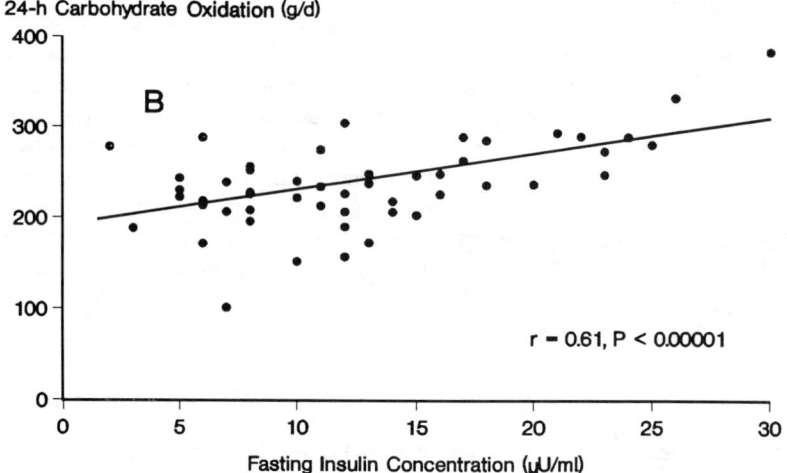

FIGURE 2. (A) Relation between 24-h lipid oxidation and waist circumference. (B) Relation between 24-h carbohydrate oxidation and fasting insulin concentration. Oxidations were measured in respiratory chambers on a fixed diet and physical activity program (in 50 premenopausal women). Reproduced with permission.[42]

genetic influence that is independent of dietary macronutrient selection. Zurlo *et al.* measured 24-h RQ in a calorimeter in Pima Indian siblings fed a controlled weight-maintenance diet while on the metabolic ward.[18] After adjustments for any prior change in body weight, 24-h energy balance, gender, and body fat, 24-h RQ was seen to be a family trait where membership explained 28% of the variation between individuals. In a study of Caucasians, 24-h RQ was measured in 71 adult subjects from 32 nuclear families on a standardized diet without control of antecedent food composition.[8] The habitual diet, however, was recorded by a 7-day food record during the period immediately before the measurement. The theoretically calculated RQ of the habitual diet (FQ) was found to be a family trait with a heritability of 45% (unpublished observations). Likewise, 24-h RQ, adjusted for age, gender, and 24-h energy balance, also aggregated within families, with a heritability of 32%. FQ and RQ were also positively correlated. However, when 24-h RQ was adjusted for differences in habitual FQ, the heritability estimate did not change. This suggests that RQ is a family trait, and the familial resemblance cannot be attributed to shared dietary composition. It cannot be ruled out, though, that shared environment, for example, physical fitness, is responsible for the family effect, but a genetic determinant is likely.

Fat Partitioning, Weight Gain, and Obesity

Individuals with the genetic predisposition gain weight when exposed to a high-fat diet over extended periods of time. Weight gain after a 6-year follow-up was generally found only weakly associated with dietary fat content at baseline, but the association was 7 times greater in overweight and obese subjects and 15 times greater in those at familial risk.[19] Rolls *et al.* have demonstrated that, in obesity-prone restrained eaters, high-fat foods suppress further intake less than in normal unrestrained subjects.[20] The physiological mechanisms behind this susceptibility in predisposed individuals may be related to their lower ability to oxidize fat. This hypothesis is supported by a study that demonstrated that restrained eaters managed to oxidize the fat content of the diet following exposure to low-fat and medium-fat diets, while on a high-fat diet (50% fat energy) their 24-h fat oxidation was impaired as compared to unrestrained subjects.[21]

There are also prospective studies to suggest a combination of a deficient fat oxidation and an enhanced fat deposition in subjects with the predisposition to weight gain and obesity. In a study in Pima Indians, 24-h RQ was measured at baseline after being on a controlled weight-maintenance diet during a stay at a metabolic ward. Those with a low fat oxidation (high RQ) were more likely to gain weight over a 3-year period than those with a low RQ (high fat oxidation).[18] Seidell *et al.* found in the Baltimore Longitudinal Study on Aging a significant positive association between resting RQ and subsequent weight gain.[22] The adjusted relative risk of gaining >5 kg in initially nonobese men with an RQ > 0.85 was 2.42 (95% confidence interval: 1.10–5.32) compared to those with an RQ < 0.76. Energy balance and antecedent diet composition were much less rigorously controlled than in the study by Zurlo *et al.*,[18] which poses the risk that the high RQ values were mainly produced by subjects who were already gaining weight and that their positive energy balance confounded the interpretation of the high RQ values. Taken together, however, the studies on "preobese" subjects support that low fat oxidation is a risk factor for weight gain when exposure to high-fat diets and low levels of physical fitness are common. The importance of the gene environment interaction is emphasized by comparing body fatness of the genetically identical Pima Indians living in very different environments in Mexico and Arizona. Pima Indians matched for age and gender to those living in the United States were 25 kg heavier than their Mexican relatives.[23]

Fat Partition in Formerly Obese Subjects

The importance of studying fat oxidation in subjects with normal sizes of fat stores is underlined by the fact that obesity brings about an increased proportion of fat utilized as fuel both in the fasting state and on a 24-h basis.[24] It is not clear whether the higher proportion of total oxidation covered by fat oxidation in obese subjects is simply brought about by their habitual consumption of a high-fat diet or whether it is rather determined by the increased supply of fat substrates by the enlarged fat mass. It is probably a combination of these two factors. Nevertheless, the finding makes obese subjects less suitable for causative studies of fat oxidation.

Apart from larger prospective studies linking RQ with subsequent weight gain and experiments using normal-weight restrained eaters, formerly obese subjects (so-called postobese), who intentionally have reduced body weight and composition compared to their normal values, are used as a more-appropriate model.[5] The first study showing that postobese subjects had impaired fat oxidation in response to high-fat diets was reported by Lean and James.[25] They reported striking differences in substrate handling by obese, postobese, and control subjects. While the mean 24-h RQ was similar in the obese and in the control group, it was significantly higher in the postobese group, both during fasting and following a high-fat diet.[25] This indicates that the postobese group utilized relatively less fat and relatively more carbohydrate than the control group during fasting and on the high-fat days.

In a more rigorously controlled dietary study, we examined the ability of postobese women to adjust their macronutrient oxidation in response to three isocaloric diets: a low-fat (20%), a medium-fat (30%), and a high-fat diet (50% fat energy) using 24-h calorimetry.[26] No differences were found between the groups on the low-fat and medium-fat diets.[26] On the high-fat diet, however, the postobese women failed to increase the ratio of fat to carbohydrate oxidation appropriately, which caused a positive fat balance (FIG. 3). The preferential storage of fat in the postobese group on the high-fat diet was caused by a failure to increase fat oxidation sufficiently to match the consumed fat. The phenomenon of deficient fat partitioning postprandially was also found after a high-fat meal.[27] Whereas the thermic effect of the meal was found to be similar in postobese and control subjects, postprandial fat oxidation was 2.5 times lower in the postobese group. The lower fat oxidation following high-fat meals and diets observed in these studies may have been subject to a confounding effect of differences in the habitual diets between the postobese and controls. If the postobese in days or weeks prior to the measurements had consumed a diet with lower fat and higher carbohydrate content, this could lead to a more-imprecise autoregulation of fat oxidation, which is likely to be seen only following high-fat challenges. There is evidence to support that postobese indeed choose low-fat, high-carbohydrate diets as compared with never-obese, otherwise healthy subjects.[28]

To avoid the confounding effect of antecedent diet composition, we undertook a 5-day calorimeter study where diet and energy balance were strictly controlled.[10] On days 0 and 1, postobese and matched controls received a diet providing 30% energy from fat, 55% from carbohydrate, and 15% from protein. On days 2, 3, and 4, they received a high-fat diet (55% fat, 30% carbohydrate, and 15% protein). While 24-h

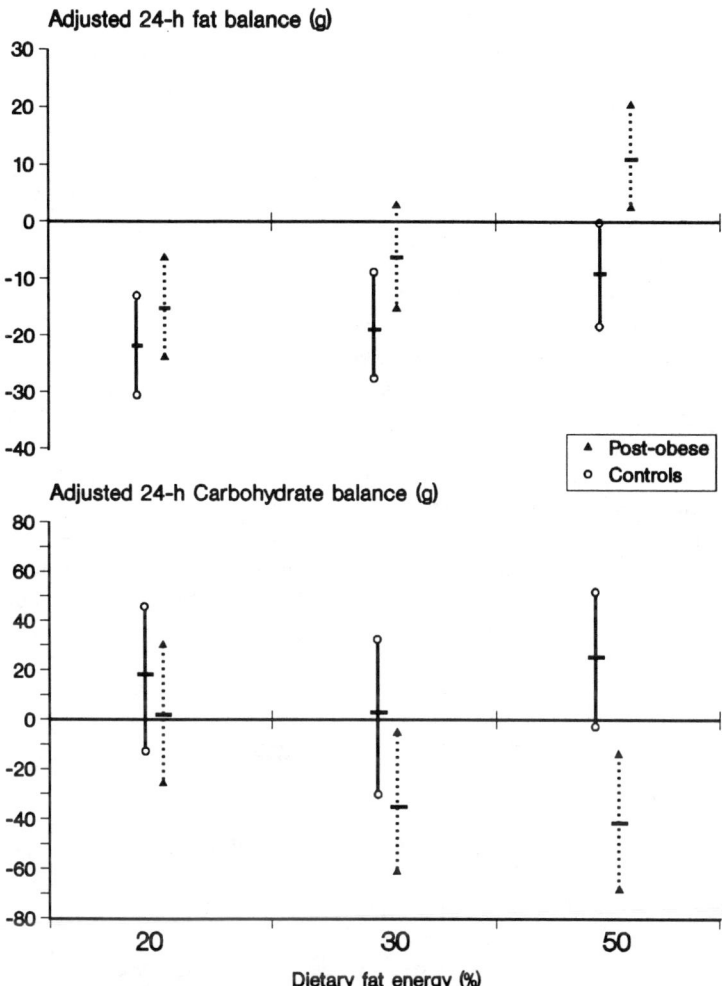

FIGURE 3. Adjusted 24-h fat balance (top) and adjusted 24-h carbohydrate balance (bottom) versus dietary fat energy in postobese women and controls measured in respiration chambers following 3 days of consumption of three different diets in random order (20%, 30%, and 50% fat in % of energy). The abnormal fat and carbohydrate partitioning was only observed in the postobese group during intake of the high-fat diet. The negative carbohydrate diet weakens the normal suppression on the subsequent day's energy intake. Reproduced with permission.[26]

RQ values changed similarly in the postobese and control subjects after introduction of the high-fat diet, the postprandial fat oxidation was consistently suppressed in the postobese group.[10] This study confirms that the postprandial fat oxidation is particularly suppressed in the postobese, but it does not exclude that the adjustment of 24-h RQ following introduction of the high-fat diet would have been impaired if the subjects were allowed to eat unrestricted.

It is possible that the accompanying negative carbohydrate balance (FIG. 3) would tend to reduce glycogen stores, which is thought to be a signal for decreased satiety and increased hunger. There is evidence to support the concept that changes in food intake from day to day are influenced by negative feedback exerted by previous carbohydrate and protein balances, and, to a minor degree, by previous fat imbalances. In a set of experiments, Stubbs *et al.* studied normal subjects without predisposition to obesity while consuming diets of constant composition for 7 consecutive days.[28] Negative correlations between changes in the energy balance from day to day and the previous day's carbohydrate balance were demonstrated, accounting for 5–10% of the variation in the energy balance of the subjects. By contrast, energy intakes were found to be correlated positively with the previous day's fat balance, which could account for 20% of the variation in energy intake (TABLE 1). Even in individuals without the predisposition to obesity, the fat content of the diet markedly influenced energy intake and balance, without eliciting counterregulatory influences on food intake that could tend to limit weight gain. These studies support the hypothesis that the macronutrients whose balance is most tightly regulated exert a suppressive effect on the subsequent day's energy intake, whereas fat does not have this effect, but may, on the contrary, stimulate appetite. Therefore, there is support for the existence of physiological mechanisms linking macronutrient balances with subsequent food intake, and it is plausible that alterations in fat oxidation and partitioning in obesity-prone individuals have consequences for their energy intake (see FIG. 3).

PHYSIOLOGICAL MECHANISMS FOR ALTERED FAT METABOLISM

Impaired Fat Mobilization or Utilization?

The impaired fat oxidation in the preobese and postobese state may be due to a number of different steps in fat and glucose metabolism, and there is the risk that the disorder is heterogeneous (i.e., reduced capacities of different pathways of the fat oxidative system). We have undertaken studies in order to assess whether fat mobilization from the adipose tissue stores through lipolysis is a limiting factor for fat oxidation in the postobese during exercise or if it rather is uptake/oxidation in skeletal muscle that is impaired. Lipolysis was measured by a combination of Xe-clearance technique and microdialysis in the abdominal subcutaneous adipose tissue.[29] Postobese and controls were matched for body composition, age, gender, and aerobic capacity. During a 60-min submaximal exercise test, the increase in lipolysis was intact in the postobese, but fat oxidation was lower in the postobese during rest and recovery and also during exercise when fat oxidation was adjusted for differences in plasma FFA. This study clearly suggests that skeletal muscle fat oxidation was low in the postobese group and that the findings cannot be explained by differences in aerobic capacity.

Fat Oxidation in Skeletal Muscle

Skeletal muscle accounts for 20–30% of the body's energy expenditure at rest and up to 90% during exercise. At rest, more than 80% of the muscle oxidation is covered by fat, and accumulating evidence points to skeletal muscle as the site of the limited capacity for uptake and oxidation of fat substrates; a number of possible mechanisms have been studied. The major fat substrates for skeletal muscle oxidation are plasma free fatty acids and triglycerides in lipoproteins, that is, VLDL and chylomicrons, which require hydrolysis by the rate-limiting enzyme, lipoprotein lipase (LPL), at the endothelial or luminal interface of muscle capillaries. A low muscle LPL activity might therefore represent a mechanism responsible for a lower fat oxidation in predisposed individuals. In a cross-sectional study, Ferraro et al. found skeletal muscle LPL activity to be inversely correlated with 24-h RQ ($r = -0.57$) in subjects who had been on a controlled diet for at least 3 days.[13] So far, however, it remains to be determined whether a low skeletal muscle LPL activity is a risk factor for weight gain. In postobese subjects, muscle LPL activity was not found to be lower than in matched controls.[30]

Similarly, insulin sensitivity of skeletal muscle has been demonstrated to be an important determinant of the local partitioning of fat and glucose substrates. Whereas insulin insensitivity is a feature of obesity, Swinburn et al. found that insulin-sensitive subjects were more likely to gain weight over 3.5 years than insulin-resistant subjects (7.6 versus 3.1 kg).[31] In postobese subjects, insulin sensitivity has also been evaluated. Using the insulin/glucose clamp technique, we were unable to detect any significant difference in insulin sensitivity and glucose oxidation in postobese and well-matched controls.[32] By contrast, two other studies with control of antecedent diet found greater insulin sensitivity in postobese subjects than in controls.[29,30] The studies in postobese subjects should not be regarded as discordant; an enhanced insulin sensitivity should rather be viewed as a risk factor, not present in all predisposed subjects, but a mechanism that may promote the uptake and oxidation of glucose at the expense of fat.

It is a commonly held belief that type I muscle fibers have a higher capacity for substrate oxidation than type II and it has been suggested that a lower ratio of type I to type II fibers may be responsible for low fat oxidation and a propensity to obesity. Based on a very small amount of material, Wade et al. found an inverse relation between fatness and the proportion of type I fibers of the vastus lateralis muscle.[33] In a larger study, this was confirmed by Lillioja et al.[34] In better controlled and larger studies, however, no significant relation was found between muscle fiber type and body fatness.[35]

The interindividual variability in fat oxidation may be due to differences in the activities of key enzymes in the β-oxidation pathway. In a number of studies, biochemical characteristics of skeletal muscle have been assessed in the vastus lateralis muscle by muscle fiber histochemistry and by measuring the activities of six key enzymes involved in the energy-generating pathways. Zurlo et al. found that 24-h RQ correlated negatively with β-hydroxy-acyl-CoA dehydrogenase (HADH) and slightly weaker with adenylokinase and creatine kinase.[14] Another interesting enzyme is malate dehydrogenase, MDH, which is an activity marker of enzymes of the Krebs cycle. Inverse correlations have been found between body fatness and MDH activity, and it has been shown that fatter men also had lower MDH activities than lean controls matched for physical fitness.[35] More convincing, however, was the finding that those individuals with a low MDH activity gained more body fat than those with high levels during 100 days of 1000-kcal/day overfeeding.[36] In postobese subjects, muscle histochemistry and biochemistry have been recently studied and compared with subjects matched with respect to aerobic capacity. The distribution of muscle fiber types and LPL activity did not differ between postobese and controls,[30] but HADH activity was 20% lower in postobese than in controls (Raben *et al.*, unpublished data). Thus, it is quite likely that several different mechanisms are involved in the lower fat oxidation of skeletal muscle in predisposed individuals. Conceivably neurohormonal influences on fat oxidation might also be responsible, such as lower thyroid hormone status[37] and lower sympathetic nervous system activity.[38] There are studies showing that both low free T_3 and low sympathetic activity are risk factors for weight gain, and both systems are likely candidates responsible for the lower fat oxidation capacities in skeletal muscle (TABLE 1).[15]

The Confounding Effect of Altered Dietary Macronutrient Selection in Obese Individuals

There is reliable evidence that obese subjects habitually consume a diet with a higher fat content than normal-weight subjects. The taste preference for fat was found to be increased in obese and postobese individuals,[39] and this finding has been confirmed.[40] This could be the mechanism that precipitates the gene expression that physiologically is seen as an inappropriate low fat oxidation, thus stimulating appetite via impaired postprandial fat oxidation. It is, therefore, of interest to assess cross-sectional and longitudinal studies that have examined the relationship between dietary fat/carbohydrate and body fatness. In the valid cross-sectional studies, case-control analysis of dietary composition in obese versus nonobese subjects consistently showed that obese individuals consume a diet with a higher fat and a lower carbohydrate content than nonobese.[41] The diet of the obese groups has been found to be 5–8 fat energy-% higher than in the control groups. Clearly, there is a need for a biological marker of habitual macronutrient intake. Such a marker can be indirectly obtained by measurement of substrate oxidation because the oxidative pattern seems to be only slightly disturbed by alterations in dietary fat content during one day. Using 24-h calorimetry, we found that oxidative fat energy in overweight and obese subjects is higher than in normal-weight controls (40.2% versus 36.0%, $P < 0.02$).[24] This supports that obese subjects consume a diet characterized by a higher fat content than nonobese subjects, but it cannot be ruled out that the higher oxidative fat energy is stimulated by the enlarged body fat stores. It stresses, however, that studies aimed at obtaining insight into the physiological and biochemical phenotypic expressions of the obesity genes and into how they interact with dietary fat should be done with caution and should probably be avoided due to the confounding effect of the increased body fat stores and their impact on glucose and fat metabolism.

ACKNOWLEDGMENTS

We thank the staff of the Energy Metabolism and Obesity Group of the Research Department of Human Nutrition, John Lind, Ole J. Victor, Inge Timmermann, Bente Knap, Helle Thomassen, and Charlotte Nielsen.

REFERENCES

1. BOUCHARD, C. & G. A. BRAY. 1996. Introduction. In Regulation of Body Weight: Biological and Behavioral Mechanisms, p. 1-13. Wiley. New York.
2. FLATT, J. P. 1996. Carbohydrate balance and body-weight regulation. Proc. Nutr. Soc. 55: 449-465.
3. PRENTICE, A. M. 1995. Alcohol and obesity. Int. J. Obes. 19(suppl. 5): S44-S50.
4. CONSIDINE, R. V., M. K. SINHA, M. L. HEIMAN, A. KRIAUCINAS, T. W. STEPHENS, M. R. NYCE et al. 1996. Serum immunoreactive-leptin concentrations in normal-weight and obese humans. N. Engl. J. Med. 334: 292-295.
5. ASTRUP, A. & J. P. FLATT. 1996. Metabolic determinants of body weight regulation. In Regulation of Body Weight: Biological and Behavioral Mechanisms, p. 193-210. Wiley. New York.
6. FLATT, J. P. 1987. Dietary fat, carbohydrate balance, and weight maintenance: effects of exercise. Am. J. Clin. Nutr. 45: 296-306.
7. ASTRUP, A. & A. RABEN. 1996. Glucostatic control of intake and obesity. Proc. Nutr. Soc. 55: 417-427.
8. TOUBRO, S., N. J. CHRISTENSEN & A. ASTRUP. 1995. Reproducibility of 24-h energy expenditure, substrate utilization, and spontaneous physical activity in obesity measured in a respiration chamber. Int. J. Obes. 19: 544-549.
9. TOUBRO, S., P. WESTERN, J. BÜLOW, I. MACDONALD, A. RABEN, N. J. CHRISTENSEN, J. MADSEN & A. ASTRUP. 1994. Insulin sensitivity in post-obese women. Clin. Sci. 84: 407-413.
10. BUEMANN, B., S. TOUBRO, A. RABEN & A. ASTRUP. 1994. Substrate oxidations in postobese women on 72-h high fat diet: a possible abnormal postprandial response? Int. J. Obes. 18(suppl. 2): 97.
11. DYCK, D. J., C. T. PUTMAN, G. J. F. HEIGHENHAUSER, E. HULTMAN & L. L. SPRIET. 1993. Regulation of fat-carbohydrate interaction in skeletal muscle during intense aerobic cycling. Am. J. Physiol. 265: E852-E859.
12. RANDLE, P. J., P. B. GARLAND, C. N. HALES & E. A. NEWSHOLME. 1963. The glucose fatty acid cycle: its role in insulin sensitivity and the metabolic disturbances of diabetes mellitus. Lancet 1: 785-789.
13. FERRARO, R. T., R. H. ECKEL, D. E. LARSON, A. M. FONTVIEILLE, R. RISING, D. R. JENSEN & E. RAVUSSIN. 1993. Relationship between skeletal muscle lipoprotein lipase activity and 24-hour macronutrient oxidation. J. Clin. Invest. 92: 441-445.
14. ZURLO, F., P. M. NEMETH, R. M. CHOKSI, S. SESODIA & E. RAVUSSIN. 1994. Whole-body energy metabolism and skeletal muscle biochemical characteristics. Metabolism 43: 481-486.
15. TREMBLAY, A. 1995. Differences in fat balance underlying obesity. Int. J. Obes. 19: S10-S14.
16. DÉRIAZ, O., F. DIONNE, L. PÉRUSSE, A. TREMBLAY, M. C. VOHOL, G. CÔTE & C. BOUCHARD. 1994. DNA variation in the genes of the Na,K-adenosine triphosphatase and its relation with resting metabolic rate, respiratory quotient, and body fat. J. Clin. Invest. 93: 838-843.
17. BOUCHARD, C., O. DÉRIAZ, L. PÉRUSSE & A. TREMBLAY. 1994. Genetics of energy expenditure in humans. In The Genetics of Obesity, p. 135-145. CRC Press. Boca Raton, Florida.
18. ZURLO, F., S. LILLIOJA, A. ESPOSITO-DEL PUENTE, B. L. NYOMBA, I. RAZ, M. F. SAAD, B. SWINBURN, W. C. KNOWLER, C. BOGARDUS & E. RAVUSSIN. 1990. Low ratio of fat to carbohydrate oxidation as predictor of weight gain: study of 24-h RQ. Am. J. Physiol. 259: E650-E657.
19. HEITMANN, B. L., L. LISSNER, T. I. A. SØRENSEN & C. BENGTSSON. 1995. Dietary fat intake and weight gain in women genetically predisposed for obesity. Am. J. Clin. Nutr. 61: 1213-1217.
20. ROLLS, B. J., S. KIM-HARRIS, M. W. FISCHMAN, R. W. FOLTIN, T. H. MORAN & S. A. STONER. 1994. Satiety after preloads with different amounts of fat and carbohydrate: implications for obesity. Am. J. Clin. Nutr. 60: 476-487.

21. VERBOEKET-VAN DE VENNE, W. P. H. G., K. R. WESTERTERP & F. TEN HOOR. 1994. Substrate utilization in man: effects of dietary fat and carbohydrate. Metabolism 43: 152-156.
22. SEIDELL, J. C., D. C. MULLER, J. D. SORKIN & R. ANDRES. 1992. Fasting respiratory exchange ratio and resting metabolic rate as predictors of weight gain: the Baltimore Longitudinal Study on Aging. Int. J. Obes. 16: 667-674.
23. RAVUSSIN, E. 1995. Metabolic differences and the development of obesity. Metabolism 44(suppl. 3): 12-14.
24. ASTRUP, A., B. BUEMANN, P. WESTERN, S. TOUBRO, A. RABEN & N. J. CHRISTENSEN. 1994. Obesity as an adaptation to a high-fat diet: evidence from a cross-sectional study. Am. J. Clin. Nutr. 59: 350-355.
25. LEAN, M. E. J. & W. P. T. JAMES. 1988. Metabolic effects of isoenergetic nutrient exchange over 24 hours in relation to obesity in women. Int. J. Obes. 12: 15-27.
26. ASTRUP, A., B. BUEMANN, N. J. CHRISTENSEN & S. TOUBRO. 1994. Failure to increase lipid oxidation in response to increasing dietary fat content in formerly obese women. Am. J. Physiol. 266: E592-E599.
27. RABEN, A., H. B. ANDERSEN, N. J. CHRISTENSEN, J. MADSEN, J. J. HOLST & A. ASTRUP. 1994. Evidence for an abnormal postprandial response to a high-fat meal in women predisposed to obesity. Am. J. Physiol. 267: E549-E559.
28. STUBBS, R. J., C. G. HABRON, P. R. MURGATROYD & A. M. PRENTICE. 1995. Covert manipulation of dietary fat and energy density: effect on substrate flux and food intake in men eating ad libitum. Am. J. Clin. Nutr. 62: 316-329.
29. ASTRUP, A., C. SØRENSEN, J. BÜLOW, B. BUEMANN, J. MADSEN & N. J. CHRISTENSEN. 1994. Fat mobilization and utilization during rest and exercise in postobese women. Int. J. Obes. 18(suppl. 2): 23.
30. RABEN, A., S. TOUBRO & A. ASTRUP. 1996. Improved insulin sensitivity after 14-d ad libitum sucrose-rich diet compared with a fat-rich or starch-rich diet. Int. J. Obes. 20(suppl. 4): 67.
31. SWINBURN, B., B. L. NYOMBA, M. F. SAAD, F. ZURLO & I. RAZ. 1991. Insulin resistance associated with lower rates of weight gain in Pima Indians. J. Clin. Invest. 88: 168-173.
32. TOUBRO, S., T. I. A. SØRENSEN, B. RØNN, N. J. CHRISTENSEN & A. ASTRUP. 1996. 24-h energy expenditure: the role of body composition, thyroid status, sympathetic activity, and family membership. J. Clin. Endocrinol. Metab. 81: 2670-2674.
33. WADE, A. J., M. M. MARBUT & J. M. ROUND. 1990. Muscle fiber type and aetiology of obesity. Lancet 335: 805-808.
34. LILLIOJA, S., A. A. YOUNG, C. L. CULTER et al. 1987. Skeletal muscle capillary density and fiber type are possible determinants of in vivo insulin resistance in man. J. Clin. Invest. 80: 415-424.
35. SIMONEAU, J.-A. & C. BOUCHARD. 1995. Skeletal muscle metabolism and body fat content in men and women. Obes. Res. 3: 23-29.
36. SIMONEAU, J.-A., A. TREMBLAY, G. THÉRIAULT & C. BOUCHARD. 1994. Relationships between the metabolic profile of skeletal muscle and body fat gain in response to overfeeding. Med. Sci. Sports Exercise 5: S159.
37. ASTRUP, A., B. BUEMANN, S. TOUBRO, C. RANNERIES & A. RABEN. 1996. Low resting metabolic rate in subjects predisposed to obesity: a role for thyroid status. Am. J. Clin. Nutr. 63: 879-883.
38. ASTRUP, A. & I. MACDONALD. 1996. Sympathoadrenal system and metabolism. In Handbook of Obesity. In press.
39. DREWNOWSKI, A., J. D. BRUNZELL, K. SANDE, P. H. IVERIUS & M. R. C. GREENWOOD. 1985. Sweet tooth reconsidered: taste responsiveness in human obesity. Physiol. Behav. 35: 617-622.
40. DREWNOWSKI, A., C. L. KURTH & J. E. RAHAIM. 1991. Taste preferences in human obesity: environmental and familial factors. Am. J. Clin. Nutr. 54: 635-641.
41. ASTRUP, A. & A. RABEN. 1996. Mono- and disaccharides: nutritional aspects. In Carbohydrates in Food, p. 159-189. Dekker. New York.
42. ASTRUP, A., B. BUEMANN, N. J. CHRISTENSEN, J. MADSEN, C. GLUUD, P. BENNETT & B. SVENSTRUP. 1992. The contribution of body composition, substrates, and hormones to the variability in energy expenditure and substrate utilization in premenopausal women. J. Clin. Endocrinol. 74: 279-286.

Fat and Energy Balance

MICHAEL J. PAGLIASSOTTI, ELLIS C. GAYLES,
AND JAMES O. HILL[a]

Department of Pediatrics
and
Center for Human Nutrition
University of Colorado Health Sciences Center
Denver, Colorado 80262

INTRODUCTION

Obesity is a major public-health problem and its prevalence is increasing.[1] It is generally agreed that obesity is a complex disorder involving an interaction between genes and environment[2,3] and between behavior and metabolism.[4-6] The high prevalence of obesity is likely due to an environment that facilitates positive energy balance coupled with an underlying genetic susceptibility. Thus, an understanding of obesity, and the factors leading to its development, requires elucidation of how the genotype and the environment interact to affect body weight and body composition. It is our view that the genotype defines the capacity of the system that regulates body weight, and nongenetic factors (e.g., environment) determine where this system operates. In short, the genotype determines what can happen and nongenetic factors determine what does happen.

When viewed in this way, obesity becomes both a behavioral and a metabolic problem and therefore cannot be understood without studying both behavior (i.e., energy intake and physical activity) and metabolism (i.e., nutrient partitioning). Whether an individual becomes obese under a given set of environmental and nutritional circumstances will be determined by the interaction between the behavioral response to these circumstances and the metabolic response to the behavior. This interaction ultimately will determine how much and what type of energy enters the body, and the fate of that energy within the body.

THE DIETARY MODEL OF OBESITY

To study specific behavioral and metabolic factors that contribute to obesity development, we use a rat model of dietary obesity that has many correlates to human obesity.[7-12] In this model, male Wistar rats are provided free access to a low-fat diet (LFD; 12% of energy from corn oil, 68% from corn starch, 20% from protein) for

---

[a] To whom all correspondence should be addressed.

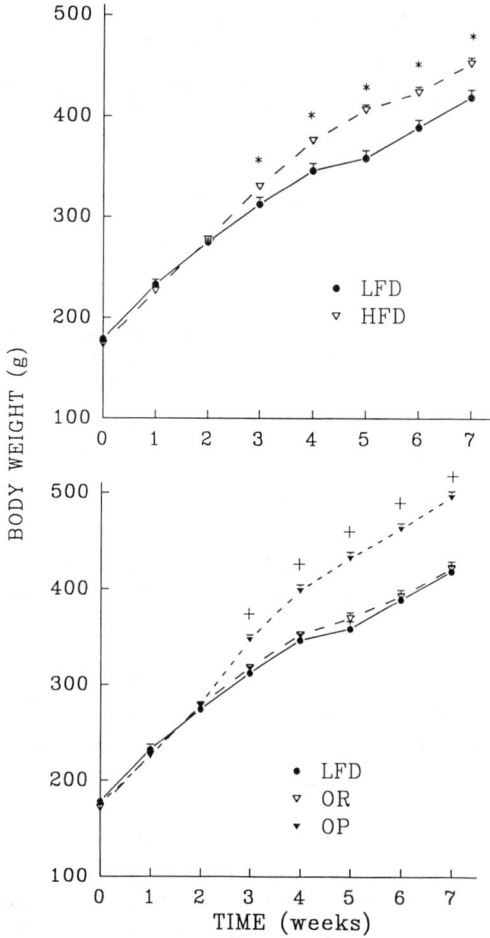

FIGURE 1. (Top) Weekly body weights of rats provided free access to either a low-fat diet (LFD) or a high-fat diet (HFD). Weight gain is shown during the initial 2-week baseline period during which both groups received the same LFD and the subsequent 5 weeks of either LFD or HFD feeding. (Bottom) Weekly body weights of rats, as described above, except that HFD rats are shown after separation into obesity-prone (OP; upper tertile of body weight gain) and obesity-resistant (OR; lower tertile of body weight gain). Values are means ± SEM. Symbols represent significant ($p < 0.05$) differences between LFD and HFD (∗) or between OP and both LFD and OR (+). Adapted from reference 9 with permission.

a 2-week baseline period. Following the baseline period, rats either remain on the LFD or are provided free access to a high-fat diet (HFD; 45% of energy from corn oil, 35% from corn starch, 20% from protein) for varying periods of time (typically for a 4–5-week period). This dietary regimen produces increased body weight and obesity when compared to LFD controls (FIG. 1, top). It also results in a heterogeneous distribution of body weight gain in the HFD group (FIG. 1, bottom). When the distribution of weight gain on the HFD is separated into tertiles, the lowest tertile (defined as obesity-resistant, OR) gains weight at a rate equivalent to LFD controls, whereas weight gain in the upper tertile (obesity-prone, OP) is significantly greater than in the lowest tertile (FIG. 1, bottom). Thus, this model allows us to characterize the behavioral (energy intake) and metabolic (nutrient partitioning) factors that contribute to the apparent differences in susceptibility to body weight gain on the HFD.

Characterization of Body Weight

Neither body weight nor body weight gain during the LFD baseline period is a good predictor of subsequent body weight gain during the HFD feeding period.[11,12] However, significant differences in the distribution of body weight gain on the HFD can be observed after only 1 week of HFD feeding, and thus separation of the HFD group into tertiles of body weight gain can be achieved by this 1-week time point (FIG. 2). The correlation between weight gain after 1 week and weight gain after 5 weeks on the HFD is strong, $r = 0.85$, and highly significant, $p < 0.001$ (FIG. 3). Thus, rats defined as OP or OR after 1 week typically remain within their classifications for the entire HFD feeding period. Analysis of body weight gain on the HFD has demonstrated that its distribution is Gaussian (FIG. 4). It therefore appears that the HFD promotes not only weight gain, but also variability in weight gain. This variability in weight gain is characterized by a normal distribution and probably results from multiple behavioral and metabolic inputs.

Similar to weight gain, there are significant differences in fat mass and % fat in the HFD rats (FIG. 2). It is important to note that the fat pad mass in FIGURE 2 represents the sum of the retroperitoneal, epididymal, and mesenteric fat pads only. Although this sum correlates with total carcass lipid ($r = 0.93$), it will underestimate the total contribution of fat pad mass gain to total weight gain. Thus, despite access to the same HFD, there is a large variability in net nutrient storage among rats of the same strain. An important question for future research is whether these differences in net nutrient storage require the HFD (i.e., environmental induction) or are simply magnified by the presence of the HFD (i.e., environmental-genetic interaction).

Characterization of Energy Intake

Although energy intake during the baseline period is correlated with subsequent energy intake on the HFD (FIG. 5, top), it does not predict the increase in energy intake observed in those rats that gain the most weight on the HFD (i.e., OP rats). Thus, energy intake during the baseline period is only weakly correlated with subsequent body weight gain on the HFD (FIG. 5, middle). In contrast, energy intake on the HFD may account for up to 50% of the variance in body weight gain on the HFD (FIG. 5, bottom). Thus, there is a definite and reproducible difference in energy intake between the three tertiles (OR; OI, obesity-intermediate or middle tertile; and OP) of HFD-fed rats. It is interesting that the greatest difference in energy intake is observed during the first week of HFD feeding, where it is typically 20–25% greater

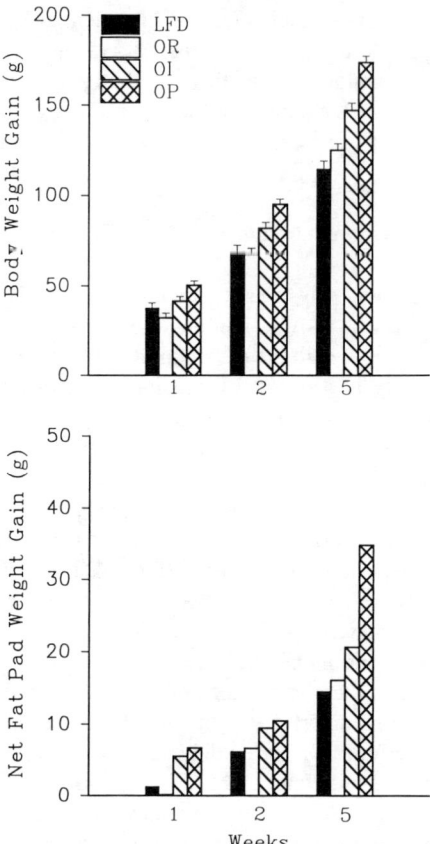

FIGURE 2. (Top) Body weight gain in low-fat-fed (LFD) and high-fat-fed obesity-resistant (OR), obesity-intermediate (OI), and obesity-prone (OP) rats after 1, 2, or 5 weeks on their respective diets. (Bottom) Fat pad weight gain in the same groups as described for the top panel. Fat pad weight gain represents the change in fat pad weight (retroperitoneal, mesenteric, and epididymal fat pads) from baseline. The baseline fat pad weights were determined on separate groups of animals. Values are means ± SEM. Adapted from reference 12 with permission.

in OP versus OR rats. During weeks 2 through 5, the difference in energy intake between OP and OR rats averages 10–15%. A similar time course is also observed for the rate of weight gain between OP and OR rats. Energy intake, like weight gain, is characterized by a normal, Gaussian distribution.

Characterization of Energy Expenditure

A reduction in energy expenditure has been implicated as an important component of obesity development in both humans[13] and animals.[14] We have previously measured

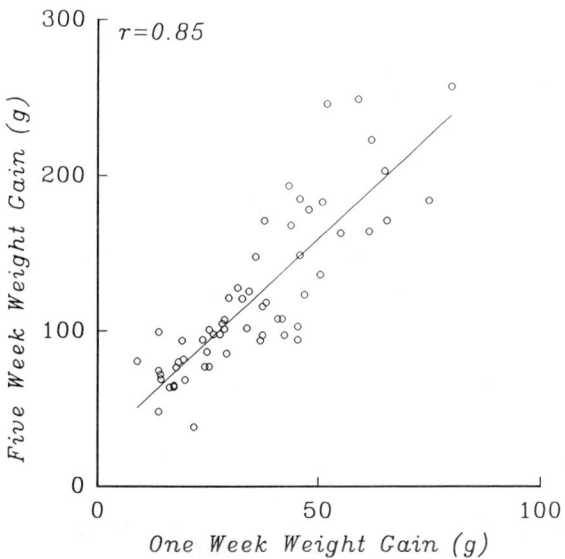

FIGURE 3. Correlation between weight gain after 1 week on the high-fat diet and 5 weeks on the high-fat diet. Adapted from reference 10 with permission.

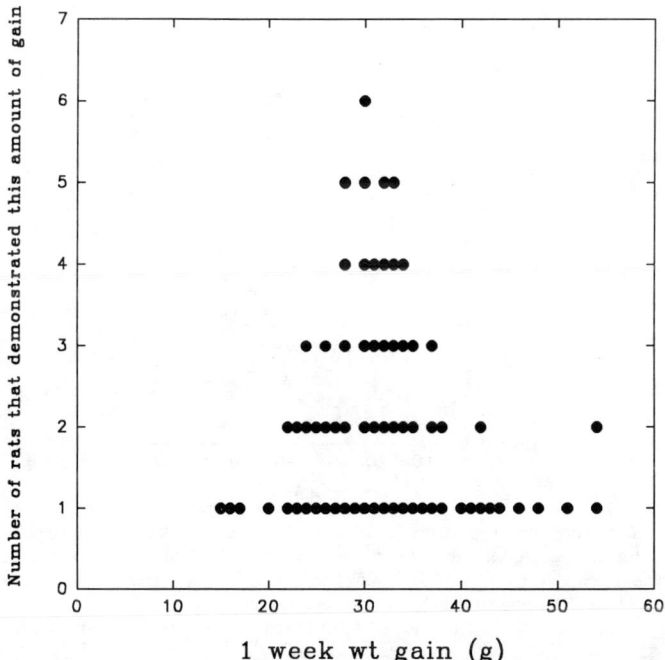

FIGURE 4. Distribution of body weight gain after 1 week on the high-fat diet. Adapted from reference 12 with permission.

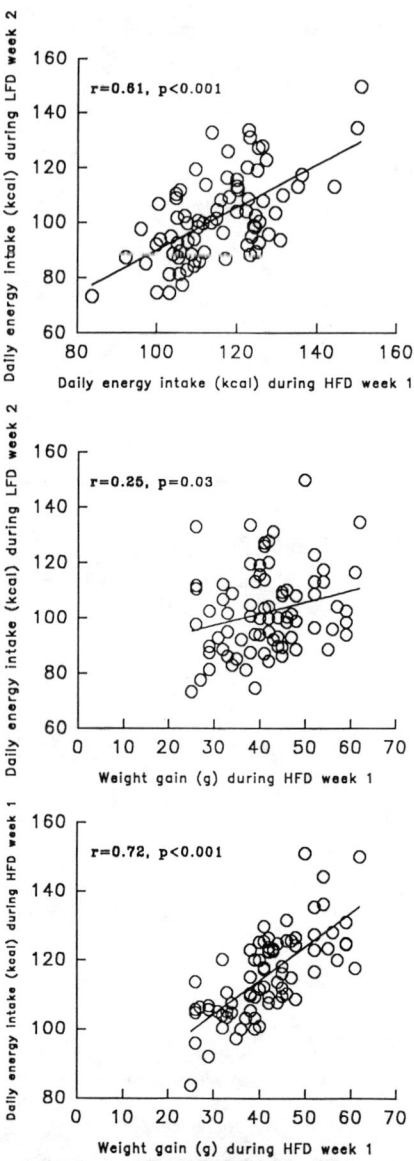

FIGURE 5. Correlations between energy intake during the second week of the LFD feeding baseline period and subsequent energy intake during the first week of HFD feeding (top), between energy intake during the second week of the LFD feeding baseline period and subsequent weight gain during the first week of HFD feeding (middle), and between energy intake during the first week of HFD feeding and weight gain during the first week of HFD feeding (bottom). Adapted from reference 12 with permission.

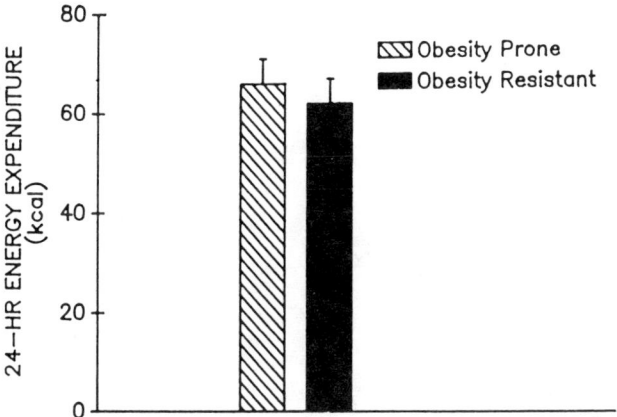

FIGURE 6. Twenty-four-hour energy expenditure in obesity-prone and obesity-resistant rats after 3–5 weeks on the HFD. Values are means ± SEM. Adapted from Chang et al.[7]

24-h energy expenditure in OP and OR rats following 3–5 weeks of HFD feeding.[7] Under these conditions, no significant differences in energy expenditure were evident between groups when expressed as total kcal (FIG. 6) or as kcal/fat free mass. [In this study, the middle tertile of body weight gain (OI rat) was not studied.] This was unexpected, especially when one considers that energy intake and energy expenditure determine energy balance, and the former only accounts for approximately 50% of the variance in total body weight gain in this model.

To better understand the contribution of energy expenditure to dietary obesity development, we are presently measuring gas exchange continuously (23 h/day) during the last week of the LFD baseline period and for 4 consecutive weeks of HFD feeding. The findings presented below are preliminary and reflect unpublished observations. During the first week of HFD feeding, there appears to be a small, but reproducible decrease in energy expenditure in rats that gain the most weight. This decrease represents a difference of approximately 25 kcal between OP and OR rats (FIG. 7). OP rats continue to display a lower energy expenditure throughout the subsequent 3-week HFD feeding period (FIG. 7). This appears to be due to a compensatory increase in energy expenditure in the OR rats. A conservative estimate of the contribution of this energetic difference to the difference in weight gain between OP and OR rats would be that all of the difference (∼ 150 kcal) was in fat weight. Based on this assumption, the contribution of energy expenditure to the difference in weight gain would be at least 30%. The energy expenditure data are not currently normalized for body mass or fat free mass. One would anticipate that this would, if anything, increase the difference between OP and OR rats. (Note that, for purposes of presentation, FIGURE 7 only contains the upper and lower tertiles of weight gain. The middle tertile, thus far, has displayed an intermediate energy expenditure during the HFD feeding period.)

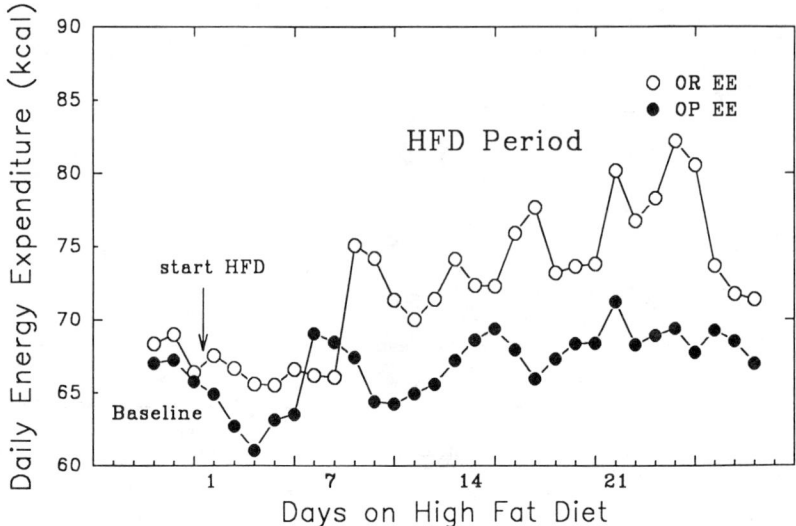

FIGURE 7. Daily energy expenditure (EE) (averaged by tertiles of weight gain) in obesity-resistant (OR) and obesity-prone (OP) rats during the last 3 days of LFD feeding (baseline) and during 28 days of HFD feeding. Data represent unpublished observations and are the average of three rats in each of the groups.

More studies are under way to determine the reproducibility of this pattern of energy expenditure. It will also be important to determine which component of total energy expenditure (i.e., resting energy expenditure or physical activity) is responsible for the observed difference between OR and OP rats.

Characterization of Respiratory Quotient (RQ)

The fundamental measurement provided by indirect calorimetry is the disappearance rate of a nutrient, regardless of any metabolic interconversions that the nutrient undergoes prior to its disappearance.[15,16] Thus, under most circumstances, RQ will provide an estimate of the net disappearance rate of a nutrient. Any presumed equivalence between net nutrient disappearance and net nutrient oxidation is disrupted when rates of gluconeogenesis, ketogenesis, or lipogenesis are elevated. Therefore, direct knowledge of the contribution and physiologic variance of these metabolic processes is required in order to equate RQ with net oxidation. In the following discussion, these limitations should be kept in mind.

A low ratio of fat to carbohydrate oxidation (high RQ) has been associated with subsequent weight gain, independent of low energy expenditure, in Pima Indians.[17] This correlation was not particularly strong ($r = 0.27$), but was highly significant ($p < 0.01$). We have also observed a higher 24-h RQ in OP versus OR rats following

3–5 weeks on a high-fat diet.[7] This difference in RQ, although statistically significant ($p < 0.05$), was small (0.79 in OP versus 0.77 in OR rats), with considerable overlap between groups. Previous interpretations regarding the physiologic basis for the relationship between RQ and susceptibility to weight gain (i.e., obesity) are based on the rationale that a lower RQ, that is, a higher net rate of fat oxidation, will restrict fat gain. In addition, a lower RQ in the presence of a high-fat diet implies that net fuel oxidation has adjusted to the fuel mixture in the diet. In fact, it has been proposed that this is a necessary condition for maintenance of stable body weight and composition.[18,19]

Energy intake must also be considered when evaluating the role of RQ in body weight gain. When diets of fixed composition are consumed, net carbohydrate oxidation (based on RQ) is proportional to energy intake, whereas net fat oxidation is negatively correlated with energy intake.[19] While carbohydrates appear to stimulate their own oxidation (and inhibit fat oxidation), fat has little impact, over the short term, in altering either fat or carbohydrate oxidation. In our dietary model of obesity, OP rats eat more energy during the HFD feeding period than OR rats. Therefore, OP rats also consume more carbohydrate, despite the fact that they are provided exactly the same HFD as OR rats. This suggests that some portion of the higher 24-h RQ observed in OP rats may be due to the greater energy and carbohydrate intake in these rats.

Recent preliminary data from our laboratory suggest that the higher RQ in OP versus OR rats may be present during the LFD baseline period (FIG. 8). This is a potentially significant observation since the difference in RQ appears to be present under conditions in which energy intake is not significantly different between groups. These data suggest that the differences in RQ between OP and OR rats may be preexisting and thus are not adaptations to the HFD. If true, RQ during the LFD baseline period will be a useful predictor of susceptibility to body weight and body fat gain. Further studies are necessary to confirm these preliminary results and suggestions.

From a metabolic standpoint, weight maintenance depends on the net oxidation of a fuel mix (RQ) that matches the composition of the diet (food quotient or FQ).[18,19] In general, if the average RQ exceeds the FQ, the net oxidation of the fuel mix may contain less fat than the nutrient entering from the diet. Therefore, a sustained period characterized by an RQ in excess of FQ implies accumulation of fat.

Recently, we have begun to investigate the relationship between RQ and FQ during the transition from an LFD to an HFD in our model of dietary obesity. We previously hypothesized that the rapidity with which RQ approached FQ (the slope of the line describing the change in steady-state RQ on the LFD from steady-state RQ on the HFD) when animals were switched from an LFD to an HFD would be inversely related to the amount of weight gain. In FIGURE 8, RQ is plotted during the last 3 days of the LFD feeding baseline period and the subsequent 28 days of HFD feeding in rats that gained the most (OP, upper tertile) or least (OR, lowest tertile) body weight over this period of time. (The middle tertile has been excluded for clarity, but was intermediate to the two other groups.) Included in the figure is the estimated FQ of the LFD (0.93) and the HFD (0.825). There are several interesting findings from these

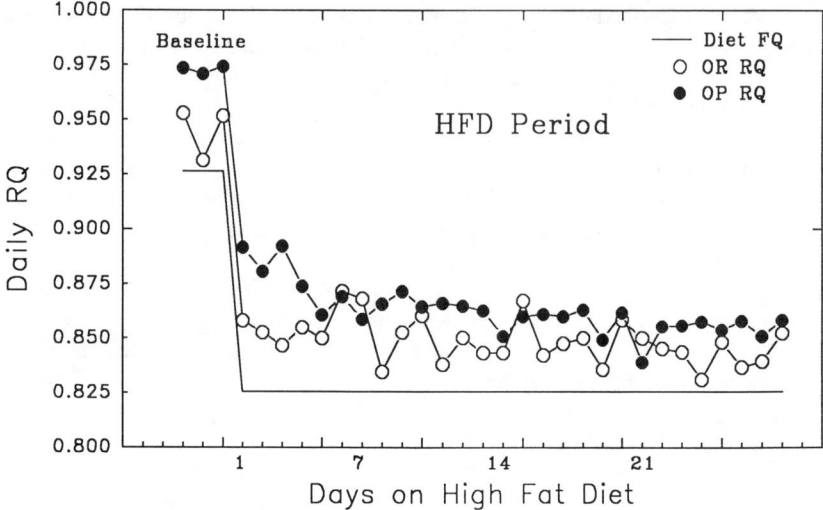

FIGURE 8. Daily respiratory quotient (RQ) (averaged by tertiles of weight gain) in obesity-resistant (OR) and obesity-prone (OP) rats during the last 3 days of LFD feeding (baseline) and during 28 days of HFD feeding. The solid line represents the estimated food quotient (FQ) of the LFD and the HFD. Data represent unpublished observations and are the average of three rats in each of the groups.

preliminary data. First, the change in RQ from the last day of the baseline period to the first day of the HFD feeding is similar between OR and OP rats. Second, the steady-state RQ of ~0.85 in OR rats is achieved after only 1 day of HFD feeding. In contrast, it takes about 5 days to achieve the steady-state RQ of ~0.86 in OP rats. Finally, since all animals are growing in this model of dietary obesity, it is important to note that RQ always exceeded FQ in both groups. These findings imply that the higher RQ observed in OP versus OR rats is largely due to preexisting differences. Whether the slower adjustment to the steady-state RQ during the HFD feeding period is due to the higher RQ in the baseline period (i.e., OP rats ultimately make a larger adjustment in the RQ), to increased energy intake, or to metabolic factors that delay the adjustment is impossible to determine from these data.

Summary of Characteristics of the Model

This dietary model of obesity results in a heterogeneous distribution of both body weight gain and body fat gain. Differences in weight and fat gain on the high-fat diet are due to differences in energy intake, energy expenditure, and whole body nutrient metabolism. We will now consider some of the potential reasons for these differences in energy intake and nutrient metabolism.

REASONS FOR DIFFERENCES IN ENERGY INTAKE

Multiple metabolic signals can regulate energy intake.[20] In this dietary model of obesity, we have been unable to find any single hormone or metabolite that can account for a significant portion of the variance in energy intake during the HFD feeding period. The variables that have been measured include fasting insulin, glucose, nonesterified fatty acids, glycerol, β-hydroxybutyrate, liver glycogen, muscle glycogen, and corticosterone. We are currently investigating the potential roles of nutrients and hormones in the fed state, palatability, leptin, and neural regulation in the differences in energy intake in this model.

In a recent study, we observed that the rate of endogenous glucose production (Ra) during a hyperinsulinemic, euglycemic clamp measured prior to the introduction of the HFD (i.e., at the end of the LFD baseline period) was strongly correlated ($r = 0.64$, $p < 0.001$) with subsequent energy intake on the HFD.[21] Therefore, reduced suppression of Ra by insulin is one metabolic variable that appears to account for a significant portion of the variance in energy intake, and thus body weight gain, on the HFD. It would appear then that a higher RQ and reduced insulin suppression of Ra are two variables that can predict obesity susceptibility in this dietary model. It is interesting that recently obese children, in contrast to obese adults, have displayed reduced insulin suppression of hepatic glucose production, but normal insulin action on adipose tissue.[22]

There are at least two possible scenarios that can account for the relationship between the ability of insulin to suppress Ra prior to the introduction of the HFD and subsequent energy intake on the HFD. In the first, reduced insulin suppression of Ra may result in greater reliance on hepatic glycogenolysis over the course of a day. In this case, increased energy intake on the HFD may be required to ensure that enough carbohydrate is eaten to replenish liver glycogen levels. This may not be required when animals are eating the LFD since it contains approximately twice the carbohydrate as the HFD. The second scenario assumes a link between metabolism and energy intake. Reduced insulin suppression of Ra should increase systemic concentrations of glucose following a meal. Increased glucose concentrations would stimulate insulin secretion. The combination of increased glucose and insulin will stimulate glucose uptake by muscle and adipose tissue. Increased glucose uptake in muscle will reduce fat uptake and result in shunting of fat to adipose tissue for acute storage. Long-term retention of this fat, that is, increased adiposity, will require a reduction in energy expenditure, increased energy intake, or both to compensate for this acute net retention of energy. Our current data suggest that both energy intake and expenditure may be used for compensation in this dietary model.

REASONS FOR DIFFERENCES IN ENERGY UTILIZATION: ROLE OF NUTRIENT PARTITIONING

The term, nutrient partitioning, is used to describe the metabolic fate of ingested nutrients. From an energy balance standpoint, nutrient partitioning represents the contribution of oxidation and storage to the disposal of excess energy. Our preliminary data suggest that obesity susceptibility is, in part, determined by the ability to partition proportionally more excess energy into storage (i.e., energy intake in excess of energy expenditure). Differences in nutrient partitioning can also be related to differences in the fuel mixture that is oxidized. The differences in RQ between OP and OR rats suggest that there are differences between these groups in the composition of the fuel mixture oxidized. These differences may be driven, in part, by differences in energy intake, but may also reflect inherent or diet-induced differences at the tissue and/or protein level.

Among the many factors suggested to contribute to differences in the composition of the fuel mixture oxidized are tissue enzymatic profile (e.g., lipoprotein lipase, citrate synthase, etc.),[23–25] skeletal muscle fiber type,[26–28] the fatty acid profile of muscle membranes,[28–30] and insulin action.[28,31–35] We have examined these variables in our dietary model of obesity.

In earlier work, we had demonstrated that the amount of type I fibers in the gastrocnemius muscle was lower in OP versus OR rats prior to and following 5 weeks of HFD feeding.[8] In contrast, oxidative capacity, based on citrate synthase activity in skeletal muscle, and skeletal muscle membrane fatty acid composition were not significantly different between OP and OR rats after 5 weeks on the HFD.[11] These findings lead to the conclusion that susceptibility to dietary obesity, in this rodent model, cannot be explained by differences in tissue oxidative capacity or muscle membrane fatty acid composition. These studies, however, do not go far enough in defining the metabolic characteristics of tissues that may contribute to differences in nutrient partitioning and obesity development. There are at least two reasons for this. First, under resting conditions, several tissues, including heart, liver, and skeletal muscle, contribute significantly to energy expenditure. This suggests that multiple tissues must be investigated in order to understand the contribution of the tissue metabolic profile to whole body nutrient partitioning. Second, metabolic pathways, and the enzymes controlling them, function as coordinated systems. Almost 30 years ago, Pette and coworkers[36,37] demonstrated that the ratios of enzymes involved in nutrient metabolism represent estimates of the relationship between different metabolic pathways.

In recent studies, we have characterized the enzymatic profile of multiple tissues using the approach of Bass et al.[36] in our dietary model of obesity. An additional advantage of these studies was our ability to study animals prior to, during, and after the development of obesity. Skeletal muscle, liver, and heart tissues taken from overnight fasted rats following 1, 2, or 5 weeks of either LFD or HFD feeding were analyzed for β-hydroxyacyl-CoA dehydrogenase activity (HADH, the key enzyme of β-oxidation), phosphofructokinase activity (PFK, the key enzyme of glycolysis), and citrate synthase activity (CS, the key enzyme of the Krebs cycle). There were no significant changes in individual enzyme activities between the LFD and HFD groups. In addition, the relationships between enzymes (i.e., HADH versus CS; PFK versus HADH) were similar in both the LFD and HFD groups. Taken together, these data suggest that the HFD did not produce an adaptation in these enzymes.

Despite this apparent lack of adaptation in individual enzymes, a significant proportion of the variance in body weight gain on the HFD was explained by the enzyme

TABLE 1. Correlations between Enzyme Ratios and Body Weight Gain for HFD Rats $(n = 31-36/\text{Week})^a$

	Week 1	Week 2	Week 5
Gastrocnemius muscle			
PFK/HADH	$r = 0.40^*$	$r = 0.48^{**}$	NS
HADH/CS	$r = -0.56^{**}$	$r = 0.36^*$	NS
Heart			
PFK/HADH	NS	$r = 0.37^*$	NS
HADH/CS	NS	$r = -0.41^*$	NS
Liver			
PFK/HADH	NA	NA	NA
HADH/CS	NS	$r = -0.55^{**}$	NS

[a] Adapted from reference 12 with permission. Terms: PFK, phosphofructokinase; HADH, β-hydroxyacyl-CoA dehydrogenase; CS, citrate synthase; NS, not significant ($p > 0.05$); NA, liver PFK activities were too low to determine group differences; (*) $p < 0.05$; (**) $p < 0.01$.

ratios PFK/HADH (an estimate of the relative capacity for glycolysis versus β-oxidation) and HADH/CS (an estimate of the relative capacity for β-oxidation versus total oxidative capacity) in the gastrocnemius muscle, heart, and liver (TABLE 1). This suggests that small opposing differences in two enzymes from different metabolic pathways can potentially make large contributions to whole body nutrient partitioning. It is important to point out that significant correlations between these enzyme ratios and weight gain on the HFD were only observed after 1 and 2 weeks of HFD feeding. No significant relationships were evident after 5 weeks on the HFD (TABLE 1). The early contribution of the tissue enzymatic profile to differences in weight gain on the HFD is also true for skeletal muscle (decreased in OP versus OR rats) and adipose tissue (increased in OP versus OR rats) lipoprotein lipase activity and mRNA (TABLE 2). It is clear that measurements made at multiple time points during a period of weight gain are necessary in order to better determine possible markers of obesity susceptibility and nutrient partitioning. These results suggest that obesity-prone animals are characterized by an enzymatic profile that is geared to utilize carbohydrate for oxidation (increased PFK/HADH ratio, decreased HADH/CS ratio) and lipid for storage (increased LPL in adipose tissue, decreased LPL in muscle). It seems likely that these enzymatic differences, as with RQ, may be preexisting and interact with other factors, such as the HFD-induced increase in energy intake and decrease in energy expenditure, to produce the large, early differences in weight and fat gain between OP and OR rats.

INSULIN ACTION AND DIETARY OBESITY

Obesity is typically characterized by widespread insulin resistance, that is, a decrease in insulin suppression of glucose production and lipolysis and a decrease in insulin stimulation of glucose uptake in muscle and adipose tissue.[31] In both Pima

TABLE 2. Epididymal Fat and Gastrocnemius Muscle Lipoprotein Lipase mRNA and Activity in LFD Controls and in HFD OR and OP Rats[a]

	Week 1	Week 2	Week 3
	mRNA (arbitrary units)		
Epididymal fat			
LFD	7.55 ± 0.65*	7.36 ± 0.18*	6.13 ± 0.48*
OR	1.57 ± 0.31	3.10 ± 1.26	1.26 ± 0.56
OP	5.16 ± 1.07+	4.16 ± 0.48	3.08 ± 0.57
Gastrocnemius muscle			
LFD	3.28 ± 0.34*	1.34 ± 0.42	2.27 ± 0.38
OR	1.70 ± 0.28	1.52 ± 0.14	1.80 ± 0.36
OP	0.31 ± 0.10+	0.36 ± 0.19*	1.39 ± 0.55
	Activity (nmol/g/min)		
Epididymal fat			
LFD	4.7 ± 0.3	5.1 ± 0.3	5.2 ± 0.2
OR	2.3 ± 0.3*	4.4 ± 0.3	4.3 ± 0.4
OP	4.6 ± 0.2	5.2 ± 0.2	5.6 ± 0.3
Gastrocnemius muscle			
LFD	2.5 ± 0.2	2.5 ± 0.3	2.6 ± 0.2
OR	2.1 ± 0.1	2.6 ± 0.4	2.6 ± 0.3
OP	1.1 ± 0.2*	1.8 ± 0.3	2.1 ± 0.3

[a] Values are means ± SEM for $n = 6$/group/time point. Terms: LFD, low-fat-diet control rats; OR, obesity-resistant HFD-fed rats; OP, obesity-prone HFD-fed rats. (*) Significantly ($p < 0.05$) different from the other two groups. (+) Significantly ($p < 0.05$) different from OR rats. Note that the differences in lipoprotein lipase activity in the epididymal fat pad among the groups after 1 week are also present when the data are expressed per cell. Adapted from reference 10 with permission.

Indians[32] and Mexican Americans,[33] lower rates of weight gain have been associated with widespread insulin resistance. Thus, it has been postulated that widespread insulin resistance may function as a feedback mechanism to attenuate weight gain.[34]

We have previously characterized whole body (glucose infusion rate during a euglycemic, hyperinsulinemic clamp) and skeletal muscle (hindquarter perfusion) insulin action on glucose metabolism in our model of dietary obesity. In these studies, OP rats were characterized by lower glucose infusion rates and lower insulin-stimulated skeletal muscle glucose uptake when compared to OR rats after 5 weeks on the HFD.[7,9] Interestingly, insulin-stimulated muscle glucose uptake in OR rats was significantly lower than in LFD control rats, despite the lack of a significant difference in body composition between these two groups. This suggests that a high-fat diet per se can influence insulin action on skeletal muscle glucose metabolism.

We have not yet characterized the time course of induction of tissue-specific insulin resistance in this model of obesity. Based on the data described above, we would predict that obesity-prone rats would have reduced insulin suppression of Ra (relative hepatic insulin resistance) prior to the introduction of the high-fat diet. In addition, Kraegen et al.[38] have demonstrated that an HFD produces insulin resistance in liver prior to skeletal muscle. It is therefore likely that hepatic insulin resistance during the early period of HFD feeding would not only be worse in OP versus OR rats, but also would be accompanied by normal or even augmented insulin action in adipose tissue.[35,39] This metabolic environment may facilitate acute nutrient retention during the early period of HFD feeding (as described above in the energy intake section). It is important to note that we have demonstrated that the induction of selective hepatic insulin resistance prior to the introduction of the HFD (by using high-sucrose-diet feeding for 1 week) can increase subsequent weight and fat gain on the HFD.[21] Thus, the role of insulin action in the development of obesity may depend on the quality of insulin action among insulin-sensitive tissues. Presumably, the induction of insulin resistance in adipose tissue may be one mechanism to restrain the continued weight and fat gain during HFD feeding. It is also likely that any contribution of insulin action to obesity development will be linked to changes in energy intake and/or energy expenditure.

SUMMARY AND CONCLUSIONS

In summary, an imbalance between energy intake and energy expenditure can explain approximately 80% of the variance in body weight gain in this dietary model of obesity. Several metabolic variables appear to contribute to differences in energy balance. A high RQ and an inappropriate suppression of glucose production by insulin appear to be linked to the increase in energy intake that occurs when obesity-prone rats are provided with the high-fat diet. In addition, early tissue enzymatic differences in obesity-prone versus obesity-resistant rats may contribute to differences in energy expenditure and/or to differences in nutrient partitioning.

In this dietary model, susceptibility to dietary obesity involves a metabolic environment that includes a high RQ and a reduced ability of insulin to suppress glucose appearance (FIG. 9). However, this environment does not lead to obesity nor to a measurable difference in body weight gain when the susceptible rats are eating a low-fat diet. The high-fat diet is a necessary catalyst for the observed variability in body weight gain and the development of obesity.

As a catalyst, the high-fat diet results in an imbalance between energy intake and energy expenditure in some, but not all, rats. This imbalance interacts with the permissive metabolic environment (tissue enzymatic profile favoring carbohydrate utilization and lipid storage) to produce obesity on the high-fat diet. Later, in the HFD feeding period, the rate of weight gain is not significantly different between OP and OR rats, although net fat accumulation remains greater in the former group. It is interesting that this later period is characterized by a reduction in the difference in both RQ and energy intake between OP and OR rats. Thus, during the later stages of HFD feeding, the discrepancy in both energy balance and nutrient balance between OP and OR rats is reduced.

This dietary model of obesity is relevant to human obesity. While the prevalence of obesity is high, the majority of people are not obese. The high prevalence of obesity may be due to environmental catalysts that interact with inherent behavioral and metabolic characteristics that favor nutrient retention. Resistance to obesity can be achieved

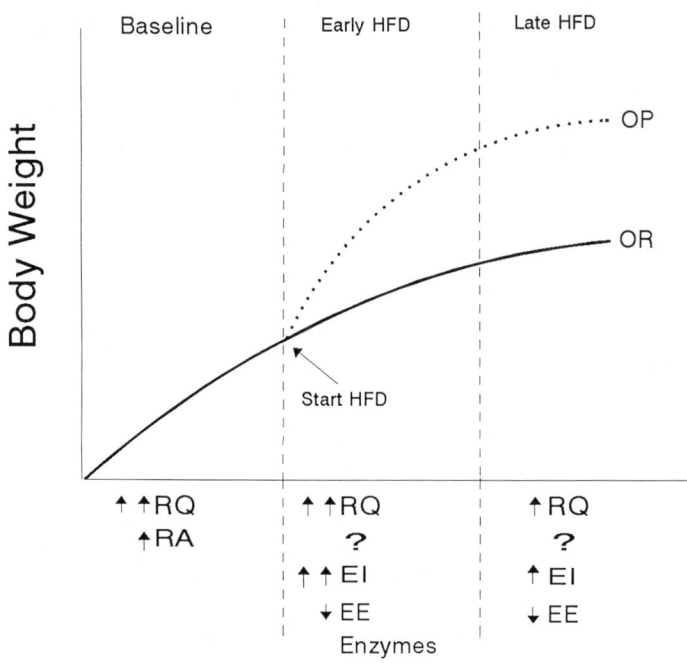

FIGURE 9. Summary of variables that may contribute to the distribution of body weight gain in high-fat-diet (HFD)-fed rats. Terms: RQ, respiratory quotient; RA, rate of glucose appearance during a euglycemic, hyperinsulinemic clamp; EI, energy intake; EE, energy expenditure; Enzymes, tissue metabolic profile as discussed in the text.

by avoiding these environmental catalysts, by having inherent characteristics that prevent nutrient retention, or both.

Our work suggests that the complete understanding of obesity will require not only the identification and functional significance of the genes that determine the inherent capacity of the behavioral and metabolic systems, but also the role of environmental catalysts in determining where and how these systems operate.

REFERENCES

1. KUCZMARSKI, R. J. 1992. Prevalence of overweight and weight gain in the United States. Am. J. Clin. Nutr. **55**: 495S–508S.
2. MUELLER, W. H. 1983. The genetics of human fatness. Yearb. Phys. Anthropol. **26**: 215–230.
3. BOUCHARD, C. 1991. Current understanding of the etiology of obesity: genetic and nongenetic factors. Am. J. Clin. Nutr. **53**: 1561S–1565S.
4. KLESGES, R. C., L. M. KLESGES, C. K. HADDOCK & L. H. ECK. 1992. A longitudinal analysis of the impact of dietary intake and physical activity on weight change in adults. Am. J. Clin. Nutr. **55**: 818–822.
5. BOUCHARD, C. & L. PERUSSE. 1993. Genetics of obesity. Annu. Rev. Nutr. **13**: 337–354; YORK, D. A. 1990. Metabolic regulation of food intake. Nutr. Rev. **48**: 64–70.
6. HILL, J. O., M. J. PAGLIASSOTTI & J. C. PETERS. 1994. Nongenetic determinants of obesity and

body fat topography. *In* Genetic Determinants of Obesity, p. 35-48. CRC Press. Boca Raton, Florida.
7. CHANG, S., B. GRAHAM, F. YAKUBU, D. LIN, J. C. PETERS & J. O. HILL. 1990. Metabolic differences between obesity-prone and obesity-resistant rats. Am. J. Physiol. **259:** R1103-R1110.
8. ABOU MRAD, J., F. YAKUBU, D. LIN, J. C. PETERS, J. B. ATKINSON & J. O. HILL. 1992. Skeletal muscle composition in dietary obesity-susceptible and dietary obesity-resistant rats. Am. J. Physiol. **262:** R684-R688.
9. PAGLIASSOTTI, M. J., K. A. SHAHROKHI & J. O. HILL. 1993. Skeletal muscle glucose metabolism in obesity-prone and obesity-resistant rats. Am. J. Physiol. **264:** R1224-R1228.
10. PAGLIASSOTTI, M. J., S. M. KNOBEL, K. A. SHAHROKHI, A. M. MANZO & J. O. HILL. 1994. Time course of adaptation to a high-fat diet in obesity-resistant and obesity-prone rats. Am. J. Physiol. **267:** R659-R664.
11. PAGLIASSOTTI, M. J., D. A. PAN, P. A. PRACH, T. A. KOPPENHAFER, L. H. STORLIEN & J. O. HILL. 1995. Tissue oxidative capacity, fuel stores, and skeletal muscle fatty acid composition in obesity-prone and obesity-resistant rats. Obes. Res. **3:** 459-464.
12. GAYLES, E. C., M. J. PAGLIASSOTTI, P. A. PRACH, T. A. KOPPENHAFER & J. O. HILL. 1997. Contribution of energy intake and tissue enzymatic profile to body weight gain in high-fat-fed rats. Am. J. Physiol. **272:** R188-R194.
13. BOGARDUS, C., S. LILLIOJA & E. RAVUSSIN. 1990. The pathogenesis of obesity in man: results of studies on Pima Indians. Int. J. Obes. **14:** 5-15.
14. STORLIEN, L. H., D. E. JAMES, K. M. BURLEIGH, D. J. CHISHOLM & E. W. KRAEGEN. 1986. Fat feeding causes widespread *in vivo* insulin resistance, decreased energy expenditure, and obesity in rats. Am. J. Physiol. **251:** E576-E583.
15. FERRANNINI, E. 1988. The theoretical basis of indirect calorimetry: a review. Metabolism **37:** 287-301.
16. SIMONSON, D. C. & R. A. DEFRONZO. 1990. Indirect calorimetry: methodological and interpretative problems. Am. J. Physiol. **258:** E399-E412.
17. ZURLO, F., S. LILLIOJA, A. ESPOSITO-DEL PUENTE, B. L. NYOMBA, I. RAZ, M. F. SAAD, B. A. SWINBURN, W. C. KNOWLER, C. BOGARDUS & E. RAVUSSIN. 1990. Low ratio of fat to carbohydrate oxidation as predictor of weight gain: study of 24-h RQ. Am. J. Physiol. **259:** E650-E657.
18. FLATT, J. P. 1993. Dietary fat, carbohydrate balance, and weight maintenance. Ann. N.Y. Acad. Sci. **683:** 122-140.
19. FLATT, J. P. 1988. Importance of nutrient balance in body weight regulation. Diabetes Metab. Rev. **4:** 571-581.
20. FRIEDMAN, M. I., M. G. TORDOFF & I. RAMIREZ. 1986. Integrated metabolic control of food intake. Brain Res. Bull. **17:** 855-859.
21. PAGLIASSOTTI, M. J., T. J. HORTON, E. C. GAYLES, T. A. KOPPENHAFER, T. D. ROSENZWEIG & J. O. HILL. 1997. Reduced insulin suppression of glucose appearance is related to susceptibility to dietary obesity in rats. Am. J. Physiol. **272:** R1264-R1270.
22. BOUGNERES, P.-F., E. ARTAVIA-LORIA, S. HENRY, A. BASDEVANT & L. CASTANO. 1989. Increased basal glucose production and utilization in children with recent obesity versus adults with long-term obesity. Diabetes **38:** 477-483.
23. COLBERG, S. R., J. A. SIMONEAU, F. L. THAETE & D. E. KELLEY. 1995. Skeletal muscle utilization of free fatty acids in women with visceral obesity. J. Clin. Invest. **95:** 1846-1853.
24. SIMONEAU, J. A. & C. BOUCHARD. 1995. Skeletal muscle metabolism and body fat content in men and women. Obes. Res. **3:** 23-29.
25. ZURLO, F., P. M. NEMETH, R. M. CHOKSI, S. SESODIA & E. RAVUSSIN. 1994. Whole body energy metabolism and skeletal muscle biochemical characteristics. Metabolism **43:** 481-486.
26. WADE, A. J., M. M. MARBUT & J. M. ROUND. 1990. Muscle fiber type and aetiology of obesity. Lancet **335:** 805-808.
27. HICKEY, M. S., J. O. CAREY, J. L. AZEVEDO, J. A. HOUMARD, W. J. PORIES, R. G. ISRAEL & G. L. DOHM. 1995. Skeletal muscle fiber composition is related to adiposity and *in vitro* glucose transport rate in humans. Am. J. Physiol. **268:** E453-E457.
28. KRIKETOS, A. D., D. A. PAN, S. LILLIOJA, G. J. COONEY, L. A. BAUR, M. R. MILNER, J. R. SUTTON, A. B. JENKINS, C. BOGARDUS & L. H. STORLIEN. 1996. Interrelationships between muscle morphology, insulin action, and adiposity. Am. J. Physiol. **270:** R1332-R1339.

29. PAN, D. A., S. LILLIOJA, M. R. MILNER, A. D. KRIKETOS, L. A. BAUR, C. BOGARDUS & L. H. STORLIEN. 1995. Skeletal muscle membrane lipid composition is related to adiposity and insulin action. J. Clin. Invest. **96:** 2802–2808.
30. PAN, D. A. & L. H. STORLIEN. 1993. Dietary lipid profile is a determinant of tissue phospholipid fatty acid composition and rate of weight gain in rats. J. Nutr. **123:** 512–519.
31. CARO, J. F., G. L. DOHM, W. J. PORIES & M. K. SINHA. 1989. Cellular alterations in liver, skeletal muscle, and adipose tissue responsible for insulin resistance in obesity and type II diabetes. Diabetes Metab. Rev. **5:** 665–689.
32. SWINBURN, B. A., B. L. NYOMBA, M. F. SAAD, F. ZURLO, I. RAZ, W. C. KNOWLER, S. LILLIOJA, C. BOGARDUS & E. RAVUSSIN. 1991. Insulin resistance associated with lower rates of weight gain in Pima Indians. J. Clin. Invest. **88:** 168–173.
33. VALDEZ, R., B. D. MITCHELL, S. M. HAFFNER, H. P. HAZUDA, P. A. MORALES, A. MONTERROSA & M. P. STERN. 1994. Predictors of weight change in a bi-ethnic population: the San Antonio Heart Study. Int. J. Obes. **18:** 85–91.
34. ECKEL, R. 1992. Insulin resistance: an adaptation for weight maintenance. Lancet **340:** 1452–1453.
35. EBERHART, G. P., D. B. WEST, C. N. BOOZER & R. L. ATKINSON. 1994. Insulin sensitivity of adipocytes from inbred mouse strains resistant or sensitive to diet-induced obesity. Am. J. Physiol. **266:** R1423–R1428.
36. BASS, A., D. BRDICZKA, P. EYER, S. HOFER & D. PETTE. 1969. Metabolic differentiation of distinct muscle types at the level of enzymatic organization. Eur. J. Biochem. **10:** 198–206.
37. STAUDTE, H. W. & D. PETTE. 1972. Correlations between enzymes of energy-supplying metabolism as a basic pattern of organization in muscle. Comp. Biochem. Physiol. **41B:** 533–540.
38. KRAEGEN, E. W., P. W. CLARK, A. B. JENKINS, E. A. DALEY, D. J. CHISHOLM & L. H. STORLIEN. 1991. Development of muscle insulin resistance after liver insulin resistance in high-fat-fed rats. Diabetes **40:** 1397–1403.
39. ROSELLA, G., J. D. ZAJAC, L. BAKER, S. J. KACZMARCZYK, S. ANDRIKOPOULOS, T. E. ADAMS & J. PROIETTO. 1995. Impaired glucose tolerance and increased weight gain in transgenic rats overexpressing a non-insulin-responsive phosphoenolpyruvate carboxykinase gene. Mol. Endocrinol. **9:** 1396–1404.

Recent Advances in the Pharmacological Control of Energy Balance and Body Weight[a]

RICHARD L. ATKINSON

Departments of Medicine and Nutritional Sciences
University of Wisconsin
Madison, Wisconsin 53706-1571

INTRODUCTION

Obesity has become epidemic in the United States and much of the world. Recent studies by the United States government demonstrate that one-third of the adult population of the United States is obese, and the prevalence of obesity increased by more than 30% in the decade from 1980 to 1990.[1] Obesity produces numerous comorbidities such as diabetes mellitus, hypertension, hyperlipoproteinemia, sleep apnea, and gallbladder disease, with the end result that mortality is increased, particularly in younger individuals.[2-5] Genetic factors contribute 20% to 50% of the variance in a population,[6,7] and biochemical and metabolic differences between obese and lean humans have been well documented.[2-7]

Although there is ample evidence that obesity is a disorder of altered physiology, psychological aspects of obesity have outweighed the physiological and biochemical factors when the treatment of obesity is undertaken. Obesity treatment has consisted predominantly of the behavioral techniques of alteration of diet, by voluntarily reducing calorie and fat intake, and of activity, predominantly by increasing exercise. Behavior modification techniques have been used to achieve these changes. Although initially successful, this form of therapy is poorly successful over the long term. In the best hands, only about two-thirds of the weight lost is maintained at 1 year, and the 5-year success rate is less than 5%.[8-12]

Despite the failures of "standard" therapy for obesity, there has been a surprising reluctance to embrace pharmacologic therapy.[13] The addictive potential of the early drugs developed for the treatment of obesity, such as dexamphetamine and methamphetamine, was amply demonstrated. Later drugs had less or much less abuse potential. Continued use of the first-generation drugs by physicians, often prescribed in combinations with excessive doses of thyroid hormone, diuretics, and even digitalis preparations, prompted restrictions on the use of these drugs by the United States government and many other governments of the world. The negative publicity gen-

---

[a] This work was supported in part by funds from the Beers-Murphy Clinical Nutrition Center, University of Wisconsin, Madison, Wisconsin.

erated a marked suspicion of obesity drugs on the part of physicians and other healthcare professionals, as well as insurance companies and the governments. Almost no third-party insurance payers in the United States pay for obesity drugs at this time. The impression persists that obesity drugs are ineffective and dangerous.

Little research into new drugs for obesity was performed in the 1970s and 1980s, and only one new drug was approved by the United States Food and Drug Administration (US FDA) from 1973 to 1996. The perception of drug treatment of obesity changed dramatically in 1992, with the publication by Weintraub[14] of a long-term study using the combination of phentermine and fenfluramine. In 1984, Weintraub et al.[15] had published a study comparing half doses of the combination to full doses of either drug alone. This study demonstrated that side effects with the combination were not different from placebo and were lower than with full doses of either single agent. This study generated little publicity, but the 1992 study received a great deal of publicity and revolutionized the field. A total of 121 subjects were treated for up to 3.5 years, with reasonable weight loss and modest side effects. Weight loss persisted for as long as drugs were given, but weight regain was rapid upon discontinuation of drugs. These studies altered the perception of the usefulness of obesity drugs and, coupled with new discoveries of the genetic basis for several animal models of obesity,[16-20] the groundwork has been laid for the current rapid expansion of research and treatment with obesity drugs.

MECHANISMS OF ACTION OF OBESITY DRUGS

Obesity drugs may act to reduce food intake or increase energy expenditure, and the consequent negative energy balance leads to weight loss. Energy intake may be reduced by decreasing appetite, by increasing satiety, or by altering the mixture of nutrients. If the total quantity of food remains unchanged, reduction of the proportion of calories as fat will reduce energy intake. There is evidence that obesity drugs may act by any or all of these three mechanisms.

Another potential method of reducing energy intake is by reduction of absorption of nutrients. As discussed below, two drugs are available that block absorption from the GI tract by binding or inactivating enzymes that digest fat or carbohydrate. The subsequent malabsorption leads to a decreased energy intake.

Obesity drugs may increase energy expenditure by stimulating an increase in activity levels or by increasing the metabolic rate directly. The literature is controversial, but animal and human studies suggest that some obesity drugs may increase energy expenditure by increasing dietary-induced thermogenesis.

Finally, it appears that obesity drugs may reduce the level at which body weight is defended. Both animals and humans reduce body weight when treated with obesity drugs; then, they tend to maintain the lower weight. The mechanisms by which this is accomplished are unclear. Levitsky and Troiano[21] showed that animals initially decreased food intake and lost weight with fenfluramine or dexamphetamine, but that food intake returned to baseline without a regain in body weight.

TABLE 1. Categories of Drugs Used to Treat Obesity

Drugs Currently Approved by the US FDA:
(A) Centrally active catecholaminergic agents:
   (1) DEA schedule II: amphetamine, methamphetamine, phenmetrazine
   (2) DEA schedule III: benzphetamine, chlorphentermine, chlortermine, phendimetrazine
   (3) DEA schedule IV: diethylpropion, mazindol, phentermine
   (4) Over-the-counter: phenylpropanolamine, ephedrine[a]
(B) Centrally active serotoninergic agents:
   (1) DEA schedule IV: d,l-fenfluramine, d-fenfluramine
   (2) Not scheduled: fluoxetine,[a] sertraline[a]
(C) Drugs affecting absorption: acarbose[a]

Experimental Drugs or Drugs Not Currently Approved by the US FDA:
(A) Drugs affecting absorption: orlistat
(B) Combined adrenergic and serotonergic mechanisms: sibutramine
(C) Currently experimental:
   (1) Gut peptides: e.g., cholecystokinin, enterostatin, GLP-1
   (2) Gut peptide antagonists: e.g., anti-NPY, anti-galanin
   (3) Growth hormone, growth factors
   (4) Opioid antagonists: e.g., naltrexone
   (5) Thermogenic agents: e.g., BRL 26380A
   (6) Lipid-oxidizing agents: e.g., RO-2-0654
   (7) Adipose tissue proteins: e.g., leptin

[a] Not currently approved for obesity treatment.

CATEGORIES OF OBESITY DRUGS

Obesity drugs currently approved for use in the United States are sparse and fall into two categories: centrally active catecholaminergic agents and centrally active serotonergic agents (TABLE 1). A number of other drugs are not currently approved for obesity or are still experimental (TABLE 1). Obesity drugs that are scheduled by the United States Drug Enforcement Agency (DEA) are presumed to have abuse or addictive potential. The drugs shown in TABLE 1 in DEA schedules II and III are not routinely used or are used only in extremely rare circumstances by responsible physicians. Catecholaminergic drugs in DEA schedule IV include phentermine, diethylpropion, and mazindol. All three of these drugs have minimal addiction or abuse potential, and the popularity of phentermine over the other two is due primarily to the publicity of the Weintraub regimen[14,15] and to high prices of the others. Also, Griffiths et al.[22] demonstrated in nonhuman primates that diethylpropion had somewhat higher reinforcement potential than did phentermine.

The two serotonergic agents approved for obesity treatment, d-fenfluramine and d,l-fenfluramine, are currently scheduled in DEA category IV, but the FDA and DEA have initiated steps to remove them from the scheduled categories as there appears to be no evidence that they produce physiological or psychological addiction. The mechanisms of action of these drugs are to increase the secretion of serotonin in nerve terminals and to block its reuptake. About 15 different serotonin receptors have been identified,[23] and the differences in effectiveness of the fenfluramines versus other serotonergic agonists may be determined by which receptors are bound.

Fluoxetine and sertraline are antidepressant agents that are not approved for obesity, but they have been shown to produce weight loss and have been used by clinicians for obesity. Weight loss with fluoxetine is approximately similar to that produced by d-fenfluramine; however, in contrast to d-fenfluramine, the loss is only temporary.[24] Weight loss reaches a maximum at about 6 months, and then there is a gradual regain so that 1-year loss is not significantly different from placebo.[24]

Sibutramine is an interesting drug that blocks the reuptake of both norepinephrine and serotonin in nerve terminals. Clinical trials show weight losses of about 7–10 kg.[25] Sibutramine is associated with minimal decreases in blood pressure, or a slight rise, despite the weight loss.[25]

Orlistat is a lipase inhibitor that binds to lipase in the intestine, inactivating it.[26,27] Orlistat is expected to have the advantage over adrenergic and serotonergic agents because it acts peripherally and is not expected to have any adverse effects on cardiovascular function. However, the side effects of intestinal gas, cramping, and diarrhea may limit its use by some patients.

Acarbose blocks intestinal amylase and inhibits digestion of complex carbohydrates.[27] Acarbose is approved in the United States for treatment of diabetes, but its effects on obesity are disappointing and it is not an adequate agent used alone. There is speculation that it may prove useful for additional weight loss when used in conjunction with other obesity agents, but there are no published studies.

A host of other potential obesity agents is under investigation. Two of the most-exciting categories are gut peptides and thermogenic agents. Numerous gut peptides have been shown to inhibit food intake and are candidates as obesity drugs. However, as peptides, they may not withstand digestion in the GI tract and may need to be given by injection. Thermogenic agents are thought to bind to beta-3 adrenergic receptors or atypical adrenergic receptors. Studies in animals were very promising until it was shown that their beta-3 receptors are different from human beta-3 receptors and that animal beta-3 agonists are not very effective in humans.

The intense publicity that surrounded the discovery of the gene defect responsible for obesity in ob/ob and db/db mice and in Zucker obese rats has stimulated hope that the gene product, leptin, will be useful for the treatment of obesity.[17-20] Leptin promotes weight loss in ob/ob mice, in normal mice, and in mice made obese by a high-fat diet.[18-20] The pessimism engendered by the findings that most obese humans have elevated levels of leptin[28] does not seem warranted since leptin causes weight loss in dietary-induced obesity, and much human obesity is exacerbated by increased dietary intake.

RESULTS OF DRUG TREATMENT OF OBESITY

Single-Drug Studies

Most studies involving drug treatment of obesity have used single agents and short-term treatment periods, usually 12 weeks or less. Scoville[29] reviewed over 200 studies

TABLE 2. Long-Term Studies with Obesity Drugs[a]

Drug Evaluated	One-Year Weight Loss (kg)		Reference
	Placebo	Obesity Drug	
Diethylpropion	−10.5	−8.9	53
Mazindol	−10.2	−14.2	54
	−	−12.0	55
Fenfluramine	−4.5	−8.7	56
Dexfenfluramine	−7.2	−9.8	31
	−2.7	−5.7	57
	−4.6	−5.2	58
Fluoxetine	+0.6	−13.9	34
	−4.5	−8.2	35
	−1.5	−2.3	24

[a] From reference 36.

of obesity drugs and reported that the average weight loss on drugs was about a half-pound a week greater than on placebo. He concluded that obesity drugs were poorly effective. Silverstone[30] compared several drugs with placebo and noted weight losses that were statistically significantly different with all drugs as compared to placebo, but again the differences were small.

Guy-Grand et al.[31] reported the results of a year-long study of dexfenfluramine in several centers in Europe. They found that dexfenfluramine produced a significantly greater weight loss than placebo at 1 year. Mean weight loss at 1 year was about 10 kg, or about 11% of initial body weight. More than twice as many subjects lost at least 10% of initial body weight on dexfenfluramine as did on placebo. About 60% of the subjects who started the year completed 1 year of drug treatment versus only about 50% of subjects on placebo. Side effects were modest.

Dexfenfluramine has been shown to selectively reduce visceral adipose tissue stores, although such selective reduction is also seen in other forms of weight reduction such as exercise.[32] Dexfenfluramine alters insulin sensitivity, even without weight loss, and thus has been recommended for use in obese patients with non-insulin-dependent diabetes.[33]

Fluoxetine given for 1 year produced modest increases in weight loss compared to placebo in some studies, but not in others.[24,34,35] The studies of Marcus et al.[34] and Darga et al.[35] were the only 2 of 10 that showed significant weight loss at 1 year.[24] Both of these studies had intensive intervention with behavioral techniques. In all studies, weight loss was maximum at about 6 months and weight regain occurred thereafter.

TABLE 2 is adapted from the review by Goldstein and Potvin[36] on long-term drug treatment of obesity and illustrates that all of the obesity drugs produce weight loss at 1 year, although for some agents there is little difference from placebo. Average weight losses at 1 year ranged from 2.3 to 14.2 kg (TABLE 2). Weight losses of 14.2 and 12.0 kg occurred with mazindol. Fluoxetine produced weight losses of 2.3 to 13.9 kg, but as noted above only 2 of 10 studies produced significant weight loss at 1 year compared to placebo. Fenfluramine and dexfenfluramine produced weight losses of 5.2 to 9.8 kg.

Many of the studies reviewed by Goldstein and Potvin[36] demonstrated quite good weight loss in the placebo groups. In these studies, both experimental and placebo groups underwent education in diet, exercise, and behavioral modification of lifestyle. The addition of these interventions has been shown to wash out the effects of an experimental perturbation;[37] thus, most of these studies are not a pure test of drug versus placebo.

Combination Drug Treatment of Obesity

The number of trials with combinations of obesity drugs is small. The only combinations of drugs that have been reported in full publications are phenylpropanolamine-benzocaine, ephedrine-methylxanthines-aspirin, and fenfluramine-phentermine.[14,15,38-41] The use of fluoxetine-phentermine for obesity has been reported in abstract form only. The experience with these combinations is described below.

First, phenylpropanolamine-benzocaine: There is only one report of the combination of phenylpropanolamine (PPA) and benzocaine.[38] PPA is an adrenergic agent with reportedly no reinforcement potential.[22] Used alone, it produces modest weight loss.[38] Benzocaine is a local anesthetic agent contained in some over-the-counter weight-reduction aids. The rationale for its use is to anesthetize the taste buds and thus reduce food intake. This combination produced similar weight loss as placebo in an 8-week trial, and Greenway[38] concluded that it is ineffective.

Second, ephedrine-caffeine-aspirin combinations: Ephedrine is an adrenergic agent that enhances norepinephrine secretion in nerve terminals.[40-44] Data in animals and to a limited extent in humans suggest that ephedrine administration results in stimulation of beta-1, -2, and -3 receptors.[42] Stimulation of beta-1 and -2 receptors may increase heart rate and blood pressure, but tachyphylaxis occurs and these responses return to baseline within the first month of treatment. There appears to be little or no tachyphylaxis of beta-3 receptor stimulation and this results in enhanced thermogenesis that persists.[42,44] Methylxanthines (caffeine, theophylline, theobromine, etc.) and aspirin reduce the metabolism of norepinephrine and thus potentiate the effects on thermogenesis.

Enhanced beta-3 stimulation produces lipolysis and increased lean body mass, and the combination of ephedrine and caffeine has been reported to reduce fat mass and increase lean body mass, even with dietary restriction and weight loss.[40-44]

Toubro *et al.*[40] compared placebo, ephedrine alone, caffeine alone, and the combination of ephedrine and caffeine over a period of 24 weeks in 180 subjects. The combination of ephedrine and caffeine produced weight loss of about 16 kg at 24 weeks. Of the initial 180 subjects, 99 were followed for another 26 weeks in an open-label study. Weight loss persisted for as long as the drugs were given. Daly *et al.*[41] treated 6 subjects in an open-label study for periods of up to 26 months and noted a persistent, modest weight loss.

Third, fenfluramine-phentermine combinations: This combination of a serotonin agonist and an adrenergic agonist was first reported in 1984 by Weintraub *et al.*[15] This group compared full doses of fenfluramine (120 mg/day) or phentermine (30 mg/day) alone with the combination of these two agents in half-strength each (60 mg/day and 15 mg/day, respectively). Weight loss was similar in all groups; however, full doses of the drugs produced side effects significantly greater than placebo, whereas side effects with the combination were not different from placebo. In a follow-up study, Weintraub[14] treated 121 subjects with the combination. All subjects received diet, exercise, and behavior modification, and subjects were randomized to experimental drugs or placebos for the first 34 weeks. Weight loss reached a plateau at about 25 weeks. All patients were then treated with the combination and, at 60 weeks, weight loss continued at 15.8 kg. Of note was the fact that subjects who lost little weight during the first 34 weeks had an increase in dose to full levels of each, but this did not increase weight loss. This suggests that responders and nonresponders may be biochemically different and that, if patients fail to lose weight initially, continued treatment is not indicated.

Weintraub[14] continued the trial for 3.5 years, with periods of intermittent treatment for some and with randomization to drugs versus placebo late in the study. When subjects were cycled off drugs or were randomized to placebo, they gained weight. When drugs were reinstituted, the subjects lost weight. Subjects were followed for 6 months off drugs at the end of the study and virtually all returned back to or near their baseline weights.

Side effects included dry mouth, fatigue, sweating, insomnia, and other sleep disturbances. In general, the side effects were tolerable and few subjects discontinued drugs for this reason. There were significant decreases in blood pressure at 34 weeks in this protocol, and serum lipid levels were assessed at 2 years. Serum cholesterol and triglycerides decreased by 34 weeks and continued to be lower at 2 years.

We[39] treated over 1300 subjects with the Weintraub regimen for periods of up to 2 years. Weight losses of about 16 kg persisted for 2 years in subjects who continued on drugs. In patients with hypertension (systolic blood pressure $\geq$ 140 mmHg or diastolic BP $\geq$ 90 mmHg), systolic BP fell by 28 mmHg and diastolic BP fell by 17 mmHg. In patients with hypercholesterolemia (serum cholesterol $\geq$ 200 mg/dL) or hypertriglyceridemia (serum triglycerides $\geq$ 150 mg/dL), cholesterol levels decreased by 27 mg/dL and triglycerides decreased by 79 mg/dL, respectively.

The dropout rate of this series of patients was 39% at 1 year and 58% at 2 years. A retrospective control group, treated only with diet, exercise, and behavior modification of lifestyle, was randomly selected from patients who had been treated in the obesity program before drugs were used. The dropout rate of the control group was 92% at 1 year.[39]

Side effects reported by this group were somewhat more disturbing than in the study of Weintraub.[14] Dry mouth was the most-frequent complaint, occurring in over 60% of patients. Fatigue, constipation, and sleep disturbances, including insomnia and vivid dreams, occurred in 10–20% of patients. Short-term memory loss, an adverse event not reported by Weintraub,[14] occurred in about 13% of patients. Seven patients discontinued the drugs or were taken off the drugs because of short-term memory loss or decreased mental acuity. Function returned to normal in all subjects. Weintraub noted in oral presentations that his subjects also had reported decreased mental acuity, but testing during a period of randomization to drugs or placebo revealed no differences (personal communication, Michael Weintraub). Our studies[39] were open label, so there was no control group. Patients were given a follow-up sheet with a list of complaints, including short-term memory loss, so the frequency with which all side effects or adverse events were reported was likely to be higher. However, we were convinced that changes in mental acuity do occur as a side effect of this combination of drugs. Randomized, placebo-controlled studies with sensitive measures of mental acuity are needed to resolve this question and to determine the true incidence of these changes.

Dhurandhar et al.[45] reported preliminary studies in abstract form demonstrating that age does not affect the outcome of treatment and that African-Americans and Caucasians respond similarly to the combination of fenfluramine and phentermine.[46] Patients aged 50–60 years lose similar amounts of weight and have similar changes in blood pressure and lipids as do patients aged 20–30 years.[45] Side effects also were similar. Blacks lost similar amounts of weight as Caucasians, although there was a trend towards lower weight loss.[46]

Heuberger et al.[47] reported in abstract form that treatment with fenfluramine and phentermine following a period of very low calorie dieting (VLCD) resulted in continued weight loss after 1 year of treatment. The dropout rate at 1 year was 23%, and 56% remained below their weight at the end of the VLCD.

Hartley et al.[48] used doses of fenfluramine up to 120 mg/day in combination with doses of phentermine resin up to 30 mg/day and found weight losses of about 16 kg. However, the frequency of side effects was greater than that found by Weintraub et al.[14,15] and by our studies.[39]

Fourth, and last, fluoxetine-phentermine: Dhurandhar et al.[49] reported in abstract form that the combination of fluoxetine (20–60 mg/day) and phentermine hydrochloride (18.75–37.5 mg/day) produced significant weight losses that were similar to those produced by the combination of fenfluramine (20–60 mg/day) and phentermine HCl (18.75–37.5 mg/day). This study extended only to 6 months, so it is not possible to determine if the weight regain that occurs with fluoxetine after 6 months will occur with the combination.

PROBLEMS AND CONCERNS ABOUT DRUG TREATMENT OF OBESITY

With the exception of dexfenfluramine, which was approved by the US FDA in 1996, all of the obesity drugs available in the United States and most of the rest of the world were developed by 1970 and approved by the US FDA by 1973. All of these drugs were approved for use for only 12 weeks as it was felt that this would limit the addiction potential and that patients needed only temporary help in learning how to eat properly to maintain weight loss. This resulted in very few studies that were conducted for longer than 12 weeks. In their comprehensive review, Goldstein and Potvin[36] were able to find only nine studies that went as long as a year. This lack of experience, coupled with several disturbing findings in the medical literature described below, is the basis for warnings to physicians to use great caution in their use of obesity drugs for the long term, and to limit these drugs only to patients with medically significant obesity. Findings in the literature that require additional research include the following:

(1) Changes in mental acuity and short-term memory loss: These symptoms have been reported in the literature predominantly as anecdotes or in uncontrolled studies.[39,47] However, some of the anecdotes are compelling and the frequency with which they are reported suggests that they have some basis in fact. Studies in animals raise questions about changes in central nervous system (CNS) biochemistry. Ricaurte et al.[50,51] reported that squirrel monkeys and rodents treated with doses of dexfenfluramine from 4 to 40 times the postulated comparable human dose had depletion of serotonin in specific CNS areas as long as 18 months after cessation of drug use. Human studies reported to the FDA by Interneuron, Incorporated, suggest that, with doses of dexfenfluramine typically used by humans, the brain levels of dexfenfluramine as assessed by CT scan are much lower than those seen in animals treated by Ricaurte et al.[50,51] In addition, Interneuron reported that the metabolites of dexfenfluramine in humans differ from those of animals, and the metabolites may also appear in the brain. The monkeys and rodents described by Ricaurte et al.[50,51] did not demonstrate any changes in function, including tests of memory and mental acuity.

(2) Primary pulmonary hypertension: Primary pulmonary hypertension (PPH) has been reported in patients treated with d-fenfluramine and other obesity drugs in a case-control study.[52] Primary pulmonary hypertension is said to be more common in obese people, so the reference group is not clear; however, it cannot be the general population. It is clear that restrictive pulmonary disease and sleep apnea, which may cause pulmonary hypertension, are more common in obese people and this may confound the interpretation of studies in the literature. In the 1960s, aminorex fumarate, a drug used for weight control in Europe, was reported to increase the risk of PPH.[52] Because there is a precedent for an obesity drug to produce PPH, there is concern that current drugs may have the same side effect. However, the history of PPH due to an obesity drug may produce a greater awareness on the part of physicians when symptoms suggestive of PPH occur, thus biasing the reporting of the disease. This may result in overreporting of PPH with obesity drugs and may give a false impression of the prevalence and the risk. Warnings by the FDA and by experts in the field suggest that patients should be followed very carefully, with a high degree of suspicion regarding any pulmonary or cardiac complaints. The actual frequency is not known, but estimates of the risk of PPH in people on obesity drugs range from 1 per 500,000 to 1 per 20,000.

CONCLUSIONS

Since obesity is a chronic disease, long-term drug treatment almost certainly will be needed to control its progression and to prevent the appearance of complications such as diabetes, cardiac disease, strokes, etc. Short-term studies suggest that obesity drugs produce moderate weight loss and reduction of risk factors associated with obesity and that these effects persist for 2 to 3 years. At this time, side effects appear to be modest, but there are insufficient data to assume that treatment with these drugs for the long term is safe or efficacious. Two areas of major concern are the possibility that obesity drugs may produce decreases in mental acuity, presumably from biochemical changes in the CNS, and primary pulmonary hypertension, which has been linked to obesity drugs by retrospective studies or anecdotal reports that have not been or cannot be adequately evaluated with the current data. More research is needed to establish safety and long-term efficacy and, until such studies are done, physicians must provide very careful follow-up and continuous assessment for the appearance of side effects. Most obesity experts agree that obesity drugs are an adjunct to diet, increased activity, and a healthier lifestyle. Obesity drugs must be used with caution and only in patients with medically significant obesity.

REFERENCES

1. KUCZMARSKI, R. J., K. M. FLEGAL, S. M. CAMPBELL & C. L. JOHNSON. 1994. Increasing prevalence of overweight among US adults. JAMA 272: 205-211.
2. BRAY, G. A. 1976. The risks and disadvantages of obesity. In The Obese Patient, p. 215-251. Saunders. Philadelphia.
3. GRUNDY, S. M. & J. P. BARNETT. 1990. Metabolic and health complications of obesity. Dis. Mon. 36(12): 641-731.
4. SIMOPOULOS, A. P. & T. B. VAN ITALLIE. 1984. Body weight, health, and longevity. Ann. Intern. Med. 100: 285-295.
5. PI-SUNYER, F. X. 1993. Medical hazards of obesity. Ann. Intern. Med. 119: 655-660.
6. BOUCHARD, C. & L. PERUSSE. 1993. Genetics of obesity. Annu. Rev. Nutr. 13: 337-354.
7. BOUCHARD, C. & L. PERUSSE. 1996. Current status of the human obesity gene map. Obes. Res. 4: 81-90.
8. ANDERSEN, T., K. H. STOKHOLM, O. G. BACKER & F. QUAADE. 1988. Long-term (5-year) results after either horizontal gastroplasty or very-low calorie diet for morbid obesity. Int. J. Obes. 12: 277-284.
9. KRAMER, F. M., R. W. JEFFERY, J. L. FORSTER & M. K. SNELL. 1989. Long-term follow-up of behavioural treatment for obesity: patterns of regain among men and women. Int. J. Obes. 13: 123-136.
10. WADDEN, T. A., J. A. STERNBERG, K. A. LETIZIA, A. J. STUNKARD & G. D. FOSTER. 1989. Treatment of obesity by very low calorie diet, behavior therapy, and their combination: a five-year perspective. Int. J. Obes. 13(suppl. 2): 39-46.
11. PERRI, M. G. 1992. Improving maintenance of weight loss following treatment by diet and lifestyle modification. In Treatment of the Seriously Obese Patient, p. 456-477. Guilford Press. New York.
12. WILSON, G. T. 1993. Behavioral treatment of obesity: thirty years and counting. Adv. Behav. Res. Ther. 16: 31-75.
13. BRAY, G. A. 1991. Barriers to the treatment of obesity. Ann. Intern. Med. 115(2): 152-153.
14. WEINTRAUB, M. 1992. Long-term weight control: the National Heart, Lung, and Blood Institute funded multimodal intervention study. Clin. Pharmacol. Ther. 51: 581-646.
15. WEINTRAUB, M., J. D. HASDAY, A. I. MUSHLIN & D. H. LOCKWOOD. 1984. A double-blind clinical trial in weight control: use of fenfluramine and phentermine alone and in combination. Arch. Intern. Med. 144: 1143-1148.
16. COLEMAN, D. L. 1978. Genetics of obesity in rodents. In Recent Advances in Obesity Research: II, p. 142-152. Newman. London.
17. ZHANG, Y., R. PROENCA, M. MAFFEI, M. BARONE, L. LEOPOLD & J. M. FRIEDMAN. 1994. Positional cloning of the mouse obese gene and its human homologue. Nature 372: 425-431.

18. PELLEYMOUNTER, M. A., M. J. CULLEN, M. B. BAKER, R. HECHT, D. WINTERS, T. BOONE & F. COLLINS. 1995. Effects of the *obese* gene product on body weight regulation in ob/ob mice. Science **269:** 540-543.
19. HALAAS, J. L., K. S. GAJIWALA, M. MAFFEI, S. L. COHEN, B. T. CHAIT, D. RABINOWITZ, R. L. LALLONE, S. K. BURLEY & J. M. FRIEDMAN. 1995. Weight-reducing effects of the plasma protein encoded by the *obese* gene. Science **269:** 543-546.
20. CAMPFIELD, L. A., F. J. SMITH, Y. GUISEZ, R. DEVOS & P. BURN. 1995. Recombinant mouse OB protein: evidence for a peripheral signal linking adiposity and central neural networks. Science **269:** 546-549.
21. LEVITSKY, D. A. & R. TROIANO. 1992. Metabolic consequences of fenfluramine for the control of body weight. Am. J. Clin. Nutr. **55:** 167S-172S.
22. GRIFFITHS, R. R., J. V. BRADY & L. D. BRADFORD. 1979. Predicting the abuse liability of drugs with animal drug self-administration procedures: psychomotor stimulants and hallucinogens. Adv. Behav. Pharmacol. **2:** 163-208.
23. BAEZ, M., J. D. KURSAR, L. A. HELTON, D. B. WAINSCOTT & D. L. G. NELSON. 1995. Molecular biology of serotonin receptors. Obes. Res. **3**(suppl. 4): 441S-447S.
24. GOLDSTEIN, D. J., A. H. RAMPEY, JR., B. E. DORNSEIF, L. R. LEVINE, J. H. POTVIN & L. A. FLUDZINSKI. 1993. Fluoxetine: a randomized clinical trial in the maintenance of weight loss. Obes. Res. **1:** 92-98.
25. RYAN, D. H., P. KAISER & G. A. BRAY. 1995. Sibutramine: a novel new agent for obesity treatment. Obes. Res. **3**(suppl. 4): 553S-559S.
26. DRENT, M. L. & E. A. VAN DER VEEN. 1995. First clinical studies with orlistat: a short review. Obes. Res. **3**(suppl. 4): 623S-625S.
27. BERGER, M. 1992. Pharmacological treatment of obesity: digestion and absorption inhibitors—clinical perspective. Am. J. Clin. Nutr. **55:** 318S-319S.
28. CONSIDINE, R. V., M. K. SINHA, M. L. HEIMAN, A. KRIAUCIUNAS, T. W. STEPHENS, M. R. NYCE, J. P. OHANNESIAN, C. C. MARCO, L. J. MCKEE, T. L. BAUER & J. F. CARO. 1996. Serum immunoreactive-leptin concentrations in normal-weight and obese humans. N. Engl. J. Med. **334:** 292-295.
29. SCOVILLE, B. A. 1976. Review of amphetamine-like drugs by the Food and Drug Administration. *In* Obesity in Perspective. Fogarty International Center for Advanced Studies in the Health Sciences, Series on Preventive Medicine, Vol. II, p. 441-443. U.S. Govt. Printing Office. Washington, District of Columbia.
30. SILVERSTONE, T. 1992. Appetite suppressants: a review. Drugs **43:** 820-836.
31. GUY-GRAND, B., M. APFELBAUM, G. CREPALDI, A. GRIES, P. LEFEBVRE & P. TURNER. 1989. International trial of long-term dexfenfluramine in obesity. Lancet **2:** 1142-1145.
32. MARKS, S. J., N. R. MOORE, M. L. CLARK, B. J. G. STRAUSS & T. D. R. HOCKADAY. 1996. Reduction of visceral adipose tissue and improvement of metabolic indices: effect of dexfenfluramine in NIDDM. Obes. Res. **4:** 1-7.
33. WILLEY, K. A., L. M. MOLYNEAUX, J. E. OVERLAND & D. K. YUE. 1992. The effects of dexfenfluramine on blood glucose control in patients with type 2 diabetes. Diabetes Med. **9:** 341-343.
34. MARCUS, M. D., R. R. WING, L. EWING, E. KERN, M. MCDERMOTT & W. GOODING. 1990. A double-blind, placebo-controlled trial of fluoxetine plus behavior modification in the treatment of obese binge-eaters and non-binge-eaters. Am. J. Psychiatry **147:** 876-881.
35. DARGA, L. L., L. CARROLL-MICHALS, S. J. BOTSFORD & C. P. LUCAS. 1991. Fluoxetine's effect on weight loss in obese subjects. Am. J. Clin. Nutr. **54:** 321-325.
36. GOLDSTEIN, D. J. & J. H. POTVIN. 1994. Long-term weight loss: the effect of pharmacologic agents. Am. J. Clin. Nutr. **60:** 647-657.
37. ATKINSON, R. L., F. L. GREENWAY, G. A. BRAY, W. T. DAHMS, M. E. MOLITCH, K. HAMILTON & J. RODIN. 1977. Treatment of obesity: comparison of physician and nonphysician therapists using placebo and anorectic drugs in a double-blind trial. Int. J. Obes. **1:** 113-120.
38. GREENWAY, F. L. 1992. Clinical studies with phenylpropanolamine: a meta-analysis. Am. J. Clin. Nutr. **55**(suppl. 1): 203S-205S.
39. ATKINSON, R. L., R. C. BLANK, J. F. LOPER, D. SCHUMACHER & R. A. LUTES. 1995. Combined treatment of obesity. Obes. Res. **3**(suppl. 4): 497S-500S.
40. TOUBRO, S., A. V. ASTRUP, L. BREUM & F. QUAADE. 1993. Safety and efficacy of long-term treatment with ephedrine, caffeine, and an ephedrine/caffeine mixture. Int. J. Obes. **17**(suppl. 1): S69-S72.

41. DALY, P. A., D. R. KRIEGER, A. G. DULLOO, J. B. YOUNG & L. LANDSBERG. 1993. Ephedrine, caffeine, and aspirin: safety and efficacy for treatment of human obesity. Int. J. Obes. 17(suppl. 1): S73–S78.
42. DULLOO, A. G. 1993. Ephedrine, xanthines, and prostaglandin-inhibitors: actions and interactions in the stimulation of thermogenesis. Int. J. Obes. 17(suppl. 1): S35–S40.
43. ARNER, P. 1993. Adenosine, prostaglandins, and phosphodiesterase as targets for obesity pharmacotherapy. Int. J. Obes. 17(suppl. 1): S57–S59.
44. LIU, Y. L., S. TOUBRO, A. ASTRUP & M. J. STOCK. 1995. Contributions of β_3-adrenoceptor activation to ephedrine-induced thermogenesis in humans. Int. J. Obes. 19: 678–685.
45. DHURANDHAR, N. V., R. C. BLANK, D. SCHUMACHER, D. L. RITCH, E. CHAN, T. S. REIG & R. L. ATKINSON. 1995. Combination drug treatment of obesity in women of different ages. Obes. Res. 3(suppl. 3): 341S.
46. DHURANDHAR, N. V., R. C. BLANK, D. SCHUMACHER, D. L. RITCH, E. CHAN, T. S. RIEG & R. L. ATKINSON. 1995. Racial differences in response to combination drug treatment of obesity. Obes. Res. 3(suppl. 3): 405S.
47. HEUBERGER, R. A., J. LOPER, E. J. BALTES, R. LUTES, R. K. MAY, J. T. BROYLES & R. L. ATKINSON. 1995. VLCD followed by combination obesity drugs: results at one year. Obes. Res. 3(suppl. 3): 405S.
48. HARTLEY, G. G., S. NICOL, C. HALSTENSON, M. KHAN & A. PHELEY. 1995. Phentermine, fenfluramine, diet, behavior modification, and exercise for treatment of obesity. Obes. Res. 3(suppl. 3): 340S.
49. DHURANDHAR, N. V. & R. L. ATKINSON. 1996. Comparison of serotonin agonists in combination with phentermine for treatment of obesity. FASEB J. 10: A561.
50. RICAURTE, G. A., M. E. MOLLIVER, M. B. MARTELLO, J. L. KATZ, M. A. WILSON & A. L. MARTELLO. 1991. Dexfenfluramine neurotoxicity in brains of non-human primates. Lancet 338: 1487–1488.
51. MCCANN, U., G. HATZIDIMITRIOU, A. RIDENOUR, C. FISCHER, J. YUAN, J. KATZ & G. RICAURTE. 1994. Dexfenfluramine and serotonin neurotoxicity: further preclinical evidence that clinical caution is indicated. Clin. Pharmacol. Exp. Ther. 269: 792–798.
52. BRENOT, F., P. HERVE, P. PETITPRETZ, F. PARENT, P. DUROUX & G. SIMONNEAU. 1993. Primary pulmonary hypertension and fenfluramine use. Br. Heart J. 70: 537–541.
53. SILVERSTONE, J. T. & T. SOLOMON. 1965. The long-term management of obesity in general practice. Br. J. Clin. Pract. 19: 395–398.
54. ENZI, G., A. BARITUSSIO, E. MARCHIORI & G. CREPALDI. 1976. Short-term and long-term clinical evaluation of a non-amphetamine anorexiant (mazindol) in the treatment of obesity. J. Int. Med. Res. 4: 305–317.
55. INOUE, S., M. EGAWA, S. SATOH, M. SAITO, H. SUZUKI, Y. KUMAHARA, M. ABE, A. KUMAGAI, Y. GOTO, K. SHIZUME, N. SHIMIZU, C. NAITO & T. ONISHI. 1992. Clinical and basic aspects of an anorexiant, mazindol, as an antiobesity agent in Japan. Am. J. Clin. Nutr. 55: 199S–202S.
56. HUDSON, K. D. 1977. The anorectic and hypotensive effect of fenfluramine in obesity. J. R. Coll. Gen. Pract. 27: 497–501.
57. TAUBER-LASSEN, E., P. DAMSBO, J. E. HENRIKSEN, B. PALMVIG & H. BECK-NIELSEN. 1990. Improvement of glycemic control and weight loss in type 2 (non-insulin-dependent) diabetics after one year of dexfenfluramine treatment. Diabetologia 33(suppl.): A124.
58. MATHUS-VLIEGEN, E. M. H., K. VAN DE VOORE, A. M. E. KOK & A. A. RES. 1992. Dexfenfluramine in the treatment of severe obesity: a placebo-controlled investigation of the effects on weight loss, cardiovascular risk factors, food intake, and eating behavior. J. Intern. Med. 232: 119–127.

Fat Substitutes and Energy Balance

JOHN C. PETERS[a]

The Procter and Gamble Company
Cincinnati, Ohio 45224

Despite having an increased awareness of the health risks associated with excessive fat ingestion and obesity, the average American has not reduced fat intake to meet recommended dietary levels. Dietary fat is particularly conducive to weight gain because it increases the caloric density of food (which encourages passive overconsumption), is not immediately burned after a meal, and is substantially oxidized only after carbohydrate and protein are used as fuel. This lower metabolic priority for fat utilization may underlie the observation in some studies that fat is less satiating than other macronutrients. Because humans tend to eat for volume in the short term and there is a weak relationship between fat intake and utilization, many subjects consuming low-fat diets do not increase food intake sufficiently to compensate for the loss of fat calories. Therefore, many studies have found that subjects maintained on a low-fat/high-carbohydrate diet lose weight. Based on evidence that low-fat diets are beneficial in controlling caloric intake and body weight, a number of dietary fat substitutes have been developed to help consumers reduce fat and calorie intakes. These fat substitutes may be classified as carbohydrate (noncaloric or caloric)-, protein-, fat-, or synthetic fat (noncaloric or caloric)-based.

One of the most-extensively studied and well known of the synthetic noncaloric fat substitutes is olestra. Olestra is neither digested nor absorbed; thus, it contributes no energy to the diet. Results of numerous studies in animals and humans indicate that it is not toxic and produces no harmful side effects. The effect of olestra on energy intake and diet selection has been studied in lean and obese adults and in children. Replacement of fat with olestra consistently reduces dietary fat intake and leads to reduced total energy intake in some cases. In subjects who compensate for the replacement of fat with olestra by increasing their energy intake, there is no evidence of "fat-specific" compensation. Collectively, these data indicate that substituting dietary fat with olestra is effective at reducing fat intake even in subjects who compensate. Therefore, long-term use of foods containing fat substitutes such as olestra in lieu of fat may be beneficial in helping people control energy intake and body weight, and thus reduce the incidence of obesity.

---

[a] J. C. Peters is an employee of the Procter and Gamble Company, producers of food and beverage products, including the fat substitute, olestra.

FAT IN THE DIET

In recent years, it has been recognized that high intakes of dietary fat are associated with an increased incidence of a variety of chronic diseases or conditions, including cardiovascular disease, cancer, diabetes, and obesity. The most-recent national health survey shows fully one-third of American adults are obese, as are over 20% of children 12-19 years of age.[1] The awareness of these associations has led to recommendations that dietary fat intake should be reduced to no more than 30% of dietary energy.[2-5] However, despite this scientific and public-health consensus that dietary fat reduction is a desirable goal and despite considerable effort to inform the public, fat intake remains high. Dietary fat provides approximately 34-37% of the energy intake in the average American diet and the amount of fat (in grams) consumed daily has not changed in the last 20 years.[6,7] How does this high-fat intake affect energy balance and body weight from an individual and population perspective?

FAT PROMOTES POSITIVE ENERGY BALANCE

Review of the available research led the 1990 WHO Report on Diet, Nutrition, and the Prevention of Chronic Diseases[8] to state that "there is increasing evidence from experimental animal studies, human physiological measures of energy metabolism, and bioenergetic considerations that dietary fat is particularly conducive to weight gain." An extensive review of laboratory studies[9] has shown that humans or animals, when allowed to eat *ad libitum* on high-fat diets, consume similar amounts of food, but have more energy intake and accompanying weight gain than when fed on lower fat diets. This association of dietary fat and positive energy balance is also supported in a recent review of epidemiologic studies.[10] Most of the secular and cross-sectional studies reviewed concluded that the percentage of fat in the diet correlated positively with energy intake and obesity.

Why does fat promote positive energy balance and consequent obesity? First and foremost, it increases the caloric density of food. Studies have shown that people tend to eat for volume,[11] at least in the short term. Thus, if a food has a high-caloric density, people will tend to "passively" overconsume energy as they select and eat a standard volume. This concept of "passive" overconsumption is probably critical to the rising prevalence of obesity. It is unlikely that people are suddenly consciously increasing the amount of food consumed; rather, with increasingly busy lifestyles, the amount of food consumed outside the home is increasing and foods served in the home are more likely to be prepackaged convenience-type foods. Both outside and inside the home, these convenience foods have a high likelihood of being high-fat and calorically dense. Of course, all this is occurring in the presence of decreased physical activity as well.

Data from Lissner *et al.*[12] demonstrate the effect of caloric density in a group of young women allowed to eat *ad libitum* diets having low-, medium-, or high-energy density achieved by modifying the dietary fat content. Subjects overate energy when only an assortment of high-fat foods was provided for three weeks and underate (relative to the medium-fat diet) when only low-fat foods were available. This suggests that the body does not monitor fat intake precisely. If you modify the fat content in food, energy intake tends to follow along with it.

Increasing the caloric density of foods and thus encouraging "passive" overconsumption is certainly not the only mechanism by which dietary fat contributes to positive energy balance. Both animal[13] and human[14-16] studies show that, even at similar energy intakes, more body fat is gained when high-fat diets are consumed than when low-fat, high-carbohydrate diets are eaten. Fat is the fuel that the body stores most efficiently. When extra fat is ingested, very little energy needs to be expended in putting it into fat stores. In contrast, if carbohydrate and protein are eaten in excess, the body needs to expend significant energy to convert them into storage fat. Furthermore, fat oxidation is not tied to fat intake. Excess carbohydrate or protein that is eaten is burned almost immediately because the body has a limited capacity for storage of these nutrients. Fat, however, has essentially a limitless storage pool. Immediately after a mixed meal containing protein, carbohydrate, and fat, most of the fat ingested is stored, while the postingestive fuel mixture burned is dominated by carbohydrate and protein oxidation.[17] The weak relationship between fat burned in the postprandial period and fat ingested means that in order for individuals to maintain fat balance (i.e., not gain or lose body fat) the fat stored right after the meal must be oxidized at some later time. Several hours after a meal, when the exogenous fuel (protein and carbohydrate from the meal) has been metabolized, the body is faced with a choice: either burn the fat (and glycogen) that was stored right after the meal or go and get something more to eat. It would appear that all too often the latter path is chosen, which leads to positive energy balance and the inevitable consequence, obesity.

One final means by which fat contributes to overconsumption is by its effect, or comparative lack of effect, on satiation (bringing a "meal" or "episode of eating" to a close) and satiety (inhibition of further consumption). Some studies have concluded that, calorie for calorie, fat may be less satiating than other macronutrients. However, studies examining the satiating effect of fat are confounded by the difficulty of separating the effects of dietary fat, per se, from those of energy density. Despite this, meals with a high fat:carbohydrate ratio suppress hunger less than meals with a low fat:carbohydrate ratio.[18] Intravenous infusion of a fixed amount of energy as fat suppressed subsequent oral energy intake less than equicaloric infusion of amino acids and carbohydrate.[19]

LOW-FAT DIETS DECREASE ENERGY INTAKE

Given these mechanisms by which dietary fat contributes to positive energy balance, can low-fat diets lead to reduced energy intake and zero (or negative) energy balance? Based on the observation of a relatively weak relationship between the consumption of fat and its subsequent oxidation, one might hypothesize that reducing fat intake would not elicit potent compensatory mechanisms designed to maintain a given level of fat intake. Studies with low-fat diets support this hypothesis.

There is evidence that the compensation for a reduction in dietary fat energy is incomplete in normal-weight subjects[20] and virtually absent in overweight subjects.[12]

A recent 12-week study with Danish workers showed that intervention at the work site with an *ad libitum* low-fat/high-carbohydrate diet caused decreases in weight among both normal-weight and overweight subjects, but greater losses for the overweight subjects.[21] Most encouragingly, the weight loss was not regained up to a year after the end of the intervention.

The more-marked effect on overweight subjects is consistent with observations comparing sedentary and more-active people. Dietary fat reduction appears to show the greatest benefit in people with low-activity levels.[22,23] In two parallel studies, subjects were fed *ad libitum* low-, medium-, and high-fat diets. In one study, the six male subjects were in a whole body calorimeter for one week and therefore very sedentary. In the other study, the subjects were resident in a hotel metabolic suite, which allowed greater activity. While energy intake increased with the percentage of dietary fat, it was virtually identical between the two activity levels. Average energy expenditure was 2.8 MJ/day higher in the active group. Subjects did not increase energy intake to compensate for the increased energy expenditure. Conversely, subjects in the calorimeter did not decrease energy intake when their energy expenditure was low. Thus, there was a more-negative energy balance for the active group at each fat level. The contrast in energy balance between the two groups was marked. For the sedentary group, even at the moderate fat level in the diet, positive energy balance was observed. Thus, when physical activity is limited, even at moderate fat intakes a reduction in dietary fat is needed to prevent weight gain.

Given that low-fat diets appear to be at least one strategy to approach the problem of increasing obesity in the population, how can we implement this strategy? The evidence shows that there is a strong effect of environment (e.g., diet, physical activity) on the stable body weight that is reached by different individuals. About 50–75% of the variance in body weight is apparently due to environmental influences or environmental interaction with genotype.[24] Using fat substitutes may help create an environment where gaining weight is more difficult.

Dietary fat substitutes represent one tool that individuals may use to help limit excess fat and energy intake in populations exposed to an abundance of high-fat/energy-dense food.

FAT SUBSTITUTE TECHNOLOGIES

A variety of fat substitutes have been developed. They may be organized according to their structural base. An exhaustive list of these fat substitutes, along with a discussion of the mechanisms involved in gaining regulatory approval for their use, is provided in the 1996 edition of *Present Knowledge in Nutrition*.[25]

Carbohydrate-based Fat Substitutes

(a) Noncaloric—These fat substitutes are not digestible, being based on cellulose, seaweed, and gums. Examples are carrageenan, cellulose gel, guar gum, and

xanthan gum. The soluble fibers may provide some energy via absorption of products of colonic fermentation.
(b) Caloric—Dextrins and modified starches constitute most of this category. These fat substitutes are digestible, yielding an energy value of about 4 kcal/g. However, since most of these products are used in a hydrated form, their caloric contribution as fat replacers is lowered. Amalean®, Leanmaker®, Litese®, Maltrin®, N-oil®, Oatrim®, and Sta-Slim® are some examples of commercially available carbohydrate-based fat substitutes based on corn starch, oat bran, polydextrose, maltodextrin, starch, oat dextrin, and potato starch, respectively.

The carbohydrate-based fat substitutes can replace all or part of the fat in salad dressings, frozen desserts, chips, spreads, baked goods, meat products, frostings, and soups, but cannot be used for frying. At high temperatures, the water that hydrates these ingredients is driven off, destroying their fat-mimetic functionality.

Protein-based Fat Substitutes

Protein microparticulation techniques produce 1–1.5-µm micelles that give the slippery and smooth-mouth feel associated with fat. These fat substitutes cannot be used in cooking or frying because the protein denatures at these temperatures. The fat substitute in this category that has received the most attention is Simplesse® (NutraSweet Company), a product formulated from milk and egg albumin protein. As a fat replacer in ice cream, 1 g of Simplesse® (4 kcal) can replace 3 g of fat (27 kcal). Other protein-based fat substitutes developed for use in frozen desserts and dairy products include Dairylight® and Dairy-lo® (whey), Enrich 301® (milk), Lita® (zein), and Ultra-freeze® (egg white and milk).

Fat-based Fat Substitutes

These are true triacylglycerols formulated to yield <9 kcal/g. Commercial examples are Caprenin® (Procter & Gamble) and Salatrim® (Nabisco). Caprenin® is a triglyceride containing caprylic, capric, and behenic acids. As behenic acid is poorly absorbed, this triglyceride provides only 5 kcal/g rather than 9 kcal/g. Caprenin® is designed as a cocoa butter substitute in confections.

Salatrim® is a group of structured triglycerides containing short- and long-chain fatty acids. The short-chain fatty acids can be acetic, propionic, and butyric acids, and the long-chain fatty acid is predominately stearic acid. The short-chain fatty acids provide fewer calories and stearic acid and the longer-chain saturated fatty acids are poorly absorbed, yielding the reduced caloric value of 5 kcal/g. Its commercial application is currently in baked and filled dairy products.

Synthetic Fat Substitutes

These fat substitutes are physically similar to fats and oils and can theoretically replace fat on a one-to-one weight basis in foods. They are heat-stable and noncaloric or low-calorie:

(a) Noncaloric—One of the most-extensively studied and well known of the synthetic noncaloric fat substitutes is olestra, a specific composition of lipids from the category of molecules referred to as sucrose polyesters (SPE). Olestra is the common name for the mixture of hexa-, hepta-, and octa-esters formed from the reaction of fatty acids with sucrose. Although olestra has physical and organoleptic properties similar to conventional triglycerides, it is not hydrolyzed by gastric or pancreatic lipases.[26] As a result, olestra is neither digested nor absorbed, and it contributes no energy to the diet. Olestra has physical and sensory properties identical to regular fat, making it possible to examine the effects of a pure fat manipulation on food intake and selection. Fat-derived energy can be varied while preserving the identical bulk, appearance, texture, and sensory characteristics of conventional food. Since olestra can be heated, it may be used for a wide range of cooked and uncooked foods. In regard to taste, foods made with olestra are virtually indistinguishable from their ordinary fat counterparts. Olestra has recently been approved by the United States Food and Drug Administration (US FDA) (January 30, 1996) as an ingredient to replace fat in savory snack foods. Olestra is currently marketed under the trade name, Olean®.

Other noncaloric synthetic fat substitutes have been developed including dialkyl dihexadecylmalonate (DDM), synthesized from malonic acid, hexadecane, and fatty acids; esterified propoxylated glycerol (EPG), which contains oxypropylene incorporated between the glycerol and fatty acids; and trialkoxytricarballate (TATCA), with tricarballylic acid replacing the glycerol and saturated and unsaturated alcohols replacing the fatty acids. These compounds are not digested nor absorbed by the body, and thus provide no caloric value.

(b) Caloric—A low-calorie (values ranging from 1 to 3 kcal/g) synthetic fat substitute has been developed as an outgrowth of the TATCA technology, carboxy/carboxylate esters. The backbones of these molecules are malonic or tartaric acids with fatty acids esterified to the hydroxyl groups and fatty alcohols esterified to the acid groups.

SAFETY OF FAT SUBSTITUTES

Because the safety of olestra has been extensively studied and published safety data concerning other fat substitutes are limited, only the safety of olestra will be discussed here. Results of short- and long-term studies in rats, dogs, mice, and monkeys indicate that, at levels as high as 15% (w/w) of the diet, olestra is not toxic or carcinogenic.[27-34] It does not affect mating, conception, embryonic development, viability, or growth of rats or mice.[35,36] Results of standard genotoxicity assays show that it does not produce mutations, cause chromosomal aberrations, or affect DNA repair.[37] Studies in animals and in humans have established that olestra does not affect absorption of orally dosed drugs.[38,39]

A number of studies have been performed in pigs and in humans to assess the effect of olestra on absorption of nutrients from the GI tract.[40-50] Both the pig and human studies used controlled diets and exaggerated olestra doses and eating patterns to define the maximum effect of olestra on fat-soluble vitamins and water-soluble micronutrients and macronutrients. Results of these studies show that olestra affects only absorption of highly lipophilic dietary components (e.g., cholesterol; vitamins A, D, E, K; and carotenoids) and that olestra's effect on absorption of these lipophiles can be negated by adding these back to olestra-containing foods. Thus, the essential vitamins A, D, E, and K are added to olestra foods available to consumers to maintain adequate status of these vitamins. The amounts of vitamins added back pose no safety concerns because they simply maintain tissue concentrations at the levels that they would be at if olestra were not being consumed.[40]

No medically significant, olestra-related adverse events or effects have been observed in human studies utilizing normal and obese subjects, diabetic adults, patients with inflammatory bowel disease, or hypercholesterolemic outpatients.[40,41,51-54]

FAT SUBSTITUTES AND ENERGY BALANCE

Comparisons between the types of fat substitutes and their effects on energy balance are not possible as virtually all the research available on energy balance has focused on the sucrose polyesters. Among the various fat replacers that have been developed, olestra has been most-extensively studied for its effects on energy intake and energy balance. When considering the potential impact of fat substitutes on energy intake and diet selection, several questions should be addressed. Do individuals compensate for changes in dietary energy density when dietary fat is replaced with a fat substitute and, if so, is the compensation "fat-specific"? Does fat replacement reduce the proportion of energy derived from fat? Is there a net decrease in total energy intake? Do individuals feel hungrier after consuming foods containing fat substitutes and do they respond by increasing the amount of food consumed? These questions have been addressed in a series of short-term studies in both adults and children using the sucrose polyester, olestra.

There have been numerous published studies in which olestra was substituted for fat and the effect on subsequent energy and fat intake was determined.[55-65] Studies have been conducted with lean[56-59,61-65] and obese[55,63] adults and with children.[60] All studies examined the effects of diluting the energy density and reducing the fat content of the diet (by substituting fat with olestra) on aspects of feeding behavior and energy balance. In most of these studies, the substitution was covert, although two studies examined the effect of overt substitution.[61,63]

Short-Term Studies

Lean adults showed generally incomplete compensation for a decrease in the energy density of the diet by replacement of fat with olestra. In a series of two-day studies, where subjects were allowed to eat *ad libitum* from a variety of familiar foods following a mandatory intake of either SPE-containing or full-fat foods, the maximal compensation observed for the energy deficit produced by SPE was 27%.[61] This series of studies also indicated that the ability of the fat replacer to provide similar sensory properties to real fat may have been related to the poor compensation observed. The effect of differing amounts of SPE, fat, and water added to the required food intake was investigated using the same two-day format. While energy compensation was not significant when fat was replaced by SPE, men compensated by 66% for the energy deficit when fat was substituted by water.[61] This suggests that SPE may have a greater ability to "fool" the body's appetite system, at least in the short term, than when fat is substituted with water. The ability of SPE to provide similar taste properties as fat while decreasing the energy density offers hope for greater compliance with low-fat regimens.

In another series of two-day studies, lean male subjects were fed a fixed diet, containing either full-fat or olestra, throughout day 1 and then on day 2 were allowed *ad libitum* intake. No energy compensation was seen and fat energy was reduced from 40% to about 30%.[58] However, when a more-severe reduction was imposed (from 32% fat kcal to 20%), the subjects compensated for 74% of the energy deficit.[59] The explanation for this stronger compensatory response under these particular dietary conditions is unclear.

In contrast to these studies in which relatively poor energy compensation was observed, two replicate short-term studies using lean male subjects found complete energy compensation when the entire olestra intake was provided solely at the breakfast meal. Using an identical protocol, two separate, blinded, dose-response studies were carried out by independent investigators.[22,57] At each site, 24 lean young men were studied in a randomized three-period crossover design. Each subject was studied for three separate 24-hour periods, during which olestra was used to replace 0, 20, or 36 grams of fat at a mandatory breakfast meal. Following the breakfast, food intake and ratings of hunger were monitored over the subsequent 24 hours. At lunch and dinner, subjects were provided with an array of foods from which to choose, including foods high in fat, protein, and carbohydrate. In addition, snacks were also available. Food intake at these occasions was *ad libitum*. Subjects rated feelings of hunger and fullness just prior to each meal and at intervals throughout the day. The results at both study sites showed that subjects compensated nearly quantitatively for the energy dilution at the breakfast meal by consuming more food at other meals during the day, primarily at dinner and evening snacks. In these studies, the subjects compensated almost perfectly for the energy deficit. Fat intake was decreased (FIG. 1), but not total caloric intake. This observation suggests that there was no "fat-specific" compensation, but that subjects compensated for energy by consuming more of the total diet. There were no differences across fat levels in ratings of hunger or fullness throughout the test days, despite the differences in actual measured intake. Hulshof[61] also showed in his two-

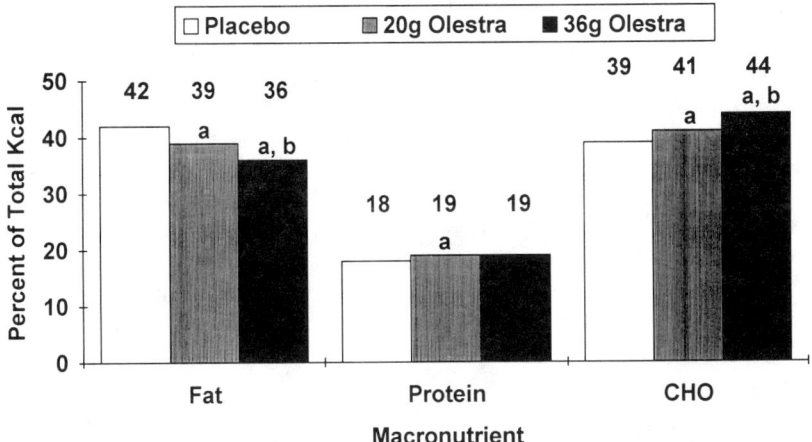

FIGURE 1. Percentage of total energy intake contributed by protein, carbohydrate, and fat in lean young men over 24 hours following replacement of 0, 20, or 36 grams of fat with olestra at breakfast (data plotted from references 56 and 57). Numbers above the bars represent the percentages of total kcal derived from each macronutrient. Significance: (a) different from placebo treatment, $p < 0.05$; (b) different from 20-g olestra treatment, $p < 0.05$.

day studies that there was a tendency for compensation to be greater after a breakfast substitution than when the substitution occurred at lunch time or during midafternoon. It may be that manipulation at breakfast allows a maximal time for the caloric difference to be detected and for compensation to occur before bedtime. This would be consistent with a satiety signal linked to the metabolism of diet-derived fat, which is typically prolonged over several hours after a meal. For example, peak plasma concentration of triglycerides occurs about three to four hours after a meal, compared to about one hour for glucose.

The effect of olestra on energy intake and diet selection in 29 boys and girls, ages 2–5 years, was studied by Birch et al.[60] It was considered that children would compensate better than adults for caloric dilution of the diet because per kg of body weight their energy requirements are higher than those of adults. Using a previously developed approach[66] for studying energy intake in young children, food selection and energy intake were measured on four separate occasions, with each occasion consisting of 48 consecutive hours. Individual children were studied on two occasions in which olestra was incorporated into food items given at breakfast, morning snack, and lunch on the first day of each 48-hour block, and on two occasions at which conventional fat foods were provided at the same three eating occasions. All other food provided for the remainder of day 1 and throughout day 2 was prepared with conventional fat. Children were provided with several different foods of different composition at each meal and were given *ad libitum* access to the food so that both food selection and free-living energy intake could be monitored. The total replacement of fat with olestra achieved over the three eating occasions on day 1 was about 120 kcal, which represented approximately 10% of the average energy intake of these children.

The children compensated almost completely for the reduced energy intakes when olestra was incorporated into the diet compared to the full-fat diet. Total energy intakes at the end of the 48-hour test period were within 20 kcal of the intakes observed on placebo days. However, as in the adult studies, despite compensation for energy in response to dilution of dietary energy with the fat substitute, there was no evidence of "fat-specific" compensation. The proportion of total energy consumed from fat, when olestra was incorporated, was lower than the fat intake on placebo days.

These results demonstrate that short-term energy intake in young children is well regulated and substitution of fat with olestra does not affect normal regulation. In addition, incorporation of olestra into the diet produced a reduction in total fat intake, suggesting that there is no "fat-specific" appetite in children.

Longer-Term Studies

Studies lasting from 12 to 14 days in both lean and obese men and women have shown that compensation for both fat and energy is incomplete when dietary fat is replaced with olestra in food consumed throughout the day.[61,62,64,65] The highest energy compensation reported in any of these studies was 46% by lean young men in studies by Hulshof.[61] In addition, a 20-day study in obese subjects enrolled in a weight-loss program demonstrated that olestra can accelerate weight loss and improve diet palatability in individuals making a concerted effort at losing weight.[55]

"Real-World" Situations

Miller et al.[63] examined the effects of consuming full-fat- or olestra-containing potato chips on feeding behavior in 96 subjects (men and women, lean and obese, restrained and unrestrained) who were habitual potato chip–eaters. Potato chips are one of the product forms for which olestra is presently approved for use in the United States. In half of the subjects the manipulation was overt and in half it was covert. In all groups, the olestra-chip eaters consumed less fat and energy during the 10-day study period. Informing subjects of what they ate (providing product labels and nutrition information) increased the intake of the olestra chips in restrained eaters (by about 15%), but their intake was still significantly lower in fat and energy than subjects fed full-fat potato chips.

A brief report by De Graaf et al.[67] can give us some idea of what effect on energy balance to expect when low-fat products are used to replace full-fat products over a longer time period. In this study, 103 subjects with free access to 50 commercially available full-fat products and 117 subjects with access to the 50 reduced-fat equivalents were followed for six months. Compared to baseline measurements, energy intake increased in the control group, while it remained constant in the reduced-fat group. The investigators concluded that "long-term consumption of reduced-fat products leads to a lower energy and fat intake, compared to the consumption of full-fat products."

GASTROINTESTINAL EFFECTS OF FAT SUBSTITUTES

Among the various fat substitutes, olestra has been most-extensively studied for its potential to affect the gastrointestinal (GI) tract. Studies performed in animals indicate that, even when the intestinal mucosa is injured, olestra is essentially not absorbed from the GI tract.[68-71] Based on this fact, numerous studies have been performed to assess if olestra has any adverse effects on GI structure or function. Results of long-term studies in animals indicate that, at doses up to 100 times expected for human consumption, olestra does not alter the morphology or physiology of the GI tract.[40] Furthermore, in animals and humans, olestra has no effect on gastric emptying, gastrointestinal transit, pancreatic enzymes, immune function, enteric microflora, or bile acid synthesis, function, or excretion in the gut.[40,72]

Although not considered medically significant, symptoms such as loose or soft stools, flatulence, bloating, and nausea have been reported by some subjects consuming either placebo (triglyceride) or olestra in controlled clinical studies, generally conducted under exaggerated dosing conditions. Although some subjects have also reported experiencing diarrhea, studies indicate that olestra does not increase the water content of the stool or the number of bowel movements per day, change gastrointestinal transit, or cause fluid loss or any other clinical signs of dehydration.[27,45,73-75] Therefore, the FDA and numerous experts have concluded that the diarrhea-like stools reported by some subjects ingesting large amounts of olestra do not represent pathological diarrhea, but are extremely loose stools caused by the physical effect of olestra on stool consistency.[71]

In free-living consumer tests involving adults and children, where there is *ad libitum* snack consumption over periods of several months, reports of GI symptoms were similar for subjects consuming olestra-containing or triglyceride snacks.[75]

CONCLUSIONS

Studies exploring olestra's effect on energy balance indicate that substituting dietary fat with olestra is effective in reducing fat and energy intake over the time periods studied. These effects occur in both lean and obese men and women under a variety of conditions ranging from the laboratory to free-living. Whether or not fat and energy reductions will be compensated over longer periods remains to be tested, but unless complete compensation occurred benefits would still accrue. Data from children, although limited, indicate that olestra consumption will likely not compromise the energy intakes of children.

The studies by Miller *et al.*[63] with potato chips suggest that fat substitutes like olestra can be of value in producing significant reductions in fat and energy derived from high-fat foods like salted snacks. This results in modest reductions in the total fat intake of free-living people who consume these products.

Collectively, these results suggest that, within the range of fat intakes typically encountered in many developed countries, replacement of fat with fat substitutes can help reduce daily fat intake, even among individuals who regulate energy intake well. Furthermore, under some circumstances, total energy intake may also be reduced. The presence and use of fat substitutes in the food supply should not reduce our efforts to improve eating habits and increase physical activity in our population. However, fat substitutes can serve as one tool to help manage fat and energy intake, especially in populations with easy access to convenient, energy-dense, high-fat foods. Because fat substitutes can substantially reduce the energy density of food, they hold promise as a means to reduce "passive" excess energy intake, and hence reduce the incidence of positive energy balance and obesity, especially among sedentary populations. Olestra is the first versatile fat substitute to be approved by the FDA and may serve as an example of a broader range of products that may become available. However, specific studies designed to explore the effectiveness and safety of these other fat substitutes are still needed.

REFERENCES

1. KUCZMARSKI, R. J. et al. 1994. Increasing prevalence of overweight among U.S. adults: the National Health and Nutrition Examination Surveys 1960-1991. JAMA 272: 205-211.
2. SELECT COMMITTEE ON NUTRITION AND HUMAN NEEDS. 1977. U.S. Senate Dietary Goals for the United States. U.S. Govt. Printing Office. Washington, D.C.
3. DHHS. 1988. The Surgeon General's Report on Nutrition and Health. DHHS (PHS) Publication No. 88-50210. U.S. Govt. Printing Office. Washington, D.C.
4. NATIONAL RESEARCH COUNCIL, COMMITTEE ON DIET AND HEALTH. 1989. Diet and Health: Implications for Reducing Chronic Disease Risk. National Academy Press. Washington, D.C.
5. DHHS. 1990. Dietary Guidelines for Americans. Third edition. Home and Garden Bulletin No. 232. U.S. Govt. Printing Office. Washington, D.C.
6. DHHS. 1983. Dietary Intake Source Data: United States, 1976-80. DHHS (PHS) Publication No. 83-1681. U.S. Govt. Printing Office. Washington, D.C.
7. THIRD NATIONAL HEALTH AND NUTRITION EXAMINATION SURVEY (PHASE 1, 1988-91). 1992. Daily dietary fat and total food-energy intakes. Morb. Mortal. Wkly. Rep. 43(7): 116-117 and 123-125.
8. WHO. 1990. Report on Diet, Nutrition, and the Prevention of Chronic Diseases. Technical Report Series No. 797.
9. WARWICK, Z. S. & S. S. SCHIFFMAN. 1992. Role of dietary fat in calorie intake and weight gain. Neurosci. Biobehav. Rev. 16: 585-596.
10. LISSNER, L. & B. L. HEITMANN. 1995. Dietary fat and obesity: evidence from epidemiology. Eur. J. Clin. Nutr. 49: 79-90.
11. ROLLS, B. J. & D. SHIDE. 1994. Dietary fat and control of food intake. In Appetite and Body Weight Regulation: Sugar, Fat, and Macronutrient Substitutes. Chapter 12. CRC Press. Boca Raton, Florida.
12. LISSNER, L. et al. 1987. Dietary fat and the regulation of energy intake in human subjects. Am. J. Clin. Nutr. 46: 886-892.
13. DONATO, K. A. & D. M. HEGSTED. 1985. Efficiency of utilization of various energy sources for growth. Proc. Natl. Acad. Sci. U.S.A. 82: 4866-4870.
14. TURNER, L. A. & M. J. KANO. 1992. Dietary fat and body fat: a multivariate study of 205 adult females. Am. J. Clin. Nutr. 56: 616-622.
15. ASTRUP, A. 1993. Dietary composition, substrate balances, and body fat in subjects with a predisposition to obesity. Int. J. Obes. 17: S32-S36.
16. PRENTICE, A. M. et al. 1992. Overview: energy requirements and energy storage. In Energy Metabolism: Tissue Determinants and Cellular Corollaries. Raven Press. New York.
17. FLATT, J. P. et al. 1985. Effects of dietary fat on postprandial substrate oxidation and on carbohydrate and fat balances. J. Clin. Invest. 76: 1019-1024.

18. VAN AMELSVOORT, J. M. M. et al. 1989. Effects of varying the carbohydrate:fat ratio in a hot lunch on postprandial variables in male volunteers. Br. J. Nutr. **61:** 267-283.
19. FRIEDMAN, M. I. 1990. Body fat and the metabolic control of food intake. Int. J. Obes. **14**(suppl. 3): 53-57.
20. HILL, A. J., P. D. LEATHWOOD & J. E. BLUNDELL. 1987. Some evidence for short-term caloric compensation in normal-weight human subjects: the effects of high- and low-energy meals on hunger, food preference, and food intake. Hum. Nutr. Appl. Nutr. **41A:** 244-257.
21. SIGGAARD, R., A. RABEN & A. ASTRUP. 1996. Weight loss during 12 weeks' *ad libitum* carbohydrate-rich diet in overweight and normal-weight subjects at a Danish work site. Obes. Res. **4:** 347-356.
22. STUBBS, R. J. et al. 1995. Covert manipulation of dietary fat and energy density: effect on substrate flux and food intake in men feeding *ad libitum*. Am. J. Clin. Nutr. **62:** 316-330.
23. STUBBS, R. J. et al. 1995. Covert manipulation of the ratio of dietary fat to energy density: effect on food intake and energy balance in free-living men feeding *ad libitum*. Am. J. Clin. Nutr. **62:** 330-338.
24. BOUCHARD, C. et al. 1988. Inheritance of the amount and distribution of human body fat. Int. J. Obes. **12:** 205-215.
25. FINLEY, J. W. & G. A. LEVEILLE. 1996. Macronutrient substitutes. *In* Present Knowledge in Nutrition. Chapter 59. ILSI Press. Washington, D.C.
26. MATTSON, F. H. & R. A. VOLPENHEIN. 1972. Rate and extent of absorption of the fatty acids of fully esterified glycerol, erythritol, and sucrose as measured in thoracic duct cannulated rats. J. Nutr. **102:** 1177-1180.
27. FOOD ADDITIVE PETITION 7A 3997. 1987. Submitted to the U.S. Food and Drug Administration. The Procter and Gamble Company, Cincinnati, Ohio.
28. WOOD, F. E., B. R. DEMARK, E. J. HOLENBACH, M. C. SARGENT & K. C. TRIEBWASSER. 1991. Analysis of liver tissue for olestra following long-term feeding to rats and monkeys. Food Chem. Toxicol. **29:** 231-236.
29. MILLER, K. W. & P. H. LONG. 1990. A 91-day feeding study in rats with heated olestra/vegetable oil blends. Food Chem. Toxicol. **28:** 307-315.
30. WOOD, F. E., W. J. TIERNEY, A. L. KNEZEVICH, H. F. BOLTE, J. K. MAURER & R. D. BRUCE. 1991. Chronic toxicity and carcinogenicity studies of olestra in Fischer 344 rats. Food Chem. Toxicol. **29:** 223-230.
31. MILLER, K. W., F. E. WOOD, S. B. STAURD & C. L. ALDEN. 1991. A 20-month olestra feeding study in dogs. Food Chem. Toxicol. **29:** 427-435.
32. LAFRANCONI, W. M. 1991. Ninety-one day feeding study of olestra in the mouse. Thirtieth Annual Meeting of the Society of Toxicology, New Orleans, Louisiana.
33. LAFRANCONI, W. M., P. H. LONG, J. E. ATKINSON, A. L. KNEZEVICH & W. L. WOODING. 1994. Chronic toxicity and carcinogenicity of olestra in Swiss CD-1 mice. Food Chem. Toxicol. **32:** 789-798.
34. ADAMS, M., M. R. MCMAHAN, F. H. MATTSON & T. B. CLARKSON. 1981. The long-term effects of dietary sucrose polyesters on African green monkeys. Proc. Soc. Exp. Biol. Med. **167:** 346-353.
35. NOLEN, G. A., F. WOOD & T. A. DIERCKMAN. 1987. A two-generation reproductive and developmental toxicity study of sucrose polymer. Food Chem. Toxicol. **25:** 1-8.
36. DENINE, E. P. & R. E. SCHROEDER. 1993. A segment II teratology study in rabbits dosed with olestra which had been used for deep frying potatoes. Toxicologist **13:** 80.
37. SKARE, K. L., J. A. SKARE & E. D. THOMPSON. 1990. Evaluation of olestra in short-term genotoxicity assays. Food Chem. Toxicol. **28:** 69-73.
38. ROBERTS, R. J. & R. D. LEFF. 1989. Influence of absorbable and nonabsorbable lipids and lipidlike substances on drug bioavailability. Clin. Pharmacol. Ther. **45:** 299-304.
39. MILLER, K. W., S. D. WILLIAMS, S. B. CARTER, M. B. JONES & D. R. MISHELL. 1990. The effect of olestra on systemic levels of oral contraceptives. Clin. Pharmacol. Ther. **48:** 34-40.
40. LAWSON, K. D., S. J. MIDDLETON & C. D. HASSALL. 1997. Olestra, a nonabsorbed, noncaloric replacement for dietary fat: a review. Drug Metab. Rev. In press.
41. MELLIES, M. J., R. J. JANDACEK *et al.* 1983. A double-blind placebo-controlled study of sucrose polyester in hypercholesterolemic outpatients. Am. J. Clin. Nutr. **37:** 339-346.
42. MILLER, K. W. & G. S. ALGOOD. 1993. Nutritional assessment of olestra, a non-caloric fat substitute. Int. J. Food Sci. Nutr. **44:** S77-S82.
43. JONES, D. Y., K. W. MILLER, B. P. KOONSVITSKY, M. L. EBERT, P. Y. T. LIN, M. B. JONES

& H. F. DELUCA. 1991. Serum 25-hydroxyvitamin D concentrations of free-living subjects consuming olestra. Am. J. Clin. Nutr. **53:** 1281-1287.
44. JONES, D. Y., B. P. KOONSVITSKY, M. L. EBERT, M. B. JONES, P. Y. T. LIN, B. H. WILL & J. W. SUTTIE. 1991. Vitamin K status of free-living subjects consuming olestra. Am. J. Clin. Nutr. **53:** 943-946.
45. KOONSVITSKY, B. P., D. Y. JONES, D. A. BERRY & M. B. JONES. 1991. Status of vitamins E, D, and K in free-living subjects consuming olestra. Am. J. Clin. Nutr. **53:** 21.
46. SCHLAGHECK, T., J. MCEDWARDS, K. RICCARDI *et al.* 1994. Effect of olestra on nutritional status in man. FASEB J. **8:** A933.
47. ZORICH, N. L., G. DAHER, D. COOPER *et al.* 1994. The effect of olestra on triglyceride absorption. FASEB J. **8:** A735.
48. DAHER, G., D. COOPER, K. RICCARDI *et al.* 1994. The effect of olestra on retinyl palmitate absorption in man. FASEB J. **8:** A443.
49. COOPER, D. A., D. BERRY, M. JONES *et al.* 1994. An assessment of the nutritional effects of olestra in the domestic pig. FASEB J. **8:** A191.
50. PETERS, J. C., D. A. COOPER, G. C. DAHER *et al.* 1994. The domestic pig as a model in which to evaluate olestra nutritional effects. FASEB J. **8:** A937.
51. BERGHOLZ, C. M. 1992. Safety evaluation of olestra, a nonabsorbed, fatlike fat replacement. Crit. Rev. Food Sci. Nutr. **32:** 141-146.
52. ZORICH, N. L., S. B. CARTER, J. M. MCEDWARDS & T. M. BAYLESS. 1994. Safety of olestra in patients with inflammatory bowel disease. Gastroenterology **106:** A796.
53. GRUNDY, S. M., J. V. ANASTASIA, A. KESANIEMI & J. ABRAMS. 1986. Influence of sucrose polyester on plasma lipoproteins and cholesterol metabolism in obese patients with and without diabetes mellitus. Am. J. Clin. Nutr. **44:** 620-629.
54. GEIL, P. B., S. R. LAWRENCE & J. W. ANDERSON. 1989. Effects of sucrose polyester (SPE) in patients with type II diabetes mellitus. Annual Meeting of the American Dietetic Association, Kansas City, Missouri.
55. GLUECK, C. J. *et al.* 1982. Sucrose polyester and covert caloric dilution. Am. J. Clin. Nutr. **35:** 1352-1359.
56. BLUNDELL, J. E., V. J. BURLEY & J. C. PETERS. 1991. Dietary fat and human appetite: effects of non-absorbable fat on energy and nutrient intakes. *In* Obesity: Dietary Factors and Control. Japan Sci. Soc. Press. Tokyo.
57. ROLLS, B. J. *et al.* 1992. Effects of olestra in noncaloric fat substitute on daily energy and fat intakes in lean men. Am. J. Clin. Nutr. **56:** 84-92.
58. COTTON, J. R. *et al.* 1996. Fat substitution and food intake: effect of replacing fat with sucrose polyester at lunch or evening meals. Br. J. Nutr. **75:** 545-556.
59. COTTON, J. R., J. A. WESTSTRATE & J. E. BLUNDELL. 1996. Replacement of dietary fat with sucrose polyester: effects on energy intake and appetite control in nonobese males. Am. J. Clin. Nutr. **63:** 891-896.
60. BIRCH, L. L. *et al.* 1993. Effects of a nonenergy fat substitute on children's energy and macronutrient intake. Am. J. Clin. Nutr. **58:** 326-333.
61. HULSHOF, T. 1994. Fat and non-absorbable fat and the regulation of food intake. Ph.D. thesis. Wageningen, the Netherlands.
62. JOHNSON, S. L. *et al.* 1995. Effects of covert substitution of olestra on self-selected food intake. FASEB J. **10:** A550 (abstract).
63. MILLER, D. L. *et al.* 1995. Effects of substituting fat-free (olestra) potato chips on 24-hour fat and energy intake. Obes. Res. 3(suppl. 3): O10 (abstract).
64. BRAY, G. A. *et al.* 1995. Effect of two weeks fat replacement by olestra on food intake and energy metabolism. FASEB J. **9:** A459 (abstract).
65. SPARTI, A. *et al.* 1995. Subjects eat for carbohydrate, not calories, after dietary fat replacement with olestra. Am. J. Clin. Nutr. **61:** 902.
66. BIRCH, L. L. *et al.* 1991. The variability of young children's energy intake. N. Engl. J. Med. **324:** 232-235.
67. DE GRAAF, C. *et al.* 1995. Energy consumption during long-term consumption of reduced fat foods. Int. J. Obes. 19(suppl. 2): 27.
68. MILLER, K. W., K. D. LAWSON, D. H. TALLMADGE *et al.* 1995. Disposition of ingested olestra in the Fischer 344 rat. Fundam. Appl. Toxicol. **24:** 229-237.

69. MILLER, K., K. LAWSON, B. MADISON et al. 1992. Absorption, distribution, and elimination of olestra after oral administration in rats. Toxicologist **12:** 145.
70. DAHER, G. C., K. D. LAWSON, D. H. TALLMADGE et al. 1996. Disposition of ingested olestra in weanling mini-pigs. Food Chem. Toxicol. **34:** 693–699.
71. FEDERAL REGISTER. January 30, 1996. **61**(20): 3118–3173.
72. ALGOOD, G. 1996. Research behind the US FDA approval of olestra. BNF Nutr. Bull. **21:** 174–182.
73. AGGARWAL, A. M., M. CAMILLERI, S. F. PHILLIPS et al. 1993. Olestra, a nondigestible nonabsorbable fat: effects on gastrointestinal function and colonic transit. Dig. Dis. Sci. **38:** 1009–1014.
74. SIIGUR, U., K. E. NORIN, G. S. ALGOOD, T. G. SCHLAGHECK & T. MIDTVEDT. 1996. Concentrations and correlations of faecal short-chain fatty acids and faecal water content in man. Microb. Ecol. Health Dis. **7:** 9–17.
75. FRESTON, J. W., D. J. AHNEN, S. J. CZININ et al. 1997. Review and analysis of the effects of olestra, a dietary fat substitute, on gastrointestinal function and symptoms. JAMA. Submitted.

Nonesterified Fatty Acid Regulation of Lipid and Glucose Oxidation in the Obese

SARAH E. KING, JANET M. BRYSON,
LOUISE A. BAUR, SOJI SWARAJ, AND
IAN D. CATERSON

Department of Endocrinology
Royal Prince Alfred Hospital
Camperdown, NSW 2050, Australia

In the "glucose–fatty acid" cycle (Randle, 1963),[1] overproduction and utilization of nonesterified fatty acids (NEFAs) inhibits glucose uptake and disposal. This may contribute to the insulin resistance seen in obesity and non-insulin-dependent diabetes mellitus (NIDDM).[1] Elevated plasma NEFA levels have been associated with body fatness,[2,3] particularly in those with excess abdominal fat deposition,[4] impaired glucose tolerance,[5] and decreased whole-body glucose uptake in NIDDM[6] and in nondiabetic normal-weight subjects.[7-9] Artificial reduction of NEFA availability improves glucose oxidation and decreases lipid utilization, while heparin-induced rises in serum NEFA enhance lipid oxidation and reduce glucose oxidation in healthy NIDDM subjects.[10] In a recent study involving normal subjects, circulating NEFA levels were directly associated with rates of glucose and lipid oxidation and glycogen synthesis, suggesting a regulatory role for NEFA in both oxidative and nonoxidative pathways.[11] As part of an ongoing study of metabolic changes in obesity, we studied the interrelationships between circulating levels of NEFAs and different parameters of glucose and lipid metabolism in obese subjects, both basally and following insulin stimulation.

METHODS

Twenty-three clinically obese, but otherwise healthy subjects (11 males, 12 females) were recruited from the Metabolism and Obesity Services at Royal Prince Alfred Hospital (BMI: 40.1 ± 0.9 kg/m^2; age: 21–62 years; weight: 113.1 ± 2.6 kg). Fasting blood samples were collected for measurement of serum insulin, NEFAs, and lipids. Participants underwent indirect calorimetry (Datex, Deltatrac, Finland) for determination of respiratory quotient (RQ) and % glucose (% GOX) and % lipid (% LOX) oxidation ($n = 23$). A euglycemic hyperinsulinemic clamp (40 mU/m^2/min) ($n = 21$) was then performed for a minimum of 2 hours and indirect calorimetry was repeated during the final hour. Total glucose disposal (GIR) was determined from

TABLE 1. Relationship between Fasting Serum NEFA Levels and Indirect Calorimetry Measurements

	Range	r	p
Basal (n = 23)			
respiratory quotient	0.75–0.92	−0.655	0.0007
% lipid oxidation	18.5–69.6	0.669	0.0005
% glucose oxidation	17.8–68.8	−0.640	0.0010
Insulin-stimulated (n = 21)			
respiratory quotient	0.80–0.99	−0.642	0.0023
% lipid oxidation	17.8–57.2	0.624	0.0032
% glucose oxidation	28.6–69.3	−0.650	0.0019

the glucose infusion rate during this final hour and nonoxidative glucose disposal (NOX) was calculated by subtracting the oxidative glucose disposal (GOX) during the clamp from the measured GIR.

RESULTS

Fasting serum NEFA levels ranged from 0.35 to 1.27 mmol/L. There was no association between the fasting NEFA levels and body weight (r = −0.012, p = 0.955), waist/hip ratio (r = −0.257, p = 0.237), and fasting insulin (r = 0.204, p = 0.351); however, NEFAs were positively correlated with BMI (r = 0.418, p = 0.047). There was no relationship between fasting serum NEFAs and basal levels of total cholesterol (r = −0.065, p = 0.7674), LDL cholesterol (r = −0.228, p = 0.295), HDL cholesterol (r = 0.142, p = 0.5191), and triglycerides (r = 0.192, p = 0.3812). In the basal state, there was a strong negative correlation between NEFA levels and RQ and % GOX, and a reciprocal association with % LOX (TABLE 1). Insulin stimulation during the euglycemic clamp produced a significant increase in RQ and GOX, and a fall in LOX (TABLE 2). A strong association was seen between fasting NEFA levels and GIR during the clamp (FIG. 1a). This was mainly due to an association between NEFA levels and the rate of nonoxidative glucose disposal (FIG. 1b), rather than with oxidative glucose disposal (FIG. 1c).

TABLE 2. Indirect Calorimetry in the Basal and Insulin-stimulated State in Obese Subjects[a]

	Basal	Insulin-stimulated
respiratory quotient	0.82 ± 0.01	0.87 ± 0.01[b]
% glucose oxidation	35.0 ± 13.3	48.9 ± 2.3[b]
% lipid oxidation	47.9 ± 13.4	34.1 ± 2.4[b]

[a] Results are mean ± SEM.
[b] p < 0.01 compared to basal values.

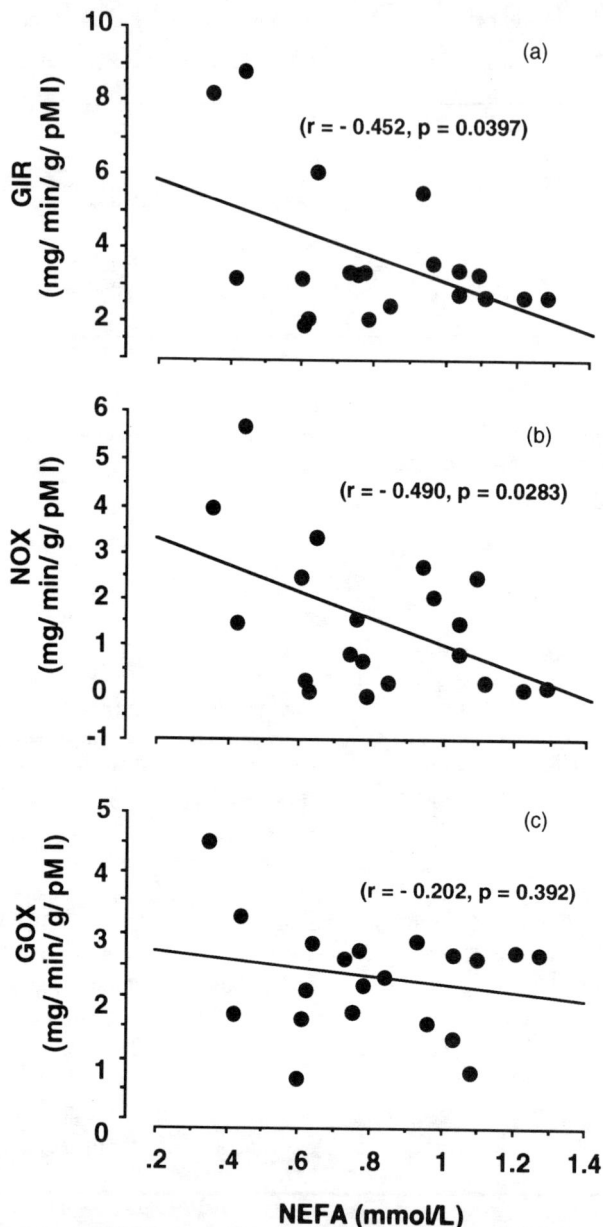

FIGURE 1. Relationship between fasting serum NEFA concentration and total glucose disposal rate (GIR) (a), nonoxidative glucose disposal (NOX) (b), and oxidative glucose disposal (GOX) (c). GIR, NOX, and GOX are expressed as milligrams of glucose infused per minute per gram of body weight per pM of insulin during the clamp.

DISCUSSION

The strong relationship between NEFA levels and different aspects of glucose and lipid metabolism in this study demonstrates that Randle's glucose–fatty acid cycle is operating in obesity. The associations found in this study are similar to those previously reported in normal and NIDDM subjects.[6–9] The strong inverse relationship between NEFAs and nonoxidative glucose disposal is consistent with the reported suppressive effects of NEFAs on glycogen synthesis.[11] Whereas the effects of NEFA on glucose oxidation can be explained by its inhibitory effect on the activity of the pyruvate dehydrogenase complex, the key enzyme for glucose oxidation,[1] the mechanisms for NEFAs suppressive effects on nonoxidative glucose disposal are not as obvious and require further investigation. In conclusion, this study indicates that circulating NEFA levels play an important regulatory role in determining the fate of glucose in these insulin-resistant subjects, both in the basal state and following insulin stimulation, affecting both oxidative and nonoxidative disposal.

REFERENCES

1. RANDLE, P. J., P. B. GARLAND, C. N. HALES & E. A. NEWSHOLME. 1963. The glucose–fatty acid cycle: its role in insulin sensitivity and the metabolic disturbances of diabetes mellitus. Lancet I: 785–789.
2. BYRNE, C. D., N. J. WAREHAM, D. C. BROWN, P. M. S. CLARK, L. J. COX, N. E. DAY, C. R. PALMER, T. W. M. WANG, D. R. R. WILLIAMS & C. N. HALES. 1994. Hypertriglyceridaemia in subjects with normal and abnormal glucose tolerance: relative contributions of insulin secretion, insulin resistance, and suppression of plasma non-esterified fatty acids. Diabetologia 37: 889–896.
3. KISSEBAH, A. H. & A. N. PEIRIS. 1989. Biology of regional body fat distribution: relationship to non-insulin dependent diabetes mellitus. Diabetes Metab. Rev. 5(2): 83–109.
4. KISSEBAH, A. H. & G. R. KRAKOWER. 1994. Regional adiposity and morbidity. Physiol. Rev. 74(4): 761–811.
5. BYRNE, C. D., N. J. WAREHAM, N. E. DAY, R. MCLEISH, D. R. R. WILLIAMS & C. N. HALES. 1995. Decreased non-esterified fatty acid suppression and features of the insulin resistance syndrome occur in a subgroup of individuals with normal glucose tolerance. Diabetologia 38: 1358–1366.
6. REAVEN, G. M. 1988. Role of insulin resistance in human disease. Diabetes 37: 1595–1607.
7. FERRANNINI, E., E. J. BARRETT, S. BEVILACQUA & R. A. DEFRONZO. 1983. Effect of fatty acids on glucose production and utilization in man. J. Clin. Invest. 72: 1737–1747.
8. WALKER, M., G. R. FULCHER, C. CATALANO, G. PETRANYI, H. ORSKOV & K. G. M. M. ALBERTI. 1990. Physiological levels of plasma non-esterified fatty acids impair forearm glucose uptake in normal man. Clin. Sci. 79: 167–174.
9. WOLFE, B. M., S. KLEIN, E. J. PETERS, B. F. SCHMIDT & R. R. WOLFE. 1988. Effect of elevated free fatty acids on glucose oxidation in normal humans. Metabolism 37(4): 323–329.
10. ABAYOMI, O. A., E. OSIFO, M. KIRK & T. D. R. HOCKADAY. 1993. The effects of changes in plasma nonesterified fatty acid levels on oxidative metabolism during moderate exercise in patients with non-insulin dependent diabetes mellitus. Metabolism 42(4): 426–434.
11. EBLING, P. & V. A. KOIVISTO. 1994. Non-esterified fatty acids regulate lipid and glucose oxidation and glycogen synthesis in healthy man. Diabetologia 37: 202–209.

Nuclear All-*trans* Retinoic Acid Receptors in Liver of Rats with Diet-induced Insulin Resistance[a]

J. BRTKO,[b] E. ŠEBÖKOVÁ,[b] D. GAŠPERÍKOVÁ,[b]
I. KLIMEŠ,[b] S. HUDECOVÁ,[c] AND J. BRANSOVÁ[b]

[b]*Institute of Experimental Endocrinology*
Slovak Academy of Sciences
833 06 Bratislava, Slovak Republic
[c]*Department of Molecular Biology*
Faculty of Natural Sciences
Comenius University
842 15 Bratislava, Slovak Republic

In vertebrates, thyroid hormones, retinoids, and their cognate nuclear receptors are involved in a complex arrangement of physiological and developmental responses. Two principal classes of nuclear retinoic acid receptors have been identified—the all-*trans* retinoic acid receptors (RAR) and the 9-*cis* retinoic acid receptors (RXR). Both of them are ligand-inducible transcription factors and are related in structure to steroid and thyroid hormone receptors.[1] Since all-*trans* retinoic acid (RA), a principal circulating regulatory retinoid, has been found to stimulate glucose transport in rat muscle cells,[2] the RA-all-*trans* retinoic acid receptor complex may play a role in insulin action. Thus, all-*trans* retinoic acid receptor α (RARα) expression and all-*trans*-retinoic acid receptor (RAR) specific binding characteristics were investigated in the liver of rats with high-sucrose (HS) diet–induced insulin resistance[3] and compared to those of rats fed diets enriched with fish oil. Thyroid hormone receptor α1 mRNA accumulation in the liver of rats with HS diet–induced insulin resistance was also evaluated.

MATERIALS AND METHODS

Male Wistar rats (VELAZ, Prague) weighing approximately 250 g were randomized into four groups of ten animals each and were fed a basal (B) or HS (63 cal%) diet with or without fish oil (FO) supplement (Activepa 30-TG, Martens, Norway, 30% w/w of total fatty acids) for two weeks.

Highly purified rat liver nuclei and the fraction of nuclear proteins containing nuclear receptors of the steroid/thyroid receptor family were prepared by procedures

---

[a] This work was supported in part by VEGA Grant No. 2/3015/96 and GAV Grant No. 2/544/94.

according to DeGroot and Torresani.[4,5] Nuclear all-*trans* retinoic acid receptors (RAR) were determined in the dark in a high-ionic-strength buffer (0.3 M KCl, 1 mM $MgCl_2$, 10 mM Tris-HCl buffer, pH 7.0) according to a method recently developed in our laboratory.[6] Total cytoplasmic RNA was isolated from the liver of rats according to the method of Chirgwin et al.,[7] and polyadenylated mRNA was isolated using an oligo(dT) cellulose column. The Pvu II fragment from plasmid rbe A12-P500 that contained the neuronal c-erbA (thyroid hormone receptor α1) cDNA (nucleotide position 607–1113) and the KpnI/SacI 503 base-pair fragment from Lambda gt 11 that contained the RARα cDNA were used as the random prime-labeled probes. Hybridizations were performed in formamide (50%) at 42°C, and Hybond N+ membranes after autoradiography were scanned by laser densitometry (ULTROSCAN, LKB, Bromma, Sweden). All data are presented as the mean ± SD and ANOVA was used to compare groups.

RESULTS

As shown in FIGURE 1, we found a significant augmentation ($p < 0.01$) in RARα mRNA accumulation in rats fed HS diet when compared to rats fed B diet. In comparison with rats fed B + FO diet, a significant increase ($p < 0.005$) in the RARα expression was found in rats fed HS + FO diet. In [^{3}H]-RA binding studies, Scatchard plots confirmed a significant increase ($p < 0.05$) in the RAR maximal binding capacity (B_{max}) only in rats fed HS + FO diet when compared to rats fed B diet (FIG. 2). No significant changes were observed in the equilibrium association constant (K_a) in rats fed HS, HS + FO, or B + FO diets when compared to rats fed B diet.

In contrast to RARα, a significant decrease ($p < 0.005$) in nuclear thyroid hormone receptor α1 (TRα1) expression was found in rats fed HS diet when compared to rats fed B diet (FIG. 1). As shown in FIGURE 1, a significant decrease ($p < 0.05$) in the TRα1 expression was also detected in rats fed HS + FO diet in comparison with rats fed B + FO diet. In addition, an analogous pattern in the expression of the TR isoform (TRα2) was observed as well. Fish oil did not have any effect on retinoic acid receptor α expression; however, it decreased the expression of the thyroid hormone receptor α1 in rat liver in comparison with rats fed B diet.

DISCUSSION

RA binds specifically to its cognate nuclear receptors (RAR), and the RA-RAR complex acts as a transcriptional activator by binding to hexameric nucleotide sequences (AGGTCA) in the responsive elements of target genes that are also well recognized by thyroid hormone receptors as well as by dihydroxy-vitamin D_3 receptors (VDR). RAR homodimerization or heterodimerization with RXR, TR, or VDR was found to be important for stable interaction of nuclear receptors with their cognate responsive elements.[8]

Recent data demonstrating a rapid stimulation of glucose transport and glucose

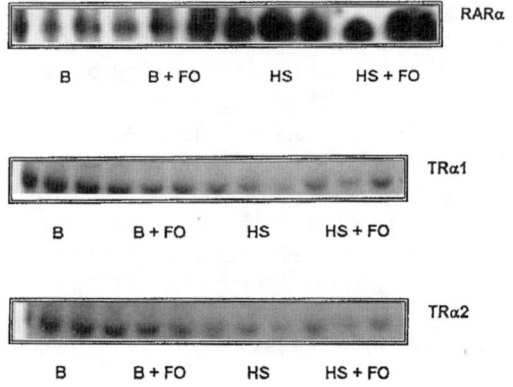

FIGURE 1. The effect of basal (B) or high-sucrose (HS) diet with or without fish oil (FO) on the expression of the nuclear retinoic acid receptor α (RARα), the thyroid hormone receptor α1 (TRα1), and the variant TRα2 in rat liver. Quantitative evaluation of autoradiograms from Northern blots of polyadenylated mRNA was accomplished by laser densitometry. B versus HS, B + FO versus HS + FO: (∗) $p < 0.05$, (∗∗) $p < 0.01$, (∗∗∗) $p < 0.005$; B versus B + FO: (†) $p < 0.05$.

transporter expression in both aligned myoblasts and mature myocytes by RA[2] suggested that the RA-RAR complex might play a role in insulin action. Our results on liver RAR and TR status in rats with the high-sucrose diet–induced insulin resistance showed an increase in both RARα expression and RAR concentration. The affinity of RAR remained unchanged and, in contrast, TRα1 and TRα2 mRNA accumulation decreased. A possible explanation for these results is that stimulation of the RAR may represent an early event in RA action in diet-induced insulin resistance.

SUMMARY

Retinoic acid receptor α (RARα) expression and RAR binding characteristics were investigated in the liver of rats with high-sucrose (HS) diet–induced insulin resistance. Animals were fed a basal (B) or HS (63 cal%) diet with or without fish oil (FO) (30% w/w of total fatty acids) for two weeks. A significant augmentation ($p < 0.01$) in the RARα mRNA accumulation in rats fed HS diet when compared to rats fed B diet

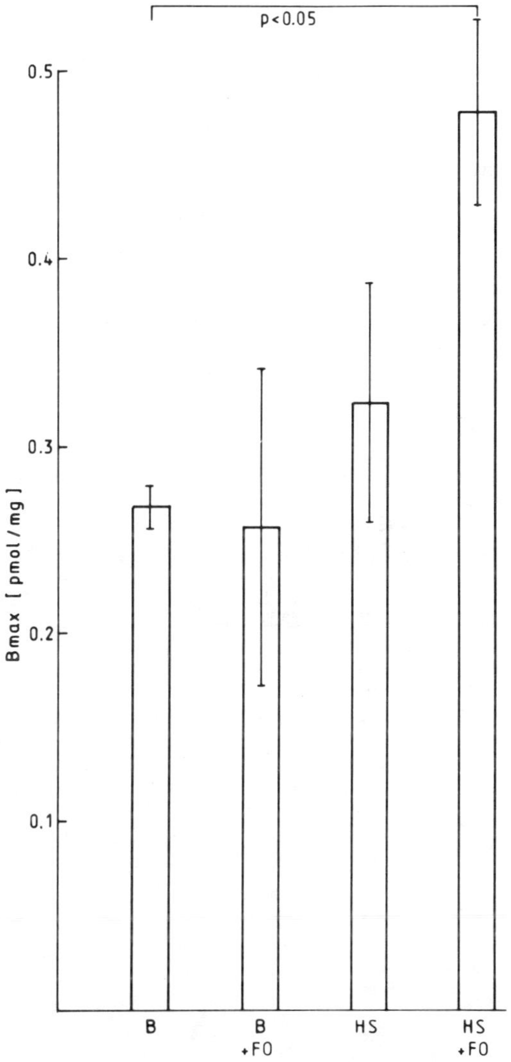

FIGURE 2. Maximal binding capacity (B_{max}) of liver nuclear all-*trans* retinoic acid receptors (RAR) in rats fed basal (B) diet (K_a = 1.32 ± 0.85 × 10^9 L/mol), B + fish oil (FO) diet (K_a = 1.38 ± 1.10 × 10^9 L/mol), high-sucrose (HS) diet (K_a = 1.83 ± 1.04 × 10^9 L/mol), and HS + FO diet (K_a = 0.87 ± 0.43 × 10^9 L/mol).

was demonstrated. In comparison with rats fed B + FO diet, a significant increase ($p < 0.005$) in the RARα expression was found in rats fed HS + FO diet. In [^{3}H]-retinoic acid (RA) binding studies, Scatchard plots confirmed a significant increase ($p < 0.05$) in the RAR maximal binding capacity (B_{max}) only in rats fed HS + FO diet when compared to rats fed B diet. No significant changes in the association constant (K_a) were found among the groups when compared to rats fed B diet. In contrast to RARα, a significant decrease ($p < 0.005$) in nuclear thyroid hormone receptor α1 (TRα1) expression was found in rats fed HS diet when compared to rats fed B diet.

A significant decrease ($p < 0.05$) in the TRα1 expression was also detected in rats fed HS + FO diet in comparison with rats fed B + FO diet. In addition, an analogous pattern in the expression of the TR isoform (TRα2) was evaluated as well.

In conclusion, the high-sucrose diet–induced insulin resistance might be associated with an increased RARα expression and RAR population, and also with a decreased TRα1 and TRα2 mRNA accumulation.

ACKNOWLEDGMENT

We wish to acknowledge the generosity of Ronald M. Evans for providing the specific cDNA probes.

REFERENCES

1. EVANS, R. M. 1988. The steroid and thyroid hormone receptor superfamily. Science **240:** 889–895.
2. SLEEMAN, M. W., H. ZHOU, S. ROGERS, K. W. NG & J. D. BEST. 1995. Retinoic acid stimulates glucose transporter expression in L6 muscle cells. Mol. Cell. Endocrinol. **108:** 161–167.
3. KLIMEŠ, I., E. ŠEBÖKOVÁ, A. VRÁNA & L. KAZDOVÁ. 1993. Raised dietary intake of N-3 polyunsaturated fatty acids in high sucrose-induced insulin resistance: animal studies. Ann. N.Y. Acad. Sci. **683:** 69–81.
4. DEGROOT, L. J. & J. TORRESANI. 1975. Triiodothyronine binding to isolated liver cell nuclei. Endocrinology **96:** 357–369.
5. TORRESANI, J. & L. J. DEGROOT. 1975. Triiodothyronine binding to liver nuclear solubilized proteins *in vitro*. Endocrinology **96:** 1201–1209.
6. BRTKO, J. 1994. Accurate determination and physicochemical properties of rat liver nuclear retinoic acid (RA) receptors. Biochem. Biophys. Res. Commun. **204:** 439–445.
7. CHIRGWIN, J. M., A. E. PRZYBYLA, R. J. MACDONALD & W. J. RUTTER. 1979. Isolation of biologically active ribonucleic acid from sources enriched in ribonuclease. Biochemistry **13:** 5294–5299.
8. BUGGE, T. H., J. POHL, O. LONNOY & H. G. STUNNENBERG. 1992. RXRα, a promiscuous partner of retinoic acid and thyroid hormone receptors. EMBO J. **11:** 1409–1418.

Diet-induced Insulin Resistance Is Associated with Decreased Activity of Type I Iodothyronine 5'-Deiodinase in Rat Liver[a]

J. BRANSOVÁ,[b] A. BRTKOVÁ,[c] E. ŠEBÖKOVÁ,[b]
I. KLIMEŠ,[b] P. LANGER,[b] AND J. BRTKO[b]

[b]Institute of Experimental Endocrinology
Slovak Academy of Sciences
833 06 Bratislava, Slovak Republic
[c]Research Institute of Nutrition
833 37 Bratislava, Slovak Republic

Dietary manipulations leading to the development of insulin resistance[1,2] are believed to be involved in the regulation of type I iodothyronine 5'-deiodinase (5'-DI) activity.[3] 5'-DI, an enzyme consisting of two identical 27-kDa subunits with one atom of selenium per molecule, catalyzes the conversion of the main secretory product of the thyroid, the prohormone thyroxine, to the active 3,5,3'-triiodothyronine (T_3), predominantly in liver, kidney, and thyroid.[4] Since T_3, originating from the monodeiodination of thyroxine (T_4) by 5'-DI, is involved in a complex array of developmental responses and metabolic functions,[4] the present study was undertaken to investigate the effects of feeding Wistar rats with high-sucrose or high-fat diets for two weeks on serum selenium concentration, T_3 and T_4 levels, and 5'-DI activity in the liver.

MATERIALS AND METHODS

Male Wistar rats (VELAZ, Prague) weighing approximately 250 g were randomized into three groups of ten animals each and were fed a basal (B), high-sucrose (HS) (63 cal% of sucrose), or high-fat (HF) (30 cal% of fat) diet for two weeks. Selenium in serum was determined by graphite-furnace atomic absorption spectrometry (VARIAN SpectrAA-30, Australia) with deuterium background and a reduced palladium modifier, according to Jacobson and Lockitch.[5] T_3 and T_4 were determined by radioimmunoassay using specific antisera prepared at our institute.[6]

---

[a] This work was supported in part by VEGA Grant No. 2/3015/96 and GAV Grant No. 2/544/94.

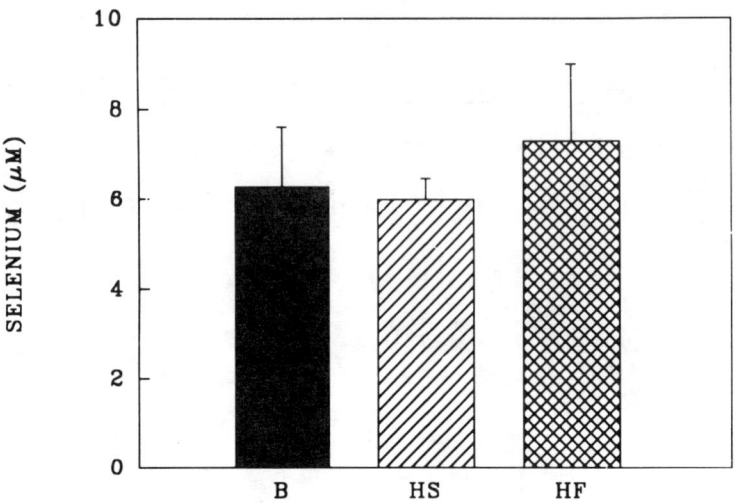

FIGURE 1. Serum selenium concentration in rats fed basal (B), high-sucrose (HS), or high-fat (HF) diets for two weeks. Data are presented as the mean ± SD.

The livers of Wistar rats fed the B, HS, or HF diet were homogenized by sonication (three times for 5 s) in ice-cold homogenization buffer containing 0.25 M sucrose, 20 mM Hepes (pH 7.4), 1 mM ethylenediaminetetraacetic acid (EDTA), and 1 mM D,L-dithiothreitol (DTT). 5'-DI activity was determined according to Leonard and Rosenberg[7] by the release of $^{125}I^-$ from [^{125}I]-3,3',5'-triiodo-L-thyronine (reverse triiodo-L-thyronine, rT$_3$) using 2 µM nonradioactive rT$_3$ and 40 mM DTT in the absence (total iodothyronine deiodinase activity) or presence of 0.1 mM 6-n-propyl-2-thiouracil (PTU). The fraction of iodide release blocked by PTU was assigned to the 5'-DI activity. Specific activity of the 5'-DI was expressed as pmol of $^{125}I^-$ released per min and per mg of protein. Protein concentration was determined by the method of Lowry using human albumin as a standard.[8] All data are presented as the mean ± SD and ANOVA was used to evaluate the significance of the data.

RESULTS

No significant differences in serum selenium concentration were found among the groups of rats fed B, HS, or HF diets (FIG. 1). On the other hand, a significant reduction of the 5'-DI activity ($p < 0.005-0.05$) was found in groups of rats fed either HS or HF diet in comparison with rats fed B diet (FIG. 2).

No differences in serum T$_3$ level were detected between rats fed B or HS diet. However, a statistically significant increase in serum T$_4$ concentration ($p < 0.001$) was observed in rats fed HS diet when compared to rats fed B diet (TABLE 1).

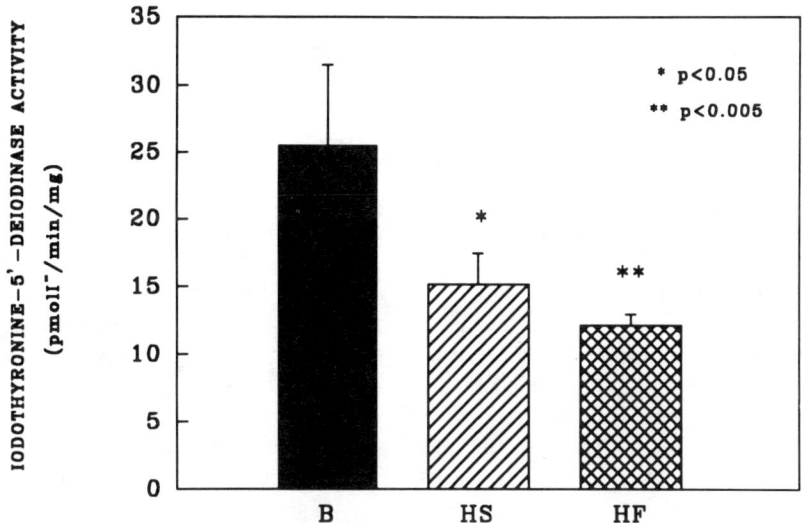

FIGURE 2. The effect of feeding Wistar rats with basal (B), high-sucrose (HS), or high-fat (HF) diets (for two weeks) on the activity of type I iodothyronine 5'-deiodinase in the liver. The activity of the 5'-DI is expressed as pmol of $^{125}I^-$ released per min and per mg of protein. Data are presented as the mean ± SD and are compared by ANOVA.

DISCUSSION

High 5'-DI activities occur in liver, kidney, and thyroid, as well as in euthyroid pituitary, CNS, placental membranes, lactating mammary glands, heart, skeletal muscle, lung, pancreas, spleen, intestine, and skin. Thus, 5'-DI appears to play an important role as "gatekeeper" for the nuclear 3,5,3'-triiodo-L-thyronine receptors that mediate most of the thyroid hormone action.[3] *In vivo* and *in vitro* studies explicitly demonstrate that gene expression of 5'-DI is regulated by selenium status.[9] Carbohydrate-dependent metabolic signals are also believed to be involved in regulation of the 5'-DI activity.[3] Several mechanisms of regulation of 5'-DI activity have been proposed, but none have been clearly proven. Moreover, genetic defects of 5'-DI have also been suggested to contribute to the phenotype of obese Zucker rats.[10] Our results suggest that

TABLE 1. Serum T_3 and T_4 Levels (ng/mL) in Rats Fed Basal or High-Sucrose Diets[a]

	Basal Diet		High-Sucrose Diet	
n	T_3	T_4	T_3	T_4
8	1.16 ± 0.3	24.97 ± 2.9	–	–
6	–	–	1.49 ± 0.2	42.10 ± 10.5

[a] Data are presented as the mean ± SD and are compared by ANOVA (T_4: B versus HS, $p < 0.001$).

decreased 5'-DI activity in high-sucrose or high-fat diet-induced insulin resistance accompanied by increased serum T_4 is not due to selenium status, but it may involve other diet-related factors.

SUMMARY

The effect of feeding Wistar rats with high-sucrose (63 wt% of sucrose, HS) or high-fat (30 wt% of fat, HF) diets for two weeks on serum selenium concentration and type I iodothyronine 5'-deiodinase (5'-DI) activity in liver was investigated. No significant differences in serum selenium concentration (as determined by graphite-furnace atomic absorption spectrometry) were found among the groups of rats fed basal, HS, or HF diets. A significant reduction of the 5'-DI activity ($p < 0.005-0.05$) was found in groups of rats fed either HS or HF diet in comparison with rats fed B diet.

In conclusion, it is suggested that decreased 5'-DI activity in HS or HF diet-induced insulin resistance is not due to selenium status, but it may involve other dietary-related factors.

REFERENCES

1. KLIMEŠ, I., E. ŠEBÖKOVÁ, A. VRÁNA & L. KAZDOVÁ. 1993. Raised dietary intake of N-3 polyunsaturated fatty acids in high sucrose-induced insulin resistance: animal studies. Ann. N.Y. Acad. Sci. **683**: 69-81.
2. STORLIEN, L. H., D. A. PAN, A. D. KRIKETOS & L. A. BAUR. 1993. High fat diet-induced insulin resistance: lessons and implications from animal studies. Ann. N.Y. Acad. Sci. **683**: 82-90.
3. KÖHRLE, J. 1994. Thyroid hormone deiodination in target tissues—a regulatory role for trace element selenium? Exp. Clin. Endocrinol. **102**: 63-89.
4. KÖHRLE, J., L. SCHOMBURG, S. DRESCHER, E. FEKETE & K. BAUER. 1995. Rapid stimulation of type I 5'-deiodinase in rat pituitaries by 3,3',5-triiodo-L-thyronine. Mol. Cell. Endocrinol. **108**: 17-21.
5. JACOBSON, B. J. & B. LOCKITCH. 1988. Direct determination of selenium in serum by graphite-furnace atomic absorption spectrometry with deuterium background correction and reduced palladium modifier. Clin. Chem. **34**: 709.
6. FÖLDES, O., P. LANGER, K. STRAUSSOVÁ, H. BROZMANOVÁ & K. GSCHWENDTOVÁ. 1982. Direct quantitative estimation of several iodothyronines in rat bile by radioimmunoassay and basal data on their biliary excretion. Biochim. Biophys. Acta **716**: 383-390.
7. LEONARD, J. L. & I. N. ROSENBERG. 1980. Iodothyronine 5'-deiodinase from rat kidney: substrate specificity and the 5'-deiodination of reverse triiodothyronine. Endocrinology **107**: 1376-1383.
8. LOWRY, O. H., N. J. ROSENBROUGH, A. L. FARR & R. J. RANDALL. 1951. Protein measurement with Folin reagent. J. Biol. Chem. **193**: 265.
9. KÖHRLE, J. 1996. Thyroid hormone deiodinases—a selenoenzyme family acting as gatekeepers to thyroid hormone action. Acta Med. Austriaca **23**: 17-30.
10. KATZEFF, H. L. & C. SELGRAD. 1993. Impaired peripheral thyroid hormone metabolism in genetic obesity. Endocrinology **132**: 989-995.

Insulin and Catecholamines Act at Different Stages of Rat Liver Regeneration[a]

J. KNOPP, L. MACHO, M. FICKOVÁ, Š. ZÓRAD,
R. KVETŇANSKÝ, AND I. JAROŠČÁKOVÁ

Institute of Experimental Endocrinology
Slovak Academy of Sciences
833 06 Bratislava, Slovak Republic

INTRODUCTION

Both insulin and catecholamines are required for optimal stimulation of mitogenesis in the liver in partially hepatectomized rats. The signals (e.g., second messengers) initiated by these hormones that control rat liver regeneration are not clear. In studying the significant role of insulin in the stimulation of liver growth, it was demonstrated[1] (using a membrane fraction of liver cells) that there are no changes in insulin binding during the early period (2–24 h) of regeneration after partial hepatectomy (PH). However, no further studies were performed on the effect of insulin administration on insulin receptor binding capacity during rat liver regeneration.

Circulating catecholamines could be involved in the promotion of the rat liver regeneration. In spite of several suggestions of the importance of the sympathoadrenal system in this phenomenon, there is no direct evidence for the secretion of catecholamines in the rat after partial hepatectomy. Recent published data showed that plasma catecholamine levels increased at various times after PH.[2] However, in this study, plasma catecholamine levels were measured in blood collected from abdominal aorta and vena cava under ether anesthesia, and ether anesthesia is known to be followed by a substantial rise of plasma epinephrine and norepinephrine, even without surgery.[3]

In the present study, we examined plasma catecholamine levels in conscious cannulated rats during 1–5 days of regeneration under carefully controlled conditions. Further insulin binding parameters were investigated using a whole cell system instead of a membrane fraction of rat liver.

METHODS

Adult male Wistar rats (300–350 g) were subjected to 60% partial hepatectomy.[4] Sham-operated controls were subjected to a laparotomy. For insulin binding studies,

[a] This research was supported by the Slovak Grant Science Agency: GAV Grant Nos. 2/1289/96 and 2/541/96.

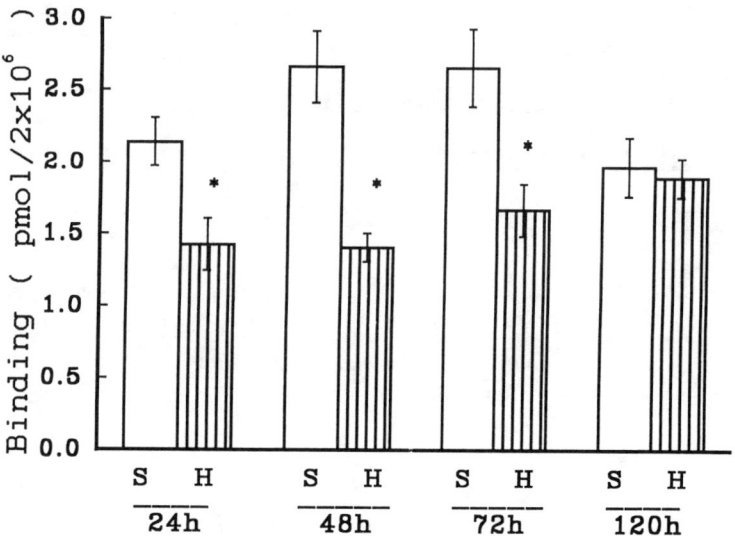

FIGURE 1. Insulin binding capacity to isolated hepatocytes after partial hepatectomy: (*) sham-operated (S) versus partially hepatectomized (H) ($p < 0.05$); period after surgery in hours.

hepatocytes were isolated by the *in situ* method and receptor binding capacity was determined by the method described previously.[5] Insulin was administered subcutaneously (2.5 IU/kg body weight) to rats at 24 h after sham operation or partial hepatectomy.

Plasma catecholamine levels were measured in cannulated rats.[6] In these animals, after 20–24 h, a control blood sample was taken from each rat. Then, partial hepatectomy or a sham operation was performed. Further blood samples (0.5–0.8 mL) were collected at 20, 60, and 240 min and 24 h after surgery. Plasma catecholamines were measured in 50-μL aliquots of plasma by a radioenzymatic assay. All data are presented as the mean ± SEM. One-way analysis of variance was used to analyze the data.

RESULTS

Significant decreases in the insulin receptor binding capacity and in the number of insulin receptors were found in hepatocytes of rats at 1, 2, and 3 days after partial hepatectomy; after 5 days, no differences were noted (FIG. 1). Plasma insulin and glucose levels were not significantly different. Insulin injection did not influence the changes that occurred in insulin receptor number. An increase of membrane microviscosity in regenerating liver cells was also observed.

Plasma epinephrine and norepinephrine levels increased rapidly at 20 min after both laparotomy and partial hepatectomy (FIG. 2). After 60 min, however, plasma norepinephrine levels returned to control values, while epinephrine levels in partially hepatectomized rats remained significantly elevated as compared to rats with laparotomy.

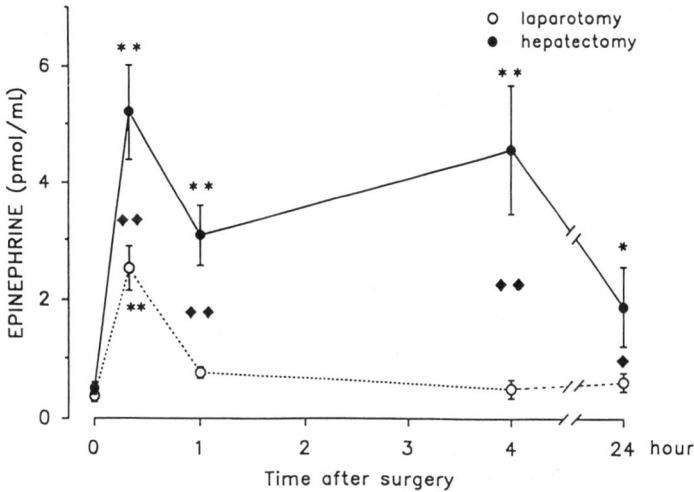

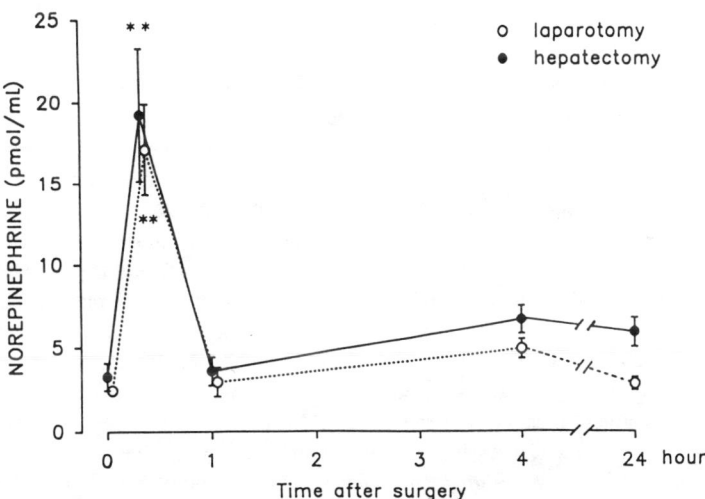

FIGURE 2. Plasma levels of epinephrine (top) and norepinephrine (bottom) in rats after laparotomy and partial hepatectomy. Data are mean ± SEM values for eight rats: (∗) $p < 0.05$, (∗∗) $p < 0.01$ compared to controls; (♦) $p < 0.05$, (♦♦) $p < 0.01$ compared to the appropriate group after laparotomy.

DISCUSSION

The results of our experiments demonstrate a significant decrease in the binding capacity of insulin membrane receptors in hepatocytes during liver regeneration after partial hepatectomy. Insulin receptor binding capacity is regulated by plasma insulin levels,[7] by the physicochemical properties of the plasma membrane,[8] and by the action of several other hormones.[9] In our experiments, no differences in the plasma levels of insulin and glucose were found. However, the transfer of all of the insulin produced in the pancreas by portal venous blood into only one-third of the liver could result in the downregulation of the insulin receptors. The increased microviscosity of the cell membrane and the increased plasma corticosterone levels after partial hepatectomy suggest that plasma corticosterone and plasma membrane properties may also play a role in the insulin receptor alteration.

One of the pathways involved in the control of proliferation of the rat liver after partial hepatectomy is the cyclic AMP system.[10] Early in regeneration, there is an increase in the concentration of cAMP in the rat liver. Only an indirect effect of agents that interfere with catecholamine receptors and cAMP has been demonstrated. Plasma epinephrine levels in our studies strongly support the idea that the peaks of cAMP concentration at 4 h and 12 h after partial hepatectomy might be due to epinephrine. These results suggest that epinephrine plays a role in stimulating insulin in rat liver regeneration after partial hepatectomy.

SUMMARY

Insulin and catecholamines are known to exert effects on hepatocyte growth and metabolism. The binding of insulin, the plasma levels of insulin (INS), and the plasma catecholamine levels of epinephrine (EPI) and norepinephrine (NE) were measured during liver regeneration after partial hepatectomy (PH). A significant decrease ($p < 0.05$) of INS receptor binding capacity was found at 1, 2, and 3 days after operation. A single insulin injection (2.5 IU/kg body weight) at 24 h after sham operation or partial hepatectomy did not affect these changes of INS binding to hepatocytes. The plasma insulin and glucose levels were similar in both hepatectomized and sham-operated rats. Within 20 min after liver resection or sham operation, plasma NE and EPI concentrations increased rapidly. Then, a significant decrease was observed in plasma catecholamine levels at 1 h after laparotomy and PH. In both groups, laparotomized and partially hepatectomized plasma levels of NE at 4 h reached control values and remained unchanged at the 4- and 24-h periods. After PH, the levels of EPI remained elevated at 4 h in comparison with laparotomy. Adrenal tyrosine hydroxylase mRNA levels were significantly elevated at 4 h in both PH and sham-operated groups. These results suggest that signals that are initiated by catecholamines and transduced through second messengers presumably participate in the trigger mechanism of liver regeneration, while insulin (considered as a secondary mitogen) enhances a stimulus for liver regeneration.

ACKNOWLEDGMENTS

We thank H. Farkasova, D. Janovova, I. Szalayova, M. Skultetyova, and E. Visztova for their technical assistance.

REFERENCES

1. LEFERT, H., N. M. ALEXANDER, G. FALOONA, B. RUBALCAVA & R. UNGER. 1975. Specific endocrine and hormonal receptor changes associated with liver regeneration in adult rats. Proc. Natl. Acad. Sci. U.S.A. **72:** 4033-4038.
2. CRUISE, J. L., S. J. KNECHTLE, R. R. BOLLINGER, C. KHAN & G. MICHALOPOULOS. 1987. Alpha-adrenergic effects of liver regeneration. Hepatology **7:** 1189-1194.
3. MICHALIKOVÁ, S., H. BALAZOVÁ, D. JEŽOVÁ & R. KVETŇANSKÝ. 1991. Changes in circulating catecholamine levels in old rats under basal conditions and during stress. Bratisl. Lek. Listy **91:** 689-693.
4. HIGGINS, G. M. & R. M. ANDERSON. 1933. Experimental pathology of the liver. I. Restoration of the liver of the white rat following partial hepatectomy. Cancer Res. **46:** 1318-1323.
5. ZÓRAD, Š., M. FICKOVÁ, I. KLIMEŠ & L. MACHO. 1986. Comparison of goodness of fit of two mathematical models for the estimation of insulin binding. Diabetol. Croat. **15:** 183-191.
6. KVETŇANSKÝ, R., C. SUN, C. R. LAKE, N. B. THOA, T. TORDA & I. J. KOPIN. 1978. Effects of handling and forced immobilization on rat plasma levels of epinephrine, norepinephrine, and dopamine beta-hydroxylase. Endocrinology **103:** 1868-1874.
7. YOKONO, K., Y. IMAMURA, K. SHII, H. SAKAI & S. BABA. 1982. Insulin binding and degradation in liver of fed and fasted rats: effect of antiserum to insulin-degrading enzyme on insulin binding and degradation. Endocrinol. Jpn. **29:** 299-306.
8. GINSBERG, B. H., P. CHATERJEE & M. YOREK. 1990. Increased membrane fluidity is associated with greater sensitivity to insulin. *In* Insulin and Cell Membrane, p. 313-329. Gordon & Breach. New York.
9. MACHO, L. & M. FICKOVÁ. 1992. *In vivo* role of corticosterone in regulation of insulin receptors in rat adipocytes during hypokinesia. Endocrinol. Exp. **26:** 183-187.
10. THROWER, S. & M. G. ORD. 1974. Hormonal control of liver regeneration. Biochem. J. **144:** 361-369.

Fatty Acid Composition in Fractions of Structural and Storage Lipids in Liver and Skeletal Muscle of Hereditary Hypertriglyceridemic Rats[a]

PAVOL BOHOV, ELENA ŠEBÖKOVÁ,
DANIELA GAŠPERÍKOVÁ, PAVEL LANGER,
AND IWAR KLIMEŠ

Diabetes and Nutrition Research Group
Institute of Experimental Endocrinology
Slovak Academy of Sciences
SK-833 06 Bratislava, Slovak Republic

It has been demonstrated both in rats[1,2] and in humans[3,4] that insulin action is associated with the fatty acid (FA) profile, mainly in skeletal muscle phospholipids. A shift in FA composition towards a higher proportion of saturated fatty acids (SFA) and conversely a lower percentage of polyunsaturated fatty acid (PUFA) n-3 series or PUFA n-6 and n-3 series with 20 and more carbon atoms (C20-22 PUFA) may play a critical role in insulin resistance.[1-4]

It is well documented that diet influences the FA profile in structural and storage lipids[5] and induces changes in plasma membrane fluidity.[6] Because changes in the lipid environment in the close vicinity of membrane proteins may regulate their activity, dietary manipulation may have functional consequences.[7]

The recently selected nonobese hereditary hypertriglyceridemic (HTG) rat[8] was shown to manifest a majority of the metabolic abnormalities known as the "Syndrome X,"[9] including hypertriglyceridemia, insulin resistance,[10,11] and hypertension.[12] Little is known, however, about whether abnormal lipid metabolism in these animals[11] is accompanied also by changes in FA metabolism.

This study was undertaken to examine the influence of genetic background and the effect of a diet containing high sucrose and/or highly polyunsaturated fat on the fatty acid composition of polar and storage lipids in liver and skeletal muscle of HTG rats.

---

[a] This work was supported by research grants of the Institute of Experimental Endocrinology, Slovak Academy of Sciences, Bratislava, provided by the Slovak Grant Agency for Science (GAV) (Grant No. 2/544/92-95), the Slovak Diabetes Association (Grant No. 1/95), and a COST B5 grant.

MATERIALS AND METHODS

Animals and Diets

Adult, male HTG rats bred in our facility with initial weights of 260–310 g were housed in wire-mesh cages in a temperature and light-controlled room. Eight rats per group were fed *ad libitum* four types of diets for 14 days. The commercially available standard diet for laboratory rats (Velaz, Prague, Czech Republic) was supplemented (10 wt%) with partially hydrogenated beef tallow (B) or with fish oil (BFO) (Activepa 30 TG, Martens, Norway). Partially purified high-sucrose diet was also supplemented (10 wt%) with partially hydrogenated beef tallow (HS) or with fish oil (HSFO) as reported earlier.[11] The diets were prepared weekly and stored at −20°C. The fatty acid composition of these diets is shown in TABLE 1. Control (NTG) male Wistar rats (Velaz, Prague, Czech Republic) were fed the same experimental diets. All experiments reported herein were approved by the Institute of Experimental Endocrinology Animal House Ethics Committee.

Biochemical Analyses

Plasma triglycerides were examined in the fed state after dietary treatment (Kit No. 1301132, Lachema, Brno, Czech Republic). Plasma from blood samples of overnight (16–18 h) fasted rats obtained after decapitation was used for assays of insulin (RIA kit from Novo Nordisk, Copenhagen, Denmark) and glucose (Beckman glucose analyzer, Fullerton, California). Liver and skeletal muscle (quadriceps femoris) were removed after decapitation and immediately frozen in liquid nitrogen.

Lipid Analysis

Lipids from the tissues of four HTG and four NTG animals were extracted with chloroform-methanol (1:2 v/v)[13] and separated by thin-layer chromatography on precoated silica gel plates (Merck No. 11844, 10 × 20 cm, 0.25-mm layer, Darmstadt, Germany). The developing solvent system for the separation of total phospholipids from neutral lipids was a mixture of hexane–diethyl ether–acetic acid (80:20:2, v/v/v). Lipid fractions (phosphatidylcholine, phosphatidylserine, phosphatidylinositol, phosphatidylethanolamine, cardiolipin, free fatty acids, diglycerides, triglycerides, and cholesterol esters) were separated first with solvent system hexane–diethyl ether (1:1, v/v) followed with chloroform-ethanol-triethylamine-acetone-water (15:20:22:2:3, all v/v). Spots were detected under ultraviolet light after spraying with Rhodamine 6G (0.05% w/v in methanol). The fractions of phosphatidylcholine and triglycerides were scraped off into a test tube and eluted from silica gel with chloroform-methanol (1:2).[14] The results of FA analyses of total phospholipids, phosphatidylcholine, and triglycerides are presented in this work.

TABLE 1. Fatty Acid Composition of Experimental Diets (wt%)[a]

Fatty Acids[b]	B	BFO	HS	HSFO
14:0	1.88	1.07	4.09	3.19
16:0	20.24	10.89	22.32	12.28
16:1n-7	1.61	2.87	1.82	3.13
18:0	16.07	4.94	18.57	5.57
18:1n-9t	5.14	0.24	6.19	0.22
18:1n-7t	2.79	–	3.05	–
18:1n-9	27.22	14.13	28.04	12.55
18:1n-7	2.19	2.06	2.17	1.79
18:2n-6	14.84	12.34	5.59	2.53
18:3n-6	–	0.35	–	0.42
18:3n-3	2.16	1.92	1.25	0.99
18:4n-3	–	2.61	–	3.05
20:3n-6	–	0.31	–	0.37
20:4n-6	–	1.64	–	1.94
20:5n-3	0.73	20.76	–	23.71
22:5n-3	0.12	3.47	–	4.18
22:6n-3	0.49	16.09	–	17.79
SFA	40.80	18.20	49.92	24.08
MUFA	32.65	21.27	33.80	19.40
PUFAn-6	14.93	15.43	5.59	6.48
PUFAn-3	3.50	44.86	1.25	49.71
trans FA	8.12	0.24	9.44	0.33
PUFAn-6M	–	3.08	–	3.96
PUFAn-3M	1.34	42.94	–	48.72
PUFA/SFA	0.45	3.31	0.14	2.33
PUFAn-6/PUFAn-3	4.28	0.34	4.49	0.13
18:2n-6/SFA	0.36	0.68	0.11	0.10
C20-22 PUFA	1.43	42.92	–	49.07
DBI	0.85	2.92	0.58	3.02

[a] Abbreviations: B = basal diet; BFO = basal diet supplemented with fish oil; HS = high-sucrose diet; HSFO = high-sucrose diet supplemented with fish oil; SFA = total saturated fatty acids; MUFA = total monounsaturated fatty acids; PUFA = total polyunsaturated fatty acids; M = total metabolites; C20-22 PUFA = 20:3n-6 + 20:4n-6 + 20:5n-3 + 22:4n-6 + 22:5n-6 + 22:5n-3 + 22:6n-3; DBI = double-bond index, which is the sum of the percentage of each fatty acid multiplied by the number of double bonds in that fatty acid divided by 100; t = trans-unsaturated fatty acids.
[b] Minor fatty acids are not shown in the table, so columns may not add up to 100%.

Fatty Acid Analysis

The fatty acid methyl esters were prepared after hydrolysis of lipids with 0.5 M KOH in ethanol-water (9:1, v/v) by esterification with diazomethane. Fatty acid determination was performed by gas-liquid chromatography (GC) (Finnigan, Model 9001, Austin, Texas) with flame ionization detection using an SP 2340 fused silica capillary column (60 m × 0.32 mm × 0.20 µm) (Supelco, Bellefonte, Pennsylvania). Details of FA preparation and GC analysis are described elsewhere.[15] Data were quantified based on heneicosanoic acid (21:0) as an internal standard with the aid of the CSW 1.6 chromatography station (Data Apex, Prague, Czech Republic).

Statistical Analysis

The effect of rat strain and dietary treatment on biochemical parameters and FA composition was examined by the two-way analysis of variance (ANOVA) procedure. The Bonferroni multiple-range test was used to test the significance of differences ($p < 0.05$ at least) between the means of individual groups with the effect of diet included in the model.

RESULTS

The initial body weight of the HTG rats was about 15% lower than that of the age-matched control animals (TABLE 2). Weight gain after dietary treatment was significantly lower in all HTG groups in comparison to NTG controls.

Plasma triglyceride levels were significantly elevated in HTG rats on B and HS diets. In both rat strains, the plasma triglyceride levels after fish oil dietary treatment decreased below the values obtained in control animals on B diet. Plasma glucose levels did not differ substantially either between the strains or as an effect of dietary treatment. Plasma insulin was reduced in rats fed the fish oil–supplemented basal diet (TABLE 2).

The composition of major fatty acids present in liver and skeletal muscle phospholipids is illustrated in TABLE 3. When HTG rats on B and HS diets were compared with NTG controls on the same diets, a decreased percentage of total PUFA n-6 metabolites (PUFAn-6M, the sum of n-6 FA derived from 18:2n-6) and C20-22 PUFA and a lower double-bond index (DBI, defined as the number of double bonds per fatty acyl group) were observed in the liver of HTG rats. These effects were largely due to greatly decreased arachidonic acid (20:4n-6) and docosahexaenoic acid (22:6n-3) levels. On the other hand, the amounts of oleic acid (18:1n-9), linoleic acid (18:2n-6), and eicosapentaenoic acid (20:5n-3) were higher in the liver of HTG rats. This pattern was not observed in skeletal muscle. In this tissue, the decreased levels of 18:2n-6 and alpha-linolenic acid (18:3n-3) with the concomitant increased amounts of stearic acid (18:0) and total PUFA n-3 metabolites were especially noticeable. The percentage of *trans*-unsaturated FA was increased in both tissues of HTG rats.

The effect of HS diet was evident mainly in the increased level of 18:1n-9 and total monounsaturated FA (MUFA) in both tissues of both strains. In liver, the level of 18:2n-6 decreased in both rat strains and the amount of 20:4n-6 also decreased in NTG rats. On the other hand, the amount of dihomo-gamma-linolenic acid (20:3n-6) rose in the tissues of HTG rats. Thus, a lower 20:4n-6/20:3n-6 ratio was found in liver phospholipids of HTG rats after HS feeding. A similar effect of HS diet on the 20:4n-6/20:3n-6 ratio was also observed in skeletal muscle of HTG rats.

TABLE 2. Effect of Diet on Body Weight, Weight Gain, Plasma Triglycerides, Glucose, and Insulin Levels of Hereditary Hypertriglyceridemic (HTG) and Control (NTG) Wistar Rats (Mean ± SEM)

	NTG				Effect of Strain	S × D	HTG			
	B	BFO	HS	HSFO			B	BFO	HS	HSFO
Initial body weight (g)	297.5 ± 5.7^d	298.1 ± 3.3^d	290.0 ± 5.3^{b,c,d}	292.5 ± 7.2^{c,d}	***↓	NA	260.5 ± 7.8^{a,b}	263.3 ± 11.0^{a,b,c}	256.3 ± 5.2^a	261.3 ± 9.7^{a,b}
Gain on diet (g)	80.6 ± 5.4^{b,c}	66.9 ± 4.6^b	85.0 ± 3.7^c	71.3 ± 4.9^{b,c}	***↓	NS	15.5 ± 1.3^a	11.4 ± 1.5^a	19.6 ± 2.6^a	16.8 ± 2.7^a
Triglycerides (mM)	2.55 ± 0.27^b	0.89 ± 0.16^a	4.44 ± 0.36^c	1.29 ± 0.19^a	***↑	***	6.05 ± 0.29^d	1.27 ± 0.15^a	7.32 ± 0.05^e	1.82 ± 0.23^{a,b}
Glucose (mM)	5.3 ± 0.16	5.7 ± 0.17	6.1 ± 0.23	6.1 ± 0.20	NS	NS	5.5 ± 0.29	5.6 ± 0.15	5.3 ± 0.37	5.5 ± 0.17
Insulin (μU/mL)	37.4 ± 2.8^{b,c}	25.5 ± 1.5^a	42.8 ± 3.5^c	31.9 ± 2.7^{a,b,c}	NS	NS	36.8 ± 4.0^{a,b,c}	27.5 ± 2.0^{a,b}	33.3 ± 1.8^{a,b,c}	31.6 ± 2.7^{a,b,c}

Note: B = basal diet; BFO = basal diet supplemented with fish oil; HS = high-sucrose diet; HSFO = high-sucrose diet supplemented with fish oil. Values within a line without a common superscript are significantly different, $p < 0.05$. Gain on diet is after 14 days of dietary treatment. S × D, strain × diet interaction; NA, not applicable; NS, not significant; (***) $p < 0.001$; ↑, higher in HTG rats; ↓, lower in HTG rats. A diet main effect was significant (at least $p < 0.01$) for all parameters except glucose (for the initial body weight, this effect was not applicable).

Feeding fish oil increased the percentage of PUFA n-3, C20–22 PUFA, and DBI and decreased the percentage of PUFA n-6 and the PUFA n-6/PUFA n-3 ratio in liver and skeletal muscle of both rat strains. The increase of PUFA n-3 was due to the higher levels of 20:5n-3 and docosahexaenoic acid (22:6n-3) in both tissues. The lower levels of PUFA n-6 were caused particularly by the lower amounts of 18:2n-6 and 20:4n-6. Despite an increase of C20-22 PUFA, the level of these FA was still lower in liver phospholipids of HTG rats than in their NTG counterparts. In contrast, the PUFA n-6 and DBI levels were similar in HTG and NTG rats after fish oil feeding. Feeding fish oil also increased the amount of SFA, especially in the tissues of NTG rats.

Similar trends in FA composition as seen in total phospholipids were observed in the phosphatidylcholine fraction of liver and skeletal muscle (TABLE 4). Some differences, however, were more or less expressed. Thus, in liver, 18:2n-6 levels were higher and total PUFA n-6 metabolite and 20:4n-6 levels were lower in all HTG groups in comparison with the controls. In skeletal muscle, the decrease in 18:3n-3 was significant only in HTG rats fed HS diet.

In triglycerides, total SFA rose in liver of HTG rats on all diets when compared with the NTG strain. This resulted from an increase of palmitic acid (16:0) and 18:0 (TABLE 5). In skeletal muscle, among saturated FA only, the 18:0 level rose in HTG rats fed the B and HS diets. The levels of 20:4n-6 and total PUFA n-6 metabolites were lower in liver of HTG rats on the B and HS diets. In both tissues, the levels of 18:1n-9 and total MUFA were lower in HTG rats. The increased percentage of *trans* FA in HTG rats is also remarkable.

High-sucrose feeding increased the levels of 18:1n-9 and total MUFA in both rat strains, being more pronounced in liver than in skeletal muscle. On the other hand, PUFA n-6 declined after HS diet as a result of decreased levels of 18:2n-6, gamma-linolenic acid (18:3n-6), and 20:4n-6 in liver and 18:2n-6 in skeletal muscle. The levels of total PUFA n-3 also decreased in liver of both strains. This was caused by the decline in the percentage of both 18:3n-3 and 22:6n-3.

The replacement of beef tallow with fish oil substantially increased the levels of 20:5n-3, 22:6n-3, total PUFA n-3, and PUFA n-3 metabolites in triglycerides in both tissues of NTG and HTG rats at the expense of MUFA and, in liver, also at the expense of SFA. Higher values of the PUFA/SFA ratio and DBI were also found along with a lower PUFA n-6/PUFA n-3 ratio. The level of 18:2n-6 decreased in liver and remained unaltered in skeletal muscle. This is in contrast to PUFA n-6 metabolites, which increased in skeletal muscle of both rat strains and in liver of HTG rats.

DISCUSSION

Fatty acid analysis in tissue polar and storage lipids revealed major differences in FA composition between the two strains of rats; however, the overall pattern of change in response to dietary manipulation was similar. Liver and skeletal muscle were particular tissues of interest in our study because of the known high lipid and carbohydrate metabolic rate in liver, whereas skeletal muscle is the most-important tissue for insulin-stimulated glucose metabolism.

Differences between the HTG and NTG strains can be summarized as follows.

TABLE 3. Fatty Acid Composition (wt%) in Liver and Skeletal Muscle Phospholipids of Hereditary Hypertriglyceridemic (HTG) and Control (NTG) Wistar Rats (Mean ± SEM)

Fatty Acids	NTG				Effect of			HTG			
	B	BFO	HS	HSFO	Strain	S × D		B	BFO	HS	HSFO
Liver											
16:0	12.76 ± 0.56^a	16.38 ± 0.66^d	14.75 ± 0.24^{b,c}	17.82 ± 0.71^e	*↓	NS		13.25 ± 0.25^a	15.69 ± 0.35^{c,d}	13.44 ± 0.23^{a,b}	16.21 ± 0.45^d
18:0	24.32 ± 0.40^b	23.56 ± 0.27^{a,b}	22.52 ± 0.38^a	22.08 ± 0.58^a	****↓	*		24.84 ± 0.32^b	23.65 ± 0.13^{a,b}	24.61 ± 0.49^b	23.68 ± 0.24^{a,b}
18:1n-9	4.82 ± 0.08^b	3.37 ± 0.10^a	5.43 ± 0.11^{c,d}	5.58 ± 0.26^d	****↓	***		5.16 ± 0.03^c	4.78 ± 0.03^b	6.24 ± 0.07^e	6.04 ± 0.01^e
18:2n-6	12.21 ± 0.47^{d,e}	9.44 ± 0.38^c	9.37 ± 0.30^c	5.50 ± 0.41^a	****↓	***		17.45 ± 0.19^f	10.70 ± 0.27^{c,d}	12.87 ± 0.44^e	7.23 ± 0.25^b
18:3n-6	0.08 ± 0.004^a	0.04 ± 0.003^a	0.05 ± 0.001^{a,b,c}	0.04 ± 0.003^{a,b}	****↓	***		0.09 ± 0.005^{e,f}	0.06 ± 0.006^{c,d}	0.11 ± 0.007^f	0.06 ± 0.001^{b,c}
20:3n-6	1.57 ± 0.02^{c,d}	1.46 ± 0.10^{b,c}	1.72 ± 0.04^d	1.48 ± 0.03^{b,c,d}	NS	***		1.35 ± 0.06^{a,b,c}	1.25 ± 0.20^{a,b}	2.25 ± 0.10^e	1.17 ± 0.01^a
20:4n-6	26.18 ± 0.62^d	18.28 ± 0.44^b	23.83 ± 0.97^c	17.53 ± 0.27^a	****↓	**		21.83 ± 0.04^{b,c}	17.88 ± 0.20^a	21.02 ± 0.34^b	16.29 ± 0.20^a
18:3n-3	0.06 ± 0.006^{c,d}	0.08 ± 0.009^e	0.05 ± 0.003^{a,b}	0.08 ± 0.005^{d,e}	NS	NS		0.06 ± 0.001^{b,c}	0.07 ± 0.003^{c,d,e}	0.04 ± 0.003^a	0.07 ± 0.001^{c,d,e}
20:5n-3	0.86 ± 0.07^{a,b}	3.96 ± 0.14^d	0.63 ± 0.08^a	5.07 ± 0.39^e	****↓	**		1.76 ± 0.02^c	5.67 ± 0.19^e	1.44 ± 0.06^{b,c}	6.89 ± 0.12^f
22:6n-3	9.70 ± 0.76^b	15.85 ± 0.55^d	13.00 ± 0.98^c	16.11 ± 0.70^d	****↓	NS		6.84 ± 0.12^a	12.98 ± 0.21^c	9.44 ± 0.50^{a,b}	13.51 ± 0.48^{c,d}
SFA	37.99 ± 0.17^a	41.00 ± 0.41^c	38.15 ± 0.46^a	40.87 ± 0.30^c	****↓	***		39.03 ± 0.09^{a,b}	40.47 ± 0.34^{b,c}	39.03 ± 0.30^{a,b}	40.94 ± 0.69^c
MUFA	7.14 ± 0.06^b	6.30 ± 0.18^a	8.67 ± 0.09^d	9.42 ± 0.24^e	****↓	***		7.43 ± 0.07^{b,c}	7.90 ± 0.02^c	9.58 ± 0.18^e	10.21 ± 0.08^f
PUFAn-6	40.63 ± 0.88^d	29.77 ± 0.69^b	35.73 ± 1.22^c	25.12 ± 0.41^a	****↓	NS		41.10 ± 0.11^d	30.41 ± 0.45^b	36.67 ± 0.65^c	25.23 ± 0.39^a
PUFAn-3	12.29 ± 0.80^a	22.63 ± 0.47^{c,d}	15.36 ± 1.11^b	24.31 ± 0.77^d	****↓	NS		10.44 ± 0.13^a	20.54 ± 0.45^c	12.28 ± 0.45^a	22.92 ± 0.64^{c,d}
trans FA	1.60 ± 0.04^d	0.20 ± 0.02^a	1.28 ± 0.02^c	0.22 ± 0.01^a	****↓	**		1.80 ± 0.03^e	0.62 ± 0.02^b	1.70 ± 0.04^{d,e}	0.65 ± 0.04^b
PUFAn-6M	28.42 ± 0.63^d	20.33 ± 0.49^b	26.36 ± 0.98^d	19.63 ± 0.27^{a,b}	****↓	**		23.65 ± 0.09^c	19.71 ± 0.19^{a,b}	23.80 ± 0.27^c	18.00 ± 0.18^a
PUFAn-3M	12.23 ± 0.81^a	22.55 ± 0.48^{c,d}	15.32 ± 1.11^b	24.24 ± 0.77^d	****↓	NS		10.38 ± 0.13^a	20.47 ± 0.45^c	12.24 ± 0.45^a	22.85 ± 0.64^{c,d}
PUFA/SFA	1.39 ± 0.01^c	1.28 ± 0.02^{b,c,d}	1.34 ± 0.03^{d,e}	1.21 ± 0.02^{a,b}	**↓	NS		1.32 ± 0.01^{c,d,e}	1.26 ± 0.02^{b,c}	1.26 ± 0.01^{b,c}	1.18 ± 0.04^a
PUFAn-6/ PUFAn-3	3.37 ± 0.32^{c,d}	1.32 ± 0.06^a	2.37 ± 0.23^b	1.04 ± 0.05^a	***↓	NS		3.94 ± 0.06^d	1.48 ± 0.05^a	3.01 ± 0.17^{b,c}	1.11 ± 0.04^a
C20-22 PUFA	40.09 ± 0.60^b	42.67 ± 0.29^{c,d}	41.05 ± 0.73^{b,c}	43.69 ± 0.73^d	****↓	**		33.60 ± 0.22^a	39.95 ± 0.29^b	35.56 ± 0.35^a	40.68 ± 0.65^{b,c}
20:4n-6/ 20:3n-6	16.72 ± 0.25^e	12.67 ± 0.83^{b,c}	13.87 ± 0.49^{b,c}	11.83 ± 0.11^b	NS	***		16.26 ± 0.77^{d,e}	14.33 ± 0.39^{c,d}	9.43 ± 0.59^a	13.91 ± 0.29^{b,c}
DBI	2.16 ± 0.03^b	2.34 ± 0.01^{c,d}	2.23 ± 0.04^{b,c}	2.35 ± 0.03^d	****↓	NS		1.96 ± 0.01^a	2.23 ± 0.02^b	2.02 ± 0.01^a	2.24 ± 0.04^{b,c}

Skeletal Muscle

16:0	18.03 ± 0.43	20.47 ± 0.41	18.82 ± 0.60	20.25 ± 1.05	NS	NS	18.70 ± 0.43	20.37 ± 0.37	19.86 ± 0.34	20.19 ± 0.69
18:0	13.41 ± 0.25[a,b]	13.54 ± 0.45[a,b,c]	13.04 ± 0.07[a]	13.29 ± 0.29[a,b]	*↑	NS	14.12 ± 0.20[c]	13.62 ± 0.17[a,b,c]	13.32 ± 0.09[a,b]	13.82 ± 0.12[b,c]
18:1n-9	5.47 ± 0.19[b]	3.94 ± 0.15[a]	7.38 ± 0.26[c]	4.24 ± 0.07[a]	**↓	NS	5.10 ± 0.15[b]	3.89 ± 0.11[a]	6.66 ± 0.24[c]	3.70 ± 0.08[a]
18:2n-6	20.45 ± 0.51[e]	12.09 ± 0.45[b]	18.55 ± 0.28[d]	9.12 ± 0.64[a]	***↓	*	17.34 ± 0.10[c,d]	9.90 ± 0.45[a]	16.91 ± 0.28[c]	8.68 ± 0.20[a]
18:3n-6	0.07 ± 0.009[a,b]	0.06 ± 0.001[a,b]	0.07 ± 0.005[b]	0.09 ± 0.006[c]	**↓	***	0.05 ± 0.005[a]	0.06 ± 0.001[a,b]	0.07 ± 0.001[b,c]	0.07 ± 0.004[b,c]
20:3n-6	1.23 ± 0.04[c]	0.87 ± 0.01[a]	1.21 ± 0.01[c]	1.01 ± 0.02[b]	NS	*	1.18 ± 0.04[c]	0.85 ± 0.02[a]	1.41 ± 0.02[d]	0.99 ± 0.02[b]
20:4n-6	11.55 ± 0.06[e]	9.21 ± 0.43[a]	10.96 ± 0.20[d,e]	9.43 ± 0.12[a,b]	*↑	***	11.61 ± 0.19[e]	10.47 ± 0.12[b,c,d]	10.84 ± 0.29[c,d,e]	9.87 ± 0.29[a,b,c]
18:3n-3	0.11 ± 0.005[d,e]	0.13 ± 0.005[e]	0.10 ± 0.008[c,d]	0.09 ± 0.003[b,c]	***↓	NS	0.09 ± 0.005[b,c]	0.11 ± 0.004[c,d,e]	0.08 ± 0.001[a,b]	0.07 ± 0.004[a]
20:5n-3	0.84 ± 0.06[a]	3.17 ± 0.39[b]	0.63 ± 0.03[a]	5.76 ± 0.16[d]	**↑	**	0.99 ± 0.01[a]	4.50 ± 0.29[c]	0.76 ± 0.04[a]	5.74 ± 0.14[d]
22:6n-3	16.62 ± 0.81[a]	25.46 ± 0.62[b]	16.08 ± 0.73[a]	25.44 ± 1.23[b]	NS	NS	18.07 ± 0.42[a]	25.36 ± 0.20[b]	16.68 ± 0.18[a]	25.24 ± 0.91[b]
SFA	32.44 ± 0.40[a]	35.36 ± 0.29[b]	32.78 ± 0.68[a]	34.70 ± 0.98[a,b]	*↑	NS	33.99 ± 0.31[a,b]	35.33 ± 0.33[b]	34.24 ± 0.29[a,b]	35.33 ± 0.72[b]
MUFA	8.98 ± 0.20[b]	8.49 ± 0.30[a,b]	11.72 ± 0.23[c]	8.79 ± 0.10[a,b]	***↓	**	8.59 ± 0.08[a,b]	8.16 ± 0.22[a,b]	11.18 ± 0.32[c]	7.96 ± 0.08[a]
PUFAn-6	34.41 ± 0.56[d]	23.08 ± 0.49[c]	31.98 ± 0.20[e]	20.48 ± 0.53[a,b]	***↓	***	30.82 ± 0.21[d,e]	21.93 ± 0.43[b,c]	29.92 ± 0.14[d]	20.22 ± 0.33[a]
PUFAn-3	22.10 ± 0.96[a]	32.67 ± 0.40[c]	21.75 ± 0.65[a]	35.65 ± 1.39[d]	*↑	NS	24.70 ± 0.51[b]	33.99 ± 0.26[c,d]	22.73 ± 0.16[a,b]	35.85 ± 0.96[d]
trans FA	1.88 ± 0.07[e]	0.32 ± 0.01[a]	1.42 ± 0.05[c]	0.32 ± 0.01[a]	***↓	***	1.78 ± 0.02[d,e]	0.53 ± 0.02[b]	1.61 ± 0.07[d]	0.56 ± 0.02[b]
PUFAn-6M	13.96 ± 0.12[d]	10.99 ± 0.41[a]	13.43 ± 0.21[d]	11.36 ± 0.13[a,b]	NS	*	13.49 ± 0.13[d]	12.03 ± 0.12[b,c]	13.02 ± 0.27[c,d]	11.54 ± 0.29[a,b]
PUFAn-3M	21.99 ± 0.96[a]	32.55 ± 0.40[c]	21.65 ± 0.66[a]	35.55 ± 1.40[d]	*↑	NS	24.60 ± 0.51[b]	33.88 ± 0.25[c,d]	22.65 ± 0.16[a,b]	35.78 ± 0.96[d]
PUFA/SFA	1.75 ± 0.03	1.58 ± 0.01	1.64 ± 0.06	1.63 ± 0.08	NS	NS	1.64 ± 0.02	1.59 ± 0.01	1.54 ± 0.01	1.59 ± 0.05
PUFAn-6/PUFAn-3	1.57 ± 0.09[d]	0.71 ± 0.02[a]	1.48 ± 0.04[c,d]	0.58 ± 0.03[a]	***↓	**	1.25 ± 0.03[b]	0.65 ± 0.02[a]	1.32 ± 0.01[b,c]	0.57 ± 0.02[a]
C20-22 PUFA	35.39 ± 0.96[a]	43.20 ± 0.42[b]	34.53 ± 0.82[a]	46.56 ± 1.51[b,c]	*↑	NS	37.76 ± 0.33[a]	45.63 ± 0.29[b,c]	35.30 ± 0.44[a]	47.05 ± 0.66[c]
20:4n-6/20:3n-6	9.41 ± 0.23[b,c]	10.57 ± 0.56[c]	9.08 ± 0.16[b]	9.34 ± 0.31[b,c]	NS	***	9.88 ± 0.42[b,c]	12.40 ± 0.18[d]	7.69 ± 0.20[a]	9.94 ± 0.12[b,c]
DBI	2.33 ± 0.04[b]	2.65 ± 0.01[b]	2.28 ± 0.08[b]	2.75 ± 0.07[b]	NS	NS	2.39 ± 0.02[a]	2.71 ± 0.01[b]	2.28 ± 0.01[a]	2.76 ± 0.05[b]

Note: Terms and abbreviations as described in TABLES 1 and 2. (*) $p < 0.05$; (**) $p < 0.01$; (***) $p < 0.001$. A diet main effect was significant (at least $p < 0.01$) for all fatty acids except 18:0 and the PUFA/SFA ratio in skeletal muscle.

TABLE 4. Fatty Acid Composition (wt%) in Liver and Skeletal Muscle Phosphatidylcholine of Hereditary Hypertriglyceridemic (HTG) and Control (NTG) Wistar Rats (Mean ± SEM)

Fatty Acids	NTG				Effect of			HTG			
	B	BFO	HS	HSFO	Strain	S × D	B	BFO	HS	HSFO
Liver											
16:0	17.18 ± 0.52^a	21.71 ± 0.65^b	19.87 ± 0.12^b	24.48 ± 1.10^c	***↑	**	16.78 ± 0.29^a	20.42 ± 0.09^b	16.23 ± 0.16^a	20.68 ± 0.45^b
18:0	22.54 ± 0.40^{b,c}	20.88 ± 0.09^b	20.90 ± 0.75^b	18.29 ± 0.70^a	***↓	**	24.05 ± 0.42^c	21.91 ± 0.16^b	24.44 ± 0.31^c	21.87 ± 0.03^b
18:1n-9	5.92 ± 0.14^b	4.00 ± 0.10^a	6.85 ± 0.24^{c,d}	6.81 ± 0.37^{c,d}	***↑	**	6.59 ± 0.06^c	5.84 ± 0.01^b	7.64 ± 0.02^e	7.42 ± 0.13^{d,e}
18:2n-6	11.14 ± 0.57^c	9.09 ± 0.45^b	9.22 ± 0.38^b	5.90 ± 0.42^a	***↓	***	19.32 ± 0.25^e	11.79 ± 0.33^c	14.34 ± 0.46^d	8.72 ± 0.31^b
18:3n-6	0.12 ± 0.005^c	0.05 ± 0.003^a	0.07 ± 0.003^{a,b}	0.05 ± 0.001^a	***↓	***	0.13 ± 0.005^{c,d}	0.08 ± 0.008^b	0.15 ± 0.009^d	0.07 ± 0.003^{a,b}
20:3n-6	1.28 ± 0.04^{a,b,c}	1.32 ± 0.10^{b,c}	1.52 ± 0.03^c	1.28 ± 0.04^{a,b,c}	NS	***	1.15 ± 0.06^{a,b}	1.04 ± 0.03^a	1.97 ± 0.12^d	1.01 ± 0.03^a
20:4n-6	27.03 ± 0.85^e	20.27 ± 0.66^c	23.51 ± 0.83^d	19.01 ± 0.16^{b,c}	***↓	***	18.79 ± 0.16^{b,c}	17.60 ± 0.28^{a,b}	19.23 ± 0.23^{b,c}	16.31 ± 0.14^a
18:3n-3	0.06 ± 0.006^b	0.06 ± 0.010^b	0.04 ± 0.003^{a,b}	0.04 ± 0.001^{a,b}	NS	NS	0.04 ± 0.003^{a,b}	0.05 ± 0.007^{a,b}	0.03 ± 0.003^a	0.04 ± 0.003^{a,b}
20:5n-3	0.80 ± 0.08^{a,b}	4.45 ± 0.19^d	0.60 ± 0.11^a	5.77 ± 0.47^e	***↑	*	1.59 ± 0.01^c	6.06 ± 0.21^e	1.33 ± 0.07^{b,c}	7.52 ± 0.11^f
22:6n-3	7.47 ± 0.60^{b,c}	12.22 ± 0.42^e	9.87 ± 1.09^{c,d,e}	11.76 ± 0.92^{d,e}	***↑	NS	4.88 ± 0.08^a	8.98 ± 0.21^{b,c}	7.26 ± 0.33^{a,b}	9.33 ± 0.37^{b,c,d}
SFA	40.56 ± 0.13^a	43.66 ± 0.67^{b,c,d}	41.59 ± 0.84^{a,b,c}	43.77 ± 0.59^d	***↑	NS	41.78 ± 0.15^{a,b,c,d}	43.57 ± 0.28^{b,c,d}	41.54 ± 0.38^{a,b}	43.70 ± 0.40^{c,d}
MUFA	7.95 ± 0.18^b	6.49 ± 0.19^a	9.66 ± 0.21^d	9.90 ± 0.38^{d,e}	***↑	**	8.76 ± 0.10^c	8.74 ± 0.04^c	10.42 ± 0.04^{e,f}	10.81 ± 0.10^f
PUFAn-6	39.93 ± 0.93^d	31.11 ± 0.75^b	34.88 ± 0.94^c	26.62 ± 0.25^a	***↓	NS	39.61 ± 0.10^d	30.84 ± 0.30^b	35.95 ± 0.65^c	26.43 ± 0.33^a
PUFAn-3	9.57 ± 0.66^{a,b}	18.46 ± 0.28^{c,d}	11.86 ± 1.25^b	19.41 ± 1.04^d	***↑	NS	7.66 ± 0.10^a	16.13 ± 0.46^c	9.66 ± 0.35^{a,b}	18.30 ± 0.52^{c,d}
trans FA	1.83 ± 0.05^d	0.22 ± 0.02^a	1.53 ± 0.03^c	0.25 ± 0.01^a	***↓	NS	2.10 ± 0.04^e	0.70 ± 0.02^b	1.98 ± 0.02^e	0.72 ± 0.04^b
PUFAn-6M	28.80 ± 0.83^e	22.03 ± 0.63^c	25.66 ± 0.78^d	20.72 ± 0.18^{b,c}	***↓	***	20.30 ± 0.21^{b,c}	19.05 ± 0.31^{a,b}	21.62 ± 0.18^c	17.71 ± 0.14^a
PUFAn-3M	9.51 ± 0.67^{a,b}	18.40 ± 0.29^{c,d}	11.82 ± 1.25^b	19.37 ± 1.04^d	***↑	NS	7.62 ± 0.09^a	16.08 ± 0.46^c	9.62 ± 0.35^{a,b}	18.26 ± 0.52^{c,d}
PUFA/SFA	1.22 ± 0.01^b	1.13 ± 0.03^{a,b}	1.13 ± 0.05^{a,b}	1.05 ± 0.04^a	*↓	NS	1.13 ± 0.01^{a,b}	1.01 ± 0.01^a	1.10 ± 0.02^a	1.03 ± 0.02^a
PUFAn-6/PUFAn-3	4.26 ± 0.41^{c,d}	1.69 ± 0.07^a	3.05 ± 0.35^b	1.38 ± 0.08^a	***↓	NS	5.17 ± 0.07^d	1.92 ± 0.07^a	3.75 ± 0.20^{b,c}	1.45 ± 0.05^a
C20-22 PUFA	37.67 ± 0.52^{c,d,e}	40.19 ± 0.54^e	36.73 ± 1.30^{c,d}	39.90 ± 1.09^{d,e}	***↑	**	27.44 ± 0.32^a	34.92 ± 0.61^c	30.67 ± 0.24^b	35.79 ± 0.61^c
20:4n-6/20:3n-6	21.25 ± 1.13^c	15.67 ± 1.41^b	15.51 ± 0.80^b	14.95 ± 0.39^b	***↓	***	16.42 ± 0.72^b	16.94 ± 0.22^b	9.92 ± 0.78^a	16.14 ± 0.43^b
DBI	2.00 ± 0.01^b	2.16 ± 0.02^c	2.00 ± 0.06^b	2.12 ± 0.05^{b,c}	***↑	NS	1.72 ± 0.01^a	1.98 ± 0.02^b	1.81 ± 0.01^a	2.00 ± 0.03^b

Skeletal Muscle

Fatty acid	1	2	3	4	Sig	5	6	7	8
16:0	30.95 ± 0.50ᵃ	33.49 ± 0.04ᵇ	33.47 ± 0.33ᵇ	35.32 ± 0.68ᶜ	**↓	30.61 ± 0.34ᵃ	32.76 ± 0.22ᵇ	32.89 ± 0.20ᵇ	33.49 ± 0.20ᵇ
18:0	7.07 ± 0.10ᶜ	5.44 ± 0.50ᵃ	6.02 ± 0.10ᵃ,ᵇ	5.39 ± 0.34ᵃ	***↑	7.75 ± 0.21ᵈ	6.28 ± 0.23ᵇ	6.66 ± 0.22ᵇ,ᶜ	6.60 ± 0.12ᵇ,ᶜ
18:1n-9	6.26 ± 0.13ᵇ	4.23 ± 0.16ᵃ	9.65 ± 0.34ᵈ	4.64 ± 0.08ᵃ	***↓	5.97 ± 0.10ᵇ	4.09 ± 0.12ᵃ	8.33 ± 0.41ᶜ	3.95 ± 0.02ᵃ
18:2n-6	22.08 ± 0.41ᵈ	10.07 ± 0.48ᵇ	19.37 ± 0.61ᶜ	6.96 ± 0.80ᵃ	*	18.88 ± 0.30ᶜ	7.92 ± 0.26ᵈ	17.81 ± 0.15ᶜ	6.78 ± 0.38ᵇ
18:3n-6	0.10 ± 0.008ᵃ,ᵇ,ᶜ	0.09 ± 0.003ᵃ,ᵇ	0.11 ± 0.008ᵇ,ᶜ	0.13 ± 0.014ᶜ	***↓	0.08 ± 0.005ᵃ	0.08 ± 0.004ᵃ	0.12 ± 0.003ᶜ	0.10 ± 0.003ᵃ,ᵇ,ᶜ
20:3n-6	1.40 ± 0.06ᵇ	0.75 ± 0.02ᵃ	1.37 ± 0.01ᵇ	0.85 ± 0.09ᵃ	NS	1.33 ± 0.04ᵇ	0.71 ± 0.02ᵃ	1.65 ± 0.02ᶜ	0.86 ± 0.06ᵃ
20:4n-6	11.64 ± 0.20ᵃ,ᵇ	11.05 ± 0.76ᵇ	10.90 ± 0.39ᵃ	11.54 ± 0.10ᵃ,ᵇ	***↓	13.08 ± 0.21ᵇ	12.76 ± 0.16ᵇ	11.96 ± 0.40ᵃ,ᵇ	12.29 ± 0.26ᵃ,ᵇ
18:3n-3	0.13 ± 0.009ᶜ,ᵈ	0.13 ± 0.003ᵈ	0.11 ± 0.007ᵇ,ᶜ	0.09 ± 0.003ᵃ,ᵇ	***↓	0.11 ± 0.003ᵇ,ᶜ	0.12 ± 0.003ᶜ,ᵈ	0.09 ± 0.003ᵃ	0.07 ± 0.005ᵃ
20:5n-3	0.97 ± 0.08ᵃ	3.92 ± 0.56ᵇ	0.73 ± 0.05ᵃ	7.91 ± 0.26ᵈ	***↑	1.34 ± 0.02ᵃ	6.27 ± 0.46ᶜ	1.03 ± 0.07ᵃ	8.52 ± 0.21ᵈ
22:6n-3	8.88 ± 0.60ᵃ,ᵇ	20.08 ± 0.89ᵈ	6.84 ± 0.46ᵃ	16.68 ± 0.68ᶜ	NS	9.28 ± 0.37ᵇ	18.51 ± 0.17ᶜ,ᵈ	7.38 ± 0.10ᵃ,ᵇ	16.37 ± 0.55ᶜ
SFA	39.23 ± 0.41ᵃ	40.53 ± 0.45ᵇ,ᶜ	40.64 ± 0.21ᶜ	42.21 ± 0.43ᵉ	NS	39.65 ± 0.13ᵃ,ᵇ	40.63 ± 0.17ᶜ	40.81 ± 0.12ᶜ	41.67 ± 0.14ᵉ
MUFA	10.12 ± 0.05ᵇ	9.34 ± 0.37ᵃ,ᵇ	14.74 ± 0.26ᵈ	9.64 ± 0.11ᵃ,ᵇ	***↓	9.93 ± 0.03ᵇ	8.77 ± 0.18ᵃ	13.55 ± 0.44ᶜ	8.61 ± 0.11ᵃ
PUFAn-6	35.82 ± 0.31ᵉ	22.51 ± 0.41ᵃ	32.48 ± 0.25ᶜ	20.00 ± 0.81ᵃ	*↓	33.74 ± 0.34ᵈ	21.96 ± 0.38ᵇ	31.97 ± 0.26ᶜ	20.46 ± 0.60ᵃ
PUFAn-3	13.01 ± 0.78ᵇ,ᶜ	27.28 ± 0.52ᵈ	10.55 ± 0.47ᵃ	27.81 ± 0.51ᵈ	**↑	14.76 ± 0.39ᶜ	28.18 ± 0.52ᵈ	11.84 ± 0.26ᵃ,ᵇ	28.76 ± 0.58ᵈ
trans FA	1.68 ± 0.03ᶜ,ᵈ	0.31 ± 0.01ᵃ	1.22 ± 0.02ᵇ	0.29 ± 0.003ᵃ	***↓	1.82 ± 0.10ᵈ	0.43 ± 0.01ᵃ	1.51 ± 0.03ᶜ	0.45 ± 0.004ᵃ
PUFAn-6M	13.74 ± 0.19ᵃ,ᵇ,ᶜ	12.44 ± 0.76ᵇ	13.11 ± 0.37ᵃ,ᵇ	13.05 ± 0.02ᵃ,ᵇ	***↑	14.86 ± 0.19ᶜ	14.04 ± 0.19ᵃ,ᵇ,ᶜ	14.16 ± 0.40ᵇ,ᶜ	13.69 ± 0.32ᵃ,ᵇ,ᶜ
PUFAn-3M	12.89 ± 0.78ᵇ,ᶜ	27.15 ± 0.52ᵈ	10.44 ± 0.47ᵃ	27.72 ± 0.51ᵈ	**↑	14.65 ± 0.39ᶜ	28.06 ± 0.52ᵈ	11.75 ± 0.26ᵃ,ᵇ	28.69 ± 0.58ᵈ
PUFA/SFA	1.25 ± 0.03ᵈ	1.23 ± 0.02ᵈ	1.06 ± 0.01ᵃ	1.13 ± 0.02ᵇ,ᶜ	NS	1.23 ± 0.01ᵈ	1.07 ± 0.01ᵃ,ᵇ	1.18 ± 0.01ᶜ,ᵈ	
PUFAn-6/PUFAn-3	2.79 ± 0.18ᶜ	0.83 ± 0.03ᵃ	3.10 ± 0.16ᶜ	0.72 ± 0.04ᵃ	*↓	2.29 ± 0.08ᵇ	0.78 ± 0.02ᵃ	2.71 ± 0.04ᵇ,ᶜ	0.71 ± 0.04ᵃ
C20–22 PUFA	26.03 ± 0.89ᵇ	39.24 ± 0.53ᵈ	22.86 ± 0.82ᵃ	40.38 ± 0.52ᵈ,ᵉ	***↑	29.15 ± 0.35ᶜ	41.82 ± 0.52ᵈ,ᵉ	25.43 ± 0.67ᵃ,ᵇ	42.09 ± 0.35ᵉ
20:4n-6/20:3n-6	8.35 ± 0.38ᵃ,ᵇ	14.89 ± 1.35ᶜ	7.96 ± 0.33ᵃ,ᵇ	14.05 ± 1.55ᶜ	NS	9.90 ± 0.40ᵇ	18.01 ± 0.37ᵈ	7.26 ± 0.23ᵃ	14.37 ± 0.63ᶜ
DBI	1.83 ± 0.04ᵇ,ᶜ	2.35 ± 0.02ᵈ	1.65 ± 0.03ᵃ	2.31 ± 0.02ᵈ	**↑	1.90 ± 0.02ᶜ	2.39 ± 0.02ᵈ	1.73 ± 0.02ᵃ,ᵇ	2.37 ± 0.02ᵈ

Note: Terms and abbreviations as described in TABLES 1 and 2. (*) $p < 0.05$; (**) $p < 0.01$; (***) $p < 0.001$. A diet main effect was significant (at least $p < 0.01$) for all fatty acids except 20:4n-6 in skeletal muscle.

TABLE 5. Fatty Acid Composition (wt%) in Liver and Skeletal Muscle Triglycerides of Hereditary Hypertriglyceridemic (HTG) and Control (NTG) Wistar Rats (Mean ± SEM)

Fatty Acids	NTG				Effect of Strain	S × D	HTG			
	B	BFO	HS	HSFO			B	BFO	HS	HSFO
Liver										
16:0	25.86 ± 0.45^c	15.52 ± 1.69^a	31.94 ± 0.82^d	17.40 ± 0.66^a	***↑	NS	32.72 ± 1.18^d	21.66 ± 0.99^b	36.08 ± 0.30^e	25.21 ± 0.46^c
18:0	3.14 ± 0.13^{c,d,e}	1.90 ± 0.13^a	3.61 ± 0.04^{e,f}	2.40 ± 0.11^{a,b}	***↑	**	4.20 ± 0.28^f	2.78 ± 0.23^{b,c,d}	3.43 ± 0.14^{d,e}	2.61 ± 0.11^{b,c}
18:1n-9	28.91 ± 0.72^d	13.62 ± 0.62^a	35.37 ± 1.26^e	19.16 ± 1.05^b	NS	*	25.04 ± 0.60^c	14.33 ± 0.19^a	35.41 ± 0.98^e	18.47 ± 0.16^b
18:2n-6	19.97 ± 0.31^d	15.19 ± 0.12^c	10.81 ± 0.23^b	7.35 ± 0.83^a	NS	*	19.68 ± 0.48^d	14.85 ± 0.46^c	10.49 ± 0.60^b	9.45 ± 0.25^b
18:3n-6	0.41 ± 0.033^c	0.13 ± 0.007^a	0.13 ± 0.004^a	0.11 ± 0.009^a	NS	***	0.28 ± 0.019^b	0.17 ± 0.009^a	0.17 ± 0.017^a	0.12 ± 0.003^a
20:3n-6	0.24 ± 0.01^{c,d}	0.30 ± 0.02^e	0.20 ± 0.01^{b,c}	0.28 ± 0.01^{d,e}	***↓	NS	0.19 ± 0.01^b	0.27 ± 0.01^{d,e}	0.14 ± 0.01^a	0.22 ± 0.01^{b,c}
20:4n-6	1.47 ± 0.02^d	1.07 ± 0.09^{b,c}	0.98 ± 0.03^b	1.08 ± 0.08^{b,c}	***↓	***	0.85 ± 0.03^b	1.28 ± 0.03^{c,d}	0.46 ± 0.07^a	0.95 ± 0.02^b
18:3n-3	1.50 ± 0.02^e	1.37 ± 0.12^e	0.74 ± 0.03^b	1.05 ± 0.07^{c,d}	***↓	NS	1.09 ± 0.05^{c,d}	1.15 ± 0.02^d	0.56 ± 0.02^a	0.90 ± 0.01^{b,c}
20:5n-3	1.58 ± 0.09^a	9.50 ± 1.09^b	0.49 ± 0.08^a	10.00 ± 0.24^b	*↑	*	1.45 ± 0.06^a	12.51 ± 0.80^c	0.40 ± 0.02^a	10.59 ± 0.40^{b,c}
22:6n-3	5.39 ± 0.06^c	27.12 ± 1.17^d	3.02 ± 0.17^b	25.20 ± 1.09^e	***↓	***	3.22 ± 0.14^b	19.18 ± 0.56^d	1.13 ± 0.06^a	18.04 ± 0.25^d
SFA	30.48 ± 0.53^c	18.53 ± 1.88^d	37.47 ± 0.84^d	21.10 ± 0.82^b	***↓	NS	38.78 ± 1.41^{d,e}	26.03 ± 1.22^b	41.35 ± 0.47^e	29.50 ± 0.42^{b,c}
MUFA	33.69 ± 0.77^d	18.59 ± 1.01^a	42.02 ± 1.43^e	25.77 ± 1.12^{b,c}	NS	NS	29.33 ± 0.64^c	19.19 ± 0.19^a	41.86 ± 1.17^e	24.59 ± 0.26^b
PUFAn-6	22.64 ± 0.29^d	17.49 ± 0.24^c	12.65 ± 0.32^b	9.55 ± 0.93^a	NS	*	21.29 ± 0.47^d	17.20 ± 0.48^c	11.41 ± 0.69^{a,b}	11.27 ± 0.27^{a,b}
PUFAn-3	10.29 ± 0.09^c	45.21 ± 2.37^e	5.42 ± 0.42^{a,b}	43.35 ± 1.54^e	***↓	*	7.40 ± 0.33^{b,c}	37.07 ± 1.55^d	2.55 ± 0.02^a	34.14 ± 0.52^d
trans FA	2.68 ± 0.15^d	0.15 ± 0.01^a	2.10 ± 0.12^c	0.20 ± 0.01^{a,b}	***↑	NS	3.11 ± 0.07^e	0.49 ± 0.05^b	2.65 ± 0.05^d	0.48 ± 0.03^b
PUFAn-6M	2.68 ± 0.05^d	2.29 ± 0.13^c	1.83 ± 0.08^b	2.20 ± 0.11^c	***↓	***	1.62 ± 0.01^b	2.35 ± 0.03^{c,d}	0.92 ± 0.09^a	1.82 ± 0.04^b
PUFAn-3M	8.80 ± 0.07^c	43.83 ± 2.32^e	4.68 ± 0.39^{a,b}	42.30 ± 1.50^e	***↓	*	6.31 ± 0.28^{b,c}	35.92 ± 1.54^d	1.99 ± 0.04^a	33.24 ± 0.52^d
PUFA/SFA	1.08 ± 0.01^{a,b}	3.56 ± 0.59^e	0.48 ± 0.01^a	2.53 ± 0.16^d	***↓	*	0.75 ± 0.04^{a,b}	2.11 ± 0.14^{c,d}	0.34 ± 0.01^a	1.54 ± 0.03^{b,c}
PUFAn-6/ PUFAn-3	2.20 ± 0.05^b	0.39 ± 0.01^a	2.37 ± 0.13^{b,c}	0.22 ± 0.02^a	***↑	***	2.89 ± 0.08^c	0.47 ± 0.03^a	4.49 ± 0.30^a	0.33 ± 0.01^a
C20-22 PUFA	10.72 ± 0.04^c	45.66 ± 2.39^e	6.22 ± 0.43^b	44.02 ± 1.47^e	***↓	*	7.45 ± 0.24^b	37.62 ± 1.51^d	2.65 ± 0.08^a	34.55 ± 0.50^d
20:4n-6/ 20:3n-6	6.16 ± 0.25^d	3.58 ± 0.26^{a,b}	4.93 ± 0.21^c	3.87 ± 0.20^{a,b}	**↓	***	4.43 ± 0.17^{b,c}	4.76 ± 0.07^c	3.26 ± 0.26^a	4.40 ± 0.07^{b,c}
DBI	1.40 ± 0.01^c	3.09 ± 0.13^g	1.02 ± 0.01^b	2.89 ± 0.08^f	***↓	*	1.16 ± 0.03^b	2.61 ± 0.08^e	0.82 ± 0.01^a	2.38 ± 0.02^d

Skeletal Muscle

16:0	27.31 ± 0.79[a,b,c]	25.88 ± 1.60[a,b]	30.52 ± 1.32[b,c]	30.11 ± 0.40[b,c]	NS	NS	27.28 ± 1.00[a,b,c]	25.06 ± 0.81[a]	29.67 ± 0.59[a,b,c]	31.25 ± 1.59[c]
18:0	6.18 ± 0.43[c]	5.12 ± 0.17[a,b]	4.97 ± 0.27[a,b]	4.61 ± 0.17[a]	***†	NS	7.93 ± 0.24[d]	5.86 ± 0.14[b,c]	6.43 ± 0.22[c]	5.53 ± 0.07[a,b,c]
18:1n-9	33.97 ± 0.55[c,d]	22.50 ± 1.09[a]	36.92 ± 1.27[d]	25.81 ± 0.95[a,b]	*†	**	31.43 ± 0.84[c]	23.75 ± 0.29[a,b]	32.22 ± 0.38[c]	26.15 ± 0.54[b]
18:2n-6	11.93 ± 0.68[c,d]	13.93 ± 0.53[d]	6.04 ± 0.50[a]	7.55 ± 0.59[a,b]	***†	*	13.85 ± 0.51[d]	14.25 ± 0.65[d]	10.56 ± 0.58[c]	9.69 ± 0.92[b,c]
18:3n-6	0.05 ± 0.005[b]	0.14 ± 0.005[d]	0.03 ± 0.003[a]	0.13 ± 0.010[d]	NS	***	0.05 ± 0.003[b]	0.14 ± 0.005[d]	0.05 ± 0.003[b]	0.10 ± 0.003[c]
20:3n-6	0.08 ± 0.008[a,b]	0.24 ± 0.026[f]	0.04 ± 0.005[a]	0.16 ± 0.010[d,e]	NS	***	0.11 ± 0.007[b,c]	0.20 ± 0.006[e,f]	0.10 ± 0.007[b,c]	0.13 ± 0.003[c]
20:4n-6	0.32 ± 0.01[a,b]	0.92 ± 0.08[c]	0.18 ± 0.03[a]	0.68 ± 0.06[d]	*†	*	0.46 ± 0.05[b,c]	0.92 ± 0.06[c]	0.43 ± 0.04[b]	0.66 ± 0.02[c,d]
18:3n-3	1.09 ± 0.07[c,d]	1.25 ± 0.09[d]	0.58 ± 0.04[a]	0.86 ± 0.06[b,c]	*†	**	0.86 ± 0.03[b,c]	0.99 ± 0.04[b,c,d]	0.75 ± 0.04[a,b]	0.74 ± 0.09[a,b]
20:5n-3	0.10 ± 0.01[a]	4.38 ± 0.56[c]	0.05 ± 0.01[a]	4.85 ± 0.24[c,d]	NS	***	0.20 ± 0.05[a]	5.56 ± 0.23[d]	0.25 ± 0.13[a]	3.18 ± 0.25[b]
22:6n-3	0.76 ± 0.06[a]	7.56 ± 0.34[e]	0.41 ± 0.03[a]	5.16 ± 0.34[c]	***†	****	0.62 ± 0.08[a]	6.29 ± 0.38[d]	0.57 ± 0.07[a]	3.34 ± 0.21[b]
SFA	37.10 ± 0.85[a,b]	34.79 ± 1.96[a]	39.66 ± 1.56[a,b]	39.11 ± 0.36[a,b]	NS	NS	39.22 ± 0.95[a,b]	34.83 ± 1.16[a]	40.81 ± 0.46[b]	41.72 ± 1.56[b]
MUFA	43.45 ± 1.08[d]	32.12 ± 1.38[e]	49.82 ± 1.18[e]	37.95 ± 1.22[b]	***†	**	39.68 ± 0.75[b,c]	32.39 ± 0.22[a]	42.58 ± 0.49[c,d]	37.07 ± 0.30[b]
PUFAn-6	12.63 ± 0.71[c,d]	15.88 ± 0.41[e]	6.43 ± 0.46[a]	8.94 ± 0.70[a,b]	***↓	***	14.76 ± 0.49[d,e]	16.05 ± 0.58[e]	11.39 ± 0.52[b,c]	10.92 ± 0.91[b,c]
PUFAn-3	2.48 ± 0.15[a]	16.66 ± 0.83[d]	1.37 ± 0.09[a]	13.58 ± 0.72[c]	**↑	***	2.33 ± 0.15[a]	15.70 ± 0.75[c,d]	2.11 ± 0.26[a]	9.14 ± 0.57[b]
trans FA	4.29 ± 0.36[d]	0.50 ± 0.07[a,b]	2.68 ± 0.08[c]	0.39 ± 0.07[a]	**↑	*	3.98 ± 0.22[d]	1.00 ± 0.07[a,b]	3.05 ± 0.09[c]	1.14 ± 0.11[b]
PUFAn-6M	0.71 ± 0.04[b]	1.95 ± 0.12[d]	0.40 ± 0.05[a]	1.39 ± 0.11[c]	NS	**	0.91 ± 0.08[b]	1.81 ± 0.07[d]	0.83 ± 0.07[b]	1.23 ± 0.01[c]
PUFAn-3M	1.39 ± 0.09[a]	15.42 ± 0.80[d]	0.79 ± 0.10[a]	12.72 ± 0.66[c]	**↓	***	1.47 ± 0.18[a]	14.72 ± 0.75[c,d]	1.35 ± 0.23[a]	8.40 ± 0.54[b]
PUFA/SFA	0.41 ± 0.03[b,c]	0.95 ± 0.08[d]	0.20 ± 0.02[a]	0.57 ± 0.04[c]	NS	NS	0.44 ± 0.02[b,c]	0.92 ± 0.06[d]	0.33 ± 0.02[a,b]	0.49 ± 0.05[b,c]
PUFAn-6/ PUFAn-3	5.10 ± 0.12[b,c]	0.96 ± 0.06[a]	4.76 ± 0.44[b]	0.66 ± 0.02[a]	**↑	NS	6.45 ± 0.60[c]	1.03 ± 0.06[a]	5.59 ± 0.56[b,c]	1.20 ± 0.10[a]
C20-22 PUFA	1.87 ± 0.09[a]	16.55 ± 0.79[a]	1.06 ± 0.13[a]	13.28 ± 0.71[c]	**↓	***	2.13 ± 0.23[a]	15.62 ± 0.72[d]	1.95 ± 0.22[a]	9.07 ± 0.51[b]
20:4n-6/ 20:3n-6	3.92 ± 0.25[a]	3.96 ± 0.19[a]	4.62 ± 0.17[a,b]	4.22 ± 0.11[a,b]	*↑	NS	4.10 ± 0.21[a,b]	4.63 ± 0.18[a,b]	4.31 ± 0.31[a,b]	5.01 ± 0.23[b]
DBI	0.86 ± 0.02[a]	1.56 ± 0.05[d]	0.72 ± 0.02[a]	1.30 ± 0.04[c]	NS	**	0.85 ± 0.01[a]	1.51 ± 0.05[d]	0.80 ± 0.02[a]	1.09 ± 0.04[b]

Note: Terms and abbreviations as described in TABLES 1 and 2. (*) $p < 0.05$; (**) $p < 0.01$; (***) $p < 0.001$. A diet main effect was significant (at least $p < 0.01$) for all fatty acids.

Higher levels of 18:2n-6 with a concomitant decrease in 20:4n-6, total long-chain PUFA n-6, and C20-22 PUFA in the liver phospholipids and phosphatidylcholine of HTG rats in comparison with controls were found. Our findings are consistent with those reported in streptozotocin-induced[16] and spontaneously[17] diabetic rats and may result from an overall decrease in fatty acid desaturase enzyme activities.[17-19] The decreased level of long-chain PUFA n-6, especially 20:4n-6, in the adipocyte plasma membranes of streptozotocin-induced diabetic rats has also been suggested to be associated with decreased insulin binding in isolated adipocytes.[18] Higher levels of *trans* FA, found in HTG liver lipid fractions, might further contribute to the lowering of the PUFA n-6 levels in this tissue via inhibition of delta–6 desaturase activity.[20] However, several other factors, such as FA chain elongation, membrane lipid turnover, and FA oxidation, may also contribute to these alterations. We also detected lower levels of 20:4n-6 and other long-chain PUFA n-6 in liver triglycerides of HTG rats fed basal or HS diets. This is in contrast with the results of type 1 diabetic rats, where the amounts of these desaturation and elongation products of 18:2n-6 in liver triglycerides were increased.[21]

Unlike the changes in PUFA n-6 in liver, a decreased percentage in skeletal muscle of 18:2n-6 and total PUFA n-6 in polar lipids of the HTG rats fed basal diet was detected.

In our study, the amount of PUFA n-3 in lipid fractions differed considerably between the HTG and NTG strains. These changes were tissue-specific. While the levels of these FA were lower in the liver of HTG rats, the opposite trend was detected in skeletal muscle. Unlike the changes in PUFA n-6 in liver, a decreased percentage in skeletal muscle of 18:2n-6 and total PUFA n-6 in polar lipids of the HTG rats fed basal diet was also detected. Skeletal muscle is the major tissue for insulin-stimulated glucose uptake.[22] It has been shown that long-chain PUFA n-3[1] or C20-22 PUFA[23] in phospholipids of skeletal muscles may be important for efficient insulin action. However, differences in the phospholipid FA composition between rat muscle types have been reported.[24] Thus, substantially higher amounts of 18:2n-6 and 22:6n-3 in red quadriceps than in white quadriceps were found.[24] In our study, we have analyzed FA composition in mixed quadriceps. This may be one cause of the differences in our finding in insulin-resistant HTG rats compared to other reports in the literature.[1,23] Another reason may be due to the fact that we used much lower dietary lipid amounts than some other investigators.[1,25] Thus, our diet may evoke less-profound alterations in skeletal muscles than in the liver. However, the slight increase of total SFA in skeletal muscle phospholipids in our HTG rats is in accordance with the data shown in the dietary-induced insulin-resistant rats.[1,23,25]

It has been shown that high-carbohydrate diet accelerates the conversion of SFA to MUFA.[26] This is in accordance with our results in rats fed high-sucrose diet demonstrating the increased level of 18:1n-9 in all lipid fractions in both strains except of skeletal muscle triglycerides. The observed decrease in the level of 18:2n-6, more profound in liver lipids than in skeletal muscle polar lipids, may be due to the lower amount of this FA in the HS diet and is accompanied by lower levels of 20:4n-6 in liver triglycerides of both strains and in liver polar lipids of NTG rats. Similar depletion of PUFA n-6 in liver polar lipids as a result of the low level of dietary 18:2n-6 was recently reported.[27]

The substantial increase in liver phospholipid long-chain PUFA n-3 in spite of the low dietary amount of 18:3n-3 (without long-chain PUFA n-3 in the diet) may reflect the lower 18:2n-6/SFA ratio in the HS diet, when the formation of long-chain PUFA n-3 from precursor 18:3n-3 is increased at the cost of 20:4n-6.[28] In skeletal muscle phospholipids, this increase in phospholipid long-chain PUFA n-3 was not detected.

The present study demonstrates that FA composition in structural and storage lipids is markedly sensitive to dietary long-chain PUFA n-3 supplied as fish oil. The replacement of dietary SFA and partially MUFA with long-chain PUFA n-3 increased the amounts of individual long-chain PUFA n-3 by 1.5- to 20-fold, mainly at the cost of 18:2n-6 in structural lipids and of 18:1n-9 in triglycerides. This increase was higher in liver than in skeletal muscle and higher in triglycerides than in polar lipids. The magnitude of incorporation of individual long-chain PUFA n-3 (especially 20:5n-3 and 22:6n-3) was more proportional in lipid fractions of skeletal muscle of both rat strains, while a tendency to higher increments of 20:5n-3 and lower 22:6n-3 was observed in the liver lipids in the HTG rats in comparison with controls. Feeding fish oil also increased C20-22 PUFA in all lipid fractions of both rat strains and partially restored the decreased levels of these long-chain FA in the liver of HTG rats. The higher levels of C20-22 PUFA in skeletal muscle phospholipids of HTG rats were associated with a higher glucose infusion rate, that is, with improved insulin action, as we are reporting elsewhere in this volume.[29] This finding confirms the results of Storlien et al.[1,23] demonstrating that skeletal long-chain PUFA n-3 can positively influence insulin action.

In spite of the tremendous increase of long-chain PUFA n-3, the PUFA/SFA ratio was quite stable in polar lipid fractions. However, a significant increase of this ratio was noticed in triglycerides. This finding further substantiates the suggestion[30,31] that FA composition in storage lipids is more susceptible to dietary treatment than that in structural lipids.

In addition to the increase in the content of long-chain PUFA n-3, there was a concomitant decrease in the level of PUFA n-6 metabolites, mainly in 18:3n-6 and 20:4n-6, in structural liver and to a smaller extent in skeletal muscle lipids. This is in agreement with earlier studies[2,28] suggesting that 20:5n-3 inhibits desaturation and chain elongation of 18:2n-6 by acting as an analogue of 20:4n-6.

However, the increased amount of 20:4n-6 in liver triglycerides of HTG rats and in skeletal muscle triglycerides of both rat strains after fish oil diets would support the suggestion that the decrease in 20:4n-6 content of membrane and tissue phospholipids may be, at least in part, due to a shift of 20:4n-6 from phospholipids to storage lipid pools in serum or in the same tissue.[32] This shift may result from a greater specificity of plasma lecithin:cholesterol acyltransferase and liver acyltransferase for 20:5n-3 than for 20:4n-6. Nonpolar lipid esters, therefore, may play a buffering role in the homeostatic maintenance of tissue phospholipid levels of 20:4n-6.[32]

In conclusion, (1) there were major differences between the HTG and control rats in FA composition of liver and skeletal muscle lipid fractions, indicating the impaired FA metabolism in HTG rats; (2) the differences were tissue-specific, being more pronounced in liver than in skeletal muscle, and included the decreased levels of long-chain PUFA metabolites; these changes indicate decreased desaturation in HTG rats; (3) diet had a major effect on FA composition in tissue lipids with similar trends across strains; and (4) dietary treatment with long-chain PUFA n-3 as fish oil was able to shift the long-chain PUFA levels of tissue lipids toward the control values. Thus, the HTG rat seems to provide an interesting animal model for study of impaired fatty acid metabolism.

SUMMARY

The fatty acid (FA) compositions of liver and skeletal muscle structural lipids, overall phospholipids and phosphatidylcholine, and triglycerides (TG) were determined in the hereditary hypertriglyceridemic (HTG) rat, a nonobese animal model of the insulin resistance syndrome. Four groups of HTG rats and four groups of control animals were fed equal-energy diets for two weeks: basal (B), high-sucrose (HS), or fish oil–supplemented basal (BFO) or high-sucrose (HSFO) diets.

In the liver of HTG rats, a decrease of n-6 long-chain polyunsaturated FA (PUFA), especially in 20:4n-6, in comparison with controls was found. Moreover, a concomitant accumulation of 18:2n-6 in structural lipids was observed. These differences were more pronounced in liver than in skeletal muscle. HS feeding raised the proportion of 18:1n-9 and decreased 18:2n-6 in lipid fractions. In both tissues and in both strains, the amounts of long-chain n-3 PUFA, as well as the level of total C20-22 PUFA, went up after fish oil feeding. However, the effects were somewhat less pronounced in the HTG rats. The increase in n-3 PUFA occurred mainly at the expense of reduced levels of 18:2n-6 in structural lipids and of 18:1n-9 in triglycerides. These changes were associated, in companion studies reported in this volume, with improved insulin action in HTG rats.

In conclusion, the FA composition in lipid subclasses of HTG rats differs significantly from the controls mainly in liver structural lipids, suggesting the impairment of PUFA desaturation. Dietary change effected a similar modulation of FA profile across both strains, with fish oil increasing the levels of long-chain PUFA toward control values in the NTG rats. The HTG rat thus provides an interesting animal model for the study of impaired fatty acid metabolism.

ACKNOWLEDGMENTS

The skillful technical assistance of Alica Mitková, Silvia Kuklová, and Marta Bohovová is greatly appreciated.

REFERENCES

1. STORLIEN, L. H., A. B. JENKINS, D. J. CHISHOLM, W. S. PASCOE, S. KHOURI & E. W. KRAEGEN. 1991. Diabetes **40**: 280–289.
2. LIU, S., V. E. BARACOS, H. A. QUINNEY & M. T. CLANDININ. 1994. Biochem. J. **299**: 831–837.
3. BORKMAN, M., L. H. STORLIEN, D. A. PAN, A. B. JENKINS, D. J. CHISHOLM & L. V. CAMPBELL. 1993. N. Engl. J. Med. **328**: 238–244.

4. VESSBY, B., S. TENGBLAD & H. LITHELL. 1994. Diabetologia **37**: 1044-1050.
5. CLANDININ, M. T., S. CHEEMA, C. J. FIELD, M. L. GARG, J. VENKATRAMAN & T. R. CLANDININ. 1991. FASEB J. **5**: 2761-2769.
6. STUBBS, C. D. & A. D. SMITH. 1984. Biochim. Biophys. Acta **779**: 89-137.
7. MURPHY, M. G. 1990. J. Nutr. Biochem. **1**: 68-79.
8. VRÁNA, A. & L. KAZDOVÁ. 1990. Transplant. Proc. **22**: 2579.
9. REAVEN, G. M. 1988. Diabetes **37**: 1595-1607.
10. ŠTOLBA, P., H. OPLTOVÁ, P. HUŠEK, J. NEDVÍDKOVÁ, J. KUNEŠ, Z. DOBEŠOVÁ, J. NEDVÍDEK & A. VRÁNA. 1993. Ann. N.Y. Acad. Sci. **683**: 281-288.
11. ŠEBÖKOVÁ, E., I. KLIMEŠ, D. GAŠPERÍKOVÁ, P. BOHOV, P. LANGER, M. LAVAU & M. T. CLANDININ. 1996. Biochim. Biophys. Acta **1303**: 56-62.
12. KLIMEŠ, I., E. ŠEBÖKOVÁ, A. VRÁNA, P. ŠTOLBA, L. KAZDOVÁ, J. KUNEŠ, P. BOHOV, M. FICKOVÁ, J. ZICHA, D. RAUČINOVÁ & V. KŘEN. 1995. *In* Lessons from Animal Diabetes, p. 271-283. Smith-Gordon. London.
13. BLIGH, E. G. & W. J. DYER. 1959. Can. J. Biochem. Physiol. **37**: 911-917.
14. ADEYEMI, A., S. C. GARNER & S. H. ZEISEL. 1993. J. Nutr. Biochem. **4**: 123-128.
15. BOHOV, P., V. BALÁŽ & J. HRIVŇÁK. 1984. J. Chromatogr. **286**: 247-252.
16. HUANG, Y. S., D. F. HORROBIN, M. S. MANKU, J. MITCHELL & M. A. RYAN. 1984. Lipids **19**: 367-370.
17. MIMOUNI, V., M. NARCE, Y. S. HUANG, D. F. HORROBIN & J. P. POISSON. 1994. Prostaglandins Leukotrienes Essent. Fatty Acids **50**: 43-47.
18. FIELD, C. J., E. A. RYAN, A. B. R. THOMSON & M. T. CLANDININ. 1988. Biochem. J. **253**: 417-424.
19. MIMOUNI, V. & J. P. POISSON. 1992. Biochim. Biophys. Acta **1123**: 296-302.
20. SIMOPOULOS, A. P. 1994. Free Radical Biol. Med. **17**: 367-372.
21. TAKAHASHI, R., D. F. HORROBIN, Y. WATANABE, V. KYTE & V. BILLARD. 1987. Biochim. Biophys. Acta **921**: 151-153.
22. DEFRONZO, R. A., E. JACOT, E. JEQUIER, J. WAHREN & J. P. FELBER. 1981. Diabetes **30**: 1000-1007.
23. STORLIEN, L. H., L. A. BAUR, A. D. KRIKETOS, D. A. PAN, G. J. COONEY, A. B. JENKINS, G. D. CALVERT & L. V. CAMPBELL. 1996. Diabetologia **39**: 621-631.
24. KRIKETOS, A. D., D. A. PAN, J. R. SUTTON, J. F. Y. HOH, L. A. BAUR, G. J. COONEY, A. B. JENKINS & L. H. STORLIEN. 1995. Am. J. Physiol. **269**: R1154-R1162.
25. KRAEGEN, E. W., P. W. CLARK, A. B. JENKINS, E. A. DALEY, D. J. CHISHOLM & L. H. STORLIEN. 1991. Diabetes **40**: 1397-1403.
26. JEFCOAT, R. & A. T. JAMES. 1977. Lipids **12**: 469-474.
27. LEE, J. H., I. IKEDA & M. SUGANO. 1992. Nutrition **8**: 162-166.
28. GARG, M. L., A. B. R. THOMSON & M. T. CLANDININ. 1990. J. Lipid Res. **31**: 271-277.
29. GAŠPERÍKOVÁ, D., I. KLIMEŠ, T. KOLTER, P. BOHOV, A. MAASSEN, J. ECKEL, M. T. CLANDININ & E. ŠEBÖKOVÁ. 1997. This volume.
30. FIELD, C. J., A. ANGEL & M. T. CLANDININ. 1985. Am. J. Clin. Nutr. **42**: 1206-1220.
31. GARG, M. L., A. A. WIERZBICKI, A. B. R. THOMSON & M. T. CLANDININ. 1988. Biochim. Biophys. Acta **962**: 337-344.
32. GARG, M. L., A. A. WIERZBICKI, A. B. R. THOMSON & M. T. CLANDININ. 1989. Biochem. J. **261**: 11-15.

Increased Adipose Tissue Lipolysis in a Hypertriglyceridemic Rat Line[a]

A. VRÁNA AND L. KAZDOVÁ

Department of Metabolic Research
Institute for Clinical and Experimental Medicine
140 00 Prague 4, Czech Republic

It has been demonstrated recently that elevated levels of circulating free fatty acids (FFA) are a risk factor for an accelerated manifestation of diabetes and cardiovascular diseases.[1] We previously selected, from rats of the Wistar strain, a line (HTg) exhibiting elevated triglyceridemia, especially when fed a simple carbohydrate-rich diet.[2] In this HTg line, a number of other earlier recognized facets of metabolic syndrome X,[3] such as hyperinsulinemia, resistance to insulin action, impaired glucose tolerance, and elevated blood pressure, were observed.[4] VLDL transport from the liver into the plasma compartment was increased in the line, and enhanced hepatic lipogenesis was also noted.[1] In addition, increased serum FFA concentrations have been repeatedly observed;[5,6] however, the mechanisms underlying the elevation of FFA in this animal model are not clear.

Experiments, whose results are briefly reported here, were therefore designed to investigate lipolysis of adipose tissue triglycerides and stimulation of this process by epinephrine, as well as to monitor glycerol and FFA release from isolated adipose tissue of HTg rats. Commercially available normotriglyceridemic Wistar rats were used as controls.

MATERIALS AND METHODS

In all experiments, adult males of the hypertriglyceridemic (HTg) line bred in our laboratory and controls (Wistar strain, Academy of Sciences of the Czech Republic), weighing 244 ± 14 and 240 ± 12 g, respectively, were used. Rats were fed ad libitum a high-carbohydrate diet containing 70 cal% carbohydrate (sucrose), 20 cal% protein, and 10 cal% fat.[7] The rats were killed by decapitation in the postprandial state, mixed blood from the neck vessels was collected, and serum was separated at 4°C.

To determine FFA and glycerol release from adipose tissue *in vitro*, the distal parts of the epididymal fat pad (250 ± 10 mg) were incubated in vials containing 3 mL of Krebs-Ringer phosphate buffer containing half of the recommended amount of calcium with 3% bovine serum albumin (Armour, fraction V) at pH 7.4 (gas phase: air).

---

[a] This study was supported by Grant Nos. 2024-2 and 3097-2 from the Internal Grant Agency of the Ministry of Health of the Czech Republic.

TABLE 1. *In Vitro* Glycerol and FFA Release from Isolated Epididymal Adipose Tissue of Controls and HTg Rats

	Epinephrine Concentration[a]			
	0 μg/mL	0.10 μg/mL	0.25 μg/mL	1.0 μg/mL
	Glycerol Release (μmol/mg protein/2 h)[b]			
controls	0.19 ± 0.01	0.49 ± 0.03	0.88 ± 0.03	1.34 ± 0.06
HTg	0.36 ± 0.02	0.76 ± 0.10	1.19 ± 0.09	1.77 ± 0.15
p	<0.001	<0.02	<0.01	<0.01
	FFA Release (μmol/mg protein/2 h)[b]			
controls	0.55 ± 0.19	1.67 ± 0.20	2.27 ± 0.17	3.91 ± 0.14
HTg	1.02 ± 0.33	2.20 ± 0.23	3.40 ± 0.36	5.28 ± 0.56
p	NS	NS	<0.01	<0.02

[a] Epinephrine concentration in incubation medium (μg/mL).
[b] Mean values ± SEM, $n = 6$–10; p: p value, HTg versus controls; NS: statistically nonsignificant.

The concentrations of epinephrine (E) in the incubation medium are given in TABLE 1. The tissues were incubated for 2 h at 37°C in a metabolic shaker. FFA concentrations in the incubation buffer and in serum were determined by titration,[8] and glycerol and serum triglycerides were measured using enzymatic kits (Lachema, Brno, Czech Republic). Adipose tissue protein content was determined according to Lowry *et al.*[9] Statistical significance between the average values of the compared groups was calculated using Student's *t*-test.

RESULTS AND DISCUSSION

TABLE 2 gives the serum concentrations of triglycerides, glycerol, and FFA in HTg and control rats. Both serum triglyceride and glycerol levels were markedly elevated in HTg rats; a less-pronounced increase was found in serum FFA concentrations.

Glycerol release from adipose tissue was markedly higher in HTg rats compared with controls. The difference was apparent in the absence as well as in the presence of all epinephrine (E) concentrations used (TABLE 1).

FFA release from isolated adipose tissue was also elevated in the HTg line in the absence and in the presence of E (TABLE 1). The increased availability of FFA rep-

TABLE 2. Serum Triglycerides (Tg), FFA, and Glycerol Concentrations in Controls and HTg Rats[a]

	Serum Tg (mmol/L)	Serum FFA (mmol/L)	Serum Glycerol (mmol/L)
controls	1.89 ± 0.14	0.58 ± 0.02	1.21 ± 0.21
HTg	4.79 ± 0.42	0.73 ± 0.04	2.11 ± 0.20
p	<0.001	<0.01	<0.001

[a] Mean values ± SEM, $n = 6$–10; p: p values, HTg versus controls.

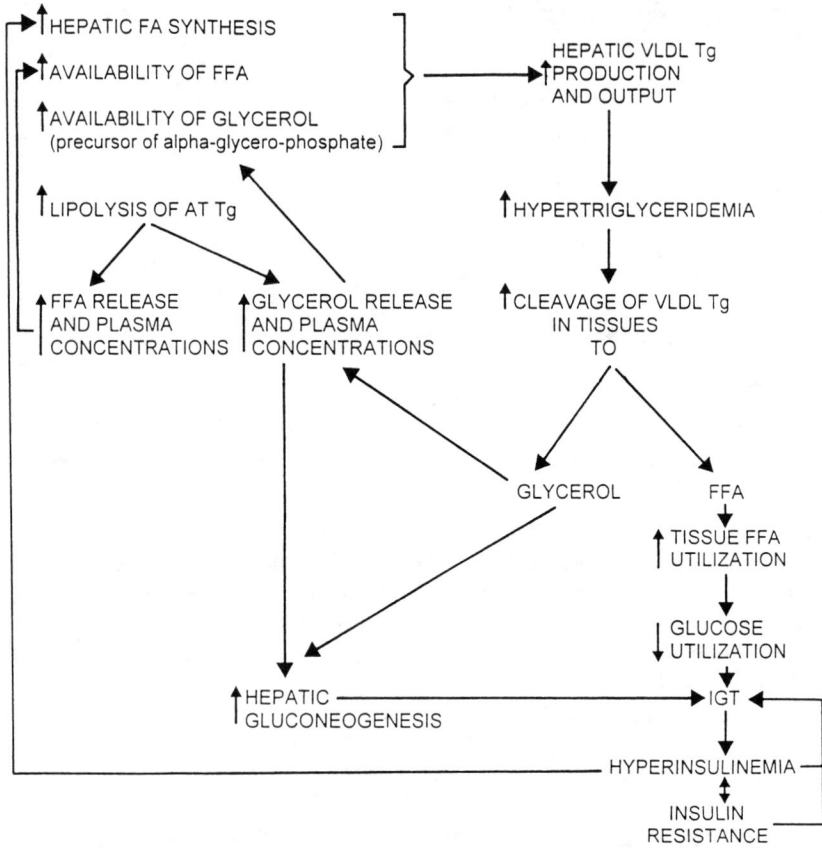

FIGURE 1. The possible mechanisms by which increased adipose tissue (AT) lipolysis, together with other metabolic changes, might be involved in some alterations of lipid and carbohydrate metabolism in the studied model of insulin resistance syndrome.

resents a precursor for hepatic VLDL triglyceride synthesis and, in addition, may compete with glucose utilization in tissues (for review, see reference 10).

A similar dual role may be played by the increased availability of circulating glycerol, the substrate for liver gluconeogenesis and for alpha-glycerol-phosphate synthesis, thereby contributing to hepatic triglyceride synthesis. The more-pronounced increase in serum glycerol concentrations in the HTg line, compared with the mild elevation in serum FFA, may reflect a higher flux of glycerol due to increased cleavage of plasma VLDL Tg in tissues by lipoprotein lipase or increased turnover of FFA compared with that of glycerol (see FIG. 1). Because of the complexity of adipose tissue lipolysis regulation, mediated especially by the balance between insulin concentration and action inhibiting such a process on the one hand and the concentration and action of lipolytic enzymes on the other, further studies are needed to elucidate the present results.

REFERENCES

1. Sowers, J. R. & M. Epstein. 1995. Hypertension **26**(part 1): 869–879.
2. Vrána, A. & L. Kazdová. 1990. Transplant. Proc. **22**: 2579.
3. Reaven, G. M. 1994. J. Intern. Med. **236**(suppl. 736): 13–22.
4. Vrána, A., L. Kazdová, Z. Dobešová, J. Kuneš, V. Křen, V. Bílá, P. Štolba & I. Klimeš. 1993. Ann. N.Y. Acad. Sci. **683**: 57–68.
5. Štolba, P., Z. Dobešová, P. Hušek, H. Opltová, J. Zicha, A. Vrána & J. Kuneš. 1992. Life Sci. **51**: 733–740.
6. Vrána, A., L. Kazdová, V. Nováková & M. Matějčková. 1996. *In* Natural Antioxidants and Food Quality in Atherosclerosis and Cancer Prevention, p. 141–144. Thomas Graham House. Cambridge, United Kingdom.
7. Fábry, P., R. Poledne, L. Kazdová & T. Braun. 1968. Nutr. Dieta **10**: 81–90.
8. Dole, V. P. 1957. J. Clin. Invest. **35**: 150–154.
9. Lowry, O. H., N. J. Rosebrough, A. L. Farr & R. J. Randall. 1955. J. Biol. Chem. **193**: 265–275.
10. Saloranta, C. & G. Leif. 1996. Diabetes Metab. Rev. **12**: 15–36.

Structural Changes in the Aorta of the Hereditary Hypertriglyceridemic Rat[a]

FRANTIŠEK KRISTEK,[b] SOŇA EDELSTEINOVÁ,[c]
ELENA ŠEBÖKOVÁ,[d] JÁN KYSELOVIČ,[c]
AND IWAR KLIMEŠ[d]

[b]Institute of Normal and Pathological Physiology
Slovak Academy of Sciences
SK-813 71 Bratislava, Slovak Republic
[c]Department of Pharmacology and Toxicology
Pharmaceutical Faculty
Comenius University
Bratislava, Slovak Republic
[d]Diabetes and Nutrition Research Group
Institute of Experimental Endocrinology
Slovak Academy of Sciences
Bratislava, Slovak Republic

The hereditary hypertriglyceridemic (hHTG) rat is an experimental model of the human metabolic syndrome X.[1] Hypertriglyceridemia, insulin resistance, glucose intolerance, and elevated blood pressure were observed in hHTG rats.[2] Also, plasma norepinephrine and epinephrine concentrations were increased.[3] Measurements of vascular reactivity of the isolated aortic rings of hHTG rats revealed changes in the response curves of aorta to contractile stimuli.[4] Thus, it has been suggested that some of the aforementioned findings could be related to changes in structure of the vascular wall.

Therefore, we studied structural changes in ascending thoracic aorta of hHTG rats using standard techniques for transmission electron microscopy.

METHODS

All experiments reported herein were approved by the Animal House Ethical Committee of the Institute of Normal and Pathological Physiology. Hereditary hypertriglyceridemic rats were obtained from Velaz (Prague, Czech Republic) and housed

---

[a] This study was supported by Slovakofarma, Joint Stock Company, Hlohovec, Slovak Republic, and by Research Grant No. GAV 425/1992 from the Slovak Grant Agency.

in a temperature-controlled room (24°C) under a 12-hour light, 12-hour dark cycle on standard diet.

Seven hHTG rats and seven male Wistar rats (used as a control) were taken for the study. The animals were sacrificed by an overdose of pentobarbital (100 mg/kg body weight, ip). In both the control and experimental animals, the chest was opened and the ascending part of the thoracic aorta was excised and fixed in 3% glutaraldehyde in 0.1 M phosphate buffer. One-mm-long segments were postfixed with 2% osmium tetroxide in 0.1 M phosphate buffer. After double fixation, the specimens were stained en bloc with 2% uranyl acetate, dehydrated through ascending concentrations of alcohols, and embedded in Durcupan ACM. Randomly selected blocks of each artery were cut perpendicularly to the long axis and examined in electron microscope Tesla BS 500 (Brno, Czech Republic).

RESULTS

The most-remarkable ultrastructural changes (when compared to aorta of control rats) were observed in the aortic tunica intima of hHTG rats (FIG. 1A–C). Endothelial cells formed a continual layer along the circumference of the artery. Only rare breaks in endothelial lining were observed. In these "gaps," subendothelial matrix faced the lumina. In the majority of endothelial cells, big electron-lucent lipid droplets were present and occupied the main part of the cytoplasm (FIG. 1C). They were intimately associated with an enlarged number of vesicles of Golgi apparatus and with dilated cisternae of endoplasmic reticulum. Degenerative microvesicular and membranous material was present in the cytoplasm of endothelial cells.

Subendothelial space was enlarged and contained lipid droplets and degenerative microvesicular components with membranous debris and abundant amorphous material (FIG. 1B). Protrusion of the endothelial cells penetrated deep into the subendothelial space. Parts of the endothelial cells were delaminated from the subendothelium by big lipid droplets present in the subendothelial space and endothelial cells were bulged out towards the lumina (FIG. 1C).

The organization and orientation of smooth muscle cells in hereditary hypertriglyceridemic rats were essentially intact. Morphological changes in smooth muscle cells were similar as in endothelial cells, but in a less-expressed form. Increased numbers of cisternae of the sarcoplasmic reticulum and vesicles of Golgi apparatus were found in the majority of the smooth muscle cells. The ultrastructural composition of these cells resembled the synthetic type of muscle cells (FIG. 2A). Big lipid droplets were in close contact with dilated cisternae of the sarcoplasmic reticulum and vesicles of Golgi apparatus, or they were localized inside of them (FIG. 2B). Cytoplasm of the smooth muscle cells often contained destructive vesicular and membranous material—myelin figures (FIG. 2C).

Extracellular matrix among the smooth muscle cells sporadically contained lipid droplets and electron-dark membranous and destructive vesicular material. Lipid droplets and lamellar structures were usually in close contact with smooth muscle cells.

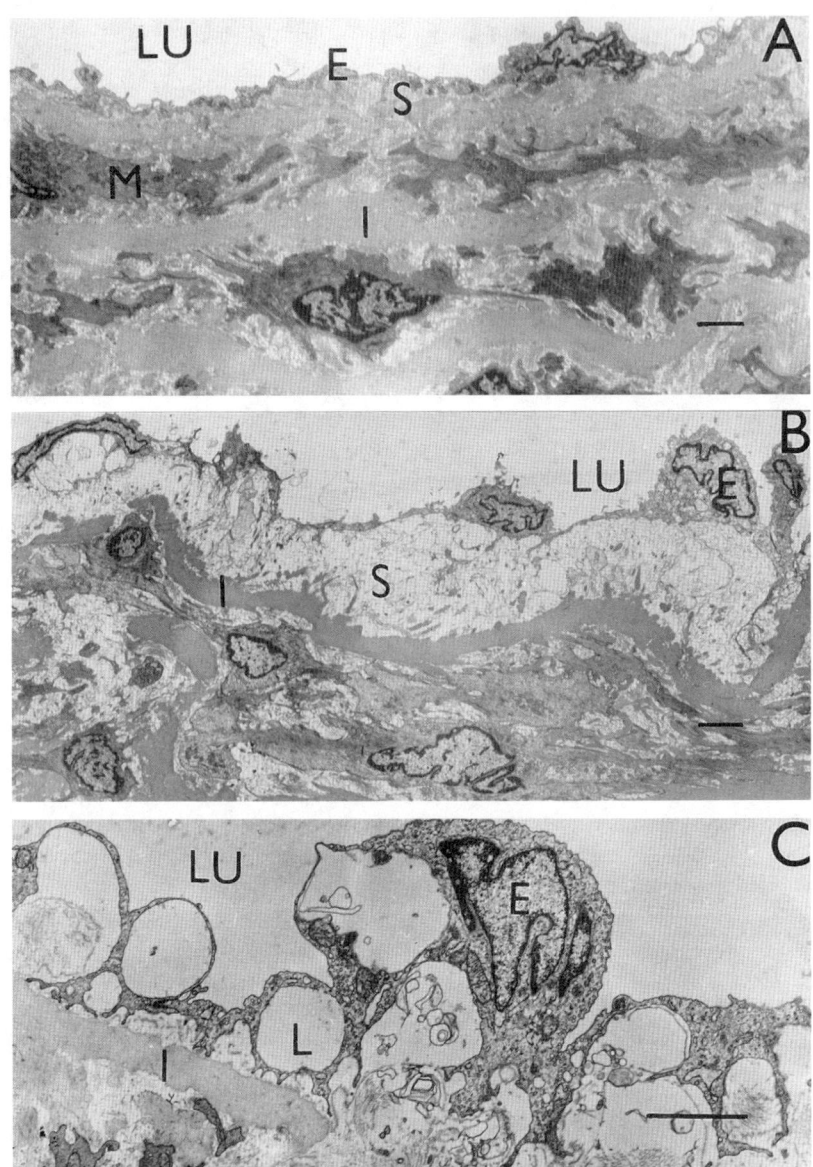

FIGURE 1. Transmission electron photomicrographs of the ascending part of the thoracic aorta. (A) Intimal part of control aorta. Bar: 2 µm. (B & C) Intimal part of hHTG aorta. Bars: 2 µm. Terms: LU, lumen of the aorta; E, endothelial cell; M, smooth muscle cell; I, elastic lamina; L, lipid droplets; S, subendothelial space.

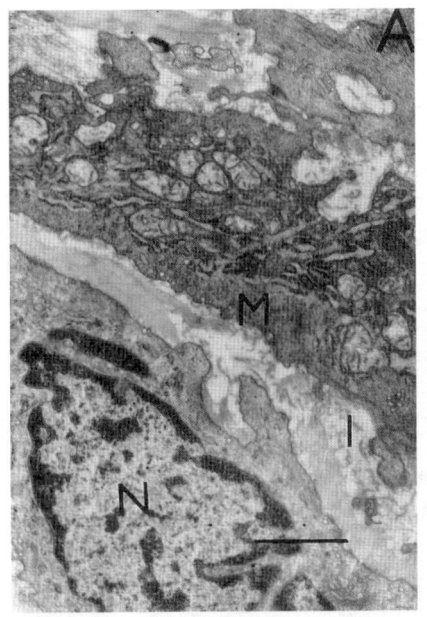
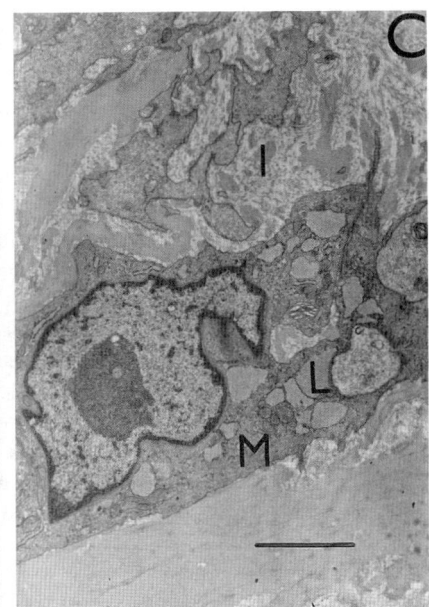
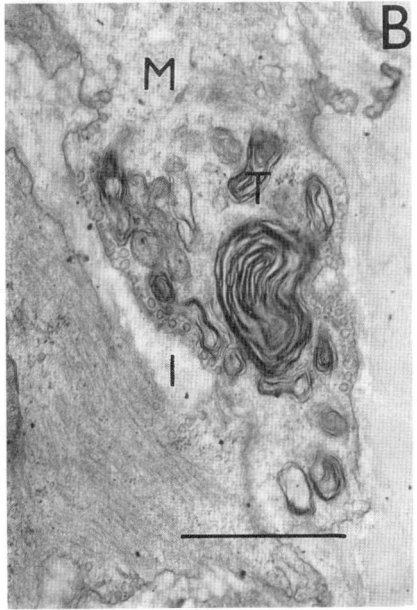

FIGURE 2. Transmission electron photomicrographs of the ascending part of the thoracic aorta. (A) Some smooth muscle cells of the tunica media of hHTG rats had a raised number of cytoplasmic organelles. Bar: 1 μm (B & C) Smooth muscle cells of hHTG rats contained lipid droplets and lamellar structures. Bars: (B) 2 μm; (C) 1 μm. Terms: M, smooth muscle cell; L, lipid droplets; I, intercellular space; T, lamellar structures; N, nucleus.

DISCUSSION

Changes in aortic wall structure can be viewed as both the consequence and the cause of at least some of the abnormalities accompanying the hereditary hypertriglyceridemic rats. Since accumulation of lipids into the arterial wall and increased volume of the extracellular matrix are the most-prominent alterations in the arterial wall, it might be speculated that hypertriglyceridemia and/or hypertension may contribute together or independently to the aforementioned alteration of the arterial structure.

Hypertriglyceridemia could increase the accumulation of lipids into the cellular and intracellular spaces of the artery. The most-pronounced accumulation of lipids was observed in endothelial cells and in subendothelial space, less in smooth muscle cells, and sporadically in intercellular space among smooth muscle cells. There was a gradient in lipid deposition in the arterial wall. It seems that the relatively compact layer of internal elastic lamina acted as a protective coat, preventing lipid penetration deep into the media. Scow and Blanchette-Mackie[5] suggested that the transport of lipids from the lumen to the arterial wall is realized by lateral movement. In this case, myoendothelial contacts that were observed between endothelial and smooth muscle cells[6] could have played an important role in this transfer. Due to an increase of subendothelial space, the number of myoendothelial contacts could have been altered.

Accumulation of lipids and increased production of the extracellular matrix resulted in the increase of subendothelial space. Raised production of extracellular matrix and concomitant changes of the smooth muscle phenotype to the synthetic type could be connected probably with the development of hypertension in hHTG rats since they were also observed in other types of hypertension (spontaneously hypertensive rats[7] or NO-deficient hypertension[8,9]). Accumulation of lipid droplets to the cellular and extracellular space was also observed in hereditary hyperlipidemic rabbits.[10] Contrary to this study, we did not observe adhesion and penetration of leukocytes into the arterial wall.

Thus, ultrastructural changes in endothelial cells and smooth muscle cells could be at least partly responsible for changes in the vascular responsiveness of hHTG rats. In particular, in an earlier study from our labs,[4] we observed a decrease of endothelium-dependent relaxation by acetylcholine and a decrease of relative force of smooth muscle cells to noradrenaline in hHTG rats.

Besides the lipid accumulation in the arterial wall, the most-pronounced ultrastructural signs in the hHTG arterial wall were the lamellar structures present in cellular and extracellular space. Lamellar structures were found also in capillaries, extracellular space, and myocytes in the myocardium of young rats injected with chylomicrons.[5,11] The lamellar structures are most probably products of lipolysis in endothelial and muscle cells.

In conclusion, important ultrastructural changes in the hHTG aortic wall were observed. Accumulation of lipids and end products of lipolysis were seen mainly in endothelial cells, in the subendothelial space, and less in smooth muscle cells. The number of vesicles of Golgi apparatus was increased, and cisternae of the endoplasmic reticulum in both endothelial and smooth muscle cells were dilated. Finally, transformation of the contractile type of muscle cells toward the synthetic type and an increased volume of extracellular matrix were found in the aorta of hhTG rats.

SUMMARY

Structural changes in the ascending thoracic aorta of hereditary hypertriglyceridemic (hhTG), insulin-resistant, and hypertensive rats were studied using transmission electron microscopy. Normotensive Wistar rats were used as controls. The most-pronounced morphological changes were observed in the tunica intima. Endothelial cells of hhTG rats formed a continual layer around the whole circumference. Subendothelial space was enlarged. Some endothelial cells were delaminated from the subendothelial space by big lipid droplets that were often present in the subendothelial space, and the endothelial cells bulged out towards the lumen. Big electron-lucent lipid droplets were present in the majority of the endothelial cells and occupied the main part of the cytoplasm. Degenerative microvesicular and membranous material was present in the cytoplasm. Increased numbers of vesicles of Golgi apparatus and cisternae of endoplasmic reticulum were found. Similar morphological alterations, but in less-extended form, were observed in smooth muscle cells. The organization and orientation of smooth muscle cells were essentially intact. In muscle cells, lipid droplets were localized in close relation to Golgi complex and in dilated cisternae of the sarcoplasmic reticulum. Lipid droplets, degenerative material, myelin figures, myelinoid membranes, and vesicular components were also sporadically found in the intercellular space among muscle cells.

This pilot morphological investigation provides further arguments for a thorough and more-focused electron microscopy study of conductance arteries of the hhTG rats.

REFERENCES

1. VRÁNA, A. & L. KAZDOVÁ. 1990. The hereditary hypertriglyceridemic nonobese rat: an experimental model of human hypertriglyceridemia. Transplant. Proc. **22:** 2579.
2. KLIMEŠ, I., A. VRÁNA, J. KUNEŠ, E. ŠEBÖKOVÁ, Z. DOBEŠOVÁ, P. ŠTOLBA & J. ZICHA. 1995. Hereditary hypertriglyceridemic rat: a new animal model of metabolic alterations in hypertension. Blood Pressure **4:** 137–142.
3. ŠTOLBA, P., Z. DOBEŠOVÁ, P. HUŠEK, H. OPLTOVÁ, J. ZICHA, A. VRÁNA & J. KUNEŠ. 1992. The hypertriglyceridemic rat as a genetic model of hypertension and diabetes. Life Sci. **51:** 733–740.
4. EDELSTEINOVÁ, S., J. KYSELOVIČ, I. KLIMEŠ, E. ŠEBÖKOVÁ, B. KOVÁCSOVÁ, F. KRISTEK, A. MITKOVÁ, A. VRÁNA & P. ŠVEC. 1993. Effects of marine fish oil on blood pressure and vascular reactivity in the hereditary hypertriglyceridemic rat. Ann. N.Y. Acad. Sci. **683:** 353–356.
5. SCOW, R. O. & E. J. BLANCHETTE-MACKIE. 1992. Endothelium, the dynamic interface in cardiac lipid transport. Mol. Cell. Biochem. **116:** 181–191.
6. KRISTEK, F. & M. GEROVÁ. 1992. Myoendothelial relations in the conduit coronary artery of the dog and rabbit. J. Vasc. Res. **29:** 29–32.
7. WIENER, J. & F. GIACOMELLI. 1983. Structural characterization of coronary arteries and myocardium in renal hypertensive hypertrophy. *In* Perspectives in Cardiovascular Research. Vol. 8, p. 59–72. Raven Press. New York.
8. DELACRETAZ, E., D. HAYOZ, M. C. OSTERHELD & C. Y. GENTON. 1994. Long-term nitric oxide

synthase inhibition and distensibility of carotid artery in intact rats. Hypertension **23**(part 2): 967–970.
9. KRISTEK, F. & M. GEROVÁ. 1996. Long-term NO-synthase inhibition affects heart weight and geometry of coronary and carotid artery. Physiol. Res. **45:** 1–9.
10. ROSENFELD, M. E., T. TSUKADA, T. GOWN & A. M. ROSS. 1987. Fatty streak initiation in Watanabe hereditable hyperlipemic and comparably hypercholesterolemic fat-fed rabbits. Arteriosclerosis **7:** 9–23.
11. WETZEL, M. G. & R. O. SCOW. 1984. Lipolysis and fatty acid transport in rat heart: electron microscopic study. Am. J. Physiol. **246:** C467–C485.

Increased Lipoprotein Oxidability and Aortic Lipid Peroxidation in an Experimental Model of Insulin Resistance Syndrome[a]

LUDMILA KAZDOVÁ,[b] ALEŠ ŽÁK,[c]
AND ANTONÍN VRÁNA[c]

[b]Department of Metabolic Research
Institute for Clinical and Experimental Medicine
140 00 Prague 4, Czech Republic

[c]Fourth Department of Internal Medicine
First Medical Faculty
Charles University
Prague, Czech Republic

Recent studies have suggested that insulin resistance, which may affect up to 25% of the adult population, is probably the most-prevalent risk factor of cardiovascular disease (CVD) in a population with normal or moderately increased cholesterolemia.[1,2] Resistance to insulin action is associated with a number of simultaneously occurring metabolic disturbances, such as hypertriglyceridemia, hyperinsulinemia, increased availability of free fatty acids (FFA), and hypertension, which increase the risk for CVD. The pathophysiological mechanisms by which the clustering of these metabolic abnormalities may promote atherogenic processes in the arterial wall are not understood well as yet.

There is growing evidence that the free radical process of lipid peroxidation is involved in many of the key events in the pathogenesis of CVD.[3] Not only low-density lipoproteins (LDL),[4] but also very low density lipoproteins (VLDL), which are predominantly elevated in insulin resistance, may be retained in the arterial wall,[5] oxidatively modified via similar mechanisms as LDL,[6] and these promote foam cell formation,[7] thereby directly contributing to atherogenesis. Moreover, triglyceride-rich VLDL have been identified in the human atherosclerotic plaque.[8]

Lipid peroxidation has not been tested in the insulin resistance syndrome yet. One of the reasons for this was the unavailability of a suitable animal model. Recently, a rat model of insulin resistance syndrome has been developed in our laboratory. The line of hypertriglyceridemic (HTg) rats, originating from the Wistar strain, exhibits insulin resistance, hyperinsulinemia, and elevated blood pressure.[9] Therefore, in this

---

[a] This study was supported by Grant Nos. 2016-3 and 3102-3 from the Internal Grant Agency of the Ministry of Health of the Czech Republic.

study, the susceptibility of the LDL + VLDL fractions to *in vitro* oxidative modification and to *in vivo* lipid peroxidation in this rat model of the insulin resistance syndrome was examined.

MATERIALS AND METHODS

Adult males of a hereditary HTg line, weighing 280–310 g, were used. Randombred males of the Wistar strain of a similar weight served as controls. All rats were fed a high-carbohydrate diet (70 cal% sucrose, 20 cal% protein, 10 cal% fat) ad libitum. After 3 weeks, the rats were killed by decapitation and nonfasted blood samples were collected into tubes containing Na_2EDTA, centrifuged, aliquoted, and immediately assayed for conjugated diene concentration. Lipoproteins (LDL + VLDL) were isolated by ultracentrifugation at 112,000g for 20 h at 10°C with a final density of 1.055 g/mL.[10,11] LDL + VLDL samples were dialyzed against oxygen-free buffer at 4°C in the dark for 24 h. The susceptibility of the lipoprotein fractions to Cu^{++}-induced oxidation *in vitro* was measured by continuous monitoring of the conjugated diene (CD) formation as described by Kleinveld *et al.*,[12] with the difference that the incubation of lipoproteins was performed at 37°C, and by measuring the production of thiobarbituric-reactive substances (TBARS) as described by Thomas *et al.*[13] Oxidative modification of the protein moiety of the lipoproteins was measured by the emission fluorescence spectra at 430 nm (excitation, 360 nm) according to Esterbauer *et al.*[14] Protein concentration of the lipoproteins was determined by the method of Lowry modified to include sodium dodecyl sulfate.[15] Plasma lipid peroxide concentration was assessed by measurement of CD[16] and by fluorometric measurement of TBARS[17] with fluorometric detection (515–553 nm). Alpha-tocopherol was estimated by HPLC analysis.[18] The serum levels of triglycerides and FFA were measured by enzymatic kits (Lachema, Czech Republic, and Boehringer-Mannheim, Germany, respectively). Aortic triglycerides, cholesterol (measured enzymatically with kits), and CD concentrations were determined in lipid extracts of the aorta.

Data are reported as the mean ± SEM. Student's t test for unpaired data was performed for all statistical analyses.

RESULTS AND DISCUSSION

As has been previously reported, HTg rats exhibited serum triglyceride levels in the fasting and postprandial stages above those of control rats. The fasting and postprandial serum levels of FFA were likewise markedly elevated in HTg rats. Plasma cholesterol was not significantly different in the two groups (TABLE 1).

The oxidability of lipoproteins was assessed using three methods: by CD formation, which is a measure of the early events of lipid peroxidation reactions; by measuring the production of TBARS as an indicator of the end products of lipid peroxidation; and by measuring fluorescent chromophores as an indicator of apoB modification.[14] As displayed in TABLE 2, the VLDL + LDL fraction of HTg rats,

TABLE 1. The Levels of Lipids, α-Tocopherol, Conjugated Dienes, and TBARS in Serum and Aorta of Controls (C) and HTg Rats

	C	HTg	p Value[a]
Serum			
triglycerides (mmol/L)	1.63 ± 0.09[b]	4.17 ± 0.43	<0.001
FFA (mmol/L)	1.05 ± 0.12	2.51 ± 0.30	<0.01
α-tocopherol (μmol/L)	17.45 ± 1.95	20.32 ± 2.38	NS
μmol/mmol triglycerides	16.62 ± 2.80	8.09 ± 1.62	<0.01
conjugated dienes (abs. U)	0.18 ± 0.02	0.28 ± 0.03	<0.05
TBARS (nmol/L)	2.15 ± 0.15	2.95 ± 0.25	<0.05
Aorta			
conjugated dienes (nmol/g wet wt)	5.55 ± 0.03	7.52 ± 0.07	<0.05
triglycerides (μmol/g)	0.49 ± 0.07	0.53 ± 0.09	NS
cholesterol (μmol/g)	2.95 ± 0.15	3.41 ± 0.25	NS

[a] Significance levels (p value) for the difference between C and HTg rats; NS = not statistically significant.
[b] Mean values ± SE (n = 8–10).

compared with that of controls, exhibited enhanced susceptibility to oxidative modification as shown by a decrease in lag time before initiation of lipoprotein oxidation, elevated total diene production, and elevated TBARS formation. Oxidative modification of the protein moiety of the lipoproteins, measured according to the formation of fluorescent chromophores due to the reaction of aldehydic lipid products of lipid peroxidation with free amino groups of the lipoproteins, was also increased in HTg rats.

There is increasing evidence that oxidative modification of lipoproteins is important for the development of atherosclerosis,[4,7] and the association between coronary atherosclerosis and susceptibility of lipoproteins to oxidative modification supports this hypothesis.[19]

The factors determining the susceptibility of lipoproteins to oxidative modification are still poorly understood. Recent studies have shown that the susceptibility of lipoproteins to oxidation (as opposed to the rate of oxidation) does not only appear to be related to vitamin E or fatty acid composition;[20] it may also be influenced by the size and density of lipoprotein particles. Small, dense LDL particles are more susceptible to oxidative modification and are associated with atherosclerosis.[21] It has

TABLE 2. Susceptibility to in Vitro Copper-induced Oxidation of Lipoproteins (VLDL + LDL Fraction) from Controls (C) and HTg Rats

	C	HTg	p Value[a]
Lag time (min)	79 ± 9[b]	56 ± 6	<0.05
Diene production (nmol/mg protein)	302 ± 14	358 ± 18	<0.05
TBARS (nmol MDA/mg protein)[c]	37.6 ± 6.8	66.2 ± 12.4	<0.001
Fluorescence intensity[c]	61 ± 5	84 ± 6	<0.05

[a] Significance levels (p value) for the difference between C and HTg rats.
[b] Mean values ± SE (n = 5–7).
[c] Measured in oxidized lipoproteins.

been suggested that size reduction of LDL particles, an integral feature of the insulin resistance syndrome,[22] is associated with an increased triglyceride content of precursor lipoproteins and that the FFA released during lipolysis of triglycerides from VLDL may be directly responsible for the observed size reduction of LDL particles.[23]

To get information about the oxidative processes that occur *in vivo* in our animal model of insulin resistance, the concentrations of lipid peroxidation products in serum and in the aorta were determined (TABLE 1). The serum concentrations of conjugated dienes and TBARS were significantly increased in HTg rats compared with controls. Lipid peroxidation in the aorta, estimated as conjugated dienes, was higher in HTg rats than in controls. On the other hand, no differences in aortic triglyceride and cholesterol levels were observed between the compared groups. These results indicate that the hypertriglyceridemia associated with insulin resistance enhances lipid peroxidation *in vivo*. The finding of autoantibodies against epitopes of oxidized lipoproteins in the aorta and serum, detected in humans and animals,[4,8] and their association with progression of atherosclerosis[24] support the concept that oxidation of lipoproteins occurs *in vivo* and that this process may play a role in the pathogenesis of atherosclerosis.

The enhanced oxidative modification of lipoproteins by free radicals under *in vivo* conditions results from a dysbalance between pro-oxidant stimuli and antioxidant defense or from a modification in the lipid composition. The serum concentration of the major antioxidant, alpha-tocopherol, when adjusted to the triglyceride level, was lower in HTg rats (TABLE 1). Evidence of the low level of antioxidant defense in HTg rats can include the results of our experiments showing a beneficial effect of antioxidant therapy on lipid peroxidation in this experimental model.[25] The production of free radicals in insulin resistance may also be increased. Enhanced plasma FFA concentrations[26] or higher free radical production by mononuclear cells, which has been observed in hypertriglyceridemia,[27] might be responsible for the increased production of free radicals. Another possibility is that hypertriglyceridemia favors the formation of small, dense lipoproteins, which are more susceptible to oxidation.[21] Their plasma residence time may be longer because of a lower affinity to the LDL receptor.[28] As a result, these lipoprotein particles may more likely enter into the arterial wall, and undergo oxidation there, to reenter the circulation again. Further studies are required to clarify the mechanism by which lipoprotein oxidation increases in insulin resistance.

In conclusion, our results indicate that the metabolic disorders accompanying the insulin resistance syndrome are associated with increased lipoprotein susceptibility to *in vitro* oxidative modification and *in vivo* lipid peroxidation. Both these processes may contribute to the accelerated atherosclerosis frequently observed with the insulin resistance states.

REFERENCES

1. REAVEN, G. M. 1988. Diabetes **37**: 1595–1607.
2. DESPRÈS, J. P. 1993. Can. Med. Assoc. J. **148**: 1339–1340.
3. WITZTUM, J. L. & D. STEINBERG. 1991. J. Clin. Invest. **88**: 1985–1992.
4. YLÄ-HERTTUALA, S. 1991. Ann. Med. **23**: 561–567.
5. NORDESTGAARD, B. G., R. WOOTTON & B. LEWIS. 1995. Arterioscler. Thromb. Vasc. Biol. **15**: 534–542.

6. MOHR, D. & R. STOCKER. 1994. Arterioscler. Thromb. **14:** 1182-1192.
7. PARTHASARANTHY, S., M. T. QUINN, D. C. SCHWENKE, T. E. CAREW & D. STEINBERG. 1989. Arteriosclerosis **9:** 398-404.
8. RAPP, J. P., A. LESPINE, R. L. HAMILTON, N. COLYVAS, A. H. CHAUMETON, J. TWEEDIE-HARDMAN, L. KOTITE, S. T. KUNITAKE, R. J. HAVEL & J. P. KANE. 1994. Arterioscler. Thromb. **14:** 1767-1774.
9. VRÁNA, A. & L. KAZDOVÁ. 1990. Transplant. Proc. **22:** 2579.
10. SCHUMACKER, D. L. & D. L. PUPPIONE. 1986. Methods Enzymol. **128:** 155-170.
11. NARUSZEWICZ, M., E. SELINGER & J. DAVIGNON. 1992. Metabolism **41:** 1215-1224.
12. KLEINVELD, H. A., H. L. M. HAK-LEMMERS, A. F. H. STELENHOEF & P. N. M. DEMACKER. 1992. Clin. Chem. **38:** 2066-2072.
13. THOMAS, C. E., R. J. JACKSON, D. F. OHLWEILER & G. KU. 1994. J. Lipid Res. **35:** 417-427.
14. ESTERBAUER, H., G. JURGENS, O. QUEHENBERGER & E. KOLLER. 1987. J. Lipid Res. **28:** 495-509.
15. MARKWELL, M. A. K., S. M. HAAS, L. L. BIEBER & N. E. TOLBERT. 1978. Anal. Biochem. **87:** 206-210.
16. WARD, P. J., G. O. PILL, J. R. HATHERILL, H. ANNENSLEY & R. G. KUKEL. 1985. J. Clin. Invest. **76:** 517-527.
17. NAITO, C., M. KAWAMURA & Y. YAMAMOTO. 1993. Ann. N.Y. Acad. Sci. **676:** 27-45.
18. CATIGNANI, G. L. 1986. Methods Enzymol. **123:** 215-219.
19. ANDREWS, B., K. BURNAND, G. PAGANGA, N. BROWSE, C. RICE-EVANS, K. SOMMERVILLE, D. LEAKE & N. TAUB. 1995. Atherosclerosis **112:** 77-84.
20. CROFT, K. D., P. WILLIAMS, S. DIMMITT, R. ABU-AMSHA & J. BEILIN. 1995. Biochim. Biophys. Acta **1254:** 250-256.
21. TRIBBLE, D. L., R. M. KRAUSS, M. G. LANSBERG, P. M. THIEL & J. J. M. VAN DEN BERG. 1995. J. Lipid Res. **36:** 662-671.
22. AUSTIN, M. A. & K. L. EDWARDS. 1996. Curr. Opin. Lipidol. **7:** 167-171.
23. KRAUS, R. M. 1987. Am. Heart J. **113:** 578-582.
24. SALONEN, J. T., S. YLÄ-HERTTUALA, R. YAMAMOTO, S. BUTLER, H. KORPELA, R. SALONEN, K. NYSSSONEN, W. PALINSKI & J. L. WITZTUM. 1992. Lancet **339:** 883-887.
25. KAZDOVÁ, L., A. VRÁNA, M. MATĚJČKOVÁ & V. NOVÁKOVÁ. 1996. *In* Natural Antioxidants and Food Quality in Atherosclerosis and Cancer Prevention, p. 31-35. Thomas Graham House. Cambridge, United Kingdom.
26. TOBOREK, M. & B. HENNING. 1994. Am. J. Clin. Nutr. **59:** 60-65.
27. HIRAMATSU, K. & S. AMORI. 1988. Diabetes **37:** 832-837.
28. NIGON, F., P. LESNICK, M. ROUIS & M. J. CHAPMAN. 1991. J. Lipid Res. **32:** 1741-1753.

Phenotype and Genotype Comparison of Hereditary Hypertriglyceridemic (hHTG) and Brown-Norway (BN) Rats

Identification of Quantitative Trait Loci (QTLs) for the Insulin Resistance Syndrome[a]

PETER KOVÁCS,[b] NILESH J. SAMANI,[c]
ELENA ŠEBÖKOVÁ,[d] BIRGER VOIGT,[b]
DANIELA GAŠPERÍKOVÁ,[d] DANIELA JEŽOVÁ,[d]
RICHARD KVETŇANSKÝ,[d] DAVID LODWICK,[c]
INGRID KLÖTING,[b] AND IWAR KLIMEŠ[d]

[b]"Gerhardt Katsch" Institute of Diabetes
University of Greifswald
17495 Karlsburg, Germany

[c]Department of Cardiology
Glenfield Hospital
Leicester LE3 9QP, United Kingdom

[d]Institute of Experimental Endocrinology
Slovak Academy of Sciences
833 06 Bratislava, Slovak Republic

INTRODUCTION

The hereditary hypertriglyceridemic (hHTG) rat manifesting hypertriglyceridemia, hyperinsulinemia,[1] insulin resistance,[2] glucose intolerance, mild hyperuricemia,[3] and hypertension[4] provides a unique animal model for the human insulin resistance syndrome[5] in which it may be possible to identify genetic factors predisposing to the syndrome. The classical method for identifying quantitative trait loci (QTLs) that influence traits in animal models is by cosegregation analysis. In this approach, the disease-prone strain is crossed with a disease-resistant strain and F_2 (or backcross) progeny are examined for cosegregation of genetic markers with quantitative differences in

---

[a] This study was supported by research grants of the Slovak Grant Agency for Science (Grant Nos. GAV 425/1991–1992 and GAV 2/544/1993–1996); by a research grant of the Slovak Grant Agency for Technic (Grant No. GAT PEKO 931 004) within the frame of the EURHYPGEN Concerted Action of the EU (Contract No. ERBBMH1CT920869); by a British Council (Slovak Branch Office) scholarship to P. Kovács; and by Grant No. Kl 771/1-3 of Deutsche Forschungsgemeinschaft.

traits. For the approach to be most successful, the two parental strains not only need to differ in the phenotypes being investigated, but they also must be highly genetically polymorphic to allow assessment of the maximum amount of the genome. In this study, we have investigated whether the Brown-Norway (BN) strain provides a suitable control for the hHTG rat. For phenotype comparisons, the strains had direct conscious unrestrained blood pressures (BP), catecholamines, triglycerides, and glucose tolerance measured. During a genome mapping, the variation between the two animal strains for 372 microsatellite primers covering the whole genome was investigated using PCR-amplified microsatellites.

EXPERIMENTAL METHODS

Phenotype

Ten hHTG and 10 BN male rats used for this study were housed before the surgery in groups of 5 (and separately after the surgery) in wire-mesh cages on 12-hour light-dark cycles and were fed semipurified standard rat diet (ST 1, VELAZ, Prague, Czech Republic).

Surgical Procedure

For insertion of a permanent cannula in the carotid artery, the modification of Koopmans' method was used.[6] After an overnight fast, a sterile silicone cannula (Silastic, medical-grade tubing, o.d. = 0.037 inches, i.d. = 0.020 inches, Dow Corning Corporation, Midland, Michigan) was placed in the carotid artery under complete anesthesia (ketamine hydrochloride = 70 mg/kg and xylazine hydrochloride = 10 mg/kg). The animals were allowed to recover from the surgery for at least two days before the next testing.

Blood Pressure

Two days after surgery, systolic and diastolic blood pressure values were obtained from conscious unrestrained rats using a computer-assisted system, DBP 1001, from Kent Scientific Corporation (Litchfield, Connecticut). All blood pressure values are the means of 30 measurements within a 30-min period, which was initiated one hour after connection of the carotid artery to the tonometer.

Determination of Plasma Adrenaline and Noradrenaline

Catecholamines were measured in 50-μL aliquots of plasma using a modification of the method of Peuler and Johnson.[7]

Immobilization Stress

Two days after the blood pressure measurements, immobilization stress[8] was begun. Blood for measurement of the catecholamines was sampled before and 20 min after stress.

Plasma Triglycerides

Plasma triglycerides were determined before surgery in the fed state using a commercially available kit from Lachema (Brno, Czech Republic).

Intravenous Glucose Tolerance Test

Blood samples were taken 0, 1, 2.5, 5, 10, 20, and 50 min after an iv glucose load (1 g glucose/kg body weight). Blood was collected and centrifuged in heparinized tubes and plasma was stored at $-20°C$ until use.

Statistical Analysis

Data are calculated as the mean ± SEM; the significance of the differences was calculated by ANOVA with the appropriate post hoc test.

Genetic Typing by PCR

To genotype hHTG and BN rats, PCR reactions were performed. Genomic DNA (5 µL, 10 ng/µL), isolated from liver tissue using the Genomix kit (Labortechnik Fröbel, Berlin, Germany), was amplified in 20-µL PCR reaction using Taq DNA polymerase (Boehringer Mannheim, Germany). The PCR primer concentrations of each of the two primers were 250 nM and the concentrations of each of the dNTPs were 65.5 µM. Primer sequences were provided by Jacob et al.[9] The reactions were overlaid with 40 µL of light mineral oil (Sigma). Reactions were performed in 96-well plates and amplified using an MJ Research Thermal Cycler. Programming of the temperature and time cycles were as follows: 3 min at 94°C, 35 cycles of 1 min at 94°C, 1 min at 55 or 63°C, and 1 min at 72°C, followed by a final extension period at 72°C for 3 min. Products were resolved by 2% agarose gel electrophoresis (FMC Bioproducts), stained with ethidium bromide, and visualized with UV fluorescence.

TABLE 1. Systolic and Diastolic Blood Pressure (BP), Plasma Adrenaline and Noradrenaline without and after Stress Exposure, and Plasma Triglycerides of Hereditary Hypertriglyceridemic (hHTG) and Brown-Norway (BN) Rat Strains[a]

	hHTG (10)	p	BN (10)
Systolic BP (mmHg)	156.7 ± 5.5	**	114.9 ± 3.4
Diastolic BP (mmHg)	130.5 ± 3.9	**	95.1 ± 3.9
Adrenaline (pg/mL)	312 ± 95	**	62 ± 10
Noradrenaline (pg/mL)	437 ± 140	**	232 ± 50
Adrenaline (after stress) (pg/mL)	2286 ± 410	**	1107 ± 183
Noradrenaline (after stress) (pg/mL)	2987 ± 610	*	1287 ± 218
Plasma triglycerides (mmol/L)	1.54 ± 0.06	*	1.24 ± 0.04

[a] Values are means ± SEM; number of animals given in parentheses; (*) $p < 0.05$, (**) $p < 0.01$.

TABLE 2. Polymorphism for PCR-analyzed Microsatellites between Hereditary Hypertriglyceridemic (hHTG) and Brown-Norway (BN) Rat Strains

Chromo-some	Product of PCR			Number of Polymorphic Markers (hHTG vs. BN)	Rate of Polymorphism (%)
	Good Product	Little Product[a]	No Product		
1	30	1	3	15	50.00
2	23	2	3	15	65.22
3	30	2	2	23	76.67
4	31	1	4	16	51.61
5	24	1	3	15	62.50
6	13	5	2	7	53.85
7	19	2	5	13	68.42
8	14	1	1	7	50.00
9	8	0	2	5	62.50
10	21	0	0	9	42.86
11	6	0	4	3	50.00
12	11	1	1	3	27.27
13	12	1	1	3	25.00
14	10	0	1	4	40.00
15	10	0	2	7	70.00
16	8	0	0	6	75.00
17	9	1	2	4	44.44
18	9	0	2	5	55.55
19	11	0	2	3	27.27
20	5	1	0	2	40.00
X	7	1	1	1	14.29
Total	311	20	41	166	53.38

[a] Difficult to detect the product on the agarose gel.

RESULTS AND DISCUSSION

As shown in TABLE 1, hHTG rats had significantly higher systolic and diastolic blood pressure and plasma adrenaline (A) and noradrenaline (NA) levels. hHTG rats responded to stress by much higher A and NA plasma levels. The levels for triglycerides were significantly higher in hHTG versus BN rats—hHTG: 1.54 ± 0.06 vs. BN: 1.24 ± 0.04 mmol/L, $p < 0.05$. There were no significant differences in the iv glucose tolerance tests.

A relatively high level of polymorphism between hHTG and BN rats was found: 311 out of 372 (83.6%) loci studied provided a PCR product and 166 out of 311 (53.4%) were polymorphic between hHTG and BN rats. The highest rate (76.7%) was observed on chromosome three with 30 loci and the lowest rate (14.3%) was observed on chromosome X with 7 loci studied (TABLE 2).

Agarose electrophoresis is very efficient in the visualization of allelic differences of between 6 and 10 base pairs (93.6%), but it is inefficient at detection of allelic differences of fewer than 5 base pairs (66.2%).[10] Taking this fact into account, it is likely that the rate of polymorphism found between hHTG and BN rats in our study (53.4%) is an underestimate of the actual degree of polymorphism. Although higher rates of polymorphism were previously described for some other strains,[11] the results of the genetic typing suggest that hHTG and BN rats are very good candidates for mapping QTLs. The differences in systolic and diastolic blood pressure and in the levels of plasma constituents (catecholamines, triglycerides) between the two strains indicate that alleles causing hypertension and hypertriglyceridemia could be identified in the hHTG rat strain.

Our results demonstrate that the BN rat provides a suitable phenotypic and genetic control to cross with the hHTG rat to identify QTLs influencing several traits of the insulin resistance syndrome. The genetic dissection of these traits according to Lander and Botstein[12] is a challenge and, therefore, an F_2 population from a cross of hHTG and BN rats is currently being developed.

ACKNOWLEDGMENTS

We thank Alica Mitková for performing the vascular surgery, Ildiko Szalayová for assistance in measuring blood pressures, and Dagmar Janovová and Daniela Chalupková for assistance with catecholamine measurements.

REFERENCES

1. VRÁNA, A. & L. KAZDOVÁ. 1990. Transpl. Proc. **22:** 2579.
2. ŠTOLBA, P., H. OPLTOVÁ, P. HUŠEK, J. NEDVÍDKOVÁ, J. KUNEŠ, Z. DOBEŠOVÁ, J. NEDVÍDEK & A. VRÁNA. 1993. Ann. N.Y. Acad. Sci. **683:** 281–288.
3. VRÁNA, A., L. KAZDOVÁ, Z. DOBEŠOVÁ, J. KUNEŠ, V. KŘEN, V. BÍLÁ, P. ŠTOLBA & I. KLIMEŠ. 1993. Ann. N.Y. Acad. Sci. **683:** 57–68.
4. LICHARDUS, B., E. ŠEBÖKOVÁ, D. JEŽOVÁ, A. MITKOVÁ, A. ZEMÁNKOVÁ, O. FÖLDEŠ, A. VRÁNA & I. KLIMEŠ. 1993. Ann. N.Y. Acad. Sci. **683:** 289–294.

5. KLIMEŠ, I., A. VRÁNA, J. KUNEŠ, E. ŠEBÖKOVÁ, Z. DOBEŠOVÁ, P. ŠTOLBA & J. ZICHA. 1995. Blood Pressure **4**: 137–142.
6. KOOPMANS, S. J., J. A. MAASEN, J. K. RADDER, M. FRÖLICH & H. M. J. KRANS. 1992. Biochim. Biophys. Acta **1115**: 230–238.
7. PEULER, Y. D. & G. A. JOHNSON. 1977. Life Sci. **21**: 625–636.
8. KVETŇANSKÝ, R. & I. MIKULAJ. 1970. Endocrinology **87**: 738–743.
9. JACOB, H. J., D. M. BROWN, R. K. BUNKER, M. J. DALY, V. J. DZAU, A. GOODMAN, G. KOIKE, V. KŘEN, T. KURTZ, A. LERNMARK, G. LEVAN, Y. MAO, A. PETTERSSON, M. PRAVENEC, J. S. SIMON, C. SZPIRER, J. SZPIRER, M. R. TROLLIET, E. S. WINER & E. S. LANDER. 1995. Nat. Genet. **9**: 63–69.
10. ROUTMAN, E. J. & J. M. CHEVERUD. 1995. Mamm. Genome **6**: 401–404.
11. KLÖTING, I., B. VOIGT & L. VOGT. 1995. J. Exp. Anim. Sci. **37**: 42–47.
12. LANDER, E. S. & D. BOTSTEIN. 1989. Genetics **121**: 185–199.

Insulin Resistance in Adipocytes of SHR Rats[a]

M. FICKOVÁ,[b] Š. ZÓRAD,[b] J. KUNEŠ,[c]
M. RUSNÁK,[b] AND L. MACHO[b]

[b]Institute of Experimental Endocrinology
Slovak Academy of Sciences
833 06 Bratislava, Slovak Republic
[c]Institute of Physiology
Czech Academy of Sciences
Prague, Czech Republic

Spontaneously hypertensive rats (SHR) represent an animal model generally accepted as the best-available model of genetic hypertension because of its numerous similarities with human essential hypertension. In addition to increased blood pressure, this model is characterized by elevated fasting concentrations of glucose, insulin, and triacylglycerols. The metabolic disturbances reflect the presence of insulin resistance.[1] Although muscle tissue mass contributes substantially to insulin-stimulated glucose utilization, no abnormalities were found in insulin receptor binding characteristics, kinase activity, or glucose transporter (GLUT4) protein level in skeletal muscle of SHR rats.[1] However, there is attenuated insulin-stimulated glucose utilization in skeletal muscle as the primary site of insulin resistance in SHR.[2] At the level of adipose tissue, Reaven and Chang described *in vitro* insulin resistance of glucose uptake, but the antilipolytic effect of insulin was augmented in SHR adipocytes. Insulin receptors in SHR fat cells display similar binding properties as in matched control rats. Thus, the attenuated insulin-stimulated glucose uptake could result from the alterations of the glucose transporter and should result in metabolic changes after glucose enters the cell.

This study was undertaken to study lipogenesis and the expression of GLUT4 in connection with insulin resistance in adipose cells of SHR rats.

---

[a] This work was supported by a grant of the Slovak Academy of Sciences (Grant No. GAV 2-541-93 to M. Ficková) and partially by Grant No. SSVT-SR95/5305/043.

MATERIALS AND METHODS

Animals

Male SHR rats and Wistar-Kyoto rats (WKY) as the normotensive control from the Prague colony at the age of 120 days (250–280 g) were used. One week before the experiment, systolic blood pressure (SBP) was measured by the noninvasive, tail-cuff method.

Lipogenic Studies

Adipocytes isolated from epididymal fat tissue by collagenase digestion were used. Fat cells were incubated for 2 h under an atmosphere of CO_2 and O_2 in Krebs-Ringer buffer (pH 7.4) supplemented with albumin (2%) and 5 mM glucose. Incorporation of U-^{14}C-glucose into total lipids under basal and insulin-stimulated conditions (1 × 10^{-9} and 1 × 10^{-6} M, respectively) was determined after lipid extraction.

Insulin Binding Studies

Binding of mono-(Tyr-A-14)-^{125}I-insulin alone and with an excess (1 × 10^{-6} M) of monocomponent pork insulin (NOVO, Denmark) was performed. Specific insulin binding (%) represents the total binding corrected for unspecific binding.

GLUT4

Plasma membranes were isolated from intact fat cells and, after solubilization, were subjected to SDS gel electrophoresis. Proteins from the gel were transferred by electroblotting to Hybond-C membrane. Western blotting with rabbit anti-GLUT4 (East Acres, United States) was followed by labeling of immunocomplexes with ^{125}I-protein A. The labeled proteins were detected by autoradiography.

Biochemical Measurements

Serum glucose (oxochrome method) and triacylglycerols were determined by commercial kits (Lachema Diagnostica, Brno, Czech Republic). Serum immunoreactive insulin was assayed using an RIA/Sax kit (Germany).

RESULTS

Blood pressure and serum parameters of nonstarved rats are shown in TABLE 1. SBP of SHR was significantly higher than that of matched controls. SHR and WKY

TABLE 1. Biochemical Characterization of Animals[a]

	SHR	WKY	p Value
SBP (mmHg)	161.0 ± 3.5	104.5 ± 2.5	<0.005
glucose (mmol/L)	8.6 ± 0.2	8.2 ± 0.1	NS
insulin (nmol/L)	0.51 ± 0.04	0.35 ± 0.01	<0.05
TG (mmol/L)	0.64 ± 0.04	0.55 ± 0.05	NS

[a] Terms—SBP: systolic blood pressure; TG: triacylglycerols; NS: not significant.

did not differ either in glycemia or in triacylglycerolemia, but a significantly higher concentration of insulin was present in the SHR group.

Insulin specific binding in isolated adipocytes was the same in SHR as in control rats (FIG. 1). Basal lipogenesis was significantly lower in SHR adipocytes and the responsiveness to the stimulatory effect of both insulin concentrations was significantly attenuated in fat cells of the SHR group (FIG. 1). Glucose transporter protein in plasma membranes of nonstimulated SHR fat cells was reduced (FIG. 2). There was no correlation between insulinemia and the extent of basal lipogenesis and a significant negative correlation between the insulin lipogenic effect (1×10^{-6} M) and serum insulin concentration ($r = -0.619$, $n = 12$, $p < 0.05$). A negative correlation ($r = -0.715$, $n = 12$, $p < 0.01$) was found between blood pressure and insulin action (1×10^{-6} M), and a positive relation between blood pressure and insulinemia ($r = 0.858$, $n = 12$, $p < 0.001$) was observed.

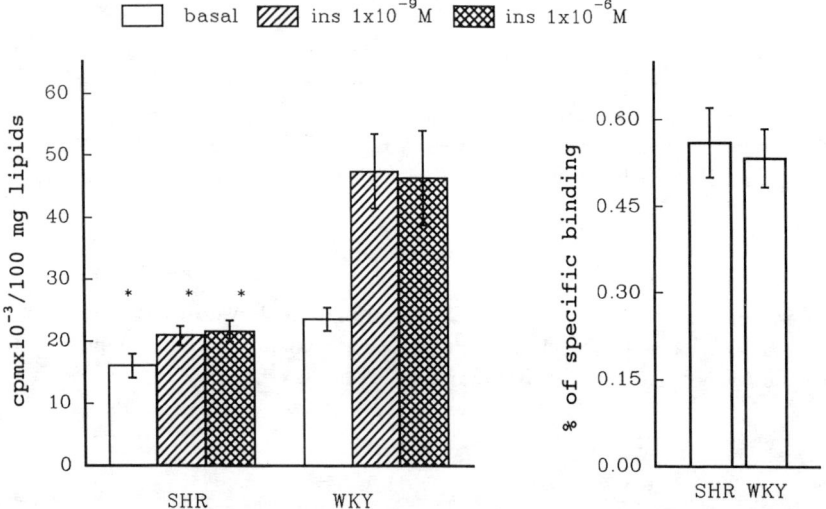

FIGURE 1. (Left) Lipogenesis: Incorporation of ^{14}C-glucose into fat cell lipids; stimulatory effect of insulin at 1×10^{-9} M and 1×10^{-6} M; (∗) $p < 0.05$, SHR vs. WKY. (Right) Insulin specific binding to isolated adipocytes. Values are expressed as the mean ± SEM and the significance of the differences was tested by Student's unpaired t test.

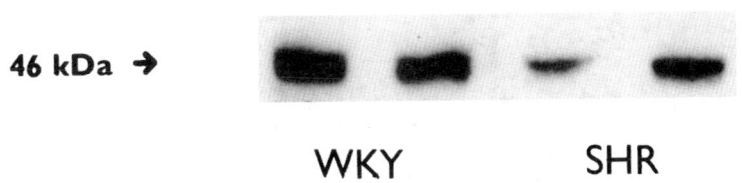

FIGURE 2. Immunoblots of GLUT4 in a plasma membrane fraction of nonstimulated fat cells.

DISCUSSION

The characterization of SHR animals of the Prague colony by values of systolic blood pressure, insulin, and triacylglycerol concentrations clearly demonstrates the similarity of these animals with those from other origins. Elevated insulinemia is an indicator of the presence of insulin resistance.

A significantly lower lipogenic effect of insulin has been found in SHR adipocytes. Moreover, a negative correlation between insulin action and serum insulin concentration clearly indicates insulin resistance in SHR adipocytes. Decreased insulin-stimulated glucose uptake into adipocytes of SHR has been described previously;[3] our observation of the reduced lipogenic effect of insulin extends the description of metabolic disturbances in SHR adipocytes. Insulin receptor binding properties in adipocytes of SHR and their WKY controls support similar results found previously.[3] This indicates that the altered glucose metabolism is not connected with the binding properties of insulin receptors, but more likely with a postreceptor mechanism. The finding of lower GLUT4 expression in fat cell plasma membranes of SHR could be related to attenuated basal lipogenic activity. The properties of GLUT4 after insulin stimulation need additional studies. The positive correlation between SBP and insulinemia and the negative relation between SBP and insulin action support the hypothesis that insulin resistance in adipose tissue and compensatory hyperinsulinemia contribute to the pathogenesis of essential hypertension.

REFERENCES

1. BADER, S., R. SCHOLZ, M. KELLERER, S. TIPPMER, K. RETT, S. MATHAEI, P. FREUND & H. U. HÄRING. 1992. Diabetologia 35: 712–718.
2. HULMAN, S., B. FALKNER & N. FREYVOGEL. 1993. Metabolism 42: 14–18.
3. REAVEN, G. M. & H. CHANG. 1991. Am. J. Hypertens. 4: 34–38.

Disproportionate Increase of Fatty Acid Binding Proteins in the Livers of Obese Diabetic *Psammomys obesus*

P. LEWANDOWSKI, D. CAMERON-SMITH,
K. MOULTON, K. WALDER, A. SANIGORSKI, AND
G. R. COLLIER[a]

*School of Nutrition and Public Health
Deakin University
Geelong, Victoria 3217, Australia*

INTRODUCTION

Fatty acid binding proteins (FABPs) are a distinct group of cytosolic transport proteins that have been shown to specifically bind and transport long-chain fatty acids (LCFA) through the aqueous environment of the cytosol.[1,2] Whether FABPs play a key role in the regulation of fat metabolism remains unclear. To examine the role of FABPs in the regulation of fat metabolism, we aim to investigate the relationship between FABP levels and lipid accumulation in the livers of a genetically heterogenous model of obesity and NIDDM, namely, the Israeli sand rat, *Psammomys obesus*. *P. obesus*, a native of North Africa and the Middle East, has been shown to develop NIDDM and obesity in a pattern that mirrors susceptible human populations.[3] Obese diabetic *P. obesus* animals also display abnormalities in lipid metabolism, including increased liver fat deposition, circulating cholesterol, and circulating triglyceride levels.[4,5]

METHODS

Animals were housed with *ad libitum* food and water under 12-h light/dark cycles at 21 ± 1 °C. The high-energy density of the diet, relative to the herbivorous diet that the sand rat subsists on in its native habitat, induces heterogenous weight gain and a distribution of glucose tolerances. At 12 weeks of age, 30 animals were bled and the blood samples were assessed for plasma glucose, insulin, and triglyceride concentrations. The animals were classified into three groups: A, normoglycemic-normoinsulinemic ($n = 11$); B, normoglycemic-hyperinsulinemic ($n = 11$); and C, hyperglycemic-hyperinsulinemic ($n = 8$). Serum insulin concentrations were deter-

---

[a] To whom all correspondence should be addressed.

mined by a radioimmunoassay (Pharmacia, Columbus, OH) using a rat insulin standard. Tissues were rapidly removed and freeze-clamped in liquid nitrogen for future analysis. Tissue FABP content was measured based on the radiochemical method of Glatz and Veerkamp.[6] Tissue homogenates were prepared in Tris-HCl buffer centrifuged for 10 min at 600g and the supernatant was centrifuged again for 90 min at 105,000g at 4°C. Albumin was removed from the cytosolic homogenate by affinity chromatography and then delipidated using a hydrophobic dextran column. The resultant FABP extract was incubated with [^{14}C]-palmitate at 37°C for 10 min and unbound [^{14}C] was removed using hydrophobic dextran and centrifugation (20,000g for 2 min at 4°C). The resulting supernatant containing the bound [^{14}C]-palmitate was assayed for radioactivity by liquid scintillation counting. Tissue lipid content was measured by chloroform methanol extraction.[7] Results were analyzed for analysis of variance (ANOVA) with Tukey's comparisons; p values of less than 0.05 were considered to be significant.

RESULTS

Animals in group A were lean, normoglycemic, and normoinsulinemic (TABLE 1). Group B animals were obese, although they maintained normoglycemia despite hyperinsulinemia (TABLE 1), and group C animals were both obese and diabetic (TABLE 1).

Interscapular fat mass was used as an indicator of whole body adipose tissue lipid accumulation. Group B and group C animals demonstrated a significant increase in adipose tissue mass when compared with lean group A animals (TABLE 1). Total FABP present (FIG. 1) in the interscapular fat of group B (26.8 ± 8.3 mg) and group C animals (28.2 ± 4.2 mg) was significantly greater than that in group A (7.5 ± 0.9 mg, $p <$ 0.05). When interscapular fat FABP content was expressed as the ratio of FABP to interscapular fat mass (FIG. 2), it was found that all groups had a similar ratio.

Hepatic lipid accumulation was measured in all animals (TABLE 1). Both obese diabetic group C animals (5.8 ± 0.56%) and obese nondiabetic group B animals (6.5 ± 0.78%) had a significantly increased amount of hepatic lipid compared with healthy lean group A animals (4.2 ± 0.34%, $p <$ 0.05). Total FABP present (FIG. 1) in the liver of group C animals (251.4 ± 55 mg) was significantly increased when compared to group A (101.3 ± 18.2 mg). Liver FABP levels did not differ in group B (151.8

TABLE 1. Body Weight, Glucose, Insulin, Interscapular Adipose Tissue Weight, and Percentage of Fat in Livers of Group A ($n = 11$), Group B ($n = 11$), and Group C ($n = 8$) *Psammomys obesus*

Group	Body Weight (g)	Glucose (mmol/L)	Insulin (μU/mL)	Interscapular Adipose (g)	Hepatic Lipids (% of Tissue Weight)
A	189.6 ± 10.9	3.5 ± 0.15	52.8 ± 7.1	1.93 ± 0.45	4.2 ± 0.34
B	225.9 ± 15.8[a]	4.2 ± 0.24	311.2 ± 42.7[a]	4.16 ± 0.73[a]	6.5 ± 0.78[a]
C	235.1 ± 9.5[a]	10.6 ± 0.58[a]	479.7 ± 118.5[a]	3.93 ± 0.56[a]	5.8 ± 0.56[a]

[a] $p <$ 0.05 compared to group A.

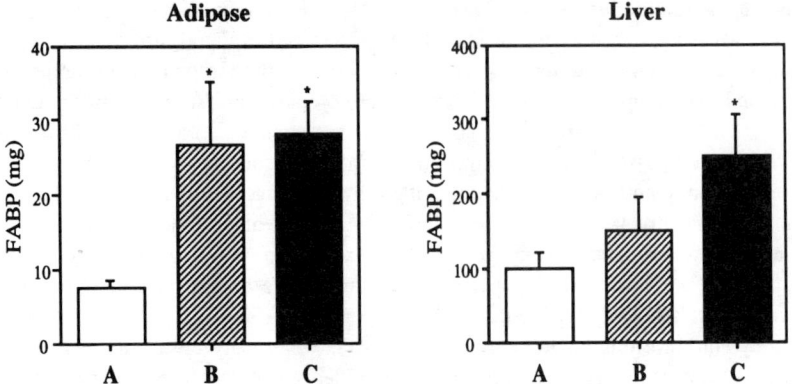

FIGURE 1. Total amount of cytosolic FABP in adipose and liver tissue in group A ($n = 11$), group B ($n = 11$), and group C ($n = 8$) *Psammomys obesus*: (∗) $p < 0.05$ compared to group A.

± 43.3 mg) when compared to the control group. When liver FABP content was expressed as the ratio of the FABP levels to fat accumulation in the liver (FIG. 2), group A (0.37 ± 0.07) and group B (0.31 ± 0.05) had similar ratios. However, group C animals (0.71 ± 0.11) demonstrated an approximate doubling of the ratio of FABP to fat accumulation in the liver compared to the other study groups.

DISCUSSION

Adipose tissue FABP concentrations were significantly increased in both obese and obese diabetic *P. obesus*. However, the ratio of FABP to interscapular fat mass

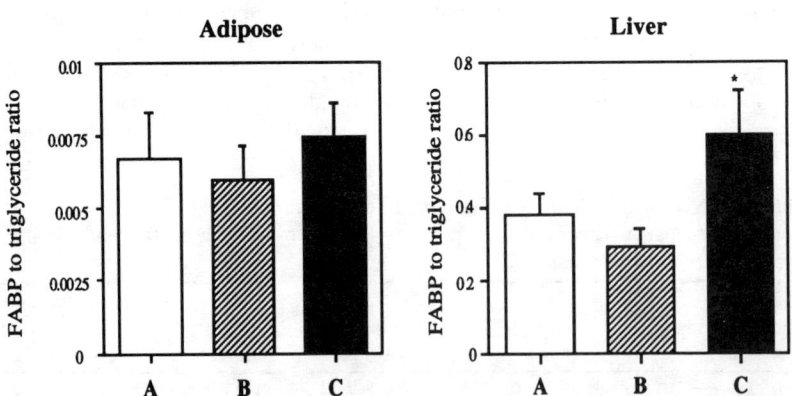

FIGURE 2. Ratio of cytosolic FABP to adipose tissue mass triglyceride or ratio of FABP to liver triglyceride content in group A ($n = 11$), group B ($n = 11$), and group C ($n = 8$) *Psammomys obesus*: (∗) $p < 0.05$ compared to group A.

was unchanged, indicating that adipose FABP levels reflected the level of fat storage. In contrast, obese diabetic animals demonstrated a significant increase in hepatic cytosolic FABPs and exhibited a disproportionate elevation in the ratio of FABP to fat in the same tissue. This suggests that there is a dysregulation of FABP levels associated with obesity and NIDDM.

The glucose and insulin concentrations in the fed state of the animals analyzed in the current study were comparable with previous studies in *P. obesus*,[8] demonstrating a range of values from normal through to the elevated concentrations seen in the diabetic state. This range of responses mirrors those seen in susceptible human populations.[3] The increased body weight observed in both group B and group C was consistent with our previous studies in *P. obesus*.[8]

The accumulation of interscapular fat in all three groups was used as an indicator of obesity in these animals. This clearly demonstrated that the increased body weight present in both group B and group C could be largely attributed to an increase in adipose tissue mass. The increase in total adipose FABP in the interscapular fat and the elevated FABP to fat mass ratio in the same tissue suggest that FABP levels in adipose tissue are linked with adipose tissue mass. As previously reported[1] and confirmed in this study, adipose tissue cytosolic FABP levels are linked with fat storage.

With increasing body weight and onset of diabetes, elevated levels of fat accumulated in the livers of *P. obesus*. However, the total amount of FABP present and the ratio of FABP to hepatic lipid accumulation in the liver were markedly and disproportionately elevated in the obese diabetic group C animals. Thus, the changes in hepatic FABP content are not linearly correlated with hepatic lipid accumulation. This would suggest that the increased FABP concentrations in the liver of obese and diabetic *P. obesus* may act to promote excessive liver fat oxidation. Elevated fat oxidation increases hepatic glucose production with suppressed glucose oxidation and elevated rates of gluconeogenesis. Evidence for this hypothesis is found in previous studies in our laboratory, where it was demonstrated that obese diabetic *P. obesus* animals exhibit elevated hepatic glucose production that was reduced following free fatty acid inhibition.[9,10] In further studies, we aim to investigate if genetic alterations in the FABP gene region could lead to increased expression in the livers of obese diabetic *P. obesus*.

In conclusion, obese and obese diabetic *P. obesus* have significantly elevated body weight, adipose tissue mass, and hepatic lipid accumulation. Adipose tissue FABPs are elevated in obesity and reflect the adipose tissue mass. In contrast, liver cytosolic FABPs exhibit a disproportionate elevation in the ratio of FABP to fat in obese diabetic *P. obesus*.

REFERENCES

1. VEERKAMP, J. H., R. A. PEETERS & R. G. H. J. MAATMAN. 1991. Structural and functional features of different types of cytoplasmic fatty acid–binding proteins. Biochim. Biophys. Acta **1081:** 1–24.
2. SWEETSER, D. A., R. O. HEUCKEROTH & J. I. GORDON. 1987. The metabolic significance of mammalian fatty-acid-binding proteins: abundant proteins in search of a function. Annu. Rev. Nutr. **7:** 337–359.

3. SHAFRIR, E. & A. GUTMAN. 1993. *Psammomys obesus* of the Jerusalem colony: a model for nutritionally induced, non-insulin-dependent diabetes. J. Basic Clin. Physiol. Pharmacol. **4**(1-2): 83-99.
4. LEE, S., A. SANIGORSKI, A. YAMAMOTO & G. COLLIER. 1995. Glucose tolerance and insulin secretion in *Psammomys obesus*: an animal model of NIDDM. Proceedings of the Australian Diabetes Society, Melbourne, p. 108.
5. REAVEN, G. 1988. Role of insulin resistance in human disease. Diabetes **37**: 1595-1607.
6. GLATZ, J. F. C. & J. H. VEERKAMP. 1983. A radiochemical procedure for the assay of fatty acid binding by proteins. Anal. Biochem. **132**: 89-95.
7. FOLCH, J., M. LEES & G. H. SLOANE STANLEY. 1957. A simple method for the isolation and purification of total lipids from animal tissues. J. Biol. Chem. **226**: 497-509.
8. BARNETT, M., G. R. COLLIER, F. M. COLLIER, P. ZIMMET & K. O'DEA. 1994. A cross-sectional and short-term longitudinal characterisation of NIDDM in *Psammomys obesus*. Diabetologia **37**: 671-676.
9. COLLIER, G., F. COLLIER, A. SINCLAIR & K. O'DEA. 1992. Phospholipid fatty acid composition and glucose tolerance in the Israeli sand rat. Diabetologia **35**(suppl. 1): A78.
10. HABITO, R., M. BARNETT, A. YAMAMOTO, D. CAMERON-SMITH, K. O'DEA, P. ZIMMET & G. COLLIER. 1995. Basal glucose turnover in *Psammomys obesus*. Acta Diabetol. **32**: 187-192.

Partial Characterization of Insulin Resistance in Adipose Tissue of Monosodium Glutamate-induced Obese Rats[a]

Š. ZÓRAD, L. MACHO, D. JEŽOVÁ,
AND M. FICKOVÁ

Institute of Experimental Endocrinology
Slovak Academy of Sciences
833 06 Bratislava, Slovak Republic

Treatment of neonatal rats with monosodium glutamate (MSG) induces a massive increase of adipose fat tissue mass.[1] This model of chemically induced obesity is characterized by normophagia, normoglycemia, fat tissue hypertrophy, and reduced muscle mass.[2] In addition, in young MSG-treated animals, the plasma insulin levels are normal despite developing obesity. In the adults, peripheral hyperinsulinemia is present, suggesting an impaired insulin sensitivity.[2] Insulin receptors and insulin action have not been studied intensively in MSG-induced obese rats; thus, the assumed tissue insulin resistance is still a matter of debate. The aim of our study was to determine *in vitro* the status of insulin receptors and to investigate insulin stimulation of glucose transport and lipogenesis in epididymal fat tissue.

MATERIALS AND METHODS

Animals

Newborn pups of Sprague-Dawley rats were subjected to MSG treatment.[3] The animals were used for experiments at the age of 3 months. The rats were starved 12 h before sacrificing for insulin receptor studies and determination of serum parameters.

Analytical Methods

Serum glucose and triacylglycerol levels were determined by Lachema Diagnostica kits (Czech Republic). Serum insulin was estimated by a radioimmunoassay kit (NOVO, Denmark).

[a] This work was supported by grants of the Slovak Academy of Sciences (Grant Nos. GA-SAV 2-541-93 to Š. Zórad and VEGA 2/1288 to D. Ježová) as well as by the EC (Grant No. CIPACT930277 to D. Ježová).

Determination of Fat Cell Size and Cellularity

Adipocytes were isolated from epididymal fat tissue by the collagenase digestion method of Rodbell.[4] Fat cell size was determined by measurement of the cell diameter under light microscopy after staining with crystal violet.[5] Cellularity of adipose tissue was determined by dividing the lipidic extract of fat tissue by the calculated mean mass of one fat cell based on the mean fat density (0.915 g/mL) and on an approximately spherical shape of the cells. Results are expressed as the number of fat cells in 1 g of fat tissue.

Assessment of Insulin Receptors

Insulin binding studies were performed in isolated plasma membrane fractions of fat cells as described elsewhere.[6] Competition curves were analyzed by using the Ligand computer program.[7]

Determination of Glucose Transport and Lipogenesis

The glucose transport assay was performed by the 2-deoxy-D-H^3-glucose method of Cherqui et al.[8] All data were corrected for extracellular trapping and passive diffusion by measuring transport in the presence of 0.25 µM of cytochalasin B.

Lipogenic activity was assessed by incubating isolated fat cells for 1 h in KRB buffer (pH 7.4) supplemented with 2% albumin and 1 mM glucose. Incorporation of ^{14}C-U-glucose into total lipids under basal and insulin-stimulated conditions was determined after lipid extraction.

Statistical Analysis

All data are expressed as the mean ± SEM. The significance of the differences was determined by Student's t test, Mann-Whitney test, and ANOVA where appropriate.

RESULTS

TABLE 1 shows that the MSG-treated rats had a clearly increased mass of epididymal fat tissue caused by enlarged fat cell size. Hypertrophic obesity in MSG-induced obese rats was further confirmed by the significantly lower cellularity of adipose tissue. Three-month-old obese rats were normoglycemic, hyperinsulinemic, and hypertriglyceridemic despite no significant weight differences (controls: 485 ± 22 g; MSG-treated: 438 ± 27 g; $n = 10$).

A Scatchard curve of insulin binding to isolated plasma membranes of epididymal

TABLE 1. Basic Serum and Epididymal Fat Tissue Parameters of the Experimental Groups

	Control	MSG
Insulin (μU/mL)	19.2 ± 4.4	38.7 ± 5.0[a]
Glucose (mmol/L)	5.8 ± 0.1	6.1 ± 0.1
Triglycerides (mmol/L)	0.7 ± 0.1	1.5 ± 0.3[a]
% of epididymal fat based on body weight	1.5 ± 0.1	2.5 ± 0.2[a]
Fat cell size (diameter, μm)	78.7 ± 2.0	103.7 ± 2.9[a]
Cellularity (10^6 cells/g tissue)	3.1 ± 0.2	1.4 ± 0.2[a]

[a] Statistically significant differences at $p < 0.05$.

fat tissue was curvilinear in both experimental groups. Specific binding was significantly higher in membranes from lean rats. Analysis of competition curves using the two-binding-site model revealed a significantly lower binding capacity [R1: 0.31 ± 0.02 vs. 0.55 ± 0.07 (pmol/mg), $p < 0.005$, $n = 10$] of high-affinity insulin receptors in MSG-treated rats. Significant differences were not noticed between MSG-treated and control rats either in low-affinity receptor capacity [R2: 3.6 ± 0.7 vs. 10.3 ± 6.1 (pmol/mg)] or in receptor affinities [K11: 2.0 ± 0.4 vs. 2.1 ± 0.2 ($\times 10^9$ M^{-1}); K12: 0.41 ± 0.10 vs. 0.41 ± 0.16 ($\times 10^8$ M^{-1})].

FIGURE 1a shows the insulin stimulation of glucose transport in isolated adipocytes. In control rats the insulin effect was dose-dependent, while in MSG-treated animals this hormone was significantly less effective. Basal glucose transport, without insulin, was also lower in obese rats (1.6 ± 0.9 vs. 4.21 ± 1.3 fmol × $10^{-6}/\mu m^2$, $n = 6$, $p < 0.002$).

Insulin stimulated the incorporation of glucose into lipids in both experimental groups dose-dependently with the exception of the highest concentration of insulin (10^{-7} M) in the MSG group (FIG. 1b). The basal and 10^{-11} and 10^{-9} M insulin-stimulated incorporation rates were significantly higher in the MSG-treated animals (117 ± 12 vs. 216 ± 30; 141 ± 16 vs. 262 ± 30; 191 ± 33 vs. 332 ± 33 (cpm/mg lipids); $n = 6$; $p < 0.05$, ANOVA).

DISCUSSION

Neonatal treatment with MSG produces lesions of the hypothalamic arcuate nucleus including growth hormone–releasing hormone neurons.[1] This causes the reduction of growth hormone (GH) secretion and also lowers the levels of insulin-like growth factor I (IGF-I).[9] Since GH is responsible *in vivo* for normal glucose transport in rat adipocytes[10] and IGF-I displays potent insulin-like activity, it seems reasonable to assume disturbed glucose metabolism in MSG-treated animals. Our results show lower basal and insulin-stimulated glucose transport per μm^2 of adipocyte surface of obese rats. This is in accordance with a decreased content of glucose transporter (GLUT 4) found in epididymal fat tissue of MSG-treated mice.[11] On the other hand, Marmo et al.[12] did not find any significant difference in glucose transport per μm^2 between

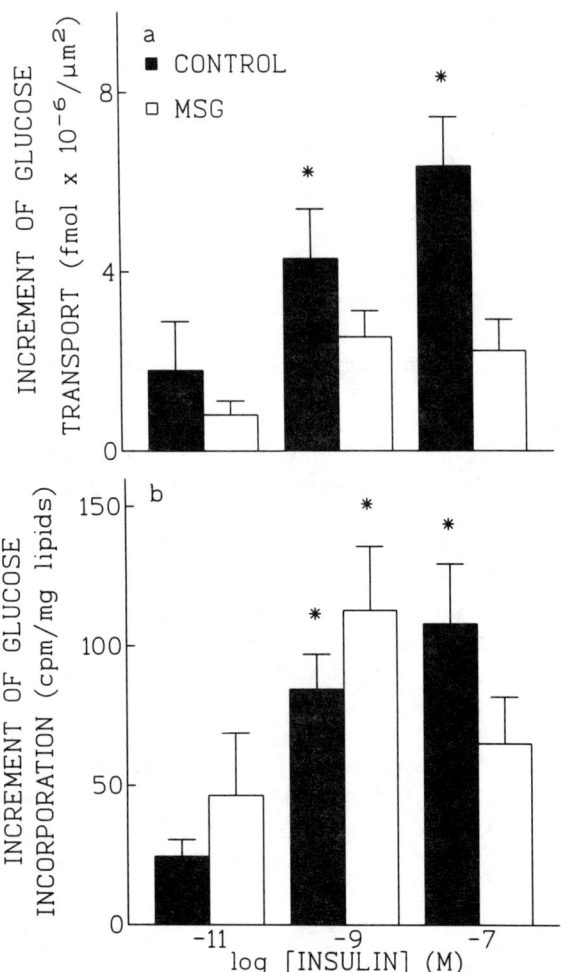

FIGURE 1. (a) Insulin-stimulated glucose transport in isolated adipocytes: (*) significant difference between control and MSG group as well as against zero increment, $n = 6$. (b) Insulin-stimulated incorporation into lipids: (*) significant difference against zero increment, $n = 6$.

control and MSG-obese rats. The discrepancy can be due to a preincubation step included in the standard glucose transport assay used by the investigators.

Nevertheless, Marmo's as well as our experiments showed increased *in vitro* lipogenesis in adipose tissue of MSG-treated rats, suggesting a shift in glucose metabolism towards lipid synthesis. In addition, our results indicated an impaired insulin action by demonstrating the presence of insulin receptor downregulation in adipocytes of obese rats. Nenoff *et al.*,[2] employing an *in vivo* radioreceptor assay, did not detect any insulin receptor changes in different tissues of MSG animals despite the presence of hyperinsulinemia. The results of the *in vivo* method, however, can be considerably influenced by uncontrolled insulin degradation and the presence of endogenous hormone. In conclusion, our data suggest the presence of insulin resistance in epididymal fat tissue of MSG-induced obese rats at the level of insulin receptors and glucose transport.

REFERENCES

1. OLNEY, J. W. 1969. Science **164:** 719-721.
2. NENOFF, P., H. REMKE, F. MULLER, T. ARNDT & T. MOTHES. 1993. Exp. Clin. Endocrinol. **101:** 215-221.
3. TOKAREV, D., V. KRISTOVÁ, M. KRIŠKA & D. JEŽOVÁ. 1997. Physiol. Res. **46:** 165-171.
4. RODBELL, M. 1964. J. Biol. Chem. **239:** 375-380.
5. BELZUNG, F., T. RACLOT & R. GROSCOLAS. 1993. Am. J. Physiol. **264:** R1111-R1118.
6. ZÓRAD, Š., E. ŠVABOVÁ, I. KLIMEŠ & L. MACHO. 1985. Endocrinol. Exp. **19:** 267-276.
7. MUNSON, P. J. & D. RODBARD. 1980. Anal. Biochem. **107:** 220-239.
8. CHERQUI, G., M. CARON, D. WICEK, J. CAPEAU & J. PICARD. 1989. Mol. Cell. Endocrinol. **65:** 13-25.
9. KUBOTA, A., Y. NAKAGAWA & Y. IGARASHI. 1994. Horm. Metab. Res. **26:** 497-503.
10. SCHOENLE, E., J. ZAPF & R. FROESCH. 1983. Endocrinology **112:** 384-385.
11. MACHADO, U. F., Y. SHIMIZU & M. SAITO. 1993. Horm. Metab. Res. **25:** 462-465.
12. MARMO, M. R., M. S. DOLNIKOFF, I. C. KETTELHUT, D. M. MATSUSHITA, N. S. HELL & F. B. LIMA. 1994. Braz. J. Med. Biol. Res. **27:** 1249-1253.

Biguanide Effects on Insulin Signaling[a]

EMMANUELLE MEUILLET,[b]
NICOLAS WIERNSPERGER,[c] PIERRE HUBERT,[b]
AND GÉRARD CRÉMEL[b,d]

[b]Centre de Neurochimie
INSERM U. 338
67084 Strasbourg, France

[c]LIPHA S.A.
69379 Lyon, France

INTRODUCTION

Metformin is a powerful antihyperglycemic agent that has been used for a long time in Europe for the treatment of diabetic patients with non-insulin-dependent diabetes mellitus. Because metformin is a positively charged molecule and nondiffusible across biomembranes, the first target of metformin could be the membrane itself or some lipid component. Recent reports have demonstrated that the major effect of metformin was either to potentiate or to mimic the action of insulin (for a review, see reference 1). These latter data suggest that metformin acts on insulin signaling pathways. Studies of modifications of membrane lipid composition have shown that lipids can affect insulin signaling.[2]

By supplementing the culture medium of HepG2 cells with a derivative of cholesterol, we have developed a cellular model with modified insulin sensitivity, where metformin action could be tested and its mode of action on insulin signaling eventually deciphered. Thus, metformin action on receptor autophosphorylation kinase activity was evaluated in whole cells.

METHODS

Cell Culture and Metformin Pretreatment

Human hepatocarcinoma HepG2 cells were grown in Dulbecco's Modified Eagle's Medium (DMEM) supplemented with 10% fetal calf serum and containing 4.5 g/L glucose. In order to change the sterol composition of the cell membrane, HepG2 cells were adapted for two passages to a culture medium supplemented with 10^{-4} M cholesterol hemisuccinate (CHS). Metformin action was evaluated by incubation of cells

---

[a] This work was supported in part by a grant from LIPHA, Lyon, France.
[d] To whom all correspondence should be addressed.

with 100 µM metformin in DMEM for 16 h. Control cells were obtained after incubation of DMEM alone during this time.

Lipid Analysis and Membrane Fluidity

Cholesterol content was estimated by a colorimetric method using $FeCl_3$ and H_2SO_4.[3] Membrane fluidity was estimated with two fluorescent probes: trimethylammonium diphenylhexatriene (TMA-DPH) (1 µM) and diphenylhexatriene (DPH) (1 µM).[4]

Immunoblotting with Antiphosphotyrosine Antibodies

HepG2 cells were treated for the indicated times with 10^{-7} M insulin, in the presence or absence of 10^{-4} M metformin. Cellular extracts were electrophoresed and transferred to nitrocellulose, and tyrosine-phosphorylated proteins were detected with antiphosphotyrosine antibodies and a peroxidase-coupled method. The band at 95 kDa corresponding to insulin receptors was quantitated after scanning.

RESULTS AND DISCUSSION

Lipid Composition and Membrane Fluidity of HepG2 Cells

In order to establish the validity of the cellular model, lipid composition of control and CHS-treated cells was determined. Culture of HepG2 cells in the presence of CHS induced a significant increase in cholesterol content (33.6 ± 3.5 µg/mg protein for control cells vs. 55.3 ± 4.7 µg/mg protein for CHS-treated cells). Total lipid content was not modified (4.0 ± 0.07 vs. 3.93 ± 0.47 mg/mg protein for control and CHS-treated cells, respectively).

Membrane fluidity of intact normal and CHS-treated cells was determined as fluorescence anisotropy for two probes (TMA-DPH and DPH). CHS treatment induced an increase in fluorescence anisotropy of both probes, according to the well-known role of cholesterol in increasing membrane rigidity: fluorescence anisotropy rose from 0.224 ± 0.002 to 0.229 ± 0.002 for control and CHS-treated cells in the case of TMA-DPH and from 0.123 ± 0.003 to 0.145 ± 0.002 in the case of DPH.

Effect of Metformin on Insulin Receptor Autophosphorylation

We incubated modified and control HepG2 cells with insulin for different times. Tyrosine-phosphorylated proteins were then visualized after electrophoresis and immunoblotting with specific antiphosphotyrosine antibodies. Insulin stimulated the

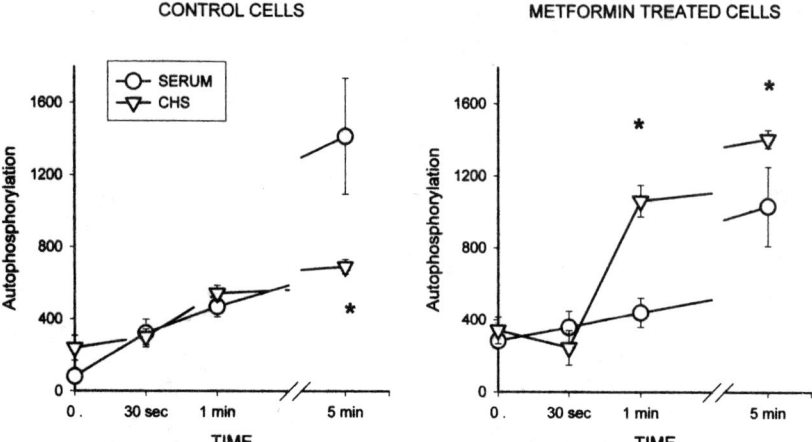

FIGURE 1. Metformin effect on autophosphorylation of insulin receptors revealed by electrophoresis and immunoblotting. Control (○) or CHS-treated (▽) HepG2 cells were stimulated for the indicated times with 10^{-7} M insulin, in the presence or absence of 10^{-4} M metformin. The band at 95 kDa corresponding to insulin receptors was quantitated after scanning. Results are depicted as the mean ± SD of four separate experiments. An asterisk denotes significant ($p < 0.05$) differences between metformin treatment and the corresponding control situation without metformin.

phosphorylation of a major band at ~95 kDa, which was identified as the insulin receptor beta subunit by immunoblotting and immunoprecipitation with anti–insulin receptor antibodies. FIGURE 1 shows the effect of 100 μM metformin on the insulin receptor autophosphorylation, which was greatly diminished for CHS-treated cells. In the presence of metformin, after 1 minute of stimulation, phosphorylation of the insulin receptor beta subunit already occurred at a higher level in CHS-treated cells than in control cells. Thus, for CHS-treated cells, where receptor autophosphorylation was greatly diminished, metformin increased receptor autophosphorylation to levels similar to those of control cells and was able to restore insulin responsiveness.

The present study demonstrates that metformin pretreatment of cholesterol-modified HepG2 hepatoma cells is able to restore their sensitivity to insulin. Metformin action is measurable at early steps of the insulin intracellular signaling cascade, namely, autophosphorylation of the insulin receptor activity, and also at the PI3 kinase or MAP kinase level. These results indicate that one site of action of metformin possibly resides within some membrane component of a signaling complex associated with the insulin receptor.

ACKNOWLEDGMENT

We thank M. M. Maine for revising the manuscript.

REFERENCES

1. WIERNSPERGER, N. & J. R. RAPIN. 1995. Diabetes Metab. Rev. **11:** s3–s12.
2. CRÉMEL, G., M. FICKOVÁ, I. KLIMEŠ, C. LERAY, V. LERAY, E. MEUILLET, M. ROQUES, C. STAEDEL & P. HUBERT. 1993. Ann. N.Y. Acad. Sci. **683:** 164–171.
3. HUANG, H., J. W. KUAN & G. D. GUIBAULT. 1975. Clin. Chem. **21:** 1605–1608.
4. BRUNEAU, C., P. HUBERT, A. WAKSMAN, J. P. BECK & C. STAEDEL-FLAIG. 1987. Biochim. Biophys. Acta **928:** 297–304.

Effect of the High-Fat Diet on the Calcium Channels in Rat Myocardium[a]

I. MINAROVIČ,[b] R. VOJTKO,[b] E. ŠEBÖKOVÁ,[c]
I. KLIMEŠ,[c] AND I. ZAHRADNÍK[b,d]

[b]*Institute of Molecular Physiology and Genetics*
[c]*Institute of Experimental Endocrinology*
Slovak Academy of Sciences
Bratislava, Slovak Republic

INTRODUCTION

The activity of ion channels depends on the composition of the surrounding lipid phase.[1] Lipid composition of the cell membrane is influenced by the content and lipid composition of the fat in the diet.[2] Therefore, modification of a diet may result in alteration of membrane properties and hence in modification of the ion channel functions. These functions can be assessed in three principal ways: first, by direct measurements of transport activity of ion channels; second, by observing changes in their metabolic regulation; and third, by assessing the binding characteristics of their receptors.

The question addressed in this study was which functions, if any, of the cardiac L-type calcium channel, which is of utmost importance for transforming the excitation signal to contraction of the myocyte,[3,4] can be altered by a diet designed to modify membrane properties. Here, we inspected their voltage-dependent and antagonist-binding characteristics with the use of electrophysiological techniques.

MATERIALS AND METHODS

Male Wistar rats weighing 120–150 g were divided into two groups of 10 animals per group. Each group of rats was fed 14 days either "fish oil" diet (FO; 900 g standard laboratory chow supplemented with 100 g fish oil) or "high-fat" diet (HF; 600 g standard laboratory chow supplemented with 270 g beef tallow, 30 g safflower oil, and 20 g cholesterol). The fatty acid composition of these diets is given in TABLE 1. Diets were prepared in advance and stored at $-20°C$. Food and water were available ad libitum. As a control, we used male Wistar rats of the same age (weighing 240–290 g) fed standard laboratory chow (VELAZ, Prague, Czech Republic).

[a] This work was supported by Grant Nos. VEGA 1203 and GAV 543.
[d] To whom all correspondence should be addressed.

TABLE 1. Fatty Acid Composition of the Experimental Diets

Fatty Acids	FO Diet [wt%]	HF Diet [wt%]
Σ saturated	13.0	47.0
Σ monounsaturated	29.4	39.7
Σ polyunsaturated	57.6	13.3

Rat ventricular myocytes were isolated using a collagenase digestion procedure,[5,6] providing about 50% yield of calcium-tolerant cells. Cells were kept at room temperature in a "KB" medium (in mmol/L: 106 CH_3SO_3K, 1 EGTA, 22 taurine, 22 glucose, 2.4 $MgSO_4$, 8 K_2HPO_4, pH 7.3) until use on the same day. Calcium currents were recorded from single myocytes using a standard whole-cell patch-clamp technique.[7] Patch pipettes with an inner-tip diameter of 1–2 μm filled with the internal solution (in mmol/L: 135 $CsCH_3SO_3$, 10 CsCl, 10 Hepes, 1 EGTA, 3 $MgSO_4$, 3 ATP, 0.05 cAMP, pH 7.3) had a resistance of 1.5–3 MΩ. The patch-clamped myocyte, held at a potential of −50 mV against grounded bath, was placed into the perfusion channel (volume 50 μL), allowing complete exchange of external solutions within about 1 s. The external solution, used as a vehicle for the drugs (in mmol/L: 135 NaCl, 5.4 CsCl, 10 Hepes, 5 $MgCl_2$, 0.33 NaH_2PO_4, 1 $CaCl_2$, pH 7.3), was supplemented with 20 μM tetrodotoxin to block the sodium current and with 10 μM isobutylmethylxanthine to inhibit the activity of the cAMP-splitting phosphodiesterases. Calcium current measurements were carried out at room temperature.

RESULTS

Effect of Diets on the Electrophysiological Characteristics of the Calcium Channel

Under our recording conditions, the current elicited by membrane depolarization is carried by Ca^{2+} ions through the L-type calcium channels. In accordance with the physiological convention, Ca^{2+} current (I_{Ca}) is displayed as a downward (negative) deflection of the trace, corresponding to a flow of Ca^{2+} ions to the cell. Generally, under voltage-clamp conditions, the amplitude of the current at an imposed membrane potential is proportional to the number of open channels at a given time, and the rate of the current changes is proportional to the kinetics of the intramolecular transitions of the channel protein.[8]

Traces of the Ca^{2+} current (see FIG. 1A) were characterized by a transient response to a depolarizing voltage pulse. Activation kinetics of the current were faster with higher amplitudes of depolarization. The biphasic inactivation of I_{Ca}, not complete within the 70-ms duration of the pulses, was fastest at voltage steps to about 0 mV. The small exponentially decaying current flowing after the end of the voltage pulse reflected the deactivation of calcium channels that remained open at the end of the pulse.

In FIGURE 1B, the current-voltage characteristics, constructed from the peak calcium current amplitudes, were compared for myocytes of rats fed control, FO (fish

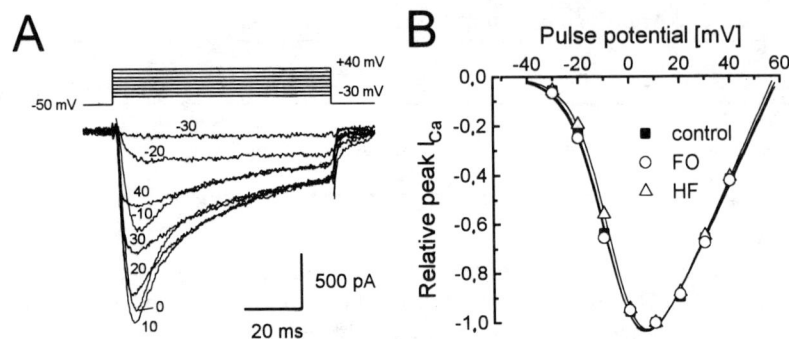

FIGURE 1. Effect of the diets on the Ca^{2+} current-voltage characteristics. (A) Superimposed records of I_{Ca} (noisy traces with numbers referring to the amplitudes of the corresponding voltage pulses in mV; control group) and the corresponding voltage pulses (above traces). (B) Normalized I_{Ca}-V relationships for each group, constructed from the peak I_{Ca} values of records such as in part A. The lines represent the best fits of the theoretical equation to the data.[8]

oil), or HF (high-fat) diets. For obvious reasons, this comparison cannot be made on the same myocyte. Therefore, we compared the normalized *I*-V characteristics independent of the real current amplitude, naturally varying among myocytes according to their membrane surface area. The parameters of the theoretical equation[8] fitted into experimental data were found to be the same for all groups. Similar observations were made for the kinetics of the I_{Ca}. The rates of activation and inactivation of the Ca^{2+} current were not significantly different among myocytes isolated from rats fed different diets (not shown).

In other words, the time course and voltage dependence of the I_{Ca} recorded from each group were similar to records shown in FIGURE 1A. It can therefore be concluded that the basic characteristics of calcium channels in cardiac myocytes are not changed by the specific diets used.

More advanced analysis of the calcium current has revealed that the inactivation of the calcium channels can be influenced by the intake of fat in the food. FIGURE 2A shows the stimulation protocol used to measure the steady-state inactivation curves of the Ca^{2+} current together with typical traces of I_{Ca} recorded on the test pulse. Evaluation of all experiments (5–8 cells per group) given in FIGURE 2B showed a small, but highly significant, difference in the half-inactivation potential between groups. It is concluded that calcium channels of myocytes obtained from rats fed the high-fat diet are more prone to inactivation (inactivate at less depolarizing potentials) than calcium channels in hearts of the fish oil–fed rats.

Effect of Diets on the Pharmacological Characteristics of the Calcium Channel

The L-type calcium channels have three well-described types of receptors: the dihydropyridine (hence the name DHP receptor channel), the phenylalkylamine (PAA),

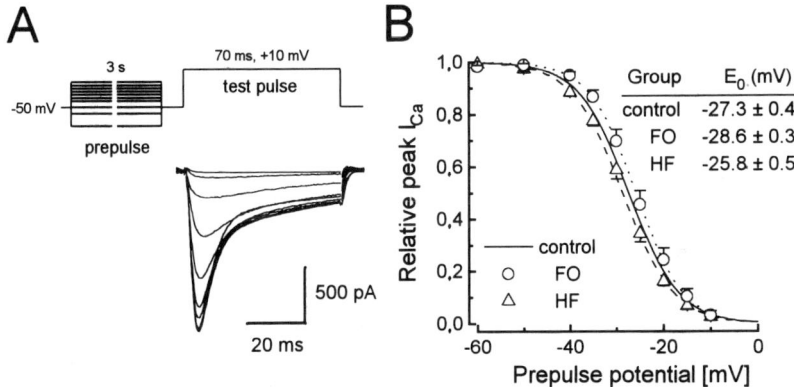

FIGURE 2. Effect of the diets on the steady-state inactivation of the Ca current. (A) An example of the inactivation experiment. The peak amplitude of the I_{Ca} recorded on the test pulse was progressively depressed (inactivated) by increasing amplitude of the prepulse voltage. (B) The inactivation curves for the control (solid line; measured data not shown for clarity), FO (dotted line), and HF (dashed line) groups, constructed from the peak I_{Ca} recorded like in part A and plotted against the prepulse voltage. The lines represent the corresponding best fits of the theoretical equation to the data.[8] E_0 = half-inactivation potential.

and the benzothiazepine (BT) receptors. In this study, we have assessed the binding characteristics of the PAA receptor with verapamil and those of the BT receptor with diltiazem. The effects of the drugs on Ca^{2+} current-voltage characteristics were evaluated at concentrations inhibiting I_{Ca} by about 50%.

It was found that in both the HF diet– and FO diet–fed rats, like in the control group, neither verapamil nor diltiazem had an effect on the parameters of the I_{Ca}-V curves other than reducing the amplitude of the current by the same factor at all voltages (data not shown). This finding indicates that the diet-induced changes in the membrane of myocytes did not lead to qualitative changes in the mechanism of interaction between the drugs and the channel receptor.

Binding characteristics of the calcium channel blockers, verapamil and diltiazem, in all three groups were inspected with the use of a constant-pulse voltage protocol. The cells were repeatedly stimulated at a frequency of 0.3 Hz with 70-ms voltage pulses from a holding potential of -50 mV to 0 mV. When a stable reading of the I_{Ca} was observed, the pertinent drug at the desired concentration was applied to the cell. The effect of the drug on the current amplitude stabilized usually within 1 minute, after which a new increased concentration of the drug was applied until full inhibition of the I_{Ca} was reached. The steady amplitude of the current in the presence of the drug was related to the amplitude measured before drug addition.

The relative I_{Ca} obtained for every drug concentration from 3 to 5 cells of every group (each cell from a different animal) were averaged, and the means ± SEM were plotted as a function of the drug concentration (FIG. 3). Fitting the dose-response data with the inhibitory equation provided, in all three diet groups and for both drugs, binding curves with the Hill coefficient $n = 1$. This finding provides further evidence for a lack of effect of the diets used on the mechanism of the drug-channel interaction.

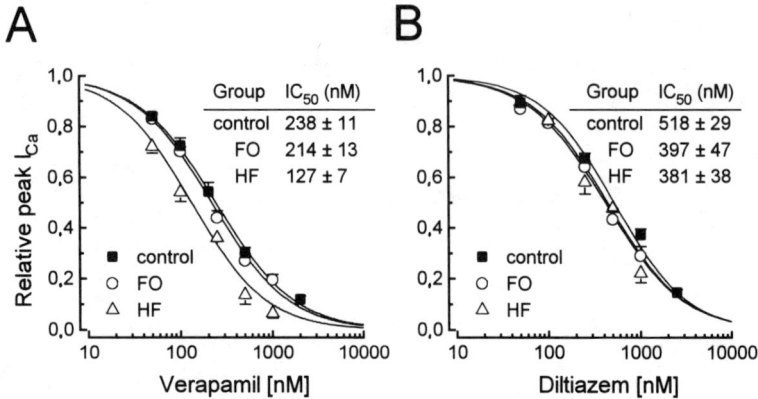

FIGURE 3. The dose-response curves of the Ca current inhibition by verapamil (A) and diltiazem (B). The lines represent the best fits of the theoretical equation to the data.[8]

A significant shift of the verapamil dose-response curve to the left was found in the HF group, but not in the FO group, relative to the control group. The binding constants of diltiazem did not differ significantly among the diet groups. Therefore, the effect of the HF diet on increasing the affinity of the calcium channel to verapamil can be regarded as specific, as the FO diet had no effect on the binding of either drug, and the effect of the HF diet on diltiazem binding was not significant.

DISCUSSION

To summarize, some functions of the calcium channels were found to be independent of the dietary fat amount and type: specifically, (i) the rate of activation, (ii) the rate of the fast component of inactivation, (iii) the voltage dependence of the peak I_{Ca} amplitude, and (iv) the mechanism of antagonists interacting with the channel's phenylalkylamine and benzothiazepine receptors. All these functions reflect dynamic behavior of the calcium channel. On the other hand, we have found two functions of the calcium channels that were sensitive to the dietary changes, namely, the shift of the steady-state inactivation curve and the increase in affinity of the agonist to the receptor. The affected functions reflect equilibrium characteristics of the channel.

Explanation of these findings will need more-detailed studies. We do not know much about changes in lipid composition within the membrane of myocytes. Nevertheless, changes in membrane thickness by short-chain phospholipids were reported to inhibit voltage-gated sodium channels[9] and to reduce the activity of acetylcholine-activated channels.[10] These substances, however, also strongly perturb membrane structural and dynamic characteristics.[11] Changes of membrane-phase characteristics by synthetic detergents and cholesterol shifted inactivation, but not activation, of the N-type Ca channels.[12] The results of our study, therefore, can be interpreted as a consequence of changes in the properties of the membrane environment.

Calcium channel blockers, besides their strong binding to the corresponding receptors of the calcium channel, show potent physicochemical interaction with the membrane lipid,[13] which strongly depends on the membrane lipid composition.[14] We have observed a remarkable change in the affinity of verapamil to the PAA receptor, but not that of diltiazem to the BT receptor. Although both drugs have comparable affinities to their receptors, verapamil is a much-stronger lipophilic agent than diltiazem, with log P values of 5.5 and 2.6, respectively. Therefore, in light of the drug-receptor interaction theory,[15] it is not surprising that the lipophilic core of the membrane has a greater effect in the case of verapamil. A similar observation was made for the interaction between pentobarbital and voltage-dependent sodium channels,[16] where increasing content of cholesterol within the membrane had no direct effect on the channel activity, but significantly reduced the effectiveness of pentobarbital.

A general conclusion, which can be drawn from this study, is that there are particular diets that can have influence on some functions of the calcium channels. The function of calcium channels prone to modification by a diet may result in malfunction of the myocardium and changes in its pharmacology.

ACKNOWLEDGMENTS

We gratefully acknowledge D. Gašperíková, P. Bohov, and E. Danišová for assistance with the experiments.

REFERENCES

1. SANDERMAN, H. 1978. Biochim. Biophys. Acta **515**: 209–237.
2. STORLIEN, L. H., A. B. JENKINS, D. J. CHISHOLM, W. S. PASCOE, S. KHOURI & E. W. KRAEGEN. 1991. Diabetes **40**: 280–289.
3. BERS, D. M. 1993. Excitation-Contraction Coupling and Cardiac Contractile Force. Kluwer. Dordrecht.
4. KATZ, A. M. 1992. Physiology of the Heart. Raven Press. New York.
5. ZAHRADNÍK, I. & A. ZAHRADNÍKOVÁ. 1989. Gen. Physiol. Biophys. **8**: 119–132.
6. ZAHRADNÍK, I. & P. PALADE. 1993. Pflügers Arch. **424**: 129–136.
7. HAMILL, O. P., A. MARTY, E. NEHER, B. SAKMANN & F. J. SIGWORTH. 1981. Pflügers Arch. **391**: 85–100.
8. HILLE, B. 1984. Ionic Channels of Excitable Membranes. Sinauer Assoc. Sunderland, Massachusetts.
9. HENDRY, B. M., J. R. ELLIOTT & D. A. HAYDON. 1985. Biophys. J. **47**: 841–845.
10. BRAUN, M. S. & D. A. HAYDON. 1991. Pflügers Arch. **418**: 62–67.
11. ONDRIAŠ, K. & A. STAŠKO. 1992. Chem. Biol. Interactions **84**: 143–151.
12. LUNDBAEK, J. A., P. BIRN, J. GIRSHMAN, A. J. HANSEN & O. S. ANDERSEN. 1996. Biochemistry **26**: 3825–3830.
13. MASON, R. P. & M. W. TRUMBORE. 1996. Biochem. Pharmacol. **51**: 653–660.
14. ONDRIAŠ, K., A. STAŠKO, V. MIŠÍK, J. REGULI & E. ŠVAJDLENKA. 1991. Chem. Biol. Interactions **79**: 197–206.
15. HILLE, B. 1977. J. Gen. Physiol. **69**: 497–515.
16. REHBERG, B., B. W. URBAN & D. S. DUCH. 1995. Anesthesiology **82**: 749–758.

The Effect of Fasting and Vitamin E on Insulin Action in Obese Type 2 Diabetes Mellitus

J. ŠKRHA, G. ŠINDELKA, AND J. HILGERTOVÁ

Department of Internal Medicine 3
Faculty of Medicine 1
Charles University
128 21 Prague 2, Czech Republic

An impaired insulin action has been repeatedly reported in obese Type 2 diabetic patients as well as in obese subjects without diabetes.[1,2] Insulin resistance accompanied by hyperinsulinemia is a common finding in these individuals. The impaired insulin action reflects one of the most-important consequences in metabolic syndromes.

The improvement of insulin action may have beneficial effect not only for glucose utilization, but for the whole metabolism regulated by insulin. A decrease of body weight may improve insulin sensitivity in both obese nondiabetic individuals and obese diabetic patients.[3] A positive effect of high doses of vitamin E on insulin action has been reported earlier.[4] It induced us to evaluate the influence of short-time fasting as well as of vitamin E administration on insulin sensitivity and insulin receptors in obese Type 2 diabetic patients.

PATIENTS AND METHODS

Two groups of obese Type 2 diabetic patients were studied in the present investigation. Group A consisted of 12 patients (mean age: 50 ± 8 years; body mass index, BMI: 40.3 ± 8.9 kg/m^2)—7 of them were treated by oral agents and the remaining 5 were on dietary regimen. All patients underwent fasting during 7 days after initial examination and the tests were repeated on the eighth day. Blood glucose was examined daily and oral agents could be omitted in all patients because of improvement of blood glucose control during fasting. Mild acidosis developed within the first 4 days, but pH was not lower than 7.35 at the end of fasting. Nine Type 2 diabetic patients were examined in group B (mean age: 46 ± 9 years; BMI: 31.4 ± 3.7 kg/m^2)—5 patients were on dietary regimen and 4 were on oral agents. They were examined before and after 3 months of treatment with daily doses of 600 mg of vitamin E administered orally. No changes of previous medication were done within the testing period. The control group consisted of 12 healthy nonobese persons (mean age: 42 ± 8 years; BMI: 24.1 ± 2.1 kg/m^2).

All tests were performed after an overnight fast. Blood glucose for biochemical analysis was collected immediately after cannulation, and hyperinsulinemic isoglycemic clamp was continued on Biostator GCIIS (Miles, Elkhart, Indiana), mode 7:1, during 90 min after a 30-min stabilization period with an insulin infusion rate of 1 mU/kg/min as previously described.[5] Blood samples for insulin determination were collected at the beginning and twice at the end of the clamp. Blood glucose concentration was repeatedly controlled by Glucose Analyzer ESAT 6660-2 (Medingen, Germany).

The following variables were analyzed. Glucose disposal rate was calculated from the amount of glucose maintaining blood glucose concentration at a constant level (M), metabolic clearance rate of glucose was expressed as a ratio between M and the blood glucose concentration in the last 10 min of the clamp (MCR_G), and an index of insulin sensitivity was calculated as a ratio between the glucose disposal rate and insulin concentration (M/I).

Heparinized whole blood was collected for the evaluation of insulin receptors on erythrocytes.[6] Specific insulin binding, insulin binding capacity, and receptor affinity were determined in all patients.

Serum insulin concentration was evaluated by radioimmunoassay technique[7] and glycated hemoglobin HbA_{1C} was evaluated by HPLC method.[8]

Statistical evaluation was performed by using a t test for paired and unpaired values. ANOVA was used to compare the differences between separate groups, and Pearson's correlation was done to evaluate the relationship between variables. The results are expressed as the means with SD range.

RESULTS

Significant reduction of body mass index (BMI) was observed in group A at the end of fasting (40.3 ± 8.9 vs. 37.0 ± 6.9 kg/m^2, $p < 0.01$). This was accompanied by a decrease of the fasting plasma glucose (11.4 ± 3.6 vs. 8.4 ± 3.0 mmol/L, $p < 0.001$), whereas basal serum insulin concentration did not change. Insulin resistance was documented in both A and B groups by decreased glucose disposal rate ($p < 0.01$), decreased metabolic clearance rate of glucose ($p < 0.001$), and decreased insulin sensitivity index ($p < 0.001$) as compared with control persons (TABLE 1).

In group A, an increase of the glucose disposal rate together with an increase of the metabolic clearance rate of glucose (both $p < 0.01$) were observed at the end of fasting. Insulin sensitivity index was not significantly improved. A decrease of insulin receptor number was counterbalanced by an increase of affinity so that the resulting insulin binding was not changed.

In group B, the administration of vitamin E induced significant decreases of fasting serum insulin concentration (29 ± 16 vs. 20 ± 8 mU/L, $p < 0.01$), glucose disposal rate ($p < 0.02$), and metabolic clearance rate of glucose ($p < 0.05$). These changes were accompanied by a downregulation of insulin receptors with an increase of their affinity (TABLE 1). A mild worsening of diabetes control was found after treatment by vitamin E (HbA_{1C}: $6.52 \pm 1.94\%$ vs. $7.11 \pm 1.29\%$, $p < 0.01$).

An inverse relationship was observed between body mass index and insulin sensitivity index in a total cohort of examined subjects (FIG. 1).

TABLE 1. Metabolic Variables of Glucose Metabolism before and after Treatment by Fasting (Group A) and Vitamin E (Group B) as Compared with Controls

	Group A		Group B		
	Before Fasting	After Fasting	Before Vitamin E	After Vitamin E	Controls
HbA$_{1C}$ (%)	7.0 ± 1.3^a	–	6.5 ± 1.4^a	7.1 ± 1.3^a	4.9 ± 0.4
G$_o$ (mmol/L)	11.4 ± 3.9^a	8.4 ± 3.0^{a,x}	7.9 ± 2.7^a	7.4 ± 2.0^a	4.4 ± 0.2
M (μmol/kg/min)	24.0 ± 7.5^b	29.5 ± 8.9^x	26.6 ± 9.8	21.3 ± 8.5^{b,y}	34.0 ± 11.7
I (mU/L)	147 ± 45^a	173 ± 71^a	121 ± 29^a	104 ± 22^b	79 ± 12
MCR$_G$ (mL/kg/min)	2.3 ± 0.9^a	4.0 ± 2.5^{a,x}	3.7 ± 1.7^a	2.9 ± 0.8^{a,z}	7.9 ± 3.5
M/I (μmol/kg/min per mU/L × 100)	16.3 ± 7.4^a	18.0 ± 10.5^a	23.0 ± 9.1^a	21.7 ± 9.4^a	40.3 ± 13.3
B (%)	14.7 ± 3.0	14.5 ± 2.5	14.4 ± 4.7	13.7 ± 3.4	14.4 ± 3.0
R$_o$ (pmol/L)	245 ± 66^b	192 ± 61^{a,y}	282 ± 95	150 ± 22^{a,x}	319 ± 78
K$_a$ (10^8 L/mol)	12.8 ± 3.9	17.3 ± 5.4^{b,x}	12.5 ± 7.1	20.6 ± 6.1^{a,x}	10.0 ± 3.6

Note: Glucose disposal rate (M), serum insulin concentration at the end of the clamp (I), metabolic clearance rate of glucose (MCR$_G$), and insulin sensitivity index (M/I) are compared with insulin receptor characteristics: B = insulin binding (%), R_o = insulin binding capacity (pmol/L), and K_a = receptor affinity (in 10^8 L/mol). Statistical significance as compared to control values ($^a p < 0.001$, $^b p < 0.01$) and to pretreated results ($^x p < 0.01$, $^y p < 0.02$, $^z p < 0.05$).

DISCUSSION

We demonstrated that short fasting may improve the insulin action, especially at the postreceptor level. It was documented by an increase of glucose disposal rate and of metabolic clearance rate of glucose. However, fasting has been suggested earlier as a cause of insulin insensitivity, at least partly due to developed acidosis. We did not observe serious acidosis in our group and this could explain our finding.

In contrast to previous observations,[4] we found a worsening of insulin action after vitamin E administration. We used a slightly lower dose of vitamin E as in the above study (600 vs. 900 mg daily), but it was still more than usually administered. A decrease of insulin action on postbinding level was accompanied by a downregulation of insulin receptors. Although the erythrocytes are not target cells for insulin like muscle cells or adipocytes, they may offer useful information on insulin binding.[5]

In conclusion, a diverse effect of fasting and vitamin E administration on insulin action was demonstrated in the present study. The fasting positively influenced the insulin action in a very early phase. However, vitamin E caused a worsening of insulin action at both receptor and postreceptor levels. Further work would be useful to evaluate the mechanism by which vitamin E interferes with insulin action.

SUMMARY

The influence of either short-term fasting or vitamin E administration on insulin action was studied in two groups of obese Type 2 diabetic patients. Twelve patients

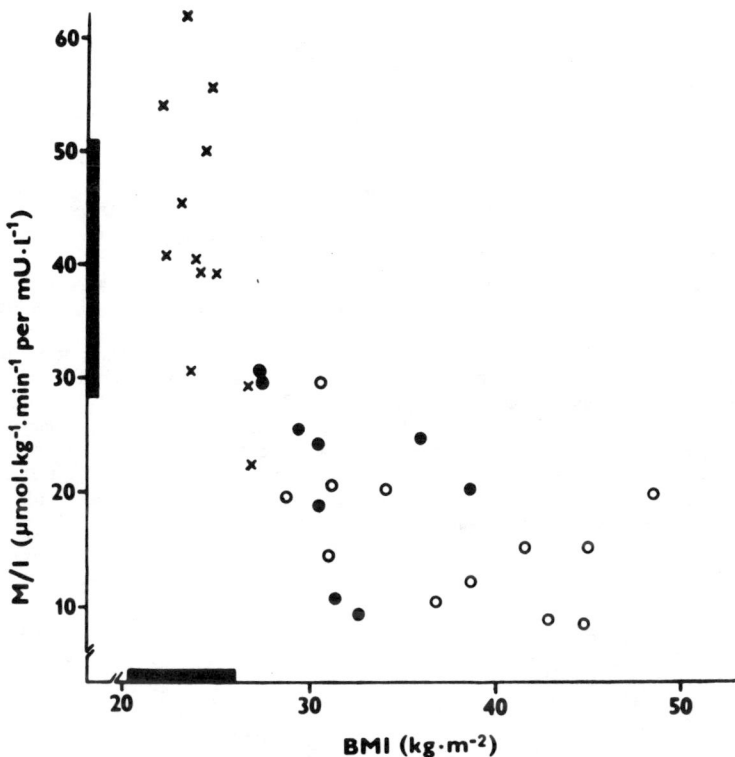

FIGURE 1. Relationship of insulin sensitivity index (M/I) and body mass index (BMI) in a total cohort of examined persons: healthy nonobese controls ($\times$); obese Type 2 diabetic patients treated by vitamin E ($\bullet$) and by fasting ($\circ$) ($\log = 10.4 \log x^{-2.1}$, $r = -0.80$, $p < 0.001$). The thick bars indicate the normal values (2SD range).

underwent 7 days of fasting (group A), whereas 600 mg of vitamin E was administered daily during 3 months in 9 diabetic patients (group B). Insulin action was examined by using hyperinsulinemic isoglycemic clamps (insulin infusion rate, 1.0 mU/kg/min) and insulin receptors on erythrocytes before and after respective regimens. An increase of glucose disposal rate (29.5 ± 8.9 vs. 24.0 ± 7.5 µmol/kg/min, $p < 0.01$) and an increase of metabolic clearance rate of glucose (4.0 ± 2.5 vs. 2.3 ± 0.9 mL/kg/min, $p < 0.01$) were observed in group A after fasting. On the contrary, decreases of glucose disposal rate (21.3 ± 8.5 vs. 26.6 ± 9.8 µmol/kg/min, $p < 0.02$), metabolic clearance rate of glucose (2.9 ± 0.8 vs. 3.7 ± 1.7 mL/kg/min, $p < 0.05$), and insulin receptor number ($p < 0.01$) were found after vitamin E administration as compared with pretreated values. A worsening of diabetes control as observed by an increase of HbA_{1C} ($p < 0.01$) was present in the latter group. In summary, we found an improvement of insulin action after short-term fasting in contrast with the worsening of metabolic parameters after vitamin E administration in obese Type 2 diabetic patients.

REFERENCES

1. FERRANNINI, E. 1995. Physiological and metabolic consequences of obesity. Metabolism **44**(suppl. 3): 15-17.
2. OLEFSKY, M. J. 1995. Insulin resistance in non-insulin-dependent diabetes mellitus. Curr. Opin. Endocrinol. Diabetes **2**: 290-299.
3. GOLAY, A., J. P. FELBER & M. DUSMET. 1985. Effect of weight loss on glucose disposal in obese and obese diabetic patients. Int. J. Obes. **9**: 181-190.
4. PAOLISSO, G., A. D'AMORE, D. GIUGLIANO, A. CERIELLO, M. VARRICCHIO & F. D'ONOFRIO. 1993. Pharmacologic doses of vitamin E improve insulin action in healthy subjects and NIDDM patients. Am. J. Clin. Nutr. **57**(5): 650-656.
5. ŠKRHA, J., G. ŠINDELKA, T. HAAS, J. HILGERTOVÁ & V. JUSTOVÁ. 1996. Comparison of insulin sensitivity in patients with insulinoma and obese Type 2 diabetes mellitus. Horm. Metab. Res. **28**: 595-598.
6. HOVORKA, R. & J. HILGERTOVÁ. 1991. Identification of insulin receptor systems: assessing the impact of model selection and measurement error on precision of parameter estimates using Monte Carlo study. J. Theor. Biol. **151**: 367-383.
7. ŠRÁMKOVÁ, J., J. PÁV & O. ENGELBERTH. 1975. Inordinately high levels of serum immunoreactive insulin in monoclonal immunoglobulinemia. Diabetes **24**: 214-224.
8. JEPPSON, J., P. JERNTORP, G. SUNDKVIST, H. ENGLUND & V. NYLUND. 1986. Measurement of hemoglobin A_{IC} by a new liquid-chromatographic assay: methodology, clinical utility, and relation to glucose tolerance evaluated. Clin. Chem. **32**: 1867-1872.

The Effect of Hyperlipidemia on Serum Fatty Acid Composition in Type 2 Diabetics[a]

PAVOL BOHOV,[b] VILIAM BALÁŽ,[c]
ELENA ŠEBÖKOVÁ,[b] AND IWAR KLIMEŠ[b]

[b]*Diabetes and Nutrition Research Group*
Institute of Experimental Endocrinology
Slovak Academy of Sciences
SK-833 06 Bratislava, Slovak Republic
[c]*Research Institute of Gerontology*
901 01 Malacky, Slovak Republic

The prevalence of atherosclerosis is increased in Type 2 (non-insulin-dependent, NIDDM) diabetes mellitus and causes death in three-quarters of the diabetic population.[1] It is quite likely that the impaired lipid metabolism contributes significantly to the increased mortality from coronary heart disease (CHD) in NIDDM. This type of diabetes is characterized by higher serum triglyceride concentration, generally no difference in the total cholesterol or low-density lipoprotein concentration, and variable patterns in the high-density lipoproteins (HDL).[2] However, the differences in these classical atherogenic factors explain the excessive risk only partly and, thus, other mechanisms may contribute to the atherogenesis in NIDDM.

The extensive studies in the previous decades revealed that, besides the serum lipids, the process of atherogenesis is seriously influenced by abnormal serum and tissue fatty acid (FA) composition. While some saturated fatty acids (SFA) are directly correlated with CHD, linoleic acid and the long-chain n-3 polyunsaturated fatty acids (PUFA) exhibit the opposite effect.[3] The increased levels of SFA and monounsaturated fatty acids (MUFA) and the decreased amount of linoleic acid were found in serum lipid fractions of NIDDM patients[4] and in men who developed NIDDM.[5] Little is known, however, about whether the changes in serum FA composition in NIDDM may be influenced by impaired lipid metabolism.

To address the aforementioned question, the serum FA composition of groups of randomly selected Type 2 diabetics with diverse types of hyperlipidemia were compared with data obtained in a group of nondiabetic subjects.

---

[a] This work was supported by a research grant of the Institute of Experimental Endocrinology, Slovak Academy of Sciences, Bratislava, provided by the Slovak Grant Agency for Science (GAV) (Grant No. 2/544/92-96).

SUBJECTS AND METHODS

The patients were randomly selected from Type 2 diabetics with known duration of diabetes of >5 years, treated with oral antidiabetic drugs for >1 year (sulfonylurea derivatives), who were attending two diabetes outpatient clinics in a rural area. The patients were instructed not to change their recommended diet (weight-maintaining diet consisting of about 50 cal% carbohydrate, 35 cal% fat, and 15 cal% protein, respecting the local dietary habits) at least 30 days before blood withdrawal. None of them were on any special lipid-altering drugs. According to the degree of hyperlipidemia and/or dyslipidemia, patients were divided into four groups. Out of the total of 114 NIDDM patients, 21 (10 men) had normal serum lipids (DM-NLP), 11 (4 men) had hypercholesterolemia (DM-HCHOL), 43 (14 men) had hypertriglyceridemia (DM-HTG), and 39 (16 men) had combined hypercholesterolemia and hypertriglyceridemia (DM-HLP). Thirty-two (14 men) age-matched healthy subjects without a family history of diabetes mellitus served as controls.

Blood samples were collected after overnight fasting. Serum cholesterol, HDL-cholesterol, triglycerides, and glucose were measured using enzymatic methods with the aid of an Impact 400 Analyzer (Gilford, United States). Lipids were extracted from serum with chloroform-methanol.[6] After alkaline hydrolysis of lipids, FA were esterified with diazomethane and were determined by gas chromatography (GC) on a capillary column coated with SP 2340 stationary phase as previously described.[7] Internal standard C21:0 was used for FA quantification.

Significance of differences was evaluated by ANOVA using a Tukey multiple-range test. The relationships between variables were analyzed by linear correlation (Pearson or Spearman coefficient) and confirmed by linear regression analysis.

RESULTS

As shown in TABLE 1, no differences were observed in the age of subjects and in the duration of diabetes. Although most of the patients were free of pronounced obesity, about a quarter of the diabetics with hyperlipidemia had a body mass index in the range of 30–35 kg/m². Fasting serum glucose level was significantly higher in all groups of diabetics compared with controls, which shows that oral antidiabetic treatment ensured only poor control for these patients. Serum cholesterol and triglyceride levels were the highest in the corresponding hyperlipidemic patients. The lowest level of HDL-cholesterol, as found in the DM-HTG group, was significantly different from the groups with normal serum lipids.

Fatty acid composition of total serum lipids revealed some significant differences between the hyperlipidemic diabetics and normolipidemic subjects (TABLE 2). Among them, the most remarkable seems to be the increased percentage of palmitic and oleic acids and, conversely, the decreased proportion of linoleic acid in patients with higher triglyceride levels (DM-HTG, DM-HLP). Decreased proportions of some PUFA n-6 metabolites were also detected in DM-HLP patients, including dihomo-gamma-linolenic and arachidonic acids.

TABLE 1. Clinical and Serum Biochemical Characteristics of Subjects (Mean ± SEM)

	Controls (n = 32)	DM-NLP (n = 21)	DM-HCHOL (n = 11)	DM-HTG (n = 43)	DM-HLP (n = 39)
Age (years)	65.5 ± 2.24	64.7 ± 1.47	67.5 ± 2.17	62.1 ± 1.13	61.2 ± 1.16
DM duration (years)	NA	10.6 ± 0.91	10.7 ± 1.19	9.4 ± 0.67	10.4 ± 0.74
Body mass index (kg/m^2)	24.6 ± 0.20^a	26.3 ± 0.66^{a,b}	28.7 ± 0.96^{b,c}	29.1 ± 0.66^c	28.6 ± 0.58^{b,c}
Fasting glucose (mM)	5.0 ± 0.15^a	12.2 ± 1.15^b	15.0 ± 1.37^{b,c}	15.1 ± 0.61^c	15.5 ± 0.70^c
Triglycerides (mM)	1.20 ± 0.07^a	1.35 ± 0.08^a	1.45 ± 0.09^a	3.32 ± 0.19^b	4.99 ± 0.56^c
Total cholesterol (mM)	5.5 ± 0.10^a	5.3 ± 0.14^a	7.4 ± 0.20^b	5.7 ± 0.11^a	7.5 ± 0.14^b
HDL-cholesterol (mM)	1.44 ± 0.04^a	1.43 ± 0.08^a	1.38 ± 0.14^{a,b}	1.16 ± 0.04^b	1.27 ± 0.06^{a,b}

Note: DM = diabetes mellitus; HDL = high-density lipoprotein; DM-NLP = diabetes mellitus with normal serum lipids; DM-HCHOL = diabetes mellitus with hypercholesterolemia; DM-HTG = diabetes mellitus with hypertriglyceridemia; DM-HLP = diabetes mellitus with hypercholesterolemia and hypertriglyceridemia; NA = not applicable. Values within a line without a common superscript are significantly different; p values for significant differences ranged from 0.01 to 0.05.

TABLE 2. Fatty Acid Composition (wt%) of Total Serum Lipids in Non-Insulin-dependent Diabetic Patients and Control Subjects (Mean ± SEM)

Fatty Acids	Controls (n = 32)	DM-NLP (n = 21)	DM-HCHOL (n = 11)	DM-HTG (n = 43)	DM-HLP (n = 39)
14:0	1.30 ± 0.18	0.93 ± 0.07	0.95 ± 0.08	1.08 ± 0.06	1.09 ± 0.05
16:0	20.26 ± 0.27^a	21.34 ± 0.44^a	21.69 ± 0.31^{a,b}	23.00 ± 0.31^{b,c}	23.62 ± 0.30^c
16:1n-7	2.09 ± 0.08^{a,b}	1.73 ± 0.08^a	2.02 ± 0.14^{a,b}	2.21 ± 0.08^b	2.45 ± 0.13^b
18:0	7.04 ± 0.12^a	6.89 ± 0.14^{a,b}	6.73 ± 0.27^{a,b}	6.62 ± 0.10^{a,b}	6.49 ± 0.13^b
18:1n-9t	0.35 ± 0.03	0.33 ± 0.04	0.43 ± 0.06	0.45 ± 0.04	0.36 ± 0.04
18:1n-9	21.42 ± 0.35^a	22.06 ± 0.68^a	21.84 ± 1.00^a	24.79 ± 0.41^b	25.43 ± 0.54^b
18:1n-7	2.20 ± 0.09	2.12 ± 0.07	2.13 ± 0.10	2.21 ± 0.05	2.34 ± 0.05
18:2n-6	27.71 ± 0.34^a	27.19 ± 0.94^a	27.12 ± 1.16^a	23.05 ± 0.45^b	23.07 ± 0.52^b
18:3n-6	0.45 ± 0.03	0.38 ± 0.03	0.41 ± 0.05	0.44 ± 0.02	0.40 ± 0.02
18:3n-3	0.59 ± 0.04^a	0.34 ± 0.02^b	0.37 ± 0.02^b	0.43 ± 0.03^b	0.47 ± 0.03^{a,b}
20:3n-6	1.87 ± 0.07^a	1.75 ± 0.07^{a,b}	1.67 ± 0.16^{a,b}	1.70 ± 0.07^{a,b}	1.46 ± 0.63^b
20:4n-6	7.08 ± 0.31^a	7.19 ± 0.26^a	6.81 ± 0.37^{a,b}	6.68 ± 0.19^{a,b}	6.03 ± 0.27^b
20:5n-3	0.64 ± 0.05	0.63 ± 0.06	0.81 ± 0.13	0.66 ± 0.05	0.62 ± 0.04
22:5n-3	0.47 ± 0.03	0.53 ± 0.07	0.57 ± 0.08	0.48 ± 0.02	0.49 ± 0.03
22:6n-3	1.79 ± 0.12	2.06 ± 0.13	2.30 ± 0.15	2.01 ± 0.09	1.86 ± 0.10
SFA	30.48 ± 0.49^a	31.15 ± 0.53^{a,b}	31.21 ± 0.64^{a,b}	32.60 ± 0.34^b	32.74 ± 0.34^b
MUFA	27.63 ± 0.44^a	27.67 ± 0.76^a	27.53 ± 1.09^a	30.80 ± 0.44^b	31.84 ± 0.59^b
PUFAn-6	37.90 ± 0.58^a	37.23 ± 1.07^a	36.72 ± 1.05^a	32.49 ± 0.52^b	31.56 ± 0.72^b
PUFAn-3	3.49 ± 0.18	3.56 ± 0.23	4.05 ± 0.32	3.59 ± 0.15	3.44 ± 0.15
PUFAn-6M	10.19 ± 0.38^a	10.04 ± 0.33^a	9.60 ± 0.49^{a,b}	9.43 ± 0.24^{a,b}	8.49 ± 0.33^b
PUFAn-3M	2.90 ± 0.17	3.22 ± 0.22	3.68 ± 0.32	3.16 ± 0.14	2.97 ± 0.14
C20-22 PUFA	12.29 ± 0.45^a	12.62 ± 0.44^a	12.63 ± 0.65^{a,b}	11.94 ± 0.28^{a,b}	10.83 ± 0.39^b
DBI	1.39 ± 0.02^a	1.39 ± 0.02^{a,b}	1.40 ± 0.02^{a,b}	1.32 ± 0.01^{b,c}	1.28 ± 0.02^c

Note: t = *trans* unsaturated double bonds; SFA = total saturated fatty acids; MUFA = total monounsaturated fatty acids; PUFA = total polyunsaturated fatty acids; M = total metabolites; DBI = double-bond index, which is the sum of the percentages of each fatty acid multiplied by the number of double bonds in that fatty acid divided by 100; C20-22 PUFA = 20:3n-6 + 20:4n-6 + 20:5n-3 + 22:4n-6 + 22:5n-6 + 22:5n-3 + 22:6n-3. Other abbreviations and definitions as described in TABLE 1. Some minor FA were omitted from the final tabulation.

As a consequence of changes in the spectrum of individual FA, higher proportions of total SFA and total MUFA and lower proportions of total PUFA n-6 were found in DM-HTG and DM-HLP patients. Double-bond index was decreased in these patients as well. In DM-HLP diabetics, the proportions of total PUFA n-6 metabolites were also decreased simultaneously with C20-C22 long-chain PUFA.

DISCUSSION

In our work, we have studied the FA composition of total serum lipids. The FA profile reflects the average fatty acid composition of all serum lipid esters and free fatty acids taken together. We have not found a similar work in the literature in the last couple of years, when improvements of GC analysis of FA enabled the separation of the detailed spectrum of FA with many positional and geometric isomers, possessing completely different biological effects. Thus, our results are compared with studies where the FA composition of defined lipid esters or free fatty acids had been analyzed.[4,5,8,9] Such an evaluation is rather limited because of the quite different FA spectrum in various lipid fractions. The question could arise of whether the differences between the hypertriglyceridemic diabetic patients on the one hand and the normolipidemic subjects on the other might not be due to the increased levels of some lipid fractions as such, rather than due to qualitative changes of FA in lipid fractions.

It is well documented that serum triglycerides contain about a half of the total serum MUFA, but substantially less PUFA, than serum phospholipids and cholesterol esters. The amount of SFA carried in the triglyceride fraction is between the amounts found in phospholipids and cholesterol esters.[8,10] Thus, a threefold to fourfold increased level of serum triglycerides in DM-HTG and DM-HLP patients could significantly contribute to the higher percentage of MUFA and lower percentage of PUFA in these patients, while SFA would not change significantly. This anticipation is supported by correlation relationships between triglyceride levels and MUFA and PUFA n-6 ($r = 0.6511$, $p < 0.001$ vs. $r = -0.6119$, $p < 0.001$), respectively, in patients of the DM-HTG and DM-HLP groups. In patients without HTG, such relationships were not seen.

However, changes in some FA cannot be explained simply by increasing the amount of triglycerides. About 80% of the gamma-linolenic acid is bound in cholesterol esters and a similar amount of docosahexaenoic acid is bound in phospholipids.[10,11] If only the quantitative parameters should contribute to the observed changes in the FA profiles, the percentage of these FA would be significantly lower in diabetics with HTG. However, this is not our case. In contrast, about half of the total serum amount of alpha-linolenic acid, carried in triglycerides, should have increased this FA in the total fatty acid spectrum. Nevertheless, we found a lower level of alpha-linolenic acid in HTG and HLP patients. These and some other changes (no change in PUFA n-3 or *trans*-fatty acids) indicate that the differences found in our work could be partly ascribed to alterations in FA profiles in lipid fractions as discussed above.

In serum of DM-HTG and DM-HLP, significantly higher amounts of fatty acids are being carried in serum lipoproteins, which enable the higher supply of FA to body tissues. In accordance with previous findings in serum and tissue lipid fractions,[2,4,5,9] the higher amounts of palmitic acid, palmitoleic acid, and total SFA and the lower proportions of linoleic acid in the above-mentioned DM groups and of C20-C22 long-chain PUFA in the total serum lipids of DM-HLP diabetics in our study may reflect a deterioration of lipid metabolism. Thus, the difference with regard to the proportions of FA in the hyperlipidemic diabetics, as compared with the normolipidemic controls, is probably to a large part dependent on the elevated lipid levels, but it cannot be excluded that the diabetic state as such has also contributed to the changes.

The lower proportions of dihomo-gamma-linolenic and arachidonic acids in our patients, however, are in contrast with some previous studies showing the opposite tendency for these FA in serum cholesterol esters[4,8] or serum phospholipids[8] of NIDDM. It is to be said that diabetics from the other studies[4,8] were much better controlled, even mostly with the diet, and they did not have severe impairments of serum lipid levels.

In our work, we did not find any differences in the long-chain PUFA n-3 between the diabetics with different levels of serum lipids or between the diabetics and controls. Such observations are consistent with findings seen in lipid fractions.[4,5,8,9] A possible explanation would lay in setting the equilibrium between the quantity and quality of dietary fat, the activity of desaturation enzymes, and the demands of the body for these specific fatty acids. On the other hand, we observed a decreased proportion of precursor alpha-linolenic acid in all groups of diabetics except DM-HLP. Similar lower proportions of alpha-linolenic acid were shown by others.[4,8]

From our results, it cannot be positively asserted if the observed changes resulted from different dietary habits of HTG and HLP diabetics or from other factors as well. Diabetic patients and healthy control subjects originated from the same area and from a similar socioeconomic group. Besides, the diabetics were instructed not to eat excessive and fatty meals. That this instruction was quite well respected can be seen from the body mass indices. These indicated that obesity in the majority of patients was lower than is usually found in such poorly controlled diabetics. Another, at least as likely, explanation is that the relatively low body weight in these very poorly controlled patients is due to a caloric deficit resulting from glucosuria, which is an obligate consequence of the very high blood glucose levels in these patients who are not insulin-treated. However, the dietary intake was not verified during the study, and local dietary distinctions had to be anticipated and accepted. Since the consumption of fish is not common among the people from the investigated area, similar levels of long-chain PUFA n-3 in controls and diabetics are probably derived from parental alpha-linolenic acid.

In conclusion, the increases of serum SFA level at the expense of PUFA, especially those of the n-6 series, in hypertriglyceridemic diabetics and patients with combined hyperlipidemia may have different courses. These changes could possibly be of some importance for atherogenesis, but more studies are needed to understand the importance of these findings.

SUMMARY

Fatty acid (FA) profiles of total serum lipids were determined by capillary gas chromatography in Type 2 diabetic patients (NIDDM) with diverse types of hyperlipidemia. In patients with hypertriglyceridemia (DM-HTG) and combined hypertriglyceridemia and hypercholesterolemia (DM-HLP), a significantly different total FA composition was found compared with healthy controls or diabetics with normal serum lipids. In particular, the proportions of saturated and monounsaturated FA were increased and the proportions of n-6 polyunsaturated FA were decreased. In DM-HLP patients, PUFA n-6 metabolites and C20-C22 PUFA were also decreased. Thus, hyperlipidemia shifts significantly the serum FA composition in NIDDM patients into an atherogenic profile. More study is needed, however, to understand if serum FA changes may contribute to the increased atherogenesis commonly found in these patients.

REFERENCES

1. JARRETT, R. J. 1984. *In* Diabetes and Heart Disease, p. 1-23. Elsevier. Amsterdam/New York.
2. STEINER, G. 1994. Atherosclerosis **110**: S27-S33.
3. SIMON, J. A., M. L. HODGKINS, W. S. BROWNER, J. M. NEUHAUS, J. T. BERNERT, JR. & S. B. HULLEY. 1995. Am. J. Epidemiol. **142**: 469-476.
4. SALOMAA, V., I. AHOLA, J. TUOMILEHTO, A. ARO, P. PIETINEN, H. J. KORHONEN & I. PENTTILÄ. 1990. Metabolism **39**: 1285-1291.
5. VESSBY, B., A. ARO, E. SKARFORS, L. BERGLUND, I. SALMINEN & H. LITHELL. 1994. Diabetes **43**: 1353-1357.
6. BLIGH, E. G. & W. J. DYER. 1959. Can. J. Biochem. Physiol. **37**: 911-917.
7. BOHOV, P., V. BALÁŽ & J. HRIVŇÁK. 1984. J. Chromatogr. **286**: 247-252.
8. PELIKÁNOVÁ, T., M. KOHOUT, J. VÁLEK, J. BAŠE & Z. STEFKA. 1991. Metabolism **40**: 175-180.
9. BORKMAN, M., L. H. STORLIEN, D. A. PAN, A. B. JENKINS, D. J. CHISHOLM & L. V. CAMPBELL. 1993. N. Engl. J. Med. **328**: 238-244.
10. MANTZIORIS, E., M. J. JAMES, R. A. GIBSON & L. G. CLELAND. 1994. Am. J. Clin. Nutr. **59**: 1304-1309.
11. ROEMEN, T. H. M., H. KEIZER & G. J. VAN DER VUSSE. 1990. J. Chromatogr. **528**: 447-452.

High Lipid Levels in Slovak Rural Population

Consequence of Thyroid Dysfunction or Nutritional Status?[a]

P. LANGER,[b] E. HANZEN,[c] M. TAJTÁKOVÁ,[d]
Z. PUTZ,[c] A. KREZE,[c] E. ŠEBÖKOVÁ,[b] P. BOHOV,[b]
D. GAŠPERÍKOVÁ,[b] M. HUČKOVÁ,[e] AND I. KLIMEŠ[b]

[b] *Institute of Experimental Endocrinology*
Slovak Academy of Sciences
833 06 Bratislava, Slovak Republic

[c] *Institute of Clinical Endocrinology*
Ľubochňa, Slovak Republic

[d] *First Clinic of Internal Medicine*
Faculty of Medicine
P. J. Šafárik University
Košice, Slovak Republic

[e] *Department of Clinical Chemistry*
Faculty Hospital
Faculty of Medicine
Comenius University
Bratislava, Slovak Republic

After World War II, a very high prevalence of endemic goiter has been found in Slovakia.[1] However, mandatory and well-controlled use of iodized salt beginning from the early 1950s resulted in satisfactory thyroid volume in schoolchildren and adolescents.[2] Nevertheless, there has been continuous evidence showing a possible high prevalence of subclinical and even undetected overt hypothyroidism, as suggested by excessive blood levels of lipids in a great majority of outpatients and inpatients in this country, apparently resulting in a high prevalence of cardiovascular diseases and by far the lowest life expectancy in Slovakia (e.g., 66 years in men and 75 years in women) as compared with other European countries.

These circumstances forced us to organize field surveys aimed to estimate the blood levels of lipids and the prevalence of overt and subclinical thyroid disorders in a group of adult women and to compare the data with those obtained from young adolescents.

---

[a] This work was supported by Grant No. GA 425/1994 from the Slovak Grant Agency for Science and partly by Grant No. 1/990549/92 from the Ministry of Education of the Slovak Republic.

SUBJECTS AND METHODS

Subjects

Subjects included 278 women (average age 54, median 54) from a mountainous village with a formerly high prevalence of endemic goiter and 120 children aged 13 years from the same mountainous area. Informed consent was obtained from all adult participants and from the parents of appropriate children.

Analytical Methods

In serum of all subjects, the following measures were made: (1) thyrotropin (TSH) by sensitive immunoassay (Abbott, United States); (2) total cholesterol (CH) and triglycerides (TG) by enzymatic oxochrome methods (Lachema, Czech Republic); (3) apolipoproteins A1 and B (APO A1 and APO B) with immunoturbidimetric methods (Immuno, Austria).

Statistical Evaluation

The homogeneity of the variations for each parameter was tested first using Cochran's, Bartlett's, and Hartley's tests. The correlations between the level of TSH and individual lipids and apolipoproteins were evaluated by Pearson's or Spearman's test for parametric or nonparametric distributions, respectively.

RESULTS

Prevalence of Overt and Subclinical Thyroid Disorders

In 3/278 (1.1%) women, overt hypothyroidism was found, with the level of TSH being in the range of 27.3–64.2 mU/L; in an additional 17/278 (6.1%), the level of TSH was in the hypothyroid range of 4.6–12.1 mU/L, thus giving 7.2% prevalence of hypothyroidism. In contrast, 7/278 women (2.5%) showed TSH levels in the hyperthyroid range (i.e., below 0.08 mU/L), but only 1 showed clinical symptoms. The remaining 251 values were in the range of 0.08–4.50 mU/L.

In children, only 4/120 (3.3%) cases with marginally increased TSH levels (4.5–7.2 mU/L) were found.

Serum Lipids and Apolipoproteins—Their Interrelations with Thyroid Status

Cholesterol

In women, the mean ± SE was 6.39 ± 0.068 mmol/L (median 6.3), with 110 values (39.5%) being in a very high range of 6.6–9.5 mmol/L and 25 values (8.9%) between

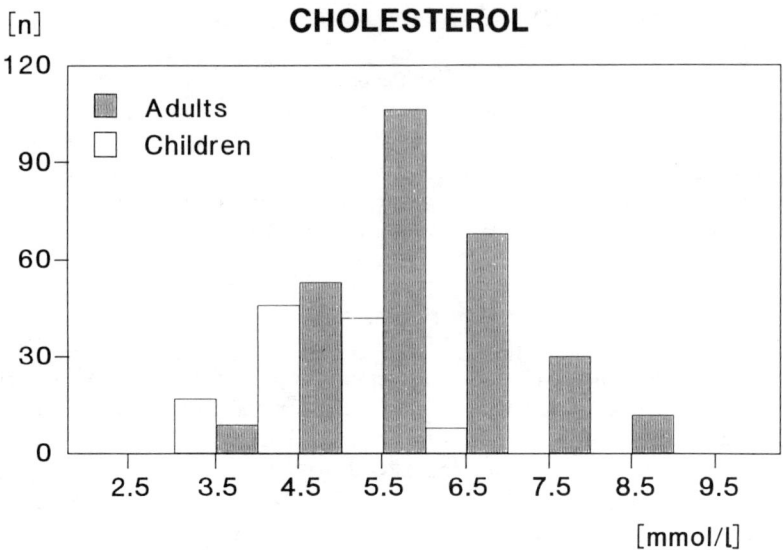

FIGURE 1. Distribution of serum cholesterol levels in adult women (gray columns) and children (white columns).

8.1 and 9.5 mmol/L. In children, the values were much lower, with the range being 2.8–5.9 mmol/L (median 4.4). (See FIG. 1.)

Triglycerides

In women, the mean ± SE was 1.93 ± 0.057 mmol/L (median 1.7), with 85 values (30.6%) being in a very high range of 2.6–6.7 mmol/L and 20 values (7.2%) in the range of 3.6–6.7 mmol/L. In children, the values were much lower, with the range being 0.41–2.17 mmol/L (median 0.92). (See FIG. 2.)

APO Al

In women, the mean ± SE was 218.9 ± 2.67 mg/dL (median 220), with 24 values (8.6%) being below 160 mg/dL and 4 values (1.4%) below 120 mg/dL. In children, the whole range (72–216 mg/dL, median 159) was lower than the mean value found in adult women. (See FIG. 3.)

APO B

In women, the mean ± SE was 138.5 ± 1.79 mmol/L (median 136), with 168 values (60.4%) being over 131 mg/dL and 61 values (21.9%) in a high range of 161–243 mg/dL. As in the case of APO Al, the whole range in children (38–134 mg/dL, median 78) was lower than the mean value in adult women. (See FIG. 4.)

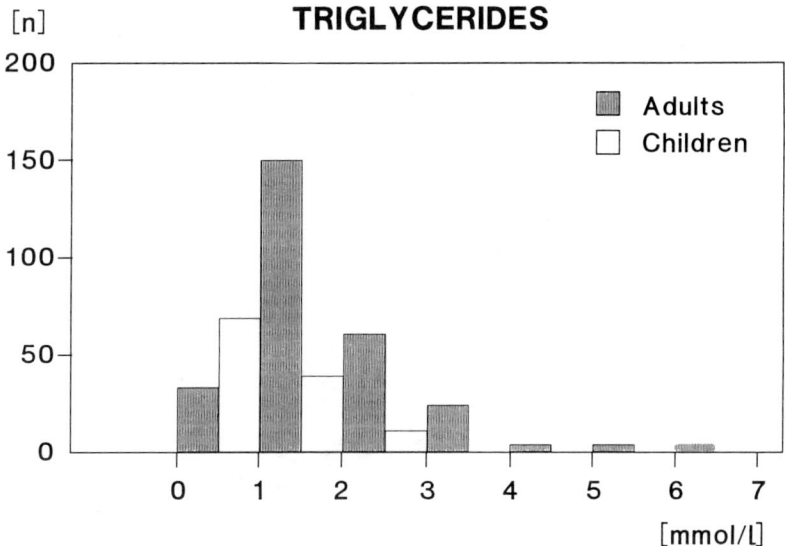

FIGURE 2. Distribution of serum triglyceride levels in adult women (gray columns) and children (white columns).

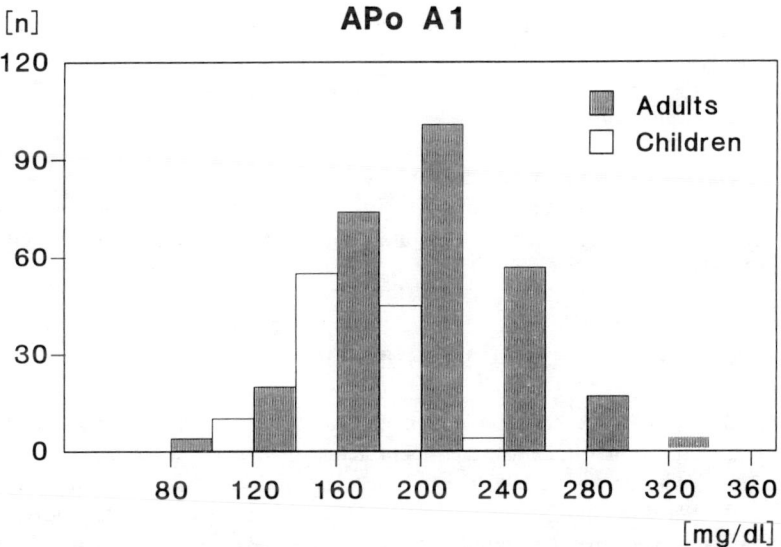

FIGURE 3. Distribution of apolipoprotein A1 levels in adult women (gray columns) and children (white columns).

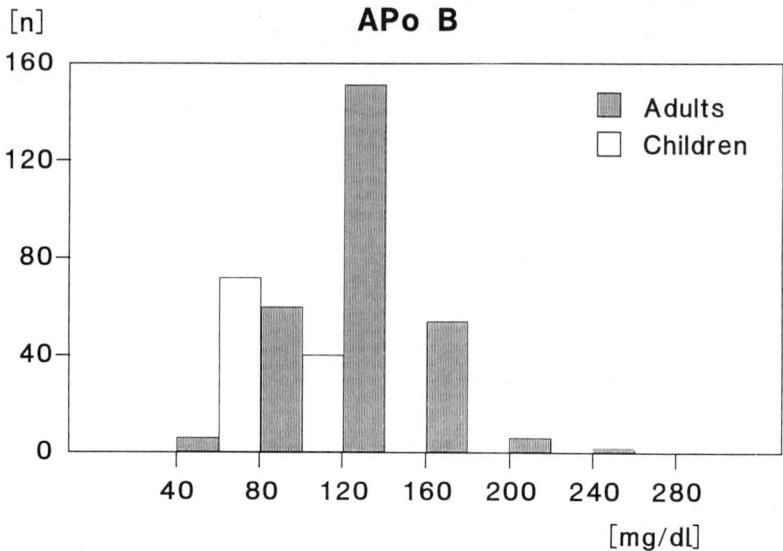

FIGURE 4. Distribution of apolipoprotein B levels in adult women (gray columns) and children (white columns).

APO B/APO Al

In women, the mean ± SE was 0.65 ± 0.01 (median 0.63, range 0.27–1.61); in children, the median was 0.51 and the range was 0.25–0.82.

Statistical Evaluation

No correlations were found between any of the individual lipid parameters and TSH in women and children. However, positive correlations ($p < 0.001$) were found between individual lipid fractions (CH and TG; CH and APO B/APO Al; TG and APO B/APO Al) in adult women.

DISCUSSION

We did not observe any significant correlation between the serum level of TSH and that of individual lipids and apolipoproteins, which appears to be in a definite discrepancy with the repeated and generally accepted findings of increased levels of CH, TG, and APO B in clinically manifested[3–7] and even subclinical hypothyroidism.[8–12] Such a discrepancy may be supported by a recent review[13] of 148 studies on hypercholesterolemia and subclinical hypothyroidism between 1976 and 1995 showing that such thyroid dysfunction was two to three times more frequent in people with elevated total plasma cholesterol.

Our findings may be explained by the traditionally very high consumption of animal fat (lard, butter, cream, high-fat milk and milk products), high-fat red meat (mainly pork), eggs, etc., by the rural population in this country. Similar circumstances apparently exist in the rural populations of surrounding countries (Poland, Czech Republic, Hungary, etc.), as recently supported by the study on changing dietary habits in East Germany after German reunification and their relations to serum lipids.[14] From this, it may be suggested that close interrelations between thyroid function and lipid levels exist perhaps mainly in developed industrialized countries in which most of the inappropriate dietetic factors may be excluded (at least in the middle and upper classes). The same may be true even for the countries (e.g., Japan) with traditionally low intake of those unhealthy foods.

CONCLUSIONS

First, the levels of CH, TG, APO A1, and APO B in women were very high and considerably higher than those in children. Second, no correlations were found between any of the individual lipids and TSH. Third, in women, positive correlations ($p < 0.001$) were found between individual lipids (CH and TG; CH and APO B/APO A1; TG and APO B/APO A1). Fourth and last, it is suggested that high lipid levels result mainly from inappropriate nutritional status of rural populations.

REFERENCES

1. PODOBA, J. 1962. Endemic Goiter in Slovakia (in Slovak). VEDA. Bratislava.
2. LANGER, P., M. TAJTÁKOVÁ, J. PODOBA, JR., L. KOŠŤÁLOVÁ & R. GUTE-KUNST. 1994. Thyroid volume and urinary iodine in schoolchildren and adolescents in Slovakia after 40 years of iodine prophylaxis. Exp. Clin. Endocrinol. **102:** 394–398.
3. KUUSI, T., M. R. TASKINEN & E. A. NIKKILA. 1988. Lipoproteins, lipolytic enzymes, and hormonal status in hypothyroid women at different levels of substitution. J. Clin. Endocrinol. Metab. **66:** 51–55.
4. O'BRIEN, T., S. F. DINNEEN, P. C. O'BRIEN & P. J. PALUMBO. 1993. Hyperlipidemia in patients with primary and secondary hypothyroidism. Mayo Clin. Proc. **68:** 860–866.
5. DE BRUIN, T. W. A., H. VAN BARLINGEN, M. VAN LINDE-SIBENIUS TRIP, A-R. VAN VUURST DE VRIES, M. J. AKVELF & D. W. ERKELENS. 1993. Lipoprotein (a) and apolipoprotein B plasma concentrations in hypothyroid, euthyroid, and hyperthyroid subjects. J. Clin. Endocrinol. Metab. **76:** 121–126.
6. PACKARD, C. J., J. SHEPHERD, G. M. LINDSAY, A. GAW & M. R. TASKINEN. 1993. Thyroid replacement therapy and its influence on postheparin plasma lipases and apolipoprotein-B metabolism in hypothyroidism. J. Clin. Endocrinol. Metab. **76:** 1209–1216.
7. WISEMAN, S. A., J. T. POWELL, S. E. HUMPHRIES & M. PRESS. 1993. The magnitude of the hypercholesterolemia of hypothyroidism is associated with variation in the low density lipoprotein receptor gene. J. Clin. Endocrinol. Metab. **77:** 108–112.
8. LITHELL, H., J. BOBERG, K. HELLSING, S. LJUNGHALL, G. LUNDQUIST, B. VESSBY & L. WIDE. 1981. Serum lipoprotein and apolipoprotein concentrations and tissue lipoprotein-lipase activity in overt and subclinical hypothyroidism: the effect of substitution therapy. Eur. J. Clin. Invest. **11:** 3–10.

9. CARON, PH., C. CALAZEL, H. J. PARRA, M. HOFF & J. P. LOUVET. 1990. Decreased HDL cholesterol in subclinical hypothyroidism: the effect of L-thyroxine therapy. Clin. Endocrinol. **33:** 519–523.
10. ELDER, J., A. McLELLAND, D. O'REILLY, C. J. PACKARD, J. J. SERIES & J. SHEPHERD. 1990. The relationship between serum cholesterol and serum thyrotropin, thyroxine, and triiodothyronine concentrations in suspected hypothyroidism. Ann. Clin. Biochem. **27:** 110–113.
11. BOGNER, U., H. R. ARNITZ, H. PETERS & H. SCHLEUSENER. 1993. Subclinical hypothyroidism and hyperlipoproteinemia: indiscriminate L-thyroxine treatment not justified. Acta Endocrinol. **126:** 202–206.
12. KUNG, A. W. C., R. W. C. PANG & E. D. JANUS. 1995. Elevated serum lipoprotein(s) in subclinical hypothyroidism. Clin. Endocrinol. **43:** 445–449.
13. TANIS, B. C., R. G. J. WESTENDORP & A. H. M. SMELT. 1996. Effect of thyroid substitution on hypercholesterolemia in patients with subclinical hypothyroidism: a reanalysis of intervention studies. Clin. Endocrinol. **44:** 643–649.
14. WAHRBURG, U., H. MARTIN, S. BERGMANN, H. SCHULTE, W. JAROSS & G. ASSMANN. 1995. Dietary habits in Eastern Germany: changes due to German reunification and their relations to serum lipids. Nutr. Metab. Cardiovasc. Dis. **5:** 201–210.

Activity of Antioxidant Enzymes during Hyperglycemia and Hypoglycemia in Healthy Subjects[a]

J. KOŠKA, D. SYROVÁ, P. BLAŽÍČEK, M. MARKO,
D. J. GRŇA, R. KVETŇANSKÝ, AND M. VIGAŠ

Institute of Experimental Endocrinology
Slovak Academy of Sciences
and
Hospital of the Ministry of Defense
Bratislava, Slovak Republic

Among the physiological states of enhanced free radical production, strenuous physical exercise,[1,2] hyperglycemia,[3,4] hyperinsulinemia,[5,6] and enhanced lipid mobilization[7] make a significant contribution. Catecholamine metabolism, which is increased in exercise and hypoglycemia, may also participate in the formation of lipid peroxides.[8] The aim of the present paper was to examine the effect of glucose, catecholamine, and insulin concentrations on the activation of antioxidant enzymes during exercise, hypoglycemia, and hyperglycemia in young healthy subjects.

SUBJECTS AND METHODS

Eight healthy males (22.3 ± 1.5 years, 179.1 ± 1.9 cm height, 70.5 ± 2.2 kg body weight) gave their informed written consent to participate in the study, which was approved by the Ethics Committee of the Institute of Experimental Endocrinology, Slovak Academy of Sciences.

Starting with the evening before the study, the subjects were asked to restrain from alcohol and tobacco and to keep their physical activity at a minimum. They then fasted overnight.

On the day of the experiment, the volunteers arrived at 8 A.M. After inserting an indwelling catheter into the cubital vein, they were allowed to rest in the sitting position for at least 30 min prior to the collection of the control blood sample. Each subject underwent both tests at least one week apart in random order.

---

[a] This work was supported by Grant No. SSVT 95/5305/043.

Hypoglycemia

Hypoglycemia was induced by short-acting insulin (0.1 IU/kg body weight; Actrapid MC, Novo) administered intravenously. The blood samples were obtained before and 30, 45, 60, and 90 min after insulin injection.

Hyperglycemia

Postprandial hyperglycemia was induced by administration of 75 g of glucose in water solution (500 mL). Blood samples were taken before glucose ingestion and at 30-min intervals up to 90 min.

Analytical Methods

Blood was collected into cooled polyethylene tubes using heparin as the anticoagulant. The samples were centrifuged at 4°C and, after separation, the aliquots of plasma for hormone determination were stored frozen ($-20°C$) until analyzed.

Antioxidant parameters were analyzed using commercial kits from Randox (United Kingdom): activity of superoxide dismutase (SOD) was determined in red blood cells and glutathione peroxidase (GPx) was measured in whole blood using the photometric method. The thiobarbituric acid (TBA) method was used for malondialdehyde (MDA) determination by HPLC.

Plasma glucose was analyzed using the glucose-oxidase method (Boehringer). Catecholamines were determined by the radioenzymatic method.[9] RIA was performed for C-peptide and insulin measurements in plasma using commercial kits from Immunotech (France).

The data were statistically evaluated by one-way analysis of variance (ANOVA) followed by pairwise comparisons according to Dunnett and Dunn. Data are shown as the mean ± SE.

RESULTS

Hypoglycemia

Insulin administration decreased blood glucose concentration with a nadir of 52% of the initial value at 30 min ($p < 0.01$). The concentration of circulating insulin rose to 503 ± 60 µU/mL at 5 min after insulin injection. Plasma C-peptide was significantly reduced ($p < 0.01$). Epinephrine concentration was elevated ($p < 0.01$), while plasma noradrenaline was not significantly increased. No changes in blood pressure and a small nonsignificant elevation of heart rate were observed.

Both the activity of SOD and of GPx decreased significantly at the nadir of hypoglycemia ($p < 0.01$ and $p < 0.001$, respectively). MDA concentration was slightly reduced ($p < 0.05$). (See TABLE 1.)

TABLE 1. Plasma Concentration of Glucose, Insulin, C-peptide, Catecholamines, and MDA and Activity of SOD and GPx after Intravenous Administration of Insulin (0.1 U/kg) (Mean ± SE)

Sample	Control	Hypoglycemia		
		30'	60'	90'
Glucose (mmol/L)	4.95 ± 0.08	2.39 ± 0.25[a]	3.93 ± 0.21[a]	4.39 ± 0.21
Insulin (μIU/mL)	7.83 ± 1.29	27.0 ± 4.11[a]	9.99 ± 3.43	5.55 ± 1.64
C-peptide (ng/mL)	1.28 ± 0.16	0.49 ± 0.04[a]	0.34 ± 0.04[a]	0.42 ± 0.07[a]
Adrenaline (pg/mL)	52 ± 9	753 ± 193[a]	214 ± 35	75 ± 23
Noradrenaline (pg/mL)	507 ± 64	601 ± 46	540 ± 49	515 ± 52
SOD (U/g Hb)	935 ± 39	796 ± 33[a]	948 ± 36	930 ± 15
GPx (U/g Hb)	46.3 ± 2.6	32.1 ± 2.0[b]	39.6 ± 2.4	42.9 ± 2.0
MDA (pg/mL)	1.27 ± 0.06	1.09 ± 0.09[c]	1.06 ± 0.11[c]	1.1 ± 0.06[c]

[a] $p < 0.01$.
[b] $p < 0.001$.
[c] $p < 0.05$.

Hyperglycemia

The maximal blood sugar increment of 70% was found at 30 min after glucose ingestion ($p < 0.01$). Hyperglycemia caused elevation of plasma insulin ($p < 0.01$) and C-peptide ($p < 0.01$) concentrations.

The activity of SOD as well as of GPx increased significantly ($p < 0.01$) at 30 min after glucose administration, corresponding with the maximal hyperglycemia. (See TABLE 2.)

DISCUSSION

Antioxidant defense mechanisms, important for protection of cells and tissues from oxidative damage, consist of nonenzymatic antioxidants and antioxidant enzymes, including SOD and GPx. A balance is assumed to exist between free radicals and pro-

TABLE 2. Plasma Concentration of Glucose, Insulin, C-peptide, and MDA and Activity of SOD and GPx after Glucose Administration (75 g Orally) (Mean ± SE)

Sample	Control	Hyperglycemia		
		30'	60'	90'
Glucose (mmol/L)	4.36 ± 0.22	7.44 ± 0.66[a]	6.06 ± 0.64[b]	4.56 ± 0.14
Insulin (μIU/mL)	7.40 ± 1.06	70.3 ± 17.4[a]	59.2 ± 12.4[a]	47.2 ± 6.8[b]
C-peptide (ng/mL)	1.81 ± 0.40	3.23 ± 0.22[a]	3.32 ± 0.23[a]	2.97 ± 0.17[b]
SOD (U/g Hb)	870 ± 16	1006 ± 50[a]	931 ± 51	872 ± 19
GPx (U/g Hb)	40.5 ± 3.3	48.6 ± 4.0[a]	46.4 ± 3.4[b]	42.0 ± 3.7
MDA (pg/mL)	0.76 ± 0.03	1.10 ± 0.08[a]	0.96 ± 0.05	0.82 ± 0.03

[a] $p < 0.01$.
[b] $p < 0.05$.

tective antioxidant reserves. When more free radicals are generated in certain physiological or pathological situations, greater activation of antioxidant enzymes and increased concentration of toxic products (e.g., MDA) are observed.[10] In our investigation, free radical production was based on this indirect evaluation, an experimental approach that has certain limitations.[10]

Different mechanisms have been suggested to be operative in oxidative stress induced by muscle exercise, hyperinsulinemia, or hyperglycemia. Increased activity of the sympathetic nervous system associated with oxidative pathways of catecholamines and mobilization of free fatty acids may induce the activation of free radical formation.[8]

Insulin administration of the dose used in our study induced a brief increase of plasma insulin concentration, which in the first few minutes exceeded 500 µU/mL. This elevated insulin concentration was followed by hypoglycemia and high elevation of plasma adrenaline. Hyperinsulinemia has also been observed to cause a rise in plasma free radical concentration.[6] Krieger-Bauer and Kather[11] demonstrated *in vitro* that insulin activated the plasma membrane–bound H_2O_2-generating system. However, in our *in vivo* study, insulin did not show any stimulatory effect on the antioxidant system. Moreover, the activity of antioxidant enzymes was inhibited and the concentration of MDA was slightly decreased. The inhibition of the activity of the antioxidant system in healthy males seems to be related to plasma glucose concentration rather than to hyperinsulinemia.

In healthy volunteers, glucose-induced hyperglycemia led to increased plasma C-peptide and insulin concentrations. However, this was significantly lower when compared to insulin-induced hypoglycemia after exogenous insulin administration. Increased activity of antioxidant enzymes and elevated concentration of MDA were observed in all subjects. This finding is in agreement with the assumption that hyperglycemia may be involved in enhancing plasma free radical production.[12] Glucose autoxidation is one pathway by which glucose itself generates free oxygen radicals. The -enediol form of glucose may be autoxidized to an -enediol radical anion. The reduced oxygen products are superoxide radical anion (O_2^-), the hydroxyl radical (OH^-), and hydrogen peroxide (H_2O_2). The hydroxyl radical produced specifically by glucose autoxidation was shown to damage proteins. Moreover, elevated glucose was also found to activate lipoxygenase enzymes[4] and to decrease the activity of NO synthase and glutathione reductase, resulting in increased susceptibility of endothelial cells to damage.[6] Remarkably, glucose ingestion was not followed by increased catecholamine release[13] and lipid mobilization[14] and thus does not participate in increasing the free radical production in hyperglycemia.

Our results indicate a clear-cut relation between plasma glucose concentration and the activity of antioxidant enzymes in healthy males. No association of antioxidant enzymes with plasma insulin, adrenaline, or noradrenaline concentration, or with the rate of glucose metabolism, was observed.

Although glucose is not supposed to be an important factor in free radical production and antioxidant enzyme activation, plasma glucose concentration should be taken into consideration as its fluctuations may bias the results of free radical generation studies.

REFERENCES

1. Aruoma, A. I. 1994. J. Nutr. Biochem. **5:** 370-381.
2. Li, L. J. 1995. Free Radical Biol. Med. **18:** 1079-1086.
3. Wolf, S. P., M. J. C. Crabbe & P. J. Thornalley. 1984. Experientia **40:** 244.
4. Nadler, J. L. & L. Winer. 1996. In Diabetes Mellitus, p. 840-847. Lippincott-Raven. Philadelphia/New York.
5. Habib, M. P., F. D. Dickerson & F. D. Mooradian. 1994. Metabolism **43:** 1442-1445.
6. Paolisso, G. & D. Giugliano. 1996. Diabetologia **39:** 357-363.
7. Toborek, M. & B. Henning. 1994. Am. J. Clin. Nutr. **59:** 60-65.
8. Singal, P. K., R. E. Beamish & N. S. Dhalla. 1983. Adv. Exp. Biol. Med. **161:** 391-401.
9. Peuler, J. D. & G. A. Johnson. 1977. Life Sci. **21:** 625-636.
10. Comminacini, L., U. Garbin & V. Lo Cassio. 1996. Diabetologia **39:** 364-366.
11. Krieger-Bauer, H. & H. Kather. 1992. J. Clin. Invest. **89:** 1006-1013.
12. Hunt, J. V., R. T. Dean & S. P. Wolf. 1988. Biochem. J. **256:** 205-212.
13. Tse, T. F., W. E. Clutter, S. D. Shah, J. P. Miller & P. E. Cryer. 1983. J. Clin. Invest. **72:** 270-277.
14. Vuorinen-Markkola, H., V. A. Koivisto & H. Yki-Jarvinen. 1992. Diabetes **41:** 571-580.

Index of Contributors

Abel, J., 353-368
Anderson, R. C., 231-245
Astrup, A., 417-430
Atkinson, R. L., 449-460

Baillie, R., 178-187
Baláž, V., 561-567
Baur, L. A., 287-301, 476-479
Bell, P. A., 231-245
Bendlová, B., 135-143
Berg, S., 64-84, 94-99
Blažíček, P., 575-579
Blundell, J. E., 392-407
Bohov, P., 144-157, 494-509, 561-567, 568-574
Bouchard, C., 408-416
Bouloux, P-M., 110-117
Bransová, J., 480-484, 485-488
Braschi, S., 100-109
Brtko, J., 480-484, 485-488
Brtková, A., 485-488
Bryson, J. M., 476-479
Buemann, B., 417-430
Burkey, B. F., 231-245

Calvert, G. D., 287-301
Cameron-Smith, D., 536-540
Caterson, I. D., 476-479
Caulfield, M., 110-117
Cavallero, E., 100-109
Cavanna, J., 100-109
Charles, A., 100-109
Cheatham, W., 163-169
Clandinin, M. T., 144-157, 188-199
Clarke, S. D., 178-187
Clifton, P. M., 369-381
Collier, G. R., 50-63, 536-540
Crémel, G., 546-549

Deems, R. O., 231-245
de Silva, A., 50-63
de Souza, C., 231-245
Domaschko, D. W., 35-49
Drevon, C. A., 158-162, 310-326
Dukát, A., 382-391
Dunning, B. E., 231-245

Eckel, J., 144-157
Edelsteinová, S., 514-520

Fauth, U., 353-368
Fenselau, S., 353-368
Ficková, M., 489-493, 532-535, 541-545
Foley, J. E., 231-245

Galton, D. J., 100-109
Gambhir, K. K., 163-169
Gašperíková, D., 118-134, 144-157, 480-484, 494-509, 526-531, 568-574
Gayles, E. C., 431-448
Gletsu, N. A., 188-199
Graham, J. J., 369-381
Grňa, D. J., 575-579

Haffner, S. M., 1-12
Halmagyi, M., 353-368
Halvorsen, B., 158-162
Hanefeld, M., 246-268, 279-286
Hanzen, E., 568-574
Heydrick, S. J., 221-230
Hilgertová, J., 556-560
Hill, J. O., 431-448
Howard, B. V., 215-220
Howe, P. R. C., 339-352, 369-381
Hsueh, W. A., 170-177
Hubert, P., 546-549
Hučková, M., 568-574
Hudecová, S., 480-484
Hughes, C. R. T., 369-381

Jacotot, B., 100-109
Jaroščáková, I., 489-493
Jeloková, J., 118-134
Jenkins, A. B., 287-301
Ježová, D., 526-531, 541-545
Jump, D. B., 178-187

Kaburagi, Y., 85-93
Kadowaki, T., 85-93
Kahle, E. B., 35-49
Kay, A., 100-109
Kazdová, L., 269-278, 510-513, 521-525
Keppler, I., 353-368
King, S. E., 476-479
Klimeš, I., xi-xii, 13-34, 118-134, 135-143, 144-157, 200-214, 480-484, 485-488, 494-509, 514-520, 526-531, 550-555, 561-567, 568-574
Klöting, I., 64-84, 94-99, 526-531
Knopp, J., 489-493

Köhler, C., 246-268
Kolter, T., 144-157
Koška, J., 575-579
Kovács, P., 64-84, 94-99, 526-531
Kreze, A., 568-574
Kriketos, A. D., 287-301
Kristek, F., 514-520
Kuneš, J., 532-535
Kurowski, T. G., 221-230
Kvetňanský, R., 118-134, 489-493, 526-531, 575-579
Kyselović, J., 514-520

Langer, P., 485-488, 494-509, 568-574
Laue, Ch., 353-368
Law, R. E., 170-177
Leibel, R. L., 35-49
Leonhardt, W., 279-286
Lewandowski, P., 536-540
Lietava, J., 382-391
Lodwick, D., 526-531
Lungershausen, Y. K., 369-381

Maassen, A., 144-157
Macdiarmid, J. I., 392-407
Macho, L., 118-134, 489-493, 532-535, 541-545
Mann, K. T., 35-49
Marko, M., 575-579
März, W., 353-368
Mazura, I., 135-143
Meuillet, E., 546-549
Minarovič, I., 550-555
Moulton, K., 536-540
Munroe, P., 110-117

Nakamura, M. T., 178-187
Nenseter, M. S., 310-326

Oates, P., 163-169
Orth, B., 353-368

Pacák, K., 118-134
Pagliassotti, M. J., 431-448
Palyzová, D., 135-143
Pan, D. A., 287-301
Pelikánová, T., 269-278
Perlemuter, L., 100-109
Perušičová, J., 135-143
Peters, J. C., 461-475
Phillips, P., 369-381
Putz, Z., 568-574

Raben, A., 417-430
Raney, S. G., 35-49
Ranheim, T., 158-162
Rivellese, A. A., 302-309

Ruderman, N. B., 221-230
Rusnák, M., 118-134, 532-535
Rustan, A. C., 158-162, 310-326

Sabban, E. L., 118-134
Saha, A. K., 221-230
Samani, N. J., 526-531
Sanigorski, A., 50-63, 536-540
Schmidt, S., 64-84
Schrezenmeir, J., 353-368
Šeböková, E., xi-xii, 13-34, 118-134, 135-143, 144-157, 200-214, 480-484, 485-488, 494-509, 514-520, 526-531, 550-555, 561-567, 568-574
Simopoulos, A. P., 327-338
Šindelka, G., 556-560
Škrha, J., 556-560
Storlien, L. H., 287-301
Stürmer, W., 353-368
Swaraj, S., 476-479
Syrová, D., 575-579

Tajtáková, M., 568-574
Tamemoto, H., 85-93
Tapsell, L. C., 287-301
Temam, S., 163-169
Temelkova-Kurktschiev, T., 246-268, 279-286
Terauchi, Y., 85-93
Thomas, D. W., 369-381
Thurzo, M., 382-391
Tobe, K., 85-93
Toubro, S., 417-430
Tremblay, A., 408-416
Tvrzická, E., 269-278

Vavvas, D., 221-230
Včelák, J., 135-143
Verma, M., 163-169
Vietor, I., 118-134
Vigaš, M., 575-579
Vogt, L., 64-84, 94-99
Voigt, B., 64-84, 94-99, 526-531
Vojtko, R., 550-555
Vrána, A., 510-513, 521-525

Walder, K., 50-63, 536-540
Wiernsperger, N., 546-549

Yamamoto, A., 50-63
Yamauchi, T., 85-93

Zahradník, I., 550-555
Žák, A., 269-278, 521-525
Zhang, Q., 100-109
Zimmet, P., 50-63
Zórad, Š., 118-134, 489-493, 532-535, 541-545